Unit Conversions (Equivalents)

Length

1 in. = 2.54 cm
1 cm = 0.394 in.
1 ft = 30.5 cm
1 m = 39.37 in. = 3.28 ft
1 mi = 5280 ft = 1.61 km
1 km = 0.621 mi
1 nautical mile (U.S.) = 1.15 mi = 6076 ft = 1.852 km
1 fermi = 1 femtometer (fm) = 10^{-15} m
1 angstrom (Å) = 10^{-10} m
1 light-year (ly) = 9.46×10^{15} m
1 parsec = 3.26 ly = 3.09×10^{16} m

Volume

1 liter (L) = 1000 mL = 1000 cm^3 = 1.0×10^{-3} m^3 =
   1.057 quart (U.S.) = 54.6 in.^{3}
1 gallon (U.S.) = 4 qt (U.S.) = 231 in.^{3} = 3.78 L =
   0.83 gal (Imperial)
1 m^3 = 35.31 ft^3

Speed

1 mi/h = 1.47 ft/s = 1.609 km/h = 0.447 m/s
1 km/h = 0.278 m/s = 0.621 mi/h
1 ft/s = 0.305 m/s = 0.682 mi/h
1 m/s = 3.28 ft/s = 3.60 km/h
1 knot = 1.151 mi/h = 0.5144 m/s

Angle

1 radian (rad) = 57.30° = 57°18′
1° = 0.01745 rad
1 rev/min (rpm) = 0.1047 rad/s

Time

1 day = 8.64×10^4 s
1 year = 3.156×10^7 s

Mass

1 atomic mass unit (u) = 1.6605×10^{-27} kg
1 kg = 0.0685 slug
[1 kg has a weight of 2.20 lb where g = 9.81 m/s^2.]

Force

1 lb = 4.45 N
1 N = 10^5 dyne = 0.225 lb

Energy and Work

1 J = 10^7 ergs = 0.738 ft·lb
1 ft·lb = 1.36 J = 1.29×10^{-3} Btu = 3.24×10^{-4} kcal
1 kcal = 4.18×10^3 J = 3.97 Btu
1 eV = 1.602×10^{-19} J
1 kWh = 3.60×10^6 J = 860 kcal

Power

1 W = 1 J/s = 0.738 ft·lb/s = 3.42 Btu/h
1 hp = 550 ft·lb/s = 746 W

Pressure

1 atm = 1.013 bar = 1.013×10^5 N/m^2
     = 14.7 lb/in.^{2} = 760 torr
1 lb/in.^{2} = 6.90×10^3 N/m^2
1 Pa = 1 N/m^2 = 1.45×10^{-4} lb/in.^{2}

SI Derived Units and Their Abbreviations

Quantity	Unit	Abbreviation	In Terms of Base Units†
Force	newton	N	kg·m/s^2
Energy and work	joule	J	kg·m^2/s^2
Power	watt	W	kg·m^2/s^3
Pressure	pascal	Pa	kg/(m·s^2)
Frequency	hertz	Hz	s^{-1}
Electric charge	coulomb	C	A·s
Electric potential	volt	V	kg·m^2/(A·s^3)
Electric resistance	ohm	Ω	kg·m^2/(A^2·s^3)
Capacitance	farad	F	A^2·s^4/(kg·m^2)
Magnetic field	tesla	T	kg/(A·s^2)
Magnetic flux	weber	Wb	kg·m^2/(A·s^2)
Inductance	henry	H	kg·m^2/(s^2·A^2)

†kg = kilogram (mass), m = meter (length), s = second (time), A = ampere (electric current).

Metric (SI) Multipliers

Prefix	Abbreviation	Value
exa	E	10^{18}
peta	P	10^{15}
tera	T	10^{12}
giga	G	10^9
mega	M	10^6
kilo	k	10^3
hecto	h	10^2
deka	da	10^1
deci	d	10^{-1}
centi	c	10^{-2}
milli	m	10^{-3}
micro	μ	10^{-6}
nano	n	10^{-9}
pico	p	10^{-12}
femto	f	10^{-15}
atto	a	10^{-18}

PHYSICS
for
SCIENTISTS & ENGINEERS

Volume I

PHYSICS

for

SCIENTISTS & ENGINEERS

Third Edition

DOUGLAS C. GIANCOLI

PRENTICE HALL
Upper Saddle River, New Jersey 07458

Library of Congress Cataloging-in-Publication Data

Giancoli, Douglas C.
    Physics: principles with applications / Douglas C. Giancoli. — 3rd ed.

    p.   cm.
    Includes index.
    ISBN 0-13-021518-X
    1. Physics.   I. Title
QC21.2.G5   2000
530—dc21

          99-40008
           CIP

Editor-in-Chief: Paul F. Corey
Production Editor: Susan Fisher
Executive Editor: Alison Reeves
Development Editor: David Chelton
Director of Marketing: John Tweedale
Senior Marketing Manager: Erik Fahlgren
Assistant Vice President of Production and Manufacturing: David W. Riccardi
Executive Managing Editor: Kathleen Schiaparelli
Manufacturing Manager: Trudy Pisciotti
Art Manager: Gus Vibal
Art Editor: Michele Giusti
Creative Director: Paul Belfanti
Art Director and Cover Designer: Amy Rosen
Advertising and Promotions Manager: Elise Schneider
Editor in Chief of Development: Ray Mullaney
Project Manager: Elizabeth Kell
Photo Research: Mary Teresa Giancoli
Photo Research Administrator: Melinda Reo
Copy Editor: Jocelyn Phillips
Editorial Assistant: Marilyn Coco
Cover photo: Onne van der Wal/Young America
Composition: Emilcomp srl / Preparé Inc.

ISBN 0-13-021518-X

Prentice-Hall International (UK) Limited, *London*
Prentice-Hall of Australia Pty. Limited, *Sydney*
Prentice-Hall Canada Inc., *Toronto*
Prentice-Hall Hispanoamericana, S.A., *Mexico City*
Prentice-Hall of India Private Limited, *New Delhi*
Prentice-Hall of Japan, Inc., *Tokyo*
Prentice-Hall (Singapore) Pte. Ltd.
Editora Prentice-Hall do Brasil, Ltda., *Rio de Janeiro*

CONTENTS—VOLUME I

1   INTRODUCTION, MEASUREMENT, ESTIMATING   1

2   DESCRIBING MOTION: KINEMATICS IN ONE DIMENSION   16

3   KINEMATICS IN TWO DIMENSIONS; VECTORS   45

4   DYNAMICS: NEWTON'S LAWS OF MOTION   77

Contents of Volume II

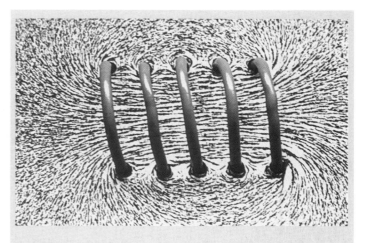

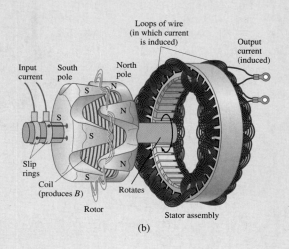

Loops of wire
(in which current
is induced)

Output
current
(induced)

Input
current

South
pole

North
pole

S

S

N

S

N

Slip
rings

Coil
(produces B)

Rotates

Rotor

Stator assembly

(b)

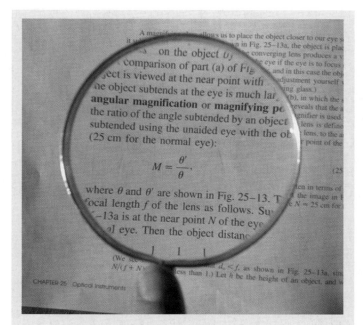

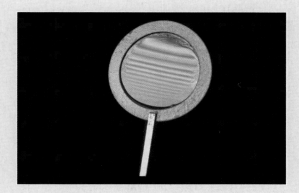

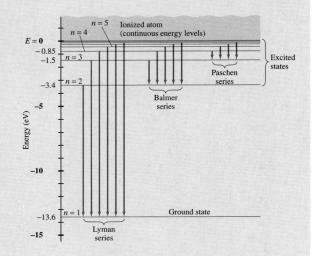

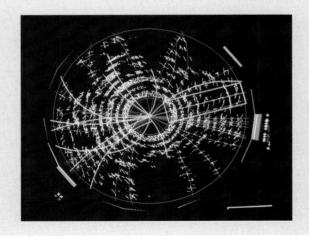

45 ASTROPHYSICS AND COSMOLOGY

APPENDICES

PREFACE

A Brand New Third Edition

It has been more than ten years since the second edition of this calculus-based introductory physics textbook was published. A lot has changed since then, not only in physics itself, but also in how physics is presented. Research in how students learn has provided textbook authors new opportunities to help students learn physics and learn it well.

This third edition comes in three versions. The standard version covers all of classical physics plus a chapter on special relativity and one on the early quantum theory. The extended version, with modern physics, contains a total of nine detailed chapters on modern physics, ending with astrophysics and cosmology. This book retains the original approach: in-depth physics, concrete and nondogmatic, readable.

This new third edition has many improvements in the physics and its applications. Before discussing those changes in detail, here is a list of some of the overall changes that will catch the eye immediately.

Full color throughout is not just cosmetic, although fine color photographs do help to attract the student readers. More important, full color diagrams allow the physics to be displayed with much greater clarity. We have not stopped at a 4-color process; this book has actually been printed in 5 pure colors (5 passes through the presses) to provide better variety and definition for illustrating vectors and other physics concepts such as rays and fields. I want to emphasize that color is used pedagogically to bring out the physics. For example, different types of vectors are given different colors—see the chart on page xxxi.

Many more diagrams, almost double the number in the previous edition, have all been done or redone carefully using full color; there are many more graphs and many more photographs throughout. See for example in optics where new photographs show lenses and the images they make.

Marginal notes have been added as an aid to students to (i) point out what is truly important, (ii) serve as a sort of outline, and (iii) help students to find details about something referred to later that they may not remember so well. Besides such "normal" marginal notes, there are also marginal notes that point out brief *problem solving* hints, and others that point out interesting *applications*.

The great laws of physics are emphasized by giving them a marginal note all in capital letters and enclosed in a rectangle. The most important equations, especially those expressing the great laws, are further emphasized by a tan-colored screen behind them.

Chapter opening photographs have been chosen to illustrate aspects of each chapter. Each was chosen with an eye to writing a caption which could serve as a kind of summary of what is in that chapter, and sometimes offer a challenge. Some chapter-opening photos have vectors or other analysis superimposed on them.

Page layout: complete derivations. Serious attention has been paid to how each page was formatted, especially for page turns. Great effort has been made to keep important derivations and arguments on facing pages. Students then don't have to turn back to check. More important, readers repeatedly see before them, on two facing pages, an important slice of physics.

New Physics

The whole idea of a new edition is to improve, to bring in new material, and to delete material that is verbose and only makes the book longer or is perhaps too advanced and not so useful. Here is a brief summary of a few of the changes involving the physics iself. These lists are selections, not complete lists.

New discoveries:
- planets revolving around distant stars
- Hubble Space Telescope
- updates in particle physics and cosmology, such as inflation and the age of the universe

New physics topics added:
- new treatment of how to make estimates (Chapter 1), including new Estimating Examples throughout (in Chapter 1, estimating the volume of a lake, and the radius of the Earth)
- symmetry used much more, including for solving problems
- new Tables illustrating the great range of lengths, time intervals, masses, voltages
- gravitation as curvature of space, and black holes (Chapter 6)
- engine efficiency (Chapter 8 as well as Chapter 20)
- rolling with and without slipping, and other useful details of rotational motion (Chapter 10)
- forces in structures including trusses, bridges, arches, and domes (Chapter 12)
- square wave (Chapter 15)
- using the Maxwell distribution (Chapter 18)
- Otto cycle (Chapter 20)
- statistical calculation of entropy change in free expansion (Chapter 20)
- effects of dielectrics on capacitor connected and not (Chapter 24)
- grounding to avoid electric hazards (Chapter 25)
- three phase ac (Chapter 31)
- equal energy in $\mathbf{E}$ and $\mathbf{B}$ of EM wave (Chapter 32)
- radiation pressure, EM wave (Chapter 32)
- photos of lenses and mirrors with their images (Chapter 33)
- detailed outlines for ray tracing with mirrors and lenses (Chapters 33, 34)
- lens combinations (Chapter 34)
- new radiation standards (Chapter 43)
- Higgs boson, supersymmetry (Chapter 44)

Modern physics. A number of modern physics topics are discussed in the framework of classical physics. Here are some highlights:
- gravitation as curvature of space, and black holes (Chapter 6)
- planets revolving around distant stars (Chapter 6)
- kinetic energy at relativistic speeds (Chapter 7)
- nuclear collisions (Chapter 9)
- star collapse (Chapter 10)
- galaxy red shift, Doppler (Chapter 16)
- atoms, theory of (Chapters 17, 18, 21)
- atomic theory of thermal expansion (Chapter 17)
- mass of hydrogen atom (Chapter 17)
- atoms and molecules in gases (Chapters 17, 18)
- molecular speeds (Chapter 18)
- equipartition of energy; molar specific heats (Chapter 19)
- star size (Chapter 19)
- molecular dipoles (Chapters 21, 23)
- cathode ray tube (Chapters 23, 27)
- electrons in a wire (Chapter 25)
- superconductivity (Chapter 25)
- discovery and properties of the electron, e/m, oil drop experiment (Chapter 27)
- Hall effect (Chapter 27)

- magnetic moment of electrons (Chapter 27)
- mass spectrometer (Chapter 27)
- velocity selector (Chapter 27)
- electron spin in magnetic materials (Chapter 28)
- light and EM wave emission (Chapter 32)
- spectroscopy (Chapter 36)

Many other examples of modern physics are found as Problems, even in early chapters. Chapters 37 and 38 contain the modern physics topics of Special Relativity, and an introduction to Quantum Theory and Models of the Atom. The longer version of this text, "with Modern Physics," contains an additional seven chapters (for a total of nine) which present a detailed and extremely up-to-date treatment of modern physics: Quantum Mechanics of Atoms (Chapters 38 to 40); Molecules and Condensed Matter (Chapter 41); Nuclear Physics (Chapter 42 and 43); Elementary Particles (Chapter 44); and finally Astrophysics, General Relativity, and Cosmology (Chapter 45).

Revised physics and reorganizations. First of all, a major effort has been made to not throw everything at the students in the first few chapters. The basics have to be learned first; many aspects can come later, when the students are more prepared. Secondly, a great part of this book has been rewritten to make it clearer and more understandable to students. Clearer does not always mean simpler or easier. Sometimes making it "easier" actually makes it harder to understand. Often a little more detail, without being verbose, can make an explanation clearer. Here are a few of the changes, big and small:

- new graphs and diagrams to clarify velocity and acceleration; deceleration carefully treated.
- unit conversion now a new Section in Chapter 1, instead of interrupting kinematics.
- circular motion: Chapter 3 now gives only the basics, with more complicated treatment coming later: non-uniform circular motion in Chapter 5, angular variables in Chapter 10.
- Newton's second law now written throughout as $ma = \Sigma F$, to emphasize inclusion of all forces acting on a body.
- Newton's third law follows the second directly, with inertial reference frames placed earlier. New careful discussions to head off confusion when using Newton's third law.
- careful rewriting of chapters on Work and Energy, especially potential energy, conservative and nonconservative forces, and the conservation of energy.
- renewed emphasis that $\Sigma\tau = I\alpha$ is not always valid: only for an axis fixed in an inertial frame or if axis is through the CM (Chapters 10 and 11).
- rolling motion introduced early in Chapter 10, with more details later, including rolling with and without slipping.
- rotating frames of reference, and Coriolis, moved later, to Chapter 11, shortened, optional, but still including why an object does not fall straight down on Earth.
- fluids reduced to a single chapter (13); some topics and details dropped or greatly shortened.
- clearer details on how an object floats (Chapter 13).
- distinction between wave interference in space, and in time (beats) (Chapter 16).
- thermodynamics reduced to four chapters; the old chapters on Heat and on the First Law of Thermodynamics have been combined into one (19), with some topics shortened and a more rational sequence of topics achieved.
- heat transfer now follows the first law of thermodynamics (Chapter 19).
- electric potential carefully rewritten for accuracy (Chapter 23).
- CRT, computer monitors, TV, treated earlier (Chapter 23).
- use of Q_{encl} and I_{encl} for Gauss's and Ampère's laws, with subscripts meaning "enclosed".
- Ohm's law and definition of resistance carefully redone (Chapter 25).
- sources of magnetic field, Chapter 28, reorganized for ease of understanding, with some new material, and deletion of the advanced topic on magnetization vector.
- circuits with L, C, and/or R now introduced via Kirchhoff's loop rule, and clarified in other ways too (Chapters 30, 31).
- streamlined Maxwell's equations, with displacement current downplayed (Chapter 32).
- optics reduced to four chapters; polarization is now placed in the same chapter as diffraction.

New Pedagogy

All of the above mentioned revisions, rewritings, and reorganizations are intended to help students learn physics better. They were done in response to contemporary research in how students learn, as well as to kind and generous input from professors who have read, reviewed, or used the previous editions. This new edition also contains some new elements, especially an increased emphasis on conceptual development:

Conceptual Examples, typically 1 or 2 per chapter, sometimes more, are each a sort of brief Socratic question and answer. It is intended that students will be stimulated by the question to think, or reflect, and come up with a response—before reading the Response given. Here are a few:

- using symmetry (Chapters 1, 44, and elsewhere)
- ball moving upward: misconceptions (Chapter 2)
- reference frames and projectile motion: where does the apple land? (Chapter 3)
- what exerts the force that makes a car move? (Chapter 4)
- Newton's third law clarification: pulling a sled (Chapter 4)
- free-body diagram for a hockey puck (Chapter 4)
- advantage of a pulley (Chapter 4), and of a lever (Chapter 12)
- to push or to pull a sled (Chapter 5)
- which object rolls down a hill faster? (Chapter 10)
- moving the axis of a spinning wheel (Chapter 11)
- tragic collapse (Chapter 12)
- finger at top of a full straw (Chapter 13)
- suction cups on a spacecraft (Chapter 13)
- doubling amplitude of SHM (Chapter 14)
- do holes expand thermally? (Chapter 17)
- simple adiabatic process: stretching a rubber band (Chapter 19)
- charge inside a conductor's cavity (Chapter 22)
- how stretching a wire changes its resistance (Chapter 25)
- series or parallel (Chapter 26)
- bulb brightness (Chapter 26)
- spiral path in magnetic field (Ch. 27)
- practice with Lenz's law (Chapter 29)
- motor overload (Chapter 29)
- emf direction in induction (Chapter 30)
- effect of inductance in simple circuits (Chapter 30)
- photo with reflection—is it upside down? (Chapter 33)
- reversible light rays (Chapter 33)
- how tall must a full-length mirror be? (Chapter 33)
- diffraction spreading (Chapter 36)

Estimating Examples, roughly 10% of all Examples, also a new feature of this edition, are intended to develop the skills for making order-of-magnitude estimates, even when the data are scarce, and even when you might never have guessed that any result was possible at all. See, for example, Section 1–6, Examples 1–5 to 1–8.

Problem Solving, with New and Improved Approaches

Learning how to approach and solve problems is a basic part of any physics course. It is a highly useful skill in itself, but is also important because the process helps bring understanding of the physics. Problem solving in this new edition has a significantly increased emphasis, including some new features.

Problem-solving boxes, about 20 of them, are new to this edition. They are more concentrated in the early chapters, but are found throughout the book. They each outline a step-by-step approach to solving problems in general, and/or specifically for the material being covered. The best students may find these separate "boxes" unnecessary (they can skip them), but many students will find it helpful to be reminded of the general approach and of steps they can take to get started; and, I think, they help to build confidence. The general problem solving box in Section 4–8 is placed there, after students have had some experience wrestling with problems, and so may be strongly motivated to read it with close attention. Section 4–8 can, of course, be covered earlier if desired.

Problem-solving Sections occur in many chapters, and are intended to provide extra drill in areas where solving problems is especially important or detailed.

Examples. This new edition has many more worked-out Examples, and they all now have titles for interest and for easy reference. There are even two new categories of Example: Conceptual, and Estimates, as described above. Regular Examples serve as "practice problems". Many new ones have been added, some of the old ones have been dropped, and many have been reworked to provide greater clarity and detail: more steps are spelled out, more of "why we do it this way", and more discussion of the reasoning and approach. In sum, the idea is "to think aloud with the students", leading them to develop insight. The total number of worked-out Examples is about 30% greater than in the previous edition, for an average of 12 to 15 per chapter. There is a significantly higher concentration of Examples in the early chapters, where drill is especially important for developing skills and a variety of approaches. The level of the worked-out Examples for most topics increases gradually, with the more complicated ones being on a par with the most difficult Problems at the end of each chapter, so that students can see how to approach complex problems. Many of the new Examples, and improvements to old ones, provide relevant applications to engineering, other related fields, and to everyday life.

Problems at the end of each chapter have been greatly increased in quality and quantity. There are over 30% more Problems than in the second edition. Many of the old ones have been replaced, or rewritten to make them clearer, and/or have had their numerical values changed. Each chapter contains a large group of Problems arranged by Section and graded according to difficulty: level I Problems are simple, designed to give students confidence; level II are "normal" Problems, providing more of a challenge and often the combination of two different concepts; level III are the most complex, typically combining different issues, and will challenge even superior students. The arrangement by Section number means only that those Problems depend on material up to and including that Section: earlier material may also be relied upon. The ranking of Problems by difficulty (I, II, III) is intended only as a guide.

General Problems. About 70% of Problems are ranked by level of difficulty (I, II, III) and arranged by Section. New to this edition are General Problems that are unranked and grouped together at the end of each chapter, and account for about 30% of all problems. The average total number of Problems per chapter is about 90. Answers to odd-numbered Problems are given at the back of the book.

Complete Physics Coverage, with Options

This book is intended to give students the opportunity to obtain a thorough background in all areas of basic physics. There is great flexibility in choice of topics so that instructors can choose which topics they cover and which they omit. Sections marked with an asterisk can be considered optional, as discussed more fully on p. xxv. Here I want to emphasize that topics not covered in class can still be read by serious students for their own enrichment, either immediately or later. Here is a partial list of physics topics, not the standard ones, but topics that might not usually be covered, and that represent how thorough this book is in its coverage of basic physics. Section numbers are given in parentheses.

- use of calculus; variable acceleration (2–8)
- nonuniform circular motion (5–4)
- velocity-dependent forces (5–5)
- gravitational versus inertial mass; principle of equivalence (6–8)
- gravitation as curvature of space; black holes (6–9)
- kinetic energy at very high speed (7–5)
- potential energy diagrams (8–9)
- systems of variable mass (9–10)
- rotational plus translational motion (10–11)
- using $\Sigma \tau_{CM} = I_{CM} \alpha_{CM}$ (10–11)
- derivation of $K = K_{CM} + K_{rot}$ (10–11)
- why does a rolling sphere slow down? (10–12)
- angular momentum and torque for a system (11–4)
- derivation of $d\mathbf{L}_{CM}/dt = \Sigma \tau_{CM}$ (11–4)
- rotational imbalance (11–6)
- the spinning top (11–8)
- rotating reference frames; inertial forces (11–9)
- coriolis effect (11–10)
- trusses (12–7)
- flow in tubes: Poiseuille's equation (13–11)
- surface tension and capillarity (13–12)
- physical pendulum; torsion pendulum (14–6)
- damped harmonic motion: finding the solution (14–7)
- forced vibrations; equation of motion and its solution; Q-value (14–8)
- the wave equation (15–5)
- mathematical representation of waves; pressure wave derivation (16–2)
- intensity of sound related to amplitude (16–3)
- interference in space and in time (16–6)
- atomic theory of expansion (17–4)
- thermal stresses (17–5)
- ideal gas temperature scale (17–10)
- calculations using the Maxwell distribution of molecular speeds (18–2)
- real gases (18–3)
- vapor pressure and humidity (18–4)
- van der Waals equation of state (18–5)
- mean free path (18–6)
- diffusion (18–7)
- equipartition of energy (19–8)
- energy availability; heat death (20–8)
- statistical interpretation of entropy and the second law (20–9)
- thermodynamic temperature scale; absolute zero and the third law (20–10)
- electric dipoles (21–11, 23–6)
- experimental basis of Gauss's and Coulomb's laws (22–4)
- general relation between electric potential and electric field (23–2, 23–8)
- electric fields in dielectrics (24–5)
- molecular description of dielectrics (24–6)
- Gauss's law in dielectrics (24–7)
- current density and drift velocity (25–8)
- superconductivity (25–9)
- RC circuits (26–4)
- use of voltmeters and ammeters; effects of meter resistance (26–5)
- transducers (26–6)
- magnetic dipole moment (27–5)
- Hall effect (27–8)
- operational definition of the ampere and coulomb (28–3)
- magnetic materials—ferromagnetism (28–7)
- electromagnets and solenoids (28–8)
- hysteresis (28–9)
- paramagnetism and diamagnetism (28–10)
- counter emf and torque; eddy currents (29–5)
- Faraday's law—general form (29–7)
- force due to changing $\mathbf{B}$ is nonconservative (29–7)
- LC circuits and EM oscillations (30–5)
- AC resonance; oscillators (31–6)
- impedance matching (31–7)
- three phase AC (31–8)
- changing electric fields produce magnetic fields (32–1)
- speed of light from Maxwell's equations (32–5)
- radiation pressure (32–8)
- fiber optics (33–7)
- lens combinations (34–3)
- aberrations of lenses and mirrors (34–10)
- coherence (35–4)
- intensity in double-slit pattern (35–5)
- luminous intensity (35–8)
- limits of resolution, the λ limit (36–4, 36–5)
- intensity for single-slit (36–2)
- diffraction for double–slit (36–3)
- resolution of the human eye and useful magnification (36–6)
- spectroscopy (36–8)
- peak widths and resolving power for a diffraction grating (36–9)
- x-rays and x-ray diffraction (36–10)
- scattering of light by the atmosphere (36–12)
- time–dependent Schrödinger equation (39–6)
- wave packets (39–7)
- tunneling through a barrier (39–9)
- free-electron theory of metals (41–6)
- semiconductor electronics (41–9)
- standard model, symmetry, QCD, GUT (44–9, 44–10)
- astrophysics, cosmology (Ch. 45)

New Applications

Relevant applications to everyday life, to engineering, and to other fields such as geology and medicine, provide students with motivation and offer the instructor the opportunity to show the relevance of physics. Applications are a good response to students who ask "Why study physics?" Many new applications have been added in this edition. Here are some highlights:

- airbags (Chapter 2)
- elevator and counterweight (Chapter 4)
- antilock brakes and skidding (Chapter 5)
- geosynchronous satellites (Chapter 6)
- hard drive and bit speed (Chapter 10)
- star collapse (Chapter 10)
- forces within trusses, bridges, arches, domes (Chapter 12)
- the Titanic (Chapter 12)
- Bernoulli's principle: wings, sailboats, TIA, plumbing traps and bypasses (Chapter 13)
- pumps (Chapter 13)
- car springs, shock absorbers, building dampers for earthquakes (Chapter 14)
- loudspeakers (Chapters 14, 16, 27)
- autofocusing cameras (Chapter 16)
- sonar (Chapter 16)
- ultrasound imaging (Chapter 16)
- thermal stresses (Chapter 17)
- R-values, thermal insulation (Ch. 19)
- engines (Chapter 20)
- heat pumps, refrigerators, AC; coefficient of performance (Chapter 20)
- thermal pollution (Chapter 20)
- electric shielding (Chapters 21, 28)
- photocopier (Chapter 21)
- superconducting cables (Chapter 25)
- jump starting a car (Chapter 26)
- aurora borealis (Chapter 27)
- solenoids and electromagnetics (Ch. 28)
- computer memory and digital information (Chapter 29)
- seismograph (Chapter 29)

- tape recording (Chapter 29)
- loudspeaker cross-over network (Ch. 31)
- antennas, for $\mathbf{E}$ or $\mathbf{B}$ (Chapter 32)
- TV and radio; AM and FM (Chapter 32)
- eye and corrective lenses (Chapter 34)
- mirages (Chapter 35)
- liquid crystal displays (Chapter 36)
- CAT scans, PET, MRI (Chapter 43)

Some old favorites retained (and improved):

- pressure gauges (Chapter 13)
- musical instruments (Chapter 16)
- humidity (Chapter 18)
- CRT, TV, computer monitors (Ch. 23, 27)
- electric hazards (Chapter 25)
- power in household circuits (Chapter 25)
- ammeters and voltmeters (Chapter 26)
- microphones (Chapters 26, 29)
- transducers (Chapter 26, and elsewhere)
- electric motors (Chapter 27)
- car alternator (Chapter 29)
- electric power transmission (Chapter 29)
- capacitors as filters (Chapter 31)
- impedance matching (Chapter 31)
- fiber optics (Chapter 33)
- cameras, telescopes, microscopes, other optical instruments (Chapter 34)
- lens coatings (Chapter 35)
- spectroscopy (Chapter 36)
- electron microscopes (Chapter 38)
- lasers, holography, CD players (Ch. 40)
- semiconductor electronics (Chapter 41)
- radioactivity (Chapters 42 and 43)

Deletions

Something had to go, or the book would have been too long. Lots of subjects were shortened—the detail simply isn't necessary at this level. Some topics were dropped entirely: polar coordinates; center-of-momentum reference frame; Reynolds number (now a Problem); object moving in a fluid and sedimentation; derivation of Poiseuille's equation; Stoke's equation; waveguide and transmission line analysis; electric polarization and electric displacement vectors; potentiometer (now a Problem); negative pressure; combinations of two harmonic motions; adiabatic character of sound waves; central forces.

Many topics have been shortened, often a lot, such as: velocity-dependent forces; variable acceleration; instantaneous axis; surface tension and capillarity; optics topics such as some aspects of light polarizarion. Many of the brief historical and philosophical issues have been shortened as well.

General Approach

This book offers an in-depth presentation of physics, and retains the basic approach of the earlier editions. Rather than using the common, dry, dogmatic approach of treating topics formally and abstractly first, and only later relating the material to the students' own experience, my approach is to recognize that physics is a description of reality and thus to start each topic with concrete observations and experiences that students can directly relate to. Then we move on to the generalizations and more formal treatment of the topic. Not only does this make the material more interesting and easier to understand, but it is closer to the way physics is actually practiced.

This new edition, even more than previous editions, aims to explain the physics in a readable and interesting manner that is accessible and clear. It aims to teach students by anticipating their needs and difficulties, but without oversimplifying. Physics is all about us. Indeed, it is the goal of this book to help students "see the world through eyes that know physics."

As mentioned above, this book includes of a wide range of Examples and applications from technology, engineering, architecture, earth sciences, the environment, biology, medicine, and daily life. Some applications serve only as examples of physical principles. Others are treated in depth. But applications do not dominate the text—this is, after all, a physics book. They have been carefully chosen and integrated into the text so as not to interfere with the development of the physics but rather to illuminate it. You won't find essay sidebars here. The applications are integrated right into the physics. To make it easy to spot the applications, a new *Physics Applied* marginal note is placed in the margin (except where diagrams in the margin prevent it).

It is assumed that students have started calculus or are taking it concurrently. Calculus is treated gently at first, usually in an optional Section so as not to burden students taking calculus concurrently. For example, using the integral in kinematics, Chapter 2, is an optional Section. But in Chapter 7, on work, the integral is discussed fully for all readers.

Throughout the text, *Système International* (SI) units are used. Other metric and British units are defined for informational purposes. Careful attention is paid to significant figures. When a certain value is given as, say, 3, with its units, it is meant to be 3, not assumed to be 3.0 or 3.00. When we mean 3.00 we write 3.00. It is important for students to be aware of the uncertainty in any measured value, and not to overestimate the precision of a numerical result.

Rather than start this physics book with a chapter on mathematics, I have instead incorporated many mathematical tools, such as vector addition and multiplication, directly in the text where first needed. In addition, the Appendices contain a review of many mathematical topics such as trigonometric identities, integrals, and the binomial (and other) expansions. One advanced topic is also given an Appendix: integrating to get the gravitational force due to a spherical mass distribution.

It is necessary, I feel, to pay careful attention to detail, especially when deriving an important result. I have aimed at including all steps in a derivation, and have tried to make clear which equations are general, and which are not, by explicitly stating the limitations of important equations in brackets next to the equation, such as

$$x = x_0 + v_0 t + \tfrac{1}{2} a t^2. \qquad \text{[constant acceleration]}$$

The more detailed introduction to Newton's laws and their use is of crucial pedagogic importance. The many new worked-out Examples include initially fairly simple ones that provide careful step-by-step analysis of how to proceed in solving dynamics problems. Each succeeding Example adds a new element or a new twist that introduces greater complexity. It is hoped that this strategy will enable even less-well-prepared students to acquire the tools for using Newton's laws correctly. If students don't surmount this crucial hurdle, the rest of physics may remain forever beyond their grasp.

Rotational motion is difficult for most students. As an example of attention to detail (although this is not really a "detail"), I have carefully distinguished the position vector ($\mathbf{r}$) of a point and the perpendicular distance of that point from an axis, which is called R in this book (see Fig. 10–2). This distinction, which enters particularly in connection with torque, moment of inertia, and angular momentum, is often not made clear—it is a disservice to students to use $\mathbf{r}$ or r for both without distinguishing. Also, I have made clear that it is not always true that $\Sigma\tau = I\alpha$. It depends on the axis chosen (valid if axis is fixed in an inertial reference frame, or through the CM). To not tell this to students can get them into serious trouble. (See pp. 250, 283, 284.) I have treated rotational motion by starting with the simple instance of rotation about an axis (Chapter 10), including the concepts of angular momentum and rotational kinetic energy. Only in Chapter 11 is the more general case of rotation about a point dealt with, and this slightly more advanced material can be omitted if desired (except for Sections 11–1 and 11–2 on the vector product and the torque vector). The end of Chapter 10 has an optional subsection containing three slightly more advanced Examples, using $\Sigma\tau_{CM} = I_{CM}\alpha_{CM}$: car braking distribution, a falling yo-yo, and a sphere rolling with and without slipping.

Among other special treatments is Chapter 28, Sources of Magnetic Field: here, in one chapter, are discussed the magnetic field due to currents (including Ampère's law and the law of Biot-Savart) as well as magnetic materials, ferromagnetism, paramagnetism, and diamagnetism. This presentation is clearer, briefer, and more of a whole, and all the content is there.

Organization

The general outline of this new edition retains a traditional order of topics: mechanics (Chapters 1 to 12); fluids, vibrations, waves, and sound (Chapter 13 to 16); kinetic theory and thermodynamics (Chapters 17 to 20). In the two-volume version of this text, volume I ends here, after Chapter 20. The text continues with electricity and magnetism (Chapters 21 to 32), light (Chapters 33 to 36), and modern physics (Chapters 37 and 38 in the short version, Chapters 37 to 45 in the extended version "with Modern Physics"). Nearly all topics customarily taught in introductory physics courses are included. A number of topics from modern physics are included with the classical physics chapters as discussed earlier.

The tradition of beginning with mechanics is sensible, I believe, because it was developed first, historically, and because so much else in physics depends on it. Within mechanics, there are various ways to order topics, and this book allows for considerable flexibility. I prefer, for example, to cover statics after dynamics, partly because many students have trouble working with forces without motion. Besides, statics is a special case of dynamics—we study statics so that we can prevent structures from becoming dynamic (falling down)—and that sense of being at the limit of dynamics is intuitively helpful. Nonetheless statics (Chapter 12) can be covered earlier, if desired, before dynamics, after a brief introduction to vector addition. Another option is light, which I have placed after electricity and magnetism and EM waves. But light could be treated immediately after the chapters on waves (Chapters 15 and 16). Special relativity is Chapter 37, but could instead be treated along with mechanics—say, after Chapter 9.

Not every chapter need be given equal weight. Whereas Chapter 4 might require $1\frac{1}{2}$ to 2 weeks of coverage, Chapter 16 or 22 may need only $\frac{1}{2}$ week.

Some instructors may find that this book contains more material than can be covered completely in their courses. But the text offers great flexibility in choice of topics. Sections marked with a star (asterisk) are considered optional. These Sections contain slightly more advanced physics material, or material not usually covered in typical courses, and/or interesting applications. They contain no material needed in later chapters (except perhaps in later optional Sections). This does not imply that all nonstarred Sections must be covered: there still remains considerable flexibility in the choice of material. For a brief course, all optional material could be dropped as well as major parts of Chapters 11, 13, 16, 26, 30, 31, and 36 as well as selected parts of Chapters 9, 12, 19, 20, 32, 34, and the modern physics chapters. Topics not covered in class can be a valuable resource for later study; indeed, this text can serve as a useful reference for students for years because of its wide range of coverage.

Thanks

Some 60 physics professors provided input or direct feedback on every aspect of this textbook. The reviewers and contributors to this third edition are listed below. I owe each a debt of gratitude.

Ralph Alexander, University of Missouri at Rolla

Zaven Altounian, McGill University

Charles R. Bacon, Ferris State University

Bruce Birkett, University of California, Berkeley

Art Braundmeier, Southern Illinois University at Edwardsville

Wayne Carr, Stevens Institute of Technology

Edward Chang, University of Massachusetts, Amherst

Charles Chiu, University of Texas at Austin

Lucien Crimaldi, University of Mississippi

Robert Creel, University of Akron

Alexandra Cowley, Community College of Philadelphia

Timir Datta, University of South Carolina

Gary DeLeo, Lehigh University

John Dinardo, Drexel University

Paul Draper, University of Texas, Arlington

Alex Dzierba, Indiana University

William Fickinger, Case Western University

Jerome Finkelstein, San Jose State University

Donald Foster, Wichita State University

Gregory E. Frances, Montana State University

Lothar Frommhold, University of Texas at Austin

Thomas Furtak, Colorado School of Mines

Edward Gibson, California State University, Sacramento

Christopher Gould, University of Southern California

John Gruber, San Jose State University

Martin den Boer, Hunter College

Greg Hassold, General Motors Institute

Joseph Hemsky, Wright State University

Laurent Hodges, Iowa State University

Mark Holtz, Texas Tech University

James P. Jacobs, University of Montana

James Kettler, Ohio University Eastern Campus

Jean Krisch, University of Michigan

Mark Lindsay, University of Louisville

Eugene Livingston, University of Notre Dame

Bryan Long, Columbia State Community College

Daniel Mavlow, Princeton University

Pete Markowitz, Florida International University

John McCullen, University of Arizona, Tucson

Peter Nemeth, New York University

Hon-Kie Ng, Florida State University

Eugene Patroni, Georgia Institute of Technology

Robert Pelcovits, Brown University

William Pollard, Valdosta State University

Joseph Priest, Miami University

Carl Rotter, West Virginia University

Lawrence Rees, Brigham Young University

Peter Riley, University of Texas at Austin

Roy Rubins, University of Texas at Arlington

Mark Semon, Bates College

Robert Simpson, University of New Hampshire

Mano Singham, Case Western University

Harold Slusher, University of Texas at El Paso

Don Sparks, Los Angeles Pierce Community College

Michael Strauss, University of Oklahoma

Joseph Strecker, Wichita State University

William Sturrus, Youngstown State University

Arthur Swift, University of Massachusetts, Amherst

Leo Takahasi, The Pennsylvania State University

Edward Thomas, Georgia Institute of Technology

Som Tyagi, Drexel University

John Wahr, University of Colorado

Robert Webb, Texas A & M University

James Whitmore, The Pennsylvania State University

W. Steve Quon, Ventura College

I owe special thanks to Irv Miller, not only for many helpful physics discussions, but for having worked out all the Problems and managed the team that also worked out the Problems, each checking the other, and finally for producing the Solutions Manual and all the answers to the odd-numbered Problems at the end of this book. He was ably assisted by Zaven Altounian and Anand Batra.

I am particularly grateful to Robert Pelcovits and Peter Riley, as well as to Paul Draper and James Jacobs, who inspired many of the new Examples, Conceptual Examples, and Problems.

Crucial for rooting out errors, as well as providing excellent suggestions, were the perspicacious Edward Gibson and Michael Strauss, both of whom carefully checked all aspects of the physics in page proof.

Special thanks to Bruce Birkett for input of every kind, from illuminating discussions on pedagogy to a careful checking of details in many sections of this book. I wish also to thank Professors Howard Shugart, Joe Cerny, Roger Falcone and Buford Price for helpful discussions, and for hospitality at the University of California, Berkeley. Many thanks also to Prof. Tito Arecchi at the Istituto Nazionale di Ottica, Florence, Italy, and to the staff of the Institute and Museum for the History of Science, Florence, for their hospitality.

Finally, I wish to thank the superb editorial and production work provided by all those with whom I worked directly at Prentice Hall: Susan Fisher, Marilyn Coco, David Chelton, Kathleen Schiaparelli, Michele Giusti, Gus Vibal, Mary Teresa Giancoli, and Jocelyn Phillips.

The biggest thanks of all goes to Paul Corey, whose constant encouragement and astute ability to get things done, provided the single strongest catalyst.

The final responsibility for all errors lies with me, of course. I welcome comments and corrections.

D.C.G.

AVAILABLE SUPPLEMENTS

For the Student

Student Study Guide and Solutions Manual
Douglas Brandt, Eastern Illinois University. (0-13-021475-2)
Contains chapter objectives, summaries with additional examples, self-study quizzes, key mathematical equations, and complete worked-out solutions to alternate odd problems in the text.

Doing Physics with Spreadsheets: A Workbook
Gordon Aubrecht, T. Kenneth Bolland, and Michael Ziegler, all of The Ohio State University.
(0-13-021474-4)
Designed to introduce students to the use of spreadsheets for solving simple and complex physics problems. Students are either provided with spreadsheets or must construct their own, then use the model to most closely approximate natural behavior. The amount of spreadsheet construction and the complexity of the spreadsheet increases as the student gains experience.

Science on the Internet: A Student's Guide, 1999
Andrew Stull and Carl Adler (0-13-021308-X)
The perfect tool to help students take advantage of the *Physics for Scientists and Engineers, Third Edition* Web page. This useful resource gives clear steps to access Prentice Hall's regularly updated physics resources, along with an overview of general World Wide Web navigation strategies. Available FREE for students when packaged with the text.

Prentice Hall/*New York Times* Themes of the Times — Physics
This unique newspaper supplement brings together a collection of the latest physics-related articles from the pages of *The New York Times*. Updated twice per year and available FREE to students when packaged with the text.

For the Instructor

Instructor's Solutions Manual
Irvin A. Miller, Drexel University.
Print version (0-13-021381-0); Electronic (CD-ROM) version (0-13-021481-7)
Contains detailed worked solutions to every problem in the text. Electronic versions are available in CD-ROM (dual platform for both Windows and Macintosh systems) for instructors with Microsoft Word or Word-compatible software.

Test Item File
Robert Pelcovits, Brown University; David Curott, University of North Alabama; et al.
(0-13-021482-5)
Contains over 2200 multiple choice questions, about 25% conceptual in nature. All are referenced to the corresponding Section in the text and ranked by difficulty.

Prentice Hall Custom Test Windows (0-13-021477-9); Macintosh (0-13-021476-0)
Based on the powerful testing technology developed by Engineering Software Associates, Inc. (ESA), Prentice Hall Custom Test includes all questions from the Test Item File and allows instructors to create and tailor exams to their own needs. With the Online Testing Program, exams can also be administered on line and data can then be automatically transferred for evaluation. A comprehensive desk reference guide is included along with online assistance.

Transparency Pack (0-13-021470-1)
Includes approximately 400 full color transparencies of images from the text.

Media Supplements

Physics for Scientists and Engineers Web Site www.prenhall.com/giancoli
A FREE innovative online resource that provides students with a wealth of activities and exercises for each text chapter. Features on the site include:

- Practice Questions, Destinations (links to related sites), NetSearch keywords and algorithmically generated numeric Practice Problems by Carl Adler of East Carolina University.
- Physlet Problems (Java-applet simulations) by Wolfgang Christian of Davidson College.
- Warmups and Puzzles essay questions and Applications from Gregor Novak and Andrew Gavrin at Indiana University-Purdue University, Indianapolis.
- Ranking Task Exercises edited by Tom O'Kuma of Lee College, Curtis Hieggelke of Joliet Junior College and David Maloney of Indiana University-Purdue University, Fort Wayne.

Using Prentice Hall CW '99 technology, the website grades and scores all objective questions, and results can be automatically e-mailed directly to the instructors if so desired. Instructors can also create customized syllabi online and link directly to activities on the Giancoli website.

Presentation Manager CD-ROM
Dual Platform (Windows/Macintosh; 0-13-214479-5)
This CD-ROM enables instructors to build custom sequences of Giancoli text images and Prentice Hall digital media for playback in lecture presentations. The CD-ROM contains all text illustrations, digitized segments from the Prentice Hall *Physics You Can See* videotape as well as additional lab and demonstration videos and animations from the Prentice Hall *Interactive Journey Through Physics* CD-ROM. Easy to navigate with Prentice Hall Presentation Manager software, instructors can preview, sequence, and play back images, as well as perform keyword searches, add lecture notes, and incorporate their own digital resources.

Physics You Can See *Video*
(0-205-12393-7)
Contains eleven two- to five-minute demonstrations of classical physics experiments. It includes segments such as "Coin and Feather" (acceleration due to gravity), "Monkey and Gun" (projectile motion), "Swivel Hips" (force pairs), and "Collapse a Can" (atmospheric pressure).

CAPA: A Computer-Assisted Personalized Approach to Assignments, Quizzes, and Exams
CAPA is an on-line homework system developed at Michigan State University that instructors can use to deliver problem sets with randomized variables for each student. The system gives students immediate feedback on their answers to problems, and records their participation and performance. Prentice Hall has arranged to have half of the even-numbered problems of Giancoli, *Physics for Scientists and Engineers, Third Edition*, coded for use with the CAPA system. For additional information about the CAPA system, please visit the web site at http://www.pa.msu.edu/educ/CAPA/.

WebAssign
WebAssign is a web-based homework delivery, collection, grading, and recording service developed and hosted by North Carolina State University. Prentice Hall will arrange for end-of-chapter problems from Giancoli, *Physics for Scientists and Engineers, Third Edition* to be coded for use with the *WebAssign* system for instructors who wish to take advantage of this service. For more information on the *WebAssign* system and its features, please visit http://webassign.net/info or e-mail webassign@ncsu.edu.

NOTES TO STUDENTS AND INSTRUCTORS ON THE FORMAT

1. Sections marked with a star (*) are considered optional. They can be omitted without interrupting the main flow of topics. No later material depends on them except possibly later starred sections. They may be fun to read though.

2. The customary conventions are used: symbols for quantities (such as m for mass) are italicized, whereas units (such as m for meter) are not italicized. Boldface ($\mathbf{F}$) is used for vectors.

3. Few equations are valid in all situations. Where practical, the limitations of important equations are stated in square brackets next to the equation. The equations that represent the great laws of physics are displayed with a tan background, as are a few other equations that are so useful that they are indispensable.

4. The number of significant figures (see Section 1–3) should not be assumed to be greater than given: if a number is stated as (say) 6, with its units, it is meant to be 6 and not 6.0 or 6.00.

5. At the end of each chapter is a set of Questions that students should attempt to answer (to themselves at least). These are followed by Problems which are ranked as level I, II, or III, according to estimated difficulty, with level I Problems being easiest. These Problems are arranged by Section, but Problems for a given Section may depend on earlier material as well. There follows a group of General Problems, which are not arranged by Section nor ranked as to difficulty. Questions and Problems that relate to optional Sections are starred.

6. Being able to solve problems is a crucial part of learning physics, and provides a powerful means for understanding the concepts and principles. This book contains many aids to problem solving: (a) worked-out Examples and their solutions in the text, which are set off with a vertical blue line in the margin, and should be studied as an integral part of the text; (b) special "Problem-solving boxes" placed throughout the text to suggest ways to approach problem solving for a particular topic—but don't get the idea that every topic has its own "techniques," because the basics remain the same; (c) special problem-solving Sections (marked in blue in the Table of Contents); (d) "Problem solving" marginal notes (see point 8 below) which refer to hints for solving problems within the text; (e) some of the worked-out Examples are Estimation Examples, which show how rough or approximate results can be obtained even if the given data are sparse (see Section 1–6); and finally (f) the Problems themselves at the end of each chapter (point 5 above).

7. Conceptual Examples look like ordinary Examples but are conceptual rather than numerical. Each proposes a question or two, which hopefully starts you to think and come up with a response. Give yourself a little time to come up with your own response before reading the Response given.

8. Marginal notes: brief notes in the margin of almost every page are printed in blue and are of four types: (a) ordinary notes (the majority) that serve as a sort of outline of the text and can help you later locate important concepts and equations; (b) notes that refer to the great laws and principles of physics, and these are in capital letters and in a box for emphasis; (c) notes that refer to a problem-solving hint or technique treated in the text, and these say "Problem Solving"; (d) notes that refer to an application of physics, in the text or an Example, and these say "Physics Applied."

9. This book is printed in full color. But not simply to make it more attractive. The color is used above all in the Figures, to give them greater clarity for our analysis, and to provide easier learning of the physical principles involved. The Table on the next page is a summary of how color is used in the Figures, and shows which colors are used for the different kinds of vectors, for field lines, and for other symbols and objects. These colors are used consistently throughout the book.

10. Appendices include useful mathematical formulas (such as derivatives and integrals, trigonometric identities, areas and volumes, expansions), and a table of isotopes with atomic masses and other data. Tables of useful data are located inside the front and back covers.

USE OF COLOR

Vectors

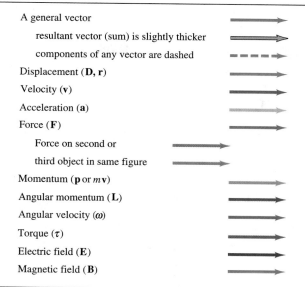

A general vector

    resultant vector (sum) is slightly thicker

    components of any vector are dashed

Displacement ($\mathbf{D}$, $\mathbf{r}$)

Velocity ($\mathbf{v}$)

Acceleration ($\mathbf{a}$)

Force ($\mathbf{F}$)

    Force on second or

    third object in same figure

Momentum ($\mathbf{p}$ or $m\mathbf{v}$)

Angular momentum ($\mathbf{L}$)

Angular velocity (ω)

Torque (τ)

Electric field ($\mathbf{E}$)

Magnetic field ($\mathbf{B}$)

Electricity and magnetism

Electric field lines

Equipotential lines

Magnetic field lines

Electric charge (+)   +   or   • +

Electric charge (−)   −   or   • −

Electric circuit symbols

Wire

Resistor

Capacitor

Inductor

Battery

Optics

Light rays

Object

Real image (dashed)

Virtual image (dashed and paler)

Other

Energy level (atom, etc.)

Measurement lines |—1.0 m—|

Path of a moving object

Direction of motion or current

Besides learning some basics about science and its theories, and about measurement and units, you will also learn in this Chapter how to make quick estimates—such as estimating the radius of the Earth using everyday observations. The photo shown here of Earth was taken from about 36,000 km away. North and South America are clearly visible below the clouds in this computer-enhanced photo.

CHAPTER 1

Introduction, Measurement, Estimating

Physics is the most basic of the sciences. It deals with the behavior and structure of matter. The field of physics is usually divided into the areas of motion, fluids, heat, sound, light, electricity and magnetism, which is referred to as classical physics; plus the modern physics topics of relativity, atomic structure, condensed matter, nuclear physics, elementary particles, and astrophysics.

The fundamentals of physics need to be understood by anyone who hopes to make a career in the sciences or technology: physicists, engineers, chemists, astronomers, mathematicians, geologists, biologists. All the sciences use physics as a fundamental basis, and so does engineering. Engineers, for example, must understand how to use the laws of thermodynamics to design heating systems; they must understand optics and electromagnetism to design medical imaging systems; and they must know how to calculate the forces within a structure in order to design a structure that will remain standing (Fig. 1–1). Indeed, in Chapter 12 we will see a worked-out example of how a simple physics calculation—or even intuition based on understanding the physics of forces—would have saved the lives of hundreds of people. We will see many examples in this book of how physics is useful in other fields, and in everyday life.

(a)

(c)

(b)

FIGURE 1–1 (a) This Roman aqueduct was built 2000 years ago and still stands. (b) So does the Golden Gate Bridge, built in 1937. (c) Collapse of the Hartford Civic Center in 1978, just two years after it was built.

The principal aim of all sciences, including physics, is generally considered to be the search for order in our observations of the world around us. Many people think that science is a mechanical process of collecting facts and devising theories. But it is not so simple. Science is a creative activity that in many respects resembles other creative activities of the human mind.

The Nature of Science

Observation One important aspect of science is **observation** of events, which includes the design and carrying out of experiments. But observation and experiment require imagination, for scientists can never include everything in a description of what they observe. Hence, scientists must make judgments about what is relevant in their observations and experiments. Consider, for example, how two great minds, Aristotle (384–322 B.C.) and Galileo (1564–1642), interpreted motion along a horizontal surface. Aristotle noted that objects given an initial push along the ground (or on a tabletop) always slow down and stop. Consequently, Aristotle argued that the natural state of an object is to be at rest. Galileo, in his reexamination of horizontal motion almost 2000 years later, chose rather to study the idealized case of motion free from resistance. Galileo imagined that if friction could be eliminated, an object given an initial push along a horizontal surface would continue to move indefinitely without stopping. He concluded that for an object to be in motion was just as natural as for it to be at rest. By inventing a new approach, Galileo founded our modern view of motion (Chapters 2, 3, and 4), and he did so with a leap of the imagination. Galileo made this leap conceptually, without actually eliminating friction.

Observation, and careful experimentation and measurement, are one side of the scientific process. The other side is the invention or creation of **theories** to explain and order the observations. Theories, it must be emphasized, are not derived directly from observations. Observations may help inspire a theory, and theories are accepted or rejected based on observation and experiment.

Theories

The great theories of science may be compared, as creative achievements, with great works of art or literature. But how does science differ from these other creative activities? One important difference is that science requires **testing** of its ideas or theories to see if their predictions are borne out by experiment.

Testing (which can never be exhaustive)

Although the testing of theories distinguishes science from other creative fields, it should not be assumed that a theory is "proved" by testing. First of all, no measuring instrument is perfect, so exact confirmation cannot be possible. Furthermore, it is not possible to test a theory in every single possible circumstance. Hence a theory cannot be absolutely verified. Indeed, the history of science tells us that long-held theories can be replaced by new ones.

1-2 | Models, Theories, and Laws

When scientists are trying to understand a particular set of phenomena, they often make use of a **model**. A model, in the scientist's sense, is a kind of analogy or mental image of the phenomena in terms of something we are familiar with. One example is the wave model of light. We cannot see waves of light as we can water waves; but it is valuable to think of light as if it were made up of waves because experiments indicate that light behaves in many respects as water waves do.

Models

The purpose of a model is to give us an approximate mental or visual picture—something to hold on to—when we cannot see what actually is happening. Models often give us a deeper understanding: the analogy to a known system (for instance, water waves in the above example) can suggest new experiments to perform and can provide ideas about what other related phenomena might occur.

You may wonder what the difference is between a theory and a model. Sometimes the words are used interchangeably. Usually, however, a model is relatively simple and provides a structural similarity to the phenomena being studied. A **theory**, on the other hand, is broader, more detailed, and attempts to solve a set of problems, often with great precision. Sometimes, as a model is developed and modified and corresponds more closely to experiment over a wide range of phenomena, it may come to be referred to as a theory. The atomic theory of matter is an example, as is the wave theory of light.

Theories (vs. models)

Models can be very helpful, and they often lead to important theories. But it is important not to confuse a model, or a theory, with the real system or the phenomena themselves.

Scientists give the title **law** to certain concise but general statements about how nature behaves (that energy is conserved, for example). Sometimes the statement takes the form of a relationship or equation between quantities (such as Newton's second law, $F = ma$.)

Laws

Scientific laws are different from political laws in that the latter are *prescriptive*: they tell us how we ought to behave. Scientific laws are *descriptive*: they do not say how nature *should* behave, but rather are meant to describe how nature *does* behave. As with theories, laws cannot be tested in the infinite variety of cases possible. So we cannot be sure that any law is absolutely true. We use the term "law" when its validity has been tested over a wide range of cases, and when any limitations and the range of validity are clearly understood. Even then, as new information comes in, certain laws may have to be modified or discarded.

Scientists normally do their work as if the accepted laws and theories were true. But they are obliged to keep an open mind in case new information should alter the validity of any given law or theory.

1–3 Measurement and Uncertainty; Significant Figures

In the quest to understand the world around us, scientists seek to find relationships among physical quantities that can be measured.

Uncertainty

Every measurement has an uncertainty.

Accurate measurements are an important part of physics. But no measurement is absolutely precise. There is an uncertainty associated with every measurement. Uncertainty arises from different sources. Among the most important, other than blunders, are the limited accuracy of every measuring instrument and the inability to read an instrument beyond some fraction of the smallest division shown. For example, if you were to use a centimeter ruler to measure the width of a board (Fig. 1–2), the result could be claimed to be precise to about 0.1 cm, the smallest division on the ruler, although half of this value might be a valid claim as well. The reason for this is that it is difficult for the observer to interpolate between the smallest divisions. Furthermore, the ruler itself may not have been manufactured to an accuracy very much better than this.[†]

FIGURE 1–2 Measuring the width of a board with a centimeter ruler. Accuracy is about ±1 mm.

When giving the result of a measurement, it is important to state the precision, or **estimated uncertainty**, in the measurement. For example, the width of a board might be written as 8.8 ± 0.1 cm. The ±0.1 cm ("plus or minus 0.1 cm") represents the estimated uncertainty in the measurement, so that the actual width most likely lies between 8.7 and 8.9 cm. The **percent uncertainty** is simply the ratio of the uncertainty to the measured value, multiplied by 100. For example, if the measurement is 8.8 and the uncertainty about 0.1 cm, the percent uncertainty is

$$\frac{0.1}{8.8} \times 100\% \approx 1\%$$

where ≈ means "is roughly equal to."

Assumed uncertainty

Often the uncertainty in a measured value is not specified explicitly. In such cases, the uncertainty is generally assumed to be one or two (or even three) units in the last digit specified. For example, if a length is given as 8.8 cm, the uncertainty is assumed to be about 0.1 cm (or perhaps 0.2 cm). It is important in this case that you do not write 8.80 cm, for this implies an uncertainty on the order of 0.01 cm; it assumes that the length is probably between 8.79 cm and 8.81 cm, when actually you believe it is between 8.7 and 8.9 cm.

Significant Figures

Which digits are significant?

The number of reliably known digits in a number is called the number of **significant figures**. Thus there are four significant figures in the number 23.21 cm and two in the number 0.062 cm (the zeros in the latter are merely place holders that show where the decimal point goes). The number of significant figures may

[†] There is a technical difference between "precision" and "accuracy." Precision in a strict sense refers to the repeatability of the measurement using a given instrument. For example, if you measure the width of a board many times, getting results like 8.81 cm, 8.85 cm, 8.78 cm, 8.82 cm (interpolating between marks as a best estimate each time), you could say the measurements give a *precision* a bit better than 0.1 cm. *Accuracy* refers to how close a measurement is to the true value. For example, if the ruler shown in Fig. 1–2 was manufactured with a 2% error, the accuracy of its measurement of the board's width (about 8.8 cm) would be about 2% of 8.8 cm or about ±0.2 cm. Estimated uncertainty is meant to take both accuracy and precision into account.

not always be clear. Take, for example, the number 80. Are there one or two significant figures? If we say it is *about* 80 km between two cities, there is only one significant figure (the 8) since the zero is merely a place holder. If it is *exactly* 80 km within an accuracy of 1 or 2 km, then the 80 has two significant figures.[†] If it is precisely 80 km, to within ± 0.1 km, then we write 80.0 km.

When making measurements, or when doing calculations, you should avoid the temptation to keep more digits in the final answer than is justified. For example, to calculate the area of a rectangle 11.3 cm by 6.8 cm, the result of multiplication would be 76.84 cm². But this answer is clearly not accurate to 0.01 cm², since (using the outer limits of the assumed uncertainty for each measurement) the result could be between $11.2 \times 6.7 = 75.04$ cm² and $11.4 \times 6.9 = 78.66$ cm². At best, we can quote the answer as 77 cm², which implies an uncertainty of about 1 or 2 cm². The other two digits (in the number 76.84 cm²) must be dropped since they are not significant. As a rough general rule, (i.e., in the absence of a detailed consideration of uncertainties), we can say that: *the final result of a multiplication or division should have only as many digits as the number with the least number of significant figures used in the calculation.* In our example, 6.8 cm has the least number of significant figures, namely two. Thus the result 76.84 cm² needs to be rounded off to 77 cm².

➥ **P R O B L E M   S O L V I N G**

Report only the proper number of significant figures in the final result. An extra digit or two can be kept during the calculation.

Similarly, when adding or subtracting numbers, the final result is no more accurate than the least accurate number used. For example, the result of subtracting 0.57 from 3.6 is 3.0 (and not 3.03).

Keep in mind when you use a calculator that all the digits it produces may not be significant. When you divide 2.0 by 3.0, the proper answer is 0.67, and not some such thing as 0.66666666. Digits should not be quoted (or written down) in a result, unless they are truly significant figures. However, to obtain the most accurate result, you should normally keep an extra significant figure or two throughout a calculation, and round off only in the final result. Note also that calculators sometimes give too few significant figures. For example, when you multiply 2.5×3.2, a calculator may give the answer as simply 8. But the answer is good to two significant figures, so the proper answer is 8.0.

In answers, keep only significant figures

In calculations, keep an extra digit or two

Scientific Notation

We commonly write numbers in "powers of ten," or "scientific" notation—for instance 36,900 as 3.69×10^4, or 0.0021 as 2.1×10^{-3}. One advantage of scientific notation is that it allows the number of significant figures to be clearly expressed. For example, it is not clear whether 36,900 has three, four, or five significant figures. With powers of ten notation the ambiguity can be avoided: if the number is known to an accuracy of three significant figures, we write 3.69×10^4, but if it is known to four, we write 3.690×10^4.

Powers of ten

CONCEPTUAL EXAMPLE 1–1 **Is the diamond yours?** A friend asks to borrow your precious diamond for a day to show her family. You are a bit worried, so you carefully have your diamond weighed on a scale which reads 8.17 grams. The scale's accuracy is claimed to be ± 0.05 grams. The next day you weigh the returned diamond again, getting 8.09 grams. Is this your diamond?

RESPONSE The scale readings are measurements and do not necessarily give the "true" value of the mass. Each measurement could have been high or low by up to 0.05 gram or so. The actual mass of your diamond lies most likely between 8.12 grams and 8.22 grams. The actual mass of the returned diamond is most likely between 8.04 grams and 8.14 grams. These two ranges overlap, so there is not a strong reason to doubt that the returned diamond is yours, at least based on the scale readings.

[†] If the 80 has two significant figures, some people prefer to write it 80., with a decimal point. This is not always (or even usually) done, so the number of significant figures in 80 can be ambiguous unless something is said about it, such as "about" or "precisely."

1–4 Units, Standards, and the SI System

The measurement of any quantity is made relative to a particular standard or **unit**, and this unit must be specified along with the numerical value of the quantity. For example, we can measure length in units such as inches, feet, or miles, or in the metric system in centimeters, meters, or kilometers. To specify that the length of a particular object is 18.6 is meaningless. The unit *must* be given; for clearly, 18.6 meters is very different from 18.6 inches or 18.6 millimeters.

Length

The first real international standard was the **meter** (abbreviated m) established as the standard of **length** by the French Academy of Sciences in the 1790s. In a spirit of rationality, the standard meter was originally chosen to be one ten-millionth of the distance from the Earth's equator to either pole,[†] and a platinum rod to represent this length was made. (This turns out to be, very roughly, the distance from the tip of your nose to the tip of your finger, with arm and hand outstretched.) In 1889, the meter was defined more precisely as the distance between two finely engraved marks on a particular bar of platinum–iridium alloy. In 1960, to provide greater precision and reproducibility, the meter was redefined as 1,650,763.73 wavelengths of a particular orange light emitted by the gas krypton 86. In 1983 the meter was again redefined, this time in terms of the speed of light (whose best measured value in terms of the older definition of the meter was 299,792,458 m/s, with an uncertainty of 1 m/s). The new definition reads: "The meter is the length of path traveled by light in vacuum during a time interval of 1/299,792,458 of a second."[‡]

British units of length (inch, foot, mile) are now defined in terms of the meter. The inch (in.) is defined as precisely 2.54 centimeters (cm; 1 cm = 0.01 m). Other conversion factors are given in the table on the inside of the front cover of this book. Table 1–1 presents some characteristic lengths, from very small to very large. See also Fig. 1–3.

FIGURE 1–3 Some lengths: (a) viruses (about 10^{-7} m long) attacking a cell; (b) Mt. Everest's height is on the order of 10^4 m (8848 m to be precise).

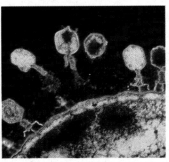

(a)

(b)

TABLE 1–1 Some typical Lengths or Distances (order of magnitude)	
Length (or distance)	**Meters (approximate)**
Neutron or proton (radius)	10^{-15} m
Atom	10^{-10} m
Virus [see Fig. 1–3]	10^{-7} m
Sheet of paper (thickness)	10^{-4} m
Finger width	10^{-2} m
Football field length	10^{2} m
Mt. Everest height [see Fig. 1–3]	10^{4} m
Earth diameter	10^{7} m
Earth to Sun	10^{11} m
Nearest star, distance	10^{16} m
Nearest galaxy	10^{22} m
Farthest galaxy visible	10^{26} m

[†] Modern measurements of the Earth's circumference reveal that the intended length is off by about one-fiftieth of 1 percent. Not bad!

[‡] The new definition of the meter has the effect of giving the speed of light the exact value of 299,792,458 m/s.

TABLE 1–2 Some typical Time Intervals	
Time interval	**Seconds (approximate)**
Lifetime of very unstable subatomic particle	10^{-23} s
Lifetime of radioactive elements	10^{-22} s to 10^{28} s
Lifetime of muon	10^{-6} s
Time between human heartbeats	10^{0} s (= 1 s)
One day	10^{5} s
One year	3×10^{7} s
Human life span	2×10^{9} s
Length of recorded history	10^{11} s
Humans on Earth	10^{14} s
Life on Earth	10^{17} s
Age of Universe	10^{18} s

TABLE 1–3 Some Masses	
Object	**Kilograms (approx.)**
Electron	10^{-30} kg
Proton, neutron	10^{-27} kg
DNA molecule	10^{-17} kg
Bacterium	10^{-15} kg
Mosquito	10^{-5} kg
Plum	10^{-1} kg
Person	10^{2} kg
Ship	10^{8} kg
Earth	6×10^{24} kg
Sun	2×10^{30} kg
Galaxy	10^{41} kg

Time

The standard unit of **time** is the **second** (s). For many years, the second was defined as 1/86,400 of a mean solar day. The standard second is now defined more precisely in terms of the frequency of radiation emitted by cesium atoms when they pass between two particular states. [Specifically, one second is defined as the time required for 9,192,631,770 periods of this radiation.] There are, by definition, 60 s in one minute (min) and 60 minutes in one hour (h). Table 1–2 presents a range of measured time intervals.

Mass

The standard unit of **mass** is the **kilogram** (kg). The standard mass is a particular platinum-iridium cylinder (Fig. 1–4), kept at the International Bureau of Weights and Measures near Paris, France, whose mass is defined as exactly 1 kg. A range of masses is presented in Table 1–3. [For practical purposes, 1 kg weighs about 2.2 pounds on Earth.]

When dealing with atoms and molecules, the **unified atomic mass unit** (u) is usually used. In terms of the kilogram

$$1 \text{ u} = 1.6605 \times 10^{-27} \text{ kg}.$$

The definitions of other standard units for other quantities will be given as we encounter them in later chapters.

FIGURE 1–4   The standard kilogram.

Unit Prefixes

In the metric system, the larger and smaller units are defined in multiples of 10 from the standard unit, and this makes calculation particularly easy. Thus 1 kilometer (km) is 1000 m, 1 centimeter is $\frac{1}{100}$ m, 1 millimeter (mm) is $\frac{1}{1000}$ m or $\frac{1}{10}$ cm, and so on. The prefixes "centi-," "kilo-," and others are listed in Table 1–4 and can be applied not only to units of length, but to units of volume, mass, or any other metric unit. For example, a centiliter (cL) is $\frac{1}{100}$ liter (L) and a kilogram (kg) is 1000 grams (g).

TABLE 1–4 Metric (SI) Prefixes		
Prefix	**Abbreviation**	**Value**
exa	E	10^{18}
peta	P	10^{15}
tera	T	10^{12}
giga	G	10^{9}
mega	M	10^{6}
kilo	k	10^{3}
hecto	h	10^{2}
deka	da	10^{1}
deci	d	10^{-1}
centi	c	10^{-2}
milli	m	10^{-3}
micro†	μ	10^{-6}
nano	n	10^{-9}
pico	p	10^{-12}
femto	f	10^{-15}
atto	a	10^{-18}

† μ is the Greek letter "mu."

Systems of Units

When dealing with the laws and equations of physics it is very important to use a consistent set of units. Several systems of units have been in use over the years. Today the most important is the **Système International** (French for International System), which is abbreviated SI. In SI units, the standard of length is the meter, the standard for time is the second, and the standard for mass is the kilogram. This system used to be called the MKS (meter-kilogram-second) system.

TABLE 1–5
SI Base Quantities and Units

Quantity	Unit	Unit Abbreviation
Length	meter	m
Time	second	s
Mass	kilogram	kg
Electric current	ampere	A
Temperature	kelvin	K
Amount of substance	mole	mol
Luminous intensity	candela	cd

A second metric system is the **cgs system**, in which the centimeter, gram, and second are the standard units of length, mass, and time, as abbreviated in the title. The **British engineering system** takes as its standards the foot for length, the pound for force, and the second for time.

SI units are the principal ones used today in scientific work. We will therefore use SI units almost exclusively in this book, although we will give the cgs and British units for various quantities when introduced.

* Base vs. Derived Units (optional)

Physical quantities can be divided into two categories: *base quantities* and *derived quantities*. The corresponding units for these quantities are called *base units* and *derived units*. A **base quantity** must be defined in terms of a standard. Scientists, in the interest of simplicity, want the smallest number of base quantities possible consistent with a full description of the physical world. This number turns out to be seven, and those used in the SI are given in Table 1–5. All other quantities can be defined in terms of these seven base quantities,[†] and hence are referred to as **derived quantities**. An example of a derived quantity is speed, which is defined as distance traveled divided by the time it takes to travel that distance. To define any quantity, whether base or derived, we can specify a rule or procedure, and this is called an **operational definition**.

1–5 Converting Units

Any quantity we measure, such as a length, a speed, or an electric current, consists of a number *and* a unit. Often we are given a quantity in one set of units, but we want it expressed in another set of units. For example, suppose we measure that a table is 21.5 inches wide, and we want to express this in centimeters. We must use a **conversion factor** which in this case is

$$1 \text{ in.} = 2.54 \text{ cm}$$

or, written another way,

$$1 = 2.54 \text{ cm/in.}$$

Since multiplying by one does not change anything, the width of our table, in cm, is

$$21.5 \text{ inches} = (21.5 \text{ in.}) \times \left(2.54 \frac{\text{cm}}{\text{in.}}\right) = 54.6 \text{ cm.}$$

Note how the units (inches in this case) cancelled out. A table containing many unit conversions is found inside the front cover of this book. Let's take some Examples.

➡ P H Y S I C S   A P P L I E D

How many yards in the 100-m dash

EXAMPLE 1–2 **The 100-m dash.** What is the length of the 100-m dash expressed in yards?

SOLUTION Let us assume the distance is accurately known to four significant figures, 100.0 m. One yard (yd) is precisely 3 feet (36 inches), so we can write

$$1 \text{ yd} = 3 \text{ ft} = 36 \text{ in.} = (36 \text{ in.})\left(2.540 \frac{\text{cm}}{\text{in.}}\right) = 91.44 \text{ cm} = 0.9144 \text{ m.}$$

We can rewrite this result as

$$1 \text{ m} = \frac{1 \text{ yd}}{0.9144} = 1.094 \text{ yd.}$$

Then

$$100 \text{ m} = (100 \text{ m})\left(1.094 \frac{\text{yd}}{\text{m}}\right) = 109.4 \text{ yd,}$$

so a 100-m dash is 9.4 yards longer than a 100-yard dash.

[†] The only exceptions are for angle (radians—see Chapter 10) and solid angle (steradian). No general agreement has been reached as to whether these are base or derived quantities.

We could have done this conversion all in one line:

$$100.0 \text{ m} = (100.0 \text{ m})\left(\frac{100 \text{ cm}}{1 \text{ m}}\right)\left(\frac{1 \text{ in.}}{2.54 \text{ cm}}\right)\left(\frac{1 \text{ yd}}{36 \text{ in.}}\right) = 109.4 \text{ yd.}$$

EXAMPLE 1–3 **Area of a semiconductor chip.** A silicon chip has an area of 1.25 square inches. Express this in square centimeters.

SOLUTION Because 1 in. = 2.54 cm, then 1 in.^{2} = (2.54 cm)^2 = 6.45 cm^2. So

$$1.25 \text{ in.}^2 = (1.25 \text{ in.}^2)\left(2.54 \frac{\text{cm}}{\text{in.}}\right)^2 = (1.25 \text{ in.}^2)\left(6.45 \frac{\text{cm}^2}{\text{in.}^2}\right) = 8.06 \text{ cm}^2.$$

EXAMPLE 1–4 **Speeds.** Where the posted speed limit is 55 miles per hour (mi/h or mph), what is this speed (*a*) in meters per second (m/s) and (*b*) in kilometers per hour (km/h)?

SOLUTION (*a*) We can write 1 mile as

$$1 \text{ mi} = (5280 \text{ ft})\left(12 \frac{\text{in.}}{\text{ft}}\right)\left(2.54 \frac{\text{cm}}{\text{in.}}\right)\left(\frac{1 \text{ m}}{100 \text{ cm}}\right) = 1609 \text{ m.}$$

Note that each conversion factor is equal to one. We also know that 1 hour equals (60 min/h) × (60 s/min) = 3600 s/h, so

Conversion factors = 1

$$55 \frac{\text{mi}}{\text{h}} = \left(55 \frac{\text{mi}}{\text{h}}\right)\left(1609 \frac{\text{m}}{\text{mi}}\right)\left(\frac{1 \text{ h}}{3600 \text{ s}}\right) = 25 \frac{\text{m}}{\text{s}}.$$

(*b*) Now we use 1 mi = 1609 m = 1.609 km; then

$$55 \frac{\text{mi}}{\text{h}} = \left(55 \frac{\text{mi}}{\text{h}}\right)\left(1.609 \frac{\text{km}}{\text{mi}}\right) = 88 \frac{\text{km}}{\text{h}}.$$

When changing units, you can avoid making an error in the use of conversion factors by checking that units cancel out properly. For example, in our conversion of 1 mi to 1609 m in Example 1–4(*a*), if we had incorrectly used the factor $\left(\frac{100 \text{ cm}}{1 \text{ m}}\right)$ instead of $\left(\frac{1 \text{ m}}{100 \text{ cm}}\right)$, the meter units would not have cancelled out; we would not have ended up with meters.

➡ **PROBLEM SOLVING**

Unit conversion is wrong if units do not cancel

1–6 Order of Magnitude: Rapid Estimating

We are sometimes interested only in an approximate value for a quantity. This might be because an accurate calculation would take more time than it is worth or would require additional data that are not available. In other cases, we may want to make a rough estimate in order to check an accurate calculation made on a calculator, to make sure that no blunders were made when entering the numbers.

A rough estimate is made by rounding off all numbers to one significant figure and its power of 10, and after the calculation is made, again only one significant figure is kept. Such an estimate is called an **order-of-magnitude estimate** and can be accurate within a factor of 10, and often better. In fact, the phrase "order of magnitude" is sometimes used to refer simply to the power of 10.

To give you some idea of how useful and powerful rough estimates can be, let us do a few "worked-out Examples."

➡ **PROBLEM SOLVING**

How to make a rough estimate

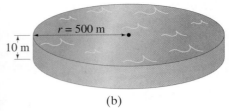

FIGURE 1–5 Example 1–5. (a) How much water is in this lake? (Photo is of one of the Rae Lakes in the Sierra Nevada of California.) (b) Model of the lake as a cylinder. [We could go one step further and estimate the mass or weight of this lake. We will see later that water has a density of 1000 kg/m^3, so this lake has a mass of about $(10^3 \text{ kg/m}^3)(10^7 \text{ m}^3) \approx 10^{10}$ kg, which is about 10 billion kg or 10 million metric tons. (A metric ton is 1000 kg, about 2200 lbs, slightly larger than a British ton, 2000 lbs.)]

➥ **PHYSICS APPLIED**

Estimating the volume (or mass) of a lake; see also Fig. 1–5

EXAMPLE 1–5 **ESTIMATE** **Volume of a lake.** Estimate how much water there is in a particular lake, Fig. 1–5a, which is roughly circular, about 1 km across, and you guess it to have an average depth of about 10 m.

SOLUTION No lake is a perfect circle, nor can lakes be expected to have a perfectly flat bottom. We are only estimating here. To estimate the volume, we use a simple model of the lake as a cylinder: we multiply the average depth of the lake times its roughly circular surface area, as if the lake were a cylinder (Fig. 1–5b). The volume V of a cylinder is the product of its height h times the area of its base: $V = h\pi r^2$, where r is the radius of the circular base. The radius r is $\frac{1}{2}$ km = 500 m, so the volume is approximately

$$V = h\pi r^2 \approx (10 \text{ m}) \times (3) \times (5 \times 10^2 \text{ m})^2 \approx 8 \times 10^6 \text{ m}^3 \approx 10^7 \text{ m}^3,$$

where π was rounded off to 3. So the volume is on the order of 10^7 m^3, ten million cubic meters. Because of all the estimates that went into this calculation, the order-of-magnitude estimate (10^7 m^3) is probably better to quote than the 8×10^6 m^3 figure.

EXAMPLE 1–6 **ESTIMATE** **Thickness of a page.** Estimate the thickness of a page of this book.

➥ **PROBLEM SOLVING**

Use symmetry when possible

FIGURE 1–6 A micrometer, which is used for measuring small thicknesses.

SOLUTION At first you might think that a special measuring device, a micrometer (Fig. 1–6), is needed to measure the thickness of one page since an ordinary ruler clearly won't do. But we can use a trick or, to put it in physics terms, make use of a *symmetry*: we can make the reasonable assumption that all the pages of this book are equal in thickness. Thus, we can use a ruler to measure hundreds of pages at once. If you measure the thickness of the first 500 pages of this book (page 1 to page 500) you might get something like 1.5 cm. Note that 500 pages counted front and back is 250 separate pieces of paper. So one page must have a thickness of about

$$\frac{1.5 \text{ cm}}{250 \text{ pages}} \approx 6 \times 10^{-3} \text{ cm} = 6 \times 10^{-2} \text{ mm}$$

or less than a tenth of a millimeter (0.1 mm).

Now let's take a simple Example of how a diagram can be useful for making an estimate. It cannot be emphasized enough how important it is to draw a diagram when trying to solve a physics problem.

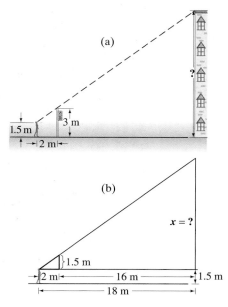

EXAMPLE 1–7 ESTIMATE Height by triangulation. Estimate the height of the building shown in Fig. 1–7a, by "triangulation," with the help of a bus-stop pole and a friend.

SOLUTION By standing your friend next to the pole, you estimate the height of the pole to be 3 m. You next step away from the pole until the top of the pole is in line with the top of the building, Fig. 1–7a. You are 5 ft 6 in. tall, so your eyes are about 1.5 m above the ground. Your friend is taller, and when she stretches out her arms, one hand touches you, and the other touches the pole, so you estimate that distance as 2 m (Fig. 1–7a). You then pace off the distance from the pole to the base of the building with big, 1-m-long, steps, and you get a total of 16 steps or 16 m. Now you draw, to scale, the diagram shown in Fig. 1–7b using these measurements. You can measure, right on the diagram, the last side of the triangle to be about $x = 13$ m. Alternatively, you can use similar triangles to obtain the height x:

$$\frac{1.5\,\text{m}}{2\,\text{m}} = \frac{x}{18\,\text{m}}, \qquad \text{so} \qquad x \approx 13\tfrac{1}{2}\,\text{m}.$$

Finally you add in your eye height of 1.5 m above the ground to get your final result: the building is about 15 m tall.

FIGURE 1–7 Example 1–7. Diagrams are really useful!

EXAMPLE 1–8 ESTIMATE Estimating the radius of the Earth. An easy way to convince yourself that the Earth is round is to watch a ship at sea disappear over the horizon on a calm day. In fact, believe it or not, you can estimate the radius of the Earth without having to go into space (see chapter opening photo). To see how, consider the following data. Suppose you measure the deck of a large sailboat moored on a lake or bay to be 2.0 m above water level. Then you go to the far side of the lake, where you are 4.4 km from the sailboat. Now you lie down right at the water's edge and you estimate that you can see only the upper $\tfrac{1}{4}$ of the sailboat's hull—that is, $\tfrac{3}{4}$ of the hull, or 1.5 m, is below the horizon (hidden behind water). Using Fig. 1–8, where $h = 1.5$ m, estimate the radius of the Earth.

FIGURE 1–8 Boat disappearing over the horizon, Example 1–8. R is the radius of the Earth. You are a distance $d = 4.4$ km from the sailboat when you can see only $\tfrac{1}{4}$ of its hull. Because of the curvature of the Earth, the water "bulges out" between you and the boat.

SOLUTION We can use the Pythagorean theorem on the large right triangle in Fig. 1–8, where R is the radius of the Earth, $h + R$ is the hypotenuse, $d = 4.4$ km, and $h = 1.5$ m:

$$R^2 + d^2 = (R + h)^2$$
$$= R^2 + 2hR + h^2$$

so

$$R \approx \frac{d^2 - h^2}{2h} = \frac{(4400\,\text{m})^2 - (1.5\,\text{m})^2}{3.0\,\text{m}} \approx 6500\,\text{km}.$$

Precise measurements give 6380 km. But look at your achievement! With a few simple rough measurements, and simple geometry, you can make a good estimate of the Earth's radius. You didn't need to go out in space, nor did you need a very long measuring stick.

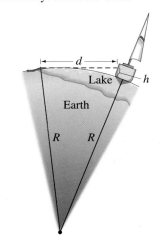

Another technique for estimating, this one made famous by Enrico Fermi to his physics students, is to estimate the number of piano tuners in a city, say, Chicago or San Francisco. To get a rough order-of-magnitude estimate of the number of piano tuners today in San Francisco, a city of about 700,000 inhabitants, we can proceed by estimating the number of functioning pianos, how often each piano is tuned, and how many pianos each tuner can tune. To estimate the number of pianos in San Francisco, we note that certainly not everyone has a piano. A guess of 1 family in 3 having a piano would correspond to 1 piano per 12 persons, assuming an average family of 4 persons. As an order of magnitude, let's say 1 piano per 10 people. This is certainly more reasonable than 1 per 100 people, or 1 per every person, so let's proceed with the estimate that 1 person in 10 has a piano, or about 70,000 pianos in San Francisco. Now a piano tuner needs an hour or two to tune a piano. So let's estimate that a tuner can tune 4 or 5 pianos a day. A piano ought to be tuned every 6 months or a year—let's say once each year. A piano tuner tuning 4 pianos a day, 5 days a week, 50 weeks a year can tune about 1000 pianos a year. So San Francisco, with its (very) roughly 70,000 pianos needs about 70 piano tuners. This is, of course, only a rough estimate.[†] It tells us that there must be many more than 10 piano tuners, and surely not as many as 1000. If you were estimating the number of car mechanics, on the other hand, your estimate would be rather different!

* 1–7 Dimensions and Dimensional Analysis

When we speak of the **dimensions** of a quantity, we are referring to the type of units or base quantities that make it up. The dimensions of area, for example, are always length squared, abbreviated $[L^2]$, using square brackets; the units can be square meters, square feet, and so on. Velocity, on the other hand, can be measured in units of km/h, m/s and mi/h, but the dimensions are always a length $[L]$ divided by a time $[T]$; that is, $[L/T]$. The formula for a quantity may be different in different cases, but the dimensions remain the same. For example, the area of a triangle of base b and height h is $A = \frac{1}{2}bh$, whereas the area of a circle of radius r is $A = \pi r^2$. The formulas are different in the two cases, but the dimensions in both cases are the same: $[L^2]$.

When we specify the dimensions of a quantity, we usually do so in terms of base quantities, not derived quantities. For example, force, which we will see later has the same units as mass $[M]$ times acceleration $[L/T^2]$, has dimensions of $[ML/T^2]$.

Dimensions can be used as a help in working out relationships, and such a procedure is referred to as **dimensional analysis**.[‡] One useful technique is the use of dimensions to check if a relationship is *incorrect*. A simple rule applies here: we add or subtract quantities only if they have the same dimensions (we don't add centimeters and pounds). This implies that the quantities on each side of an equals sign must have the same dimensions.

For example, suppose you derived the equation $v = v_0 + \frac{1}{2}at^2$, where v is the speed of an object after a time t, when it starts with an initial speed v_0 and undergoes an acceleration a. Let's do a dimensional check to see if this equation is correct. We write a dimensional equation as follows, remembering that the dimensions of speed are $[L/T]$ and (as we shall see in Chapter 2) the dimensions of acceleration are $[L/T^2]$:

$$\left[\frac{L}{T}\right] \stackrel{?}{=} \left[\frac{L}{T}\right] + \left[\frac{L}{T^2}\right][T^2]$$

$$\stackrel{?}{=} \left[\frac{L}{T}\right] + [L].$$

The dimensions are incorrect: on the right side, we have the sum of quantities whose dimensions are not the same. Thus we conclude that an error was made in the derivation of the original equation.

[†] A check of the San Francisco Yellow Pages (done after this calculation) reveals about 50 listings. Each of these listings may employ more than one tuner, but on the other hand, each may also do repairs as well as tuning. In any case, our estimate is reasonable.

[‡] The techniques described in the next few paragraphs may seem more meaningful after you have studied a few chapters of this book. Reading this Section now will give you an overview of the subject, and you can then return to it later as needed.

If such a dimensional check does come out correct, it does not prove that the equation is correct. For example, a dimensionless numerical factor (such as $\frac{1}{2}$ or 2π) could be wrong. Thus a dimensional check can only tell you when a relationship is wrong. It can't tell you if it is completely right.

Dimensional analysis can also be used as a quick check on an equation you are not sure about. For example, suppose that you can't remember whether the equation for the period T (the time to make one back-and-forth swing) of a simple pendulum of length l is $T = 2\pi\sqrt{l/g}$ or $T = 2\pi\sqrt{g/l}$, where g is the acceleration due to gravity and, like all accelerations, has dimensions $[L/T^2]$. (Do not worry about these formulas—the correct one will be derived in Chapter 14; what we are concerned about here is a person's forgetting whether it contains l/g or g/l.) A dimensional check shows that the former is correct:

$$[T] = \sqrt{\frac{[L]}{[L/T^2]}} = \sqrt{[T^2]} = [T],$$

whereas the latter is not:

$$[T] \neq \sqrt{\frac{[L/T^2]}{[L]}} = \sqrt{\frac{1}{[T^2]}} = \frac{1}{[T]}.$$

Note that the constant 2π has no dimensions and so can't be checked using dimensions.

☐ Summary

Scientists often devise models of physical phenomena. A **model** is a kind of picture or analogy that seems to explain the phenomena. A **theory**, often developed from a model, is usually deeper and more complex than a simple model.

A scientific **law** is a concise statement, often expressed in the form of an equation, which quantitatively describes a particular range of phenomena over a wide range of cases.

Measurements play a crucial role in physics, but can never be perfectly precise. It is important to specify the **uncertainty** of a measurement either by stating it directly using the ± notation, and/or by keeping only the correct number of **significant figures**.

Physical quantities are always specified relative to a particular standard or **unit**, and the unit used should always be stated. The commonly accepted set of units today is the **Système International** (SI), in which the standard units of length, mass, and time are the **meter**, **kilogram**, and **second**.

When converting units, check all **conversion factors** for correct cancellation of units.

Making rough, **order-of-magnitude estimates** is a very useful technique in science as well as in everyday life.

The **dimensions** of a quantity refer to the combination of base quantities that comprise it. Velocity, for example, has dimensions of [length/time] or $[L/T]$. Working with only the dimensions of the various quantities in a given relationship (this technique is called **dimensional analysis**) it is possible to check a relationship for correct form.

☐ Questions

1. It is advantageous that fundamental standards, such as those for length and time, be accessible (easy to compare to), invariable (do not change), indestructible, and reproducible. Discuss why these are advantages and whether any of these criteria can be incompatible with others.

2. What are the merits and drawbacks of using a person's foot as a standard? Discuss in terms of the criteria mentioned in Question 1. Consider both (a) a particular person's foot, and (b) any person's foot.

3. When traveling a highway in the mountains, you may see elevation signs that read "914 m (3000 ft)." Critics of the metric system claim that such numbers show the metric system is more complicated. How would you alter such signs to be more consistent with a switch to the metric system?

4. Suggest a way to measure the distance from Earth to the Sun.

5. What is wrong with this road sign: Memphis 7 mi (11.263 km)?

*6. Can you set up a complete set of base quantities, as in Table 1–5, that does not include length as one of them?

7. List assumptions useful to estimate the number of car mechanics in (a) San Francisco, (b) your hometown, and then make the estimates.

8. Estimate the number of hours you have spent in school thus far in your life.

9. Discuss how the notion of symmetry could be used to estimate the number of marbles in a one-liter jar.

10. You measure the radius of a wheel to be 4.16 cm. If you multiply by 2 to get the diameter, should you write the result as 8 cm or as 8.32 cm? Justify your answer.

Problems

[The problems at the end of each chapter are ranked I, II, or III according to estimated difficulty, with I problems being easiest. The problems are arranged by Sections, meaning that the reader should have read up to and including that Section, but not only that Section—problems often depend on earlier material. Each chapter also has a group of General Problems that are not arranged by Section and not ranked.]

Section 1–3

1. (I) The age of the universe is thought to be somewhere around 10 billion years. Assuming one significant figure, write this in powers of ten in (a) years, (b) seconds.

2. (I) How many significant figures do each of the following numbers have: (a) 2142, (b) 81.60, (c) 7.63, (d) 0.03, (e) 0.0086, (f) 3236, and (g) 8700?

3. (I) Write the following numbers in powers of ten notation: (a) 1,156, (b) 21.8, (c) 0.0068, (d) 27.635, (e) 0.219, and (f) 22.

4. (I) Write out the following numbers in full with a decimal point and correct number of zeros: (a) 8.69×10^4, (b) 7.1×10^3, (c) 6.6×10^{-1}, (d) 8.76×10^2, and (e) 8.62×10^{-5}.

5. (I) What is the percent uncertainty in the measurement 3.26 ± 0.25 m?

6. (I) What, approximately, is the percent uncertainty for the measurement given as 1.28 m?

7. (I) Time intervals measured with a stopwatch typically have an uncertainty of about 0.2 s, due to human reaction time at the start and stop moments. What is the percent uncertainty of a handtimed measurement of (a) 5 s, (b) 50 s, (c) 5 min?

8. (II) Multiply 2.079×10^2 m by 0.072×10^{-1}, taking into account significant figures.

9. (II) Add 9.2×10^3 s $+ 8.3 \times 10^4$ s $+ 0.008 \times 10^6$ s.

10. (II) What is the area, and its approximate uncertainty, of a circle of radius 3.8×10^4 cm?

11. (II) What is the percent uncertainty in the volume of a spherical beach ball whose radius is $r = 2.86 \pm 0.08$ m?

Sections 1–4 and 1–5

12. (I) Express the following using the prefixes of Table 1–4: (a) 10^6 volts, (b) 10^{-6} meters, (c) 6×10^3 days, (d) 18×10^2 bucks, and (e) 8×10^{-9} pieces.

13. (I) Write the following as full (decimal) numbers with standard units: (a) 286.6 mm, (b) 85 μV, (c) 760 mg, (d) 60.0 picoseconds, (e) 22.5 femtometers (f) 2.50 gigavolts.

14. (I) How many cars is 50 hectocars? What would you be if you earned a megabuck a year?

15. (I) Determine your height in meters.

16. (I) The Sun, on average, is 93 million miles from the Earth. How many meters is this? Express (a) using powers of ten, and (b) using a metric prefix.

17. (I) What is the conversion factor between (a) ft^2 and yd^2, (b) m^2 and ft^2?

18. (II) The Concorde airplane travels at about 2300 km/h. How long does it take to travel 1.0 mile?

19. (II) A typical atom has a diameter of about 1.0×10^{-10} m. (a) What is this in inches? (b) Approximately how many atoms are there along a 1.0-cm line?

20. (II) Express the following sum with the correct number of significant figures: 2.00 m $+$ 142.5 cm $+ 7.24 \times 10^5$ μm.

21. (II) Determine the conversion factor between (a) km/h and mi/h, (b) m/s and ft/s, and (c) km/h and m/s.

22. (II) How much longer (percentage) is a one-mile race than a 1500-m race ("the metric mile")?

23. (II) A *light-year* is the distance light (speed $= 2.998 \times 10^8$ m/s) travels in one year. (a) How many meters are there in 1.00 light-year? (b) An astronomical unit (AU) is the average distance from the Sun to Earth, 1.50×10^8 km. How many AU are there in 1.00 light-year? (c) What is the speed of light in AU/h?

24. (II) The diameter of the moon is 3480 km. What is the surface area, and how does it compare to the surface area of the Earth?

Section 1–6

(*Note*: Remember that for rough estimates, only round numbers are needed both as input to calculations and as final results.)

25. (I) Estimate the order of magnitude (power of ten) of: (a) 2800, (b) 86.30×10^2, (c) 0.0076, and (d) 15.0×10^8.

26. (II) Estimate how long it would take a good runner to run across the United States from New York to California.

27. (II) Make a rough estimate of the percentage of a house's outside wall area that consists of window area.

28. (II) Estimate the number of times a human heart beats in a lifetime.

29. (II) Make a rough estimate of the volume of your body (in cm^3).

30. (II) Estimate the time to drive from Beijing (Peking) to Paris (a) today, and (b) in 1906 when a great car race was run between those two cities.

31. (II) Estimate the number of dentists (a) in San Francisco and (b) in your town or city.

32. (II) Estimate how long it would take one person to mow a football field using an ordinary home lawn mower.

33. (II) The rubber worn from tires mostly enters the atmosphere as particulate pollution. Estimate how much rubber (in kg) is put into the air in the United States every year. To get you started, a good estimate for a tire tread's depth is 1 cm when new, and the density of rubber is about 1200 kg/m^3.

34. (II) Estimate how many books can be shelved in a college library with 1500 square meters of floor space.

35. (III) You are in a hot air balloon, 200 m above the flat Texas plains. You look out toward the horizon. How far out can you see—that is, how far is your horizon? The Earth's radius is about 6400 km.

* Section 1–7

* 36. (I) The speed, v, of a body is given by the equation $v = At^3 - Bt$, where t refers to time. What are the dimensions of A and B?

* 37. (I) What are the SI units for the constants A and B in Problem 36?

* 38. (II) Three students derive the following equations in which x refers to distance traveled, v the speed, a the acceleration (m/s^2), t the time, and the subscript $(_0)$ means a quantity at time $t = 0$: (a) $x = vt^2 + 2at$, (b) $x = v_0t + \frac{1}{2}at^2$, and (c) $x = v_0t + 2at^2$. Which of these could possibly be correct according to a dimensional check?

General Problems

39. An angstrom (symbol Å) is an older unit of length, defined as 10^{-10} m. (a) How many nanometers are in 1.0 angstrom? (b) How many femtometers or fermis (the common unit of length in nuclear physics) are in 1.0 angstrom? (c) How many angstroms are in 1.0 meter? (d) How many angstroms are in 1.0 light-year (see Problem 23)?

40. Use Table 1–3 to estimate the total number of protons or neutrons in (a) a bacterium, (b) a DNA molecule, (c) the human body, (d) our Galaxy.

41. (a) How many seconds are there in 1.00 year? (b) How many nanoseconds are there in 1.00 year? (c) How many years are there in 1.00 second?

42. One hectare is defined as $10^4\,\text{m}^2$. One acre is $4 \times 10^4\,\text{ft}^2$. How many acres are in one hectare?

43. Estimate the number of bus drivers (a) in Washington, D.C., and (b) in your town.

44. Computer chips (Fig. 1–9) are etched on circular silicon wafers of thickness 0.60 mm that are sliced from a solid cylindrical silicon crystal of length 30 cm. If each wafer can hold 100 chips, what is the maximum number of chips that can be produced from one entire cylinder?

45. Estimate the number of gallons of gasoline consumed by the total of all automobile drivers in the United States, per year.

46. Estimate the number of gumballs in the machine of Fig. 1–10.

FIGURE 1–9 Problem 44. The wafer held by the hand (above) is shown below, enlarged and illuminated by colored light. Visible are rows of integrated circuits (chips).

FIGURE 1–10 Problem 46. Estimate the number of gumballs in the machine.

47. An average family of four uses roughly 1200 liters (about 300 gallons) of water per day. (One liter $= 1000\,\text{cm}^3$.) How much depth would a lake lose per year if it uniformly covered an area of 50 square kilometers and supplied a local town with a population of 40,000 people? Consider only population uses, and neglect evaporation and so on.

48. How big is a ton? That is, what is the volume of something that weighs a ton? To be specific, estimate the diameter of a 1-ton rock, but first make a wild guess: will it be 1 ft across, 3 ft, or the size of a car? [*Hint:* Rock has mass per volume about 3 times that of water, which is 1 kg per liter ($10^3\,\text{cm}^3$) or 62 lb per cubic foot.]

49. A heavy rainstorm dumps 1.0 cm of rain on a city 5 km wide and 8 km long in a 2-h period. How many metric tons (1 ton $= 10^3$ kg) of water fell on the city? [1 cm^3 of water has a mass of 1 gram $= 10^{-3}$ kg.]

50. Hold a pencil in front of your eye at a position where its tip just blocks out the Moon (Fig. 1–11). Make appropriate measurements to estimate the diameter of the Moon, given that the Earth–Moon distance is 3.8×10^5 km.

FIGURE 1–11 Problem 50. How big is the Moon?

51. Estimate how long it would take to walk around the world.

52. Noah's ark was ordered to be 300 cubits long, 50 cubits wide, and 30 cubits high. The cubit was a unit of measure equal to the length of a human forearm, elbow to the tip of the longest finger. Express the dimensions of Noah's ark in meters.

53. Jean camps beside a wide river and wonders how wide it is. She spots a large rock on the bank directly across from her. She then walks upstream until she judges that the angle between her and the rock, which she can still see clearly, is now at an angle of 30° downstream (Fig. 1–12). Jean measures her stride to be about one yard long. The distance back to her camp is 120 strides. About how far across, both in yards and in meters, is the river?

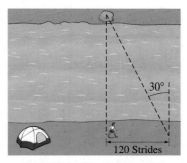

FIGURE 1–12 Problem 53.

54. One liter (1000 cm^3) of oil is spilled onto a smooth lake. If the oil spreads out uniformly until it makes an oil slick just one molecule thick, with adjacent molecules just touching, estimate the diameter of the oil slick. Assume the oil molecules have a diameter of 2×10^{-10} m.

55. Compare the percent uncertainty in θ and in $\sin\theta$ when (a) $\theta = 15.0° \pm 0.5°$, (b) $\theta = 75.0° \pm 0.5°$.

56. You are lying on the sand at the edge of the sea, watching a sailboat. If you know (or measured) that the distance from the water to the top of the boat hull is 2.5 m, estimate how far away the boat is when (using binoculars) you can no longer see the hull. The radius of the Earth is 6.38×10^6 m.

A high speed car has released a parachute to reduce its speed quickly. The directions of the car's velocity and acceleration are shown by the green (**v**) and gold (**a**) arrows. Note that **v** and **a** point in opposite directions. Motion is described using the concepts of velocity and acceleration. We see here that the acceleration **a** can sometimes be in the opposite direction from the velocity **v**. We will also examine in detail motion with constant acceleration, including the vertical motion of objects falling under gravity.

CHAPTER 2

Describing Motion: Kinematics in One Dimension

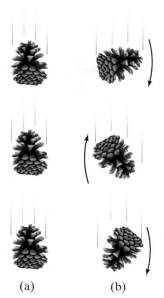

(a)   **(b)**

FIGURE 2–1 The pinecone in (a) undergoes pure translation as it falls, whereas in (b) it is rotating as well as translating.

The motion of objects—baseballs, automobiles, joggers, and even the Sun and Moon—is an obvious part of everyday life. It was not until the sixteenth and seventeenth centuries that our modern understanding of motion was established. Many contributed to this understanding, particularly Galileo Galilei (1564–1642) and Isaac Newton (1642–1727).

The study of the motion of objects, and the related concepts of force and energy, form the field called **mechanics**. Mechanics is customarily divided into two parts: **kinematics**, which is the description of how objects move, and **dynamics**, which deals with force and why objects move as they do. This chapter and the next deal with kinematics.

We start by discussing objects that move without rotating (Fig. 2–1a). Such motion is called **translational motion**. In the present chapter we will be concerned with describing an object that moves along a straight-line path, which is one-dimensional motion. In Chapter 3 we will study how to describe translational motion in two (or three) dimensions.

We will often make use of the concept, or model, of an idealized **particle** which is considered to be a mathematical point and to have no spatial extent (no size). A particle can undergo only translational motion. The particle model is useful in many real situations where we are interested only in translational motion and the object's size is not significant. For example, we might consider a billiard ball, or even a spacecraft traveling toward the moon, as a particle for many purposes.

16

2-1 | Reference Frames and Displacement

Any measurement of position, distance, or speed must be made with respect to a **frame of reference**. For example, while you are on a train traveling at 80 km/h, you might notice a person who walks past you toward the front of the train at a speed of, say, 5 km/h (Fig. 2–2). Of course this is the person's speed with respect to the train as frame of reference. With respect to the ground that person is moving at a speed of 80 km/h + 5 km/h = 85 km/h. It is always important to specify the frame of reference when stating a speed. In everyday life, we usually mean "with respect to the Earth" without even thinking about it, but the reference frame must be specified whenever there might be confusion.

All measurements are made relative to a frame of reference

FIGURE 2–2 A person walks toward the front of a train at 5 km/h. The train is moving 80 km/h with respect to the ground, so the walking person's speed, relative to the ground, is 85 km/h.

Even distances depend on the frame of reference. For example, it may not be very useful to tell you that Yosemite National Park is 300 km away unless I specify 300 km from where. Furthermore, when specifying the motion of an object, it is important to specify not only the speed but also the direction of motion. Often we can specify a direction by using the cardinal points, north, east, south, and west, and by "up" and "down." In physics, we often draw a set of **coordinate axes**, as shown in Fig. 2–3, to represent a frame of reference. We can always place the origin 0, and the directions of the x and y axes, as we like for convenience. The x and y axes are always perpendicular to each other. Objects positioned to the right of the origin of coordinates (0) on the x axis have an x coordinate which we usually choose to be positive; then points to the left of 0 have a negative x coordinate. The position along the y axis is usually considered positive when above 0, and negative when below 0, although the reverse convention can be used if convenient. Any point on the plane can be specified by giving its x and y coordinates. In three dimensions, a z axis perpendicular to the x and y axes is added.

For one-dimensional motion, we often choose the x axis as the line along which the motion takes place. Then the **position** of an object at any moment is given by its x coordinate.

We need to make a distinction between the *distance* an object has traveled, and its **displacement**, which is defined as the *change in position* of the object. That is, displacement is how far the object is from its starting point. To see the distinction between total distance and displacement, imagine a person walking 70 m to the east and then turning around and walking back (west) a distance of 30 m (see Fig. 2–4). The total *distance* traveled is 100 m, but the *displacement* is only 40 m since the person is now only 40 m from the starting point.

Displacement is a quantity that has both magnitude and direction. Such quantities are called **vectors**, and are represented by arrows in diagrams. For example, in Fig. 2–4, the blue arrow represents the displacement whose magnitude is 40 m and whose direction is to the right.

We will deal with vectors more fully in Chapter 3. For now, we deal only with motion in one dimension, along a line, and in this case vectors which point in one direction will have a positive sign, whereas vectors that point in the opposite direction will have a negative sign.

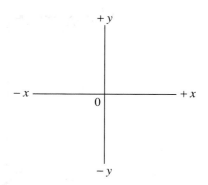

FIGURE 2–3 Standard set of xy coordinate axes.

Displacement

FIGURE 2–4 A person walks 70 m east, then 30 m west. The total distance traveled is 100 m (path is shown in black); but the displacement, shown as a blue arrow, is 40 m to the east.

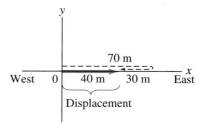

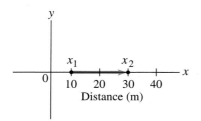

FIGURE 2–5 The arrow represents the displacement $x_2 - x_1$. Distances are in meters.

FIGURE 2–6 For the displacement $\Delta x = x_2 - x_1 = 10.0\,\text{m} - 30.0\,\text{m}$, the displacement vector points to the left.

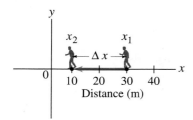

Consider the motion of an object over a particular time interval. Suppose that at some initial moment in time, call it t_1, the object is on the x axis at the point x_1 in the coordinate system shown in Fig. 2–5. At some later time, t_2, suppose the object is at point x_2. The displacement of our object is $x_2 - x_1$, and is represented by the arrow pointing to the right in Fig. 2–5. It is convenient to write

$$\Delta x = x_2 - x_1$$

where the symbol Δ (Greek letter delta) means "change in." Then Δx means "the change in x," which is the displacement. Note that the "change in" any quantity means the final value of that quantity, minus the initial value.

To be concrete, suppose $x_1 = 10.0\,\text{m}$ and $x_2 = 30.0\,\text{m}$. Then

$$\Delta x = x_2 - x_1 = 30.0\,\text{m} - 10.0\,\text{m} = 20.0\,\text{m}.$$

See Fig. 2–5.

Now consider an object moving to the left as shown in Fig. 2–6. Here the object, say a person, starts at $x_1 = 30.0\,\text{m}$ and walks to the left to the point $x_2 = 10.0\,\text{m}$. In this case

$$\Delta x = x_2 - x_1 = 10.0\,\text{m} - 30.0\,\text{m} = -20.0\,\text{m}$$

and the blue arrow representing the vector displacement points to the left. This example illustrates that when dealing with one-dimensional motion, a vector pointing to the right has a positive value, whereas one pointing to the left has a negative value.

2–2 Average Velocity

The most obvious aspect of the motion of a moving object is how fast it is moving—its speed or velocity.

The term "speed" refers to how far an object travels in a given time interval, regardless of direction. If a car travels 240 kilometers (km) in 3 hours, we say its average speed was 80 km/h. In general, the **average speed** of an object is defined as *the total distance traveled along its path divided by the time it takes to travel this distance*:

Average speed

$$\text{average speed} = \frac{\text{distance traveled}}{\text{time elapsed}}. \tag{2–1}$$

The terms velocity and speed are often used interchangeably in ordinary language. But in physics we make a distinction between the two. Speed is simply a positive number, with units. **Velocity**, on the other hand, is used to signify both the *magnitude* (numerical value) of how fast an object is moving and the *direction* in which it is moving. (Velocity is therefore a vector.) There is a second difference between speed and velocity: namely, the **average velocity** is defined in terms of *displacement*, rather than total distance traveled:

Velocity

$$\text{average velocity} = \frac{\text{displacement}}{\text{time elapsed}} = \frac{\text{final position} - \text{initial position}}{\text{time elapsed}}.$$

Average speed and average velocity have the same magnitude when the motion is all in one direction. In other cases, they may differ: recall the walk we described earlier, in Fig. 2–4, where a person walked 70 m east and then 30 m west.

The total distance traveled was 70 m + 30 m = 100 m, but the displacement was 40 m. Suppose this walk took 70 s to complete. Then the average speed was:

$$\frac{\text{distance}}{\text{time}} = \frac{100 \text{ m}}{70 \text{ s}} = 1.4 \text{ m/s}.$$

The magnitude of the average velocity, on the other hand, was:

$$\frac{\text{displacement}}{\text{time}} = \frac{40 \text{ m}}{70 \text{ s}} = 0.57 \text{ m/s}.$$

This difference between the speed and the magnitude of the velocity occurs in some cases, but only for the *average* values.

To discuss one-dimensional motion of an object in general, suppose that at some moment in time, call it t_1, the object is on the x axis at point x_1 in a coordinate system, and at some later time, t_2, suppose it is at point x_2. The elapsed time is $t_2 - t_1$, and during this time interval the displacement of our object was $\Delta x = x_2 - x_1$. Then the average velocity, defined as *the displacement divided by the elapsed time*, can be written

$$\bar{v} = \frac{x_2 - x_1}{t_2 - t_1} = \frac{\Delta x}{\Delta t}, \qquad \text{(2–2)} \qquad \textit{Average velocity}$$

where v stands for velocity and the bar ($^-$) over the v is a standard symbol meaning "average."

For the usual case of the $+x$ axis to the right, note that if x_2 is less than x_1, the object is moving to the left, and then $\Delta x = x_2 - x_1$ is less than zero. The sign of the displacement, and thus of the velocity, indicates the direction: the average velocity is positive for an object moving to the right along the $+x$ axis and negative when the object moves to the left. The direction of the average velocity is always the same as the direction of the displacement.

➡ **PROBLEM SOLVING**

+ or − sign can signify the direction for linear motion

EXAMPLE 2–1 **Runner's average velocity.** The position of a runner as a function of time is plotted as moving along the x axis of a coordinate system. During a 3.00-s time interval, the runner's position changes from $x_1 = 50.0$ m to $x_2 = 30.5$ m, as shown in Fig. 2–7. What was the runner's average velocity?

SOLUTION Average velocity is the displacement divided by the elapsed time. The displacement is $\Delta x = x_2 - x_1 = 30.5 \text{ m} - 50.0 \text{ m} = -19.5 \text{ m}$. The time interval is $\Delta t = 3.00$ s. Therefore the average velocity is

$$\bar{v} = \frac{\Delta x}{\Delta t} = \frac{-19.5 \text{ m}}{3.00 \text{ s}} = -6.50 \text{ m/s}.$$

The displacement and average velocity are negative, which tells us (if we didn't already know it) that the runner is moving to the left along the x axis, as indicated by the arrow in Fig. 2–7. Thus we can say that the runner's average velocity is 6.50 m/s to the left.

FIGURE 2–7 Example 2–1. A person runs from $x_1 = 50.0$ m to $x_2 = 30.5$ m. The displacement is −19.5 m.

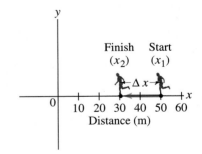

EXAMPLE 2–2 **Distance a cyclist travels.** How far can a cyclist travel in 2.5 h along a straight road if her average speed is 18 km/h?

SOLUTION We want to find the distance traveled, so we use Eq. 2–2 letting Δx be the distance and $\bar{v}$ be the average speed, and then rewrite it as

$$\Delta x = \bar{v} \, \Delta t = (18 \text{ km/h})(2.5 \text{ h}) = 45 \text{ km}.$$

If you drive a car along a straight road for 150 km in 2.0 h, the magnitude of your average velocity is 75 km/h. It is unlikely, though, that you were moving at precisely 75 km/h at every instant. To describe this situation we need the concept of *instantaneous velocity*, which is the velocity at any instant of time. (This is the magnitude that a speedometer is supposed to indicate.) More precisely, the **instantaneous velocity** at any moment is defined as *the average velocity over an infinitesimally short time interval.* That is, Eq. 2–2 is to be evaluated in the limit of Δt becoming extremely small, approaching zero. We can write the definition of instantaneous velocity, v, for one-dimensional motion as

Instantaneous velocity

$$v = \lim_{\Delta t \to 0} \frac{\Delta x}{\Delta t}. \qquad (2\text{–}3)$$

The notation $\lim_{\Delta t \to 0}$ means the ratio $\Delta x / \Delta t$ is to be evaluated in the limit of Δt approaching zero. But we do not simply set $\Delta t = 0$ in this definition, for then Δx would also be zero, and we would have an undefined number. Rather, we are considering the *ratio* $\Delta x / \Delta t$, as a whole. As we let Δt approach zero, Δx approaches zero as well. But the ratio $\Delta x / \Delta t$ approaches some definite value, which is the instantaneous velocity at a given instant.

For instantaneous velocity we use the symbol v, whereas for average velocity we use $\bar{v}$, with a bar. In the rest of this book, when we use the term "velocity" it will refer to instantaneous velocity. When we want to speak of the average velocity, we will make this clear by including the word "average."

Note that the *instantaneous* speed always equals the magnitude of the instantaneous velocity. Why? Because distance and displacement become the same when they become infinitesimally small.

If an object moves at a uniform (that is, constant) velocity over a particular time interval, then its instantaneous velocity at any instant is the same as its average velocity (see Fig. 2–8a). But in many situations this is not the case. For example, a car may start from rest, speed up to 50 km/h, remain at that velocity for a time, then slow down to 20 km/h in a traffic jam, and finally stop at its destination after traveling a total of 15 km in 30 min. This trip is plotted on the graph of Fig. 2–8b. Also shown on the graph is the average velocity (dashed line), which is $\bar{v} = \Delta x / \Delta t = 15 \text{ km}/0.50 \text{ h} = 30 \text{ km/h}$.

To better understand instantaneous velocity, let us consider a graph of the position of a particular particle versus time (x vs. t), as shown in Fig. 2–9. (Note that this is different from showing the "path" of a particle on an x vs. y plot.) The particle is at position x_1 at a time t_1, and at position x_2 at time t_2. P_1 and P_2 represent these two points on the graph. A straight line drawn from point $P_1(x_1, t_1)$ to

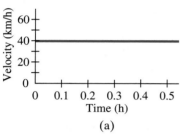

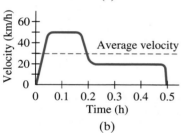

FIGURE 2–8 Velocity of a car as a function of time: (a) at constant velocity; (b) with varying velocity.

FIGURE 2–9
Graph of a particle's position x vs. time t. The slope of the straight line $P_1 P_2$ represents the average velocity of the particle during the time interval $\Delta t = t_2 - t_1$.

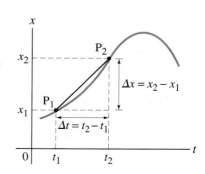

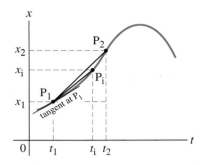

FIGURE 2–10 Same position vs. time curve as in Fig. 2–9, but note that the average velocity over the time interval $t_i - t_1$ (which is the slope of P_1P_i) is less than the average velocity over the time interval $t_2 - t_1$. The slope of the thin line tangent to the curve at point P_1 equals the instantaneous velocity at time t_1.

point $P_2(x_2, t_2)$ forms the hypotenuse of a right triangle whose sides are Δx and Δt. The ratio $\Delta x/\Delta t$ is the **slope** of the straight line P_1P_2. But $\Delta x/\Delta t$ is also the average velocity of the particle during the time interval $\Delta t = t_2 - t_1$. Therefore, we conclude that the average velocity of an object during any time interval $\Delta t = t_2 - t_1$ is equal to the slope of the straight line (or *chord*) connecting the two points (x_1, t_1) and (x_2, t_2) on an x vs. t graph.

Slope of chord joining 2 points on x vs. t graph equals average velocity

    Consider now a time t_i, intermediate between t_1 and t_2, at which time the particle is at x_i (Fig. 2–10). The slope of the straight line P_1P_i is less than the slope of P_1P_2 in this case. Thus the average velocity during the time interval $t_i - t_1$ is less than that during the time interval $t_2 - t_1$.

    Now let us imagine that we take the point P_i in Fig. 2–10 to be closer and closer to point P_1. That is, we let the internal $t_i - t_1$, which we now call Δt, to become smaller and smaller. The slope of the line connecting the two points becomes closer and closer to the slope of a line tangent to the curve at point P_1. As we take Δt smaller and smaller, the average velocity (equal to the slope of the chord) approaches the slope of the tangent at point P_1. The definition of the instantaneous velocity (Eq. 2–3) is the limiting value of the average velocity as Δt approaches zero. Thus the *instantaneous velocity equals the slope of the tangent to the curve* at that point (which we can simply call "the slope of the curve" at that point).

Slope of tangent to x vs. t curve equals instantaneous velocity

    In Eq. 2–3, the limit as $\Delta t \to 0$ is written in calculus notation as dx/dt and is called the *derivative* of x with respect to t. Thus we can write Eq. 2–3 in calculus notation as:

$$v = \lim_{\Delta t \to 0} \frac{\Delta x}{\Delta t} = \frac{dx}{dt}. \qquad (2\text{–}4)$$

Instantaneous velocity

This equation is the definition of instantaneous velocity for one-dimensional motion.

    Because the velocity at any instant equals the slope of the tangent to the x vs. t graph at that instant, we can obtain the velocity at any instant from such a graph. For example, in Fig. 2–11 (which shows the same curve as in Figs. 2–9 and 2–10), as our object moves from x_1 to x_2, the slope continually increases, so the velocity is increasing. For times after t_2, however, the slope begins to decrease and in fact reaches zero (so $v = 0$) where x has its maximum value, at point P_3 in Fig. 2–11. Beyond this point, the slope is negative, as for point P_4. The velocity is therefore negative, which makes sense since x is now decreasing—the particle is moving toward decreasing values of x, to the left on a standard xy plot.

    If an object moves with constant velocity over a particular time interval, its instantaneous velocity is equal to its average velocity. The graph of x vs. t in this case will be a straight line whose slope equals the velocity. The curve of Fig. 2–9 has no straight sections, so there are no time intervals when the velocity is constant.

FIGURE 2–11 Same x vs. t curve as in Figs. 2–9 and 2–10, but here showing the slope at four different points: At P_3, the slope is zero, so $v = 0$. At P_4 the slope is negative, so $v < 0$.

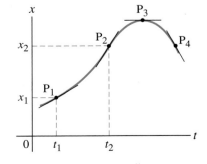

EXAMPLE 2–3 **Given x as a function of t.** A jet engine moves along an experimental track (which we call the x-axis) as shown in Fig. 2–12a. We will treat the engine as if it were a particle. Its position as a function of time is given by the equation $x = At^2 + B$, where $A = 2.10 \text{ m/s}^2$ and $B = 2.80 \text{ m}$, and this equation is plotted in Fig. 2–12b. (a) Determine the displacement of the engine during the time interval from $t_1 = 3.00 \text{ s}$ to $t_2 = 5.00 \text{ s}$. (b) Determine the average velocity during this time interval. (c) Determine the magnitude of the instantaneous velocity at $t = 5.00 \text{ s}$.

SOLUTION (a) At $t_1 = 3.00 \text{ s}$, the position (point P_1 in Fig. 2–12b) is

$$x_1 = At_1^2 + B = (2.10 \text{ m/s}^2)(3.00 \text{ s})^2 + 2.80 \text{ m} = 21.7 \text{ m}.$$

At $t_2 = 5.00 \text{ s}$, the position (P_2 in Fig. 2–12b) is

$$x_2 = (2.10 \text{ m/s}^2)(5.00 \text{ s})^2 + 2.80 \text{ m} = 55.3 \text{ m}.$$

The displacement is thus

$$x_2 - x_1 = 55.3 \text{ m} - 21.7 \text{ m} = 33.6 \text{ m}.$$

(b) The magnitude of the average velocity can then be calculated as

$$\bar{v} = \frac{x_2 - x_1}{t_2 - t_1} = \frac{33.6 \text{ m}}{2.00 \text{ s}} = 16.8 \text{ m/s}.$$

This equals the slope of the straight line joining points P_1 and P_2 shown in Fig. 2–12b.

(c) The instantaneous velocity at $t = t_2 = 5.00 \text{ s}$ equals the slope of the tangent to the curve at point P_2 shown in Fig. 2–12b, and we could measure this slope off the graph to obtain v_2. We can calculate v more precisely, and for any time t, using the given formula

$$x = At^2 + B,$$

which is the engine's position x at time t. Using the calculus formulas for derivatives

$$\frac{d}{dt}(Ct^n) = nCt^{n-1} \quad \text{and} \quad \frac{dC}{dt} = 0,$$

where C is any constant, then

$$v = \frac{dx}{dt} = \frac{d}{dt}(At^2 + B) = 2At.$$

We are given $A = 2.10 \text{ m/s}^2$, so for $t = t_2 = 5.00 \text{ s}$,

$$v_2 = 2At = 2(2.10 \text{ m/s}^2)(5.00 \text{ s}) = 21.0 \text{ m/s}.$$

FIGURE 2–12 Example 2–3. (a) Engine traveling on a straight track. (b) Graph of x vs. t: $x = At^2 + B$.

(a)

(b)

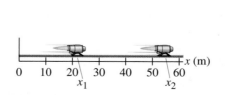

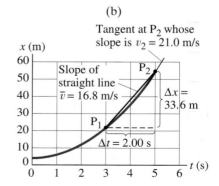

2–4 | Acceleration

An object whose velocity is changing is said to be accelerating. A car whose velocity increases in magnitude from zero to 80 km/h is accelerating. That is, acceleration specifies how rapidly the velocity of an object is changing.

Average Acceleration

Average acceleration is defined as the change in velocity divided by the time taken to make this change:

$$\text{average acceleration} = \frac{\text{change of velocity}}{\text{time elapsed}}.$$

In symbols, the average acceleration $\bar{a}$, over a time interval $\Delta t = t_2 - t_1$ during which the velocity changes by $\Delta v = v_2 - v_1$, is defined as

$$\bar{a} = \frac{v_2 - v_1}{t_2 - t_1} = \frac{\Delta v}{\Delta t}. \qquad (2\text{–}5)$$

Average acceleration

Acceleration is also a vector, but for one-dimensional motion, we need only use a plus or minus sign to indicate direction relative to a chosen coordinate system.

EXAMPLE 2–4 **Average acceleration.** A car accelerates along a straight road from rest to 75 km/h in 5.0 s, Fig. 2–13. What is the magnitude of its average acceleration?

SOLUTION The car starts from rest, so $v_1 = 0$. The final velocity is $v_2 = 75$ km/h. Then from Eq. 2–5, the average acceleration is

$$\bar{a} = \frac{75 \text{ km/h} - 0 \text{ km/h}}{5.0 \text{ s}} = 15 \frac{\text{km/h}}{\text{s}}.$$

This is read as "fifteen kilometers per hour per second" and means that, on average, the velocity changed by 15 km/h during each second. That is, assuming the acceleration was constant, during the first second the car's velocity increased from zero to 15 km/h. During the next second its velocity increased by another 15 km/h up to 30 km/h, and so on, Fig. 2–13. (Of course, if the instantaneous acceleration was not constant, these numbers could be different.)

$t_1 = 0$
$v_1 = 0$

Acceleration
$= 15 \dfrac{\text{km/h}}{\text{s}}$

at $t = 1.0$ s
$v = 15$ km/h

at $t = 2.0$ s
$v = 30$ km/h

at $t = t_2 = 5.0$ s
$v = v_2 = 75$ km/h

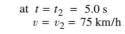

FIGURE 2–13 Example 2–4. The car is shown at the start with $v_1 = 0$ at $t_1 = 0$. It is shown three more times, at $t = 1.0$ s, at $t = 2.0$ s, and at $t = t_2 = 5.0$ s. We assume the acceleration is constant and equals 15 km/h/s. The green arrows represent the velocity vectors; the length of each represents the magnitude of the velocity at that moment. The acceleration vector is the orange arrow. Distances are not to scale.

Careful:
Do not confuse
velocity with acceleration

Note carefully that *acceleration tells us how fast the velocity changes*, whereas *velocity tells us how fast the position changes*. In this last Example, the calculated acceleration contained two different time units: hours and seconds. We usually prefer to use only seconds. To do so we can change km/h to m/s (see Section 1–5, and Example 1–4):

$$75 \text{ km/h} = \left(75 \frac{\text{km}}{\text{h}}\right)\left(\frac{1000 \text{ m}}{1 \text{ km}}\right)\left(\frac{1 \text{ h}}{3600 \text{ s}}\right) = 21 \text{ m/s}.$$

Then we get

$$\bar{a} = \frac{21 \text{ m/s} - 0.0 \text{ m/s}}{5.0 \text{ s}} = 4.2 \frac{\text{m/s}}{\text{s}} = 4.2 \frac{\text{m}}{\text{s}^2}.$$

We almost always write these units as m/s² (meters per second squared). According to the above calculation, the velocity in Example 2–4 (Fig. 2–13) changed on the average by 4.2 m/s during each second, for a total change of 21 m/s over the 5.0 s.

CONCEPTUAL EXAMPLE 2–5 | **Velocity and acceleration.** (*a*) If the velocity of an object is zero, does it mean that the acceleration is zero? (*b*) If the acceleration is zero, does it mean that the velocity is zero?

RESPONSE A zero velocity does not necessarily mean that the acceleration is zero, nor does a zero acceleration mean that the velocity is zero. (*a*) For example, when you put your foot on the gas pedal of your car which is at rest, the velocity starts from zero but the acceleration is not zero since the velocity of the car changes. (How else could your car start forward if its velocity weren't changing— that is, if the acceleration were zero?) (*b*) As you cruise along a straight highway at a constant velocity of 100 km/h, your acceleration is zero.

at $t_1 = 0$
$v_1 = 15.0$ m/s

Acceleration
$= -2.0$ m/s²

at $t_2 = 5.0$ s
$v_2 = 5.0$ m/s

FIGURE 2–14 Example 2–6, showing the position of the car at times t_1 and t_2, as well as the car's velocity represented by the green arrows. The acceleration vector (orange) points to the left.

FIGURE 2–15 The same car as in Example 2–6, but now moving to the left and decelerating. The acceleration is

$$a = \frac{v_2 - v_1}{\Delta t}$$

$$= \frac{-5.0 \text{ m/s} - (-15.0 \text{ m/s})}{5.0 \text{ s}}$$

$$= \frac{-5.0 \text{ m/s} + 15.0 \text{ m/s}}{5.0 \text{ s}} = +2.0 \text{ m/s}.$$

$v_2 = -5.0$ m/s $v_1 = -15.0$ m/s

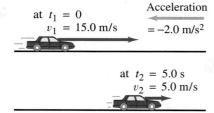

a

EXAMPLE 2–6 **Car slowing down.** An automobile is moving to the right along a straight highway, which we choose to be the positive x axis (Fig. 2–14), and the driver puts on the brakes. If the initial velocity is $v_1 = 15.0$ m/s and it takes 5.0 s to slow down to $v_2 = 5.0$ m/s, what was the car's average acceleration?

SOLUTION The average acceleration is equal to the change in velocity divided by the elapsed time, Eq. 2–5. Let us call the initial time $t_1 = 0$; then $t_2 = 5.0$ s. (Note that our choice of $t_1 = 0$ doesn't affect the calculation of $\bar{a}$ because only $\Delta t = t_2 - t_1$ appears in Eq. 2–5.) Then

$$\bar{a} = \frac{5.0 \text{ m/s} - 15.0 \text{ m/s}}{5.0 \text{ s}} = -2.0 \text{ m/s}^2.$$

The negative sign appears because the final velocity is less than the initial velocity. In this case the direction of the acceleration is to the left (in the negative x direction)—even though the velocity is always pointing to the right. We say that the acceleration is 2.0 m/s² to the left, and it is shown in Fig. 2–14 as an orange arrow.

When an object is slowing down, we sometimes say it is decelerating. But be careful: deceleration does *not* mean that the acceleration is necessarily negative. For an object moving to the right along the positive x axis and slowing down (as in Fig. 2–14), the acceleration *is* negative. But the same car moving to the left (decreasing x) and slowing down has positive acceleration that points to the right, as shown in Fig. 2–15. We have a deceleration whenever the velocity and acceleration point in opposite directions.

Instantaneous Acceleration

The **instantaneous acceleration**, a, is defined as the *limiting value of the average acceleration as we let Δt approach zero*:

$$a = \lim_{\Delta t \to 0} \frac{\Delta v}{\Delta t} = \frac{dv}{dt}. \qquad (2\text{--}6)$$

This limit, dv/dt, is the derivative of v with respect to t. We will use the term "acceleration" to refer to the instantaneous value. If we want to discuss the average acceleration, we will always include the word "average."

If we draw a graph of the velocity, v, vs. time, t, as shown in Fig. 2–16, then the average acceleration over a time interval $\Delta t = t_2 - t_1$ is represented by the slope of the straight line connecting the two points P_1 and P_2 as shown. [Compare this to the position vs. time graph of Fig. 2–9 for which the slope of the straight line represents the average velocity.] The instantaneous acceleration at any time, say t_1, is the slope of the tangent to the v vs. t curve at that time, which is also shown in Fig. 2–16. Let us use this fact for the situation graphed in Fig. 2–16; as we go from time t_1 to time t_2 the velocity continually increases but the acceleration (the rate at which the velocity changes) is decreasing since the slope of the curve is decreasing.

EXAMPLE 2–7 **Acceleration given $x(t)$.** A particle is moving in a straight line so that its position is given by the relation $x = (2.10 \text{ m/s}^2)t^2 + (2.80 \text{ m})$, as in Example 2–3. Calculate (a) its average acceleration during the time interval from $t_1 = 3.00 \text{ s}$ to $t_2 = 5.00 \text{ s}$, and (b) its instantaneous acceleration as a function of time.

SOLUTION (a) We saw in Example 2–3c that the velocity at any time t is $v = dx/dt = (4.20 \text{ m/s}^2)t$. Therefore, at $t_1 = 3.00 \text{ s}$, $v_1 = (4.20 \text{ m/s}^2)(3.00 \text{ s}) = 12.6 \text{ m/s}$ and at $t_2 = 5.00 \text{ s}$, $v_2 = 21.0 \text{ m/s}$. Therefore,

$$\bar{a} = \frac{21.0 \text{ m/s} - 12.6 \text{ m/s}}{5.00 \text{ s} - 3.00 \text{ s}} = 4.20 \text{ m/s}^2.$$

(b) With $v = (4.20 \text{ m/s}^2)t$, the instantaneous acceleration is

$$a = \frac{dv}{dt} = \frac{d}{dt}\left[(4.20 \text{ m/s}^2)t\right] = 4.20 \text{ m/s}^2.$$

The acceleration in this case is constant; it does not depend on time. Figure 2–17 shows graphs of (a) x vs. t (the same as Fig. 2–12b), (b) v vs. t, which is linearly increasing as calculated above, and (c) a vs. t, which is a horizontal straight line because a = constant.

Like velocity, acceleration is a rate. The velocity of an object is the rate at which its displacement changes with time; its acceleration, on the other hand, is the rate at which its velocity changes with time. In a sense, acceleration is a "rate of a rate." This can be expressed in equation form as follows: since $a = dv/dt$ and $v = dx/dt$, then

$$a = \frac{dv}{dt} = \frac{d}{dt}\left(\frac{dx}{dt}\right) = \frac{d^2x}{dt^2}.$$

Here d^2x/dt^2 is the *second derivative* of x with respect to time: we first take the derivative of x with respect to time (dx/dt), and then we again take the derivative with respect to time, $(d/dt)(dx/dt)$, to get the acceleration.

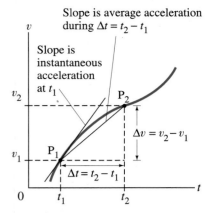

FIGURE 2–16 A graph of velocity v vs. time t. The average acceleration over a time interval $\Delta t = t_2 - t_1$ is the slope of the straight line $P_1 P_2$: $\bar{a} = \Delta v / \Delta t$. The instantaneous acceleration at time t_1 is the slope of the v vs. t curve at that instant.

FIGURE 2–17 Graphs of (a) x vs. t, (b) v vs. t, and (c) a vs. t for the motion $x = At^2 + B$. Note that v increases linearly with t and that the acceleration a is constant. Also, v is the slope of the x vs. t curve, whereas a is the slope of the v vs. t curve.

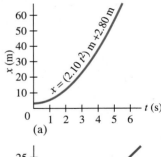

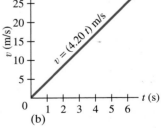

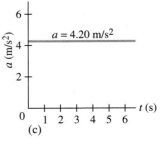

2–5 | Motion at Constant Acceleration

Many practical situations occur in which the acceleration is constant or close enough that we can assume it is constant. The acceleration due to gravity near the Earth's surface is an example. We now treat this situation when the magnitude of the acceleration is constant and the motion is in a straight line. In this case, the instantaneous and average accelerations are equal.

Let a = constant

To simplify our notation, let us take the initial time in any discussion to be zero: $t_1 = 0$. (This is effectively starting a stopwatch at t_1.) We can then let $t_2 = t$ be the elapsed time. The initial position (x_1) and initial velocity (v_1) of an object will now be represented by x_0 and v_0; and at time t the position and velocity will be called x and v (rather than x_2 and v_2). The average velocity during the time t will be (from Eq. 2–2)

$x(t=0) = x_0$
$v(t=0) = v_0$

$$\bar{v} = \frac{x - x_0}{t - t_0} = \frac{x - x_0}{t}$$

since $t_0 = 0$. And the acceleration, which is assumed constant in time, will be (from Eq. 2–5)

$$a = \frac{v - v_0}{t}.$$

A common problem is to determine the velocity of an object after a certain time, given its acceleration. We can solve such problems by solving for v in the last equation to obtain:

v related to a and t
(a = constant)

$$v = v_0 + at. \qquad \text{[constant acceleration]} \quad \textbf{(2–7)}$$

For example, it may be known that the acceleration of a particular motorcycle is $4.0 \, \text{m/s}^2$, and we wish to determine how fast it will be going after, say, 6.0 s. Assuming it starts from rest $(v_0 = 0)$, after 6.0 s the velocity will be $v = at = (4.0 \, \text{m/s}^2)(6.0 \, \text{s}) = 24 \, \text{m/s}$.

Next, let us see how to calculate the position of an object after a time t when it is undergoing constant acceleration. The definition of average velocity (Eq. 2–2) is $\bar{v} = (x - x_0)/t$, which we can rewrite as

$$x = x_0 + \bar{v}t. \qquad \textbf{(2–8)}$$

Because the velocity increases at a uniform rate, the average velocity, $\bar{v}$, will be midway between the initial and final velocities:

Average velocity
(when acceleration is constant)

$$\bar{v} = \frac{v_0 + v}{2}. \qquad \text{[constant acceleration]} \quad \textbf{(2–9)}$$

(Careful: Eq. 2–9 is not necessarily valid if the acceleration is not constant.) We combine the last two equations with Eq. 2–7 and find

$$x = x_0 + \bar{v}t$$

$$= x_0 + \left(\frac{v_0 + v}{2}\right)t$$

$$= x_0 + \left(\frac{v_0 + v_0 + at}{2}\right)t$$

or

x related to a and t
(a = constant)

$$x = x_0 + v_0 t + \tfrac{1}{2}at^2. \qquad \text{[constant acceleration]} \quad \textbf{(2–10)}$$

Equations 2–7, 2–9, and 2–10 are three of the four most useful equations for motion at constant acceleration. We now derive the fourth equation, which is useful in situations where the time t is not known. We begin with Eq. 2–8 and substitute in Eq. 2–9:

$$x = x_0 + \bar{v}t = x_0 + \left(\frac{v + v_0}{2}\right)t.$$

Next we solve Eq. 2–7 for t, obtaining

$$t = \frac{v - v_0}{a},$$

and substituting this into the equation above we have

$$x = x_0 + \left(\frac{v + v_0}{2}\right)\left(\frac{v - v_0}{a}\right) = x_0 + \frac{v^2 - v_0^2}{2a}.$$

We solve this for v^2 and obtain

$$v^2 = v_0^2 + 2a(x - x_0), \qquad \text{[constant acceleration]} \quad \textbf{(2–11)}$$

v related to a and x
(a = constant)

which is the useful equation we sought.

   We now have four equations relating position, velocity, acceleration, and time, when the acceleration a is constant. We collect them here in one place for further reference (the tan background screen is to emphasize their usefulness):

$$v = v_0 + at \qquad\qquad\qquad \text{[a = constant]} \quad \textbf{(2–12a)}$$
$$x = x_0 + v_0 t + \tfrac{1}{2}at^2 \qquad \text{[a = constant]} \quad \textbf{(2–12b)}$$
$$v^2 = v_0^2 + 2a(x - x_0) \qquad \text{[a = constant]} \quad \textbf{(2–12c)}$$
$$\bar{v} = \frac{v + v_0}{2}. \qquad\qquad\qquad \text{[a = constant]} \quad \textbf{(2–12d)}$$

Kinematic equations

for constant acceleration

(we'll use them a lot)

These useful equations are not valid unless a is a constant. In many cases we can set $x_0 = 0$, and this simplifies the above equations a bit. Note that x represents position, not distance, and $x - x_0$ is the displacement.

EXAMPLE 2–8 **Runway design.** You are designing an airport for small planes. One kind of airplane that might use this airfield must reach a speed before takeoff of at least 27.8 m/s (100 km/h), and can accelerate at 2.00 m/s². (*a*) If the runway is 150 m long, can this airplane reach the proper speed to take off? (*b*) If not, what minimum length must the runway have?

➡ **PHYSICS APPLIED**

Engineering design

SOLUTION (*a*) We are given the airplane's acceleration $(a = 2.00 \text{ m/s}^2)$, and we know the plane can travel a distance of 150 m. We want to find its velocity, to determine if it will be at least 27.8 m/s. We want to find v when we are given:

Known	Wanted
$x_0 = 0$	v
$v_0 = 0$	
$x = 150 \text{ m}$	
$a = 2.00 \text{ m/s}^2$	

Of the above four equations, Eq. 2–12c will give us v when we know v_0, a, x, and x_0:

$$v^2 = v_0^2 + 2a(x - x_0)$$
$$= 0 + 2(2.0 \text{ m/s}^2)(150 \text{ m}) = 600 \text{ m}^2/\text{s}^2$$
$$v = \sqrt{600 \text{ m}^2/\text{s}^2} = 24.5 \text{ m/s}.$$

➡ **PROBLEM SOLVING**

Equations 2–12 are valid only when the acceleration is constant, which we assume in this Example

This runway length is *not* sufficient.
(*b*) Now we want $(x - x_0)$ given $v = 27.8$ m/s and $a = 2.00$ m/s². So we use Eq. 2–12c, rewritten as

$$(x - x_0) = \frac{v^2 - v_0^2}{2a} = \frac{(27.8 \text{ m/s})^2 - 0}{2(2.0 \text{ m/s}^2)} = 193 \text{ m}.$$

2–6 Solving Problems

In this Section we do several more worked-out Examples of objects moving with constant acceleration. First let us consider generally how to approach problem solving. It is important to note that physics is *not* a collection of equations to be memorized. (In fact, rather than memorizing the very useful Eqs. 2–12, it is better to understand how to derive them from the definitions of velocity and acceleration as we did above.) Simply searching for an equation that might work can be disastrous and can lead you to a wrong result, and will surely not help you understand physics. A better approach to problem solving is to use the following (rough) procedure, which we put in a special "box":

PROBLEM SOLVING

1. **Read** *and reread* the whole problem carefully before trying to solve it.
2. **Draw** a **diagram** or picture of the situation, with coordinate axes wherever applicable. [You can choose to place the origin of coordinates and the axes wherever you like, so as to make your calculations easier. You also choose which direction is positive and which is negative. Usually we choose the x axis to the right as positive, but you could choose positive to the left.]
3. **Write down** what quantities are "known" or "given," and then what you *want* to know.
4. Think about which principles of physics apply in this problem. Then plan an approach:
5. Consider which equations (and/or definitions) relate the quantities involved. Before using equations, be sure their **range of validity** includes your problem (for example, Eqs. 2–12 are valid only when the acceleration is constant). If you find an applicable equation that involves only known quantities and one desired unknown, **solve** the equation algebraically for the unknown. In many instances several sequential calcu-

lations, or a combination of equations, may be needed. It is often advantageous to solve algebraically for the desired unknown before putting in numerical values.
6. Carry out the **calculation** if it is a numerical problem. Keep one or two extra digits during the calculations, but round off the final answers to the correct number of significant figures (Section 1–3).
7. Think carefully about the result you obtain: Is it **reasonable**? Does it make sense according to your own intuition and experience? A good check is to do a rough **estimate** using only powers of ten, as discussed in Section 1–6. Often it is preferable to do a rough estimate at the *start* of a numerical problem because it can help you focus your attention on finding a path toward a solution.
8. A very important aspect of doing problems is keeping track of **units**. An equals sign implies the units on each side must be the same, just as the numbers must. If the units do not balance, a mistake has been made. This can serve as a **check** on your solution (but it only tells you if you're wrong, not if you're right). And: always use a consistent set of units.

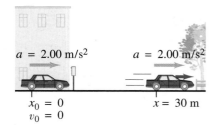

FIGURE 2–18 Example 2–9.

$x_0 = 0$
$v_0 = 0$

$x = 30\ \text{m}$

Known	Wanted
$x_0 = 0$	t
$x = 30.0\ \text{m}$	
$a = 2.00\ \text{m/s}^2$	
$v_0 = 0$	

EXAMPLE 2–9 **Acceleration of car.**   How long does it take a car to cross a 30.0-m-wide intersection after the light turns green, if it accelerates from rest at a constant 2.00 m/s²?

SOLUTION   First we make a sketch, Fig. 2–18. Next we make a table, shown in the margin, choosing $x_0 = 0$ and assume the car moves to the right along the positive x axis, and noting that "starting from rest" means $v = 0$ at $t = 0$; that is, $v_0 = 0$. Since a is constant, we can use Eqs. 2–12. Equation 2–12b is perfect since the only unknown quantity is t, which is what we are seeking. Setting $v_0 = 0$ and $x_0 = 0$, we can solve Eq. 2–12b, $x = \frac{1}{2}at^2$, for t:

$$t = \sqrt{\frac{2x}{a}} = \sqrt{\frac{2(30.0\ \text{m})}{2.00\ \text{m/s}^2}} = 5.48\ \text{s}.$$

We can check the reasonableness of our answer by calculating the final velocity v: $v = at = (2.00\ \text{m/s}^2)(5.48\ \text{s}) = 10.96\ \text{m/s}$, and then finding $x = x_0 + \bar{v}t = 0 + \frac{1}{2}(10.96\ \text{m/s} + 0)(5.48\ \text{s}) = 30.0\ \text{m}$, which is our given distance.

EXAMPLE 2–10 **ESTIMATE** **Braking distances.** Estimate the minimum stopping distances for a car, which are important for traffic safety and traffic design. The problem is best dealt with in two parts: (1) the time between the decision to apply the brakes and their actual application (the "reaction time"), during which we assume $a = 0$; and (2) the actual braking period when the vehicle slows down ($a \neq 0$). The stopping distance depends on the reaction time of the driver, the initial speed of the car (the final speed is zero), and the acceleration of the car. For a dry road and good tires, good brakes can decelerate a car at a rate of about 5 m/s^2 to 8 m/s^2. Calculate the total stopping distance for an initial velocity of 50 km/h ($14 \text{ m/s} \approx 31$ mph) and assume the acceleration of the car is -6.0 m/s^2 (the minus sign appears because the velocity is taken to be in the positive x direction and its magnitude is decreasing). Reaction time for normal drivers varies from perhaps 0.3 s to about 1.0 s; take it to be 0.50 s.

SOLUTION The car is moving to the right in the positive x direction. We take $x_0 = 0$ for the first part of the problem, in which the car travels at a constant speed of 14 m/s during the time the driver is reacting (0.50 s). See Fig. 2–19 and the Table in the margin. To find x we can use Eq. 2–12b:

$$x = v_0 t + 0 = (14 \text{ m/s})(0.50 \text{ s}) = 7.0 \text{ m}.$$

The car travels 7.0 m during the driver's reaction time, until the moment the brakes are applied.

Now for the second part, during which the brakes are applied and the car is brought to rest. We now take $x_0 = 7.0$ m (result of the first part): see Table to the right. Equation 2–12a doesn't contain x; Eq. 2–12b contains x but also the unknown t. Equation 2–12c, $v^2 - v_0^2 = 2a(x - x_0)$, is what we want; we solve for x, after setting $x_0 = 7.0$ m:

$$x = x_0 + \frac{v^2 - v_0^2}{2a}$$

$$= 7.0 \text{ m} + \frac{0 - (14 \text{ m/s})^2}{2(-6.0 \text{ m/s}^2)} = 7.0 \text{ m} + \frac{-196 \text{ m}^2/\text{s}^2}{-12 \text{ m/s}^2}$$

$$= 7.0 \text{ m} + 16 \text{ m} = 23 \text{ m}.$$

The car traveled 7.0 m while the driver was reacting and another 16 m during the braking period before coming to a stop. The total distance traveled was then 23 m. See Fig. 2–20 for graphs of v vs. t and x vs. t.

Under wet or icy conditions, the value of a may be only one third the value for a dry road since the brakes cannot be applied as hard without skidding, and hence stopping distances are much greater. Note also that the stopping distance after you hit the brakes increases with the *square* of the speed, not just linearly with speed: If you are traveling twice as fast, it takes four times the distance.

➡ P H Y S I C S A P P L I E D
Braking distances

Part 1: Reaction time

Known	Wanted
$t = 0.50$ s	x
$v_0 = 14$ m/s	
$v = 14$ m/s	
$a = 0$	
$x_0 = 0$	

Part 2: Braking

Known	Wanted
$x_0 = 7.0$ m	x
$v_0 = 14$ m/s	
$v = 0$	
$a = -6.0 \text{ m/s}^2$	

FIGURE 2–19 Example 2–10: stopping distance for a braking car.

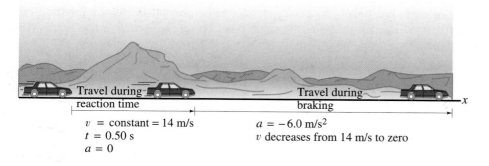

FIGURE 2–20 Graphs of v vs. t and x vs. t for Example 2–10.

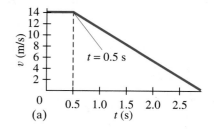

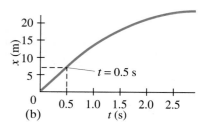

SECTION 2–6 Solving Problems **29**

EXAMPLE 2–11 ESTIMATE **Air bags.** Suppose you want to design an air-bag system that can protect the driver in a head-on collision at a speed of 100 km/h (60 mph). Estimate how fast the air bag must inflate to effectively protect the driver. Assume the car crumples upon impact over a distance of about 1 m. How does the use of a seat belt help the driver?

SOLUTION The car decelerates from 100 km/h to zero in a very short time and a very short distance (1 m). Noting that 100 km/h = 100×10^3 m/3600 s = 28 m/s, we can get the acceleration from Eq. 2–12c:

$$a = -\frac{v_0^2}{2x} = -\frac{(28 \text{ m/s})^2}{2.0 \text{ m}} = -390 \text{ m/s}^2.$$

This enormous acceleration takes place in a time given by (Eq. 2–12a):

$$t = \frac{v - v_0}{a} = \frac{0 - 28 \text{ m/s}}{-390 \text{ m/s}^2} = 0.07 \text{ s}.$$

To be effective, the air bag would need to inflate faster than this.

What does the air bag do? First, it spreads the force over a larger area of the chest. This is better than being punctured by the steering column. Also, the pressure in the bag is controlled to minimize the head's maximum deceleration. The seat belt keeps the person in the correct position against the expanding air bag.

* Two Moving Objects

We now do an Example that is a little more involved, and put it in this separate subsection marked with a * to suggest that it could be considered optional, especially on a first reading.

EXAMPLE 2–12 ESTIMATE **Catching a speeder.** A car speeding at 80 mi/h passes a still police car which immediately takes off in hot pursuit. Using simple assumptions, such as that the speeder continues at constant speed, estimate how long it takes the police car to overtake the speeder. Then estimate the police car's speed at that moment and decide if the assumptions were reasonable.

SOLUTION When the police car takes off, it accelerates, and the simplest assumption is that its acceleration is constant. This may not be reasonable, but let's see what happens. We can estimate the acceleration if we have noticed automobile ads, which claim cars can accelerate from rest to 60 mi/h in 6 seconds. So the average acceleration of the police car could be approximately

$$a_P = \frac{60 \text{ mi/h}}{6 \text{ s}} = 10 \frac{\text{mi}}{\text{h·s}}.$$

We should perhaps change units to proper SI units, but let's try to save time and work with these mixed units. We need to set up the kinematic equations to determine the unknown quantities, and since there are two moving objects, we need two separate sets of equations. We denote the speeding car's position by x_S and the police car's position by x_P. Because we are interested in solving for the time when the two vehicles arrive at the same position on the road, we use Eq. 2–12b for each car:

$$x_S = v_{0S}t = \left(80 \frac{\text{mi}}{\text{h}}\right)t$$

$$x_P = v_{0P}t + \tfrac{1}{2}a_P t^2 = \tfrac{1}{2}\left(10 \frac{\text{mi}}{\text{h·s}}\right)t^2$$

where we have set $v_{0P} = 0$ and $a_S = 0$ (speeder assumed to move at constant speed). We want the time when the cars meet, so we set $x_S = x_P$ and solve for t:

$$\left(80 \frac{\text{mi}}{\text{h}}\right)t = \left(5 \frac{\text{mi}}{\text{h} \cdot \text{s}}\right)t^2.$$

The solutions are

$$t = 0 \quad \text{and} \quad t = \frac{80 \dfrac{\text{mi}}{\text{h}}}{5 \dfrac{\text{mi}}{\text{h} \cdot \text{s}}} = 16 \, \text{s}.$$

The first solution corresponds to the moment the speeder passed the police car. The second solution tells us when the police car catches up to the speeder, 16 s later. This is our answer, but is it reasonable? The police car's speed at $t = 16 \, \text{s}$ is

$$v_P = v_{0P} + a_P t = 0 + \left(10 \frac{\text{mi}}{\text{h} \cdot \text{s}}\right)(16 \, \text{s}) = 160 \, \text{mi/h}.$$

Not reasonable, and highly dangerous. More reasonable is to give up the assumption of constant acceleration. The police car surely cannot maintain constant acceleration at those speeds. Also, the speeder, if a reasonable person, would slow down upon hearing the police siren. Figure 2–21 shows (*a*) *x* vs. *t* and (*b*) *v* vs. *t* graphs, based on the original assumption of $a_P = $ constant, whereas (*c*) shows *v* vs. *t* for more reasonable assumptions.

Careful:
Initial assumptions need to be checked out for reasonableness

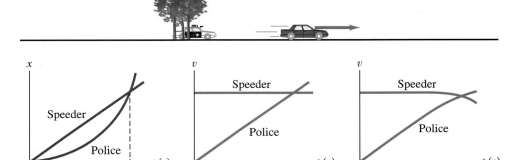

FIGURE 2–21   Example 2–12.

(a)   (b)   (c)

Falling Objects

One of the most common examples of uniformly accelerated motion is that of an object allowed to fall freely near the Earth's surface. That a falling object is accelerating may not be obvious at first. And beware of thinking, as was widely believed until the time of Galileo (Fig. 2–22), that heavier objects fall faster than lighter objects and that the speed of fall is proportional to how heavy the object is.

Galileo's analysis made use of his new and creative technique of imagining what would happen in idealized (simplified) cases. For free fall, he postulated that all objects would fall with the *same constant acceleration* in the absence of air or other resistance. He showed that this postulate predicts that for an object falling from rest, the distance traveled will be proportional to the square of the time (Fig. 2–23); that is, $d \propto t^2$. We can see this from Eq. 2–12b, but Galileo was the first to derive this mathematical relation.

To support his claim that the speed of falling objects increases as they fall, Galileo made use of a clever argument: a heavy stone dropped from a height of

FIGURE 2–22
Galileo Galilei (1564–1642).

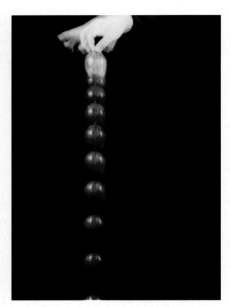

FIGURE 2–23 Multiflash photograph of a falling apple, photographed at equal time intervals. Note that the apple falls farther during each successive time interval, which means it is accelerating.

FIGURE 2–25 A rock and a feather are dropped simultaneously (a) in air, (b) in a vacuum.

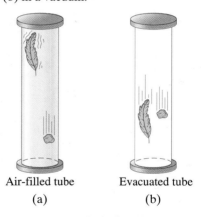

Air-filled tube
(a)

Evacuated tube
(b)

Acceleration due to gravity

(a)   (b)

FIGURE 2–24 (a) A ball and a light piece of paper are dropped at the same time. (b) Repeated, with the paper wadded up.

2 m will drive a stake into the ground much further than will the same stone dropped from a height of only 0.2 m. Clearly, the stone must be moving faster when it has fallen from a greater height.

As we saw, Galileo also claimed that *all* objects, light or heavy, fall with the *same* acceleration, at least in the absence of air. If you hold a piece of paper horizontally in one hand and a heavier object—say, a baseball—in the other, and release them at the same time as in Fig. 2–24a, the heavier object will reach the ground first. But if you repeat the experiment, this time crumpling the paper into a small wad (see Fig. 2–24b), you will find that the two objects reach the floor at nearly the same time.

Galileo was sure that air acts as a resistance to very light objects that have a large surface area. But in many ordinary circumstances this air resistance is negligible. In a chamber from which the air has been removed, even light objects like a feather or a horizontally held piece of paper will fall with the same acceleration as any other object (see Fig. 2–25). Such a demonstration in vacuum was of course not possible in Galileo's time, which makes Galileo's achievement all the greater. Galileo is often called the "father of modern science," not only for the content of his science (astronomical discoveries, inertia, free fall), but also for his approach to science (idealization and simplification, mathematization of theory, theories that have testable consequences, experiments to test theoretical predictions).

Galileo's specific contribution to our understanding of the motion of falling objects can be summarized as follows:

at a given location on the Earth and in the absence of air resistance, all objects fall with the same constant acceleration.

We call this acceleration the **acceleration due to gravity** on the Earth, and we give it the symbol g. Its magnitude is approximately

$$g = 9.80 \text{ m/s}^2.$$

In British units g is about 32 ft/s^2. Actually, g varies slightly according to latitude and elevation, but these variations are so small that we will ignore them for most purposes. The effects of air resistance are often small, and we will neglect them for the most part. However, air resistance will be noticeable even on a reasonably heavy object if the velocity becomes large.[†] Acceleration due to gravity is a vector, as is any acceleration, and its direction is downward, toward the center of the Earth.

[†] The speed of an object falling in air (or other fluid) does not increase indefinitely. If the object falls far enough, it will reach a maximum velocity called the **terminal velocity**.

When dealing with freely falling objects we can make use of Eqs. 2–12, where for a we use the value of g given above. Also, since the motion is vertical we will substitute y in place of x, and y_0 in place of x_0. We take $y_0 = 0$ unless otherwise specified. *It is arbitrary whether we choose y to be positive in the upward direction or in the downward direction; but we must be consistent about it throughout a problem's solution.*

EXAMPLE 2–13 **Falling from a tower.** Suppose that a ball is dropped from a tower 70.0 m high. How far will it have fallen after 1.00 s, 2.00 s, and 3.00 s? Assume y is positive downward. Neglect air resistance.

SOLUTION We are given the acceleration, $a = g = +9.80 \text{ m/s}^2$ which is positive because we have chosen downward as positive. Since we want to find the distance fallen given the time, t, Eq. 2–12b is the appropriate one, with $v_0 = 0$ and $y_0 = 0$. Then, after 1.00 s, the position of the ball is

$$y_1 = \tfrac{1}{2}at^2 = \tfrac{1}{2}(9.80 \text{ m/s}^2)(1.00 \text{ s})^2 = 4.90 \text{ m},$$

so the ball has fallen a distance of 4.90 m after 1.00 s. Similarly, after 2.00 s,

$$y_2 = \tfrac{1}{2}at^2 = \tfrac{1}{2}(9.80 \text{ m/s}^2)(2.00 \text{ s})^2 = 19.6 \text{ m},$$

and after 3.00 s,

$$y_3 = \tfrac{1}{2}at^2 = \tfrac{1}{2}(9.80 \text{ m/s}^2)(3.00 \text{ s})^2 = 44.1 \text{ m}.$$

See Fig. 2–26.

EXAMPLE 2–14 **Thrown down from a tower.** Suppose the ball in Example 2–13 is *thrown* downward with an initial velocity of 3.00 m/s, instead of being dropped. (*a*) What then would be its position after 1.00 s and 2.00 s? (*b*) What would its speed be after 1.00 s and 2.00 s? Compare to the speeds of a dropped ball.

SOLUTION (*a*) We can approach this in the same way as Example 2–13, using Eq. 2–12b, but this time v_0 is not zero but is $v_0 = 3.00 \text{ m/s}$. Thus, at $t = 1.00 \text{ s}$ the position of the ball is

$$y = v_0 t + \tfrac{1}{2}at^2 = (3.00 \text{ m/s})(1.00 \text{ s}) + \tfrac{1}{2}(9.80 \text{ m/s}^2)(1.00 \text{ s})^2 = 7.90 \text{ m},$$

and at $t = 2.00 \text{ s}$

$$y = v_0 t + \tfrac{1}{2}at^2 = (3.00 \text{ m/s})(2.00 \text{ s}) + \tfrac{1}{2}(9.80 \text{ m/s}^2)(2.00 \text{ s})^2 = 25.6 \text{ m}.$$

As expected, the ball falls farther each second than if it were dropped with $v_0 = 0$. (*b*) The velocity is readily obtained from Eq. 2–12a:

$$v = v_0 + at$$
$$= 3.00 \text{ m/s} + (9.80 \text{ m/s}^2)(1.00 \text{ s}) = 12.8 \text{ m/s} \quad [\text{at } t = 1.00 \text{ s}]$$
$$= 3.00 \text{ m/s} + (9.80 \text{ m/s}^2)(2.00 \text{ s}) = 22.6 \text{ m/s}. \quad [\text{at } t = 2.00 \text{ s}]$$

When the ball is dropped $(v_0 = 0)$, the first term in the above equations is zero, so

$$v = 0 + at$$
$$= (9.80 \text{ m/s}^2)(1.00 \text{ s}) = 9.80 \text{ m/s} \quad [\text{at } t = 1.00 \text{ s}]$$
$$= (9.80 \text{ m/s}^2)(2.00 \text{ s}) = 19.6 \text{ m/s}. \quad [\text{at } t = 2.00 \text{ s}]$$

We see that the speed of a dropped ball increases linearly in time. (In Example 2–13 we saw that the distance fallen increases as the *square* of the time.) The speed of the downwardly thrown ball also increases linearly ($\Delta v = 9.80$ m/s each second), but its speed at any moment is always higher than that of a falling ball by 3.0 m/s (its initial speed).

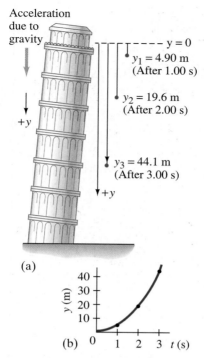

Acceleration due to gravity

$y = 0$
$y_1 = 4.90$ m (After 1.00 s)
$y_2 = 19.6$ m (After 2.00 s)
$y_3 = 44.1$ m (After 3.00 s)

$+y$

(a)

(b)

FIGURE 2–26 Example 2–13.
(a) An object dropped from a tower falls with progressively greater speed and covers greater distance with each successive second. (See also Fig. 2–23.) (b) Graph of y vs. t.

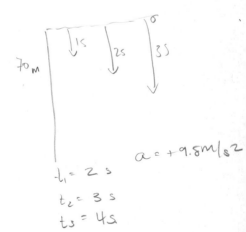

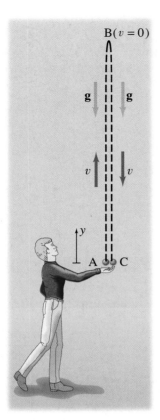

FIGURE 2–27 An object thrown into the air leaves the thrower's hand at A, reaches its maximum height at B, and returns to the original height at C. Examples 2–15, 2–16, and 2–17.

Careful:
Velocity and acceleration are not *always in the same direction*

Careful:
$a \neq 0$ *even at the highest point of a trajectory*

EXAMPLE 2–15 **Ball thrown upward.** A person throws a ball *upward* into the air with an initial velocity of 15.0 m/s. Calculate (*a*) how high it goes, and (*b*) how long the ball is in the air before it comes back to his hand. We are not concerned here with the throwing action, but only with the motion of the ball *after* it leaves the thrower's hand (Fig. 2–27).

SOLUTION Let us choose *y* to be positive in the upward direction and negative in the downward direction. (Note: This is a different convention from that used in Examples 2–13 and 2–14.) Then the acceleration due to gravity will have a negative sign, $a = -9.80 \, \text{m/s}^2$. Note that as the ball rises, its speed decreases until it reaches the highest point (B in Fig. 2–27), where its speed is zero for an instant; then it descends with increasing speed.

(*a*) To determine the maximum height, we calculate the position of the ball when its velocity equals zero ($v = 0$ at the highest point). At $t = 0$ (point A in Fig. 2–27) we have $y_0 = 0$, $v_0 = 15.0 \, \text{m/s}$, and $a = -9.80 \, \text{m/s}^2$. At time t (maximum height), $v = 0$, $a = -9.80 \, \text{m/s}^2$, and we wish to find y. We use Eq. 2–12c (replacing *x* with *y*) and solve for *y*:

$$v^2 = v_0^2 + 2ay$$
$$y = \frac{v^2 - v_0^2}{2a} = \frac{0 - (15.0 \, \text{m/s})^2}{2(-9.80 \, \text{m/s}^2)} = 11.5 \, \text{m}.$$

The ball reaches a height of 11.5 m above the hand.

(*b*) Now we need to calculate how long the ball is in the air before it returns to his hand. We could do this calculation in two parts by first determining the time required for the ball to reach its highest point, and then determining the time it takes to fall back down. However, it is simpler to consider the motion from A to B to C (Fig. 2–27) in one step and use Eq. 2–12b. We can do this because *y* (or *x*) represents position or displacement, and not the total distance traveled. Thus, at both points A and C, $y = 0$. We use Eq. 2–12b with $a = -9.80 \, \text{m/s}^2$ and find

$$y = v_0 t + \tfrac{1}{2} a t^2$$
$$0 = (15.0 \, \text{m/s})t + \tfrac{1}{2}(-9.80 \, \text{m/s}^2)t^2.$$

In this equation we can factor out one *t* to obtain

$$(15.0 \, \text{m/s} - 4.90 \, \text{m/s}^2 t)t = 0.$$

There are two solutions:

$$t = 0, \qquad \text{and} \qquad t = \frac{15.0 \, \text{m/s}}{4.90 \, \text{m/s}^2} = 3.06 \, \text{s}.$$

The first solution ($t = 0$) corresponds to the initial point (A) in Fig. 2–27, when the ball was first thrown and was also at $y = 0$. The second solution, $t = 3.06 \, \text{s}$, corresponds to point C, when the ball has returned to $y = 0$. Thus the ball is in the air for 3.06 s.

CONCEPTUAL EXAMPLE 2–16 **Two common misconceptions.** Explain the error in these two common misconceptions: (1) that acceleration and velocity are always in the same direction, and (2) that an object thrown upward has zero acceleration at the highest point (B in Fig. 2–27.)

RESPONSE Both are wrong. (1) Velocity and acceleration are *not* necessarily in the same direction. When a ball is falling downward, its velocity and acceleration are in the same direction. But when a ball is moving upward, as in Example 2–15, its velocity is upward, whereas the acceleration is downward, in the opposite direction. (2) At the highest point (B in Fig. 2–27), the ball has zero velocity for an instant. Is the acceleration also zero at this point? No. Gravity does not stop acting, so $a = -g = -9.80 \, \text{m/s}^2$ even there. Thinking that $a = 0$ at point B would lead to the conclusion that upon reaching point B, the ball would hover there. For if the acceleration (= rate of change of velocity) were zero, the velocity would remain zero, and the ball could stay up there without falling.

EXAMPLE 2-17 Ball thrown upward, II. Let us consider again the ball thrown upward of Example 2–15, and make three more calculations. Calculate (*a*) how much time it takes for the ball to reach the maximum height (point B in Fig. 2–27), (*b*) the velocity of the ball when it returns to the thrower's hand (point C), and (*c*) at what time *t* the ball passes a point 8.00 m above the person's hand.

SOLUTION Again we take *y* as positive upward. (*a*) Both Eqs. 2–12a and 2–12b contain the time *t* with other quantities known. Let us use Eq. 2–12a with $a = -9.80$ m/s², $v_0 = 15.0$ m/s, and $v = 0$:

$$v = v_0 + at,$$

so

$$t = -\frac{v_0}{a} = -\frac{15.0 \text{ m/s}}{-9.80 \text{ m/s}^2} = 1.53 \text{ s}.$$

This is just half the time it takes the ball to go up and fall back to its original position [3.06 s, calculated in part (*b*) of Example 2–15]. Thus it takes the same time to reach the maximum height as to fall back to the starting point.

(*b*) We use Eq. 2–12a with $v_0 = 15.0$ m/s and $t = 3.06$ s (the time calculated in Example 2–15 for the ball to come back to the hand):

$$v = v_0 + at = 15.0 \text{ m/s} - (9.80 \text{ m/s}^2)(3.06 \text{ s}) = -15.0 \text{ m/s}.$$

The ball has the same magnitude of velocity when it returns to the starting point as it did initially, but in the opposite direction (this is the meaning of the negative sign). Thus, as we gathered from part (*a*), we see that the motion is symmetrical about the maximum height.

Note the symmetry: the speed at any height is the same when going up as coming down

(*c*) We want *t*, given that $y = 8.00$ m, $y_0 = 0$, $v_0 = 15.0$ m/s, and $a = -9.80$ m/s². We use Eq. 2–12b:

$$y = y_0 + v_0 t + \tfrac{1}{2} a t^2$$

$$8.00 \text{ m} = 0 + (15.0 \text{ m/s})t + \tfrac{1}{2}(-9.80 \text{ m/s}^2)t^2.$$

To solve any quadratic equation of the form $at^2 + bt + c = 0$, where *a*, *b*, and *c* are constants, the **quadratic formula** gives the solutions

$$t = \frac{-b \pm \sqrt{b^2 - 4ac}}{2a}.$$

We rewrite our equation in the standard form:

$$(4.90 \text{ m/s}^2)t^2 - (15.0 \text{ m/s})t + (8.00 \text{ m}) = 0.$$

So the coefficient *a* is 4.90 m/s², *b* is −15.0 m/s, and *c* is 8.00 m. Putting these into the quadratic formula, we obtain

$$t = \frac{15.0 \text{ m/s} \pm \sqrt{(15.0 \text{ m/s})^2 - 4(4.90 \text{ m/s}^2)(8.00 \text{ m})}}{2(4.90 \text{ m/s}^2)},$$

so $t = 0.69$ s and $t = 2.37$ s. Why are there two solutions? Are they both valid? Yes, because the ball passes $y = 8.00$ m both when it goes up and when it comes down.

Figure 2–28 shows graphs of *y* vs. *t* and *v* vs. *t* for the ball thrown upward in Fig. 2–27, incorporating the results of Examples 2–15 and 2–17.

FIGURE 2–28 Graphs of (a) *y* vs. *t*, (b) *v* vs. *t* for a ball thrown upward, Examples 2–15 and 2–17.

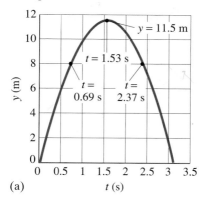

(a)

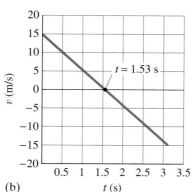

(b)

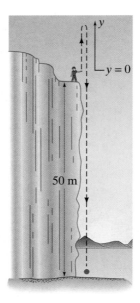

FIGURE 2–29 Example 2–18: the person in Fig. 2–27 stands on the edge of a cliff. The ball falls to the base of the cliff.

FIGURE 2–30
Example 2–18, the y vs. t graph.

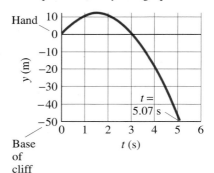

EXAMPLE 2–18 **Ball thrown upward at edge of cliff.** Suppose that the person of Examples 2–15 and 2–17 is standing on the edge of a 50.0-m-high cliff, so that the ball can fall to the base of the cliff as in Fig. 2–29. (*a*) How long does it take the ball to reach the base of the cliff? (*b*) What is the total distance traveled by the ball?

SOLUTION (*a*) Again we use Eq. 2–12b with $a = -9.80\,\text{m/s}^2$, $v_0 = 15.0\,\text{m/s}$, and $y_0 = 0$. But this time we set $y = -50.0\,\text{m}$, the bottom of the cliff, 50.0 m below the initial position ($y = 0$):

$$y = y_0 + v_0 t + \tfrac{1}{2} at^2$$

$$-50.0\,\text{m} = 0 + (15.0\,\text{m/s})t - \tfrac{1}{2}(9.80\,\text{m/s}^2)t^2.$$

Rewriting in the standard form we have

$$(4.90\,\text{m/s}^2)t^2 - (15.0\,\text{m/s})t - (50.0\,\text{m}) = 0.$$

Using the quadratic formula, we find as solutions $t = 5.07\,\text{s}$ and $t = -2.01\,\text{s}$. The first solution, $t = 5.07\,\text{s}$, is the answer to our problem. This is the time it takes the ball to rise to its highest point and then fall to the base of the cliff. To rise and fall back to the top of the cliff took 3.06 s (Example 2–15); then it took an additional 2.01 s to fall to the base. But what is the meaning of the other solution, $t = -2.01\,\text{s}$? This is a time before our problem even begins. So it isn't relevant to our problem.[†]

(*b*) From Example 2–15, the ball moves up 11.5 m, falls 11.5 m back to the top of the cliff, and then back another 50.0 m to the base of the cliff, for a total distance traveled of 73.0 m. Note that the *displacement*, however, was −50.0 m.

Figure 2–30 shows the y vs. t graph for this situation.

The acceleration of an object, particularly rockets and fast airplanes, is often given as a multiple of $g = 9.80\,\text{m/s}^2$. For example, a plane pulling out of a dive and undergoing 3.00 *g*'s or "gees" would have an acceleration of $(3.00)(9.80\,\text{m/s}^2) = 29.4\,\text{m/s}^2$. And the acceleration calculated in the collision of Example 2–11 could be expressed as $(390\,\text{m/s}^2)/(9.8\,\text{m/s}^2) = 40\,g$.

* 2–8 | Use of Calculus; Variable Acceleration

In this brief optional Section we use integral calculus to derive the kinematic equations for constant acceleration, Eqs. 2–12a and b. We also show how calculus can be used when the acceleration is not constant. If you have not yet studied simple integration in your calculus course, you may want to postpone reading this Section until you have.

Deriving the kinematic equations using calculus

First we derive Eq. 2–12a, starting with the definition of instantaneous acceleration, $a = dv/dt$, which we rewrite as

$$dv = a\,dt.$$

We take the definite integral of both sides of this equation, using the same nota-

---

[†] The solution $t = -2.01$ s could be meaningful in a different physical situation. Suppose that a person standing on top of a 50.0-m-high cliff sees a rock pass by him at $t = 0$ moving upward at 15.0 m/s; at what time did the rock leave the base of the cliff, and when did it arrive back at the base of the cliff? The equations will be precisely the same as for our original problem, and the answers $t = -2.01$ s and $t = 5.07$ s will be the correct answers. Note that we cannot put all the information for a problem into the mathematics, so we have to use common sense in interpreting results.

tion we did in Section 2–5 ($v = v_0$ at $t = 0$):

$$\int_{v=v_0}^{v} dv = \int_{t=0}^{t} a \, dt$$

which gives, since $a = $ constant,

$$v - v_0 = at.$$

Eq. 2–12a

This is Eq. 2–12a, $v = v_0 + at$.

Next we derive Eq. 2–12b starting with the definition of instantaneous velocity, Eq. 2–4, $v = dx/dt$. We rewrite this as

$$dx = v \, dt.$$

We substitute Eq. 2–12a above, $v = v_0 + at$, and integrate:

$$\int_{x=x_0}^{x} dx = \int_{t=0}^{t} (v_0 + at) \, dt$$

$$x - x_0 = \int_{t=0}^{t} v_0 \, dt + \int_{t=0}^{t} at \, dt$$

$$x - x_0 = v_0 t + \tfrac{1}{2} at^2$$

Eq. 2–12b

since v_0 and a are constants. This result is just Eq. 2–12b, $x = x_0 + v_0 t + \tfrac{1}{2} at^2$.

Finally let us use calculus to find velocity and displacement, given an acceleration that is not constant but varies in time.

EXAMPLE 2–19 **Integrating a time-varying acceleration.** An experimental vehicle starts from rest $(v_0 = 0)$ at $t = 0$ and accelerates at a rate given by $a = (7.00 \text{ m/s}^3)t$. What is (a) its velocity and (b) its displacement 2.00 s later?

SOLUTION We cannot use Eqs. 2–12 because a is not constant. Instead, we determine v as a function of t using calculus. From the definition of acceleration, $a = dv/dt$, we have

$$dv = a \, dt.$$

We take the integral of both sides from $v = 0$ at $t = 0$ to velocity v at an arbitrary time t:

$$\int_{0}^{v} dv = \int_{0}^{t} a \, dt$$

$$v = \int_{0}^{t} (7.00 \text{ m/s}^3) t \, dt$$

$$= (7.00 \text{ m/s}^3) \left(\frac{t^2}{2} \right) \Big|_{0}^{t} = (7.00 \text{ m/s}^3) \left(\frac{t^2}{2} - 0 \right) = (3.50 \text{ m/s}^3) t^2.$$

At $t = 2.00$ s, $v = (3.50 \text{ m/s}^3)(2.00 \text{ s})^2 = 14.0 \text{ m/s}$.

(b) To get the displacement, we assume $x_0 = 0$ and start with $v = dx/dt$ which we rewrite as $dx = v \, dt$. Then we integrate from $x = 0$ at $t = 0$ to position x at time t:

$$\int_{0}^{x} dx = \int_{0}^{t} v \, dt$$

$$x = \int_{0}^{2.00 \text{ s}} (3.50 \text{ m/s}^3) t^2 \, dt = (3.50 \text{ m/s}^3) \frac{t^3}{3} \Big|_{0}^{2.00 \text{ s}} = 9.33 \text{ m}.$$

In sum, at $t = 2.00$ s, $v = 14.0 \text{ m/s}$ and $x = 9.33 \text{ m}$.

Summary

[The Summary that appears at the end of each chapter in this book gives a brief overview of the main ideas of the chapter. The Summary *cannot* serve to give an understanding of the material, which can be accomplished only by a detailed reading of the chapter.]

Kinematics deals with the description of how objects move. The description of the motion of any object must always be given relative to some particular **reference frame**.

The **displacement** of an object is the change in position of the object.

Average speed is the distance traveled divided by the elapsed time. An object's **average velocity** over a particular time interval Δt is the displacement Δx divided by Δt:

$$\bar{v} = \frac{\Delta x}{\Delta t}.$$

The **instantaneous velocity**, whose magnitude is the same as the *instantaneous speed*, is the average velocity taken over an infinitesimally short time interval (Δt allowed to approach zero):

$$v = \lim_{\Delta t \to 0} \frac{\Delta x}{\Delta t} = \frac{dx}{dt},$$

where dx/dt is the derivative of x with respect to t.

On a graph of position vs. time, the *slope* is equal to the instantaneous velocity.

Acceleration is the change of velocity per unit time. An object's **average acceleration** over a time interval Δt is

$$\bar{a} = \frac{\Delta v}{\Delta t},$$

where Δv is the change of velocity during the time interval Δt.

Instantaneous acceleration is the average acceleration taken over an infinitesimally short time interval:

$$a = \lim_{\Delta t \to 0} \frac{\Delta v}{\Delta t} = \frac{dv}{dt}.$$

If an object moves in a straight line with constant acceleration, the velocity v and position x are related to the acceleration a, the elapsed time t, and the initial position x_0 and initial velocity v_0, by Eqs. 2–12:

$$v = v_0 + at, \qquad x = x_0 + v_0 t + \tfrac{1}{2}at^2,$$

$$v^2 = v_0^2 + 2a(x - x_0), \qquad \bar{v} = \frac{v + v_0}{2}.$$

Objects that move vertically near the surface of the Earth, either falling or having been projected vertically up or down, move with the constant downward **acceleration due to gravity** with magnitude of about $g = 9.80 \text{ m/s}^2$, if air resistance can be ignored.

Questions

1. Does a car speedometer measure speed, velocity, or both?
2. Can an object have a varying speed if its velocity is constant? If yes, give examples.
3. When an object moves with constant velocity, does its average velocity during any time interval differ from its instantaneous velocity at any instant?
4. In drag racing, is it possible for the car with the greatest speed crossing the finish line to lose the race? Explain.
5. If one object has a greater speed than a second object, does the first necessarily have a greater acceleration? Explain, using examples.
6. Compare the acceleration of a motorcycle that accelerates from 80 km/h to 90 km/h with the acceleration of a bicycle that accelerates from rest to 10 km/h in the same time.
7. Can an object have a northward velocity and a southward acceleration? Explain.
8. Can the velocity of an object be negative when its acceleration is positive? What about vice versa?
9. Give an example where both the velocity and acceleration are negative.
10. Two cars emerge side by side from a tunnel. Car A is traveling with a speed of 60 km/h and has an acceleration of 40 km/h/min. Car B has a speed of 40 km/h and has an acceleration of 60 km/h/min. Which car is passing the other as they come out of the tunnel? Explain your reasoning.
11. Can an object be increasing in speed as its acceleration decreases? If so, give an example. If not, explain.
12. An object that is thrown vertically upward will return to its original position with the same speed as it had initially if air resistance is negligible. If air resistance is appreciable, will this result be altered, and if so, how? [*Hint*: The acceleration due to air resistance is always in a direction opposite to the motion.]
13. As a freely falling object speeds up, what is happening to its acceleration due to gravity? Does it increase, decrease, or stay the same?
14. How would you estimate the maximum height you could throw a ball vertically upward? How would you estimate the maximum speed you could give it?
15. A rock is thrown upward with speed v from the edge of a cliff. A second rock is thrown vertically downward with the same initial speed. Which rock has greater speed when it reaches the bottom of the cliff? Ignore the effect of air resistance.
16. You travel from point A to point B in a car moving at a constant speed of 70 km/h. Then you travel the same distance to another point C, moving at a constant speed of 90 km/h. Is your average speed for the entire trip from A to C 80 km/h? Explain why or why not.

17. Describe in words the motion plotted in Fig. 2–31 in terms of v, a, etc. [*Hint:* First try to duplicate the motion plotted by walking or moving your hand.]

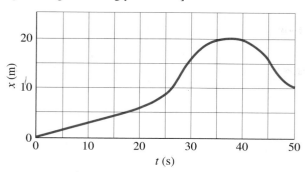

FIGURE 2–31 Question 17, Problems 11, 12, and 84.

18. Describe in words the motion of the object graphed in Fig. 2–32.

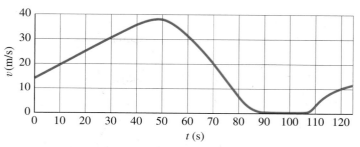

FIGURE 2–32 Question 18 and Problem 22.

Problems

[The problems at the end of each chapter are ranked I, II, or III according to estimated difficulty, with I problems being easiest. The problems are arranged by Section, meaning that the reader should have read up to and including that Section, but not only that Section—problems often depend on earlier material. Finally, there is a set of unranked "General Problems" not arranged by Section number.]

Sections 2–1 to 2–3

1. (I) A bird can fly 15 km/h. How long does it take to fly 75 km?

2. (I) What must your car's average speed be in order to travel 280 km in 3.2 h?

3. (I) If you are driving 110 km/h along a straight road and you look to the side for 2.0 s, how far do you travel during this inattentive period?

4. (I) A rolling ball moves from $x_1 = 3.4$ cm to $x_2 = -4.2$ cm during the time from $t_1 = 3.0$ s to $t_2 = 6.1$ s. What is its average velocity?

5. (I) A particle at $t_1 = -2.0$ s is at $x_1 = 3.4$ cm and at $t_2 = 4.5$ s is at $x_2 = 8.5$ cm. What is its average velocity? Can you calculate its average speed from these data?

6. (II) You are driving home from school steadily at 65 mph for 130 miles. It then begins to rain and you slow to 55 mph. You arrive home after driving 3 hours and 20 minutes. (*a*) How far is your hometown from school? (*b*) What was your average speed?

7. (II) According to a rule-of-thumb, every five seconds between a lightning flash and the following thunder gives the distance of the storm in miles. Assuming that the flash of light arrives in essentially no time at all, estimate the speed of sound in m/s from this rule.

8. (II) A person runs eight complete laps around a quarter-mile track in a total time of 12.5 min. Calculate (*a*) the average speed and (*b*) the average velocity, in m/s.

9. (II) A horse canters away from its trainer in a straight line, moving 160 m away in 17.0 s. It then turns abruptly and gallops halfway back in 6.8 s. Calculate (*a*) its average speed and (*b*) its average velocity for the entire trip, using "away from the trainer" as the positive direction.

10. (II) Two locomotives approach each other on parallel tracks. Each has a speed of 95 km/h with respect to the ground. If they are initially 8.5 km apart, how long will it be before they reach each other? (See Fig. 2–33.)

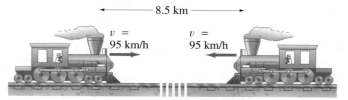

FIGURE 2–33 Problem 10.

11. (II) The position of a rabbit along a straight tunnel as a function of time is plotted in Fig. 2–31. What is its instantaneous velocity (*a*) at $t = 10.0$ s and (*b*) at $t = 30.0$ s? What is its average velocity (*c*) between $t = 0$ and $t = 5.0$ s, (*d*) between $t = 25.0$ s and $t = 30.0$ s, and (*e*) between $t = 40.0$ s and $t = 50.0$ s?

12. (II) In Fig. 2–31, (*a*) during what time periods, if any, is the rabbit's velocity constant? (*b*) At what time is its velocity the greatest? (*c*) At what time, if any, is the velocity zero? (*d*) Does the rabbit run in one direction or in both along its tunnel during the time shown?

13. (II) A dog runs 100 m away from its master in a straight line in 8.4 s, and then runs halfway back in one-third the time. Calculate (*a*) its average speed and (*b*) its average velocity.

14. (II) The position of a ball rolling in a straight line is given by $x = 2.0 - 4.6t + 1.1t^2$, where x is in meters and t in seconds. (*a*) Determine the position of the ball at $t = 1.0$, 2.0, and 3.0 s. (*b*) What is the average velocity over the interval $t = 1.0$ s to $t = 3.0$ s? (*c*) What is its instantaneous velocity at $t = 2.0$ s and at $t = 3.0$ s?

15. (II) A car traveling 90 km/h is 100 m behind a truck traveling 75 km/h. How long will it take the car to reach the truck?

16. (II) An airplane travels 2100 km at a speed of 800 km/h, and then encounters a tailwind that boosts its speed to 1000 km/h for the next 1800 km. What was the total time for the trip? What was the average speed of the plane for this trip? [*Hint:* Think carefully before using Eq. 2–12d.]

17. (II) Calculate the average speed and average velocity of a complete round-trip in which the outgoing 200 km is covered at 90 km/h, followed by a one-hour lunch break, and the return 200 km is covered at 50 km/h.

18. (II) An automobile traveling 90 km/h overtakes a 1.10-km-long train traveling in the same direction on a track parallel to the road. If the train's speed is 80 km/h, how long does it take the car to pass it, and how far will the car have traveled in this time? See Fig. 2–34. What are the results if the car and train are traveling in opposite directions?

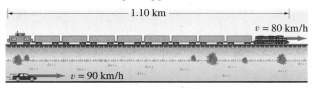

FIGURE 2–34 Problem 18.

19. (II) A bowling ball traveling with constant speed hits the pins at the end of a bowling lane 16.5 m long. The bowler hears the sound of the ball hitting the pins 2.50 s after the ball is released from his hands. What is the speed of the ball? The speed of sound is 340 m/s.

Section 2–4

20. (I) A sports car accelerates from rest to 95 km/h in 6.2 s. What is its average acceleration in m/s^2?

21. (I) At highway speeds, a particular automobile is capable of an acceleration of about $1.6 \, m/s^2$. At this rate, how long does it take to accelerate from 80 km/h to 110 km/h?

22. (I) Figure 2–32 shows the velocity of a train as a function of time. (a) At what time was its velocity greatest? (b) During what periods, if any, was the velocity constant? (c) During what periods, if any, was the acceleration constant? (d) When was the magnitude of the acceleration greatest?

23. (II) A sports car is advertised to be able to stop in a distance of 55 m from a speed of 100 km/h. What is its acceleration in m/s^2? How many g's is this $(g = 9.80 \, m/s^2)$?

24. (II) A particular automobile can accelerate approximately as shown in the velocity–time graph of Fig. 2–35. (The short flat spots in the curve represent shifting of the gears.) Estimate the average acceleration of the car in second gear and in fourth gear.

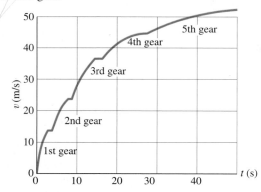

FIGURE 2–35 The velocity of a high-performance automobile as a function of time, starting from a dead stop. The flat spots in the curve represent gear shifts. Problems 24 and 25.

25. (II) Estimate the average acceleration of the car in the previous problem (Fig. 2–35) when it is in (a) first, (b) third, and (c) fifth gear. (d) What is its average acceleration through the first four gears?

26. (II) The position of a racing car, which starts from rest at $t = 0$ and moves in a straight line, has been measured as a function of time, as given in the following table. Estimate (a) its velocity and (b) its acceleration as a function of time. Display each in a table and on a graph.

$t(s)$	0	0.25	0.50	0.75	1.00	1.50	2.00	2.50
$x(m)$	0	0.11	0.46	1.06	1.94	4.62	8.55	13.79

$t(s)$	3.00	3.50	4.00	4.50	5.00	5.50	6.00
$x(m)$	20.36	28.31	37.65	48.37	60.30	73.26	87.16

27. (II) A particle moves along the x axis. Its position as a function of time is given by $x = 6.0t + 8.5t^2$, where t is in seconds and x is in meters. What is the acceleration as a function of time?

28. (II) The position of a body is given by $x = At + 6Bt^3$, where x is in meters and t is in seconds. (a) What are the units of A and B? (b) What is the acceleration as a function of time? (c) What is the velocity and acceleration at $t = 5.0 \, s$? (d) What is the velocity as a function of time if $x = At + Bt^{-3}$?

Sections 2–5 and 2–6

29. (I) A car accelerates from 12 m/s to 21 m/s in 6.0 s. What was its acceleration? How far did it travel in this time? Assume constant acceleration.

30. (I) A car slows down from 25 m/s to rest in a distance of 75 m. What was its acceleration, assumed constant?

31. (I) A light plane must reach a speed of 32 m/s for takeoff. How long a runway is needed if the (constant) acceleration is $3.0 \, m/s^2$?

32. (II) A baseball pitcher throws a baseball with a speed of 44 m/s. Estimate the average acceleration of the ball during the throwing motion. It is observed that in throwing the baseball, the pitcher accelerates the ball through a displacement of about 3.5 m, from behind the body to the point where it is released (Fig. 2–36).

FIGURE 2–36
Problem 32.

33. (II) A world-class sprinter can reach a top speed (of about 11.5 m/s) in the first 15.0 m of a race. What is the average acceleration of this sprinter and how long does it take her to reach that speed?

34. (II) Show that $\bar{v} = (v + v_0)/2$ (see Eq. 2–12d) is not valid for the case when the acceleration $a = A + Bt$, where A and B are constants.

35. (II) A car slows down uniformly from a speed of 22.0 m/s to rest in 5.00 s. How far did it travel in that time?

36. (II) In coming to a stop, a car leaves skid marks 75 m long on the highway. Assuming a deceleration of 7.00 m/s², estimate the speed of the car just before braking.

37. (II) A car traveling 55 km/h slows down at a constant 0.50 m/s² just by "letting up on the gas." Calculate (a) the distance the car coasts before it stops, (b) the time it takes to stop, and (c) the distance it travels during the first and fifth seconds.

38. (II) A car traveling at 95 km/h strikes a tree. The front end of the car compresses and the driver comes to rest after traveling 0.80 m. What was the average acceleration of the driver during the collision? Express the answer in terms of "g's," where $1.00 g = 9.80$ m/s².

39. (II) Determine the stopping distances for an automobile with an initial speed of 90 km/h and human reaction time of 1.0 s: (a) for an acceleration $a = -4.0$ m/s²; (b) for $a = -8.0$ m/s².

40. (II) A space vehicle accelerates uniformly from 65 m/s at $t = 0$ to 162 m/s at $t = 10.0$ s. How far did it move between $t = 2.0$ s and $t = 6.0$ s?

41. (II) A 75-m-long train accelerates uniformly from rest. If the front of the train passes a railway worker 140 m down the track at a speed of 25 m/s, what will be the speed of the last car as it passes the worker?

42. (II) Show that the equation for the stopping distance of a car is $d_S = v_0 t_R - v_0^2/(2a)$, where v_0 is the initial speed of the car, t_R is the driver's reaction time, and a is the constant acceleration (and is negative).

43. (II) In designing traffic signals, it is necessary to allow the yellow light to remain on long enough that a driver can either stop or pass completely through the intersection. Thus, if a driver is less than the stopping distance, d_S (calculated in Problem 42), from the intersection, the light must remain on long enough for her to travel this distance plus the width of the intersection, d_I. (a) Show that the yellow light should remain on for a time $t = t_R - v_0/(2a) + d_I/v_0$, where v_0 is a typical expected speed of a car approaching the intersection, and a and t_R are as defined in Problem 42. (b) A traffic engineer expects cars to approach a 14.4-m-wide intersection at speeds between 30.0 and 60.0 km/h. To be safe, he calculates the time for both speeds, assuming $t_R = 0.500$ s and $a = -4.00$ m/s², and chooses the longest time to be safe. What is his result?

44. (II) An unmarked police car traveling a constant 95 km/h is passed by a speeder traveling 140 km/h. Precisely 1.00 s after the speeder passes, the policeman steps on the accelerator; if the police car's acceleration is 2.00 m/s², how much time passes before the police car overtakes the speeder (assumed moving at constant speed)?

45. (III) Assume in Problem 44 that the speeder's speed is not known. If the police car accelerates uniformly as given above and overtakes the speeder after accelerating for 6.0 s, what was the speeder's speed?

46. (III) A runner hopes to complete the 10,000-m run in less than 30.0 min. After running at constant speed for exactly 27.0 min, there are still 1100 m to go. The runner must then accelerate at 0.20 m/s² for how many seconds in order to achieve the desired time?

Section 2–7 [neglect air resistance]

47. (I) If a car rolls gently $(v_0 = 0)$ off a vertical cliff, how long does it take it to reach 100 km/h?

48. (I) A stone is dropped from the top of a cliff. It is seen to hit the ground below after 2.75 s. How high is the cliff?

49. (I) Calculate (a) how long it took King Kong to fall straight down from the top of the Empire State Building (380 m high), and (b) his velocity just before "landing."

50. (II) A baseball is hit almost straight up into the air with a speed of about 20 m/s. (a) How high does it go? (b) How long is it in the air?

51. (II) A kangaroo jumps to a vertical height of 2.55 m. How long was it in the air before returning to Earth?

52. (II) A ballplayer catches a ball 3.1 s after throwing it vertically upward. With what speed did he throw it, and what height did it reach?

53. (II) Estimate the maximum speed with which you can throw an object straight up in the air. Describe your method of reaching the estimate.

54. (II) The best rebounders in basketball have a vertical leap (that is, the vertical movement of a fixed point on their body) of about 120 cm. (a) What is their initial "launch" speed off the ground? (b) How long are they in the air?

55. (II) A helicopter is ascending vertically with a speed of 5.60 m/s. At a height of 115 m above the Earth, a package is dropped from a window. How much time does it take for the package to reach the ground?

56. (II) For an object falling freely from rest, show that the distance traveled during each successive second increases in the ratio of successive odd integers (1, 3, 5, etc.). (This was first shown by Galileo.) See Figs. 2–23 and 2–26.

57. (II) If air resistance is neglected, show (algebraically) that a ball thrown vertically upward with a speed v_0 will have the same speed, v_0, when it comes back down to the starting point.

58. (II) A stone is thrown vertically upward with a speed of 23.0 m/s. (a) How fast is it moving when it reaches a height of 12.0 m? (b) How much time is required to reach this height? (c) Why are there two answers to (b)?

59. (II) Estimate the time between each photoflash of the apple in Fig. 2–23 (or number of photoflashes per second). Assume the apple is about 10 cm in diameter.

60. (II) A rocket rises vertically, from rest, with an acceleration of 3.2 m/s² until it runs out of fuel at an altitude of 1200 m. After this point, its acceleration is that of gravity, downward. (a) What is the velocity of the rocket when it runs out of fuel? (b) How long does it take to reach this point? (c) What maximum altitude does the rocket reach? (d) How much time (total) does it take to reach maximum altitude? (e) With what velocity does it strike the earth? (f) How long (total) is it in the air?

61. (II) A falling stone takes 0.30 s to travel past a window 2.2 m tall (Fig. 2–37). From what height above the top of the window did the stone fall?

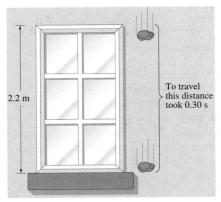

FIGURE 2–37
Problem 61.

62. (II) Suppose you adjust your garden hose nozzle for a hard stream of water. You point the nozzle vertically upward at a height of 1.5 m above the ground (Fig. 2–38). When you quickly move the nozzle away from the vertical, you hear the water striking the ground next to you for another 2.0 s. What is the water speed as it leaves the nozzle?

FIGURE 2–38
Problem 62.

63. (III) A rock is dropped from a sea cliff and the sound of it striking the ocean is heard 3.4 s later. If the speed of sound is 340 m/s, how high is the cliff?

64. (III) A rock is thrown vertically upward with a speed of 12.0 m/s. Exactly 1.00 s later, a ball is thrown up vertically along the same path with a speed of 20.0 m/s. (*a*) At what time will they strike each other? (*b*) At what height will the collision occur? (*c*) Answer (*a*) and (*b*) assuming that the order is reversed: the ball is thrown 1.00 s before the rock.

65. (III) A toy rocket passes by a 2.0-m-high window whose sill is 10.0 m above the ground. The rocket takes 0.15 s to travel the 2.0 m height of the window. What was the launch speed of the rocket, and how high will it go? Assume the propellant is burned very quickly at blastoff.

* **Section 2–8**

* **66. (II)** Given $v(t) = 25 + 18t$, where v is in m/s and t is in s, use calculus to determine the total displacement from $t_1 = 1.5$ s to $t_2 = 3.5$ s.

* **67. (III)** Air resistance acting on a falling body can be taken into account by the approximate relation for the acceleration:

$$a = \frac{dv}{dt} = g - kv,$$

where k is a constant. (*a*) Derive a formula for the velocity of the body as a function of time assuming it starts from rest ($v = 0$ at $t = 0$). [*Hint*: Change variables by setting $u = g - kv$.] (*b*) Determine an expression for the terminal velocity, which is the maximum value the velocity reaches.

* **68. (III)** The acceleration of a particle is given by $a = A\sqrt{t}$ where $A = 2.0$ m/s$^{5/2}$. At $t = 0$, $v = 10$ m/s and $x = 0$. (*a*) What is the speed as a function of time? (*b*) What is the displacement as a function of time? (*c*) What are the acceleration, speed and displacement at $t = 5.0$ s?

General Problems

69. The acceleration due to gravity on the Moon is about one sixth what it is on Earth. If an object is thrown vertically upward on the Moon, how many times higher will it go than it would on Earth, assuming the same initial velocity?

70. A person jumps from a fourth-story window 15.0 m above a firefighter's safety net. The survivor stretches the net 1.0 m before coming to rest, Fig. 2–39. (*a*) What was the average deceleration experienced by the survivor when slowed to rest by the net? (*b*) What would you do to make it "safer" (that is, generate a smaller deceleration): would you stiffen or loosen the net? Explain.

71. A person who is properly constrained by an over-the-shoulder seat belt has a good chance of surviving a car collision if the deceleration does not exceed 30 "*g*'s" $(1.00\ g = 9.80$ m/s$^2)$. Assuming uniform deceleration of this value, calculate the distance over which the front end of the car must be designed to collapse if a crash brings the car to rest from 100 km/h.

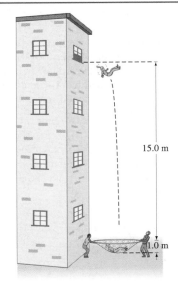

FIGURE 2–39
Problem 70.

72. A race car driver must average 200.0 km/h over the course of a time trial lasting ten laps. If the first nine laps were done at 199.0 km/h, what average speed must be maintained for the last lap?

73. A car manufacturer tests its cars for front-end collisions by hauling them up on a crane and dropping them from a certain height. (*a*) Show that the speed just before a car hits the ground, after falling from rest a vertical distance H, is given by $\sqrt{2gH}$. What height corresponds to a collision at (*b*) 50 km/h? (*c*) 100 km/h?

74. Figure 2–40 is a position versus time graph for the motion of an object along the *x* axis. As the object moves from A to B: (*a*) Is the object moving in the positive or negative direction? (*b*) Is the object speeding up or slowing down? (*c*) Is the acceleration of the object positive or negative? Next, for the time interval from D to E: (*d*) Is the object moving in the positive or negative direction? (*e*) Is the object speeding up or slowing down? (*f*) Is the acceleration of the object positive or negative? (*g*) Finally, answer these same three questions for the time interval from C to D.

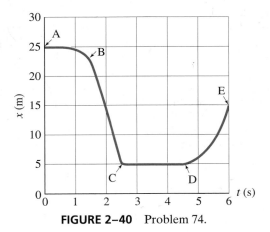

FIGURE 2–40 Problem 74.

75. Two children are playing on two trampolines. The first child can bounce up to one-and-a-half times higher than the second child. The initial speed up of the second child is 5.0 m/s. (*a*) Find the maximum height the second child reaches. (*b*) What is the initial speed of the first child? (*c*) How long was the first child in the air?

76. A 90-m-long train begins uniform acceleration from rest. The front of the train has a speed of 20 m/s when it passes a railway worker who is standing 180 m from where the front of the train started. What will be the speed of the last car as it passes the worker? (See Fig. 2–41.)

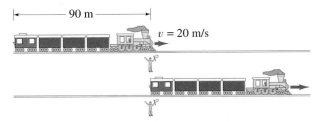

FIGURE 2–41 Problem 76.

77. A first stone is dropped from the roof of a building. 2.00 s after that, a second stone is thrown straight down with an initial speed of 30.0 m/s, and it is observed that the two stones land at the same time. (*a*) How long did it take the first stone to reach the ground? (*b*) How high is the building? (*c*) What are the speeds of the two stones just before they hit the ground?

78. A police car at rest, passed by a speeder traveling at a constant 110 km/h, takes off in hot pursuit. The police officer catches up to the speeder in 700 m, maintaining a constant acceleration. (*a*) Qualitatively plot the position versus time graph for both cars from the police car's start to the catch-up point. (*b*) Calculate how long it took the police officer to overtake the speeder, (*c*) calculate the required police car acceleration, and (*d*) calculate the speed of the police car at the overtaking point.

79. In the design of a rapid transit system, it is necessary to balance out the average speed of a train against the distance between stops. The more stops there are, the slower the train's average speed. To get an idea of this problem, calculate the time it takes a train to make a 36-km trip in two situations: (*a*) the stations at which the trains must stop are 0.80 km apart; and (*b*) the stations are 3.0 km apart. Assume that at each station the train accelerates at a rate of 1.1 m/s² until it reaches 90 km/h, then stays at this speed until its brakes are applied for arrival at the next station, at which time it decelerates at −2.0 m/s². Assume it stops at each intermediate station for 20 s.

80. Consider the street pattern as shown in Fig. 2–42. Each intersection has a traffic signal and the speed limit is 50 km/h. Suppose you are coming from the West at the speed limit and when you are 10 m from the first intersection, all the lights turn green. The lights are green for 13 s each. (*a*) Can you make it through all the lights without stopping? (*b*) Another car was stopped at the first light when all the lights turned green. It can accelerate at the rate of 2.0 m/s² to the speed limit. Can the second car clear through all the lights without stopping?

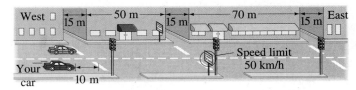

FIGURE 2–42 Problem 80.

81. A baseball is seen to pass upward by a window 25 m above the street with a vertical speed of 14 m/s. If the ball was thrown from the street, (*a*) what was its initial speed, (*b*) what altitude does it reach, (*c*) when was it thrown, and (*d*) when does it reach the street again?

82. A fugitive tries to catch a freight train traveling at a constant speed of 6.0 m/s. Just as an empty box car passes him, the fugitive starts from rest and accelerates at $a = 4.0 \text{ m/s}^2$ to his maximum speed of 8.0 m/s. (*a*) How long does it take him to reach the empty box car? (*b*) What is the distance traveled to reach the box car?

83. A stone is thrown vertically upward with a speed of 10.0 m/s from the edge of a cliff 65.0 m high (Fig. 2–43). (*a*) How much later does it reach the bottom of the cliff? (*b*) What is its speed just before hitting? (*c*) What total distance did it travel?

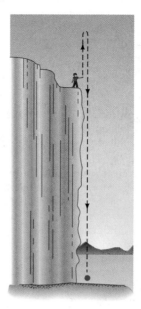

FIGURE 2–43
Problem 83.

84. Sketch the *v* vs. *t* graph for the object whose displacement as a function of time is given by Fig. 2–31.

85. A person driving her car at 50 km/h approaches an intersection just as the traffic light turns yellow. She knows that the yellow light lasts only 2.0 s before turning to red, and she is 30 m away from the near side of the intersection (Fig. 2–44). Should she try to stop, or should she make a run for it? The intersection is 15 m wide. Her car's maximum deceleration is -6.0 m/s^2, whereas it can accelerate from 50 km/h to 70 km/h in 6.0 s. Ignore the length of her car and her reaction time.

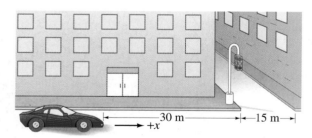

FIGURE 2–44 Problem 85.

86. Pelicans tuck their wings and free fall straight down when diving for fish. Suppose a pelican starts its dive from a height of 16.0 m and cannot change its path once committed. If it takes a fish 0.20 s to perform evasive action, at what minimum height must it spot the pelican to escape? Assume the fish is at the surface of the water.

87. In putting, the force with which a golfer strikes a ball is planned so that the ball will stop within some small distance of the cup, say 1.0 m long or short, in case the putt is missed. Accomplishing this from an uphill lie (that is, putting downhill, see Fig. 2–45) is more difficult than from a downhill lie. To see why, assume that on a particular green the ball decelerates constantly at 2.0 m/s^2 going downhill, and constantly at 3.0 m/s^2 going uphill. Suppose we have an uphill lie 7.0 m from the cup. Calculate the allowable range of initial velocities we may impart to the ball so that it stops in the range 1.0 m short to 1.0 m long of the cup. Do the same for a downhill lie 7.0 m from the cup. What in your results suggests that the downhill putt is more difficult?

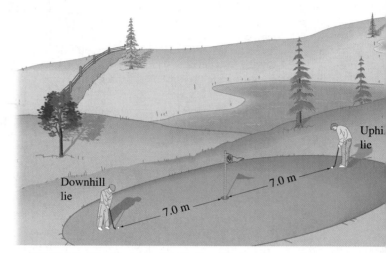

FIGURE 2–45 Problem 87. Golf on Wednesday morning.

88. A car is behind a truck going 25 m/s on the highway. The driver looks for an opportunity to pass, guessing that his car can accelerate at 1.0 m/s^2, and he gauges that he has to cover the 20-m length of the truck, plus 10 m clear room at the rear of the truck and 10 m more at the front of it. In the oncoming lane, he sees a car approaching, probably also traveling at 25 m/s. He estimates that the car is about 400 m away. Should he attempt the pass? Give details.

89. A stone is dropped from the roof of a high building. A second stone is dropped 1.50 s later. How far apart are the stones when the second one has reached a speed of 12.0 m/s?

90. Bond is standing on a bridge, 10 m above the road below, and his pursuers are getting too close for comfort. He spots a flatbed truck loaded with mattresses approaching at 30 m/s, which he measures by knowing that the telephone poles the truck is passing are 20 m apart in this country. The bed of the truck is 1.5 m above the road, and Bond quickly calculates how many poles away the truck should be when he jumps down from the bridge onto the truck, making his getaway. How many poles is it?

This multiflash photograph of a ping pong ball shows examples of motion in two dimensions. The arcs of the ping pong ball are parabolas that represent "projectile motion." Galileo analyzed projectile motion into its horizontal and vertical components, under the action of gravity (the gold arrow represents the downward acceleration of gravity, **g**). We will discuss how to manipulate vectors and how to add them. Besides analyzing projectile motion, we will also analyze uniform circular motion and how to work with relative velocity.

CHAPTER 3

Kinematics In Two Dimensions; Vectors

In Chapter 2 we dealt with motion along a straight line. We now consider the description of the motion of objects that move in paths in two (or three) dimensions. To do so we first need to discuss vectors and how they are added. Then we will examine the description of motion in general, followed by some interesting special cases including the motion of projectiles near the Earth's surface, and objects that are constrained to move in a circle.

3–1 | Vectors and Scalars

We mentioned in Chapter 2 that the term *velocity* refers not only to how fast something is moving but also to its direction. A quantity such as velocity, which has *direction* as well as *magnitude*, is a **vector** quantity. Other quantities that are also vectors are displacement, force, and momentum. However, many quantities such as mass, time, and temperature, have no direction associated with them. They are specified completely by giving a number and units. Such quantities are called **scalars**.

Drawing a diagram of a particular physical situation is always helpful in physics, and this is especially true when dealing with vectors. On a diagram, each vector is represented by an arrow. The arrow is always drawn so that it points in the direction of the vector quantity it represents. The length of the arrow is drawn proportional to the magnitude of the vector quantity. For example, in Fig. 3–1, arrows have been drawn representing the velocity of a car at various places as it rounds a curve. The magnitude of the velocity at each point can be read off this figure by measuring the length of the corresponding arrow and using the scale shown (1 cm = 90 km/h).

FIGURE 3–1 Car traveling on a road. The green arrows represent the velocity vector at each position.

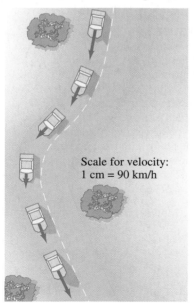

Scale for velocity:
1 cm = 90 km/h

Boldface symbols for vectors

When we write the symbol for a vector, we will always use boldface type. Thus for velocity we write **v**. (In handwritten work, the symbol for a vector can be indicated by putting an arrow over it, a $\vec{v}$ for velocity.) If we are concerned only with the magnitude of the vector, we will write simply v, in italics.

3–2 | Addition of Vectors—Graphical Methods

Because vectors are quantities that have direction as well as magnitude, they must be added in a special way. In this chapter, we will deal mainly with displacement vectors, for which we now use the symbol **D**, and velocity vectors, **v**. But the results will apply for other vectors we encounter later.

We use simple arithmetic for adding scalars. Simple arithmetic can also be used for adding vectors if they are in the same direction. For example, if a person walks 8 km east one day, and 6 km east the next day, the person will be 8 km + 6 km = 14 km east of the point of origin. We say that the *net or resultant* displacement is 14 km to the east (Fig. 3–2a). If, on the other hand, the person walks 8 km east on the first day, and 6 km west (in the reverse direction) on the second day, then the person will end up 2 km from the origin (Fig. 3–2b), so the resultant displacement is 2 km to the east. In this case, the resultant displacement is obtained by subtraction: 8 km − 6 km = 2 km.

But simple arithmetic cannot be used if the two vectors are not along the same line. For example, suppose a person walks 10.0 km east and then walks 5.0 km north. These displacements can be represented on a graph in which the positive y axis points north and the positive x axis points east, Fig. 3–3. On this graph, we draw an arrow, labeled $\mathbf{D}_1$, to represent the displacement vector of the 10.0-km displacement to the east. Then we draw a second arrow, $\mathbf{D}_2$, to represent the 5.0-km displacement to the north. Both vectors are drawn to scale, as in Fig. 3–3.

After taking this walk, the person is now 10.0 km east and 5.0 km north of the point of origin. The **resultant displacement** is represented by the arrow labeled $\mathbf{D}_R$ in Fig. 3–3. Using a ruler and a protractor, you can measure on this diagram that the person is 11.2 km from the origin at an angle of 27° north of east. In other words, the resultant displacement vector has a magnitude of 11.2 km and makes an angle $\theta = 27°$ with the positive x axis. The magnitude (length) of $\mathbf{D}_R$ can also be obtained using the theorem of Pythagoras in this case, since D_1, D_2, and D_R form a right triangle with D_R as the hypotenuse. Thus

$$D_R = \sqrt{D_1^2 + D_2^2} = \sqrt{(10.0\text{ km})^2 + (5.0\text{ km})^2} = \sqrt{125\text{ km}^2} = 11.2\text{ km}.$$

You can use the Pythagorean theorem, of course, only when the vectors are *perpendicular* to each other.

The resultant displacement vector, $\mathbf{D}_R$, is the sum of the vectors $\mathbf{D}_1$ and $\mathbf{D}_2$. That is,

$$\mathbf{D}_R = \mathbf{D}_1 + \mathbf{D}_2.$$

This is a *vector* equation. An important feature of adding two vectors that are not along the same line is that the magnitude of the resultant vector is not equal to the sum of the magnitudes of the two separate vectors, but is smaller than their sum:

$$D_R < D_1 + D_2. \qquad \text{[Vectors not along the same line]}$$

In our example (Fig. 3–3), $D_R = 11.2$ km, whereas $D_1 + D_2$ equals 15 km. We generally are not interested in $D_1 + D_2$; rather we are interested in the *vector* sum of the two vectors and its magnitude, D_R. Note also that we cannot set $\mathbf{D}_R$ equal to 11.2 km, because we have a vector equation and 11.2 km is only a part of the resultant vector, its magnitude. We could write something like this, though: $\mathbf{D}_R = \mathbf{D}_1 + \mathbf{D}_2 = (11.2\text{ km}, 27° \text{ N of E}).$

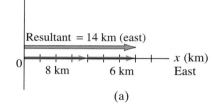

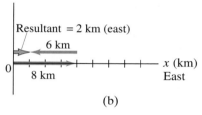

FIGURE 3–2 Combining vectors in one dimension.

FIGURE 3–3 A person walks 10.0 km east and then 5.0 km north. These two displacements are represented by the vectors $\mathbf{D}_1$ and $\mathbf{D}_2$, which are shown as arrows. The resultant displacement vector, $\mathbf{D}_R$, which is the vector sum of $\mathbf{D}_1$ and $\mathbf{D}_2$, is also shown. Measurement on the graph with ruler and protractor shows that $\mathbf{D}_R$ has a magnitude of 11.2 km and points at an angle $\theta = 27°$ north of east.

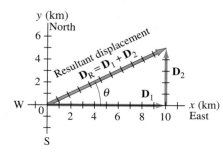

Figure 3–3 illustrates the general rules for graphically adding two vectors together, no matter what angles they make, to get their sum. The rules are as follows:

1. On a diagram, draw one of the vectors—call it $\mathbf{D}_1$—to scale.
2. Next draw the second vector, $\mathbf{D}_2$, to scale, placing its tail at the tip of the first vector and being sure its direction is correct.
3. The arrow drawn from the tail of the first vector to the tip of the second represents the *sum*, or **resultant**, of the two vectors.

Note that vectors can be translated parallel to themselves to accomplish these manipulations. The length of the resultant can be measured with a ruler and compared to the scale. Angles can be measured with a protractor. This method is known as the **tail-to-tip method of adding vectors**.

Note that it is not important in which order the vectors are added. For example, a displacement of 5.0 km north, to which is added a displacement of 10.0 km east, yields a resultant of 11.2 km and angle $\theta = 27°$ (see Fig. 3–4), the same as when they were added in reverse order (Fig. 3–3). That is,

$$\mathbf{V}_1 + \mathbf{V}_2 = \mathbf{V}_2 + \mathbf{V}_1. \qquad \text{[commutative law]} \quad \textbf{(3–1a)}$$

The tail-to-tip method can be extended to three or more vectors (Fig. 3–5) and, as indicated in that figure,

$$\mathbf{V}_1 + (\mathbf{V}_2 + \mathbf{V}_3) = (\mathbf{V}_1 + \mathbf{V}_2) + \mathbf{V}_3. \qquad \text{[associative law]} \quad \textbf{(3–1b)}$$

The left side of this equation means we first add $\mathbf{V}_2$ and $\mathbf{V}_3$ together and then add $\mathbf{V}_1$ to that sum to get the total sum. On the right side, $\mathbf{V}_1$ is added to $\mathbf{V}_2$, and this sum is added to $\mathbf{V}_3$. We see that the order in which two or more vectors are added doesn't affect the result.

A second way to add two vectors is the **parallelogram method**. It is fully equivalent to the tail-to-tip method. In this method, the two vectors are drawn starting from a common origin, and a parallelogram is constructed using these two vectors as adjacent sides as shown in Fig. 3–6b. The resultant is the diagonal drawn from the common origin. In Fig. 3–6a, the tail-to-tip method is shown, and it is clear that both methods yield the same result. It is a common error to draw the sum vector as the diagonal running between the tips of the two vectors, as in Fig. 3–6c. *This is incorrect*: it does not represent the sum of the two vectors. (In fact, it represents their difference, $\mathbf{V}_2 - \mathbf{V}_1$, as we will see in the next Section.)

FIGURE 3–6
Vector addition by two different methods, (a) and (b). Part (c) is incorrect.

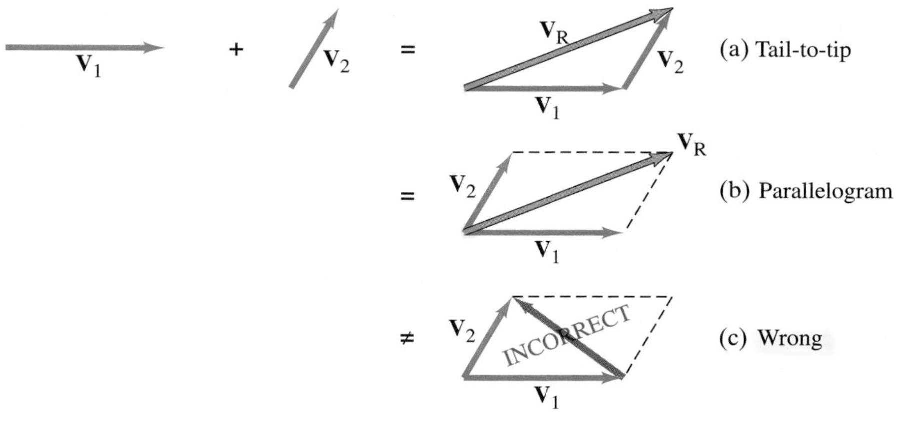

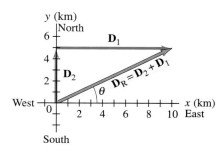

FIGURE 3–4 If the vectors are added in reverse order, the resultant is the same. (Compare Fig. 3–3.)

FIGURE 3–5 The three vectors in (a) can be added in any order with the same result, $\mathbf{V}_1 + \mathbf{V}_2 + \mathbf{V}_3$. In (b) $\mathbf{V}_R = (\mathbf{V}_1 + \mathbf{V}_2) + \mathbf{V}_3$, which clearly gives the same result as in (c) where $\mathbf{V}_R = \mathbf{V}_1 + (\mathbf{V}_2 + \mathbf{V}_3)$, which we simply write now as $\mathbf{V}_R = \mathbf{V}_1 + \mathbf{V}_2 + \mathbf{V}_3$ without parentheses.

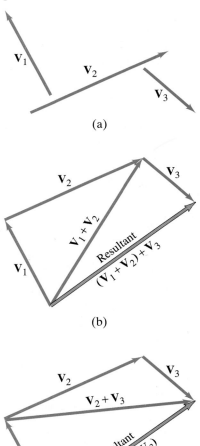

3–3 Subtraction of Vectors, and Multiplication of a Vector by a Scalar

FIGURE 3–7 The negative of a vector is a vector having the same length but opposite direction.

Given a vector **V**, we define the *negative* of this vector (−**V**) to be a vector with the same magnitude as **V** but opposite in direction, Fig. 3–7. Note, however, that no vector is ever negative in the sense of its magnitude: the magnitude of every vector is positive. A minus sign tells us about its direction.

We can now define the subtraction of one vector from another: the difference between two vectors, $\mathbf{V}_2 - \mathbf{V}_1$ is defined as

$$\mathbf{V}_2 - \mathbf{V}_1 = \mathbf{V}_2 + (-\mathbf{V}_1).$$

That is, the difference between two vectors is equal to the sum of the first plus the negative of the second. Thus our rules for addition of vectors can be applied as shown in Fig. 3–8 using the tail-to-tip method.

FIGURE 3–8 Subtracting two vectors: $\mathbf{V}_2 - \mathbf{V}_1$.

A vector **V** can be multiplied by a scalar c. We define this product so that $c\mathbf{V}$ has the same direction as **V** and has magnitude cV. That is, multiplication of a vector by a positive scalar c changes the magnitude of the vector by a factor c but doesn't alter the direction. If c is a negative scalar, the magnitude of the product $c\mathbf{V}$ is still cV (without the minus sign), but the direction is precisely opposite to that of **V**. See Fig. 3–9.

FIGURE 3–9 Multiplying a vector **V** by a scalar c gives a vector whose magnitude is c times greater and in the same direction as **V** (or opposite direction if c is negative).

3–4 Adding Vectors by Components

Adding vectors graphically using a ruler and protractor is often not sufficiently accurate and is not useful for vectors in three dimensions. We discuss now a more powerful and precise method for adding vectors.

Consider first a vector **V** that lies in a particular plane. It can be expressed as the sum of two other vectors, called the **components** of the original vector. The components are usually chosen to be along two perpendicular directions. The process of finding the components is known as **resolving the vector into its components**. An example is shown in Fig. 3–10; the vector **V** could be a displacement vector that points at an angle $\theta = 30°$ north of east, where we have chosen the

Resolving a vector into components

FIGURE 3–10 Resolving a vector **V** into its components along an arbitrarily chosen set of x and y axes. Note that the components, once found, themselves represent the vector. That is, the components contain as much information as the vector itself.

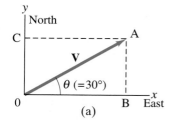

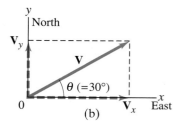

positive x axis to be to the east and the positive y axis north. This vector **V** is resolved into its x and y components by drawing dashed lines from the tip (A) of the vector and drawing these lines perpendicular to the x and y axes (lines AB and AC). Then the lines OB and OC represent the x and y components of **V**, respectively, as shown in Fig. 3–10b. These *vector components* are written $\mathbf{V}_x$ and $\mathbf{V}_y$. We generally show vector components as arrows, like vectors, but dashed. The *scalar components*, V_x and V_y, are numbers, with units, that are given a positive or negative sign depending on whether they point along the positive or negative x or y axis. As can be seen in Fig. 3–10, $\mathbf{V}_x + \mathbf{V}_y = \mathbf{V}$ by the parallelogram method of adding vectors.

Vector components

Space is made up of three dimensions, and sometimes it is necessary to resolve a vector into components along three mutually perpendicular directions. In rectangular coordinates the components are $\mathbf{V}_x$, $\mathbf{V}_y$, and $\mathbf{V}_z$, and

$$\mathbf{V} = \mathbf{V}_x + \mathbf{V}_y + \mathbf{V}_z.$$

Resolution of a vector in three dimensions is merely an extension of the above technique. We will mostly be concerned with situations in which the vectors are in a plane and two components are all that are necessary.

The use of trigonometric functions for finding the components of a vector is illustrated in Fig. 3–11, where it is seen that a vector and its two components can be thought of as making up a right triangle. We then see that the sine, cosine, and tangent are as given in the figure. Hence the x and y components of a vector V are

$$V_y = V \sin \theta \qquad \qquad \textbf{(3–2a)}$$
$$V_x = V \cos \theta. \qquad \qquad \textbf{(3–2b)}$$

Components of a vector

Note that θ is chosen (by convention) to be the angle that the vector makes with the positive x axis.

The components of a given vector will be different for different choices of coordinate axes. It is therefore crucial to specify the choice of coordinate system when giving the components.

Note that there are two ways to specify a vector in a given coordinate system:

1. We can give its components, V_x and V_y.
2. We can give its magnitude V and the angle θ it makes with the positive x axis.

Two ways to specify a vector

We can shift from one description to the other using Eqs. 3–2, and, for the reverse, by using the theorem of Pythagoras[†] and the definition of tangent (see Fig. 3–11):

$$V = \sqrt{V_x^2 + V_y^2} \qquad \qquad \textbf{(3–3a)}$$
$$\tan \theta = \frac{V_y}{V_x}. \qquad \qquad \textbf{(3–3b)}$$

Components related to magnitude and direction

[†]In three dimensions, the theorem of Pythagoras becomes $V = \sqrt{V_x^2 + V_y^2 + V_z^2}$, where V_z is the component along the third, or z, axis.

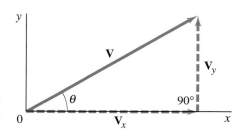

$$\sin \theta = \frac{V_y}{V}$$
$$\cos \theta = \frac{V_x}{V}$$
$$\tan \theta = \frac{V_y}{V_x}$$
$$V^2 = V_x^2 + V_y^2$$

FIGURE 3–11 Finding the components of a vector using trigonometric functions, where θ is the angle with the x axis.

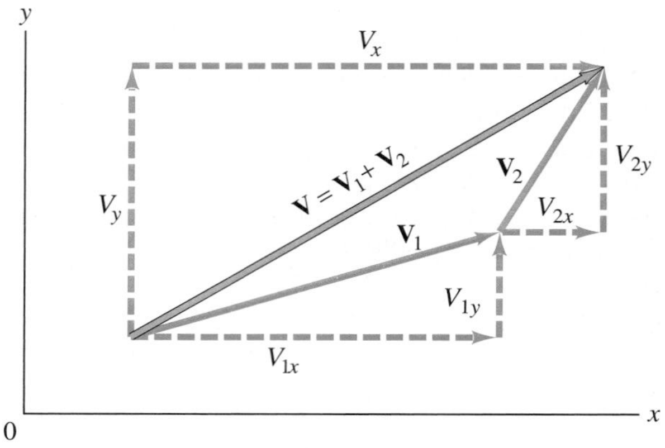

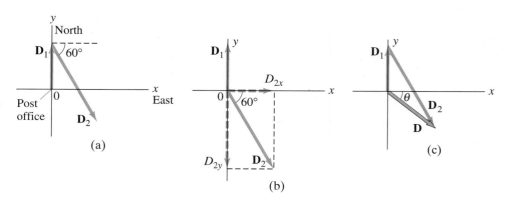

FIGURE 3–12
The components of
$\mathbf{V} = \mathbf{V}_1 + \mathbf{V}_2$ are
$V_x = V_{1x} + V_{2x}$ and
$V_y = V_{1y} + V_{2y}$.

We can now discuss how to add vectors using components. The first step is to resolve each vector into its components. Next we can see, using Fig. 3–12, that the addition of any two vectors $\mathbf{V}_1$ and $\mathbf{V}_2$ to give a resultant, $\mathbf{V} = \mathbf{V}_1 + \mathbf{V}_2$, implies that

Adding vectors analytically (by components)

$$V_x = V_{1x} + V_{2x}$$
$$V_y = V_{1y} + V_{2y}.$$

(3–4)

That is, the sum of the x components equals the x component of the resultant, and similarly for y. That this is valid can be verified by a careful examination of Fig. 3–12. But note that we add all the x components together to get the x component of the resultant; and we add all the y components together to get the y component of the resultant. We do *not* add x components to y components.

If the magnitude and direction of the resultant vector are desired, they can be obtained using Eqs. 3–3.

Choice of axes can simplify effort needed

The choice of coordinate axes is, of course, always arbitrary. You can often reduce the work involved in adding vectors by a good choice of axes—for example, by choosing one of the axes to be in the same direction as one of the vectors. Then that vector will have only one nonzero component.

EXAMPLE 3–1 **Mail carrier's displacement.** A rural mail carrier leaves the post office and drives 22.0 km in a northerly direction to the next town. She then drives in a direction 60.0° south of east for 47.0 km (Fig. 3–13a) to another town. What is her displacement from the post office?

SOLUTION We want to find her resultant displacement from the origin. We choose the positive x axis to be east and the positive y axis north, and resolve each displacement vector into its components (Fig. 3–13b). Since $\mathbf{D}_1$ has

FIGURE 3–13 Example 3–1.

magnitude 22.0 km and points north, it has only a y component:

$$D_{1x} = 0, \quad D_{1y} = 22.0 \text{ km}$$

whereas $\mathbf{D}_2$ has both x and y components:

$$D_{2x} = +(47.0 \text{ km})(\cos 60°) = +(47.0 \text{ km})(0.500) = +23.5 \text{ km}$$
$$D_{2y} = -(47.0 \text{ km})(\sin 60°) = -(47.0 \text{ km})(0.866) = -40.7 \text{ km}.$$

Notice that D_{2y} is negative because this vector component points along the negative y axis. The resultant vector, $\mathbf{D}$, has components:

$$D_x = D_{1x} + D_{2x} = \quad 0 \text{ km} + \quad 23.5 \text{ km} \quad = +23.5 \text{ km}$$
$$D_y = D_{1y} + D_{2y} = 22.0 \text{ km} + (-40.7 \text{ km}) = -18.7 \text{ km}.$$

This specifies the resultant vector completely:

$$D_x = 23.5 \text{ km}, \quad D_y = -18.7 \text{ km}.$$

We can also specify the resultant vector by giving its magnitude and angle using Eqs. 3–3:

$$D = \sqrt{D_x^2 + D_y^2} = \sqrt{(23.5 \text{ km})^2 + (-18.7 \text{ km})^2} = 30.0 \text{ km}$$

$$\tan \theta = \frac{D_y}{D_x} = \frac{-18.7 \text{ km}}{23.5 \text{ km}} = -0.796.$$

A calculator with an INV TAN or TAN^{-1} key gives $\theta = \tan^{-1}(-0.796) = -38.5°$. The negative sign means $\theta = 38.5°$ below the x axis, Fig. 3–13c.

The signs of trigonometric functions depend on which "quadrant" the angle falls in: for example, the tangent is positive in the first and third quadrants (from 0° to 90°, and 180° to 270°), but negative in the second and fourth quadrants; see Appendix A. The best way to keep track of angles, and to check any vector result, is always to draw a vector diagram. A vector diagram gives you something tangible to look at when analyzing a problem, and provides a check on the results.

➡ PROBLEM SOLVING

Check the quadrant

PROBLEM SOLVING   Adding Vectors by Components

Here is a brief summary of how to add two or more vectors using components:
1. Draw a diagram, adding the vectors graphically.
2. Choose x and y axes. Choose them in a way, if possible, that will make your work easier. (For example, choose one axis along the direction of one of the vectors so that vector will have only one component.)
3. Resolve each vector into its x and y components, showing each component along its appropriate (x or y) axis as a (dashed) arrow.
4. Calculate each component (when not given) using sines and cosines. If θ_1 is the angle vector $\mathbf{V}_1$ makes with the x axis, then:

$$V_{1x} = V_1 \cos \theta_1, \qquad V_{1y} = V_1 \sin \theta_1.$$

Pay careful attention to signs: any component that points along the negative x or y axis gets a negative sign.

5. Add the x components together to get the x component of the resultant. Similarly for y:

$$V_x = V_{1x} + V_{2x} + \text{any others.}$$
$$V_y = V_{1y} + V_{2y} + \text{any others.}$$

This is the answer: the components of the resultant vector.

6. If you want to know the magnitude and direction of the resultant vector, use Eqs. 3–3:

$$V = \sqrt{V_x^2 + V_y^2}, \qquad \tan \theta = \frac{V_y}{V_x}.$$

The vector diagram you already drew helps to obtain the correct position (quadrant) of the angle θ.

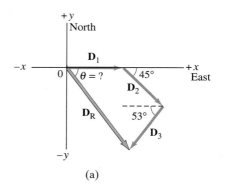

(a)

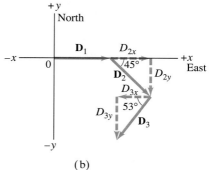

(b)

FIGURE 3–14 Example 3–2.

	Components	
Vector	**x (km)**	**y (km)**
D₁	620	0
D₂	311	−311
D₃	−331	−439
D_R	600	−750

FIGURE 3–15
Unit vectors **i, j,** and **k** along
the x, y, and z axes.

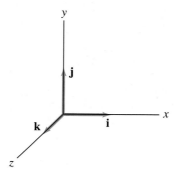

EXAMPLE 3–2 **Three short trips.** An airplane trip involves three legs, with two stopovers, as shown in Fig. 3–14a. The first leg is due east for 620 km; the second leg is southeast (45°) for 440 km; and the third leg is at 53° south of west, for 550 km, as shown. What is the plane's total displacement?

SOLUTION We follow the steps in the above Problem Solving box:
(1) and (2): Already shown in Fig. 3–14a, where we have taken the x axis as east (then **D₁** has only an x component).
(3): It is imperative to draw a good figure. The components are shown in Fig. 3–14b. Notice that instead of drawing all the vectors starting from a common origin, as we did in Fig. 3–13b, here we have drawn them "tail-to-tip" style, which is just as valid and may make it easier to see.
(4): Now we calculate the components:

$$\mathbf{D_1}: \quad D_{1x} = +D_1 \cos 0° \ = D_1 = 620 \,\text{km}$$
$$D_{1y} = +D_1 \sin 0° \ = 0 \,\text{km}$$

$$\mathbf{D_2}: \quad D_{2x} = +D_2 \cos 45° = +(440 \,\text{km})(0.707) = +311 \,\text{km}$$
$$D_{2y} = -D_2 \sin 45° = -(440 \,\text{km})(0.707) = -311 \,\text{km}$$

$$\mathbf{D_3}: \quad D_{3x} = -D_3 \cos 53° = -(550 \,\text{km})(0.602) = -331 \,\text{km}$$
$$D_{3y} = -D_3 \sin 53° = -(550 \,\text{km})(0.799) = -439 \,\text{km}.$$

Note carefully that we have given a minus sign to each component that in Fig. 3–14b points in the negative x or negative y direction. We see why a good drawing is so important. We summarize the components in the Table in the margin.
(5): This is easy:

$$D_x = D_{1x} + D_{2x} + D_{3x} = 620 \,\text{km} + 311 \,\text{km} - 331 \,\text{km} = \ \ 600 \,\text{km}$$
$$D_y = D_{1y} + D_{2y} + D_{3y} = \ \ \ 0 \,\text{km} - 311 \,\text{km} - 439 \,\text{km} = -750 \,\text{km}.$$

The x and y components are 600 km and −750 km, and point respectively to the east and south. This is one way to give the answer.
(6): We can also give the answer as

$$D_R = \sqrt{D_x^2 + D_y^2} = \sqrt{(600)^2 + (-750)^2} \,\text{km} = 960 \,\text{km}$$

$$\tan \theta = \frac{D_y}{D_x} = \frac{-750 \,\text{km}}{600 \,\text{km}} = -1.25, \qquad \text{so } \theta = -51°,$$

where we assume only two significant figures. Thus, the total displacement has magnitude 960 km and points 51° below the x axis (south of east), as was shown in our original sketch, Fig. 3–14a.

3–5 | Unit Vectors

Vectors can be conveniently written in terms of *unit vectors*. A **unit vector** is defined to have a magnitude exactly equal to one (1). It is useful to define unit vectors that point along coordinate axes, and in a rectangular coordinate system these unit vectors are called **i, j,** and **k**. They point, respectively, along the positive x, y, and z axes as shown in Fig. 3–15. Like other vectors, **i, j,** and **k** do not have to be placed at the origin, but can be placed elsewhere as long as the direction and unit length remain unchanged. Sometimes you may see unit vectors written with a "hat": **î, ĵ, k̂**.
Because of the definition of multiplication of a vector by a scalar (Section 3–3), the components of a vector **V** can be written $V_x = V_x\mathbf{i}$, $V_y = V_y\mathbf{j}$,

and $\mathbf{V}_z = V_z\mathbf{k}$. Hence any vector $\mathbf{V}$ can be written in terms of its components as

$$\mathbf{V} = V_x\mathbf{i} + V_y\mathbf{j} + V_z\mathbf{k}. \tag{3-5}$$

Unit vectors are helpful when adding vectors analytically by components. For example, Eq. 3–4 can be seen to be true by using unit vector notation for each vector (which we write for the two-dimensional case, but the extension to three dimensions is straightforward):

$$\begin{aligned}
\mathbf{V} &= (V_x)\mathbf{i} + (V_y)\mathbf{j} \\
&= \mathbf{V}_1 + \mathbf{V}_2 \\
&= (V_{1x}\mathbf{i} + V_{1y}\mathbf{j}) + (V_{2x}\mathbf{i} + V_{2y}\mathbf{j}) \\
&= (V_{1x} + V_{2x})\mathbf{i} + (V_{1y} + V_{2y})\mathbf{j}.
\end{aligned}$$

Comparing the first line to the fourth line, we get Eq. 3–4.

EXAMPLE 3–3 Using unit vectors. Write the vectors of Example 3–1 in unit vector notation, and perform the addition.

SOLUTION In Example 3–1 we resolved $\mathbf{D}_1$ and $\mathbf{D}_2$ into components and found
$$D_{1x} = 0, D_{1y} = 22.0\,\text{km}, \text{ and } D_{2x} = 23.5\,\text{km} \text{ and } D_{2y} = -40.7\,\text{km}.$$
Thus we have

$$\begin{aligned}
\mathbf{D}_1 &= 0\mathbf{i} + 22.0\,\text{km}\,\mathbf{j} \\
\mathbf{D}_2 &= 23.5\,\text{km}\,\mathbf{i} - 40.7\,\text{km}\,\mathbf{j}.
\end{aligned}$$

Then

$$\begin{aligned}
\mathbf{D}_1 + \mathbf{D}_2 &= (0 + 23.5)\,\text{km}\,\mathbf{i} + (22.0 - 40.7)\,\text{km}\,\mathbf{j} \\
&= 23.5\,\text{km}\,\mathbf{i} - 18.7\,\text{km}\,\mathbf{j}.
\end{aligned}$$

The components of the resultant displacement, $\mathbf{D}$, are $D_x = 23.5\,\text{km}$ and $D_y = -18.7\,\text{km}$.

3–6 | Vector Kinematics

We can now extend our definitions of velocity and acceleration in a formal way to two- and three-dimensional motion. Suppose a particle follows a path in the xy plane as shown in Fig. 3–16. At time t_1, the particle is at point P_1, and at time t_2, it is at point P_2. The vector $\mathbf{r}_1$ is the position vector of the particle at time t_1 (it represents the displacement of the particle from the origin of the coordinate system). And $\mathbf{r}_2$ is the position vector at time t_2.

In one dimension, we defined displacement as the *change in position* of the particle. In the more general case of two or three dimensions, the **displacement vector** is defined as the vector representing change in position. We call it $\Delta\mathbf{r}$,[†] where

$$\Delta\mathbf{r} = \mathbf{r}_2 - \mathbf{r}_1.$$

This represents the displacement during the time interval $\Delta t = t_2 - t_1$. In unit vector notation, we can write

$$\mathbf{r}_1 = x_1\mathbf{i} + y_1\mathbf{j} + z_1\mathbf{k}, \tag{3-6a}$$

where x_1, y_1, and z_1 are the coordinates of point P_1 (Fig. 3–16). Similarly,

$$\mathbf{r}_2 = x_2\mathbf{i} + y_2\mathbf{j} + z_2\mathbf{k}.$$

Hence

$$\Delta\mathbf{r} = (x_2 - x_1)\mathbf{i} + (y_2 - y_1)\mathbf{j} + (z_2 - z_1)\mathbf{k}. \tag{3-6b}$$

If the motion is along the x axis only, then $y_2 - y_1 = 0$, $z_2 - z_1 = 0$, and the magnitude of the displacement is $\Delta r = x_2 - x_1$, which is consistent with our earlier one-dimensional equation (Section 2–1). Even in one dimension, displacement is a vector, as are velocity and acceleration.

FIGURE 3–16 Path of a particle in the xy plane. At time t_1 the particle is at point P_1 given by the position vector $\mathbf{r}_1$; at t_2 the particle is at point P_2 given by the position vector $\mathbf{r}_2$. The displacement vector for the time interval $t_2 - t_1$ is $\Delta\mathbf{r} = \mathbf{r}_2 - \mathbf{r}_1$.

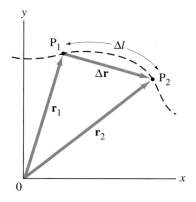

[†]We used $\mathbf{D}$ for the displacement vector earlier in the chapter for illustrating vector addition. The new notation here, $\Delta\mathbf{r}$, emphasizes that it is the difference between two position vectors.

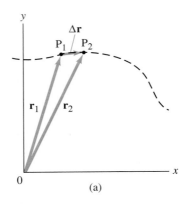
(a)

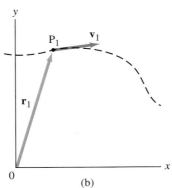
(b)

FIGURE 3–17 As we take Δt and $\Delta \mathbf{r}$ smaller and smaller [compare Fig. 3–16 and part (a) of this figure] we see that the direction of $\Delta \mathbf{r}$ and of the instantaneous velocity ($\Delta \mathbf{r}/\Delta t$, where $\Delta t \rightarrow 0$) is tangent to the curve at P_1 (part b).

FIGURE 3–18 (a) Velocity vectors $\mathbf{v}_1$ and $\mathbf{v}_2$ at instants t_1 and t_2 for the particle of Fig. 3–16. (b) Direction of the average acceleration in this case, $\bar{\mathbf{a}} = \Delta \mathbf{v}/\Delta t$.

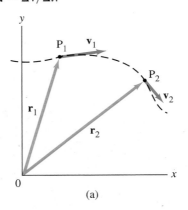
(a)

(b)

The **average velocity vector**, $\bar{\mathbf{v}}$, over the time interval $\Delta t = t_2 - t_1$ is defined as

$$\bar{\mathbf{v}} = \frac{\Delta \mathbf{r}}{\Delta t}. \tag{3–7}$$

Since $\mathbf{v}$ is a product of the vector $\Delta \mathbf{r}$ times a scalar $(1/\Delta t)$, the direction of $\mathbf{v}$ is the same as that of $\Delta \mathbf{r}$; and its magnitude is $\Delta r/\Delta t$.

We next consider shorter and shorter time intervals—that is, we let Δt approach zero so that the distance between points P_2 and P_1 also approaches zero. We define the **instantaneous velocity vector** as the limit of the average velocity as Δt approaches zero:

$$\mathbf{v} = \lim_{\Delta t \to 0} \frac{\Delta \mathbf{r}}{\Delta t} = \frac{d\mathbf{r}}{dt}. \tag{3–8}$$

The direction of $\mathbf{v}$ at any moment is along the line tangent to the path at that moment (Fig. 3–17).

Note that the magnitude of the average velocity in Fig. 3–16 is not equal to the average speed, which is the actual distance traveled, Δl, divided by Δt. In some special cases, the average speed and average velocity are equal (such as motion along a straight line in one direction), but in general they are not. However, in the limit $\Delta t \rightarrow 0$, Δr always approaches Δl, so the instantaneous speed *always* equals the magnitude of the instantaneous velocity at any time.

The instantaneous velocity (Eq. 3–8) is equal to the derivative of the position vector with respect to time. Equation 3–8 can be written in terms of components starting with Eq. 3–6a as:

$$\mathbf{v} = \frac{d\mathbf{r}}{dt} = \frac{dx}{dt}\mathbf{i} + \frac{dy}{dt}\mathbf{j} + \frac{dz}{dt}\mathbf{k}$$
$$= v_x\mathbf{i} + v_y\mathbf{j} + v_z\mathbf{k}, \tag{3–9}$$

where $v_x = dx/dt$, $v_y = dy/dt$, $v_z = dz/dt$ are the x, y, and z components of the velocity. Note that $d\mathbf{i}/dt = d\mathbf{j}/dt = d\mathbf{k}/dt = 0$ since these unit vectors are constant in both magnitude and direction.

Acceleration in two or three dimensions is treated in a similar way. The **average acceleration vector**, $\bar{\mathbf{a}}$, over a time interval $\Delta t = t_2 - t_1$ is defined as

$$\bar{\mathbf{a}} = \frac{\Delta \mathbf{v}}{\Delta t} = \frac{\mathbf{v}_2 - \mathbf{v}_1}{t_2 - t_1}, \tag{3–10}$$

where $\Delta \mathbf{v}$ is the change in the instantaneous velocity vector during that time interval: $\Delta \mathbf{v} = \mathbf{v}_2 - \mathbf{v}_1$. Note that $\mathbf{v}_2$ in many cases, such as in Fig. 3–18a, may not be in the same direction as $\mathbf{v}_1$. Hence $\bar{\mathbf{a}}$ may be in a different direction from either $\mathbf{v}_1$ or $\mathbf{v}_2$ (Fig. 3–18b). Furthermore, $\mathbf{v}_2$ and $\mathbf{v}_1$ may have the same magnitude but different directions, and the difference of two such vectors will not be zero. Hence acceleration can result from either a change in the magnitude of the velocity, or from a change in direction of the velocity, or from a change in both.

The **instantaneous acceleration vector** is defined as the limit of the average acceleration vector as the time interval Δt is allowed to approach zero:

$$\mathbf{a} = \lim_{\Delta t \to 0} \frac{\Delta \mathbf{v}}{\Delta t} = \frac{d\mathbf{v}}{dt}, \tag{3–11}$$

and is thus the derivative of $\mathbf{v}$ with respect to t. Using components gives us:

$$\mathbf{a} = \frac{d\mathbf{v}}{dt} = \frac{dv_x}{dt}\mathbf{i} + \frac{dv_y}{dt}\mathbf{j} + \frac{dv_z}{dt}\mathbf{k}$$
$$= a_x\mathbf{i} + a_y\mathbf{j} + a_z\mathbf{k}, \tag{3–12}$$

where $a_x = dv_x/dt$, etc. The instantaneous acceleration will be nonzero not only when the magnitude of the velocity changes, but also if its direction changes. For example, a person riding in a car traveling at constant speed around a curve, or a child riding on a merry-go-round, will both experience an acceleration because of a change in the direction of the velocity, even though its magnitude may be constant. (More on this later.)

In general, we will use the terms "velocity" and "acceleration" to mean the instantaneous values. If we want to discuss average values, we will use the word "average."

Constant Acceleration

In Chapter 2 we studied the important case of one-dimensional motion for which the acceleration is a constant. We now consider motion in two or three dimensions for which the acceleration vector, $\mathbf{a}$, is constant in magnitude and direction. That is, $a_x = $ constant, $a_y = $ constant, $a_z = $ constant. The average acceleration in this case is equal to the instantaneous acceleration at any moment. The equations we derived in Chapter 2 for one dimension, Eqs. 2–12a, b, and c, apply separately to each perpendicular component of two- or three-dimensional motion. In two dimensions we let $\mathbf{v}_0 = v_{x0}\mathbf{i} + v_{y0}\mathbf{j}$ be the initial velocity, and we apply Eqs. 3–6a, 3–9, and 3–12 for the position vector, $\mathbf{r}$, velocity, $\mathbf{v}$, and acceleration, $\mathbf{a}$. We can then write Eqs. 2–12a, b, and c, for two dimensions as shown in Table 3–1.

TABLE 3–1   Kinematic Equations for Constant Acceleration in 2 Dimensions

x Component (horizontal)		y Component (vertical)
$v_x = v_{x0} + a_x t$	(Eq. 2–12a)	$v_y = v_{y0} + a_y t$
$x = x_0 + v_{x0}t + \frac{1}{2}a_x t^2$	(Eq. 2–12b)	$y = y_0 + v_{y0}t + \frac{1}{2}a_y t^2$
$v_x^2 = v_{x0}^2 + 2a_x(x - x_0)$	(Eq. 2–12c)	$v_y^2 = v_{y0}^2 + 2a_y(y - y_0)$

The first two of the equations in Table 3–1 can be written more formally in vector notation (see Eqs. 3–6a, 3–9, and 3–12):

$$\mathbf{v} = \mathbf{v}_0 + \mathbf{a}t \qquad\qquad [\mathbf{a} = \text{constant}] \quad \textbf{(3–13a)}$$

$$\mathbf{r} = \mathbf{r}_0 + \mathbf{v}_0 t + \frac{1}{2}\mathbf{a}t^2. \qquad\qquad [\mathbf{a} = \text{constant}] \quad \textbf{(3–13b)}$$

Here, $\mathbf{r}$ is the position vector at any time, and $\mathbf{r}_0$ is the position vector at $t = 0$. These equations are the vector equivalent of Eqs. 2–12a and b. In practical situations, we usually use the component form given in Table 3–1.

These equations, and their use, will become clearer as we use them. We next deal with several types of motion in a plane which are common in everyday life: projectile motion, and circular motion.

3–7 Projectile Motion

In Chapter 2, we studied the motion of objects in one dimension in terms of displacement, velocity, and acceleration, including purely vertical motion of falling bodies undergoing acceleration due to gravity. Now we examine the more general motion of objects moving through the air in two dimensions near the Earth's surface, such as a golf ball, a thrown or batted baseball, kicked footballs, speeding bullets, and athletes doing the long jump or high jump. These are all examples of **projectile motion** (see Fig. 3–19), which we can describe as taking place in two dimensions. Although air resistance is often important, in many cases its effect can be ignored, and we will ignore it in the following analysis. We will not be concerned now with the process by which the object is thrown or projected. We consider only its motion *after* it has been projected and is moving freely through the air under the action of gravity alone. Thus the acceleration of the object is that

FIGURE 3–19 This strobe photograph of a ball making a series of bounces shows the characteristic "parabolic" path of projectile motion.

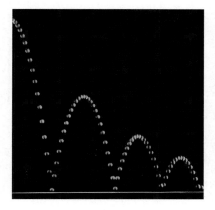

due to gravity, which acts downward with magnitude $g = 9.80 \, \text{m/s}^2$, and we assume it is constant.[†]

Horizontal and vertical motion analyzed separately

Galileo first accurately described projectile motion. He showed that it could be understood by analyzing the horizontal and vertical components of the motion separately. For convenience, we assume that the motion begins at time $t = 0$ at the origin of an xy coordinate system (so $x_0 = y_0 = 0$).

Let us look at a (tiny) ball rolling off the end of a table with an initial velocity v_{x0} in the horizontal (x) direction. See Fig. 3–20, where an object falling vertically is also shown for comparison. The velocity vector **v** at each instant points in the direction of the ball's motion at that instant and is always tangent to the path. Following Galileo's ideas, we treat the horizontal and vertical components of the velocity, v_x and v_y, separately, and we can apply the kinematic equations (Eqs. 2–12a through 2–12c) to each.

Vertical motion ($a_y = $ constant)

First we examine the vertical (y) component of the motion. Once the ball leaves the table (at $t = 0$), it experiences a vertically downward acceleration, g, the acceleration due to gravity. Thus v_y is initially zero $(v_{y0} = 0)$ but increases continually in the downward direction (until the ball hits the ground). Let us take y to be positive upwards. Then $a_y = -g$, and from Eq. 2–12a we can write $v_y = -gt$ since we set $y_0 = 0$.

Horizontal motion ($a_x = 0, v_x = $ constant)

In the horizontal direction, on the other hand, there is no acceleration. So the horizontal component of velocity, v_x, remains constant, equal to its initial value, v_{x0}, and thus has the same magnitude at each point on the path. The two vector components, **v**$_x$ and **v**$_y$, can be added vectorially to obtain the velocity **v** for each point on the path, as shown in Fig. 3–20.

One result of this analysis, which Galileo himself predicted, is that *an object projected horizontally will reach the ground in the same time as an object dropped vertically*. This is because the vertical motions are the same in both cases, as shown in Fig. 3–20. Figure 3–21 is a multiple-exposure photograph of an experiment that confirms this.

[†] This restricts us to objects whose distance traveled and maximum height above the Earth are small compared to the Earth's radius (6400 km).

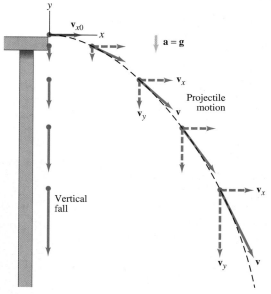

FIGURE 3–20 Projectile motion. (A vertically falling object is shown at the left for comparison.)

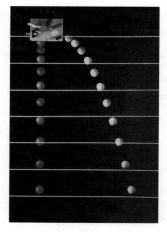

FIGURE 3–21 Multiple-exposure photograph showing positions of two balls at equal time intervals. One ball was dropped from rest at the same time the other was projected horizontally outward. The vertical position of each ball is seen to be the same.

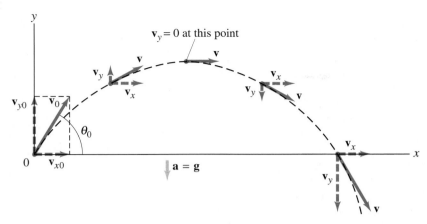

FIGURE 3–22 Path of a projectile fired with initial velocity $\mathbf{v}_0$ at angle θ to the horizontal. Path is shown in black, the velocity vectors are green arrows, and velocity components are dashed.

If an object is projected at an upward angle, as in Fig. 3–22, the analysis is similar, except that now there is an initial vertical component of velocity, v_{y0}. Because of the downward acceleration of gravity, v_y continually decreases until the object reaches the highest point on its path in Fig. 3–22, at which point $v_y = 0$. Then v_y starts to increase in the downward direction, as shown (that is, becoming negative). As before, v_x remains constant.

Galileo's analysis of almost four centuries ago is exactly equivalent to using Eqs. 2–12 separately for the horizontal (x) and vertical (y) components, as given in Table 3–1 (Section 3–6). The constant acceleration now is that of gravity alone, acting downward. As can be seen in Fig. 3–22 for an object projected at an upward angle θ, the acceleration is in one (constant) direction, whereas the velocity has two components, one of which (v_y) is continuously changing, whereas the other (v_x) stays constant.

We can simplify Eqs. 2–12 (Table 3–1) for use with projectile motion by setting $a_x = 0$. See Table 3–2, which assumes y is positive upward, so $a_y = -g = -9.80 \text{ m/s}^2$. Note also that if θ is chosen as in Fig. 3–22 then the initial velocity has components:

$$v_{x0} = v_0 \cos \theta,$$

$$v_{y0} = v_0 \sin \theta.$$

Object projected upward

TABLE 3–2 Kinematic Equations for Projectile Motion (y positive upward; $a_x = 0$, $a_y = -g = -9.80 \text{ m/s}^2$)		
Horizontal Motion $(a_x = 0, v_x = \text{constant})$		**Vertical Motion**[†] $(a_y = -g = \text{constant})$
$v_x = v_{x0}$	(Eq. 2–12a)	$v_y = v_{y0} - gt$
$x = x_0 + v_{x0}t$	(Eq. 2–12b)	$y = y_0 + v_{y0}t - \frac{1}{2}gt^2$
	(Eq. 2–12c)	$v_y^2 = v_{y0}^2 - 2gy$

[†]If y is taken positive downward, the minus ($-$) signs become $+$ signs.

3–8 Solving Problems Involving Projectile Motion

We now work through several Examples of projectile motion quantitatively. First, we give a summary of how you can approach these types of problems.

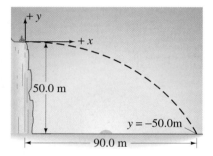

FIGURE 3–23 Example 3–4.

EXAMPLE 3–4 **Driving off a cliff.** A movie stunt driver on a motorcycle speeds horizontally off a 50.0-m-high cliff. How fast must the motorcycle leave the cliff-top if it is to land on level ground below, 90.0 m from the base of the cliff (Fig. 3–23) where the cameras are?

SOLUTION We take the y direction to be positive upward, with the top of the cliff as $y_0 = 0$, so the bottom is at $y = -50.0$ m. First, we find how long it takes the motorcycle to reach the ground below. We use Eq. 2–12b for the vertical (y) direction (Table 3–2) with $y_0 = 0$, and $v_{y0} = 0$:

$$y = -\tfrac{1}{2}gt^2$$

We solve for t and set $y = -50.0$ m:

$$t = \sqrt{\frac{2y}{-g}} = \sqrt{\frac{2(-50.0 \text{ m})}{-9.80 \text{ m/s}^2}} = 3.19 \text{ s}.$$

To calculate the initial velocity, v_{x0}, we again use Eq. 2–12b, but this time for the horizontal (x) direction, with $a_x = 0$ and $x_0 = 0$:

$$x = v_{x0}t$$

$$v_{x0} = \frac{x}{t} = \frac{90.0 \text{ m}}{3.19 \text{ s}} = 28.2 \text{ m/s},$$

which is 101 km/h (about 60 mi/h).

➡ PHYSICS APPLIED

Sports

EXAMPLE 3–5 **A kicked football.** A football is kicked at an angle $\theta_0 = 37.0°$ with a velocity of 20.0 m/s, as shown in Fig. 3–24. Calculate (a) the maximum height, (b) the time of travel before the football hits the ground, (c) how far away it hits the ground, (d) the velocity vector at the maximum height, and (e) the acceleration vector at maximum height. Assume the ball leaves the foot at ground level, and ignore air resistance (not very realistic, though).

SOLUTION We take the y direction as positive upward. The components of the initial velocity are (Fig. 3–24):

$$v_{x0} = v_0 \cos 37.0° = (20.0 \text{ m/s})(0.799) = 16.0 \text{ m/s}$$

$$v_{y0} = v_0 \sin 37.0° = (20.0 \text{ m/s})(0.602) = 12.0 \text{ m/s}.$$

(a) At the maximum height, the velocity is horizontal (Fig. 3–24), so $v_y = 0$; and this occurs (see Eq. 2–12a in Table 3–2) at time

$$t = \frac{v_{y0}}{g} = \frac{12.0 \text{ m/s}}{9.80 \text{ m/s}^2} = 1.22 \text{ s.}$$

From Eq. 2–12b, with $y_0 = 0$, we have

$$y = v_{y0}t - \tfrac{1}{2}gt^2$$
$$= (12.0 \text{ m/s})(1.22 \text{ s}) - \tfrac{1}{2}(9.80 \text{ m/s}^2)(1.22 \text{ s})^2 = 7.35 \text{ m.}$$

Alternatively, we could have used Eq. 2–12c, solved for y, and found

$$y = \frac{v_{y0}^2 - v_y^2}{2g} = \frac{(12.0 \text{ m/s})^2 - (0 \text{ m/s})^2}{2(9.80 \text{ m/s}^2)} = 7.35 \text{ m.}$$

(b) To find the time it takes for the ball to return to the ground, we use Eq. 2–12b with $y_0 = 0$ and also set $y = 0$ (ground level):

$$y = y_0 + v_{y0}t - \tfrac{1}{2}gt^2$$
$$0 = 0 + (12.0 \text{ m/s})t - \tfrac{1}{2}(9.80 \text{ m/s}^2)t^2$$

which is an equation that can be easily factored:

$$\left[\tfrac{1}{2}(9.80 \text{ m/s}^2)t - 12.0 \text{ m/s}\right]t = 0.$$

There are two solutions, $t = 0$ (which corresponds to the initial point, y_0), and

$$t = \frac{2(12.0 \text{ m/s})}{(9.80 \text{ m/s}^2)} = 2.45 \text{ s,}$$

which is the result we sought. Note that the time $t = 2.45$ s is just double the time to reach the highest point, calculated in (a). That is, the time to go up equals the time to come down to the same level—ignoring air resistance.

(c) The total distance traveled in the x direction is found by applying Eq. 2–12b with $x_0 = 0$, $a_x = 0$, $v_{x0} = 16.0$ m/s:

$$x = v_{x0}t = (16.0 \text{ m/s})(2.45 \text{ s}) = 39.2 \text{ m.}$$

(d) At the highest point, there is no vertical component to the velocity. There is only the horizontal component (which remains constant throughout the flight), so $v = v_{x0} = v_0 \cos 37.0° = 16.0$ m/s.

(e) The acceleration vector is the same at the highest point as it is throughout the flight, which is 9.80 m/s² downward.

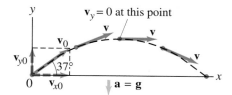

FIGURE 3–24 Example 3–5. (See also Fig. 3–22.)

Note the symmetry

CONCEPTUAL EXAMPLE 3–6 Where does the apple land? A child sits upright in a wagon which is moving to the right at constant speed as shown in Fig. 3–25. The child extends her hand and throws an apple straight upward (from her own point of view, Fig. 3–25a), while the wagon continues to travel forward at constant speed. If air resistance is neglected, will the apple land (a) behind the wagon, (b) in the wagon, or (c) in front of the wagon?

RESPONSE The child throws the apple straight up from her own point of view with initial velocity $\mathbf{v}_{0y}$ (Fig. 3–25a). But when viewed by someone on the ground, the apple also has an initial horizontal component of velocity equal to the speed of the wagon, $\mathbf{v}_{0x}$. Thus, to a person on the ground, the apple will follow the path of a projectile as shown in Fig. 3–25b. The apple experiences no horizontal acceleration, so $\mathbf{v}_{0x}$ will stay constant and equal to the speed of the wagon. As the apple follows its arc, the wagon will be directly under the apple at all times because they have the same horizontal velocity. When the apple comes down, it will drop right into the wagon, and into the outstretched hand of the child. The answer is (b).

FIGURE 3–25 Conceptual Example 3–6.

(a) Wagon reference frame

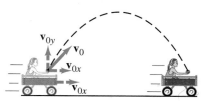

(b) Ground reference frame

CONCEPTUAL EXAMPLE 3–7 **The wrong strategy.** A boy on a small hill aims his water-balloon slingshot horizontally, straight at a second boy hanging from a tree branch a distance d away, Fig. 3–26. At the instant the water balloon is released, the second boy lets go and falls from the tree, hoping to avoid being hit. Show that he made the wrong move. (He hadn't studied physics yet.)

RESPONSE Both the water balloon and the boy in the tree start falling at the same instant, and in a time t they each fall the same vertical distance $y = \frac{1}{2}gt^2$ (see Fig. 3–21). In the time it takes the water balloon to travel the horizontal distance d, the balloon will have the same y position as the falling boy. Splat. If the boy had stayed in the tree, he'd have saved himself the humiliation.

FIGURE 3–26
Example 3–7.

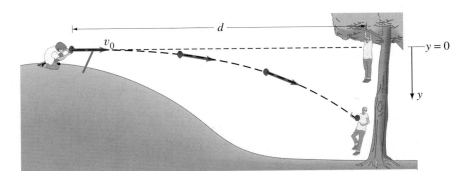

➡ **PHYSICS APPLIED**

Horizontal range of a projectile

FIGURE 3–27 Example 3–8.
(a) The range R of a projectile;
(b) shows how generally there are two angles θ_0 that will give the same range. Can you show that if one angle is θ_{01}, the other is $\theta_{02} = 90° - \theta_{01}$?

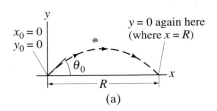

(a)

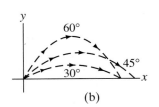

(b)

Level range formula
$[y \text{ (final)} = y_0]$

EXAMPLE 3–8 **Level horizontal range.** (*a*) Derive a formula for the horizontal range R of a projectile in terms of its initial velocity v_0 and angle θ_0. The horizontal range is defined as the horizontal distance the projectile travels before returning to its original height, which is typically the ground; that is, y (final) $= y_0$. See Fig. 3–27. (*b*) Suppose one of Napoleon's cannons had a muzzle velocity, v_0, of 60.0 m/s. At what angle should it have been aimed (ignore air resistance) to strike a target 320 m away?

SOLUTION (*a*) We set $x_0 = 0$ and $y_0 = 0$ at $t = 0$. After the projectile travels a horizontal distance R, it returns to the same level, $y = 0$, the final point. So to find a general expression for R, we set both $y = 0$ and $y_0 = 0$ in Eq. 2–12b for the vertical motion, and obtain

$$v_{y0}t - \tfrac{1}{2}gt^2 = 0.$$

We solve for t, which gives two solutions: $t = 0$ and $t = 2v_{y0}/g$. The first solution corresponds to the initial instant of projection and the second is the time when the projectile returns to $y = 0$. Then the range, R, will be equal to x at the moment t has this value, which we put into Eq. 2–12b for the horizontal motion ($x = v_{x0}t$, with $x_0 = 0$). Thus we have:

$$R = x = v_{x0}t = v_{x0}\left(\frac{2v_{y0}}{g}\right) = \frac{2v_{x0}v_{y0}}{g} = \frac{2v_0^2 \sin\theta_0 \cos\theta_0}{g} \qquad [y = y_0]$$

where we have written $v_{x0} = v_0 \cos\theta_0$ and $v_{y0} = v_0 \sin\theta_0$. This is the result we sought, and it can be rewritten, using the trigonometric identity $2\sin\theta\cos\theta = \sin 2\theta$ (Appendix A), as

$$R = \frac{v_0^2 \sin 2\theta_0}{g}. \qquad [y = y_0]$$

We see that the maximum range, for a given initial velocity v_0, is obtained when the sine takes on its maximum value of 1.0, which occurs for $2\theta_0 = 90°$; so

$$\theta_0 = 45° \text{ for maximum range, and } R_{\text{max}} = v_0^2/g.$$

[When air resistance is important, the range is less for a given v_0, and the maximum range is obtained at an angle smaller than 45°.] Note that the maximum range increases by the square of v_0, so doubling the muzzle velocity of a cannon increases its maximum range by a factor of 4.

(b) From the equation we just derived, Napoleon's cannon should be aimed (assuming, unrealistically, no air resistance) at an angle θ_0 given by

$$\sin 2\theta_0 = \frac{Rg}{v_0^2} = \frac{(320\,\text{m})(9.80\,\text{m/s}^2)}{(60.0\,\text{m/s})^2} = 0.871.$$

We want to solve for an angle θ_0 that is between 0° and 90°, which means $2\theta_0$ in this equation can be as large as 180°. Thus, $2\theta_0 = 60.6°$ is a solution, but $2\theta_0 = 180° - 60.6° = 119.4°$ is also a solution (see Appendix A). In general we will have two solutions, which in Napoleon's case are given by

$$\theta_0 = 30.3° \text{ or } 59.7°.$$

Either angle gives the same range. Only when $\sin 2\theta_0 = 1$ (so $\theta_0 = 45°$) is there a single solution (that is, both solutions are the same).

EXAMPLE 3–9 **A punt.** Suppose the football in Example 3–5 was a punt and left the punter's foot at a height of 1.00 m above the ground. How far did the football travel before hitting the ground? Set $x_0 = 0$, $y_0 = 0$.

➡ **PHYSICS APPLIED**

Sports

SOLUTION We cannot use the range formula from Example 3–8 because it is valid only if y (final) $= y_0$, which is not the case here. We have $y_0 = 0$, and the football hits the ground where $y = -1.00\,\text{m}$ (see Fig. 3–28). We can get x from Eq. 2–12b, $x = v_{x0}t$, since we know that $v_{x0} = 16.0\,\text{m/s}$. But first we must find t, the time at which the ball hits the ground. With $y = -1.00\,\text{m}$ and $v_{y0} = 12.0\,\text{m/s}$ (see Example 3–5), we use the equation

➡ **PROBLEM SOLVING**

Do not use any formula unless you are sure its range of validity fits the problem. The range formula does not apply here because $y \ne y_0$

$$y = y_0 + v_{y0}t - \tfrac{1}{2}gt^2,$$

and obtain

$$-1.00\,\text{m} = 0 + (12.0\,\text{m/s})t - (4.90\,\text{m/s}^2)t^2.$$

Rearranging this equation into standard form and using the quadratic formula gives

$$t = \frac{12.0\,\text{m/s} \pm \sqrt{(12.0\,\text{m/s})^2 - 4(4.90\,\text{m/s}^2)(-1.00\,\text{m})}}{2(4.90\,\text{m/s}^2)}$$

$$= 2.53\,\text{s} \quad \text{or} \quad -0.081\,\text{s}.$$

The second solution would correspond to a time previous to the kick, so it doesn't apply here. With $t = 2.53\,\text{s}$ for the time at which the ball touches the ground, the distance the ball traveled is (putting $v_{x0} = 16.0\,\text{m/s}$, from Example 3–5):

$$x = v_{x0}t = (16.0\,\text{m/s})(2.53\,\text{s}) = 40.5\,\text{m}.$$

Note that our assumption in Example 3–5 that the ball leaves the foot at ground level results in an underestimate of about 1.3 m in the distance traveled.

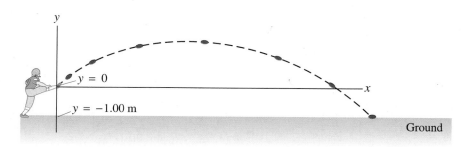

FIGURE 3–28 Example 3–9: the football leaves the punter's foot at $y = 0$, and reaches the ground where $y = -1.00\,\text{m}$.

$y = 0$

$y = -1.00\,\text{m}$

Ground

EXAMPLE 3–10 **Rescue plane drops supplies.** A rescue plane wants to drop supplies to isolated mountain climbers on a rocky ridge 200 m below. If the plane is traveling horizontally with a speed of 250 km/h (69 m/s), (a) how far in advance of the recipients (horizontal distance) must the goods be dropped (Fig. 3–29a)? (b) Suppose, instead, that the plane releases the supplies a horizontal distance of 400 m in advance of the mountain climbers. What vertical velocity should they give the supplies (up or down) so that they arrive precisely at the climbers' position (Fig. 3–29b)? (c) With what speed do the supplies land in the latter case?

SOLUTION (a) The vertical motion (take $+y$ upwards) doesn't depend on the horizontal motion, so we can find the time to reach the climbers using the vertical distance of 200 m. The supplies are "dropped" so initially they have the velocity of the airplane, $v_{x0} = 69$ m/s, $v_{y0} = 0$. Then, since $y = -\frac{1}{2}gt^2$, we have

$$t = \sqrt{\frac{-2y}{g}} = \sqrt{\frac{-2(-200\,\text{m})}{9.80\,\text{m/s}^2}} = 6.39\,\text{s}.$$

The horizontal motion of the falling supplies is at constant speed of 69 m/s. So

$$x = v_{x0}t = (69\,\text{m/s})(6.39\,\text{s}) = 440\,\text{m}.$$

(b) We are given $x = 400$ m, $v_{x0} = 69$ m/s, $y = -200$ m, and we want to find v_{y0} (see Fig. 3–29b). Like most problems, this one can be approached in various ways. Instead of searching for a formula or two, let's try to reason it out in a simple way, based on what we did in part (a). If we know t, perhaps we can get v_{y0}. Since the horizontal motion of the supplies is at constant speed (once they are released we don't care what the plane does), we have $x = v_{x0}t$, so

$$t = \frac{x}{v_{x0}} = \frac{400\,\text{m}}{69\,\text{m/s}} = 5.80\,\text{s}.$$

Now let's try to use the vertical motion to get v_{y0}: $y = y_0 + v_{y0}t - \frac{1}{2}gt^2$. Since $y_0 = 0$ and $y = -200$ m, we can solve for v_{y0}:

$$v_{y0} = \frac{y + \frac{1}{2}gt^2}{t} = \frac{-200\,\text{m} + \frac{1}{2}(9.80\,\text{m/s}^2)(5.80\,\text{s})^2}{5.80\,\text{s}} = -6.1\,\text{m/s}.$$

Thus, in order to arrive at precisely the mountain climbers' position, the supplies must be thrown *downward* from the plane with a speed of 6.1 m/s.

(c) We want to know v of the supplies at $t = 5.80$ s. The components are:

$$v_x = v_{x0} = 69\,\text{m/s}$$

$$v_y = v_{y0} - gt = -6.1\,\text{m/s} - (9.80\,\text{m/s}^2)(5.80\,\text{s}) = -63\,\text{m/s}.$$

So $v = \sqrt{(69\,\text{m/s})^2 + (-63\,\text{m/s})^2} = 93\,\text{m/s}$. (Better not to release the supplies from such an altitude or to use a parachute.)

FIGURE 3–29 Example 3–10.

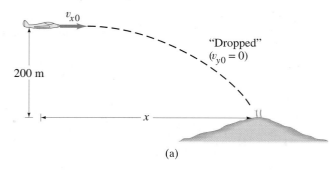

200 m

v_{x0}

"Dropped"
$(v_{y0} = 0)$

x

(a)

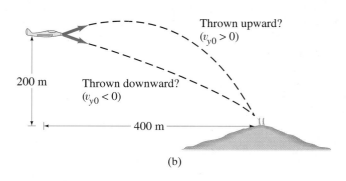

200 m

Thrown upward?
$(v_{y0} > 0)$

Thrown downward?
$(v_{y0} < 0)$

400 m

(b)

Projectile Motion Is Parabolic

We now show that the path followed by any projectile is a *parabola*, if we can ignore air resistance and can assume that **g** is constant. To do so, we need to find y as a function of x by eliminating t between the two equations for horizontal and vertical motion (Eq. 2–12b in Table 3–2), and we set $x_0 = y_0 = 0$:

$$x = v_{x0}t$$

$$y = v_{y0}t - \tfrac{1}{2}gt^2.$$

From the first equation, we have $t = x/v_{x0}$, and we substitute this into the second one to obtain

$$y = \left(\frac{v_{y0}}{v_{x0}}\right)x - \left(\frac{g}{2v_{x0}^2}\right)x^2.$$

We see that y as a function of x has the form

$$y = Ax - Bx^2,$$

Projectile motion equation is a parabola

where A and B are constants for any specific projectile motion. This is the well-known equation for a parabola. See Figs. 3–19 and 3–30.

The idea that projectile motion is parabolic was, in Galileo's day, at the forefront of physics research. Today we discuss it in Chapter 3 of introductory physics!

FIGURE 3–30 Examples of projectile motion—sparks (small hot glowing pieces of metal), water, and fireworks. All exhibit the parabolic path characteristic of projectile motion, although the effects of air resistance can be seen to significantly alter the path of some trajectories.

3–9 Uniform Circular Motion

An object that moves in a circle at constant speed v is said to undergo **uniform circular motion**. Examples are a ball on the end of a string revolved about one's head, and the nearly uniform circular motion of the Moon around the Earth. The *magnitude* of the velocity remains constant in this case, but the *direction* of the velocity is continually changing (Fig. 3–31). Since acceleration is defined as the rate of change of velocity, a change in direction of velocity means an acceleration is occurring, just as does a change in magnitude. Thus an object undergoing uniform circular motion is accelerating even if the speed remains constant $(v_1 = v_2)$. We now investigate this acceleration quantitatively.

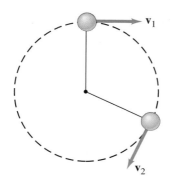

FIGURE 3–31 A particle moving in a circle, showing how the velocity changes direction. Note that at each point, the instantaneous velocity is in a direction tangent to the circular path.

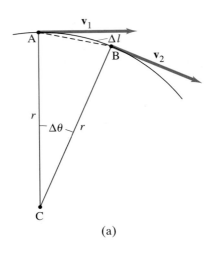

(a)

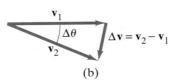

(b)

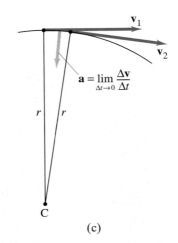

(c)

FIGURE 3–32 Determining the change in velocity, $\Delta\mathbf{v}$, for a particle moving in a circle.

Acceleration is defined by

$$\mathbf{a} = \lim_{\Delta t \to 0} \frac{\Delta\mathbf{v}}{\Delta t} = \frac{d\mathbf{v}}{dt},$$

where $\Delta\mathbf{v}$ is the change in velocity during the short time interval Δt. We will eventually consider the situation when Δt approaches zero and thus obtain the instantaneous acceleration. But for purposes of making a clear drawing we consider a nonzero time interval (Fig. 3–32). During the time Δt the particle in Fig. 3–32a moves from point A to point B, covering a small distance Δl *along the arc* which subtends a small angle $\Delta\theta$. The change in the velocity vector is $\mathbf{v}_2 - \mathbf{v}_1 = \Delta\mathbf{v}$, and is shown in Fig. 3–32b.

If we let Δt be very small (approaching zero), then Δl and $\Delta\theta$ are also very small; then $\mathbf{v}_2$ will be almost parallel to $\mathbf{v}_1$, and $\Delta\mathbf{v}$ will be essentially perpendicular to them (Fig. 3–32c). Thus $\Delta\mathbf{v}$ points toward the center of the circle. Since $\mathbf{a}$, by definition, is in the same direction as $\Delta\mathbf{v}$, it too must point toward the center of the circle. Therefore, this acceleration is called **centripetal acceleration** ("center-seeking" acceleration) or **radial acceleration** (since it is directed along the radius, toward the center of the circle), and we denote it by $\mathbf{a}_R$.

We next determine the magnitude of the centripetal (radial) acceleration, a_R. Because CA is perpendicular to $\mathbf{v}_1$, and CB is perpendicular to $\mathbf{v}_2$, it follows that the angle $\Delta\theta$, defined as the angle between CA and CB in Fig. 3–32a, is also the angle between $\mathbf{v}_1$ and $\mathbf{v}_2$. Hence the vectors $\mathbf{v}_2$, $\mathbf{v}_1$, and $\Delta\mathbf{v}$ in Fig. 3–32b form a triangle that is geometrically similar to triangle CAB in Fig. 3–32a. Taking $\Delta\theta$ small (letting Δt be very small) and setting $v = v_1 = v_2$ because the magnitude of the velocity is assumed not to change, we can write

$$\frac{\Delta v}{v} \approx \frac{\Delta l}{r}.$$

This is an exact equality when Δt approaches zero, for then the arc length Δl equals the cord length AB. Since we want to find the instantaneous acceleration, for which Δt approaches zero, we write the above expression as an equality and solve for Δv:

$$\Delta v = \frac{v}{r}\,\Delta l.$$

To get the centripetal acceleration, a_R, we divide Δv by Δt:

$$a_R = \lim_{\Delta t \to 0} \frac{\Delta v}{\Delta t} = \lim_{\Delta t \to 0} \frac{v}{r}\frac{\Delta l}{\Delta t}.$$

And since

$$\lim_{\Delta t \to 0} \frac{\Delta l}{\Delta t}$$

is the speed, v, of the object, we get

Centripetal acceleration

$$a_R = \frac{v^2}{r}. \qquad\qquad \textbf{(3–14)}$$

To summarize, *an object moving in a circle of radius r with constant speed v has an acceleration whose direction is toward the center of the circle and whose magnitude is* $a_R = v^2/r$. It is not surprising that this acceleration depends on v and r. For the greater the speed v, the faster the velocity changes direction; and the larger the radius, the less rapidly the velocity changes direction.

The acceleration vector points toward the center of the circle. But the velocity vector always points in the direction of motion, which is tangential to the circle. Thus the velocity and acceleration vectors are perpendicular to each other at every point in the path for uniform circular motion (see Fig. 3–33). This is another example that illustrates the error in thinking that acceleration and velocity are always in the same direction. For an object falling vertically, **a** and **v** are indeed parallel. But in circular motion, **a** and **v** are not parallel—nor are they in projectile motion (Section 3–7), where the acceleration **a** = **g** is always downward but the velocity vector can have various directions (Figs. 3–20 and 3–22).

Circular motion is often described in terms of the **frequency** f as so many revolutions per second. The **period** T of an object revolving in a circle is the time required for one complete revolution. Period and frequency are related by

$$T = \frac{1}{f}. \tag{3–15}$$

Acceleration and velocity are not in same direction

For example, if an object revolves at a frequency of 3 rev/s, then each revolution takes $\frac{1}{3}$ s. For an object revolving in a circle at constant speed v, we can write

$$v = \frac{2\pi r}{T}$$

since in one revolution the object travels one circumference ($= 2\pi r$).

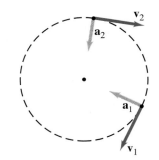

FIGURE 3–33 For uniform circular motion, **a** is always perpendicular to **v**.

EXAMPLE 3–11 **Acceleration of a revolving ball.** A 150-g ball at the end of a string is revolving uniformly in a horizontal circle of radius 0.600 m. The ball makes 2.00 revolutions in a second. What is its centripetal acceleration?

SOLUTION The centripetal acceleration is $a_R = v^2/r$. First, we determine the speed of the ball v. The ball makes two complete revolutions per second, so its period is $T = 0.500$ s. In this time it travels one circumference of the circle, $2\pi r$, so the ball has speed

$$v = \frac{2\pi r}{T} = \frac{2(3.14)(0.600\,\text{m})}{(0.500\,\text{s})} = 7.54\,\text{m/s}.$$

The centripetal acceleration is

$$a_R = \frac{v^2}{r} = \frac{(7.54\,\text{m/s})^2}{(0.600\,\text{m})} = 94.8\,\text{m/s}^2.$$

EXAMPLE 3–12 **Moon's centripetal acceleration.** The Moon's nearly circular orbit about the Earth has a radius of about 384,000 km and a period T of 27.3 days. Determine the acceleration of the Moon toward the Earth.

The Moon's acceleration toward Earth

SOLUTION In orbit around the Earth, the Moon travels a distance $2\pi r$, where $r = 3.84 \times 10^8$ m is the radius of its circular path. The speed of the Moon in its orbit about the Earth is $v = 2\pi r/T$. The period T in seconds is $T = (27.3\,\text{d})(24.0\,\text{h/d})(3600\,\text{s/h}) = 2.36 \times 10^6$ s. Therefore,

$$a_R = \frac{v^2}{r} = \frac{(2\pi r)^2}{T^2 r} = \frac{[2(3.14)(3.84 \times 10^8\,\text{m})]^2}{(2.36 \times 10^6\,\text{s})^2(3.84 \times 10^8\,\text{m})}$$
$$= 0.00272\,\text{m/s}^2 = 2.72 \times 10^{-3}\,\text{m/s}^2.$$

We can write this in terms of $g = 9.80$ m/s^2 (the acceleration of gravity at the Earth's surface) as

$$a = 2.72 \times 10^{-3}\,\text{m/s}^2\left(\frac{g}{9.80\,\text{m/s}^2}\right)$$
$$= 2.78 \times 10^{-4}\,g.$$

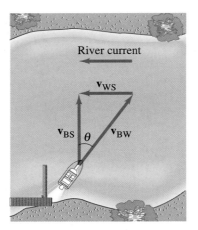

FIGURE 3–34 The boat must head upstream at an angle θ if it is to move directly across the river. Velocity vectors are shown as green arrows:

$\mathbf{v}_{BS}$ = velocity of **B**oat with respect to the **S**hore,

$\mathbf{v}_{BW}$ = velocity of **B**oat with respect to the **W**ater,

$\mathbf{v}_{WS}$ = velocity of the **W**ater with respect to the **S**hore (river current).

➥ **PROBLEM SOLVING**

Subscripts for adding velocities: first subscript for the object; second for the reference frame

3–10 | Relative Velocity

We now consider how observations made in different frames of reference are related to each other. For example, consider two trains approaching one another, each with a speed of 80 km/h with respect to the Earth. Observers on the Earth beside the train tracks will measure 80 km/hr for the speed of each of the trains. Observers on either one of the trains (a different frame of reference) will measure a speed of 160 km/h for the train approaching them. Similarly, when one car traveling 90 km/h passes a second car traveling in the same direction at 75 km/h, the first car has a speed relative to the second car of 90 km/h − 75 km/h = 15 km/h.

When the velocities are along the same line, simple addition or subtraction is sufficient to obtain the relative velocity. But if they are not along the same line, we must make use of vector addition. We emphasize, as mentioned in Section 2–1, that when specifying a velocity, it is important to specify what the reference frame is.

When determining relative velocity, it is easy to make a mistake by adding or subtracting the wrong velocities. It is useful, therefore, to use a careful labeling process that makes things clear. Each velocity is labeled by *two subscripts: the first refers to the object, the second to the reference frame in which it has this velocity.* For example, suppose a boat is to cross a river to the opposite side, as shown in Fig. 3–34. We let $\mathbf{v}_{BW}$ be the velocity of the **B**oat with respect to the **W**ater. (This is also what the boat's velocity would be relative to the shore if the water were still.) Similarly, $\mathbf{v}_{BS}$ is the velocity of the **B**oat with respect to the **S**hore, and $\mathbf{v}_{WS}$ is the velocity of the **W**ater with respect to the **S**hore (this is the river current). Note that $\mathbf{v}_{BW}$ is what the boat's motor produces (against the water), whereas $\mathbf{v}_{BS}$ is equal to $\mathbf{v}_{BW}$ plus the effect of the current. Therefore, the velocity of the boat relative to the shore is (see vector diagram, Fig. 3–34)

$$\mathbf{v}_{BS} = \mathbf{v}_{BW} + \mathbf{v}_{WS}. \tag{3–16}$$

By writing the subscripts via the convention above, we see that the inner subscripts (the two W's) on the right-hand side of Eq. 3–16 are the same, whereas the outer subscripts on the right of Eq. 3–16 (the B and the S) are the same as the two subscripts for the sum vector on the left, $\mathbf{v}_{BS}$. By following this convention (first subscript for the object, second for the reference frame), one can write down the correct equation relating velocities in different reference frames.[†] Figure 3–35 gives a derivation of Eq. 3–16, which is valid in general and can be extended to three or more velocities. For example, if a fisherman on a boat walks with a velocity $\mathbf{v}_{FB}$ relative to the boat, his velocity relative to the shore is

[†] We thus would know by inspection that (for example) the equation $\mathbf{v}_{BW} = \mathbf{v}_{BS} + \mathbf{v}_{WS}$ is wrong.

FIGURE 3–35 Derivation of relative velocity equation (Eq. 3–16), in this case for a person walking along the corridor in a train. We are looking down on the train and two reference frames are shown: xy on the Earth and $x'y'$ fixed on the train. We have

$\mathbf{r}_{PT}$ = position vector of person (P) relative to train (T)

$\mathbf{r}_{PE}$ = position vector of person (P) relative to Earth (E)

$\mathbf{r}_{TE}$ = position vector of train's coordinate system (T) relative to Earth (E)

From the diagram we see that

$$\mathbf{r}_{PE} = \mathbf{r}_{PT} + \mathbf{r}_{TE}.$$

We take the derivative with respect to time to obtain

$$\frac{d}{dt}(\mathbf{r}_{PE}) = \frac{d}{dt}(\mathbf{r}_{PT}) + \frac{d}{dt}(\mathbf{r}_{TE})$$

or, since $d\mathbf{r}/dt = \mathbf{v}$,

$$\mathbf{v}_{PE} = \mathbf{v}_{PT} + \mathbf{v}_{TE}.$$

This is the equivalent of Eq. 3–16 for the present situation (check the subscripts!).

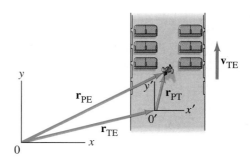

$\mathbf{v}_{FS} = \mathbf{v}_{FB} + \mathbf{v}_{BW} + \mathbf{v}_{WS}$. The equations involving relative velocity will be correct when adjacent inner subscripts are identical and when the outermost ones correspond exactly to the two on the velocity on the left of the equation. But this works only with plus signs (on the right), not minus signs.

It is often useful to remember that for any two objects or reference frames, A and B, the velocity of A relative to B has the same magnitude, but opposite direction, as the velocity of B relative to A:

$$\mathbf{v}_{BA} = -\mathbf{v}_{AB}. \qquad (3\text{--}17)$$

For example, if a train is traveling 100 km/h relative to the Earth in a certain direction, objects on the Earth (such as trees) appear to an observer on the train to be traveling 100 km/h in the opposite direction.

EXAMPLE 3–13 **Heading upstream.** A boat's speed in still water is $v_{BW} = 1.85$ m/s. If the boat is to travel northward directly across a river whose current has velocity $v_{WS} = 1.20$ m/s toward the west, at what angle must the boat head?

SOLUTION The current will drag the boat westward, so to counteract that motion the boat must head upstream in a northeasterly direction, as shown in Fig. 3–36. Figure 3–36 has been drawn with $\mathbf{v}_{BS}$, the velocity of the **B**oat relative to the **S**hore, pointing directly across the river since this is how the boat is supposed to move. (Note that $\mathbf{v}_{BS} = \mathbf{v}_{BW} + \mathbf{v}_{WS}$.) Thus, $\mathbf{v}_{BW}$ points upstream at an angle θ as shown, where

$$\sin\theta = \frac{v_{WS}}{v_{BW}} = \frac{1.20 \text{ m/s}}{1.85 \text{ m/s}} = 0.6486.$$

Thus $\theta = 40.4°$, so the boat must head upstream at a 40.4° angle.

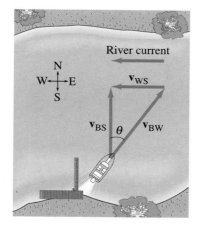

FIGURE 3–36 Example 3–13.

EXAMPLE 3–14 **Heading across the stream.** The same boat ($v_{BW} = 1.85$ m/s) now heads directly across the stream whose current is still 1.20 m/s. (*a*) What is the velocity (magnitude and direction) of the boat relative to the shore? (*b*) If the river is 110 m wide, how long will it take to cross and how far downstream will the boat be then?

SOLUTION (*a*) As shown in Fig. 3–37, the boat is pulled downstream by the current. The boat's velocity with respect to the shore, $\mathbf{v}_{BS}$, is the sum of its velocity with respect to the water, $\mathbf{v}_{BW}$, plus the velocity of the water with respect to the shore, $\mathbf{v}_{WS}$:

$$\mathbf{v}_{BS} = \mathbf{v}_{BW} + \mathbf{v}_{WS},$$

just as before. Since $\mathbf{v}_{BW}$ is directed straight across the stream, it is perpendicular to $\mathbf{v}_{WS}$, and we can get v_{BS} using the theorem of Pythagoras:

$$v_{BS} = \sqrt{v_{BW}^2 + v_{WS}^2} = \sqrt{(1.85 \text{ m/s})^2 + (1.20 \text{ m/s})^2} = 2.21 \text{ m/s}.$$

We can obtain the angle (note how θ is defined in diagram) from:

$$\tan\theta = v_{WS}/v_{BW} = (1.20 \text{ m/s})/(1.85 \text{ m/s}) = 0.6486.$$

A calculator with an INV TAN or TAN^{-1} key gives $\theta = \tan^{-1}(0.6486) = 33.0°$. Note that this angle is *not* equal to the angle calculated in Example 3–13.
(*b*) Given the river's width $D = 110$ m and using the definition of velocity, we solve for $t = D/v_{BW}$, where we use the velocity component in the direction of D; so $t = (110 \text{ m})/(1.85 \text{ m/s}) = 60$ s. The boat will have been carried downstream, in this time, a distance

$$d = v_{WS}t = (1.20 \text{ m/s})(60 \text{ s}) = 72 \text{ m}.$$

FIGURE 3–37 Example 3–14: a boat heading directly across a river whose current moves at 1.20 m/s.

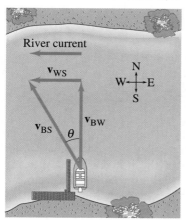

EXAMPLE 3–15 **Car velocities at 90°.** Two automobiles approach a corner at right angles to each other with the same speed of 40.0 km/h (= 11.1 m/s), as shown in Fig. 3–38a. What is the relative velocity of one car with respect to the other? That is, determine the velocity of car 1 as seen by car 2.

SOLUTION Figure 3–38a shows the situation in a reference frame fixed to the Earth. But we want to view the situation from a reference frame in which car 2 is at rest, and this is shown in Fig. 3–38b. In this reference frame (the world as seen by the driver of car 2), the Earth moves toward car 2 with velocity $\mathbf{v}_{E2}$ (speed of 40.0 km/h), which is of course equal and opposite to $\mathbf{v}_{2E}$, the velocity of car 2 with respect to the Earth (Eq. 3–17):

$$\mathbf{v}_{2E} = -\mathbf{v}_{E2}.$$

Then the velocity of car 1 as seen by car 2 is

$$\mathbf{v}_{12} = \mathbf{v}_{1E} + \mathbf{v}_{E2}$$

or (since $\mathbf{v}_{E2} = -\mathbf{v}_{2E}$)

$$\mathbf{v}_{12} = \mathbf{v}_{1E} - \mathbf{v}_{2E}.$$

That is, the velocity of car 1 as seen by car 2 is the difference of their velocities, $\mathbf{v}_{1E} - \mathbf{v}_{2E}$, both measured relative to the Earth (see Fig. 3–38c). Since the magnitudes of $\mathbf{v}_{1E}$, $\mathbf{v}_{2E}$, and $\mathbf{v}_{E2}$ are equal (40 km/h = 11.1 m/s) we see (Fig. 3–38b) that $\mathbf{v}_{12}$ points at a 45° angle toward car 2; the speed is

$$v_{12} = \sqrt{(11.1 \text{ m/s})^2 + (11.1 \text{ m/s})^2} = 15.7 \text{ m/s} \, (= 56.5 \text{ km/h}).$$

FIGURE 3–38 Example 3–15.

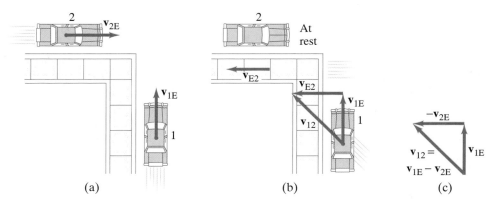

(a)   (b)   (c)

Summary

A quantity that has both a magnitude and a direction is called a **vector**. A quantity that has only a magnitude is called a **scalar**.

Addition of vectors can be done graphically by placing the tail of each successive arrow (representing each vector) at the tip of the previous one. The sum, or **resultant vector**, is the arrow drawn from the tail of the first to the tip of the last. Two vectors can also be added using the parallelogram method. Vectors can be added more accurately using the analytical method of adding their **components** along chosen axes with the aid of trigonometric functions. A vector of magnitude V making an angle θ with the x axis has components

$$V_x = V \cos \theta \qquad V_y = V \sin \theta.$$

Given the components, we can find the magnitude and direction from

$$V = \sqrt{V_x^2 + V_y^2}, \qquad \tan \theta = \frac{V_y}{V_x}.$$

It is often helpful to express a vector in terms of its components along chosen axes using **unit vectors**, which are vectors of unit length along the chosen coordinate axes; for Cartesian coordinates the unit vectors along the x, y, and z axes are called $\mathbf{i}$, $\mathbf{j}$, and $\mathbf{k}$.

The general definitions for the instantaneous **velocity**, $\mathbf{v}$, and **acceleration**, $\mathbf{a}$, of a particle (in one, two, or three dimensions) are

$$\mathbf{v} = \frac{d\mathbf{r}}{dt} \quad \text{and} \quad \mathbf{a} = \frac{d\mathbf{v}}{dt}$$

where $\mathbf{r}$ is the position vector of the particle. The kinematic equations for motion with constant acceleration can be written for each of the x, y, and z components of the motion and have the same form as for one-dimensional motion (Eqs. 2–12). Or they can be written in the more general vector form:

$$\mathbf{v} = \mathbf{v}_0 + \mathbf{a}t$$

$$\mathbf{r} = \mathbf{r}_0 + \mathbf{v}_0 t + \tfrac{1}{2}\mathbf{a}t^2.$$

Projectile motion of an object moving in the air near the Earth's surface can be analyzed as two separate motions if air resistance can be ignored. The horizontal component of the motion is at constant velocity, whereas the vertical component is at constant acceleration, g, just as for a body falling vertically under the action of gravity.

An object moving in a circle of radius r with constant speed v is said to be in **uniform circular motion** and has a **radial** or **centripetal acceleration** $\mathbf{a}_R$ toward the center of the circle of magnitude:

$$a_R = \frac{v^2}{r}.$$

The frequency f is the number of complete revolutions per second. The period T is the time required for one complete revolution, and is related to the frequency by

$$T = \frac{1}{f}.$$

The velocity of an object relative to one frame of reference can be found by vector addition if its velocity relative to a second frame of reference, and the **relative velocity** of the two reference frames, are known.

Questions

1. One car travels due east at 40 km/h, and a second car travels north at 40 km/h. Are their velocities equal? Explain.

2. Can you conclude that a car is not accelerating if its speedometer indicates a steady 60 km/h?

3. Can you give several examples of an object's motion in which a great distance is traveled but the displacement is zero?

4. Can the displacement vector for a particle moving in two dimensions ever be longer than the length of path traveled by the particle over the same time interval? Can it ever be less? Discuss.

5. During baseball practice, a batter hits a very high fly ball and then runs in a straight line and catches it. Which had the greater displacement, the player or the ball?

6. If $V = V_1 + V_2$, is V necessarily greater than V_1 and/or V_2? Discuss.

7. Two vectors have length $V_1 = 3.5$ km and $V_2 = 4.0$ km. What are the maximum and minimum magnitudes of their vector sum?

8. Can two vectors, of unequal magnitude, add up to give the zero vector? Can *three* unequal vectors? Under what conditions?

9. Can the magnitude of a vector ever (a) equal or (b) be less than one of its components?

10. Can a particle with constant speed be accelerating? What if it has constant velocity?

11. Can a vector of magnitude zero have a nonzero component?

12. Does the odometer of a car measure a scalar or a vector quantity? What about the speedometer?

13. A child wishes to determine the speed a slingshot imparts to a rock. How can this be done using only a meter stick, a rock, and the slingshot?

14. Is it ever necessary, in projectile motion, to consider the motion in three dimensions if air resistance can be neglected? What if air resistance can't be neglected? Discuss.

15. What physical factors are important for an athlete doing the long jump? What about the high jump?

16. A projectile has the least speed at what point in its path?

17. It was reported in the First World War that a French pilot flying at an altitude of 2 km caught in his bare hands a bullet fired at the plane! Using the fact that a bullet slows down considerably due to air resistance, explain how this incident occurred.

18. A car rounds a curve at a steady 50 km/h. If it rounds the same curve at a steady 70 km/h, will its acceleration be any different? Explain.

19. Will the acceleration of a car be the same if it travels around a sharp curve at 60 km/h as when it travels around a gentle curve at the same speed? Explain.

20. At some amusement parks, to get on a moving "car" the riders first hop on to a moving walkway and then on to the cars themselves. Why is this done?

21. If you are riding on a train that speeds past another train moving in the same direction on an adjacent track, it appears that the other train is moving backwards. Why?

22. If you stand motionless under an umbrella in a rainstorm where the drops fall vertically you remain relatively dry. However, if you start running, the rain begins to hit your legs even if they remain under the umbrella. Why?

23. Two cars with equal speed approach an intersection at right angles to each other. Will they necessarily collide? Show that when the relative velocity of approach is collinear (along the same line) with the relative displacement, we have the nautical maxim "constant bearing means collision."

24. A person sitting in an enclosed train car, moving at constant velocity, throws a ball straight up into the air in her reference frame. (a) Where does the ball land? What is your answer if the car (b) accelerates, (c) decelerates, (d) rounds a curve, (e) moves with constant velocity but is open to the air?

25. Two rowers, who can row at the same speed in still water, set off across a river at the same time. One heads straight across and is pulled downstream somewhat by the current. The other one heads upstream at an angle so as to arrive at a point opposite the starting point. Which rower reaches the opposite side first?

Problems

Sections 3–1 to 3–5

1. (I) A car is driven 200 km west and then 80 km southwest. What is the displacement of the car from the point of origin (magnitude and direction)? Draw a diagram.

2. (I) A delivery truck travels 18 blocks north, 10 blocks east, and 16 blocks south. What is its final displacement from the origin? Assume the blocks are equal length.

3. (I) Show that the vector labeled "wrong" in Fig. 3–6c is actually the difference of the two vectors. Is it $V_2 - V_1$, or $V_1 - V_2$?

4. (I) If $V_x = 8.80$ units and $V_y = -6.40$ units, determine the magnitude and direction of V.

5. (I) Graphically determine the resultant of the following three vector displacements: (1) 14 m, 30° north of east; (2) 18 m, 37° east of north; and (3) 20 m, 30° west of south.

6. (II) Vector V_1 is 6.0 units long and points along the negative x axis. Vector V_2 is 4.5 units long and points at +45° to the positive x axis. (a) What are the x and y components of each vector? (b) Determine the sum $V_1 + V_2$ (magnitude and angle).

7. (II) V is a vector 14.3 units in magnitude and points at an angle of 34.8° above the negative x axis. (a) Sketch this vector. (b) Find V_x and V_y. (c) Use V_x and V_y to obtain (again) the magnitude and direction of V. [*Note*: Part (c) is a good way to check if you've resolved your vector correctly.]

8. (II) Figure 3–39 shows two vectors, A and B, whose magnitudes are $A = 6.8$ units and $B = 5.5$ units. Determine C if (a) $C = A + B$, (b) $C = A - B$, (c) $C = B - A$. Give the magnitude and direction for each.

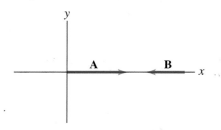

FIGURE 3–39
Problem 8.

9. (II) An airplane is traveling 635 km/h in a direction 41.5° west of north (Fig. 3–40). (a) Find the components of the velocity vector in the northerly and westerly directions. (b) How far north and how far west has the plane traveled after 3.00 h?

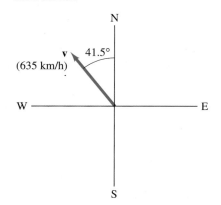

FIGURE 3–40
Problem 9.

10. (II) Let $V_1 = -6.0i + 8.0j$ and $V_2 = 4.5i - 5.0j$. Determine the magnitude and direction of (a) V_1, (b) V_2, (c) $V_1 + V_2$ and (d) $V_2 - V_1$.

11. (II) (a) Determine the magnitude and direction of the sum of the three vectors $V_1 = 4i - 8j$, $V_2 = i + j$, and $V_3 = -2i + 4j$. (b) Determine $V_1 - V_2 + V_3$.

12. (II) Three vectors are shown in Fig. 3–41. Their magnitudes are given in arbitrary units. Determine the sum of the three vectors. Give the resultant in terms of (a) components, (b) magnitude and angle with x axis.

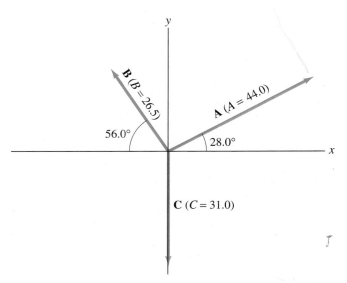

FIGURE 3–41 Problems 12, 13, 14, and 15. Vector magnitudes are given in arbitrary units.

13. (II) (a) Given the vectors A and B shown in Fig. 3–41, determine $B - A$. (b) Determine $A - B$ without using your answer in (a). Then compare your results and see if they are opposite.

14. (II) For the vectors given in Fig. 3–41, determine (a) $A - B + C$, (b) $A + B - C$, and (c) $C - A - B$.

15. (II) For the vectors shown in Fig. 3–41, determine (a) $B - 2A$, (b) $2A - 3B + 2C$.

16. (II) The summit of a mountain, 2450 m above base camp, is measured on a map to be 4580 m horizontally from the camp in a direction 32.4° west of north. What are the components of the displacement vector from camp to summit? What is its magnitude? Choose the x axis east, y axis north, and z axis up.

17. (III) You are given a vector in the xy plane that has a magnitude of 90.0 units and a y component of −35.0 units. (a) What are the two possibilities for its x component? (b) Assuming the x component is known to be positive, specify the vector which, if you add it to the original one would give a resultant vector that is 80.0 units long and points entirely in the −x direction.

18. (I) The position of a particular particle as a function of time is given by $\mathbf{r} = (7.60t\mathbf{i} + 8.85\mathbf{j} - t^2\mathbf{k})$ m. Determine the particle's velocity and acceleration as a function of time.

19. (I) What was the average velocity of the particle in Problem 18 between $t = 1.00$ s and $t = 3.00$ s? What is the magnitude of the instantaneous velocity at $t = 2.00$ s?

20. (II) What is the shape of the path of the particle of Problem 18?

21. (II) A car is moving with speed 18.0 m/s due south at one moment and 27.5 m/s due east 8.00 s later. Over this time interval, determine (a) its average velocity, (b) its average acceleration (magnitude and direction for both), and (c) its average speed. [Hint: Can you determine all these from the information given?]

22. (II) (a) A skier is accelerating down a 30.0° hill at 3.80 m/s² (Fig. 3-42). What is the vertical component of her acceleration? (b) How long will it take her to reach the bottom of the hill, assuming she starts from rest and accelerates uniformly, if the elevation change is 250 m?

$a = 3.80$ m/s²

30.0°

FIGURE 3-42 Problem 22.

23. (II) At $t = 0$, a particle starts from rest and moves in the xy plane with an acceleration $\mathbf{a} = (4.0\mathbf{i} + 3.0\mathbf{j})$ m/s². Determine (a) the x and y components of velocity, (b) the speed of the particle, and (c) the position of the particle, all as a function of time. (d) Evaluate all the above at $t = 2.0$ s.

24. (II) A particle starts from the origin at $t = 0$ with an initial velocity of 5.0 m/s along the positive x axis. If the acceleration is $(-3.0\mathbf{i} + 4.5\mathbf{j})$ m/s², determine the velocity and position of the particle at the moment it reaches its maximum x coordinate.

25. (III) The position of a particle is given by $\mathbf{r} = (6.0 \cos 3.0t\mathbf{i} + 6.0 \sin 3.0t\mathbf{j})$ meters. Determine (a) the velocity vector, $\mathbf{v}$, and (b) the acceleration vector, $\mathbf{a}$. (c) What is the path of this particle? [Hint: Determine $r = |\mathbf{r}|$.] (d) What is the relation between r and a (give a formula), and between $\mathbf{r}$ and $\mathbf{a}$ (give an angle)? (e) Show that $a = v^2/r$.

26. (I) A tiger leaps horizontally from a 6.5-m-high rock with a speed of 4.0 m/s. How far from the base of the rock will she land?

27. (I) A diver running 2.1 m/s dives out horizontally from the edge of a vertical cliff and reaches the water below 3.0 s later. How high was the cliff and how far from its base did the diver hit the water?

28. (I) Determine how much farther a person can jump on the Moon as compared to the Earth if the takeoff speed and angle are the same. The acceleration due to gravity on the Moon is one-sixth what it is on Earth.

29. (I) A fire hose held near the ground shoots water at a speed of 5.5 m/s. At what angle(s) should the nozzle point in order that the water land 3.0 m away (Fig. 3-43)? Why are there two different angles? Sketch the two trajectories.

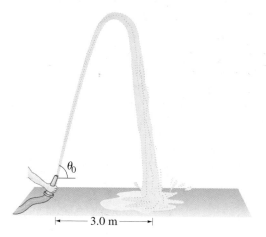

θ_0

|← 3.0 m →|

FIGURE 3-43 Problem 29.

30. (II) A ball is thrown horizontally from the roof of a building 9.0 m tall and lands 8.5 m from the base. What was the ball's initial speed?

31. (II) A football is kicked at ground level with a speed of 18.0 m/s at an angle of 32.0° to the horizontal. How much later does it hit the ground?

32. (II) A ball thrown horizontally at 22.2 m/s from the roof of a building lands 36.0 m from the base of the building. How high is the building?

33. (II) A shotputter throws the shot (mass = 7.3 kg) with an initial speed of 14 m/s at a 40° angle to the horizontal. Calculate the horizontal distance traveled by the shot if it leaves the athlete's hand at a height of 2.2 m above the ground.

34. (II) Show that the time required for a projectile to reach its highest point is equal to the time for it to return to its original height.

35. (II) A projectile is fired with an initial speed of 30.0 m/s. Plot on graph paper its trajectory for initial projection angles of $\theta = 15°, 30°, 45°, 60°, 75°$, and $90°$. Plot at least 10 points for each curve.

36. (II) William Tell must split the apple atop his son's head from a distance of 25.0 m. When he aims directly at the apple, the arrow is horizontal. At what angle must he aim it to hit the apple if the arrow travels at a speed of 22.5 m/s?

37. (II) The pilot of an airplane traveling 160 km/h wants to drop supplies to flood victims isolated on a patch of land 160 m below. The supplies should be dropped how many seconds before the plane is directly overhead?

38. (II) An athlete executing a long jump leaves the ground at a 33.0° angle and travels 7.80 m. (a) What was the takeoff speed? (b) If this speed were increased by just 5.0 percent, how much longer would the jump be?

39. (II) A projectile is fired with an initial speed of 51.2 m/s at an angle of 44.5° above the horizontal on a long flat firing range. Determine (a) the maximum height reached by the projectile, (b) the total time in the air, (c) the total horizontal distance covered (that is, the range), and (d) the velocity of the projectile 1.50 s after firing.

40. (II) A projectile is shot from the edge of a cliff 125 m above ground level with an initial speed of 65.0 m/s at an angle of 37.0° with the horizontal, as shown in Fig. 3–44. (a) Determine the time taken by the projectile to hit point P at ground level. (b) Determine the range X of the projectile as measured from the base of the cliff. At the instant just before the projectile hits point P, find (c) the horizontal and the vertical components of its velocity, (d) the magnitude of the velocity, and (e) the angle made by the velocity vector with the horizontal.

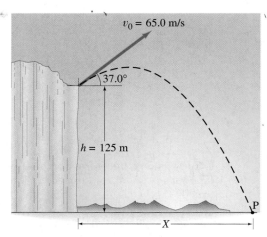

FIGURE 3–44   Problem 40.

41. (II) Revisit Conceptual Example 3–7, and assume that the boy with the slingshot is *below* the boy in the tree (Fig. 3–45) and so aims *upward*, directly at the boy in the tree. Show that again the boy in the tree makes the wrong move by letting go at the moment the water balloon is shot.

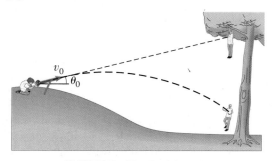

FIGURE 3–45   Problem 41.

42. (II) At what projection angle will the range of a projectile equal its maximum height?

43. (II) Suppose the kick in Example 3–5 is attempted 36.0 m from the goalposts, whose crossbar is 3.00 m above the ground. If the football is directed perfectly between the goalposts, will it pass over the bar and be a field goal? Show why or why not. If not, from what horizontal distance must this kick be made if it is to score?

44. (II) Exactly 3.0 s after a projectile is fired into the air from the ground, it is observed to have a velocity $\mathbf{v} = (7.6\mathbf{i} + 4.8\mathbf{j})$ m/s, where the x axis is horizontal and the y axis is positive upward. Determine (a) the horizontal range of the projectile, (b) its maximum height above the ground, and (c) its speed and angle of motion just before it strikes the ground.

45. (II) A high diver leaves the end of a 5.0-m-high diving board and strikes the water 1.3 s later, 3.0 m beyond the end of the board. Considering the diver as a particle, determine (a) her initial velocity, $\mathbf{v}_0$; (b) the maximum height reached; and (c) the velocity $\mathbf{v}_f$ with which she enters the water.

46. (II) A stunt driver wants to make his car jump over 8 cars parked side by side below a horizontal ramp (Fig. 3–46). (a) With what minimum speed must he drive off the horizontal ramp? The vertical height of the ramp is 1.5 m above the cars and the horizontal distance he must clear is 20 m. (b) What is the new minimum speed if the the ramp is now tilted upward, so that "takeoff angle" is 10° above the horizontal, and nothing else is changed?

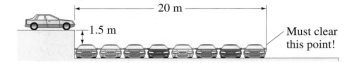

FIGURE 3–46   Problem 46.

47. (III) A ball is thrown horizontally from the top of a cliff with initial speed v_0 (at $t = 0$). At any moment, its direction of motion makes an angle θ to the horizontal (Fig. 3–47). Derive a formula for θ as a function of time, t, as the ball follows a projectile's path.

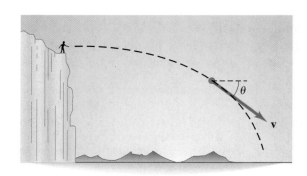

FIGURE 3–47   Problem 47.

48. (III) Derive a formula for the horizontal range R, of a projectile when it lands at a height h above its initial point. (For $h < 0$, it lands a distance $-h$ below the starting point.) Assume it is projected at an angle θ_0 with initial speed v_0.

49. (III) A person stands at the base of a hill that is a straight incline making an angle ϕ with the horizontal (Fig. 3–48). For a given initial speed v_0, at what angle θ (to the horizontal) should objects be thrown so that the distance d they land up the hill is as large as possible?

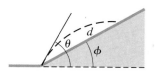

FIGURE 3–48 Problem 49. Given ϕ and v_0, determine θ to make d maximum.

50. (III) At $t = 0$ a batter hits a baseball with an initial speed of 32 m/s at a 55° angle to the horizontal. An outfielder is 85 m from the batter at $t = 0$, and, as seen from home plate, the line of sight to the outfielder makes a horizontal angle of 22° with the plane in which the ball moves (see Fig. 3–49). What speed and direction must the fielder take in order to catch the ball at the same height from which it was struck? Give angle with respect to the outfielder's line of sight to home plate.

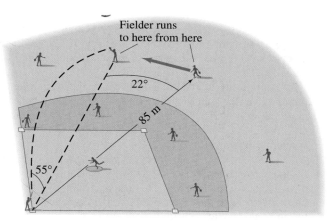

FIGURE 3–49 Problem 50.

Section 3–9

51. (I) A jet plane traveling 1800 km/h (500 m/s) pulls out of a dive by moving in an arc of radius 3.50 km. What is the plane's acceleration in "g's"?

52. (I) What is the centripetal acceleration of a child 3.6 m from the center of a merry-go-round? The child's speed is 0.85 m/s.

53. (I) Calculate the centripetal acceleration of the Earth in its orbit around the Sun. Assume the Earth's orbit is a circle of radius 1.5×10^{11} m.

54. (II) What is the magnitude of the acceleration of a speck of clay on the edge of a potter's wheel turning at 45 rpm (revolutions per minute) if its diameter is 30 cm?

55. (II) Suppose the space shuttle is in orbit 400 km from the Earth's surface, and circles the Earth about once every 90 minutes. Find the centripetal acceleration of the space shuttle in its orbit. Express your answer in terms of g, the gravitational acceleration at the Earth's surface.

56. (II) Because the Earth rotates once per day, the effective acceleration of gravity at the equator is slightly less than it would be if the Earth didn't rotate. Estimate the magnitude of this effect. What fraction of g is this?

57. (II) Use dimensional analysis (Section 1–7) to obtain the form for the centripetal acceleration, $a_R = v^2/r$.

58. (III) The position of a particle moving in the xy plane is given by $\mathbf{r} = \mathbf{i}2.0 \cos 3.0t + \mathbf{j}2.0 \sin 3.0t$, where r is in meters and t is in seconds. (a) Show that this represents circular motion of radius 2.0 m centered at the origin. (b) Determine the velocity and acceleration vectors as functions of time. (c) Determine the speed and magnitude of the acceleration. (d) Show that $a = v^2/r$. (e) Show that the acceleration vector always points toward the center of the circle.

Section 3–10

59. (I) Huck Finn walks at a speed of 1.0 m/s across his raft (that is, he walks perpendicular to the raft's motion relative to the shore). The raft is traveling down the Mississippi River at a speed of 2.5 m/s relative to the river bank (Fig. 3–50). What is the velocity (speed and direction) of Huck relative to the river bank?

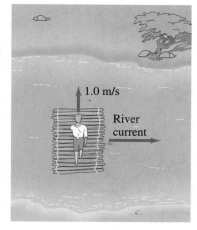

FIGURE 3–50 Problem 59.

60. (II) A boat can travel 2.20 m/s in still water. (a) If the boat points its prow directly across a stream whose current is 1.20 m/s, what is the velocity (magnitude and direction) of the boat relative to the shore? (b) What will be the position of the boat, relative to its point of origin, after 3.00 s? (See Fig. 3–37.)

61. (II) Two planes approach each other head-on. Each has a speed of 780 km/h, and they spot each other when they are initially 10.0 km apart. How much time do the pilots have to take evasive action?

62. (II) An airplane is heading due south at a speed of 550 km/h. If a wind begins blowing from the southwest at a speed of 90.0 km/h (average), calculate: (a) the velocity (magnitude and direction) of the plane, relative to the ground, and (b) how far off course it will be after 12.0 min if the pilot takes no corrective action. [*Hint*: First draw a diagram.]

63. (II) Determine the speed of the boat with respect to the shore in Example 3–13.

64. (II) A passenger on a boat moving at 1.80 m/s on a still lake walks up a flight of stairs at a speed of 0.60 m/s, Fig. 3–51. The stairs are angled at 45° pointing in the direction of motion as shown. What is the velocity of the passenger relative to the water?

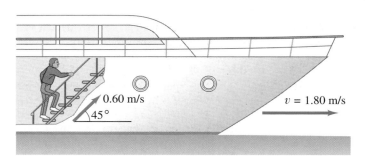

FIGURE 3–51 Problem 64.

65. (II) A motorboat whose speed in still water is 3.70 m/s must aim upstream at an angle of 29.5° (with respect to a line perpendicular to the shore) in order to travel directly across the stream. (a) What is the speed of the current? (b) What is the resultant speed of the boat with respect to the shore? (See Fig. 3–34.)

66. (II) A boat, whose speed in still water is 2.40 m/s, must cross a 280-m-wide river and arrive at a point 120 m upstream from where it starts (Fig. 3–52). To do so, the pilot must head the boat at a 45.0° upstream angle. What is the speed of the river's current?

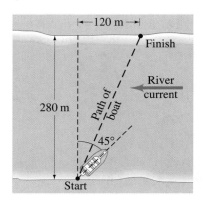

FIGURE 3–52
Problem 66.

67. (II) A swimmer is capable of swimming 1.00 m/s in still water. (a) If she aims her body directly across a 75-m-wide river whose current is 0.80 m/s, how far downstream (from a point opposite her starting point) will she land? (b) How long will it take her to reach the other side?

68. (II) At what upstream angle must the swimmer in Problem 67 aim, if she is to arrive at a point directly across the stream?

69. (II) Two cars approach a street corner at right angles to each other. Car 1 travels at 30 km/h and car 2 at 50 km/h. What is the relative velocity of car 1 as seen by car 2? What is the velocity of car 2 relative to car 1?

70. (III) An airplane, whose air speed is 680 km/h, is supposed to fly in a straight path 35.0° N of E. But a steady 120 km/h wind is blowing from the north. In what direction should the plane head?

71. (III) A motorcycle traveling 95.0 km/h approaches a car traveling in the same direction at 75.0 km/h. When the motorcycle is 60.0 m behind the car, the rider accelerates uniformly and passes the car 10.0 s later. What was the acceleration of the motorcycle?

General Problems

72. Two vectors, $\mathbf{V}_1$ and $\mathbf{V}_2$, add to a resultant $\mathbf{V} = \mathbf{V}_1 + \mathbf{V}_2$. Describe $\mathbf{V}_1$ and $\mathbf{V}_2$ if (a) $V = V_1 + V_2$, (b) $V^2 = V_1^2 + V_2^2$, (c) $V_1 + V_2 = V_1 - V_2$.

73. A plumber steps out of his truck, walks 60 m east and 35 m south, and then takes an elevator 12 m into the subbasement of a building where a bad leak is occurring. What is the displacement of the plumber relative to his truck? Give your answer in components, and also in magnitude and angle notation. Assume *x* is east, *y* is north, and *z* is up.

74. On mountainous downhill roads, escape routes are sometimes placed to the side of the road for trucks whose brakes might fail. Assuming a constant upward slope of 30°, calculate the horizontal and vertical components of the acceleration of a truck that slowed from 120 km/h to rest in 12 s. See Fig. 3–53.

FIGURE 3–53 Problem 74.

75. Romeo is chucking pebbles gently up to Juliet's window, and he wants the pebbles to hit the window with only a horizontal component of velocity. He is standing at the edge of a rose garden 8.0 m below her window and 9.0 m from the base of the wall (Fig. 3–54). How fast are the pebbles going when they hit her window?

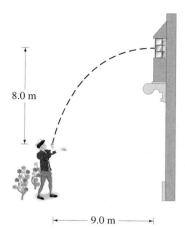

8.0 m

9.0 m

FIGURE 3–54   Problem 75.

76. A hunter aims directly at a target (on the same level) 65.0 m away. (*a*) If the bullet leaves the gun at a speed of 145 m/s, by how much will it miss the target? (*b*) At what angle should the gun be aimed so the target will be hit?

77. What is the *y* component of a vector in the *xy* plane whose magnitude is 52.8 and whose *x* component is 46.4? What is the direction of this vector (angle it makes with the *x* axis)?

θ

FIGURE 3–55   Problem 78.

78. Raindrops make an angle θ with the vertical when viewed through a moving train window (Fig. 3–55). If the speed of the train is v_T, what is the speed of the raindrops in the reference frame of the Earth in which they are assumed to fall vertically?

79. A light plane is headed due south with a speed relative to still air of 240 km/h. After 1 hour, the pilot notices that they have covered only 180 km and their direction is not south but southeast. What is the wind velocity?

80. An Olympic long jumper is capable of jumping 8.0 m. Assuming his horizontal speed is 9.2 m/s as he leaves the ground, how long is he in the air and how high does he go? Assume that he lands standing upright—that is, the same way he left the ground.

81. Apollo astronauts took a "nine iron" to the Moon and hit a golf ball about 180 m. Assuming that the swing, launch angle, and so on, were the same as on Earth where the same astronaut could hit it only 30 m, estimate the acceleration due to gravity on the surface of the Moon. (We neglect air resistance in both cases, but on the Moon there is none.)

82. When Babe Ruth hit a homer over the 12-m-high right-field fence 92 m from home plate, roughly what was the minimum speed of the ball when it left the bat? Assume the ball was hit 1.0 m above the ground and its path initially made a 40° angle with the ground.

83. The cliff divers of Acapulco push off horizontally from rock platforms about 35 m above the water, but they must clear rocky outcrops at water level that extend out into the water 5.0 m from the base of the cliff directly under their launch point. See Fig. 3–56. What minimum pushoff speed is necessary to do this? How long are they in the air?

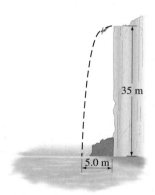

35 m

5.0 m

FIGURE 3–56   Problem 83.

84. At serve, a tennis player aims to hit the ball horizontally. What minimum speed is required for the ball to clear the 0.90-m-high net about 15.0 m from the server if the ball is "launched" from a height of 2.50 m? Where will the ball land if it just clears the net (and will it be "good" in the sense that it lands within 7.0 m of the net)? How long will it be in the air? See Fig. 3–57.

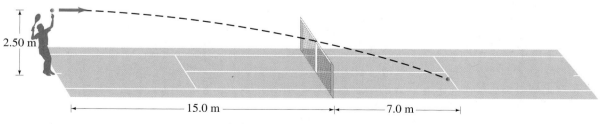

2.50 m

15.0 m

7.0 m

FIGURE 3–57   Problem 84.

85. The speed of a boat in still water is v. The boat is to make a round trip in a river whose current travels at speed u. Derive a formula for the time needed to make a round trip of total distance D if the boat makes the round trip (a) by moving upstream and back downstream, and (b) by moving directly across the river and back. We must assume $u < v$; why?

86. A plane whose airspeed is 200 km/h heads due north. But a 100-km/h northeast wind (that is, coming from the northeast) suddenly begins to blow. What is the resulting velocity of the plane with respect to the ground?

87. Spymaster Tim, flying a constant 200 km/h horizontally in a low-flying helicopter, wants to drop secret documents into his contact's open car which is traveling 150 km/h on a level highway 78.0 m below. At what angle (with the horizontal) should the car be in his sights when the packet is released (Fig. 3–58)?

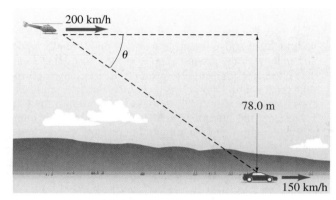

FIGURE 3–58 Problem 87.

88. A basketball leaves a player's hands at a height of 2.10 m above the floor. The basket is 2.60 m above the floor. The player likes to shoot the ball at a 38.0° angle. If the shot is made from a horizontal distance of 11.00 m and must be accurate to ±0.22 m (horizontally), what is the range of initial speeds allowed to make the basket?

89. What is the speed of a point (a) on the equator of the Earth, (b) at a latitude of 40° N, due to the Earth's daily rotation?

90. A projectile is launched from ground level to the top of a cliff which is 195 m away and 155 m high (see Fig. 3–59). If the projectile lands on top of the cliff 7.6 s after it is fired, find the initial velocity of the projectile (magnitude and direction). Neglect air resistance.

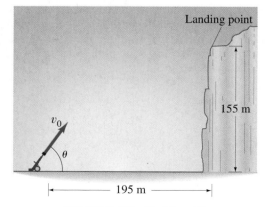

FIGURE 3–59 Problem 90.

91. In hot pursuit, Agent Logan of the FBI must get directly across a 1600-m-wide river in minimum time. The river's current is 0.80 m/s, he can row a boat at 1.50 m/s, and he can run 3.00 m/s. Describe the path he should take (rowing plus running along the shore) for the minimum crossing time, and determine the minimum time.

92. A person going for a morning jog on the deck of a cruise ship is running toward the bow (front) of the ship at 2.0 m/s while the ship is moving ahead at 8.5 m/s. What is the velocity of the jogger relative to the water? Later, the jogger is moving toward the stern (rear) of the ship. What is the jogger's velocity relative to the water now?

93. A projected space station consists of a circular tube that is set rotating about its center (like a tubular bicycle tire, Fig. 3–60). The circle formed by the tube has a diameter of about 1.1 km. What must be the rotation speed (revolutions per day) if an effect equal to gravity at the surface of the Earth (1.0 g) is to be felt?

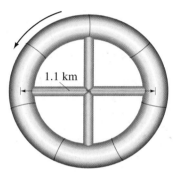

FIGURE 3–60 Problem 93.

94. A jet pilot takes his aircraft in a vertical loop (Fig. 3–61). If the jet is moving at a speed of 700 km/h at the lowest point on the loop, determine the minimum radius of the circle so that the acceleration at the lowest point does not exceed 6.0 g's.

FIGURE 3–61 Problem 94.

This airplane is taking off. It is accelerating, increasing in speed rapidly. To do so, a force must be exerted on it, according to Newton's second law, $\Sigma \mathbf{F} = m\mathbf{a}$. What exerts this force? The two jet engines of this plane exert a strong force on the gases they push out toward the rear of the plane. According to Newton's third law, these ejected gases exert an equal and opposite force on the airplane in the forward direction. The forces exerted on the **p**lane by the **g**ases, labeled $\mathbf{F}_{PG}$, are what accelerate the plane.

CHAPTER 4

Dynamics: Newton's Laws of Motion

We have discussed how motion is described in terms of velocity and acceleration. Now we deal with the question of *why* objects move as they do: What makes an object at rest begin to move? What causes a body to accelerate or decelerate? What is involved when an object moves in a circle? We can answer in each case that a force is required. In this chapter,[†] we will investigate the connection between force and motion, which is the subject called **dynamics**.

4–1 Force

Intuitively, we experience **force** as any kind of a push or a pull on an object. When you push a stalled car or a grocery cart (Fig. 4–1), you are exerting a force on it. When a motor lifts an elevator, or a hammer hits a nail, or the wind blows the leaves of a tree, a force is being exerted. We say that an object falls because of the *force of gravity*. Forces do not always give rise to motion. For example, you may push very hard on a heavy desk and it may not move.

If an object is at rest, to start it moving requires force—that is, to accelerate it from zero velocity to a non-zero velocity. For an object already moving, if you want to change its velocity—either in direction or in magnitude—again a force is required. In other words, to accelerate an object, a force is required. Indeed, we can define force as an action capable of accelerating an object. In Section 4–4 we will discuss the precise relationship between force and acceleration, which is Newton's second law.

FIGURE 4–1 A force exerted on a grocery cart—in this case exerted by a child.

[†] We treat everyday sort of objects in motion here. The submicroscopic world of atoms and molecules has to be treated separately. Also, velocities that are extremely high, close to the speed of light (3.0×10^8 m/s), must be treated using the theory of relativity (Chapter 37).

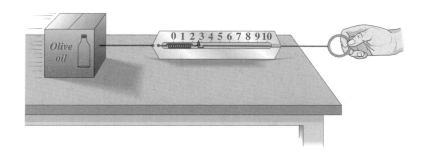

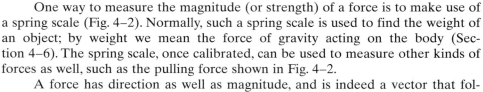

FIGURE 4–2 A spring scale used to measure a force.

One way to measure the magnitude (or strength) of a force is to make use of a spring scale (Fig. 4–2). Normally, such a spring scale is used to find the weight of an object; by weight we mean the force of gravity acting on the body (Section 4–6). The spring scale, once calibrated, can be used to measure other kinds of forces as well, such as the pulling force shown in Fig. 4–2.

Measuring force

A force has direction as well as magnitude, and is indeed a vector that follows the rules of vector addition discussed in Chapter 3. We can represent any force by an arrow, on a diagram, just as we did with velocity. The direction of the arrow is the direction of the push or pull, and its length is drawn proportional to the magnitude of the force.

4–2 | Newton's First Law of Motion

What is the exact connection between force and motion? Aristotle (384–322 B.C.) believed that a force was required to keep an object moving along a horizontal plane. To Aristotle, the natural state of a body was at rest, and a force was believed necessary to keep a body in motion. Furthermore, Aristotle argued, the greater the force on the body, the greater its speed.

Some 2000 years later, Galileo disagreed, and maintained that it is just as natural for an object to be in horizontal motion with a constant velocity as it is for it to be at rest.

To understand Galileo's idea, consider the following observations involving motion along a horizontal plane. To push an object with a rough surface along a tabletop at constant speed requires a certain amount of force. To push an equally heavy object with a very smooth surface across the table at the same speed will require less force. If a layer of oil or other lubricant is placed between the surface of the object and the table, then almost no force is required to move the object. Notice that in each successive step, less force is required. As the next step, we can imagine a situation in which the object does not rub against the table at all—or there is a perfect lubricant between the object and the table—and theorize that once started, the object would move across the table at constant speed with *no* force applied. A steel ball bearing rolling on a hard horizontal surface approaches this situation. So does a puck on an air table (Fig. 4–3), in which a thin layer of air reduces friction almost to zero.

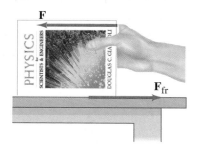

FIGURE 4–3 Photo of an air table. Air issuing from many tiny holes forms a thin air layer between the table and a puck, which, when given an initial shove, will travel at very nearly constant speed in a straight line (until it hits a wall or another puck).

It was Galileo's genius to imagine such an idealized world—in this case, one where there is no friction—and to see that it could produce a more useful view of the real world. It was this idealization that led him to his remarkable conclusion that if no force is applied to a moving object, it will continue to move with constant speed in a straight line. An object slows down only if a force is exerted on it. Galileo thus interpreted friction as a force akin to ordinary pushes and pulls.

FIGURE 4–4 **F** represents the force applied by the person and **F**$_{fr}$ represents the force of friction.

To push an object across a table at constant speed requires a force from your hand only to balance out the force of friction (Fig. 4–4). When the object moves at constant speed, your pushing force is equal in magnitude to the friction force, but these two forces are in opposite directions, so the *net* force on the object (the vector sum of the two forces) is zero. This is consistent with Galileo's viewpoint, for the object moves with constant speed when no net force is exerted on it.

Upon this foundation, Isaac Newton (Fig. 4–5) built his great theory of motion. Newton's analysis of motion is summarized in his famous "three laws of motion." In his great work, the *Principia* (published in 1687), Newton readily acknowledged his debt to Galileo. In fact, **Newton's first law of motion** is very close to Galileo's conclusions. It states that

> **Every body continues in its state of rest or of uniform speed in a straight line unless acted on by a nonzero net force.**

The tendency of a body to maintain its state of rest or of uniform motion in a straight line is called **inertia**. As a result, Newton's first law is often called the **law of inertia**.

Inertial Reference Frames

Newton's first law does not hold in every reference frame. For example, if your reference frame is fixed in a car, and the car accelerates, an object such as a cup resting on the dashboard may begin to move toward you (it stayed at rest as long as the car's velocity remained constant). The cup accelerated toward you but neither you nor anything else exerted a force on it in that direction. In such an accelerating reference frame, Newton's first law does not hold. Reference frames in which Newton's first law does hold are called **inertial reference frames**—the law of inertia is valid in them. For most purposes, we can usually assume that reference frames fixed on the Earth are inertial frames. This is not precisely true, due to the Earth's rotation, but usually it is close enough. Any reference frame that moves with constant velocity (say, a car or an airplane) relative to an inertial frame is also an inertial reference frame. Reference frames where the law of inertia does *not* hold, such as the accelerating reference frame discussed above, are called **noninertial** reference frames. How can we be sure a reference frame is inertial or not? By checking to see if Newton's first law holds. Thus Newton's first law serves as the definition of inertial reference frames.

Inertia

FIGURE 4–5
Isaac Newton (1642–1727).

4–3 | Mass

Newton's second law, which we come to in the next Section, makes use of the concept of mass. Newton used the term *mass* as a synonym for *quantity of matter*. This intuitive notion of the mass of a body is not very precise because the concept "quantity of matter" is not very well defined. More precisely, we can say that **mass** is a *measure of the inertia of a body*. The more mass a body has, the greater the force needed to give it a particular acceleration. It is harder to start it moving from rest, or to stop it when it is moving, or to change its velocity sideways out of a straight-line path. A truck has much more inertia than a baseball moving at the same speed, and it requires a much greater force to change its velocity at the same rate. It therefore has much more mass. How we define mass in terms of inertia is discussed in the next Section.

Mass as inertia

To quantify the concept of mass, we must define a standard. In SI units, the unit of mass is the **kilogram** (kg) as we discussed in Chapter 1, Section 1–4.

The terms *mass* and *weight* are often confused with one another, but it is important to distinguish between them. Mass is a property of a body itself—it is a measure of a body's inertia, or its "quantity of matter". On the other hand, *weight is a force*, the force of gravity acting on a body. To see the difference, suppose we take an object to the Moon. The object will weigh only about one sixth as much as it did on Earth, since the force of gravity is weaker, but its mass will be the same. It will have the same amount of matter and it will have just as much inertia—for in the absence of friction, it will be just as hard to start it moving or to stop it once it is moving. (More on weight in Section 4–6.)

Mass vs. weight

FIGURE 4–6 The bobsled accelerates because the team exerts a force.

4–4 | Newton's Second Law of Motion

Newton's first law states that if no net force is acting on a body, it remains at rest, or if moving, it continues moving with constant speed in a straight line. But what happens if a net force is exerted on a body? Newton perceived that the velocity will change (Fig. 4–6). A net force exerted on an object may make its speed increase, or, if the net force is in a direction opposite to the motion, it will reduce the speed. If the net force acts sideways on a moving object, the *direction* of the velocity changes (and the magnitude may as well). Since a change in speed or velocity is an acceleration (Chapter 2, Section 2–4), we can say that *a net force gives rise to acceleration*.

What precisely is the relationship between acceleration and force? Everyday experience can answer this question. Consider the force required to push a cart whose friction is ignorable. If there is friction, consider the *net* force, which is the force you exert minus the force of friction. Now if you push with a gentle but constant force for a certain period of time, you will make the cart accelerate from rest up to some speed, say 3 km/h. If you push with twice the force, you will find that the cart will reach 3 km/h in half the time. That is, the acceleration will be twice as great. If you double the force, the acceleration doubles. If you triple the force, the acceleration is tripled, and so on. Thus, the acceleration of a body is directly proportional to the net applied force.

But the acceleration depends on the mass of the object as well. If you push an empty grocery cart with the same force as you push one that is filled with groceries, you will find that the more massive cart accelerates more slowly. The greater the mass, the less the acceleration for the same net force. The mathematical relation, as Newton argued, is that the acceleration of a body is inversely proportional to its mass. These relationships are found to hold in general and can be summarized as follows:

> **The acceleration of an object is directly proportional to the net force acting on it and is inversely proportional to its mass. The direction of the acceleration is in the direction of the net force acting on the object.**

This is **Newton's second law of motion.** As an equation, it can be written

$$\mathbf{a} = \frac{\Sigma \mathbf{F}}{m},$$

where **a** stands for acceleration, m for the mass, and $\Sigma \mathbf{F}$ for the *net force*. The symbol Σ (Greek "sigma") stands for "sum of"; **F** stands for force, so $\Sigma \mathbf{F}$ means the *vector sum of all forces* acting on the body, which we define as the **net force**.

Net force

We rearrange this equation to obtain the familiar statement of Newton's second law:

$$\Sigma \mathbf{F} = m\mathbf{a}. \tag{4–1}$$

Newton's second law relates the description of motion to the cause of motion, force. It is one of the most fundamental relationships in physics. From Newton's second law we can make the definition of **force** as *an action capable of accelerating an object*.

Force defined

Every force **F** is a vector, with magnitude and direction. Equation 4–1 is a vector equation valid in any inertial reference frame. It can be written in component form in rectangular coordinates as

$$\Sigma F_x = ma_x, \qquad \Sigma F_y = ma_y, \qquad \Sigma F_z = ma_z.$$

If the motion is all along a line (one dimensional), we can leave out the subscripts and simply write $\Sigma F = ma$.

In SI units, with the mass in kilograms, the unit of force is called the **newton** (N). One newton, then, is the force required to impart an acceleration of 1 m/s² to a mass of 1 kg. Thus $1 \text{ N} = 1 \text{ kg} \cdot \text{m/s}^2$.

In cgs units, the unit of mass is the gram (g) as mentioned earlier.[†] The unit of force is the *dyne*, which is defined as the net force needed to impart an acceleration of 1 cm/s² to a mass of 1 g. Thus 1 dyne = 1 g·cm/s². It is easy to show that 1 dyne = 10^{-5} N.

In the British system, the unit of force is the *pound* (abbreviated lb), where 1 lb = 4.44822 N ≈ 4.45 N. The unit of mass is the *slug*, which is defined as that mass which will undergo an acceleration of 1 ft/s² when a force of 1 lb is applied to it. Thus 1 lb = 1 slug·ft/s². Table 4–1 summarizes the units in the different systems.

It is very important that only one set of units be used in a given calculation or problem, with the SI being preferred. If the force is given in, say, newtons, and the mass in grams, then before attempting to solve for the acceleration in SI units, the mass must be changed to kilograms. For example, if the force is given as 2.0 N along the x axis and the mass is 500 g, we change the latter to 0.50 kg, and the acceleration will then automatically come out in m/s² when Newton's second law is used:

$$a_x = \frac{\Sigma F_x}{m} = \frac{2.0 \text{ N}}{0.50 \text{ kg}} = \frac{2.0}{0.50} \frac{\text{kg} \cdot \text{m}}{\text{kg} \cdot \text{s}^2} = 4.0 \text{ m/s}^2.$$

TABLE 4–1
Units for Mass and Force

System	Mass	Force (including weight)
SI	kilogram (kg)	newton (N) (= kg·m/s²)
cgs	gram (g)	dyne (= g·cm/s²)
British	slug	pound (lb)

Conversion factors: 1 dyne = 10^5 N;
1 lb ≈ 4.45 N.

➥ **PROBLEM SOLVING**

Use a consistent set of units

EXAMPLE 4–1 ESTIMATE Force to accelerate a fast car. Estimate the net force needed to accelerate (*a*) a 1000-kg car at $\frac{1}{2}g$; (*b*) a 200-g apple at the same rate.

SOLUTION (*a*) The car's acceleration is $a = \frac{1}{2}g = \frac{1}{2}(9.8 \text{ m/s}^2) \approx 5 \text{ m/s}^2$. We use Newton's second law to get the net force needed to achieve this acceleration:

$$\Sigma F = ma \approx (1000 \text{ kg})(5 \text{ m/s}^2) = 5000 \text{ N}.$$

(If you are used to British units, to get an idea of what a 5000 N force is, you can divide by 4.45 N/lb and get a force of about 1000 lb.)
(*b*) For the apple,

$$\Sigma F = ma \approx (0.200 \text{ kg})(5 \text{ m/s}^2) = 1 \text{ N}.$$

EXAMPLE 4–2 Force to stop a car. What constant net force is required to bring a 1500-kg car to rest from a speed of 100 km/h within a distance of 55 m?

SOLUTION We use Newton's second law, $\Sigma F = ma$, but first we must determine the acceleration a, which is constant since the net force is constant. We assume the motion is along the $+x$ axis (Fig. 4–7). We are given the initial velocity $v_0 = 100 \text{ km/h} = 28 \text{ m/s}$, the final velocity $v = 0$, and the distance traveled $x - x_0 = 55 \text{ m}$. From Eq. 2–12c, we have

$$v^2 = v_0^2 + 2a(x - x_0)$$

so

$$a = \frac{v^2 - v_0^2}{2(x - x_0)} = \frac{0 - (28 \text{ m/s})^2}{2(55 \text{ m})} = -7.1 \text{ m/s}^2.$$

The net force required is then

$$\Sigma F = ma = (1500 \text{ kg})(-7.1 \text{ m/s}^2) = -1.1 \times 10^4 \text{ N}.$$

The force must be exerted in the direction *opposite* to the initial velocity, which is what the negative sign tells us.

[†] Be careful not to confuse g for gram with g for the acceleration due to gravity. The latter is always italicized (or bold face as a vector).

$v_0 = 100 \text{ km/h}$ $v = 0$

$x = 0$ $x = 55 \text{ m}$ $x(\text{m})$

FIGURE 4–7 Example 4–2.

As mentioned in Section 4–3, we can quantify the concept of mass using its definition as a measure of inertia. How to do this is evident from Eq. 4–1, where we see that the acceleration of a body is inversely proportional to its mass. If the same net force ΣF acts to accelerate each of two masses, m_1 and m_2, then the ratio of their masses can be defined as the inverse ratio of their accelerations:

Mass defined

$$\frac{m_2}{m_1} = \frac{a_1}{a_2}.$$

If one of the masses is known (it could be the standard kilogram) and the two accelerations are precisely measured, then the unknown mass is obtained from this definition. For example, if $m_1 = 1.00\,\text{kg}$, and for a particular force $a_1 = 3.00\,\text{m/s}^2$ and $a_2 = 2.00\,\text{m/s}^2$, then $m_2 = 1.50\,\text{kg}$.

Newton's second law, like the first law, is valid only in inertial reference frames. In the non-inertial reference frame of an accelerating car, for example, a cup on the dashboard starts sliding—it accelerates—even though the net force on it is zero; thus $\Sigma \mathbf{F} = m\mathbf{a}$ doesn't work in such an accelerating reference frame.

4–5 | Newton's Third Law of Motion

*A force is exerted **on** a body and is exerted **by** another body*

Newton's second law of motion describes quantitatively how forces affect motion. But where, we may ask, do forces come from? Observations suggest that a force applied to any object is always applied *by another object*. A horse pulls a wagon, a person pushes a grocery cart, a hammer pushes on a nail, a magnet attracts a paper clip. In each of these examples, a force is exerted *on* one body, and that force is exerted *by* another body. For example, the force exerted *on* the nail is exerted *by* the hammer.

But Newton realized that things are not so one-sided. True, the hammer exerts a force on the nail (Fig. 4–8). But the nail evidently exerts a force back on the hammer as well, for the hammer's speed is rapidly reduced to zero upon contact. Only a strong force could cause such a rapid deceleration of the hammer. Thus, said Newton, the two bodies must be treated on an equal basis. The hammer exerts a force on the nail, and the nail exerts a force back on the hammer. This is the essence of **Newton's third law of motion**:

NEWTON'S THIRD LAW OF MOTION

> **Whenever one object exerts a force on a second object, the second exerts an equal and opposite force on the first.**

Action and reaction act on different objects

This law is sometimes paraphrased as "to every action there is an equal and opposite reaction." This is perfectly valid. But to avoid confusion, it is very important to remember that the "action" force and the "reaction" force are acting on *different* objects.

As evidence for the validity of Newton's third law, look at your hand when you push against a grocery cart or against the edge of a desk, Fig. 4–9. Your hand's shape is distorted, clear evidence that a force is being exerted on it. You can *see* the edge of the desk pressing into your hand. You can even *feel* the desk exerting a force on your hand; it hurts! The harder you push against the desk, the harder the desk pushes back on your hand. (Note that you only feel forces exerted *on* you, not forces that you exert on other objects.)

FIGURE 4–8 Multiflash photo of a hammer striking a nail. In accordance with Newton's third law the hammer exerts a force on the nail and the nail exerts a force back on the hammer. The latter force decelerates the hammer and brings it to rest.

FIGURE 4–9 If your hand pushes against the edge of a desk (the force vector is shown in red), the desk pushes back against your hand (this force vector is shown in a different color, violet, to remind us that this force acts on a different object).

Force exerted on hand by desk

Force exerted on desk by hand

(a) (b)

FIGURE 4–10 Two examples of Newton's third law. (a) When an ice-skater pushes against the railing, the railing pushes back and this force causes her to move away. (b) The launch of a rocket. The rocket engine pushes out the gases, and the gases exert an equal and opposite force back on the rocket, accelerating it.

As another demonstration of Newton's third law, consider the ice-skater in Fig. 4–10a. Since there is very little friction between her skates and the ice, she will move freely if a force is exerted on her. She pushes against the railing; and then *she* starts moving backward. Clearly, there had to be a force exerted on her to make her move. The force she exerts on the railing cannot make *her* move, for that force acts on the railing. Something had to exert a force on her to make her start moving, and that force could only have been exerted by the railing. The force with which the railing pushes on her is, by Newton's third law, equal and opposite to the force she exerts on the railing.

When a person throws a package out of a boat (initially at rest), the boat starts moving in the opposite direction. The person exerts a force on the package. The package exerts an equal and opposite force back on the person, and this force propels the person (and the boat) backward slightly. Rocket propulsion also is explained using Newton's third law (Fig. 4–10b). A common misconception is that rockets accelerate because the gases rushing out the back of the engine push against the ground or the atmosphere. Not true. What happens, instead, is that a rocket exerts a strong force on the gases, expelling them; and the gases exert an equal and opposite force *on the rocket*. It is this latter force that propels the rocket forward. Thus, a space vehicle is maneuvered in empty space by firing its rockets in the direction opposite to that in which it needs to accelerate.

Consider how we walk. A person begins walking by pushing with the foot against the ground. The ground then exerts an equal and opposite force back on the person (Fig. 4–11) and it is this force, *on* the person, that moves the person forward. (If you doubt this, try walking on very smooth slippery ice.) In a similar way, a bird flies forward by exerting a force on the air, but it is the air pushing back on the bird's wings that propels the bird forward.

➥ PHYSICS APPLIED

How does a rocket accelerate?

FIGURE 4–11 We can walk forward because, when one foot pushes backward against the ground, the ground pushes forward on that foot. The two forces shown *act on different objects.*

Horizontal force exerted on the ground by person's foot ◄———

Horizontal force exerted on the person's foot by the ground ———►

CONCEPTUAL EXAMPLE 4–3   **What exerts the force on a car?** What makes a car go forward?

RESPONSE   A common answer is that the engine makes the car move forward. But it is not so simple. The engine makes the wheels go around. But what good is that if they are on slick ice or mud? They just spin. A car moves forward due to the friction force exerted by the ground on the tires, and this force is the reaction to the force exerted on the ground by the tires.

We tend to associate forces with active bodies such as humans, animals, engines, or a moving object like a hammer. It is often difficult to see how an inanimate object at rest, such as a wall or a desk, can exert a force. The explanation lies in the fact that every material, no matter how hard, is elastic, at least to some degree. No one can deny that a stretched rubber band can exert a force on a wad

Inanimate objects can exert a force

of paper and send it flying across the room. Other materials may not stretch as easily as rubber, but they do stretch when a force is applied to them. And just as a stretched rubber band exerts a force, so does a stretched (or compressed) wall, desk, or car fender.

From the examples discussed above, it is clear that it is quite important to remember *on* what object a given force is exerted and *by* what object that force is exerted. The point is that a force influences the motion of an object only when it is applied *on* that object. A force exerted *by* a body does not influence that body; it only influences the other body *on* which it is exerted. Thus, to avoid confusion, the two prepositions *on* and *by* must always be used—and used with care.

One way to keep clear which force acts on which object is to use double subscripts. For example the force exerted on the **P**erson by the **G**round in Fig. 4–11 can be labeled F_{PG} as shown in Fig. 4–12. And the force exerted on the ground by the person is F_{GP}. Note that we have used different colors for the force vectors when they act on different objects. By Newton's third law

$$\mathbf{F}_{GP} = -\mathbf{F}_{PG}. \tag{4–2}$$

$\mathbf{F}_{GP}$ and $\mathbf{F}_{PG}$ have the same magnitude, and the minus sign reminds us that these two forces are in opposite directions.

Note carefully that the two forces shown in Fig. 4–11 or 4–12 act on different objects—hence we used slightly different colors for the arrows representing the forces. These two forces would never appear together in a sum of forces in Newton's second law, $\Sigma\mathbf{F} = m\mathbf{a}$. Why not? Because **a** is the acceleration of one particular body and $\Sigma\mathbf{F}$ must include only the forces on that body.

CONCEPTUAL EXAMPLE 4–4 | **Third law clarification.** Michelangelo's assistant has been assigned the task of moving a block of marble using a sled (Fig. 4–13). He says to his boss, "When I exert a forward force on the sled, the sled exerts an equal and opposite force backward. So how can I ever start it moving? No matter how hard I pull, the backward reaction force always equals my forward force, so the net force must be zero. I'll never be able to move this load." Is this a case of a little knowledge being dangerous? Explain.

RESPONSE Yes. Although it is true that the action and reaction forces are equal in magnitude, the assistant has forgotten that they are exerted on different objects. The forward ("action") force is exerted by the assistant on the sled (Fig. 4–13), whereas the backward "reaction" force is exerted by the sled on the assistant. To determine if the *assistant* moves or not, we must consider only

NEWTON'S THIRD LAW OF MOTION

$\mathbf{F}_{GP}$ $\mathbf{F}_{PG}$

FIGURE 4–12 Newton's third law. Subscripts on forces remind us *on* which body a force acts and *by* which body it is exerted.

FIGURE 4–13 Seventy-year-old Michelangelo has selected a fine block of marble for his next sculpture. Shown here is his assistant pulling it on a sled away from the quarry. Forces on the assistant are shown as red (magenta) arrows. Forces on the sled are purple arrows. Forces acting on the ground are orange arrows. Action–reaction forces that are equal and opposite are labeled by the same subscripts but reversed (such as $\mathbf{F}_{GA}$ and $\mathbf{F}_{AG}$) and are of different colors because they act on different objects. [Only horizontal forces are shown.]

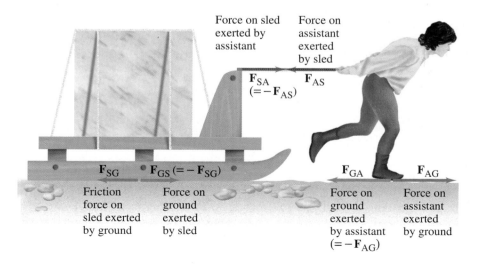

Force on sled exerted by assistant   Force on assistant exerted by sled

$\mathbf{F}_{SA}$ ($= -\mathbf{F}_{AS}$) $\mathbf{F}_{AS}$

$\mathbf{F}_{SG}$ $\mathbf{F}_{GS}$ ($= -\mathbf{F}_{SG}$) $\mathbf{F}_{GA}$ $\mathbf{F}_{AG}$

Friction force on sled exerted by ground   Force on ground exerted by sled   Force on ground exerted by assistant ($= -\mathbf{F}_{AG}$)   Force on assistant exerted by ground the forces *on the assistant* and then apply $\Sigma \mathbf{F} = m\mathbf{a}$, where $\Sigma \mathbf{F}$ is the net force *on the assistant*, $\mathbf{a}$ is the acceleration of the assistant, and m is the assistant's mass. There are two forces on the assistant that affect his forward motion and they are shown as bright red (magenta) arrows in Figs. 4–13 and 4–14: they are (1) the horizontal force $\mathbf{F}_{AG}$ exerted on the assistant by the ground (the harder he pushes backward against the ground, the harder the ground pushes forward on him— Newton's third law), and (2) the force $\mathbf{F}_{AS}$ exerted on the assistant by the sled, pulling backward on him; see Fig. 4–14. When the ground pushes forward on the assistant harder than the sled pulls backward, the assistant accelerates forward (Newton's second law). The sled, on the other hand, accelerates forward when the force on *it* exerted by the assistant is greater than the frictional force acting backward (that is, when $\mathbf{F}_{SA}$ has greater magnitude than $\mathbf{F}_{SG}$ in Fig. 4–13).

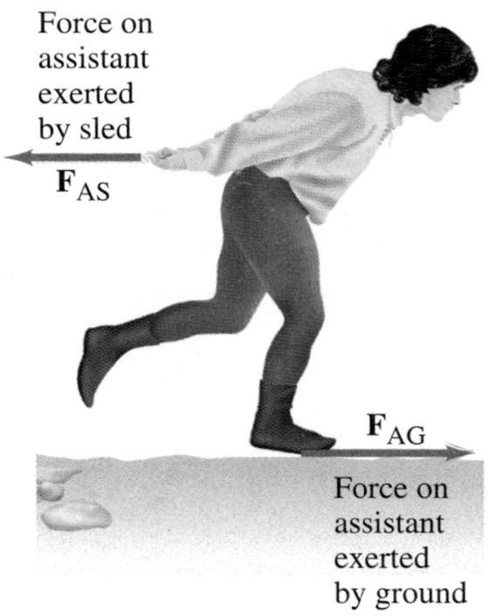

Using double subscripts to clarify Newton's third law can become cumbersome, and we won't usually use them in this way. Nevertheless, if there is any confusion in your mind about a given force, go ahead and use them to identify *on* what object and *by* what object the force is exerted.

Force on assistant exerted by sled $\mathbf{F}_{AS}$

$\mathbf{F}_{AG}$ Force on assistant exerted by ground

FIGURE 4–14 The forces on the assistant of Example 4–4.

| **CONCEPTUAL EXAMPLE 4–5** | **Collision using Newton's second and third laws.** |

A massive truck collides head-on with a small sports car. Which vehicle experiences the greater force of impact? Which experiences the greater acceleration? Which of Newton's laws is useful to obtain the correct answer?

RESPONSE The truck and the car experience the same magnitude impact force (Newton's third law). The car experiences a much greater acceleration in stopping than the truck because its mass is much less (Newton's second law).

4–6 Weight—the Force of Gravity; and the Normal Force

Galileo claimed that objects dropped near the surface of the Earth will all fall with the same acceleration, $\mathbf{g}$, if air resistance can be neglected. The force that gives rise to this acceleration is called the force of gravity. We now apply Newton's second law to the gravitational force; and for the acceleration, $\mathbf{a}$, we use the downward acceleration due to gravity, $\mathbf{g}$. Thus, the force of gravity on an object, $\mathbf{F}_G$, whose magnitude is commonly called its **weight**, can be written as

$$\mathbf{F}_G = m\mathbf{g}. \tag{4–3}$$

Weight = force of gravity

The direction of this force is down toward the center of the Earth. If the y axis is upwards, we can write $\mathbf{F}_G = -mg\mathbf{j}$.

In SI units, $g = 9.80 \text{ m/s}^2 = 9.80 \text{ N/kg}$,[†] so the weight of a 1.00-kg mass on Earth is $1.00 \text{ kg} \times 9.80 \text{ m/s}^2 = 9.80 \text{ N}$. We will mainly be concerned with the weight of objects on Earth, but we note that on the Moon, on other planets, or in space, the weight of a given mass will be different. For example, on the Moon the acceleration due to gravity is about one sixth what it is on Earth, and a 1.0 kg mass weighs only 1.7 N. Although we will not have occasion to use British units, we note that for practical purposes on the Earth, a mass of 1 kg weighs about 2.2 lb. (On the Moon, 1 kg weighs only about 0.4 lb.)

The force of gravity acts on an object when it is falling. When an object is at rest on the Earth, the gravitational force on it does not disappear, as we know if we weigh it on a spring scale. The same force, given by Eq. 4–3, continues to act. Why, then, doesn't the object move? From Newton's second law, the net force on an object that remains at rest is zero. There must be another force on the object to balance the gravitational force. For an object resting on a table, the table exerts

[†] Since $1 \text{ N} = 1 \text{ kg} \cdot \text{m/s}^2$ (Section 4–4), $1 \text{ m/s}^2 = 1 \text{ N/kg}$.

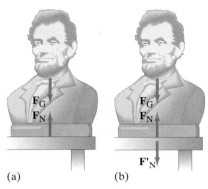

(a)   (b)

FIGURE 4–15 (a) The net force on an object at rest is zero according to Newton's second law. Therefore the downward force of gravity ($\mathbf{F}_G$) on an object must be balanced by an upward force (the normal force $\mathbf{F}_N$) exerted by the table in this case. (b) $\mathbf{F}'_N$ is the force exerted on the table by the statue and is the reaction force to $\mathbf{F}_N$ as per Newton's third law. ($\mathbf{F}'_N$ is shown in a different color to remind us it acts on a different body.) The reaction to $\mathbf{F}_G$ is not shown.

this upward force; see Fig. 4–15a. The table is compressed slightly beneath the object, and due to its elasticity, it pushes up on the object as shown. The force exerted by the table is often called a **contact force**, since it occurs when two objects are in contact. (The force of your hand pushing on a cart is also a contact force.) When a contact force acts *perpendicular* to the common surface of contact, it is usually referred to as the **normal force** ("normal" means perpendicular); hence it is labeled $\mathbf{F}_N$ in the diagram.

The two forces shown in Fig. 4–15a are both acting on the statue, which remains at rest, so the vector sum of these two forces must be zero (Newton's second law: if $\mathbf{a} = 0$, $\Sigma \mathbf{F} = 0$). Hence $\mathbf{F}_G$ and $\mathbf{F}_N$ must be of equal magnitude and in opposite directions. But they are *not* the equal and opposite forces spoken of in Newton's third law. The action and reaction forces of Newton's third law act on *different objects*, whereas the two forces shown in Fig. 4–15a act on the *same* object. For each of the forces shown in Fig. 4–15a, we can ask, "What is the reaction force?" The upward force, $\mathbf{F}_N$, on the statue is exerted by the table. The reaction to this force is a force exerted by the statue on the table. It is shown in Fig. 4–15b, where it is labeled $\mathbf{F}'_N$. This force $\mathbf{F}'_N$, exerted on the table by the statue, is the reaction force to $\mathbf{F}_N$ in accord with Newton's third law. Now, what about the other force on the statue, the force of gravity $\mathbf{F}_G$? Can you guess what the reaction is to this force? [We will see in Chapter 6 that the reaction force is also a gravitational force exerted on the Earth by the statue, and can be considered to act at the Earth's center.]

EXAMPLE 4–6 **Weight, normal force, and a box.** A friend has given you a special gift, a box of mass 10.0 kg with a mystery surprise inside. It's a reward for your fine showing on the physics final. The box, is resting on the smooth (frictionless) horizontal surface of a table, Fig. 4–16a. (a) Determine the weight of the box and the normal force acting on it. (b) Now your friend pushes down on the box with a force of 40.0 N, as in Fig. 4–16b. Again determine the normal force acting on the box. (c) If your friend pulls upward on the box with a force of 40.0 N (Fig. 4–16c), what now is the normal force on the box?

SOLUTION (a) The box is resting on the table. The weight of the box is $mg = (10.0 \text{ kg})(9.80 \text{ m/s}^2) = 98.0$ N, and this force acts downward. The only other force on the box is the normal force exerted upward on it by the table, as shown in Fig. 4–16a. We chose the upward direction as the positive y direction, and then the net force ΣF_y on the box is $\Sigma F_y = F_N - mg$. Since the box is at rest, the net force on it must be zero ($\Sigma F_y = ma_y$, and $a_y = 0$). Thus

$$\Sigma F_y = F_N - mg = 0,$$

so

$$F_N = mg.$$

The normal force on the box, exerted by the table, is 98.0 N upward, and has magnitude equal to the box's weight.

FIGURE 4–16 Example 4–6. (a) A 10-kg gift box is at rest on a table. (b) A person pushes down on the box with a force of 40.0 N. (c) A person pulls upward on the box with a force of 40.0 N. The forces are all assumed to act along a line; they are shown slightly displaced in order to be distinguishable on the diagram. Only forces acting on the box are shown.

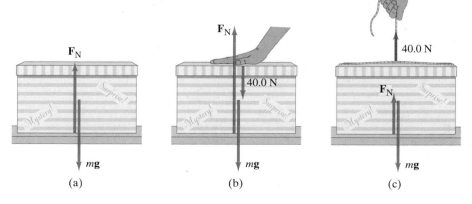

(a)   (b)   (c)

(*b*) Your friend is pushing down on the box with a force of 40.0 N. So now there are three forces acting on the box, as shown in Fig. 4–16b. The weight of the box is still $mg = 98.0$ N. The net force is $\Sigma F_y = F_N - mg - 40.0$ N, and is equal to zero since the box remains at rest. Thus, since $a = 0$, Newton's second law gives

$$\Sigma F_y = F_N - mg - 40.0 \text{ N} = 0,$$

so the normal force is now

$$F_N = mg + 40.0 \text{ N} = 98.0 \text{ N} + 40.0 \text{ N} = 138.0 \text{ N},$$

which is greater than in (*a*). The table pushes back with more force.

(*c*) The box's weight is still 98.0 N and acts downward. The force exerted by your friend and the normal force both act upward (positive direction), as shown in Fig. 4–16c. The box doesn't move since your friend's upward force is less than the weight. The net force, again set to zero, is

Normal force is not always equal in magnitude to the weight

$$\Sigma F_y = F_N - mg + 40.0 \text{ N} = 0,$$

so

$$F_N = mg - 40.0 \text{ N} = 98.0 \text{ N} - 40.0 \text{ N} = 58.0 \text{ N}.$$

The table does not push against the full weight of the box because of the upward pull exerted by your friend.

EXAMPLE 4–7 **Accelerating the box.** What happens when a person pulls upward on the box in Example 4–6 (*c*) with a force equal to, or greater than, the box's weight, say $F_P = 100.0$ N rather than the 40.0 N shown in Fig. 4–16c?

SOLUTION The net force is now

$$\Sigma F_y = F_N - mg + F_P = F_N - 98.0 \text{ N} + 100.0 \text{ N},$$

and if we set this equal to zero, we would get $F_N = -2.0$ N. This is nonsense, since the negative sign implies F_N points downward, and the table surely cannot pull down on the box (unless there's glue on the table). The least F_N can be is zero, which it will be in this case. What really happens here is clear: the box accelerates upward since the net force is not zero; it is

$$\Sigma F_y = F_P - mg = 100.0 \text{ N} - 98.0 \text{ N} = 2.0 \text{ N}$$

upward. See Fig. 4–17. So the box moves upward with an acceleration of magnitude

$$a_y = \Sigma F_y/m = 2.0 \text{ N}/10.0 \text{ kg} = 0.20 \text{ m/s}^2.$$

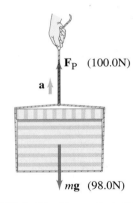

FIGURE 4–17 Example 4–7. The box accelerates upwards because $F_P > mg$.

EXAMPLE 4–8 **Unwanted weight loss?** A 65-kg woman descends in an elevator that briefly, when leaving a floor, accelerates at $0.20g$. (*a*) If she stands on a scale during this acceleration, what is her weight and what does the scale read? (*b*) What does the scale read when the elevator descends at a constant speed of 2.0 m/s?

SOLUTION (*a*) Figure 4–18 shows all the forces that act on the woman (and only those that act on her). The direction of the acceleration is downward, which we take as positive. From Newton's second law,

$$\Sigma F = ma$$

$$mg - F_N = 0.20mg$$

$$F_N = mg - 0.20mg = 0.80mg.$$

The normal force $\mathbf{F}_N$ is the force the scale exerts on the person, and is equal and opposite to the force she exerts on the scale, $\mathbf{F}'_N = -0.80\,mg$. Her weight (force of gravity on her) is still $mg = (65 \text{ kg})(9.8 \text{ m/s}^2) = 640$ N. But the scale, needing to exert a force of only $0.80mg$, will show her mass as $0.80m = 52$ kg.

(*b*) Now there is no acceleration, $a = 0$, so by Newton's second law, $mg - F_N = 0$ and $F_N = mg$. The scale reads her correct mass of 65 kg.

FIGURE 4–18 Example 4–8.

4-7 Solving Problems with Newton's Laws: Free-Body Diagrams

Newton's second law tells us that the acceleration of an object is proportional to the *net force* acting on the object. The **net force**, as mentioned earlier, is the *vector sum* of all forces acting on the object. Indeed, extensive experiments have shown that forces do add together as vectors precisely according to the rules we developed in Chapter 3. For example, in Fig. 4–19, two forces of equal magnitude (100 N each) are shown acting on an object at right angles to each other. Intuitively, we can see that the object will move at a 45° angle and thus the net force acts at a 45° angle. This is just what the rules of vector addition give. From the theorem of Pythagoras, the magnitude of the resultant force is $F_R = \sqrt{(100\,\text{N})^2 + (100\,\text{N})^2} = 141\,\text{N}$.

FIGURE 4–19 (a) Two forces, $\mathbf{F}_1$ and $\mathbf{F}_2$, act on an object. (b) The sum, or resultant, of $\mathbf{F}_1$ and $\mathbf{F}_2$ is $\mathbf{F}_R$.

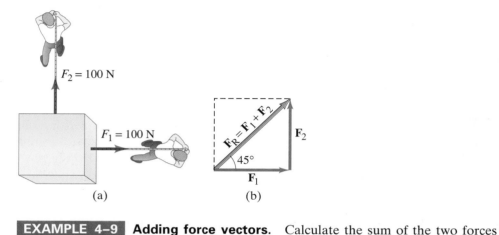

(a)　　　　(b)

FIGURE 4–20 Two force vectors act on a boat (Example 4–9).

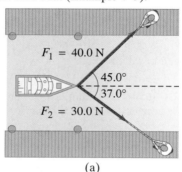

(a)

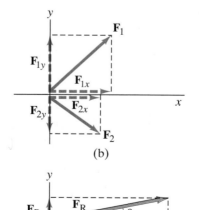

(b)

(c)

EXAMPLE 4–9 **Adding force vectors.** Calculate the sum of the two forces acting on the boat shown in Fig. 4–20a.

SOLUTION These two forces are shown resolved in Fig. 4–20b. We add the forces using the method of components. The components of $\mathbf{F}_1$ are

$$F_{1x} = F_1 \cos 45.0° = (40.0\,\text{N})(0.707) = 28.3\,\text{N},$$

$$F_{1y} = F_1 \sin 45.0° = (40.0\,\text{N})(0.707) = 28.3\,\text{N}.$$

The components of $\mathbf{F}_2$ are

$$F_{2x} = +F_2 \cos 37.0° = +(30.0\,\text{N})(0.799) = +24.0\,\text{N},$$

$$F_{2y} = -F_2 \sin 37.0° = -(30.0\,\text{N})(0.602) = -18.1\,\text{N}.$$

F_{2y} is negative because it points along the negative y axis. The components of the resultant force are (see Fig. 4–20c)

$$F_{Rx} = F_{1x} + F_{2x} = 28.3\,\text{N} + 24.0\,\text{N} = 52.3\,\text{N},$$

$$F_{Ry} = F_{1y} + F_{2y} = 28.3\,\text{N} - 18.1\,\text{N} = 10.2\,\text{N}.$$

To find the magnitude of the resultant force, we use the Pythagorean theorem:

$$F_R = \sqrt{F_{Rx}^2 + F_{Ry}^2} = \sqrt{(52.3)^2 + (10.2)^2} = 53.3\,\text{N}.$$

The only remaining question is the angle θ that the net force $\mathbf{F}_R$ makes with the x axis. We use:

$$\tan \theta = \frac{F_{Ry}}{F_{Rx}} = \frac{10.2\,\text{N}}{52.3\,\text{N}} = 0.195,$$

and $\tan^{-1}(0.195) = 11.0°$.

When solving problems involving Newton's laws and force, it is very important to draw a diagram showing all the forces acting *on* each object involved. Such a diagram is called a **free-body diagram**, or **force diagram**: draw an arrow to represent each force acting on a given body, being sure to include *every* force acting on that body. Do not show forces that the body exerts on *other* objects. When concerned only about translational motion, all the forces on a given body can be drawn as acting at the center of the object, thus treating the object as a point particle. However, when doing problems involving rotation or statics, the place *where* each force acts is also important, as we shall see.

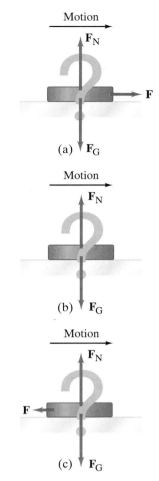

FIGURE 4–21 Which is the correct free-body diagram for a hockey puck sliding across frictionless ice (Example 4–10)?

CONCEPTUAL EXAMPLE 4–10 **The hockey puck.** A hockey puck is sliding at constant velocity across a flat horizontal ice surface that is assumed to be frictionless. Which of the sketches in Fig. 4–21 is the correct free-body diagram for this puck? What would your answer be if the puck slowed down?

RESPONSE Did you choose (a)? If so, can you answer the question: what exerts the horizontal force labeled **F**? If you say that it is the force needed to maintain the motion (as the ancient Greeks said), ask yourself: what exerts this force? Remember that another object must exert any force—and there simply isn't any possibility here. Therefore, (a) is wrong. Besides, the force **F** in Fig. 4–21a would give rise to an acceleration by Newton's second law. It is (b) that is correct, as long as there is no friction. No net force acts on the puck, and the puck slides at constant velocity across the ice.

If someone insists that we come down from the ivory tower of idealized frictionless surfaces, down to the real world where even smooth ice exerts at least a tiny friction force, then (c) is the correct answer. The tiny friction force is in the direction opposite to the motion (it ought to be labeled **F**$_{fr}$, not simply **F**), and the puck's velocity decreases, even if very slowly.

Here now is a brief summary of how to approach solving problems involving Newton's laws:

1. Draw a sketch of the situation.

2. Consider only one object (at a time), and draw a **free-body diagram** for that body, showing *all* the forces acting *on* that body, including any unknown forces that you have to solve for. Do not show any forces that the body exerts on other bodies. Draw the arrow for each force vector reasonably accurately for direction and magnitude. Label each force, including forces you must solve for, as to its source (gravity, person, friction, and so on). If several bodies are involved, draw a free-body diagram for each body *separately*, showing all the forces acting *on that body* (and *only* forces acting on that body). For each (and every) force, you

must be clear about: *on* what object that force acts; and *by* what object that force is exerted. Only forces acting *on* a given body can be included in $\Sigma F = ma$ for that body.

3. Newton's second law involves vectors, and it is usually important to resolve vectors into components. Choose an x and a y axis in a way that simplifies the calculation.

4. For each body, Newton's second law can be applied to the x and y components separately. That is, the x component of the net force on that body will be related to the x component of that body's acceleration: $\Sigma F_x = ma_x$, and similarly for the y direction.

5. Solve the equation or equations for the unknown(s).

This problem-solving box should not be considered a prescription. Rather it is a summary of things to do that will start your mind thinking and getting involved in the problem at hand.

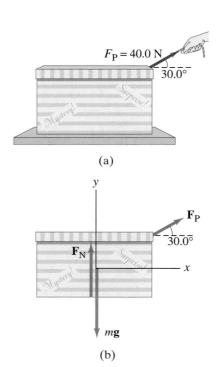

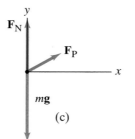

FIGURE 4–22 Example 4–11; (b) is the free-body diagram for the box, and (c) is the free-body diagram considering all the forces to act at a point (translational motion only, which is what we have here).

In the Examples that follow, we assume that all surfaces are very smooth so that friction can be ignored. (Friction, and Examples using it, are discussed in the next chapter.)

EXAMPLE 4–11 **Pulling the mystery box.** Suppose a friend asks to examine the 10.0-kg box you were given (Example 4–6, Fig. 4–16), hoping to guess what is inside; and you respond, "Sure, pull the box over to you." She then pulls the box by the attached ribbon (or string), as shown in Fig. 4–22a, along the smooth surface of the table. The magnitude of the force exerted by the person is $F_P = 40.0$ N, and it is exerted at a 30.0° angle as shown. Calculate (a) the acceleration of the box, and (b) the magnitude of the upward force F_N exerted by the table on the box. Assume that friction can be neglected.

SOLUTION Figure 4–22b shows the free-body diagram of the box, which means we show *all* the forces acting on the box and *only* the forces acting on the box. They are: the force of gravity $m\mathbf{g}$; the normal force exerted by the table $\mathbf{F_N}$; and the force exerted by the person $\mathbf{F_P}$. We are interested only in translational motion, so we can show the three forces acting at a point, Fig. 4–22c. With the y axis vertical and the x axis horizontal, the pull of 40.0 N has components

$$F_{Px} = (40.0\,\text{N})(\cos 30.0°) = (40.0\,\text{N})(0.866) = 34.6\,\text{N},$$

$$F_{Py} = (40.0\,\text{N})(\sin 30.0°) = (40.0\,\text{N})(0.500) = 20.0\,\text{N}.$$

(a) In the horizontal (x) direction, $\mathbf{F_N}$ and $m\mathbf{g}$ have zero components. Thus the horizontal component of the net force is F_{Px}. From Newton's second law, $\Sigma F_x = ma_x$, we have

$$F_{Px} = ma_x,$$

so

$$a_x = \frac{F_{Px}}{m} = \frac{(34.6\,\text{N})}{(10.0\,\text{kg})} = 3.46\,\text{m/s}^2.$$

The acceleration of the box is thus 3.46 m/s² to the right.

(b) In the vertical (y) direction, with upward as positive, again using Newton's second law we have

$$\Sigma F_y = ma_y$$

$$F_N - mg + F_{Py} = ma_y.$$

Now $mg = (10.0\,\text{kg})(9.80\,\text{m/s}^2) = 98.0$ N and $F_{Py} = 20.0$ N as we calculated above. Furthermore, since $F_{Py} < mg$, the box does not move vertically, so $a_y = 0$. Thus

$$F_N - 98.0\,\text{N} + 20.0\,\text{N} = 0$$

which tells us that the normal force is

$$F_N = 78.0\,\text{N}.$$

Notice that F_N is less than mg. The table does not push against the full weight of the box since part of the pull exerted by the person is in the upward direction. Compare this to Example 4–6, part (c).

Tension in a cord

When a flexible cord pulls on an object, the cord is said to be under **tension**, and the force it exerts on the object is the tension $\mathbf{F_T}$. If the cord has negligible mass, the force exerted at one end is transmitted undiminished to each adjacent piece of cord along the entire length to the other end. Why? Because for the cord, $\Sigma\mathbf{F} = m\mathbf{a} = 0$ no matter what $\mathbf{a}$ is if m is zero (negligible); hence the forces pulling on the cord at its two ends must add up to zero (F_T and $-F_T$). Note that flexible ropes and cords can only pull. They can't push because they bend.

FIGURE 4–23 Example 4–12; (a) Two boxes are connected by a cord. A person pulls horizontally on box 1 with force $F_P = 40.0$ N. (b) Free-body diagram for box 1. (c) Free-body diagram for box 2.

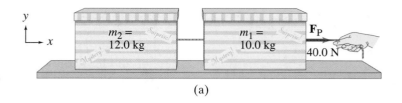

(a)

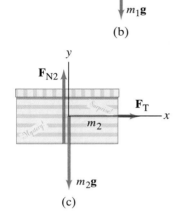

EXAMPLE 4–12 **Two boxes connected by a cord.** Two boxes are connected by a lightweight cord and are resting on a table. The boxes have masses of 12.0 kg and 10.0 kg. A horizontal force F_P of 40.0 N is applied by a person to the 10.0-kg box, as shown in Fig. 4–23a. Find (*a*) the acceleration of each box, and (*b*) the tension in the cord.

SOLUTION (*a*) The free-body diagram for each of the boxes is shown in Figs. 4–23b and c. We consider each box by itself so that Newton's second law can be applied to each. The cord is light, so we neglect its mass relative to the mass of the boxes. The force F_P acts on box 1. Box 1 exerts a force F_T on the connecting cord, and the cord exerts an opposite but equal magnitude force F_T back on box 1 (Newton's third law). These forces on box 1 are shown in Fig. 4–23b. Because the cord is considered to be massless, the tension at each end is the same as we saw just above. Hence the cord exerts a force F_T on the second box; Fig. 4–23c shows the forces on box 2. There will be only horizontal motion. We take the positive x axis to the right, and we use subscripts 1 and 2 to refer to the two boxes. Applying $\Sigma F_x = ma_x$ to box 1, we have:

$$\Sigma F_x = F_P - F_T = m_1 a_1.$$ [box 1]

For box 2, the only horizontal force is F_T, so

$$\Sigma F_x = F_T = m_2 a_2.$$ [box 2]

The boxes are connected, and if the cord remains taut and doesn't stretch, then the two boxes will have the same acceleration a. Thus $a_1 = a_2 = a$, and we are given $m_1 = 10.0$ kg and $m_2 = 12.0$ kg. We add the two equations above and obtain

$$(m_1 + m_2)a = F_P - F_T + F_T = F_P$$

or

$$a = \frac{F_P}{m_1 + m_2} = \frac{40.0 \text{ N}}{22.0 \text{ kg}} = 1.82 \text{ m/s}^2.$$

This is what we sought. Notice that we would have obtained the same result had we considered a single system, of mass $m_1 + m_2$, acted on by a net horizontal force equal to F_P. (The tension forces F_T would then be considered internal to the system as a whole, and summed together would make zero contribution to the net force on the *whole* system.)

(*b*) From the equation above for box 2 $(F_T = m_2 a_2)$, the tension in the cord is

$$F_T = m_2 a = (12.0 \text{ kg})(1.82 \text{ m/s}^2) = 21.8 \text{ N}.$$

Thus, F_T is less than F_P (= 40.0 N), as we expect, since F_T acts to accelerate only m_2.

➡ **PROBLEM SOLVING**

An alternate analysis

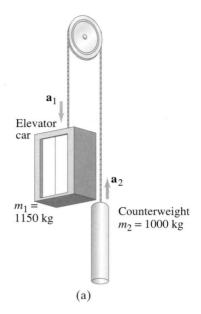

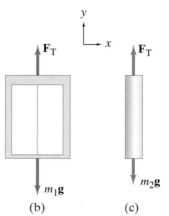

(a)

(b)   (c)

FIGURE 4–24 Example 4–13.
(a) Atwood's machine in the form of
an elevator-counterweight system.
(b) and (c) Free-body diagrams for
the two masses.

⇒ PROBLEM SOLVING

*Check your result by seeing if it
works in situations where the
answer is easily guessed*

EXAMPLE 4–13 **Elevator and counterweight (Atwood's machine).** Two masses suspended over a pulley by a cable, as shown in Fig. 4–24a, is sometimes referred to generically as an *Atwood's machine*. Consider the real-life application of an elevator (m_1) and its counterweight (m_2). To minimize the work done by the motor to raise and lower the elevator safely, m_1 and m_2 are similar in mass. We leave the motor out of the system for this calculation, and assume that the cable's mass is negligible and that the mass of the pulley,[†] as well as any friction, is small and ignorable. These assumptions assure that the tension F_T in the cord has the same magnitude on both sides of the pulley. Let the mass of the counterweight be $m_2 = 1000$ kg. Assume the mass of the empty elevator is 850 kg, and its mass when carrying four passengers is $m_1 = 1150$ kg. For the latter case $(m_1 = 1150$ kg$)$, calculate (*a*) the acceleration of the elevator and (*b*) the tension in the cable.

SOLUTION (*a*) Figures 4–24b and c show the free-body diagrams for the two masses. It is clear that m_1, being the heavier, will accelerate downward, and m_2 will accelerate upward. The magnitudes of their accelerations will be equal (we assume the cable doesn't stretch). For the counter-weight, $m_2 g = (1000$ kg$)(9.80$ m/s$^2) = 9800$ N, so F_T must be greater than 9800 N (in order that m_2 will accelerate upward). For the elevator, $m_1 g = (1150$ kg$)(9.80$ m/s$^2) = 11{,}300$ N, which must have greater magnitude than F_T so that m_1 accelerates downward. Thus our calculation must give F_T between 9800 N and 11,300 N. To find F_T as well as the acceleration a, we apply $\Sigma F = ma$ to each mass, where we take upward as the positive y direction for both masses. With this choice of axes, $a_2 = a$, and $a_1 = -a$. Thus

$$F_T - m_1 g = m_1 a_1 = -m_1 a$$

$$F_T - m_2 g = m_2 a_2 = +m_2 a.$$

We subtract the first equation from the second to get

$$(m_1 - m_2)g = (m_1 + m_2)a.$$

We solve this for a:

$$a = \frac{m_1 - m_2}{m_1 + m_2} g = \frac{1150 \text{ kg} - 1000 \text{ kg}}{1150 \text{ kg} + 1000 \text{ kg}} g = 0.070g = 0.68 \text{ m/s}^2.$$

The elevator (m_1) accelerates downward (and the counterweight m_2 upward) at $a = 0.070g = 0.68$ m/s^2.
(*b*) The tension in the cord F_T can be obtained from either of the two $\Sigma F = ma$ equations, setting $a = 0.070g = 0.68$ m/s^2:

$$F_T = m_1 g - m_1 a = m_1(g - a)$$
$$= 1150 \text{ kg} (9.80 \text{ m/s}^2 - 0.68 \text{ m/s}^2) = 10{,}500 \text{ N},$$

$$F_T = m_2 g + m_2 a = m_2(g + a)$$
$$= 1000 \text{ kg} (9.80 \text{ m/s}^2 + 0.68 \text{ m/s}^2) = 10{,}500 \text{ N},$$

which are consistent.
We can check our equation for the acceleration a in this Example by noting that if the masses were equal $(m_1 = m_2)$, then our equation above for a would give $a = 0$, as we should expect. Also, if one of the masses is zero (say, $m_1 = 0$), then the other mass $(m_2 \neq 0)$ would be predicted by our equation to accelerate at $a = g$, again as expected.

[†]We'll see how to deal with a rotating pulley with mass in Chapter 10.

CONCEPTUAL EXAMPLE 4–14 **The advantage of a pulley.** A mover is trying to lift a piano (slowly) up to a second-story apartment (Fig. 4–25). He is using a rope looped over two pulleys as shown. What force must he exert on the rope to slowly lift the piano's 2000 N weight?

RESPONSE Look at the forces acting on the lower pulley at the piano. The weight of the piano is pulling down. The tension in the rope, looped through this pulley, pulls up *twice*, once on each side of the pulley. Thus, Newton's second law gives

$$2F_T - mg = ma.$$

To move the piano with constant speed ($a = 0$) requires a tension in the cord, and hence a pull on the cord, of $F_T = mg/2$. The mover exerts a force equal to half the piano's weight. We say the pulley has given a **mechanical advantage** of 2, since without the pulley the mover would have to exert twice the force.

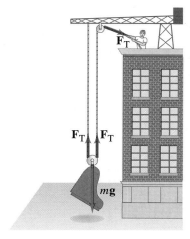

FIGURE 4–25 Example 4–14.

EXAMPLE 4–15 **Getting the car out of the mud.** Finding her car stuck in the mud, a bright graduate of a good physics course ties a strong rope to the back bumper of the car, and the other end to a tree, as shown in Fig. 4–26a. She pushes at the midpoint of the rope with her maximum effort, which she estimates to be a force $F_P \approx 300$ N. The car just begins to budge with the rope at an angle θ (see the Figure) which she estimates to be 5°. With what force is the rope pulling on the car? Neglect the mass of the rope.

How to get out of the mud

SOLUTION First, note that the tension in a rope is always along the rope. Any component perpendicular to the rope would cause the rope to bend or buckle (as it does here where $\mathbf{F}_P$ acts)—in other words, a rope can support a tension force only along its length. Let $\mathbf{F}_{T1}$ and $\mathbf{F}_{T2}$ be the forces the rope exerts on the tree and on the car, as shown in Fig. 4–26a. As our "free body," we choose the tiny section of rope where she pushes. The free-body diagram is shown in Fig. 4–26b, which shows $\mathbf{F}_P$ as well as the tensions in the rope (note that we have used Newton's third law). At the moment the car budges, the acceleration is still essentially zero, so $\mathbf{a} = 0$. For the x component of $\Sigma\mathbf{F} = m\mathbf{a} = 0$ on that small section of rope, we have

$$\Sigma F_x = F_{T2x} - F_{T1x} = 0, \quad \text{or} \quad F_{T1}\cos\theta - F_{T2}\cos\theta = 0.$$

Hence $F_{T1} = F_{T2}$, and we can write $F_T = F_{T1} = F_{T2}$. In the y direction, the forces acting are $\mathbf{F}_P$, and the components of $\mathbf{F}_{T1}$ and $\mathbf{F}_{T2}$ that point in the negative y direction (each equal to $F_T\sin\theta$). So for the y component of $\Sigma\mathbf{F} = m\mathbf{a}$, we have

$$\Sigma F_y = F_P - 2F_T\sin\theta = 0.$$

We solve this for F_T and insert $F_P \approx 300$ N, which was given:

$$F_T = \frac{F_P}{2\sin\theta} \approx \frac{300 \text{ N}}{2\sin 5°} \approx 1700 \text{ N}.$$

She was able to magnify her effort almost six times using this technique! Notice the symmetry of the problem, which ensures that $F_{T1} = F_{T2}$.

➡ **PROBLEM SOLVING**

Use any symmetry present to simplify a problem

FIGURE 4–26 Example 4–15. Getting a car out of the mud.

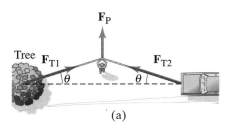

(a)

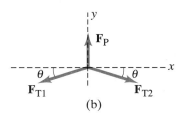

(b)

EXAMPLE 4–16 **Accelerometer.** A small mass m hangs from a thin string and can swing like a pendulum. You attach it above the window of your car as shown in Fig. 4–27a. When the car is at rest, the string hangs vertically. What angle θ does the string make (a) when the car accelerates at a constant $a = 1.20 \text{ m/s}^2$, and (b) when the car moves at constant velocity, $v = 90 \text{ km/h}$?

SOLUTION (a) Figure 4–27b shows the pendulum at some angle θ and the forces on it: mg downward and the tension $\mathbf{F_T}$ in the cord. These forces do not add up to zero if $\theta \neq 0$, and since we have an acceleration a, we therefore expect $\theta \neq 0$. Note that θ is the angle relative to the vertical. The acceleration $a = 1.20 \text{ m/s}^2$ is horizontal, so from Newton's second law,

$$ma = F_T \sin\theta$$

for the horizontal component, whereas the vertical component gives

$$0 = F_T \cos\theta - mg.$$

From these two equations, we obtain

$$\tan\theta = \frac{F_T \sin\theta}{F_T \cos\theta} = \frac{ma}{mg} = \frac{a}{g}$$

or

$$\tan\theta = \frac{1.20 \text{ m/s}^2}{9.80 \text{ m/s}^2} = 0.122$$

so

$$\theta = 7.0°.$$

(b) The velocity is constant, so $a = 0$ and $\tan\theta = 0$. Hence the pendulum hangs vertically ($\theta = 0°$). This simple device is an **accelerometer**—it can be used to measure acceleration.

FIGURE 4–27 Example 4–16.

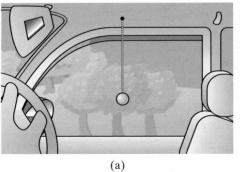

(a)

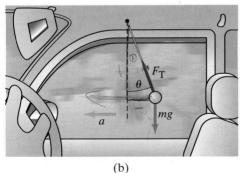

(b)

Now we consider what happens when an object slides down an incline, such as a hill or ramp. Such problems are interesting because gravity is the accelerating force, yet the acceleration is not vertical. Solving problems is usually easier if we choose the xy coordinate system so the x axis points along the incline and the y axis is perpendicular to the incline, as shown in Fig. 4–28a. Note also that the normal force is not vertical, but is perpendicular to the plane, Fig. 4–28b.

Good choice of coordinate system simplifies the calculation

EXAMPLE 4–17 **Box slides down an incline.** A box of mass m is placed on a smooth (frictionless) incline that makes an angle θ with the horizontal, as shown in Fig. 4–28a. (*a*) Determine the normal force on the box. (*b*) Determine the box's acceleration. (*c*) Evaluate for a mass $m = 10\,\mathrm{kg}$ and an incline of $\theta = 30°$.

SOLUTION The free-body diagram is shown in Fig. 4–28b. The forces on the box are its weight mg vertically downward, which is shown resolved into its components parallel and perpendicular to the incline, and the normal force F_N. The incline acts as a constraint, allowing motion along its surface. The "constraining" force is the normal force. Since we know the motion is along the incline we choose the x axis downward along the incline (the direction of motion), and the y axis perpendicular to the incline upward, as shown. There is no motion in the y direction, so $a_y = 0$. Applying Newton's second law we have

$$F_y = ma_y$$
$$F_\mathrm{N} - mg\cos\theta = 0,$$

where F_N and the y component of gravity ($mg\cos\theta$) are all the forces acting on the box in the y direction. Thus the answer to part (*a*) is that the normal force is given by

$$F_\mathrm{N} = mg\cos\theta.$$

Note carefully that unless $\theta = 0°$, F_N has magnitude less than the weight mg. (*b*) In the x direction the only force acting is the x component of $m\mathbf{g}$, which we see from the diagram is $mg\sin\theta$. The acceleration a is in the x direction so

$$F_x = ma_x$$
$$mg\sin\theta = ma,$$

and we see that the acceleration down the plane is

$$a = g\sin\theta.$$

Thus the acceleration along an incline is always less than g, except at $\theta = 90°$, for which $\sin\theta = 1$ and $a = g$. This makes sense, of course, since $\theta = 90°$ is pure vertical fall. For $\theta = 0°$, $a = 0$, which makes sense since $\theta = 0°$ means the plane is horizontal so gravity causes no acceleration. Note too that the acceleration does not depend on the mass m.
(*c*) For $\theta = 30°$, $\cos\theta = 0.866$ and $\sin\theta = 0.500$, so

$$F_\mathrm{N} = 0.866mg = 85\,\mathrm{N},$$

and

$$a = 0.500\,g = 4.9\,\mathrm{m/s^2}.$$

We will discuss more Examples of motion on an incline in the next chapter, where friction will be included.

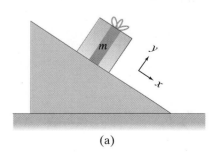
(a)

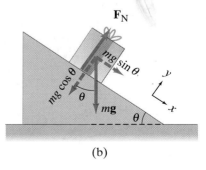
(b)

FIGURE 4–28 Example 4–17. (a) Box sliding on inclined plane. (b) Free-body diagram of box.

4–8 Problem Solving—A General Approach

A basic part of a physics course is solving problems effectively. The approach discussed here, though emphasizing Newton's laws, can be applied generally for other topics discussed throughout this book.

PROBLEM SOLVING	**In General**

1. **Read** and reread written problems carefully. A common error is to leave out a word or two when reading, which can completely change the meaning of a problem.

2. **Draw** an accurate picture or diagram of the situation. (This is probably the most overlooked, yet most crucial, part of solving a problem.) Use arrows to represent vectors such as velocity or force, and label the vectors with appropriate symbols. When dealing with forces and applying Newton's laws, make sure to include all forces on a given body, including unknown ones, and make clear what forces act on what body (otherwise you may make an error in determining the *net force* on a particular body). A separate **free-body diagram** needs to be drawn for each body involved, and it must show *all* the forces acting on a given body (and only on that body). Do not show forces that the body exerts on other bodies.

3. Choose a convenient xy **coordinate system** (choose one that makes your calculations easier). Vectors are to be resolved into components along these axes. When using Newton's second law, apply $\Sigma \mathbf{F} = m\mathbf{a}$ separately to x and y components, remembering that x direction forces are related to a_x, and similarly for y.

4. Note what the unknowns are—that is, what you are trying to determine—and decide what you need in order to find the unknowns. For problems in the present chapter, we use Newton's laws. More generally, it may help to see if there are one or more **relationships** (or **equations**) that relate the unknowns to the knowns. But be sure each relationship is applicable in the given case. It is very important to know the limitations of each formula or relationship—when it is valid and when not. In this book, the more general

equations have been given numbers, but even these can have a limited range of validity (often stated briefly, in brackets, to the right of the equation).

5. Try to solve the problem approximately, to see if it is doable (to check if enough information has been given) and reasonable. Use your intuition, and make **rough calculations**—see "Order of Magnitude Estimating" in Section 1–6. A rough calculation, or a reasonable guess about what the range of final answers might be, is very useful. And a rough calculation can be checked against the final answer to catch errors in calculation, such as in a decimal point or the powers of 10.

6. **Solve** the problem, which may include algebraic manipulation of equations and/or numerical calculations. Recall the mathematical rule that you need as many independent equations as you have unknowns; if you have three unknowns, for example, then you need three independent equations. It is usually best to work out the algebra symbolically before putting in the numbers. Why? Because (*a*) you can then solve a whole class of similar problems with different numerical values; (*b*) you can check your result for cases already understood (say, $\theta = 0°$ or $90°$); (*c*) there may be cancellations or other simplifications; (*d*) there is usually less chance for numerical error; and (*e*) because you may gain better insight into the problem.

7. Be sure to keep track of **units**, for they can serve as a check (they must balance on both sides of any equation).

8. Again consider if your answer is **reasonable**. The use of dimensional analysis, described in Section 1–7, can also serve as a check for many problems.

Summary

Newton's three laws of motion are the basic classical laws describing motion.

Newton's first law (the law of inertia) states that if the net force on an object is zero, an object originally at rest remains at rest, and an object in motion remains in motion in a straight line with constant velocity.

Newton's second law states that the acceleration of a body is directly proportional to the net force acting on it, and inversely proportional to its mass:

$$\Sigma \mathbf{F} = m\mathbf{a}.$$

Newton's second law is one of the most important and fundamental laws in classical physics.

Newton's third law states that whenever one body exerts a force on a second body, the second body always exerts a force on the first body which is equal in magnitude but opposite in direction:

$$\mathbf{F}_{12} = -\mathbf{F}_{21}.$$

The tendency of a body to resist a change in its motion is called **inertia**. **Mass** is a measure of the inertia of a body.

Weight refers to the force of gravity on a body, and is equal to the product of the body's mass m and the acceleration of gravity $\mathbf{g}$:

$$\mathbf{F}_G = m\mathbf{g}.$$

Force, which is a vector, can be considered as a push or pull; or, from Newton's second law, force can be defined as an action capable of giving rise to acceleration. The **net force** on an object is the vector sum of all forces acting on it.

For solving problems involving the forces on one or more bodies, it is essential to draw a **free-body diagram** for each body, showing all the forces acting on only that body. Newton's second law can be applied to the vector components for each body.

Questions

1. Why does a child in a wagon seem to fall backward when you give the wagon a sharp pull?

2. A box rests on the smooth bed of a truck. The truck driver starts the truck and accelerates forward. The box immediately starts to slide toward the rear of the truck bed. Discuss the motion of the box, in terms of Newton's laws, as seen (*a*) by Andrea standing on the ground beside the truck, and (*b*) by Dave who is riding on the truck (Fig. 4–29). Assume that the contact surfaces between the box and the truck bed are so smooth that you can ignore friction.

FIGURE 4–29 Question 2.

3. If the acceleration of a body is zero, are no forces acting on it?

4. Why do you push down harder on the pedals of a bicycle when first starting out than when moving at constant speed?

5. Only one force acts on an object. Can the object have zero acceleration? Can it have zero velocity?

6. When a golf ball is dropped to the pavement, it bounces back up. (*a*) Is a force needed to make it bounce back up? (*b*) If so, what exerts the force?

7. Which of the following objects weighs about 1 N: (*a*) an apple, (*b*) a mosquito, (*c*) this book, (*d*) you?

8. Why might your foot hurt if you kick a heavy desk or a wall?

9. When you are running and want to stop quickly, you must decelerate quickly. (*a*) What is the origin of the force that causes you to stop? (*b*) Estimate (using your own experience) the maximum rate of deceleration of a person running at top speed to come to rest.

10. A stone hangs by a fine thread from the ceiling, and a section of the same thread dangles from the bottom of the stone (Fig. 4–30). If a person gives a sharp pull on the dangling thread, where is the thread likely to break: below the stone or above it? What if the person gives a slow and steady pull? Explain your answers.

FIGURE 4–30 Question 10.

11. The force of gravity on a 2-kg rock is twice as great as that on a 1-kg rock. Why then doesn't the heavier rock fall faster?

12. Observe the motion of the moving disks of an air hockey game. Explain how Newton's first, second, and third laws apply. The disks float on a layer of air, ejected through tiny holes, so friction is reduced to a very small amount.

13. Compare the effort (or force) needed to lift a 10-kg object when you are on the Moon as compared to lifting it on Earth. Compare the force needed to throw a 2-kg object horizontally with a given speed when on the Moon as compared to on Earth.

14. According to Newton's third law, each team in a tug of war (Fig. 4–31) pulls with equal force on the other team. What, then, determines which team will win?

FIGURE 4–31 A tug of war. Describe the forces on each of the teams and on the rope. Question 14.

15. Whiplash sometimes results from an automobile accident when the victim's car is struck violently from the rear. Explain why the head of the victim seems to be thrown backward in this situation. Is it really?

16. If you walk on a log floating on a lake, why does the log move in the opposite direction?

17. Mary exerts an upward force of 40 N to hold a bag of groceries. Describe the "reaction" force (Newton's third law) by stating (*a*) its magnitude, (*b*) its direction, (*c*) *on* what body it is exerted, and (*d*) *by* what body it is exerted.

18. When you stand still on the ground, how large a force does the ground exert on you? Why doesn't this force make you rise up into the air?

19. A bear sling, Fig. 4–32, is used in some national parks for placing backpackers' food out of the reach of bears. Explain why the force needed to pull the backpack up increases as the backpack gets higher and higher. Is it possible to pull the rope hard enough so that it doesn't sag at all?

FIGURE 4–32 Question 19.

Problems

Sections 4–4 to 4–6

1. (I) Show that a $\frac{1}{4}$-lb cube of butter, or any other "quarter pounder" weighs about 1 N.

2. (I) A net force of 255 N accelerates a bike and rider at 2.20 m/s². What is the mass of the bike and rider?

3. (I) How much force is required to accelerate a 7.0-g object at 10,000 "*g*'s" (say, in a centrifuge)?

4. (I) How much tension must a rope withstand if it is used to accelerate a 1250-kg car horizontally at 1.30 m/s²? Ignore friction.

5. (I) What is the weight of a 58-kg astronaut (*a*) on Earth, (*b*) on the Moon ($g = 1.7$ m/s²), (*c*) on Mars ($g = 3.7$ m/s²), (*d*) in outer space traveling with constant velocity?

6. (II) What average force is required to stop a 1050-kg car in 7.0 s if it is traveling at 90 km/h?

7. (II) What average force is needed to accelerate a 6.25-gram pellet from rest to 155 m/s over a distance of 0.700 m along the barrel of a rifle?

8. (II) A 30.0-kg box rests on a table. (*a*) What is the weight of the box and the normal force acting on it? (*b*) A 20.0-kg box is placed on top of the 30.0-kg box, as shown in Fig. 4–33. Determine the normal force that the table exerts on the 30.0-kg box and the normal force that the 30.0-kg box exerts on the 20.0-kg box.

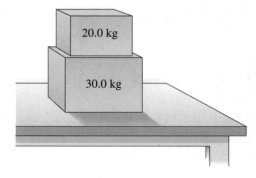

FIGURE 4–33 Problem 8.

9. (II) Superman must stop a 100-km/h train in 150 m to keep it from hitting a stalled car on the tracks. If the train's mass is 3.6×10^5 kg, how much force must he exert? Compare to the weight of the train.

10. (II) Consider a box resting on a frictionless surface. A constant force acts on the box for a given amount of time, accelerating it to some final speed. Then this process is repeated with another box, which has twice the mass of the first one. Compare the final speed of the heavier box to that of the first box.

11. (II) A fisherman yanks a fish out of the water with an acceleration of 3.5 m/s^2 using very light fishing line that has a breaking strength of 25 N. The fisherman unfortunately loses the fish as the line snaps. What can you say about the mass of the fish?

12. (II) A 0.140-kg baseball traveling 41.0 m/s strikes the catcher's mitt, which, in bringing the ball to rest, recoils backward 12.0 cm. What was the average force applied by the ball on the glove?

13. (II) Estimate the average force exerted by a shot-putter on a 7.0-kg shot if the shot is moved through a distance of 2.8 m and is released with a speed of 13 m/s.

14. (II) How much tension must a rope withstand if it is used to accelerate a 1200-kg car vertically upward at 0.80 m/s^2?

15. (II) A 7.50-kg bucket is lowered by a rope in which there is 63.0 N of tension. What is the acceleration of the bucket? Is it up or down?

16. (II) An elevator (mass 4125 kg) is to be designed so that the maximum acceleration is 0.0600 g. What are the maximum and minimum forces the motor should exert on the supporting cable?

17. (II) A 65-kg petty thief wants to escape from a third-story jail window. Unfortunately, a makeshift rope made of sheets tied together can support a mass of only 57 kg. How might the thief use this "rope" to escape? Give a quantitative answer.

18. (II) A person stands on a bathroom scale in a motionless elevator. When the elevator begins to move, the scale briefly reads only 0.75 of the person's regular weight. Calculate the acceleration of the elevator, and find the direction of acceleration.

19. (II) The cable supporting a 2100-kg elevator has a maximum strength of 21,750 N. What maximum upward acceleration can it give the elevator without breaking?

20. (II) A particular race car can cover a quarter-mile track (402 m) in 6.40 seconds, starting from a standstill. Assuming the acceleration is constant, how many "g's" does the driver experience? If the combined mass of the driver and race car is 280 kg, what horizontal force must the road exert on the tires?

21. (II) A Saturn V rocket has a mass of 2.75×10^6 kg and exerts a force of 33×10^6 N on the gases it expels. Determine (a) the initial vertical acceleration of the rocket, (b) its velocity after 8.0 s, and (c) how long it takes to reach an altitude of 9500 m. Ignore the mass of gas expelled (not realistic) and assume g remains constant.

22. (II) (a) What is the acceleration of two falling sky divers (mass 120.0 kg including parachute) when the upward force of air resistance is equal to one-fourth of their weight? (b) After popping open the parachute, the divers descend leisurely to the ground at constant speed. What now is the force of air resistance on the sky divers and their parachute? See Fig. 4–34.

FIGURE 4–34 Problem 22.

23. (II) An exceptional standing jump would raise a person 0.80 m off the ground. To do this, what force must a 61-kg person exert against the ground? Assume the person crouches a distance of 0.20 m prior to jumping, and thus the upward force has this distance to act over before he leaves the ground.

24. (III) A person jumps from the roof of a house 3.5-m high. When he strikes the ground below, he bends his knees so that his torso decelerates over an approximate distance of 0.70 m. If the mass of his torso (excluding legs) is 43 kg, find (a) his velocity just before his feet strike the ground, and (b) the average force exerted on his torso by his legs during deceleration.

25. (III) The 100-m dash can be run by the best sprinters in 10.0 s. A 62-kg sprinter accelerates uniformly for the first 45 m to reach top speed, which he maintains for the remaining 55 m. (a) What is the average horizontal component of force exerted on his feet by the ground during acceleration? (b) What is the speed of the sprinter over the last 55 m of the race (i.e., his top speed)?

Section 4–7

26. (I) A box weighing 85 N rests on a table. A rope tied to the box runs vertically upward over a pulley and a weight is hung from the other end (Fig. 4–35). Determine the force that the table exerts on the box if the weight hanging on the other side of the pulley weighs (a) 30 N, (b) 60 N, and (c) 90 N.

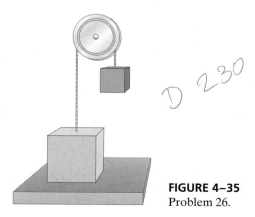

FIGURE 4–35
Problem 26.

27. (I) A 750-N force acts in a northwesterly direction. A second 750-N force must be exerted in what direction so that the resultant of the two forces points westward?

28. (I) Draw the free-body diagram for a basketball player (a) just before leaving the ground on a jump, and (b) while in the air. (Fig. 4–36).

FIGURE 4–36 Problem 28.

29. (I) Sketch the free-body diagram of a baseball (a) at the moment it is hit by the bat, and again (b) after it has left the bat and is flying toward the outfield.

30. (II) At the instant a race began, a 57-kg sprinter was found to exert a force of 80 N on the starting block at a 22° angle with respect to the ground. (a) What was the horizontal acceleration of the sprinter? (b) If the force was exerted for 0.34 s, with what speed did the sprinter leave the starting block?

31. (II) The two forces $\mathbf{F}_1$ and $\mathbf{F}_2$ shown in Fig. 4–37a and b (looking down) act on a 29.0-kg object on a frictionless tabletop. If $F_1 = 20.2$ N and $F_2 = 26.0$ N, find the net force on the object and its acceleration for each situation, (a) and (b).

32. (II) A person pushes a 13.0-kg lawn mower at constant speed with a force of 78.0 N directed along the handle, which is at an angle of 45.0° to the horizontal (Fig. 4–38). (a) Draw the free-body diagram showing all forces acting on the mower. Calculate (b) the horizontal retarding force on the mower, then (c) the normal force exerted vertically upward on the mower by the ground, and (d) the force the person must exert on the lawn mower to accelerate it from rest to 1.2 m/s in 2.0 seconds (assuming the same retarding force).

FIGURE 4–38 Problem 32.

33. (II) Redo Example 4–13 but (a) set up the equations so that the direction of the acceleration **a** of each object is in the direction of motion of that object. (In Example 4–13, we took **a** as positive upward for both masses.) (b) Solve the equations to obtain the same answers as in Example 4–13.

34. (II) A 7500-kg helicopter accelerates upward at 0.52 m/s² while lifting a 1200-kg car. (a) What is the lift force exerted by the air on the rotors? (b) What is the tension in the cable (ignore its mass) that connects the car to the helicopter?

35. (II) One 3.5-kg paint bucket is hanging by a massless cord from another 3.5-kg paint bucket, also hanging by a massless cord, as shown in Fig. 4–39. (a) If the buckets are at rest, what is the tension in each cord? (b) If the two buckets are pulled upward with an acceleration of 1.60 m/s² by the upper cord, calculate the tension in each cord.

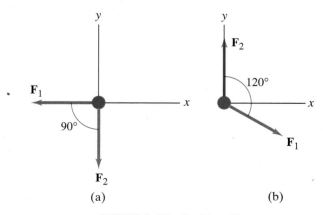

FIGURE 4–37 Problem 31.

FIGURE 4–39
Problems 35 and 44.

36. (II) A window washer pulls herself upward using the bucket-pulley apparatus shown in Fig. 4–40. (*a*) How hard must she pull downward to raise herself slowly at constant speed? (*b*) If she increases this force by 10 percent, what will her acceleration be? The mass of the person plus the bucket is 58 kg.

FIGURE 4–40 Problem 36.

37. (II) Arlene is to walk across a "high wire" strung horizontally between two buildings 10.0 m apart. The sag in the rope when she is at the mid-point should not exceed 10°, as shown in Fig. 4–41. If her mass is 50.0 kg, what must be the tension in the rope?

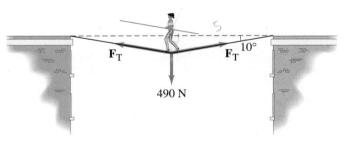

FIGURE 4–41 Problem 37.

38. (II) Tom's hang glider supports his weight using the six ropes shown in Fig. 4–42. Each rope is designed to support an equal fraction of Tom's weight. Tom's mass is 70.0 kg. What is the tension in each of the support ropes?

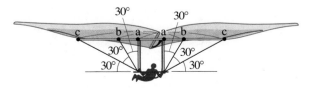

FIGURE 4–42 Problem 38.

39. (II) Two snowcats tow a housing unit to a new location at McMurdo Base, Antarctica, as shown in Fig. 4–43. The sum of the forces $\mathbf{F}_A$ and $\mathbf{F}_B$ exerted on the unit by the horizontal cables is parallel to the line L, and $F_A = 4500$ N. Determine F_B and the magnitude of $\mathbf{F}_A + \mathbf{F}_B$.

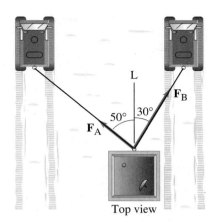

FIGURE 4–43 Problem 39.

40. (II) The block shown in Fig. 4–44 has mass $m = 7.0$ kg and lies on a smooth frictionless plane tilted at an angle $\theta = 22.0°$ to the horizontal. (*a*) Determine the acceleration of the block as it slides down the plane. (*b*) If the block starts from rest 12.0 m up the plane from its base, what will be the block's speed when it reaches the bottom of the incline?

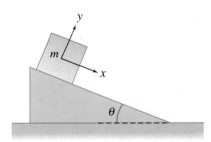

FIGURE 4–44 Block on inclined plane. Problems 40 and 41.

41. (II) A block is given an initial speed of 4.0 m/s up the 22° plane shown in Fig. 4–44. (*a*) How far up the plane will it go? (*b*) How much time elapses before it returns to its starting point? Ignore friction.

42. (II) A 27-kg chandelier hangs from a ceiling on a vertical 4.0-m-long wire. (*a*) What horizontal force would be necessary to displace its position 0.10 m to one side? (*b*) What will be the tension in the wire?

43. (II) A mass m is at rest at $t = 0$. Then a constant force F_0 acts on it for a time t_0. Suddenly the force doubles to $2F_0$ and remains constant until $t = 2t_0$. Determine the total distance traveled from $t = 0$ to $t = 2t_0$.

44. (II) The cords accelerating the buckets in Problem 35 (Fig. 4–39) are each 1.0 m long and each has a weight of 2.0 N. Determine the tension in each cord at its points of attachment (highest and lowest points).

45. (II) A train locomotive is pulling two cars of the same mass behind it. Show that the tension in the coupling between the locomotive and the first car is twice that between the first car and the second car, for any nonzero acceleration of the train.

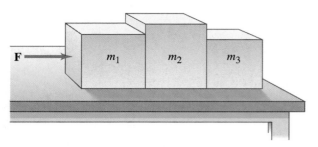

FIGURE 4–45 Problem 46.

46. (II) Three blocks on a frictionless horizontal surface are in contact with each other as shown in Fig. 4–45. A force **F** is applied to block 1 (mass m_1). (a) Draw a free-body diagram for each block. Determine (b) the acceleration of the system (in terms of m_1, m_2, and m_3), (c) the net force on each block, and (d) the force of contact that each block exerts on its neighbor. (e) If $m_1 = m_2 = m_3 = 12.0$ kg and $F = 96.0$ N, give numerical answers to (b), (c), and (d). Do your answers make sense intuitively?

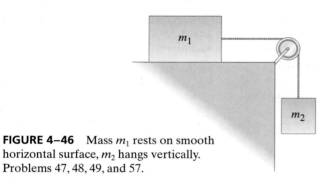

FIGURE 4–46 Mass m_1 rests on smooth horizontal surface, m_2 hangs vertically. Problems 47, 48, 49, and 57.

47. (II) Figure 4–46 shows a block (mass m_1) on a smooth horizontal surface, connected by a thin cord that passes over a pulley to a second block (m_2), which hangs vertically. (a) Draw a free-body diagram for each block, showing the force of gravity on each, the force (tension) exerted by the cord, and any normal force. (b) Determine formulas for the acceleration of the system and for the tension in the cord. Ignore friction and the masses of the pulley and cord.

48. (II) (a) If $m_1 = 13.0$ kg and $m_2 = 6.0$ kg in Fig. 4–46, determine the acceleration of each block. (b) If initially m_1 is at rest 1.250 m from the edge of the table, how long does it take to reach the edge of the table if the system is allowed to move freely? (c) If $m_2 = 1.0$ kg, how large must m_1 be if the acceleration of the system is to be kept at $\frac{1}{100}g$?

49. (III) Determine a formula for the acceleration of the system shown in Fig. 4–46 (see Problem 47) if the cord has a nonnegligible mass m_C. Specify in terms of l_1 and l_2, the lengths of cord from the respective masses to the pulley. (The total cord length is $l = l_1 + l_2$.)

50. (III) The two masses shown in Fig. 4–47 are each initially 1.80 m above the ground, and the massless frictionless pulley is fixed 4.8 m above the ground. What maximum height does the lighter object reach after the system is released? [*Hint*: First determine the acceleration of the lighter mass and then its velocity at the moment the heavier one hits the ground. This is its "launch" speed. Assume it doesn't hit the pulley.]

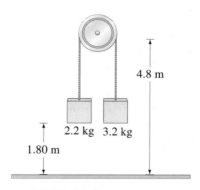

FIGURE 4–47 Problem 50.

51. (III) Suppose the cord in Example 4–12 and Fig. 4–23 is a heavy rope of mass 1.0 kg. Calculate the acceleration of each box and the tension at each end of the cord, using the free-body diagrams shown in Fig. 4–48. Assume the cord is fairly rigid so it doesn't sag.

FIGURE 4–48 Problem 51. Free-body diagrams for each of the objects of the system shown in Fig. 4–23a. Vertical forces, $\mathbf{F}_N$ and $\mathbf{F}_G$, are not shown.

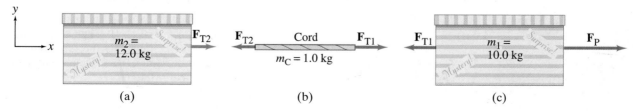

52. (III) Suppose the pulley in Fig. 4–49 is suspended by a cord C. Determine the tension in this cord after the mass is released and before it hits the ground. Ignore the mass of the pulley.

FIGURE 4–49
Problem 52.

1.2 kg 3.2 kg

53. (III) A small block of mass m rests on the sloping side of a triangular block of mass M which itself rests on a horizontal table as shown in Fig. 4–50. Assuming all surfaces are frictionless, determine the force F that must be applied to M so that m remains in a fixed position relative to M (that is, m doesn't move on the incline).

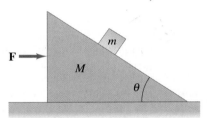

FIGURE 4–50
Problem 53.

54. (III) Determine a formula for the magnitude of the force **F** exerted on the large block (m_3) in Fig. 4–51 so that the mass m_1 does not move relative to m_3. Ignore all friction. Assume m_2 does not make contact with m_3.

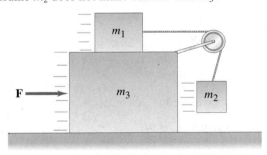

FIGURE 4–51   Problem 54.

55. (III) The double Atwood machine shown in Fig. 4–52 has frictionless, massless pulleys and cords. Determine (a) the acceleration of masses m_1, m_2, and m_3, and (b) the tensions F_{T1} and F_{T3} in the cords.

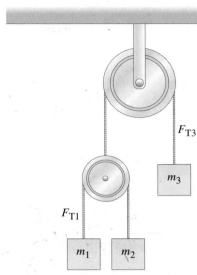

FIGURE 4–52   Problem 55.

* 56. (III) A particle of mass m, initially at rest, is accelerated by a force that increases in time as $F = Ct^2$. Determine its velocity v and position x as a function of time.

* 57. (III) Determine a formula for the speed v of the masses shown in Fig. 4–46 (see Problem 49) assuming the cord has mass m_C and is uniform. At $t = 0$, $v = 0$, and mass m_2 is at the pulley whereas mass m_1 is a distance l from the pulley. Assume the pulley is very small (ignore its diameter) and ignore friction. [*Hint:* Use the chain rule, $dv/dt = (dv/dy)(dy/dt)$, and integrate.]

* 58. (III) A heavy steel cable of length L and mass M passes over a small massless, frictionless pulley. (a) If a length y hangs on one side of the pulley (so $L - y$ hangs on the other side), calculate the acceleration of the cable as a function of y. (b) Assuming the cable starts from rest with length y_0 on one side of pulley, determine the velocity v_f at the moment the whole cable has fallen from the pulley; (c) evaluate v_f for $y_0 = \frac{2}{3}L$. [*Hint:* Use the chain rule, $dv/dt = (dv/dy)(dy/dt)$, and integrate.]

General Problems

59. According to a simplified model of a mammalian heart, at each pulse, approximately 20 g of blood is accelerated from 0.25 m/s to 0.35 m/s during a period of 0.10 s. What is the magnitude of the force exerted by the heart muscle?

60. A person has a reasonable chance of surviving an automobile crash if the deceleration is no more than 30 "g's." Cal-

culate the force on a 70-kg person accelerating at this rate. What distance is traveled if brought to rest at this rate from 90 km/h?

61. A 2.0-kg purse is dropped 55 m from the top of the Leaning Tower of Pisa and reaches the ground with a speed of 29 m/s. What was the average force of air resistance?

62. A fisherman in a boat is using a "10-lb test" fishing line. This means that the line can exert a force of 45 N without breaking (1 lb = 4.45 N). (a) How heavy a fish can the fisherman land if he pulls the fish up vertically at constant speed? (b) If he accelerates the fish upwards at 2.0 m/s², what maximum weight fish can he land?

63. An elevator in a tall building is allowed to reach a maximum speed of 3.5 m/s going down. What must the tension be in the cable to stop this elevator over a distance of 3.0 m if the elevator has a mass of 1300 kg including occupants?

64. A crane's trolley at point P in Fig. 4–53 moves to the right with constant acceleration, and the 800-kg load hangs at a 5.0° angle to the vertical as shown. What is the acceleration of the trolley and load?

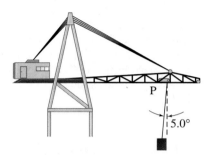

FIGURE 4–53 Problem 64.

65. A wet bar of soap (m = 150 g) slides freely down a ramp 3.0 m long inclined at 9.5°. How long does it take to reach the bottom? How would this change if the soap's mass were 300 grams?

66. A block (mass m_1) lying on a frictionless inclined plane is connected to a mass m_2 by a massless cord passing over a pulley, as shown in Fig. 4–54. (a) Determine a formula for the acceleration of the system in terms of m_1, m_2, θ, and g. (b) What conditions apply to masses m_1 and m_2 for the acceleration to be in one direction (say, m_1 down the plane), or in the opposite direction?

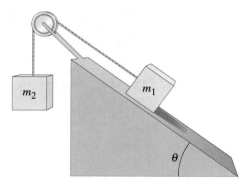

FIGURE 4–54 Problems 66 and 67.

67. (a) In Fig. 4–54, if $m_1 = m_2 = 1.00$ kg and $\theta = 30°$, what will be the acceleration of the system? (b) If $m_1 = 1.00$ kg and $\theta = 30°$, and the system remains at rest, what must the mass m_2 be? (c) Calculate the tension in the cord for (a) and (b).

68. The masses m_1 and m_2 slide on the smooth (frictionless) inclines shown in Fig. 4–55. (a) Determine a formula for the acceleration of the system in terms of m_1, m_2, θ_1, θ_2, and g. (b) If $\theta_1 = 30°$, $\theta_2 = 20°$, and $m_1 = 5.0$ kg, what value of m_2 would keep the system at rest? What would be the tension in the cord in this case?

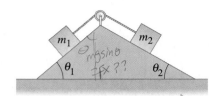

FIGURE 4–55 Problem 68.

69. If a bicyclist of mass 65 kg (including the bicycle) can coast down a 6.0° hill at a steady speed of 6.0 km/h because of air resistance, how much force must be applied to climb the hill at the same speed (and the same air resistance)?

70. A 75.0-kg person stands on a scale in an elevator. What does the scale read (in N and in kg) when (a) the elevator is at rest, (b) the elevator is climbing at a constant speed of 3.0 m/s, (c) the elevator is falling at 3.0 m/s, (d) the elevator is accelerating upward at 3.0 m/s², (e) the elevator is accelerating downward at 3.0 m/s²?

71. A city planner is working on the redesign of a hilly portion of a city. An important consideration is how steep the roads can be so that even low-powered cars can get up the hills without slowing down. It is given that a particular small car, with a mass of 1100 kg, can accelerate on a level road from rest to 21 m/s (75 km/h) in 14.0 s. Using these data, calculate the maximum steepness of a hill.

72. A bicyclist can coast down a 5.0° hill at a constant speed of 6.0 km/h. If the force of air resistance is proportional to the speed v so that $F_{\text{air}} = cv$, calculate (a) the value of the constant c, and (b) the average force that must be applied in order to descend the hill at 20.0 km/h. The mass of the cyclist plus bicycle is 80.0 kg.

73. Francesca, who likes physics experiments, dangles her watch from a thin piece of string while the jetliner she is in takes off from Dulles Airport (Fig. 4–56). She notices that the string makes an angle of 25° with respect to the vertical while the aircraft accelerates for takeoff, which takes about 18 seconds. Estimate the takeoff speed of the aircraft.

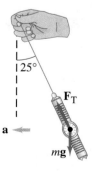

FIGURE 4–56 Problem 73.

74. In the design of a supermarket, there are to be several ramps connecting different parts of the store. Customers will have to push grocery carts up the ramps and it is obviously desirable that this not be too difficult. The engineer has done a survey and found that almost no one complains if the force required is no more than 20 N. Ignoring friction, at what maximum angle θ should the ramps be built, assuming a full 20-kg grocery cart?

75. (*a*) What minimum force F is needed to lift the piano (mass M) using the pulley apparatus shown in Fig. 4–57? (*b*) Determine the tension in each section of rope: F_{T1}, F_{T2}, F_{T3}, and F_{T4}.

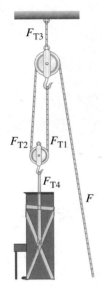

FIGURE 4–57   Problem 75.

76. A jet aircraft is climbing at an angle of 45° above the horizontal and is also accelerating at 4.5 m/s². What is the total force that the cockpit seat exerts on the 75-kg pilot?

77. In the design process for a child restraint chair, an engineer considers the following set of conditions: A 12-kg child is riding in the chair, which is securely fastened to the seat of an automobile (Fig. 4–58). Assume the automobile is involved in a head-on collision with another vehicle. The initial speed v_0 of the car is 50 km/h and this speed is reduced to zero during the collision time of 0.20 s. Assume a constant car deceleration during the collision and estimate

FIGURE 4–58   Problem 77.

the net horizontal force F that the straps of the restraint chair must exert on the child in order to keep her fixed to the chair. Treat the child as a particle and state any additional assumptions made during your analysis.

78. A helicopter is to lift a 600-kg magnesium frame at a construction site, as shown in Fig. 4–59. As the helicopter accelerates upward at 0.15g, what is the magnitude of the tension F_T in the cord?

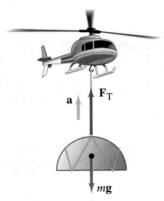

FIGURE 4–59   Problem 78.

79. A new high-speed 12-car Italian train has a mass of 660 metric tons (660,000 kg). It can exert a maximum force of 400 kN horizontally against the tracks, whereas at maximum velocity (300 km/h), it exerts a force of about 150 kN. Calculate (*a*) its maximum acceleration, and (*b*) estimate the force of air resistance at top speed.

Newton's laws are fundamental in physics, and this chapter covers some different applications. These photos show two situations of using Newton's laws which involve some new elements in addition to those discussed in the previous chapter. The downhill skier illustrates **friction** on an incline, although at this moment she is not touching the snow, and so is retarded only by air resistance which is a velocity-dependent force (an optional topic in this chapter). The people on the rotating amusement park ride on the right illustrate the dynamics of circular motion.

CHAPTER 5

Further Applications of Newton's Laws

This chapter continues our study of Newton's laws and emphasizes their fundamental importance in physics. We cover some important applications of Newton's laws, including friction and the very important subject of the dynamics of circular motion (we studied its kinematics in Chapter 3). Although some material in this chapter may seem to repeat topics covered in Chapter 4, in fact, new elements are involved.

5-1 Applications of Newton's Laws Involving Friction

FIGURE 5–1 An object moving to the right on a table or floor. The two surfaces in contact are rough, at least on a microscopic scale.

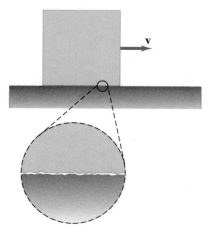

Until now we have ignored friction, but it must be taken into account in most practical situations. Friction exists between two solid surfaces because even the smoothest looking surface is quite rough on a microscopic scale, Fig. 5–1. When we try to slide an object across another surface, these microscopic bumps impede the motion. In addition, at the atomic level, the atoms on a bump of one surface come so close to the atoms of the other surface that electric forces between the atoms can form chemical bonds, as a tiny weld between the two surfaces. Sliding an object across a surface is often jerky due to the making and breaking of these bonds. Even when a body rolls across a surface, there is still some friction, called *rolling friction*, although it is generally much less than when a body slides across a surface. We will be concerned mainly with sliding friction in this section, and it is usually called **kinetic friction** (*kinetic* is from the Greek for "moving").

When a body is in motion along a rough surface, the force of kinetic friction acts opposite to the direction of the body's velocity. The magnitude of the force of kinetic friction depends on the nature of the two sliding surfaces. For given surfaces, experiment shows that the friction force is approximately proportional to the *normal force* between the two surfaces, which is the force that either object exerts on the other, perpendicular to their common surface of contact (see Fig. 5–2). The force of friction between hard surfaces depends very little on the total surface area of contact; that is, the friction force on this book is roughly the same whether it is being slid on its wide face or on its spine, assuming the surfaces have the same smoothness. We can write the proportionality as an equation by inserting a constant of proportionality, μ_k:

$$F_{fr} = \mu_k F_N.$$

This relation is not a fundamental law; it is an experimental relation between the magnitude of the friction force $\mathbf{F}_{fr}$ which acts parallel to the two surfaces, and the magnitude of the normal force $\mathbf{F}_N$ which acts perpendicular to the surfaces. It is *not* a vector equation since the two forces are perpendicular to one another. The term μ_k is called the **coefficient of kinetic friction**, and its value depends on the nature of the two surfaces. Measured values for a variety of surfaces are given in Table 5–1. These are only approximate, however, since μ depends on whether the surfaces are wet or dry, on how much they have been sanded or rubbed, if any burrs remain, and other such factors. But μ_k is roughly independent of the sliding speed.

What we have been discussing up to now is *kinetic friction*, when one object slides over another. There is also **static friction**, which refers to a force parallel to the two surfaces that can arise even when they are not sliding. Suppose an object such as a desk is resting on a horizontal floor. If no horizontal force is exerted on the desk, there also is no friction force. But now, suppose you try to push the desk, but it doesn't move. You are exerting a horizontal force, but the desk isn't moving, so there must be another force on the desk keeping it from moving (the net force is zero on an object that doesn't move). This is the force of *static friction* exerted by the floor on the desk. If you push with a greater force without moving the desk, the force of static friction also has increased. If you push hard enough, the desk will eventually start to move, and kinetic friction takes over. At this point, you have exceeded the maximum force of static friction, which is given by $F_{max} = \mu_s F_N$, where μ_s is the **coefficient of static friction** (Table 5–1). Since the force of static friction can vary from zero to this maximum value, we write

$$F_{fr} \leq \mu_s F_N.$$

Kinetic friction

$\mathbf{F}_{fr} \perp \mathbf{F}_N$

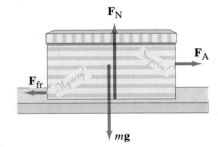

FIGURE 5–2 When an object is pulled by an applied force $\left(\mathbf{F}_A\right)$ along a surface, the force of friction $\mathbf{F}_{fr}$ opposes the motion. The magnitude of $\mathbf{F}_{fr}$ is proportional to the magnitude of the normal force F_N.

Static friction

TABLE 5–1 Coefficients of Friction†

Surfaces	Coefficient of Static Friction, μ_s	Coefficient of Kinetic Friction, μ_k
Wood on wood	0.4	0.2
Ice on ice	0.1	0.03
Metal on metal (lubricated)	0.15	0.07
Steel on steel (unlubricated)	0.7	0.6
Rubber on dry concrete	1.0	0.8
Rubber on wet concrete	0.7	0.5
Rubber on other solid surfaces	1–4	1
Teflon® on Teflon in air	0.04	0.04
Teflon on steel in air	0.04	0.04
Lubricated ball bearings	< 0.01	< 0.01
Synovial joints (in human limbs)	0.01	0.01

† Values are approximate and are intended only as a guide.

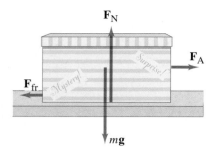

FIGURE 5–2 Repeated for Example 5–1.

FIGURE 5–3 Magnitude of the force of friction as a function of the external force applied to a body initially at rest. As the applied force is increased in magnitude, the force of static friction increases linearly to just match it, until the applied force equals $\mu_s F_N$. If the applied force increases further, the body will begin to move, and the friction force drops to a roughly constant value characteristic of kinetic friction.

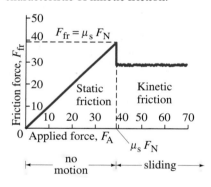

You may have noticed that it is often easier to keep a heavy object moving, such as pushing a table, than it is to start it moving in the first place. This is consistent with the fact (see Table 5–1) that μ_s is generally greater than μ_k. (It can never be less. Why?)

EXAMPLE 5–1 **Friction: static and kinetic.** Our 10.0-kg mystery box rests on a horizontal floor. The coefficient of static friction is $\mu_s = 0.40$ and the coefficient of kinetic friction is $\mu_k = 0.30$. Determine the force of friction, F_{fr}, acting on the box if a horizontal external applied force F_A is exerted on it of magnitude: (a) 0, (b) 10 N, (c) 20 N, (d) 38 N, and (e) 40 N.

SOLUTION The free-body diagram of the box is shown in Fig. 5–2. Examine it carefully. In the vertical direction there is no motion, so $\Sigma F_y = ma_y = 0$ yields $F_N - mg = 0$. Hence the normal force for all cases is

$$F_N = mg = (10.0\,\text{kg})(9.80\,\text{m/s}^2) = 98.0\,\text{N}.$$

(a) Since no force is applied in this first case, the box doesn't move, and $F_{fr} = 0$.
(b) The force of static friction will oppose any applied force up to a maximum of

$$\mu_s F_N = (0.40)(98.0\,\text{N}) = 39\,\text{N}.$$

The applied force is $F_A = 10\,\text{N}$. Thus the box will not move; since $\Sigma F_x = F_A - F_{fr} = 0$ then $F_{fr} = 10\,\text{N}$.
(c) An applied force of 20 N is also not sufficient to move the box. Thus $F_{fr} = 20\,\text{N}$ to balance the applied force.
(d) The applied force of 38 N is still not quite large enough to move the box; so the friction force has now increased to 38 N to keep the box at rest.
(e) A force of 40 N will start the box moving since it exceeds the maximum force of static friction, $\mu_s F_N = (0.40)(98.0\,\text{N}) = 39\,\text{N}$. Instead of static friction, we now have kinetic friction, and its magnitude is

$$F_{fr} = \mu_k F_N = (0.30)(98.0\,\text{N}) = 29\,\text{N}.$$

There is now a net (horizontal) force on the box of magnitude $F = 40\,\text{N} - 29\,\text{N} = 11\,\text{N}$, so the box will accelerate at a rate

$$a_x = \Sigma F/m = 11\,\text{N}/10\,\text{kg} = 1.1\,\text{m/s}^2$$

as long as the applied force is 40 N. Figure 5–3 shows a graph that summarizes this Example.

Now we look at some Examples involving kinetic friction in a variety of situations. Note that both the normal force and the friction force are forces exerted by one surface on another; one is perpendicular to the contact surfaces (the normal force), and the other is parallel (the friction force).

CONCEPTUAL EXAMPLE 5–2 **To push or to pull a sled?** Your little sister wants a ride on her sled. If you are on flat ground, will you exert less force if you push her or pull her? See Figs. 5–4a and b. Assume the same angle θ in each case.

RESPONSE Free-body diagrams are shown in Figs. 5–4c and d. If you push her, and $\theta > 0$, there is a vertically downward component to your force. Hence the normal force upward exerted by the ground will be larger than mg (where m is the mass of sister plus sled). If you pull her, your force has a vertically upward component, so the normal force F_N can be less than mg. Because the friction force is proportional to the normal force, it will be less if you pull her. So you exert less force if you pull her.

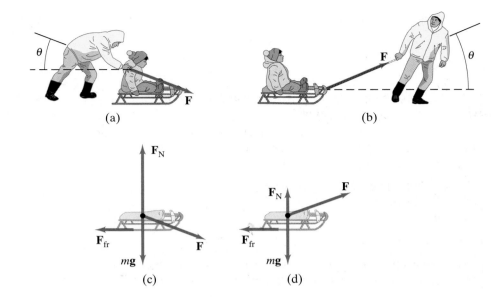

(a)   (b)

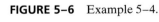

(c)   (d)

FIGURE 5–4
Example 5–2.

CONCEPTUAL EXAMPLE 5–3 | **A box against a wall.** You can hold a box against a rough wall (Fig. 5–5) and prevent it from slipping down by pressing hard horizontally. How does the application of a horizontal force keep an object from moving vertically?

RESPONSE This won't work well if the wall is slippery. You need friction. Even then, if you don't press hard enough, the box will slip. The horizontal force you apply produces a normal force on the box exerted by the wall (why?). The force of gravity, mg, acting downward on the box, can now be balanced by an upward friction force whose magnitude is proportional to the normal force. The harder you push, the greater F_N is and the greater F_{fr} can be. If you don't press hard enough, then $mg > \mu_s F_N$ and the box begins to slide down.

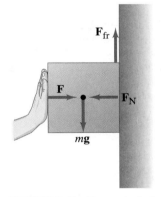

FIGURE 5–5   Example 5–3.

EXAMPLE 5–4 | **Pulling against friction.** A 10.0-kg box is pulled along a horizontal surface by a force F_P of 40.0 N which is applied at a 30.0° angle. This is like Example 4–11 in the preceding chapter except now there is friction, and we assume a coefficient of kinetic friction of 0.30. The free-body diagram is shown in Fig. 5–6. Calculate the acceleration.

SOLUTION The force of kinetic friction on the box opposes the direction of motion and is parallel to the surfaces of contact. The calculation for the vertical (y) direction is just the same as before (Example 4–11): there is no motion in the vertical direction because $F_{Py} = F_y \sin 30° = (40.0\,\text{N})(\sin 30.0°) = 20.0\,\text{N}$ is less than the weight of the box $mg = (10.0\,\text{kg})(9.80\,\text{m/s}^2) = 98.0\,\text{N}$. Taking y as positive upward, we have

$$F_N - mg + F_{Py} = ma_y$$

$$F_N - 98.0\,\text{N} + 20.0\,\text{N} = 0$$

and the normal force is $F_N = 78.0\,\text{N}$. Next we apply Newton's second law for the horizontal (x) direction (positive to the right):

$$F_{Px} - F_{fr} = ma_x.$$

FIGURE 5–6   Example 5–4.

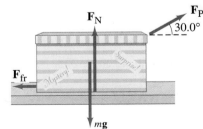

The friction force is kinetic as long as $F_{fr} = \mu_k F_N$ is less than F_{Px}, which it is since

$$F_{fr} = \mu_k F_N = (0.30)(78.0\,\text{N}) = 23.4\,\text{N}$$

and

$$F_{Px} = F_P \cos 30.0° = (40.0\,\text{N})(0.866) = 34.6\,\text{N}.$$

[If $\mu_k F_N$ were greater than F_{Px}, what would you conclude?] Hence the box does accelerate:

$$a_x = \frac{F_{Px} - F_{fr}}{m} = \frac{34.6\,\text{N} - 23.4\,\text{N}}{10.0\,\text{kg}} = 1.1\,\text{m/s}^2.$$

In the absence of friction, as we saw in Example 4–11, the acceleration would be much greater. Note that our final answer should have only two significant figures because our least significant input value has two ($\mu_k = 0.30$)

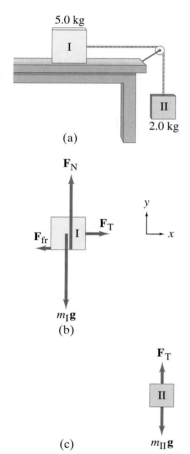

5.0 kg

I

II

2.0 kg

(a)

$\mathbf{F}_N$

$\mathbf{F}_{fr}$ I $\mathbf{F}_T$

y

x

$m_I \mathbf{g}$

(b)

$\mathbf{F}_T$

II

(c) $m_{II}\mathbf{g}$

FIGURE 5–7 Example 5–5.

EXAMPLE 5–5 **Two boxes and a pulley.** In Fig. 5–7a, two boxes are connected by a cord running over a pulley. The coefficient of kinetic friction between box I and the table is 0.20. We ignore the mass of the cord and pulley and any friction in the pulley, which means we can assume that a force applied to one end of the cord will have the same magnitude at the other end. We wish to find the acceleration, a, of the system, which will have the same magnitude for both boxes assuming the cord doesn't stretch. As box II moves down, box I moves to the right.

SOLUTION Free-body diagrams are shown for each box in Fig. 5–7b and c. Box I does not move vertically, so the normal force just balances the weight,

$$F_N = m_I g = (5.0\,\text{kg})(9.8\,\text{m/s}^2) = 49\,\text{N}.$$

In the horizontal direction, there are two forces on box I (Fig. 5–7b): F_T, the tension in the cord (whose value we don't know), and the force of friction

$$F_{fr} = \mu_k F_N = (0.20)(49\,\text{N}) = 9.8\,\text{N}.$$

The horizontal acceleration is what we wish to find; we use Newton's second law in the x direction, $\Sigma F_{Ix} = m_I a_x$, which becomes (taking the positive direction to the right and setting $a_{Ix} = a$):

$$\Sigma F_{Ix} = F_T - F_{fr} = m_I a.\qquad\qquad\text{[box I]}$$

Next consider box II. The force of gravity $F_G = m_{II} g = 19.6\,\text{N}$ pulls downward; and the cord pulls upward with a force F_T. So we can write Newton's second law for box II (taking the downward direction as positive):

$$\Sigma F_{IIy} = m_{II} g - F_T = m_{II} a.\qquad\qquad\text{[box II]}$$

[Note here that if $a \neq 0$, then F_T is not equal to $m_{II} g$.] We have two unknowns, a and F_T, and we also have two equations. We solve the box I equation for F_T:

$$F_T = F_{fr} + m_I a,$$

and substitute this into the box II equation:

$$m_{II} g - F_{fr} - m_I a = m_{II} a.$$

Now we solve for a and put in numerical values:

$$a = \frac{m_{II} g - F_{fr}}{m_I + m_{II}} = \frac{19.6\,\text{N} - 9.8\,\text{N}}{5.0\,\text{kg} + 2.0\,\text{kg}} = 1.4\,\text{m/s}^2,$$

which is the acceleration of box I to the right, and of box II down. If we wish, we can calculate F_T using the first equation:

$$F_T = F_{fr} + m_I a = 9.8\,\text{N} + (5.0\,\text{kg})(1.4\,\text{m/s}^2) = 17\,\text{N}.$$

Next we take an example of an object moving down an incline, as in Example 4–17 of Chapter 4; but now we include friction.

EXAMPLE 5–6 **The skier.** The skier in Fig. 5–8a has just begun descending the 30° slope. Assuming the coefficient of kinetic friction is 0.10, calculate (a) her acceleration and (b) the speed she will reach after 4.0 s.

➡ PHYSICS APPLIED

Sport

SOLUTION First we draw the free-body diagram showing all the forces acting on the skier, Fig. 5–8b: her weight ($\mathbf{F}_G = m\mathbf{g}$) downward, and the two forces exerted on her skis by the snow—the normal force perpendicular to the snow's surface, and the friction force parallel to the surface. These three forces are shown acting at one point in Fig. 5–8b, for convenience. Also for convenience, we choose the x axis parallel to the snow surface with positive direction downhill, and the y axis perpendicular to the surface. With this choice, we have to resolve only one vector into components, the weight, and its components are shown as dashed lines in Fig. 5–8c. They are given by

$$F_{Gx} = mg \sin\theta,$$

$$F_{Gy} = mg \cos\theta,$$

where we have stayed general, using θ rather than 30° for now.
(a) To calculate her acceleration down the hill, a_x, we apply Newton's second law to the x direction:

$$\Sigma F_x = ma_x$$

$$mg \sin\theta - \mu_k F_N = ma_x$$

where the two forces are the component of the gravity force (+x direction) and the friction force (−x direction). We want to find the value of a_x, but we don't yet know F_N in the last equation. Let's see if we can get F_N from the y component of Newton's second law:

$$\Sigma F_y = ma_y$$

$$F_N - mg \cos\theta = ma_y = 0$$

where we set $a_y = 0$ because there is no motion in the y direction (perpendicular to the slope). Thus we can solve for F_N:

$$F_N = mg \cos\theta$$

and we can substitute this into our equation above for ma_x:

$$mg \sin\theta - \mu_k(mg \cos\theta) = ma_x.$$

There is an m in each term which can be canceled out. Thus (setting $\theta = 30°$ and $\mu_k = 0.10$):

$$a_x = g \sin 30° - \mu_k g \cos 30° = 0.50\, g - (0.10)(0.866)g = 0.41\, g.$$

The skier's acceleration is 0.41 times the acceleration of gravity, which in numbers is $a = (0.41)(9.8\,\text{m/s}^2) = 4.0\,\text{m/s}^2$. It is interesting that the mass canceled out here, and so we have the useful conclusion that *the acceleration doesn't depend on the mass*. That such a cancellation sometimes occurs, and thus may give a useful conclusion as well as saving calculation, is a big advantage of working with the algebraic equations and putting in the numbers only at the end.
(b) The speed after 4.0 s is found, since the acceleration is constant, by using Eq. 2–12a:

$$v = v_0 + at = 0 + (4.0\,\text{m/s}^2)(4.0\,\text{s}) = 16\,\text{m/s},$$

where we assumed a start from rest.

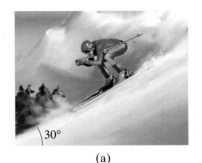

(a)

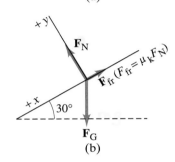

(b)

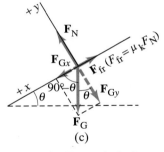

(c)

FIGURE 5–8 Example 5–6: A skier descending a slope.

➡ PROBLEM SOLVING

It is often helpful to put in numbers only at the end

EXAMPLE 5–7 **Measuring μ_k.** Suppose in Example 5–6 that the snow is slushy and the skier moves down the 30° slope at constant speed. What can you say about the coefficient of friction, μ_k?

SOLUTION Now the skier moves down the slope at constant speed, and we want to find μ_k. The free-body diagram and the $\Sigma F = ma$ equations for the x and y components will be the same as above, except that now we are given $a_x = 0$. Thus

$$\Sigma F_y = F_N - mg\cos\theta = ma_y = 0$$

$$\Sigma F_x = mg\sin\theta - \mu_k F_N = ma_x = 0.$$

From the first equation, we have $F_N = mg\cos\theta$; we substitute this into the second equation:

$$mg\sin\theta - \mu_k(mg\cos\theta) = 0.$$

Now we solve for μ_k:

$$\mu_k = \frac{mg\sin\theta}{mg\cos\theta} = \frac{\sin\theta}{\cos\theta} = \tan\theta$$

which for $\theta = 30°$ is

$$\mu_k = \tan\theta = \tan 30° = 0.58.$$

A method for determining μ_k

Notice that we could use the equation

$$\mu_k = \tan\theta$$

to determine μ_k under a variety of conditions. All we need to do is observe at what slope angle the skier descends at constant speed. Here is another reason why it is often useful to plug in numbers only at the end: we obtained a general result useful for other situations as well.

EXAMPLE 5–8 **A plane, a pulley, and two boxes.** A box of mass $m_1 = 10.0\,\text{kg}$ rests on a surface inclined at $\theta = 37°$ to the horizontal. It is connected by a lightweight cord, which passes over a massless and frictionless pulley, to a second box of mass m_2, which hangs freely as shown in Fig. 5–9a. (a) If the coefficient of static friction is $\mu_s = 0.40$, determine what range of values for mass m_2 will keep the system at rest. (b) If the coefficient of kinetic friction is $\mu_k = 0.30$, and $m_2 = 10.0\,\text{kg}$, determine the acceleration of the system.

SOLUTION (a) Figure 5–9b shows two free-body diagrams for box m_1. The tension force exerted by the cord is labeled $\mathbf{F}_T$. The force of friction can be either up or down the slope, and we show both possibilities in Fig. 5–9b: (i) if $m_2 = 0$ or is sufficiently small, m_1 would tend to slide down the incline, so $\mathbf{F}_{fr}$ would be directed up the incline; (ii) if m_2 is large enough, m_1 will tend to be pulled up the plane, so $\mathbf{F}_{fr}$ would point down the plane. For both cases, Newton's second law for the y direction (perpendicular to the plane) is the same:

$$F_N - m_1 g\cos\theta = m_1 a_y = 0$$

since there is no y motion. So

$$F_N = m_1 g\cos\theta.$$

Now for the x motion. We consider case (i) first for which $\Sigma F = ma$ gives

$$m_1 g\sin\theta - F_T - F_{fr} = m_1 a_x.$$

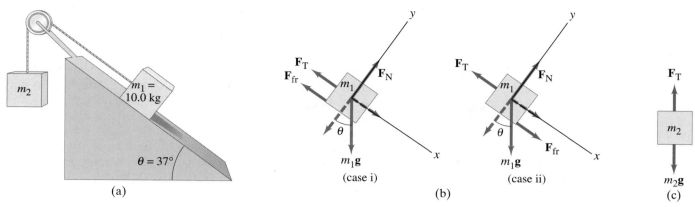

FIGURE 5–9 Example 5–8.

We want $a_x = 0$ and we solve for F_T since F_T is related to m_2 (whose value we are seeking) by $F_T = m_2 g$ (see Fig. 5–9c). Thus

$$m_1 g \sin\theta - F_{fr} = F_T = m_2 g.$$

Since F_{fr} can be at most $\mu_s F_N = \mu_s m_1 g \cos\theta$, the minimum value that m_2 can have to prevent motion $(a_x = 0)$ is, after dividing through by g,

$$m_2 = m_1 \sin\theta - \mu_s m_1 \cos\theta$$
$$= (10.0 \text{ kg})(\sin 37° - 0.40 \cos 37°) = 2.8 \text{ kg}.$$

Thus if $m_2 < 2.8$ kg, then box 1 will slide down the incline.
Now for case (ii). Newton's second law is

$$m_1 g \sin\theta + F_{fr} - F_T = ma_x = 0.$$

Then the maximum value m_2 can have without causing acceleration is given by

$$F_T = m_2 g = m_1 g \sin\theta + \mu_s m_1 g \cos\theta$$

or

$$m_2 = m_1 \sin\theta + \mu_s m_1 \cos\theta$$
$$= (10.0 \text{ kg})(\sin 37° + 0.40 \cos 37°) = 9.2 \text{ kg}.$$

Thus, to prevent motion, we have the condition

$$2.8 \text{ kg} < m_2 < 9.2 \text{ kg}.$$

(*b*) If $m_2 = 10.0$ kg and $\mu_k = 0.30$, then m_2 will fall and m_1 will rise up the plane with an acceleration a given by

$$m_1 a = F_T - m_1 g \sin\theta - \mu_k F_N.$$

Since m_2 also accelerates, $F_T = m_2 g - m_2 a$ (see Fig. 5–9c) and we substitute this into the equation above:

$$m_1 a = m_2 g - m_2 a - m_1 g \sin\theta - \mu_k F_N.$$

We solve for the acceleration a and substitute $F_N = m_1 g \cos\theta$, and then $m_1 = m_2 = 10.0$ kg, to find

$$a = \frac{m_2 g - m_1 g \sin\theta - \mu_k m_1 g \cos\theta}{m_1 + m_2}$$
$$= \frac{(9.8 \text{ m/s}^2)(10.0 \text{ kg})(1 - \sin 37° - 0.30 \cos 37°)}{20.0 \text{ kg}} = 0.079g = 0.78 \text{ m/s}^2.$$

Friction can be a hindrance. It slows down moving objects and causes heating and binding of moving parts in machinery. Friction can be reduced by using lubricants such as oil. More effective in reducing friction between two surfaces is to maintain a layer of air or other gas between them. Devices using this concept, which is not practical for most situations, include air tracks and air tables (or games) in which the layer of air is maintained by forcing air through many tiny holes. Another technique to maintain the air layer is to suspend objects in air using magnetic fields ("magnetic levitation"). On the other hand, friction can be helpful. Our ability to walk depends on friction between the soles of our shoes (or feet) and the ground. (Does walking involve static friction or kinetic friction?) The movement of a car, and also its stability, depend on friction. When friction is low, such as on ice, safe walking or driving become difficult.

5–2 | Dynamics of Uniform Circular Motion

In Section 3–9 we saw that a particle revolving in a circle of radius r with uniform speed v undergoes a radial (or centripetal) acceleration which at any moment is given by

$$a_R = \frac{v^2}{r}.$$

This acceleration, a_R, is called radial or centripetal acceleration because it is directed toward the center of the circle. For uniform circular motion (v = constant), the magnitude of the acceleration is constant, but its direction is continually changing. Hence the acceleration vector $\mathbf{a}_R$ is a variable acceleration. The direction of $\mathbf{a}_R$ is always perpendicular to the velocity $\mathbf{v}$. [The equation $a_R = v^2/r$ is valid even if v is not constant, and we will treat that case in Section 5–4.]

Force is needed to provide centripetal acceleration

An object moving uniformly in a circle, such as a ball swung horizontally on the end of a string, must have a net force applied to it to keep it moving in that circle, instead of in a straight line. That is, a net force is necessary to give it a centripetal acceleration. The magnitude of the required net force can be calculated using Newton's second law for the radial component, $\Sigma F_R = ma_R$, where $a_R = v^2/r$ is the centripetal acceleration, and ΣF_R is the total (net) force in the radial direction:

$$\Sigma F_R = ma_R = m\frac{v^2}{r}. \qquad \text{[circular motion]} \quad \textbf{(5–1)}$$

Since a_R is directed along the radius toward the center of the circle at any moment, the *net force too must be directed toward the center of the circle* for uniform circular motion (v = constant). A net force is clearly necessary because otherwise, if no net force were exerted on the object, it would not move in a circle but in a straight line, as Newton's first law tells us. To pull an object out of its "natural" straight-line path, a net force to the side is necessary. For uniform circular motion, this sideways net force must act toward the circle's center (see Fig. 5–10). The direction of the force is thus continually changing so that it is always directed toward the center of the circle. This force is sometimes called a centripetal ("aiming toward the center") force. But be aware that "centripetal force" does not indicate some new kind of force. The term merely describes the direction of the force: that the net force is directed toward the circle's center. The force *must be applied by other objects*. For example, when a person swings a ball in a circle on the end of a string, the person pulls on the string and the string exerts the force on the ball.

Careful: Centripetal force is not a new kind of force

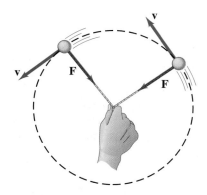

FIGURE 5–10 A force is required to keep an object moving in a circle. If the speed is constant, the force is directed toward the center of the circle.

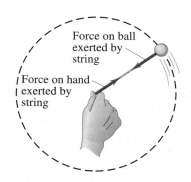

FIGURE 5–11 Swinging a ball on the end of a string.

There is a common misconception that an object moving in a circle has an outward force acting on it, a so-called centrifugal ("center-fleeing") force. Consider, for example, a person swinging a ball on the end of a string around her head (Fig. 5–11). If you have ever done this yourself, you know that you feel a force pulling outward on your hand. The misconception arises when this pull is interpreted as an outward "centrifugal" force pulling on the ball that is transmitted along the string to your hand. This is not what is happening at all. To keep the ball moving in a circle, you pull inwardly on the string, which in turn exerts the force on the ball. The ball exerts an equal and opposite force on your hand (Newton's third law), and *this* is the force your hand feels (see Fig. 5–11). The force *on the ball* is the one exerted *inwardly* on it by the string.

For even more convincing evidence that a "centrifugal force" does not act on the ball, consider what happens when you let go of the string. If a centrifugal force were acting, the ball would fly outward, as shown in Fig. 5–12a. But it doesn't; the ball flies off tangentially (Fig. 5–12b), in the direction of the velocity it had at the moment it was released, because the inward force no longer acts. Try it and see!

Careful:
Beware a misconception
for a "centrifugal" force

FIGURE 5–12 If centrifugal force existed, the ball would fly off as in (a) when released. In fact, it flies off tangentially as in (b). Similarly, sparks fly in straight lines tangentially from the edge of a rotating grinding wheel (c).

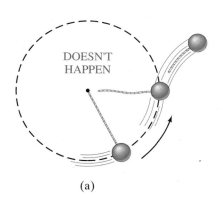

(a)

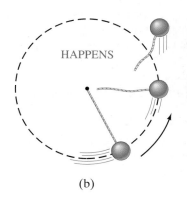

(b)

(c)

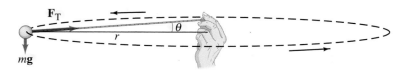

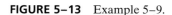

FIGURE 5–13 Example 5–9. $m\mathbf{g}$

EXAMPLE 5–9 **ESTIMATE** **Force on revolving ball (horizontal).** Estimate the force a person must exert on a string attached to a 0.150-kg ball to make the ball revolve in a horizontal circle of radius 0.600 m. The ball makes 2.00 revolutions per second ($T = 0.500$ s), as in our earlier Example 3–11.

SOLUTION First we draw the free-body diagram for the ball, Fig. 5–13, which shows the two forces acting on the ball: the force of gravity, $m\mathbf{g}$; and the tension force $\mathbf{F}_T$ that the string exerts (which occurs because the person exerts that same force on the string). The ball's weight complicates matters and makes it impossible to revolve a ball with the cord horizontal. But if the weight is small enough, we can ignore it. Then $\mathbf{F}_T$ will act nearly horizontally ($\theta \approx 0$ in Fig. 5–13) and provide the force necessary to give the ball its centripetal acceleration. We apply Newton's second law to the radial direction, which now is horizontal, so we call it x:

$$\Sigma F_x = ma_x$$

where $a_x = v^2/r$ and $v = 2\pi r/T = 2\pi(0.600 \text{ m})/(0.500 \text{ s}) = 7.54 \text{ m/s}$. Thus

Tension in cord acts to provide the centripetal acceleration

$$F_{Tx} = m\frac{v^2}{r} = (0.150 \text{ kg})\frac{(7.54 \text{ m/s})^2}{(0.600 \text{ m})} \approx 14 \text{ N},$$

where we have rounded off because our estimate ignores the ball's weight. We keep only two significant figures here in the answer because $mg = (0.150 \text{ kg})(9.80 \text{ m/s}^2) = 1.5 \text{ N}$, being about 1/10 of our result, is small, but not so small as to justify stating a more precise answer since we ignored the effect of $m\mathbf{g}$. [Note: if you want to include the effect of $m\mathbf{g}$ here, resolve $\mathbf{F}_T$ in Fig. 5–13 into components, and set the horizontal component of $\mathbf{F}_T$ equal to mv^2/r and its vertical component equal to mg—something like Example 5–10 below.]

FIGURE 5–14 Example 5–10. Conical pendulum.

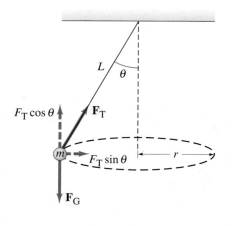

EXAMPLE 5–10 **Conical pendulum.** A small ball of mass m, suspended by a cord of length L, revolves in a circle of radius $r = L \sin\theta$, where θ is the angle the string makes with the vertical (Fig. 5–14). (*a*) In what direction is the acceleration of the ball, and what causes the acceleration? (*b*) Calculate the speed and period (time required for one revolution) of the ball in terms of L, θ, and m.

SOLUTION (*a*) The acceleration points horizontally toward the center of the ball's circular path (not along the cord). The force responsible for the acceleration is the *net* force which here is the vector sum of the forces acting on the mass m: its weight $\mathbf{F}_G$ (of magnitude $F_G = mg$) and the force exerted by the tension in the cord, $\mathbf{F}_T$. The latter has horizontal and vertical components of magnitude $F_T \sin\theta$ and $F_T \cos\theta$, respectively. We apply Newton's second law to the horizontal and vertical directions. In the vertical direction, there is no motion, so the acceleration is zero and the net force in the vertical direction is zero:

$$F_T \cos\theta - mg = 0.$$

In the horizontal direction there is only one force, of magnitude $F_T \sin\theta$, that acts toward the center of the circle and gives rise to the acceleration v^2/r. Newton's second law tells us:

$$F_T \sin\theta = m\frac{v^2}{r}.$$

(b) We solve the two equations above for v by eliminating F_T between them (and using $r = L \sin \theta$):

$$v = \sqrt{\frac{rF_T \sin \theta}{m}} = \sqrt{\frac{r}{m} \left(\frac{mg}{\cos \theta} \right) \sin \theta} = \sqrt{\frac{Lg \sin^2 \theta}{\cos \theta}}.$$

The period T is the time required to make one revolution, a distance of $2\pi r = 2\pi L \sin \theta$. The speed v can thus be written $v = 2\pi L \sin \theta / T$; then

$$T = \frac{2\pi L \sin \theta}{v} = \frac{2\pi L \sin \theta}{\sqrt{\dfrac{Lg \sin^2 \theta}{\cos \theta}}} = 2\pi \sqrt{\frac{L \cos \theta}{g}}.$$

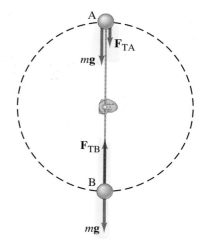

FIGURE 5–15 Example 5–11, with free-body diagrams at the two positions.

EXAMPLE 5–11 **Revolving ball (vertical circle).** A 0.150-kg ball on the end of a 1.10-m-long cord (negligible mass) is swung in a *vertical* circle. Determine the minimum speed the ball must have at the top of its arc so that it continues moving in a circle. (b) Calculate the tension in the cord at the bottom of the arc if the ball is moving at twice the speed of part (a).

SOLUTION The free-body diagram is shown in Fig. 5–15 for both situations. (a) At the top (point A), two forces can act on the ball: $m\mathbf{g}$, its weight; and $\mathbf{F}_{TA}$, the tension force the cord exerts at point A. Both act downward, and their vector sum acts to give the ball its centripetal acceleration a_R. We apply Newton's second law, for the vertical direction, choosing downward as positive (toward the center):

$$\Sigma F_R = ma_R$$

$$F_{TA} + mg = m \frac{v_A^2}{r}.$$

Gravity and cord tension together provide centripetal acceleration

From this equation we can see that the tension force F_{TA} at A will get larger if v_A (ball's speed at top of circle) is made larger, as expected. But we are asked for the *minimum* speed to keep the ball moving in a circle. The cord will remain taut as long as there is tension in it; but if the tension disappears (because v_A is too small) the cord can go limp, and the ball will fall out of its circular path. Thus, the minimum speed will occur if $F_{TA} = 0$, for which we have $mg = mv_A^2/r$. Then:

$$v_A = \sqrt{gr} = \sqrt{(9.80 \text{ m/s}^2)(1.10 \text{ m})} = 3.28 \text{ m/s}.$$

Gravity alone provides centripetal acceleration

This is the minimum speed at the top of the circle if the ball is to continue moving in a circular path.

(b) At the bottom of the circle (see Fig. 5–15) the cord exerts its tension force F_{TB} upward whereas the force of gravity, $m\mathbf{g}$, acts downward. So Newton's second law, this time choosing *upward* as positive (toward the center), gives

$$\Sigma F_R = ma_R$$

$$F_{TB} - mg = m \frac{v_B^2}{r}.$$

Cord tension and gravity acting in opposite directions provide centripetal acceleration

The speed v_B is given as twice what we found in (a), namely 6.56 m/s. [Note that the speed changes here because gravity acts on the ball all along the path, but Eq. 5–1 still remains valid, $\Sigma F_R = mv^2/r$.] We solve for F_{TB} in the last equation:

$$F_{TB} = m \frac{v_B^2}{r} + mg$$

$$= (0.150 \text{ kg}) \frac{(6.56 \text{ m/s})^2}{(1.10 \text{ m})} + (0.150 \text{ kg})(9.80 \text{ m/s}^2) = 7.34 \text{ N}.$$

Note that the cord's tension not only provides the centripetal acceleration, but must be even larger than ma_R to compensate for the downward force of gravity.

1. Draw a free-body diagram, showing all the forces acting on each object under consideration. Be sure you can identify the source of each force (tension in a cord, Earth's gravity, friction, normal force, and so on), so you don't put in something that doesn't belong (like a centrifugal force).

2. Determine which of these forces, or which of their components, act to provide the centripetal acceleration—that is, all the forces or components that act radially, toward or away from the center of the circu-

lar path. The sum of these forces (or components) provides the centripetal acceleration, $a_R = v^2/r$.

3. Choose a coordinate system, and positive and negative directions, and apply Newton's second law to the radial component:

$$\Sigma F_R = ma_R = m\frac{v^2}{r}. \qquad \text{[radial direction]}$$

Include only radial components of force.

5–3 | Highway Curves, Banked and Unbanked

An example of centripetal acceleration occurs when an automobile rounds a curve. In such a situation, you may feel that you are thrust outward. But there is not some mysterious centrifugal force pulling on you. What is happening is that you tend to move in a straight line, whereas the car has begun to follow a curved path. To make you go in the curved path, the seat (friction) or the door of the car (direct contact) exerts a force on you (Fig. 5–16). The car itself must have an inward force exerted on it if it is to move in a curve. On a flat road, this force is supplied by friction between the tires and the pavement.

If the wheels and tires of the car are rolling normally without slipping or sliding, the bottom of the tire is at rest against the road at each instant; so the friction force the road exerts on the tires is static friction. But if the static friction force is not great enough, as under icy conditions, sufficient friction force cannot be applied and the car will skid out of a circular path into a more nearly straight path. See Fig. 5–17. Once a car skids or slides, the friction force becomes kinetic friction, which is less than static friction.

FIGURE 5–16 The road exerts an inward force (friction against the tires) on a car to make it move in a circle; and the car exerts an inward force on the passenger.

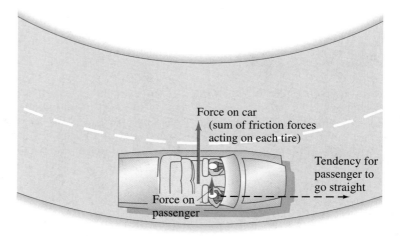

Force on car
(sum of friction forces
acting on each tire)

Tendency for
passenger to
go straight

Force on
passenger

EXAMPLE 5–12 **Skidding on a curve.** A 1000-kg car rounds a curve on a flat road of radius 50 m at a speed of 50 km/h (14 m/s). Will the car make the turn, or will it skid, if: (a) the pavement is dry and the coefficient of static friction is $\mu_s = 0.60$; (b) the pavement is icy and $\mu_s = 0.25$?

SOLUTION Figure 5–18 shows the free-body diagram for the car. The normal force, F_N, on the car is equal to the weight since the road is flat and there is no vertical acceleration:

$$F_N = mg = (1000 \, \text{kg})(9.8 \, \text{m/s}^2) = 9800 \, \text{N}.$$

In the horizontal direction the only force is friction, and we must compare it to the force needed to produce the centripetal acceleration to see if it is sufficient. The net horizontal force required to keep the car moving in a circle around the curve is

$$\Sigma F_R = ma_R = m\frac{v^2}{r} = (1000 \, \text{kg})\frac{(14 \, \text{m/s})^2}{(50 \, \text{m})} = 3900 \, \text{N}.$$

Naturally we hope the maximum total friction force (the sum of the friction forces acting on each of the four tires) will be at least this large. For (a), $\mu_s = 0.60$, and the maximum friction force attainable (recall from Section 5–1 that $F_{fr} \leq \mu_s F_N$) is

$$(F_{fr})_{max} = \mu_s F_N = (0.60)(9800 \, \text{N}) = 5900 \, \text{N}.$$

Since a force of only 3900 N is needed, and that is, in fact, how much will be exerted by the road as a static friction force, the car can make the turn fine. But in (b) the maximum friction force possible is

$$(F_{fr})_{max} = \mu_s F_N = (0.25)(9800 \, \text{N}) = 2500 \, \text{N}.$$

The car will skid because the ground cannot exert sufficient force (3900 N is needed) to keep it moving in a curve of radius 50 m.

FIGURE 5–17 Race car heading down into a curve. From the tire marks we can see that most cars experienced a sufficient friction force to give them the needed centripetal acceleration for rounding the curve safely. But, we can also see a few tire tracks of cars on which there was not sufficient force—and which followed more nearly straight-line paths.

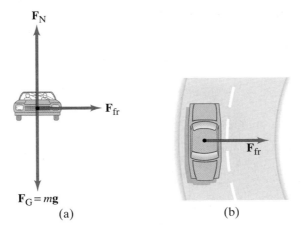

FIGURE 5–18 Forces on a car rounding a curve on a flat road. Example 5–12. (a) Front view, (b) top view.

The situation is worse if the wheels lock (stop rotating) when the brakes are applied too hard. When the tires are rolling, static friction exists. But if the wheels lock, the tires slide and the friction force, which is now kinetic friction, is less. Moreover, when the road is wet or icy, locking of the wheels occurs with less force on the brake pedal since there is less road friction to keep the wheels turning rather than sliding. Antilock brakes (ABS) are designed to limit brake pressure just before the point where sliding would occur, by means of delicate sensors and a fast computer.

➡ **PHYSICS APPLIED**

Antilock brakes (ABS)

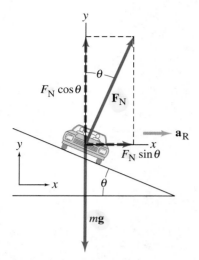

FIGURE 5–19 Normal force on a car rounding a banked curve, resolved into its horizontal and vertical components. Note that the centripetal acceleration is horizontal (and not parallel to the sloping road). The friction force on the tires is not shown. It could point up or down the slope, depending on the car's speed. The friction force will be zero for one particular speed.

Horizontal component of normal force alone acts to provide centripetal acceleration (friction is desired to be zero— otherwise it too would contribute)

The banking of curves can reduce the chance of skidding because the normal force of the road, acting perpendicular to the road, will have a component toward the center of the circle (Fig. 5–19), thus reducing the reliance on friction. For a given banking angle, θ, there will be one speed for which no friction at all is required. This will be the case when the horizontal component of the normal force toward the center of the curve, $F_N \sin \theta$ (see Fig. 5–19), is just equal to the force required to give a vehicle its centripetal acceleration—that is, when

$$F_N \sin \theta = m \frac{v^2}{r}.$$

The banking angle of a road, θ, is chosen so that this condition holds for a particular speed, called the "design speed."

EXAMPLE 5–13 Banking angle. (*a*) For a car traveling with speed v around a curve of radius r, determine a formula for the angle at which a road should be banked so that no friction is required. (*b*) What is this angle for an expressway off-ramp curve of radius 50 m at a design speed of 50 km/h?

SOLUTION We choose our x and y axes as horizontal and vertical so that a_R, which is horizontal, is along the x axis. The components of F_N are as shown in Fig. 5–19. (*a*) For the horizontal direction, $\Sigma F_R = m a_R$ gives

$$F_N \sin \theta = \frac{mv^2}{r}.$$

In the vertical direction, the forces are $F_N \cos \theta$ upward (Fig. 5–19) and the weight of the car (mg) downward. Since there is no vertical motion, the y component of the acceleration is zero, so $\Sigma F_y = m a_y$ gives us

$$F_N \cos \theta - mg = 0.$$

Thus,

$$F_N = \frac{mg}{\cos \theta}.$$

[Note in this case that $F_N \geq mg$ since $\cos \theta \leq 1$.] We substitute this relation for F_N into the equation for the horizontal motion,

$$F_N \sin \theta = m \frac{v^2}{r},$$

and obtain

$$\frac{mg}{\cos \theta} \sin \theta = m \frac{v^2}{r}$$

or

$$mg \tan \theta = m \frac{v^2}{r},$$

so

$$\tan \theta = \frac{v^2}{rg}.$$

This is the formula for the banking angle θ.
(*b*) For $r = 50$ m and $v = 50$ km/h (or 14 m/s),

$$\tan \theta = \frac{(14 \text{ m/s})^2}{(50 \text{ m})(9.8 \text{ m/s}^2)} = 0.40,$$

so $\theta = 22°$.

Nonuniform Circular Motion

If the speed of a particle revolving in a circle is changing, there will be a tangential acceleration, $\mathbf{a}_{tan}$, as well as the radial (centripetal) acceleration, $\mathbf{a}_R$. The tangential acceleration arises from the change in the magnitude of the velocity:

$$a_{tan} = \frac{dv}{dt}, \tag{5-2}$$

whereas the radial acceleration arises from the change in direction of the velocity, and as we have already seen, has magnitude

$$a_R = \frac{v^2}{r}.$$

The tangential acceleration always points in a direction tangent to the circle, and is in the direction of motion (parallel to $\mathbf{v}$) if the speed is increasing, as shown in Fig. 5–20 for a particle moving counterclockwise. If the speed is decreasing, $\mathbf{a}_{tan}$ points antiparallel to $\mathbf{v}$. In either case, $\mathbf{a}_{tan}$ and $\mathbf{a}_R$ are always perpendicular to each other, and their directions change continually as the particle moves along its circular path. The total vector acceleration, $\mathbf{a}$, is the sum of these two:

$$\mathbf{a} = \mathbf{a}_{tan} + \mathbf{a}_R. \tag{5-3}$$

Since $\mathbf{a}_R$ and $\mathbf{a}_{tan}$ are always perpendicular to each other, the magnitude of $\mathbf{a}$ at any moment is

$$a = \sqrt{a_{tan}^2 + a_R^2}.$$

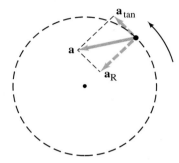

FIGURE 5–20   For nonuniform circular motion, the acceleration has a tangential ($\mathbf{a}_{tan}$) as well as a radial ($\mathbf{a}_R$) component.

EXAMPLE 5–14 **Two components of acceleration.** A racing car starts from rest in the pit area and accelerates at a uniform rate to a speed of 35 m/s in 11 s, moving on a circular track of radius 500 m. Assuming constant tangential acceleration, find (*a*) the tangential acceleration, and (*b*) the radial acceleration, at the instant when the speed is $v = 15$ m/s, and again when $v = 30$ m/s.

SOLUTION   (*a*) a_{tan} is constant, of magnitude

$$a_{tan} = \frac{dv}{dt} = \frac{(35 \text{ m/s} - 0 \text{ m/s})}{11 \text{ s}} = 3.2 \text{ m/s}^2,$$

and is the same for both instants.
(*b*) At the earlier instant,

$$a_R = \frac{v^2}{r} = \frac{(15 \text{ m/s})^2}{(500 \text{ m})} = 0.45 \text{ m/s}^2.$$

At the second instant,

$$a_R = \frac{(30 \text{ m/s})^2}{(500 \text{ m})} = 1.8 \text{ m/s}^2.$$

The radial acceleration continually increases, although the tangential acceleration stays constant.

These concepts can be used for an object moving along any curved path, such as that shown in Fig. 5–21. We can treat any portion of the curve as an arc of a circle with a radius of curvature r. The velocity at any point is always tangent to the path. The acceleration can be written, in general, as a sum of two components, the tangential component $a_{tan} = dv/dt$, and the radial (centripetal) component $a_R = v^2/r$.

FIGURE 5–21   Particle following a curved path. At point P the path has a radius of curvature r; the particle has velocity $\mathbf{v}$, tangential acceleration $\mathbf{a}_{tan}$ (it is increasing in speed), and radial (centripetal) acceleration $\mathbf{a}_R = v^2/r$ pointing toward the center of curvature C.

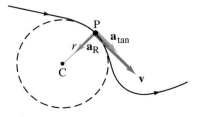

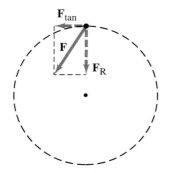

FIGURE 5–22 The speed of a particle moving in a circle changes if the net force on it has a tangential component.

Circular motion at constant speed occurs when the net force on an object is exerted toward the center of the circle. If the net force is not directed toward the center, but acts at an angle as shown in Fig. 5–22, then the force has two components. The component directed toward the center of the circle, $\mathbf{F}_R$, gives rise to the centripetal acceleration, $\mathbf{a}_R$, and keeps the object moving in a circle. The component $\mathbf{F}_{tan}$, tangential to the circle, acts to increase (or decrease) the speed and thus gives rise to the tangential acceleration (Fig. 5–20).

When you first start revolving a ball on the end of a string around your head, you must give it tangential acceleration. You do this by pulling on the string with your hand displaced from the center of the circle. In athletics, a hammer thrower accelerates the hammer tangentially in a similar way so that it reaches a high speed before release.

*5–5 Velocity-Dependent Forces; Terminal Velocity

When an object slides along a surface, the force of friction acting on the object is nearly independent of how fast the object is moving. But other types of resistive forces do depend on the object's velocity. The most important example is for an object moving through a liquid or gas, such as air. The fluid offers resistance to the motion of the object, and this resistive force, or **drag force**, depends on the velocity of the object.

The way the drag force varies with velocity is complicated in general. But for small objects at very low speeds, a good approximation can often be made by assuming that the drag force, F_D, is directly proportional to the velocity, v:

$$F_D = -bv. \tag{5–4}$$

The minus sign is necessary since the drag force opposes the motion. Here b is a constant (approximately) that depends on the viscosity of the fluid and on the size and shape of the object. This works well for small objects moving at low speed in a viscous liquid. It also works for very small objects moving in air at very low speeds, such as dust particles. For objects moving at high speeds, such as an airplane, a sky diver, a baseball, or an automobile, the force of air resistance can be better approximated as being proportional to v^2:

$$F_D \propto v^2.$$

For accurate calculations, however, more complicated forms and numerical integration generally need to be used. For objects moving through liquids, Eq. 5–4 works well for everyday objects at normal speeds (e.g., a boat in water).

Let us consider an object that falls from rest, through air or other fluid, under the action of gravity and a resistive force proportional to v. The forces acting on the object are the force of gravity, mg, acting downward, and the drag force, $-bv$, acting upward (Fig. 5–23a). Since the velocity $\mathbf{v}$ points downward, let us take the positive direction as downward. Then the net force on the object can be written

$$\Sigma F = mg + F_D$$
$$= mg - bv.$$

From Newton's second law $\Sigma F = ma$, we have

$$mg - bv = m\frac{dv}{dt}, \tag{5–5}$$

where we have written the acceleration according to its definition as rate of change of velocity, $a = dv/dt$. At $t = 0$, $v = 0$ and the acceleration $dv/dt = g$. But as the object falls and increases in speed, the resistive force increases, and this reduces the acceleration, dv/dt (see Fig. 5–23b). The velocity continues to increase, but at a slower rate. Eventually, the velocity becomes so large that the

FIGURE 5–23 (a) Forces acting on an object falling downward. (b) Graph of the velocity of a body falling in air when air resistance drag force is $F_D = -bv$. Initially, $v = 0$ and $dv/dt = g$, but as time goes on dv/dt (= slope of curve) decreases because of F_D. Eventually, v approaches a maximum value, v_T, the terminal velocity, which occurs when F_D has magnitude equal to mg.

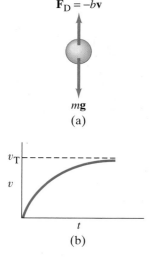

magnitude of the resistive force, $-bv$, approaches that of the gravitational force, mg; when the two are equal, we have

$$mg - bv = 0. \qquad (5-6)$$

At this point $dv/dt = 0$ and the object no longer increases in speed. It has reached its **terminal velocity** and continues to fall at this constant velocity until it hits the ground. This sequence of events is shown in the graph of Fig. 5–23b. The value of the terminal velocity v_T can be obtained from Eq. 5–6:

$$v_T = \frac{mg}{b}. \qquad (5-7)$$

If the resistive force is assumed proportional to v^2, or an even higher power of v, the sequence of events is similar and a terminal velocity reached, although it will not be given by Eq. 5–7.

EXAMPLE 5–15 **Force proportional to velocity.** Determine the velocity as a function of time for a body falling vertically from rest when there is a resistive force linearly proportional to v.

SOLUTION We start with Eq. 5–5 and write

$$\frac{dv}{dt} = g - \frac{b}{m}v.$$

There are two variables, v and t. We collect variables of the same type on one or the other side of the equation:

$$\frac{dv}{g - \dfrac{b}{m}v} = dt \qquad \text{or} \qquad \frac{dv}{v - \dfrac{mg}{b}} = -\frac{b}{m}\,dt.$$

Now we can integrate, remembering $v = 0$ at $t = 0$:

$$\int_0^v \frac{dv}{v - \dfrac{mg}{b}} = -\frac{b}{m}\int_0^t dt$$

$$\ln\left(v - \frac{mg}{b}\right) - \ln\left(-\frac{mg}{b}\right) = -\frac{b}{m}t$$

or

$$\ln\frac{v - mg/b}{-mg/b} = -\frac{b}{m}t.$$

We raise each side to the exponential [note that the natural log and the exponential are inverse operations of each other: $e^{\ln x} = x$, or $\ln(e^x) = x$] and obtain

$$v - \frac{mg}{b} = -\frac{mg}{b}e^{-\frac{b}{m}t} \qquad \text{or finally,} \qquad v = \frac{mg}{b}\left(1 - e^{-\frac{b}{m}t}\right).$$

This relation gives the velocity v as a function of time and corresponds to the graph of Fig. 5–23b. As a check, note that at $t = 0$, $v = 0$ and

$$a(t = 0) = \frac{dv}{dt} = \frac{mg}{b}\frac{d}{dt}\left(1 - e^{-\frac{b}{m}t}\right) = \frac{mg}{b}\left(\frac{b}{m}\right) = g,$$

as expected (see also Eq. 5–5). At large t, $e^{-\frac{b}{m}t}$ approaches zero, so v approaches mg/b, which is the terminal velocity, v_T, as we saw earlier. If we set $\tau = m/b$, then $v = v_T(1 - e^{-t/\tau})$. So $\tau = m/b$ is the time required for the velocity to reach 63 percent of the terminal velocity (since $e^{-1} = 0.37$).

Summary†

When two bodies slide over one another, the force of **friction** that each exerts on the other can be written approximately as $F_{fr} = \mu_k F_N$, where F_N is the **normal force** (the force each body exerts on the other perpendicular to their contact surface), and μ_k is the coefficient of **kinetic friction**. If the bodies are at rest relative to each other, then F_{fr} is just large enough to hold them at rest and satisfies the inequality $F_{fr} < \mu_s F_N$, where μ_s is the coefficient of **static friction**.

A particle revolving in a circle of radius r with constant speed v must have a net force acting on it that is directed toward the center of the circle at every moment. The magnitude of this net force must equal the product of the particle's mass m and its centripetal acceleration v^2/r.

† Optional material is generally not included in the Summary.

Questions

1. Cross-country skiers prefer their skis to have a large coefficient of static friction but a small coefficient of kinetic friction. Explain why. [*Hint*: Think of uphill and downhill.]

2. When you must brake your car very quickly, why is it safer if the wheels don't lock? When driving on slick roads, why is it advisable to apply the brakes slowly?

3. Why is the stopping distance of a truck much shorter than for a train going the same speed?

4. Can a coefficient of friction exceed 1.0?

5. A block is given a push so that it slides up a ramp. When the block reaches its highest point, it slides back down. Why is its acceleration less on the descent than on the ascent?

6. A heavy crate rests on the bed of a flatbed truck. When the truck accelerates, the crate remains where it is on the truck, so it, too, accelerates. What force causes the crate to accelerate?

7. The game of tetherball is played with a ball tied to a pole with a string. When the ball is struck, it whirls around the pole as shown in Fig. 5–24. In what direction is the acceleration of the ball, and what causes the acceleration?

8. When attempting to stop a car quickly on dry pavement, which of the following methods will stop the car in the least time? (*a*) Slam on the brakes as hard as possible, locking the wheels and *skidding* to a stop. (*b*) Press the brakes as hard as possible without locking the wheels and *rolling* to a stop. Explain.

9. Sometimes it is said that water is removed from clothes in a spin dryer by centrifugal force throwing the water outward. Is this correct? Discuss.

10. Technical reports often specify only the rpm for centrifuge experiments. Why is this inadequate?

11. Suppose a car moves at constant speed along a mountain road. At which of the following places does it exert the greatest and the least forces on the road: (*a*) At the top of a hill, (*b*) at a dip between two hills, (*c*) on a level stretch near the bottom of a hill?

12. A rider on a Ferris wheel moves in a vertical circle of radius r at constant speed v (Fig. 5–25). Is the normal force that the seat exerts on the rider at the top of the circle (*a*) less than, (*b*) more than, or (*c*) the same as, the force the seat exerts at the bottom of the circle? Explain.

FIGURE 5–24 Question 7.

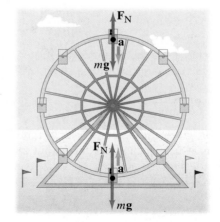

FIGURE 5–25 Question 12.

13. Describe all the forces acting on a child riding a horse on a merry-go-round. Which of these forces provides the centripetal acceleration of the child?

14. Why do bicycle riders lean in when rounding a curve at high speed?

15. A bucket of water can be whirled in a vertical circle without the water spilling out even at the top of the circle when the bucket is upside down. Explain.

16. Astronauts who spend long periods in outer space could be adversely affected by weightlessness. One way to simulate gravity is to shape the spaceship like a bicycle wheel that rotates about an axis just like a wheel, with the astronauts walking on the inside of the "tire." Explain how this simulates gravity. Consider (a) how objects fall, (b) the force we feel on our feet, and (c) any other aspects of gravity you can think of.

17. Why do airplanes bank when they turn? How would you compute the banking angle given the airspeed and radius of the turn?

Problems

Section 5–1

1. (I) If the coefficient of kinetic friction between a 12.0-kg crate and the floor is 0.30, what horizontal force is required to move the crate at a steady speed across the floor? What horizontal force is required if μ_k is zero?

2. (I) A force of 25.0 N is required to start a 6.0-kg box moving across a horizontal concrete floor. (a) What is the coefficient of static friction between the box and the floor? (b) If the 25.0-N force continues, the box accelerates at 0.50 m/s². What is the coefficient of kinetic friction?

3. (I) (a) A box sits at rest on a rough 37° inclined plane. Draw the free-body diagram, showing all the forces acting on the box. (b) How would the diagram change if the box were sliding down the plane. (c) How would it change if the box were sliding up the plane after an initial shove?

4. (I) The coefficient of friction between hard rubber and normal street pavement is about 0.8. On how steep a hill (maximum angle) can you leave a car parked?

5. (I) Suppose that you are standing on a train accelerating at 0.20 g. What minimum coefficient of static friction must exist between your feet and the floor if you are not to slide?

6. (I) What is the maximum acceleration a car can undergo if the coefficient of static friction between the tires and the ground is 0.80?

7. (II) A 15.0-kg box is released on a 30° incline and accelerates down the incline at 0.30 m/s². Find the friction force impeding its motion. What is the coefficient of kinetic friction?

8. (II) A car can decelerate at −4.80 m/s² without skidding when coming to rest on a level road. What would its deceleration be if the road were inclined at 13° uphill? Assume the same static friction force.

9. (II) For the system shown in Fig. 5–7 (Example 5–5), how large a mass must box I have to prevent any motion from occurring? Assume $\mu_s = 0.25$.

10. (II) A wet bar of soap slides freely down a ramp 9.0 m long inclined at 8.0°. How long does it take to reach the bottom? Assume $\mu_k = 0.060$.

11. (II) A box is given a push so that it slides across the floor. How far will it go, given that the coefficient of kinetic friction is 0.25 and the push imparts an initial speed of 2.5 m/s?

12. (II) Determine a formula for the acceleration of the system shown in Fig. 5–7 in terms of m_I, m_{II}, and the mass of the cord, m_C. Define any other variables needed.

13. (II) A 1000-kg car pulls a 350-kg trailer. The car exerts a horizontal force of 3.5×10^3 N against the ground in order to accelerate. What force does the car exert on the trailer? The trailer wheels are not friction-free, so estimate the net friction force on the trailer, using an effective friction coefficient of 0.15.

14. (II) (a) Show that the minimum stopping distance for an automobile traveling at speed v is equal to $v^2/2\mu_s g$, where μ_s is the coefficient of static friction between the tires and the road, and g is the acceleration of gravity. (b) What is this distance for a 1200-kg car traveling 95 km/h if $\mu_s = 0.75$? (c) What would it be if the car were on the Moon (the acceleration of gravity on the Moon is about g/6) but all else stayed the same?

15. (II) Piles of snow on slippery roofs can become dangerous projectiles as they melt. Consider a chunk of snow at the ridge of a roof with a pitch of 30°. (a) What is the minimum value of the coefficient of static friction needed to keep the snow from sliding down? (b) As the snow begins to melt the coefficient of static friction decreases and the snow finally slips. Assuming that the distance from the chunk to the edge of the roof is 5.0 m and the coefficient of kinetic friction is 0.20, calculate the speed of the snow chunk when it slides off the roof. (c) If the edge of the roof is 10.0 m above ground, what is the speed of the snow when it hits the ground?

16. (II) Police lieutenants, examining the scene of an accident involving two cars, measure the skid marks of one of the cars, which nearly came to a stop before colliding, to be 80 m long. The coefficient of kinetic friction between rubber and the pavement is about 0.8. Estimate the initial speed of that car.

17. (II) Two crates, of mass 80 kg and 210 kg, are in contact and at rest on a horizontal surface (Fig. 5–26). A 750-N force is exerted on the 80-kg crate. If the coefficient of kinetic friction is 0.12, calculate (a) the acceleration of the system, and (b) the force that each crate exerts on the other. (c) Repeat with the crates reversed.

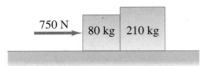

FIGURE 5–26 Problem 17.

18. (II) The block shown in Fig. 5–27 lies on a plane tilted at an angle $\theta = 22.0°$ to the horizontal, with $\mu_k = 0.17$. (a) Determine the acceleration of the block as it slides down the plane. (b) If the block starts from rest 9.3 m up the plane from its base, what will be the block's speed when it reaches the bottom of the incline?

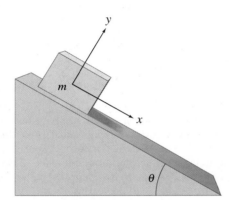

FIGURE 5–27 Block on inclined plane. Problems 18 and 19.

19. (II) A block is given an initial speed of 3.0 m/s up the 22.0° plane shown in Fig. 5–27. (a) How far up the plane will it go? (b) How much time elapses before it returns to its starting point? Assume $\mu_k = 0.17$.

20. (II) Radio engineers are erecting a communications tower that is 18 m high. During the installation they stabilize the tower with 30-m-long cables running from the top of the tower to the ground. The anchors consist of concrete blocks to which the cables can be secured. Each block weighs 1600 N. If the coefficient of static friction between a block and the ground is 0.80, what is the maximum tension that can exist in a cable before that cable's anchor is in danger of moving?

21. (II) Two blocks made of different materials connected together by a thin cord, slide down a plane ramp inclined at an angle θ to the horizontal as shown in Fig. 5–28 (block 2 is above block 1). The masses of the blocks are m_1 and m_2, and the coefficients of friction are μ_1 and μ_2. If

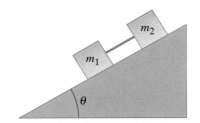

FIGURE 5–28 Problems 21, 22, and 23.

$m_1 = m_2 = 5.0$ kg, and $\mu_1 = 0.20$ and $\mu_2 = 0.30$, determine (a) the acceleration of the blocks and (b) the tension in the cord, for an angle $\theta = 30°$.

22. (II) For two blocks, connected by a cord and sliding down the incline shown in Fig. 5–28 (see Problem 21), describe the motion (a) if $\mu_1 < \mu_2$, and (b) if $\mu_1 > \mu_2$. (c) Determine a formula for the acceleration of each block and the tension F_T in the cord in terms of m_1, m_2, and θ; interpret your results in light of your answers to (a) and (b).

23. (II) Do Problem 21, but assume the two blocks sliding on the plane (Fig. 5–28) are connected by a rigid rod instead of a cord.

24. (II) In Fig. 5–29 the coefficient of static friction between mass m_1 and the table is 0.40, whereas the coefficient of kinetic friction is 0.30 (a) What minimum value of m_1 will keep the system from starting to move? (b) What value(s) of m_1 will keep the system moving at constant speed?

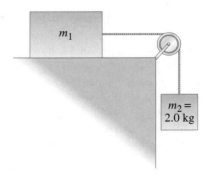

FIGURE 5–29 Problem 24.

25. (II) A small block of mass m is given an initial speed v_0 up a ramp inclined at angle θ to the horizontal. It travels a distance d up the ramp and comes to rest. (a) Determine a formula for the coefficient of kinetic friction between block and ramp. (b) What can you say about the value of the coefficient of static friction?

26. (II) On an icy day, you worry about parking your car in your driveway, which has an incline of 12°. Your neighbor Ralph's driveway has an incline of 9.0°, and Bonnie's driveway across the street has one of 6.0°. The coefficient of static friction between tire rubber and ice is 0.15. Which driveway(s) will be safe to park in?

27. (II) What is the acceleration of the system shown in Fig. 5–30 if the coefficient of kinetic friction is 0.10? Assume that the blocks start from rest and that (a) $m_1 = 5.0$ kg and (b) $m_1 = 2.0$ kg.

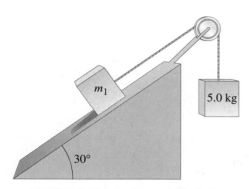

FIGURE 5–30 Problems 27 and 28.

28. (II) What are the minimum and maximum values of m_1 in Fig. 5–30 to keep the system from accelerating? Take $\mu_s = \mu_k = 0.50$.

29. (II) A child slides down a slide with a 28° incline, and at the bottom her speed is precisely half what it would have been if the slide had been frictionless. Calculate the coefficient of kinetic friction between the slide and the child.

30. (III) Boxes are moved on a conveyor belt from where they are filled to the packing station 10 m away. The belt is initially stationary and must finish with zero speed. The most rapid transit is accomplished if the belt accelerates for half the distance, then decelerates for the final half of the trip. If the coefficient of static friction between a box and the belt is 0.60, what is the minimum transit time for each box?

31. (III) A bicyclist can coast down a 7.0° hill at a steady 9.5 km/h. If the drag force is proportional to the square of the speed v, so that $F_D = cv^2$, calculate (a) the value of the constant c and (b) the average force that must be applied in order to descend the hill at 25 km/h. The mass of the cyclist plus bicycle is 80.0 kg. Ignore other types of friction.

32. (III) A 4.0-kg block is stacked on top of a 12.0-kg block, which is accelerating along a horizontal table at $a = 5.2$ m/s² (Fig. 5–31). Let $\mu_k = \mu_s = \mu$. (a) What minimum coefficient of friction μ between the two blocks will prevent the 4.0-kg block from sliding off? (b) If μ is only half this minimum value, what is the acceleration of the 4.0-kg block with respect to the table, and (c) with respect to the 12.0-kg block? (d) What is the force that must be applied to the 12.0-kg block in (a) and in (b), assuming that the table is frictionless?

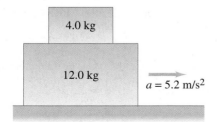

FIGURE 5–31
Problem 32.

33. (III) A small block of mass m rests on the rough, sloping side of a triangular block of mass M which itself rests on a horizontal frictionless table as shown in Fig. 5–32. If the coefficient of static friction is μ, determine the minimum horizontal force F applied to M that will cause the small block m to start moving up the incline.

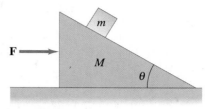

FIGURE 5–32 Problem 33.

Section 5–2 and 5–3

34. (I) Calculate the centripetal acceleration of the Earth in its orbit around the Sun and the net force exerted on the Earth. What exerts this force on the Earth? Assume the Earth's orbit is a circle of radius 1.50×10^{11} m.

35. (I) What is the maximum speed with which a 1200-kg car can round a turn of radius 80.0 m on a flat road if the coefficient of friction between tires and road is 0.55? Is this result independent of the mass of the car?

36. (I) A horizontal force of 60.0 N is exerted on a 2.00-kg discus as it is rotated uniformly in a horizontal circle (at arm's length) of radius 1.00 m. Calculate the speed of the discus.

37. (I) A child moves with a speed of 1.50 m/s when 9.0 m from the center of a merry-go-around. Calculate (a) the centripetal acceleration of the child and (b) the net horizontal force exerted on the child (mass = 25 kg).

38. (II) How fast (in rpm) must a centrifuge rotate if a particle 9.0 cm from the axis of rotation is to experience an acceleration of 100,000 g's?

39. (II) Is it possible to whirl a bucket of water fast enough in a vertical circle so that the water won't fall out? If so, what is the minimum speed? Define all quantities needed.

40. (II) At what minimum speed must a roller coaster be traveling when upside down at the top of a circle (Fig. 5–33) if the passengers are not to fall out? Assume a radius of curvature of 8.0 m.

FIGURE 5–33 Problem 40.

41. (II) A coin is placed 12.0 cm from the axis of a rotating turntable of variable speed. When the speed of the turntable is slowly increased, the coin remains fixed on the turntable until a rate of 50 rpm is reached, at which point the coin slides off. What is the coefficient of static friction between the coin and the turntable?

42. (II) The design of a new road includes a straight stretch that is horizontal and flat but that suddenly dips down a steep hill at 22°. The transition should be rounded with what minimum radius so that cars traveling 90 km/h will not leave the road (Fig. 5–34)?

FIGURE 5–34 Problem 42.

43. (II) On an ice rink two skaters of equal mass grab hands and spin in a mutual circle once every three seconds. If we assume their arms are each 0.80 m long, how hard are they pulling on one another, assuming their individual masses are 60.0 kg?

44. (II) Tarzan plans to cross a gorge by swinging in an arc from a hanging vine. If his arms are capable of exerting a force of 1400 N on the rope, what is the maximum speed he can tolerate at the lowest point of his swing? His mass is 80 kg and the vine is 4.8 m long.

45. (II) Redo Example 5–9, precisely this time, by not ignoring the weight of the ball. In particular, find the magnitude of $\mathbf{F}_T$ and the angle it makes with the horizontal.

46. (II) A 0.35-kg ball, attached to the end of a horizontal cord, is rotated in a circle of radius 1.0 m on a frictionless horizontal surface. If the cord will break when the tension in it exceeds 80 N, what is the maximum speed the ball can have? How would your answer be affected if there were friction?

47. (II) A 1000-kg sports car moving at 20 m/s crosses the rounded top of a hill (radius = 100 m). Determine (a) the normal force on the car, (b) the normal force on the 70-kg driver, and (c) the car speed at which the normal force is zero.

48. (II) Two masses, m_1 and m_2, connected to each other and to a central post by cords as shown in Fig. 5–35, rotate about the post at frequency f (revolutions per second) on a frictionless horizontal surface at distances r_1 and r_2 from the post. Derive an algebraic expression for the tension in each segment of the cord.

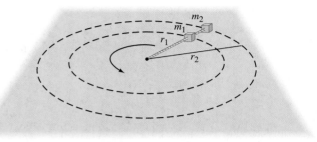

FIGURE 5–35 Problem 48.

49. (III) A thin circular horizontal hoop of mass m and radius R rotates at frequency f about a vertical axis through its center (Fig. 5–36). Determine the tension within the hoop. [*Hint*: Consider a tiny section of the hoop.]

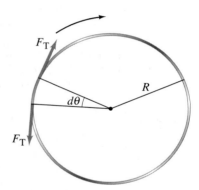

FIGURE 5–36 Problem 49.

50. (III) If a curve with a radius of 65 m is properly banked for a car traveling 70 km/h, what must be the coefficient of static friction for a car not to skid when traveling at 100 km/h?

51. (III) A curve of radius 60 m is banked for a design speed of 90 km/h. If the coefficient of static friction is 0.30 (wet pavement), at what range of speeds can a car safely make the curve?

* Section 5–4

* 52. (II) A particle starting from rest revolves with uniformly increasing speed in a clockwise circle in the xy plane. The center of the circle is at the origin of an xy coordinate system. At $t = 0$, the particle is at $x = 0.0$, $y = 2.0$ m. At $t = 2.0$ s, it has made one-quarter of a revolution and is at $x = 2.0$ m, $y = 0.0$. Determine (a) its speed at $t = 2.0$ s, (b) the average velocity vector, and (c) the average acceleration vector during this interval.

* 53. (II) In Problem 52 assume the tangential acceleration is constant and determine the components of the instantaneous acceleration at (a) $t = 0.0$, (b) $t = 1.0$ s, and (c) $t = 2.0$ s.

* 54. (II) A particle rotates in a circle of radius 3.60 m. At a particular instant its acceleration is 0.210 g in a direction that makes an angle of 28.0° to its direction of motion. Determine its speed (a) at this moment and (b) 2.00 s later, assuming constant tangential acceleration.

* 55. (II) A body moves in a circle of radius 20 m with its speed given by $v = 3.6 + 1.5t^2$, with v in meters per second and t in seconds. At $t = 3.0$ s, find (a) the tangential acceleration and (b) the radial acceleration.

* 56. (III) An object of mass m is constrained to move in a circle of radius r. Its tangential acceleration as a function of time is given by $a_{tan} = b + ct^2$, where b and c are constants. If $v = v_0$ at $t = 0$, determine the tangential and radial components of the force, F_{tan} and F_R, acting on the object at any time $t > 0$.

* Section 5–5

* 57. (I) In Example 5–15, use dimensional analysis to determine if the time constant τ is $\tau = m/b$ or $\tau = b/m$.

* 58. (II) The terminal velocity of a 3×10^{-5}-kg raindrop is about 9 m/s. Assuming a drag force $F_D = -bv$, determine (a) the value of the constant b and (b) the time required for such a drop, starting from rest, to reach 63 percent of terminal velocity.

* 59. (II) An object moving vertically has $\mathbf{v} = \mathbf{v}_0$ at $t = 0$. Determine a formula for its velocity as a function of time assuming a resistive force $F = -bv$ as well as gravity for two cases: (a) $\mathbf{v}_0$ is downward and (b) $\mathbf{v}_0$ is upward.

* 60. (II) The drag force on large objects such as cars, planes, and sky divers moving through air is more nearly $F_D = -bv^2$. (a) For this quadratic dependence on v, determine a formula for the terminal velocity v_T, of a vertically falling object. (b) A 75-kg sky diver has a terminal velocity of about 60 m/s; determine the value of the constant b. (c) Draw a curve like that of Fig. 5–23 for this case of $F_D \propto v^2$. For the same terminal velocity, would this curve lie above or below that in Fig. 5–23? Explain why.

* 61. (III) Two drag forces act on a bicycle and rider: F_{D1} due to rolling resistance, which is essentially velocity independent; and F_{D2} due to air resistance, which is proportional to v^2. For a specific bike plus rider of total mass 80 kg, $F_{D1} \approx 4.0$ N; and for a speed of 2.2 m/s, $F_{D2} \approx 1.0$ N. (a) Show that the total drag force is

$$F_D = 4.0 + 0.21v^2,$$

where v is in m/s, and F_D is in N and opposes the motion. (b) Determine at what slope angle θ the bike and rider can coast downhill at a constant speed of 10 m/s.

* 62. (III) Determine a formula for the position and acceleration of a falling object as a function of time if the object starts from rest at $t = 0$ and undergoes a resistive force $F = -bv$, as in Example 5–15.

* 63. (III) A motorboat traveling at a speed of 2.4 m/s shuts off its engines at $t = 0$. How far does it travel before coming to rest if it is noted that after 3.0 s its speed has dropped to half its original value? Assume that the drag force of the water is proportional to v.

* 64. (III) A block slides along a horizontal surface lubricated with a thick oil which provides a drag force proportional to the square root of velocity:

$$F_D = -bv^{\frac{1}{2}}.$$

If $v = v_0$ at $t = 0$, determine v and x as functions of time.

General Problems

65. A 2.0-kg silverware drawer becomes stuck. The owner gradually pulls with more and more force, and when the applied force reaches 8.0 N, the drawer suddenly opens, throwing all the utensils to the floor. What is the coefficient of static friction between the drawer and the cabinet?

66. Devise a method using an inclined plane to measure the coefficient of static friction, μ_s, between two surfaces.

67. Drag race tires in contact with an asphalt surface probably have one of the highest coefficients of static friction in the everyday world. Assuming a constant acceleration and no slipping of tires, estimate the coefficient of static friction for a drag racer that covers the quarter mile in 6.0 s.

68. A coffee cup on the dashboard of a car slides forward on the dash when the driver decelerates from 45 km/h to rest in 3.5 s or less, but not if he decelerates in a longer time. What is the coefficient of static friction between the cup and the dash?

69. An 18.0-kg box is released on a 37.0° incline and accelerates down the incline at 0.270 m/s². Find the friction force impeding its motion. How large is the coefficient of friction?

70. A flatbed truck is carrying a heavy crate. The coefficient of static friction between the crate and the bed of the truck is 0.75. What is the maximum rate at which the driver can decelerate and still avoid having the crate slide against the cab?

71. In an earthquake, the ground is found to accelerate with a maximum value of a_{max}. (a) If an object is going to "hold its place" on the ground, show that it must have a coefficient of static friction with respect to the ground of at least $\mu_s = a_{max}/g$. (b) Numerically, the famous Loma Prieta Earthquake that stopped the 1989 World Series produced maximum ground accelerations of up to 4.0 m/s² in the San Francisco Bay Area. Would a chair start to slide on your linoleum floor if the coefficient of static friction were 0.25?

72. A roller coaster reaches the top of the steepest hill with a speed of 6.0 km/h. It then descends the hill, which is at an average angle of 45° and is 45.0 m long. What will its speed be when it reaches the bottom? Assume $\mu_k = 0.12$.

73. A motorcyclist is coasting with the engine off at a steady speed of 20.0 m/s but enters a sandy stretch where the coefficient of kinetic friction is 0.80. Will the cyclist emerge from the sandy stretch without having to start the engine if the sand lasts for 15 m? If so, what will be the speed upon emerging?

74. A flat puck (mass M) is rotated in a circle on an air hockey (frictionless) table top, and is held in this orbit by a light cord which is connected to a dangling mass (mass m) through the central hole as shown in Fig. 5–37. Show that the speed of the puck is given by $v = \sqrt{\dfrac{mgR}{M}}$.

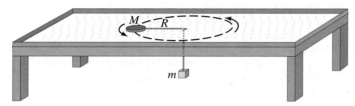

FIGURE 5–37   Problem 74.

75. A device for training astronauts and jet fighter pilots is designed to rotate the trainee in a horizontal circle of radius 10.0 m. If the force felt by the trainee is 7.75 times her own weight, how fast is she rotating? Express your answer in both m/s and rev/s.

76. In a "Rotor-ride" at a carnival, people pay money to be rotated in a vertical cylindrically walled "room." (See Fig. 5–38). If the room radius is 5.0 m, and the rotation frequency is 0.50 revolutions per second when the floor drops out, what is the minimum coefficient of static friction so that the people will not slip down? People describe this ride by saying they were being "pressed against the wall." Is this true? Is there really an outward force pressing them against the wall? If so, what is its source? If not, what is the proper description of their situation (besides nausea)?

FIGURE 5–38   Problem 76.

*77. Determine the tangential and centripetal components of the net force exerted on a car (by the ground) when its speed is 30 m/s, and it has accelerated to this speed from rest in 9.0 s on a curve of radius 450 m. The car's mass is 1000 kg.

78. A 1000-kg car rounds a curve of radius 80 m banked at an angle of 14°. If the car is traveling at 80 km/h, will a friction force be required? If so, how much and in what direction?

79. A car maintains a constant speed v as it traverses the hill and valley shown in Fig. 5–39. Both the hill and valley have a radius of curvature R. (a) How do the normal forces acting on the car at A, B, and C compare? (Which is largest? Smallest?) Explain. (b) Where would the driver feel heaviest? Lightest? Explain. (c) How fast can the car go without losing contact with the road at A?

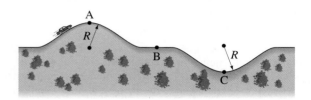

FIGURE 5–39   Problem 79.

*80. A car at the Indianapolis 500 accelerates uniformly from the pit area, going from rest to 320 km/h in a semicircular arc with a radius of 200 m. Determine the tangential and radial acceleration of the car when it is halfway through the turn, assuming constant acceleration. If the curve were flat, what would the coefficient of static friction have to be between the tires and the roadbed to provide this acceleration with no slipping or skidding?

*81. A particle revolves in a horizontal circle of radius 2.70 m. At a particular instant, its acceleration is 1.05 m/s², in a direction that makes an angle of 32.0° to its direction of motion. Determine its speed (a) at this moment, and (b) 2.00 s later, assuming constant tangential acceleration.

82. Figaro the cat (5.0 kg) is hanging on the tablecloth, pulling Cleo's fishbowl (11 kg) toward the edge of the table (Fig. 5–40). The coefficient of kinetic friction between the tablecloth (ignore its mass) under the fishbowl and the table is 0.44. (a) What is the acceleration of Figaro and the fishbowl? (b) If the fishbowl is 0.90 m from the edge of the table, how much time does it take for Figaro to pull Cleo off the table?

FIGURE 5–40   Problem 82.

83. A small mass m is set on the surface of a sphere, Fig. 5–41. If the coefficient of static friction is $\mu_s = 0.60$, at what angle ϕ would the mass start sliding?

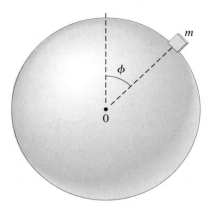

FIGURE 5–41   Problem 83.

84. The 70.0-kg climber in Fig. 5–42 is supported in the "chimney" by the friction forces exerted on his shoes and back. The static coefficients of friction between his shoes and the wall, and between his back and the wall, are 0.80 and 0.60, respectively. What is the minimum normal force he must exert? Assume the walls are vertical and that the static friction forces are both at their maximum, $F_{fr} = \mu_s F_N$.

FIGURE 5–42   Problem 84.

85. A friend throws you a baseball with an initial horizontal velocity of 30 m/s. Its path is parabolic, but at any point we can determine a *radius of curvature*. (*a*) Define a method for determining the radius of curvature. (*b*) What is this radius of curvature for the baseball immediately after it left your friend's hand?

86. A 28.0-kg block is connected to an empty 1.00-kg bucket by a cord running over a frictionless pulley (Fig. 5–43). The coefficient of static friction between the table and the block is 0.450 and the coefficient of kinetic friction between the table and the block is 0.320. Sand is gradually added to the bucket until the system just begins to move. (*a*) Calculate

28.0 kg

FIGURE 5–43   Problem 86.

the mass of sand added to the bucket. (*b*) Calculate the acceleration of the system.

87. An airplane traveling at 520 km/h needs to reverse its course. The pilot decides to accomplish this by banking the wings at an angle of 38°. (*a*) Find the time needed for this maneuver. (*b*) What additional force will the passengers experience during the turn? [*Hint*: Assume an aerodynamic "lift" force that acts perpendicularly to the flat wings; see Fig. 5–44.]

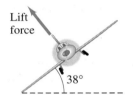

Lift force

38°

FIGURE 5–44   Problem 87.

88. A circular curve of radius R in a new highway is designed so that a car traveling at speed v_0 can negotiate the turn safely on glare ice (zero friction). If a car travels too slowly, then it will slip toward the center of the circle. If it travels too fast, then it will slip away from the center of the circle. As the coefficient of static friction increases, it becomes possible for a car to stay on the road while traveling at any speed within a range from v_{min} to v_{max}. Derive formulas for v_{min} and v_{max} as functions of μ_s, v_0, and R.

89. A train traveling at a constant speed rounds a curve of radius 275 m. A pendulum suspended from the ceiling swings out to an angle of 17.5° throughout the turn. What is the speed of the train?

∗90. The force of air resistance on a rapidly falling body has the form $F = -kv^2$, so that Newton's second law applied to such an object is

$$m \frac{dv}{dt} = mg - kv^2$$

when the downward direction is considered positive. Determine the speed and position at 2.0-s intervals, up to 20.0 s, of a 75-kg person (sky diver) who starts from rest, assuming $k = 0.22$ kg/m. Also, show that the body eventually reaches a steady speed, the *terminal speed*, and explain why this happens.

*91. Assume a net force $F = -mg - kv^2$ acts during the upward vertical motion of a 250-kg rocket, starting at the moment ($t = 0$) when the fuel has burned out and the rocket has an upward speed of 120 m/s. Let $k = 0.65$ kg/m. Calculate the maximum height reached by the rocket. Compare to free-flight conditions without air resistance ($k = 0$).

92. A small bead of mass m is constrained to slide without friction inside a circular vertical hoop of radius r which rotates about a vertical axis (Fig. 5–45) at a frequency f. (a) Determine the angle θ where the bead will be in equilibrium—that is, where it will have no tendency to move up or down along the hoop. (b) If $f = 4.0$ rev/s and $r = 20$ cm, what is θ? (c) Can the bead ride as high as the center of the circle ($\theta = 90°$)? Explain.

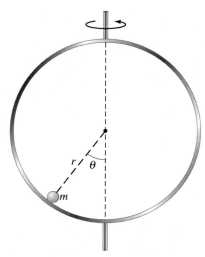

FIGURE 5–45   Problem 92.

93. The sides of a cone make an angle ϕ with the vertical. A small mass m is placed on the inside of the cone and the cone, with its point down, is revolved at a frequency f (revolutions per second) about its symmetry axis. If the coefficient of static friction is μ_s, at what positions on the cone can the mass be placed without sliding on the cone? (Give the maximum and minimum distances, r, from the axis).

*94. A ball of mass $m = 1.0$ kg at the end of a thin cord of length $r = 0.80$ m revolves in a vertical circle about point O, as shown in Fig. 5–46. During the time we observe it, the only forces acting on the ball are gravity and the tension in the cord. The motion is circular but not uniform because of the force of gravity. The ball increases in speed as it descends and decelerates as it rises on the other side of the circle. At the moment the cord makes an angle $\theta = 30°$ below the horizontal, the ball's speed is 6.0 m/s. At this point, determine the tangential acceleration, the radial acceleration, and the tension in the cord, F_T. Take θ increasing downward as shown.

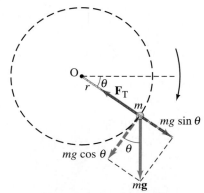

FIGURE 5–46   Problem 94.

The astronauts in the upper left of this photo are working on the space shuttle. As they orbit the Earth—at a pretty high speed—they experience apparent weightlessness. The Moon, in the background, also is orbiting the Earth at high speed. What keeps the Moon and the space shuttle (and its astronauts) from moving off in a straight line away from Earth? It is the force of gravity. According to Newton's law of universal gravitation, all objects attract all other objects with a force proportional to their masses and inversely proportional to the square of the distance between them.

Gravitation and Newton's Synthesis

S that serve as the foundation for the study of dynamics. He also conceived of another great law to describe one of the basic forces in nature, gravitation, and he applied it to understand the motion of the planets. This new law, published in 1687 in his great book *Philosophiae Naturalis Principia Mathematica* (the *Principia*, for short), is called Newton's law of universal gravitation. It was the capstone of Newton's analysis of the physical world. Indeed, Newtonian mechanics, with its three laws of motion and the law of universal gravitation, was accepted for centuries as a mechanical basis for the way the universe works.

6–1 Newton's Law of Universal Gravitation

Among his many great accomplishments, Sir Isaac Newton examined the motion of the heavenly bodies—the planets and the Moon. In particular, he wondered about the nature of the force that must act to keep the Moon in its nearly circular orbit around the Earth.

Newton was also thinking about the problem of gravity. Since falling bodies accelerate, Newton had concluded that they must have a force exerted on them, a force we call the force of gravity. Whenever a body has a force exerted *on* it, that force is exerted *by* some other body. But what *exerts* the force of gravity? Every object on the surface of the Earth feels the force of gravity, and no matter where the object is, the force is directed toward the center of the Earth (Fig. 6–1). Newton concluded that it must be the Earth itself that exerts the gravitational force on objects at its surface.

According to an early account, Newton was sitting in his garden and noticed an apple drop from a tree. He is said to have been struck with a sudden inspiration: if the effect of gravity acts at the tops of trees, and even at the tops of mountains, then perhaps it acts all the way to the Moon! Whether this story is true or not, it does seem to capture something of Newton's reasoning and inspiration. With this idea that it is terrestrial gravity that holds the Moon in its orbit, Newton developed his great theory of gravitation. [But there was controversy at the time. Many thinkers had trouble accepting the idea of a force "acting at a distance." Typical forces act through contact—your hand pushes a cart and pulls a wagon, a bat hits a ball, and so on. But gravity acts without contact, said Newton: the Earth exerts a force on a falling apple and on the Moon, even though there is no contact, and the two objects may even be very far apart.]

Newton set about determining the magnitude of the gravitational force that the Earth exerts on the Moon as compared to the gravitational force on objects at the Earth's surface. At the surface of the Earth, the force of gravity accelerates objects at 9.80 m/s^2. But what is the centripetal acceleration of the Moon? Since the Moon moves with nearly uniform circular motion, the acceleration can be calculated from $a_R = v^2/r$. We already performed this calculation in Example 3–12 and found that $a_R = 0.00272 \text{ m/s}^2$. In terms of the acceleration of gravity at the Earth's surface, g, this is equivalent to

$$a_R \approx \frac{1}{3600} g.$$

That is, the acceleration of the Moon toward the Earth is about $\frac{1}{3600}$ as great as the acceleration of objects at the Earth's surface. The Moon is 384,000 km from the Earth, which is about 60 times the Earth's radius of 6380 km. That is, the Moon is 60 times farther from the Earth's center than are objects at the Earth's surface. But $60 \times 60 = 60^2 = 3600$. Again that number 3600! Newton concluded that the gravitational force exerted by the Earth on any object decreases with the square of its distance, r, from the Earth's center:

$$\text{force of gravity} \propto \frac{1}{r^2}.$$

The Moon, being 60 Earth radii away, feels a gravitational force only $\frac{1}{60^2} = \frac{1}{3600}$ times as strong as it would if it were at the Earth's surface. Any object placed 384,000 km from the Earth would experience the same acceleration due to the Earth's gravity as the Moon experiences: 0.00272 m/s^2.

Newton realized that the force of gravity on an object depends not only on distance but also on the object's mass. In fact, it is directly proportional to its mass, as we have seen. According to Newton's third law, when the Earth exerts its gravitational force on any body, such as the Moon, that other body exerts an equal and opposite force on the Earth (Fig. 6–2). Because of this symmetry, Newton reasoned, the magnitude of the force of gravity must be proportional to *both* the masses. Thus

$$F \propto \frac{m_E m_B}{r^2}$$

where m_E is the mass of the Earth, m_B the mass of the other body, and r the distance from the Earth's center to the center of the other body.

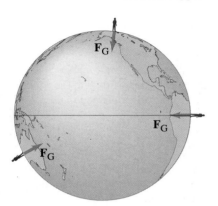

Newton's apple

FIGURE 6–1 Anywhere on Earth, whether in Alaska, Australia or Peru, the force of gravity acts downward toward the center of the Earth.

The Moon's acceleration toward Earth

FIGURE 6–2 The gravitational force one body exerts on a second body is directed toward the first body, and is equal and opposite to the force exerted by the second body on the first.

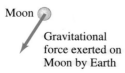

Moon

Gravitational force exerted on Moon by Earth

Earth

Gravitational force exerted on Earth by the Moon

Newton went a step further in his analysis of gravity. In his examination of the orbits of the planets, he concluded that the force required to hold the different planets in their orbits around the Sun seems to diminish as the inverse square of their distance from the Sun. This led him to believe that it is also the gravitational force that acts between the Sun and each of the planets to keep them in their orbits. And if gravity acts between these objects, why not between all objects? Thus he proposed his famous **law of universal gravitation**, which we can state as follows:

Every particle in the universe attracts every other particle with a force that is proportional to the product of their masses and inversely proportional to the square of the distance between them. This force acts along the line joining the two particles.

The magnitude of the gravitational force can be written as

$$F = G \frac{m_1 m_2}{r^2}, \tag{6-1}$$

where m_1 and m_2 are the masses of the two particles, r is the distance between them, and G is a universal constant which must be measured experimentally and has the same numerical value for all objects.

The value of G must be very small, since we are not aware of any force of attraction between ordinary-sized objects, such as between two baseballs. The force between two ordinary objects was first measured, over 100 years after Newton's publication of his law, by Henry Cavendish in 1798. To detect and measure the incredibly small force, he used an apparatus like that shown in Fig. 6–3. Cavendish confirmed Newton's hypothesis that two bodies attract one another, and that Eq. 6–1 accurately describes this force. In addition, because he could measure F, m_1, m_2, and r accurately, he was able to determine the value of the constant G as well. The accepted value today is

$$G = 6.67 \times 10^{-11}\,\text{N}\cdot\text{m}^2/\text{kg}^2.$$

Strictly speaking, Eq. 6–1 gives the magnitude of the gravitational force that one particle exerts on a second particle that is a distance r away. For an extended object (that is, not a point), we must consider how to measure the distance r. You might think that r would be the distance between the centers of the objects. This is often true and often a good approximation even when not quite true; but to do a calculation correctly, each extended body must be considered as a collection of tiny particles, and the total force is the sum of the forces due to all the particles. The sum over all these particles is often best done using integral calculus, which Newton himself invented. Newton showed that for two uniform spheres, Eq. 6–1 gives the correct force where r is the distance between their centers. (The derivation is given in Appendix C.) Also, when extended bodies are small compared to the distance between them (as for the Earth–Sun system), little inaccuracy results from considering them as point particles.

FIGURE 6–3 Schematic diagram of Cavendish's apparatus. Two spheres are attached to a light horizontal rod, which is suspended at its center by a thin fiber. When a third sphere labeled A is brought close to one of the suspended spheres, the gravitational force causes the latter to move, and this twists the fiber slightly. The tiny movement is magnified by the use of a narrow light beam directed at a mirror mounted on the fiber. The beam reflects onto a scale. Previous determination of how large a force will twist the fiber a given amount then allows one to determine the magnitude of the gravitational force between two objects.

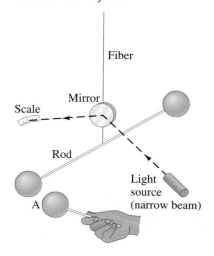

EXAMPLE 6–1 ESTIMATE Can you attract another person gravitationally? A 50-kg person and a 75-kg person are sitting on a bench so that their centers are about 50 cm apart. Estimate the magnitude of the gravitational force each exerts on the other.

SOLUTION We use Eq. 6–1, which gives

$$F = \frac{(6.67 \times 10^{-11}\,\text{N}\cdot\text{m}^2/\text{kg}^2)(50\,\text{kg})(75\,\text{kg})}{(0.50\,\text{m})^2} = 1.0 \times 10^{-6}\,\text{N},$$

which is unnoticeably small unless very delicate instruments are used.

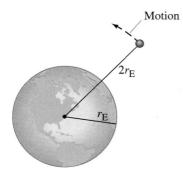

Motion

$2r_E$

r_E

FIGURE 6–4   Example 6–2.

FIGURE 6–5   Orientation of Sun (S), Earth (E), and Moon (M) for Example 6–3 (not to scale).

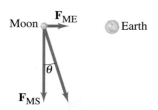

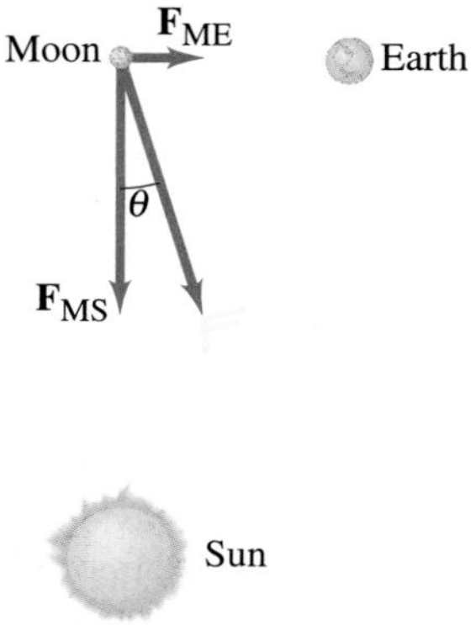

Moon $\mathbf{F}_{ME}$ Earth

θ

$\mathbf{F}_{MS}$

Sun

FIGURE 6–6   The displacement vector $\mathbf{r}_{21}$ points from particle of mass m_2 to particle of mass m_1. The unit vector shown, $\hat{\mathbf{r}}_{21}$ is in the same direction as $\mathbf{r}_{21}$, but is defined as having length one.

m_1

$\mathbf{r}_{21}$

$\hat{\mathbf{r}}_{21}$

m_2

EXAMPLE 6–2 Spacecraft at $2r_E$. What is the force of gravity acting on a 2000-kg spacecraft when it orbits two Earth radii from the Earth's center (that is, a distance $r_E = 6380$ km above the Earth's surface, Fig. 6–4)? The mass of the Earth is $M_E = 5.98 \times 10^{24}$ kg.

SOLUTION   We could plug all the numbers into Eq. 6–1, but there is a simpler approach. The spacecraft is twice as far from the Earth's center as when at the surface of the Earth. Therefore, since the force of gravity decreases as the square of the distance $\left(\text{and } \frac{1}{2^2} = \frac{1}{4}\right)$, the force of gravity on it will be only one fourth its weight at the Earth's surface:

$$F_G = \tfrac{1}{4}mg = \tfrac{1}{4}(2000 \text{ kg})(9.80 \text{ m/s}^2) = 4900 \text{ N}.$$

EXAMPLE 6–3 Force on the Moon.   Find the net force on the Moon $\left(m_M = 7.35 \times 10^{22} \text{ kg}\right)$ due to the gravitational attraction of both the Earth $\left(m_E = 5.98 \times 10^{24} \text{ kg}\right)$ and the Sun $\left(m_S = 1.99 \times 10^{30} \text{ kg}\right)$, assuming they are at right angles to each other (Fig. 6–5).

SOLUTION   We must add the two forces vectorially. First we calculate their magnitudes. The Earth is 3.84×10^5 km $= 3.84 \times 10^8$ m from the Moon, so F_{ME} (the force on the Moon due to the Earth) is

$$F_{ME} = \frac{\left(6.67 \times 10^{-11} \text{ N}\cdot\text{m}^2/\text{kg}^2\right)\left(7.35 \times 10^{22} \text{ kg}\right)\left(5.98 \times 10^{24} \text{ kg}\right)}{\left(3.84 \times 10^8 \text{ m}\right)^2}$$

$$= 1.99 \times 10^{20} \text{ N}.$$

The Sun is 1.50×10^8 km from the Earth and the Moon, so F_{MS} (the force on the Moon due to the Sun) is

$$F_{MS} = \frac{\left(6.67 \times 10^{-11} \text{ N}\cdot\text{m}^2/\text{kg}^2\right)\left(7.35 \times 10^{22} \text{ kg}\right)\left(1.99 \times 10^{30} \text{ kg}\right)}{\left(1.50 \times 10^{11} \text{ m}\right)^2}$$

$$= 4.34 \times 10^{20} \text{ N}.$$

Since the two forces act at right angles in the case we are considering (Fig. 6–5), the total force is

$$F = \sqrt{(1.99)^2 + (4.34)^2} \times 10^{20} \text{ N} = 4.77 \times 10^{20} \text{ N}$$

which acts at an angle $\theta = \tan^{-1}(1.99/4.34) = 24.6°$.

6–2 Vector Form of Newton's Law of Universal Gravitation

We can write Newton's law of universal gravitation in vector form as

$$\mathbf{F}_{12} = -G \frac{m_1 m_2}{r_{21}^2} \hat{\mathbf{r}}_{21}, \tag{6–2}$$

where $\mathbf{F}_{12}$ is the vector force on particle 1 $\left(\text{of mass } m_1\right)$ exerted by particle 2 $\left(\text{of mass } m_2\right)$, which is a distance r_{21} away; $\hat{\mathbf{r}}_{21}$ is a unit vector that points from particle 2 toward particle 1 along the line joining them so that $\hat{\mathbf{r}}_{21} = \mathbf{r}_{21}/r_{21}$, where $\mathbf{r}_{21}$ is the displacement vector as shown in Fig. 6–6. The minus sign in Eq. 6–2 is necessary because the force on particle 1 due to particle 2 points toward m_2, in the direction opposite to $\hat{\mathbf{r}}_{21}$. The displacement vector $\mathbf{r}_{12}$ is a vector of the same magnitude as $\mathbf{r}_{21}$, but it points in the opposite direction so that

$$\mathbf{r}_{12} = -\mathbf{r}_{21}.$$

By Newton's third law, the force $\mathbf{F}_{21}$ acting on m_2 exerted by m_1 must have the same magnitude as $\mathbf{F}_{12}$ but acts in the opposite direction (Fig. 6–7), so that

$$\mathbf{F}_{21} = -\mathbf{F}_{12} = G\frac{m_1 m_2}{r_{21}^2}\,\hat{\mathbf{r}}_{21}$$

$$= -G\frac{m_2 m_1}{r_{12}^2}\,\hat{\mathbf{r}}_{12}.$$

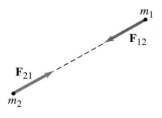

FIGURE 6–7 By Newton's third law, the gravitational force on particle 1 exerted by particle 2, $\mathbf{F}_{12}$, is equal and opposite to that on particle 2 exerted by particle 1, $\mathbf{F}_{21}$; that is $\mathbf{F}_{21} = -\mathbf{F}_{12}$.

The force of gravity exerted on one particle by a second particle is always directed toward the second particle, as in Fig. 6–6. When many particles interact, the total gravitational force on a given particle is the vector sum of the forces exerted by each of the others. For example, the total force on particle number 1 is

$$\mathbf{F}_1 = \mathbf{F}_{12} + \mathbf{F}_{13} + \mathbf{F}_{14} + \cdots + \mathbf{F}_{1n} = \sum_{i=2}^{n} \mathbf{F}_{1i} \qquad (6\text{–}3)$$

where $\mathbf{F}_{1i}$ means the force on particle 1 exerted by particle i, and n is the total number of particles.

This vector notation can be very helpful, especially when sums over many particles are needed. However, in many cases we don't have to be so formal and we can deal with directions by making careful diagrams.

6–3 Gravity Near the Earth's Surface; Geophysical Applications

When Eq. 6–1 is applied to the gravitational force between the Earth and an object at its surface, m_1 becomes the mass of the Earth, m_E, m_2 becomes the mass of the object, m, and r becomes the distance of the object from the Earth's center,[†] which is the radius of the Earth r_E. This force of gravity due to the Earth is the weight of the object, which we have been writing as mg. Thus,

$$mg = G\frac{m m_E}{r_E^2}.$$

Hence

$$g = G\frac{m_E}{r_E^2}. \qquad (6\text{–}4)$$

g in terms of G

Thus, the acceleration of gravity at the surface of the Earth, g, is determined by m_E and r_E. (Be careful not to confuse G with g; they are very different quantities, but are related by Eq. 6–4.)

Until G was measured, the mass of the Earth was not known. But once G was measured, Eq. 6–4 could be used to calculate the Earth's mass, and Cavendish was the first to do so. Since $g = 9.80\ \text{m/s}^2$ and the radius of the Earth is $r_E = 6.38 \times 10^6\ \text{m}$, then, from Eq. 6–4, we obtain

$$m_E = \frac{g r_E^2}{G} = \frac{(9.80\ \text{m/s}^2)(6.38 \times 10^6\ \text{m})^2}{6.67 \times 10^{-11}\ \text{N}\cdot\text{m}^2/\text{kg}^2} = 5.98 \times 10^{24}\ \text{kg}$$

Mass of the Earth

for the mass of the Earth.

When dealing with the weight of objects at the surface of the Earth, we can continue to use simply mg. If we wish to calculate the force of gravity on an object some distance from the Earth, or the force due to some other heavenly body, such as that exerted by the Moon or a planet, we can calculate the effective value of g from Eq. 6–4, replacing r_E (and m_E) by the appropriate distance (and mass), or we can use Eq. 6–1 directly.

[†]That the distance is measured from the Earth's center does not imply that the force of gravity somehow emanates from that one point. Rather, all parts of the Earth attract gravitationally, but the net effect is a force acting toward the Earth's center. (See Appendix C.)

EXAMPLE 6–4 **ESTIMATE** **Gravity on Everest.** Estimate the effective value of g on the top of Mt. Everest, 8848 m (29,028 ft) above the Earth's surface. That is, what is the acceleration due to gravity of objects allowed to fall freely at this altitude?

SOLUTION Let us call the acceleration of gravity at the given point g'. We use Eq. 6–4, with r_E replaced by $r = 6380 \text{ km} + 8.8 \text{ km} = 6389 \text{ km} = 6.389 \times 10^6 \text{ m}$:

$$g' = G\frac{m_E}{r^2} = \frac{(6.67 \times 10^{-11} \text{ N·m}^2/\text{kg}^2)(5.98 \times 10^{24} \text{ kg})}{(6.389 \times 10^6 \text{ m})^2} = 9.77 \text{ m/s}^2$$

which is a reduction of about 3 parts in a thousand (0.3%). However, we ignored the mass accumulated under the mountain top.

Note that Eq. 6–4 does not give precise values for g at different locations because the Earth is not a perfect sphere. The Earth not only has mountains and valleys, and bulges at the equator, but its mass is not distributed precisely uniformly. See Table 6–1. The Earth's rotation also has an effect on the value of g, as discussed in Example 6–5 below.

The value of g can vary locally on the Earth's surface because of the presence of irregularities and rocks of different densities. Such variations in g, known as "gravity anomalies," are very small—on the order of 1 part per 10^6 or 10^7 in the value of g. But they can be measured ("gravimeters" today can detect variations in g to 1 part in 10^9). Geophysicists use such measurements as part of their investigations into the structure of the Earth's crust, and in mineral and oil exploration. Mineral deposits, for example, often have a greater density than surrounding material; because of the greater mass in a given volume, g can have a slightly greater value on top of such a deposit than at its flanks. "Salt domes," under which petroleum is often found, have a lower than average density and searches for a slight reduction in the value of g in certain locales have led to the discovery of oil.

EXAMPLE 6–5 **Effect of Earth's rotation on g.** Assuming the Earth is a perfect sphere, determine how the Earth's rotation affects the value of g at the equator compared to its value at the poles.

SOLUTION Figure 6–8 shows a mass m hanging from a spring scale at two places on the Earth. At the North Pole there are two forces acting on the mass m: the force of gravity, $\mathbf{F}_G = mg$, and the force with which the spring pulls up on the mass, $\mathbf{w}$. We call this latter force w because it is what the scale reads as the weight of the object; by Newton's third law it equals the force with which the mass pulls down on the spring. Since the mass is not accelerating, Newton's second law tells us

$$mg - w = 0,$$

so $w = mg$. Thus the weight w that the spring registers equals mg, which is no surprise. Next, at the equator, there *is* an acceleration because the Earth is rotating. The same force of gravity $\mathbf{F}_G = mg$ acts downward (we are letting g represent the acceleration of gravity in the absence of rotation and we ignore the slight bulging of the equator). The spring pulls upward with a force w'; w' is also the force with which the mass pulls on the spring (Newton's third law) and hence is the weight registered on the spring scale. From Newton's second law we now have (see Fig. 6–8)

$$mg - w' = m\frac{v^2}{r_E}$$

where $r_E = 6.38 \times 10^6 \text{ m}$ is the Earth's radius and v is the speed of m due to the Earth's daily rotation [1 day = (24 h)(3600 s/h) = 8.64×10^4 s] which equals

$$v = 2\pi r_E/1 \text{ day} = (6.28)(6.38 \times 10^6 \text{ m})/(8.64 \times 10^4 \text{ s}) = 4.64 \times 10^2 \text{ m/s}.$$

TABLE 6–1
Acceleration Due to Gravity at Various Locations on Earth

Location	Elevation (m)	g (m/s²)
New York	0	9.803
San Francisco	100	9.800
Denver	1650	9.796
Pikes Peak	4300	9.789
Equator	0	9.780
North Pole (calculated)	0	9.832

➡ **PHYSICS APPLIED**

Geology—mineral and oil exploration

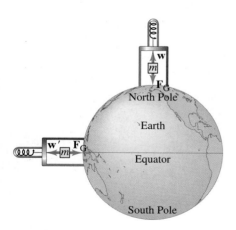

FIGURE 6–8 Example 6–5.

The effective weight is w', so the effective value of g, which we call g', is $g' = w'/m$. Solving the equation above for w', we have

$$w' = m(g - v^2/r_E),$$

so

$$g' = \frac{w'}{m} = g - \frac{v^2}{r_E}.$$

Hence

$$\Delta g = g - g' = v^2/r_E = (4.64 \times 10^2 \,\text{m/s})^2 / (6.38 \times 10^6 \,\text{m}) = 0.0337 \,\text{m/s}^2.$$

In Table 6–1 we see that the difference is actually greater than this: $(9.832 - 9.780) \,\text{m/s}^2 = 0.052 \,\text{m/s}^2$; this discrepancy is due mainly to the Earth's being slightly fatter at the equator (by 21 km) than at the poles. The calculation of the effective value of g at latitudes other than at the poles or equator is a two-dimensional problem because $\mathbf{F}_G$ acts radially toward the Earth's center whereas the centripetal acceleration is directed perpendicular to the axis of rotation, parallel to the equator and that means that a plumb line (the effective direction of g) is not precisely vertical except at the equator and the poles.

6–4 Satellites and "Weightlessness"

Artificial satellites circling the Earth are now commonplace (Fig. 6–9). A satellite is put into orbit by accelerating it to a sufficiently high tangential speed with the use of rockets, as shown in Fig. 6–10. If the speed is too high, the spacecraft will not be confined by the Earth's gravity and will escape, never to return. If the speed is too low, it will return to Earth. Satellites are usually put into circular (or nearly circular) orbits, because they require the least takeoff speed. It is sometimes asked: "What keeps a satellite up?" The answer is: its high speed. If a satellite stopped moving, it would, of course, fall directly to Earth. But at the very high speed a satellite has, it would quickly fly out into space (Fig. 6–11) if it weren't for the gravitational force of the Earth pulling it into orbit. In fact, a satellite *is* falling (accelerating toward Earth), but its high tangential speed keeps it from hitting Earth.

➡ P H Y S I C S   A P P L I E D

Artificial Earth satellites

FIGURE 6–9
A satellite circling the Earth.

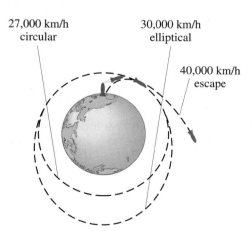

FIGURE 6–10 Artificial satellites launched at different speeds.

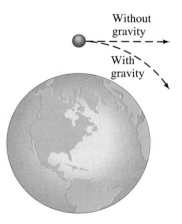

FIGURE 6–11 A moving satellite "falls" out of a straight-line path toward the Earth.

For satellites that move in a circle (at least approximately), the needed acceleration is v^2/r. The force that gives a satellite this acceleration is the force of gravity, and since a satellite may be at a considerable distance from the Earth, we must use Eq. 6–1 for the force acting on it. When we apply Newton's second law, $\Sigma F_R = ma_R$, we find

$$G\frac{mm_E}{r^2} = m\frac{v^2}{r},$$ (6–5)

where m is the mass of the satellite. This equation relates the distance of the satellite from the Earth's center, r, to its speed, v. Note that only one force—gravity—is acting on the satellite, and that r is the sum of the Earth's radius r_E plus the satellite's height h above the Earth: $r = r_E + h$.

EXAMPLE 6–6 **Geosynchronous satellite.** A *geosynchronous* satellite is one that stays above the same point on the Earth, which is possible only if it is above a point on the equator. Such satellites are used for such purposes as cable TV transmission, for weather forecasting, and as communication relays. Determine (a) the height above the Earth's surface such a satellite must orbit and (b) such a satellite's speed. (c) Compare to the speed of a satellite orbiting 200 km above Earth's surface.

SOLUTION (a) The only force on the satellite is gravity, so we apply Eq. 6–5 assuming the satellite moves in a circle:

$$G\frac{m_{Sat}m_E}{r^2} = m_{Sat}\frac{v^2}{r}.$$

This equation seems to have two unknowns, r and v. But we know that v must be such that the satellite revolves around the Earth with the same period that the Earth rotates on its axis, namely once in 24 hours. Thus the speed of the satellite must be

$$v = \frac{2\pi r}{T}$$

where $T = 1$ day $= (24\,h)(3600\,s/h) = 86,400\,s$. We put this into the first equation above and obtain (after canceling m_{Sat} on both sides):

$$G\frac{m_E}{r^2} = \frac{(2\pi r)^2}{rT^2}.$$

We solve for r:

$$r^3 = \frac{Gm_E T^2}{4\pi^2} = \frac{(6.67 \times 10^{-11}\,N\cdot m^2/kg^2)(5.98 \times 10^{24}\,kg)(86,400\,s)^2}{4\pi^2}$$

$$= 7.54 \times 10^{22}\,m^3,$$

and, taking the cube root, $r = 4.23 \times 10^7$ m, or 42,300 km from the Earth's center. We subtract the Earth's radius of 6380 km to find that the satellite must orbit about 36,000 km (about 6 r_E) above the Earth's surface.
(b) We solve Eq. 6–5 for v:

$$v = \sqrt{\frac{Gm_E}{r}} = \sqrt{\frac{(6.67 \times 10^{-11}\,N\cdot m^2/kg^2)(5.98 \times 10^{24}\,kg)}{(4.23 \times 10^7\,m)}} = 3070\,m/s.$$

We get the same result if we use $v = 2\pi r/T$.
(c) The last equation above for v shows $v \propto \sqrt{1/r}$. So for $r = r_E + h = 6380\,km + 200\,km = 6580\,km$, we get

$$v' = v\sqrt{r/r'} = (3070\,m/s)\sqrt{(42,300\,km)/(6580\,km)} = 7780\,m/s.$$

CONCEPTUAL EXAMPLE 6–7 **Catching a satellite.** You are an astronaut in the space shuttle pursuing a satellite in need of repair. You find yourself in a circular orbit of the same radius as the satellite, but 30 km behind it. How will you catch up with it?

RESPONSE We saw in Example 6–6 (or see Eq. 6–5) that the velocity is proportional to $1/\sqrt{r}$. Thus you need to aim for a smaller orbit and at the same time increase your velocity. Note that you cannot just increase your speed without changing your orbit.

People and other objects in a satellite circling the Earth are said to experience apparent weightlessness. Before tackling the case of a satellite, however, let us first look at the simpler case of a falling elevator. In Fig. 6–12a, we see an elevator at rest with a bag hanging from a spring scale. The scale reading indicates the downward force exerted on it by the bag. This force, exerted *on* the scale, is equal and opposite to the force exerted *by* the scale upward on the bag. We call this force **w**. Since the mass, m, is not accelerating, we apply $\Sigma F = ma$ to the bag and obtain

$$w - mg = 0,$$

where mg is the weight of the bag. Thus, $w = mg$, and since the scale indicates the force w exerted on it by the bag, it registers a force equal to the weight of the bag as we expect. If, now, the elevator has an acceleration, a, then applying $\Sigma F = ma$ to the bag, we have

$$w - mg = ma.$$

Solving for w, we have

$$w = mg + ma.$$

We have chosen the positive direction up. Thus, if the acceleration a is up, a is positive; and the scale, which measures w, will read more than mg. We call w the *apparent weight* of the bag, which here is greater than its actual weight (mg). If the elevator accelerates downward, a will be negative and w, the apparent weight, will be less than mg. Note that the direction of the velocity v doesn't matter. Only the direction of the acceleration **a** influences the scale reading.

If, for example, the elevator's acceleration is $\frac{1}{2}g$ upward, then we find $w = mg + m(\frac{1}{2}g) = \frac{3}{2}mg$. That is, the scale reads $1\frac{1}{2}$ times the actual weight (Fig. 6–12b). The apparent weight of the bag is $1\frac{1}{2}$ times its real weight. The same is true of the person: her apparent weight (equal to the normal force exerted on her by the elevator floor) is $1\frac{1}{2}$ times her real weight. We can say that she is experiencing $1\frac{1}{2}g$'s, just as astronauts experience so many g's at a rocket's launch.

If, instead, the elevator's acceleration is $-\frac{1}{2}g$ (downward), then $w = mg - \frac{1}{2}mg = \frac{1}{2}mg$. That is, the scale reads one half the actual weight. If the elevator is in *free fall* (for example, if the cables break), then $a = -g$ and $w = mg - mg = 0$. The scale reads zero! See Fig. 6–12c. The bag seems weightless.

"Weightlessness" in a falling elevator

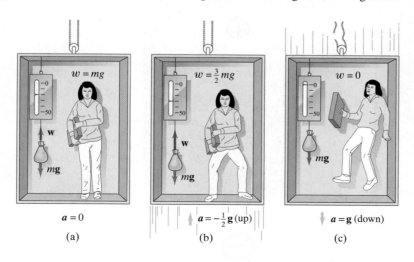

$$a = 0 \qquad\qquad a = -\frac{1}{2}\mathbf{g}\ (\text{up}) \qquad\qquad a = \mathbf{g}\ (\text{down})$$

(a)   (b)   (c)

FIGURE 6–12 (a) An object in an elevator at rest exerts a force on a spring scale equal to its weight. (b) In an elevator accelerating upward at $\frac{1}{2}g$, the object's apparent weight is $1\frac{1}{2}$ times larger. (c) In a freely falling elevator, the object experiences "weightlessness": the scale reads zero.

If the person in the elevator let go of a pencil, say, it would not fall to the floor. True, the pencil would be falling with acceleration g. But so would the floor of the elevator and the person. The pencil would hover right in front of the person. This phenomenon is called *apparent weightlessness* because, in fact, gravity is still acting on the object and its weight is still mg. The objects seem weightless only because the elevator is in free fall.

The "weightlessness" experienced by people in a satellite orbit close to the Earth (Fig. 6–13) is the same apparent weightlessness experienced in a freely falling elevator. It may seem strange, at first, to think of a satellite as freely falling. But a satellite is indeed falling toward the Earth, as was shown in Fig. 6–11. The force of gravity causes it to "fall" out of its natural straight-line path. The acceleration of the satellite must be the acceleration due to gravity at that point, since the only force acting on it is gravity. (We used this to obtain Eq. 6–5.) Thus, although the force of gravity acts on objects within the satellite, the objects experience an apparent weightlessness because they, and the satellite, are accelerating as in free fall.

Figure 6–14 shows some examples of "free fall," or apparent weightlessness, experienced by people on Earth for brief moments.

A completely different situation occurs when a spacecraft is out in space far from the Earth, the Moon, and other attracting bodies. The force of gravity due to the Earth and other heavenly bodies will then be quite small because of the distances involved, and persons in such a spacecraft will experience real weightlessness.

The effects on human beings of weightlessness (whether real or apparent makes no difference) are interesting. In ordinary circumstances, for example, people can become quite tired holding out their arms horizontally. But for a person experiencing weightlessness, no effort is needed. The arms will just "float" there, since there is no sensation of weight. This effect has many applications in athletics (Fig. 6–14). During a jump or a dive, while on a trampoline, and even between strides while running, a person is experiencing apparent weightlessness or free fall, although only for a short time. During these brief periods, limbs can be moved much more easily, since only inertia needs to be overcome. The loss of control because of lack of contact with the ground is compensated for by the increased mobility. Prolonged weightlessness in space, however, can have harmful effects on health. Red blood cells diminish in number, blood collects in the thorax, bones lose calcium and become brittle, and muscles lose their tone. These effects are being carefully studied.

"Weightlessness" in a satellite

FIGURE 6–13 This astronaut is making a repair to the Hubble Space Telescope. He must feel very free because he is experiencing apparent weightlessness.

FIGURE 6–14 Experiencing weightlessness on Earth.

(a)   (b)   (c)

6–5 Kepler's Laws and Newton's Synthesis

More than a half century before Newton proposed his three laws of motion and his law of universal gravitation, the German astronomer Johannes Kepler (1571–1630) had written a number of astronomical works in which we can find a detailed description of the motion of the planets about the Sun. Kepler's work resulted in part from the many years he spent examining data collected by Tycho Brahe (1546–1601) on the positions of the planets in their motion through the heavens. Among Kepler's writings were three empirical findings that we now refer to as **Kepler's laws of planetary motion**. These are summarized as follows, with additional explanation in Figs. 6–15 and 6–16.

Kepler's first law: The path of each planet about the Sun is an ellipse with the Sun at one focus (Fig. 6–15).

Kepler's second law: Each planet moves so that an imaginary line drawn from the Sun to the planet sweeps out equal areas in equal periods of time (Fig. 6–16).

Kepler's third law: The ratio of the squares of the periods of any two planets revolving about the Sun is equal to the ratio of the cubes of their semimajor axes. [The semimajor axis is half the long (major) axis of the orbit, as shown in Fig. 6–15, and represents the planet's average distance from the Sun.[†]] That is, if T_1 and T_2 represent the periods (the time needed for one revolution about the Sun) for any two planets, and s_1 and s_2 represent their semimajor axes, then

$$\left(\frac{T_1}{T_2}\right)^2 = \left(\frac{s_1}{s_2}\right)^3.$$

We can rewrite this as

$$\frac{s_1^3}{T_1^2} = \frac{s_2^3}{T_2^2},$$

meaning that s^3/T^2 should be the same for each planet. Present-day data are given in Table 6–2; see the last column.

[†] The semimajor axis is equal to the planet's average distance from the Sun in the sense that it equals the average of the planet's nearest and farthest distances from the Sun (points Q and R in Fig. 6–15). Most planetary orbits are close to circles, and for a circle the semimajor axis is the radius of the circle.

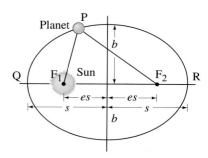

FIGURE 6–15 (a) *Kepler's first law.* An ellipse is a closed curve such that the sum of the distances from any point P on the curve to two fixed points (called the foci, F_1 and F_2) remains constant. That is, the sum of the distances, $F_1 P + F_2 P$, is the same for all points on the curve. A circle is a special case of an ellipse in which the two foci coincide, at the center of the circle. The semimajor axis is s (that is, the long axis is $2s$) and the semiminor axis is b, as shown. The *eccentricity*, e, is defined so that es is the distance from the center to either focus. The Earth and most of the other planets have nearly circular orbits. For Earth $e = 0.017$.

FIGURE 6–16 *Kepler's second law.* The two shaded regions have equal areas. The planet moves from point 1 to point 2 in the same time as it takes to move from point 3 to point 4. Planets move fastest in that part of their orbit where they are closest to the Sun. Exaggerated scale.

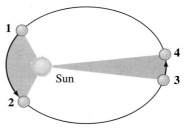

TABLE 6–2	Planetary Data Applied to Kepler's Third Law		
Planet	**Average Distance from Sun, s (10^6 km)**	**Period, T (Earth years)**	**s^3/T^2 (10^{24} km^3/y^2)**
Mercury	57.9	0.241	3.34
Venus	108.2	0.615	3.35
Earth	149.6	1.0	3.35
Mars	227.9	1.88	3.35
Jupiter	778.3	11.86	3.35
Saturn	1427	29.5	3.34
Uranus	2870	84.0	3.35
Neptune	4497	165	3.34
Pluto	5900	248	3.33

Kepler arrived at his laws through careful analysis of experimental data. Fifty years later, Newton was able to show that Kepler's laws could be derived mathematically from the law of universal gravitation and the laws of motion. He also showed that for any reasonable form for the gravitational force law, only one that depends on the inverse square of the distance is fully consistent with Kepler's laws. He thus used Kepler's laws as evidence in favor of his law of universal gravitation, Eq. 6–1.

Derivation of Kepler's third law   We will derive Kepler's second law in Chapter 11 when we study angular momentum. Here we derive Kepler's third law, and we do it for the special case of a circular orbit, in which case the semimajor axis is the radius r of the circle. (Note that most planetary orbits are close to a circle.) First, we write down Newton's second law of motion $\Sigma F = ma$. Then for F we substitute the law of universal gravitation, Eq. 6–1, and for a the centripetal acceleration, v^2/r:

$$\Sigma F = ma$$

$$G \frac{m_1 M_S}{r_1^2} = m_1 \frac{v_1^2}{r_1}.$$

Here m_1 is the mass of a particular planet, r_1 its distance from the Sun, and v_1 its average speed in orbit; M_S is the mass of the Sun, since it is the gravitational attraction of the Sun that keeps each planet in its orbit. The period T_1 of the planet is the time required for one complete orbit, a distance equal to $2\pi r_1$, the circumference of a circle, so

$$v_1 = \frac{2\pi r_1}{T_1}.$$

We substitute this formula for v_1 into the equation above:

$$G \frac{m_1 M_S}{r_1^2} = m_1 \frac{4\pi^2 r_1}{T_1^2}.$$

We rearrange this to get

$$\frac{T_1^2}{r_1^3} = \frac{4\pi^2}{GM_S}. \tag{6–6}$$

We derived this for planet 1 (say, Mars). The same derivation would apply for a second planet (say, Saturn):

$$\frac{T_2^2}{r_2^3} = \frac{4\pi^2}{GM_S},$$

where T_2 and r_2 are the period and orbit radius, respectively, for the second planet. Since the right sides of the two previous equations are equal, we have $T_1^2/r_1^3 = T_2^2/r_2^3$ or, rearranging,

Kepler's third law
$$\left(\frac{T_1}{T_2}\right)^2 = \left(\frac{r_1}{r_2}\right)^3, \tag{6–7}$$

which is Kepler's third law. Equations 6–6 and 6–7 are valid also for elliptical orbits if we replace r with the semimajor axis s.

The derivations of Eqs. 6–6 and 6–7 (Kepler's third law) are general enough to be applied to other systems. For example, we could determine the mass of the Earth from Eq. 6–6 using the period of the Moon about the Earth and the Moon's distance from the Earth, or the mass of Jupiter from the period and distance of one of its moons (this is indeed how masses are determined; see the Problems). We can also use Eqs. 6–6 and 6–7 to compare objects that orbit other attracting centers, such as the Moon and a weather satellite orbiting Earth. But be careful not to use Eq. 6–7 to compare, say, the Moon's orbit around the Earth to the orbit of Mars around the Sun because they depend on different attracting centers.

In the following examples, we assume the orbits are circles, although it is not quite true in general.

EXAMPLE 6–8 **Where is Mars?** Mars' period (its "year") was first noted by Kepler to be about 687 days (Earth-days), which is $(687\,d/365\,d) = 1.88\,\text{yr}$. Determine the distance of Mars from the Sun using the Earth as a reference.

SOLUTION The period of the Earth is $T_E = 1\,\text{yr}$, and the distance of Earth from the Sun is $r_{ES} = 1.50 \times 10^{11}\,\text{m}$. From Kepler's third law (Eq. 6–7):

$$\frac{r_{MS}}{r_{ES}} = \left(\frac{T_M}{T_E}\right)^{\frac{2}{3}} = \left(\frac{1.88\,\text{yr}}{1\,\text{yr}}\right)^{\frac{2}{3}} = 1.52.$$

So Mars is 1.52 times the Earth's distance from the Sun, or $2.28 \times 10^{11}\,\text{m}$.

EXAMPLE 6–9 **The Sun's mass determined.** Determine the mass of the Sun given the Earth's distance from the Sun as $r_{ES} = 1.5 \times 10^{11}\,\text{m}$.

➡ **PHYSICS APPLIED**

Determining the Sun's mass

SOLUTION We can use Eq. 6–6 and solve for M_S:

$$M_S = \frac{4\pi^2 r_{ES}^3}{G T_E^2} = \frac{4\pi^2 (1.5 \times 10^{11}\,\text{m})^3}{(6.67 \times 10^{-11}\,\text{N} \cdot \text{m}^2/\text{kg}^2)(3.16 \times 10^7\,\text{s})^2} = 2.0 \times 10^{30}\,\text{kg}$$

where we used the fact that

$$T_E = 1\,\text{yr} = (365\tfrac{1}{4}\,d)(24\,\text{h/d})(3600\,\text{s/h}) = 3.16 \times 10^7\,\text{s}.$$

EXAMPLE 6–10 **ESTIMATE** **Geosynchronous satellite, simplified.** A *geosynchronous* satellite of the Earth (as mentioned in Example 6–6) is one that stays above the same point on the equator of the Earth. Estimate the height above the Earth's surface needed for a geosynchronous weather satellite. (This is to be a "lunchtime" calculation done on a napkin, without calculator, as compared to our earlier calculation in Example 6–6.)

SOLUTION To use Kepler's third law we must compare the satellite to some other object that orbits Earth. The simplest choice is the Moon because we know its period and distance. The Moon's period is about $T_M \approx 27\,d$ and its distance from the Earth about $r_{ME} \approx 380,000\,\text{km}$. The period of the weather satellite needs to be $T_{Sat} = 1\,d$ so that it stays above the same place on the Earth. Hence,

$$r_{Sat} = r_{ME}\left(\frac{T_{Sat}}{T_M}\right)^{\frac{2}{3}} = r_{ME}\left(\frac{1\,d}{27\,d}\right)^{\frac{2}{3}} = r_{ME}\left(\frac{1}{3}\right)^2 = \frac{r_{ME}}{9}.$$

(How nice the Moon's approximate period turns out to be a perfect cube.) A geosynchronous satellite must be $\frac{1}{9}$ the distance to the Moon, which is 42,000 km from the center of the Earth or 36,000 km above the Earth's surface. This is about 6 Earth radii high.

Accurate measurements on the orbits of the planets indicated that they did not precisely follow Kepler's laws. For example, slight deviations from perfectly elliptical orbits were observed. Newton was aware that this was to be expected from the law of universal gravitation ("every body in the universe attracts every other body…") because each planet exerts a gravitational force on the other planets. Since the mass of the Sun is much greater than that of any planet, the force on one planet due to any other planet will be small in comparison to the force on it due to the Sun. (The derivation of perfectly elliptical orbits ignores the forces due to other planets.) But because of this small force, each planetary orbit should depart from a perfect ellipse, especially when a second planet is fairly close to it. Such deviations, or **perturbations**, as they are called, from perfect ellipses are indeed observed. In fact, Newton's recognition of perturbations in the orbit of Saturn was a hint that helped him formulate the law of universal gravitation, that all

➡ **PHYSICS APPLIED**

Perturbations and discovery of planets

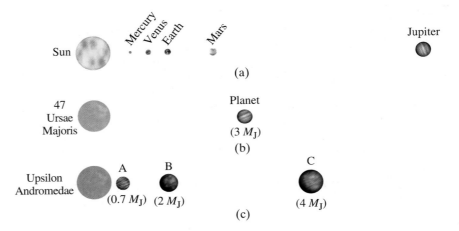

FIGURE 6–17 (a) Our solar system, compared to recently discovered planets orbiting (b) the star 47 Ursae Majoris, and (c) the star Upsilon Andromedae with at least three planets. M_J is the mass of Jupiter. (Not to scale.)

bodies attract gravitationally. Observation of other perturbations later led to the discovery of Neptune and Pluto. Deviations in the orbit of Uranus, for example, could not all be accounted for by perturbations due to the other known planets. Careful calculation in the nineteenth century indicated that these deviations could be accounted for if there were another planet farther out in the solar system. The position of this planet was predicted from the deviations in the orbit of Uranus, and telescopes focused on that region of the sky quickly found it; the new planet was called Neptune. Similar but much smaller perturbations of Neptune's orbit led to the discovery of Pluto in 1930.

More recently, in 1996, planets revolving about distant stars (Fig. 6–17) were inferred from the regular "wobble" of each star due to the gravitational attraction of the revolving planet.

The development by Newton of the law of universal gravitation and the three laws of motion was a major intellectual achievement. For with these laws, Newton was able to describe the motion of objects on Earth and in the heavens. The motions of heavenly bodies and bodies on Earth were seen to follow the same laws (something not previously recognized generally, although Galileo and Descartes had argued in its favor). For this reason, and also because Newton integrated the results of earlier workers into his system, we sometimes speak of Newton's "synthesis."

Newton's work was so encompassing that it constituted a theory of the universe, and influenced philosophy and other fields. The laws formulated by Newton are referred to as **causal laws**. By **causality** we mean the idea that one occurrence can cause another. We have repeatedly observed, for example, that when a rock strikes a window, the window almost immediately breaks. We infer that the rock *caused* the window to break. This idea of "cause and effect" took on more forceful meaning with Newton's laws. For the motion—or rather the acceleration—of any object was seen to be *caused* by the net force acting on it. As a result, the universe came to be pictured by many scientists and philosophers as a big machine whose parts move in a predetermined way—according to natural laws. However, this *deterministic* view of the universe had to be modified by scientists in the twentieth century.

6–6 Gravitational Field

Most of the forces we meet in everyday life are contact forces: you push or pull on a lawn mower, a tennis racket exerts a force on a tennis ball when they make contact, or a ball exerts a force on a window when they make contact. But the gravitational force (and even the electromagnetic force, as we shall see later) acts over a distance: there is a force even when the two objects are not in contact. The Earth, for example, exerts a force on a falling apple. It also exerts a force on the Moon, 384,000 km away. And the Sun exerts a gravitational force on the Earth. The idea of a force *acting at a distance* was a difficult one for early thinkers. Newton himself felt uneasy with this concept when he published his law of universal gravitation.

Another point of view that avoids some of these conceptual difficulties is the concept of the **field**, developed in the nineteenth century by Michael Faraday (1791–1867) to aid understanding of electromagnetism. Only later was it applied to gravity. According to the field concept, a **gravitational field** surrounds every body that has mass, and this field permeates all of space. A second body at a particular location near the first body experiences a force because of the gravitational field that exists there. Because the gravitational field at the location of the second mass is considered to act directly on this mass, we are a little closer to the idea of a contact force. *Field*

To be quantitative, we can define the **gravitational field** as the gravitational force per unit mass at any point in space. If we want to measure the gravitational field at any point, we place a small "test" mass m at that point and measure the force $\mathbf{F}$ exerted on it (making sure only gravitational forces are acting). Then the gravitational field, $\mathbf{g}$, at that point is defined as

$$\mathbf{g} = \frac{\mathbf{F}}{m}.$$ (6–8)   *Gravitational field (defined)*

The units of $\mathbf{g}$ are N/kg.

It is clear that the gravitational field which an object experiences has magnitude equal to the acceleration due to gravity at that point. (When we speak of acceleration, however, we use units m/s², which is equivalent to N/kg, since $1\,\mathrm{N} = 1\,\mathrm{kg \cdot m/s^2}$.)

If the gravitational field is due to a single body of mass M (such as when m is near the Earth's surface), then the gravitational field at a distance r from M has magnitude

$$g = \frac{1}{m} G \frac{mM}{r^2} = G \frac{M}{r^2}.$$

In vector notation we write

$$\mathbf{g} = -\frac{GM}{r^2}\,\hat{\mathbf{r}},$$ *Gravitational field due to a mass M*

where $\hat{\mathbf{r}}$ is a unit vector pointing radially outward from mass M, and the minus sign reminds us that the field points toward mass M (see Eqs. 6–1, 6–2, and 6–4). If several different bodies contribute significantly to the gravitational field, then we write $\mathbf{g}$ as the vector sum of all these contributions. In interplanetary space, for example, $\mathbf{g}$ at any point in space is the vector sum of terms due to the Earth, Sun, Moon, and other bodies that contribute. The gravitational field $\mathbf{g}$ at any point in space does not depend on the value of our test mass, m, placed at that point; $\mathbf{g}$ depends only on the masses (and locations) of the bodies that create the field there.

6–7 Types of Forces in Nature

We have already discussed that Newton's law of universal gravitation, Eq. 6–1, describes how a particular type of force—gravity—depends on the distance between, and masses of, the objects involved. Newton's second law, $\Sigma \mathbf{F} = m\mathbf{a}$, on the other hand, tells how a body will accelerate due to *any* type of force. But what are the types of forces that occur in nature besides gravity?

In the twentieth century, physicists came to recognize four different fundamental forces in nature: (1) the gravitational force; (2) the electromagnetic force (we shall see later that electric and magnetic forces are intimately related); (3) the strong nuclear force; and (4) the weak nuclear force. In this chapter, we discussed the gravitational force in detail. The nature of the electromagnetic force will be discussed in detail in Chapters 21 to 32. The strong and weak nuclear forces operate at the level of the atomic nucleus, and although they manifest themselves in such phenomena as radioactivity and nuclear energy, they are much less obvious in our daily lives.

Electroweak and GUT

Physicists have been working on theories that would unify these four forces—that is, to consider some or all of these forces as different manifestations of the same basic force. So far, the electromagnetic and weak nuclear forces have been theoretically united to form a successful *electroweak* theory, in which the electromagnetic and weak forces are seen as two different manifestations of a single *electroweak force*. Attempts to further unify the forces, such as in *grand unified theories* (GUT), are hot research topics today.

Contact forces

But where do everyday forces fit into this scheme? Ordinary forces, such as pushes, pulls, and other contact forces like the normal force and friction, are today considered to be due to the electromagnetic force acting at the atomic level. For example, the force your fingers exert on a pencil is the result of electrical repulsion between the outer electrons of the atoms of your finger and those of the pencil.

* 6–8 Gravitational Versus Inertial Mass; the Principle of Equivalence

We have dealt with two aspects of mass. In Chapter 4, we defined mass as a measure of the inertia of a body. Newton's second law relates the force acting on a body to its acceleration and its **inertial mass**, as we call it. We might say that inertial mass represents a resistance to any force. In this chapter we have dealt with mass as a property related to the gravitational force—that is, mass as a quantity that determines the strength of the gravitational force between two bodies. This we call the **gravitational mass**.

Now it is not at all obvious that the inertial mass of a body should be equal to its gravitational mass. (The force of gravity might have depended on a completely different property of a body, just as the electrical force depends on a property called electric charge.) Newton's and Cavendish's experiments indicated that these two types of mass are equal for a body, and modern experiments confirm it to a precision of about 1 part in 10^{12}.

Principle of equivalence

The experimental evidence that gravitational and inertial masses are equal (or at least proportional) is remarkable. This equivalence between gravitational and inertial masses was raised by Albert Einstein (1879–1955) to a principle of nature. Einstein called it simply the **principle of equivalence**, and he used it as a foundation for his general theory of relativity (c. 1916). The principle of equivalence, as Einstein gave it, can be stated in another way: there is no experiment observers can perform to distinguish if an acceleration arises because of a gravitational force or because their reference frame is accelerating. If, for example, you were far out in space and an apple fell to the floor of your spacecraft, you might assume a gravitational force was acting on the apple. On the other hand, it would also be possible that the apple fell because your spacecraft accelerated upward (relative to an inertial system). The effects would be indistinguishable, according to the principle of equivalence, because the apple's inertial and gravitational masses—that determine how a body "reacts" to outside influences—are indistinguishable.

FIGURE 6–18 (a) Light beam goes straight across an elevator that is not accelerating. (b) The light beam bends (exaggerated) in an elevator accelerating in an upward direction.

*6–9 Gravitation as Curvature of Space; Black Holes

The principle of equivalence can be used to show that light ought to be deflected due to the gravitational force of a massive body. Let us consider a thought experiment, in an elevator in free space where no gravity acts. If there is a hole in the side of the elevator and a beam of light enters from outside, the beam travels straight across the elevator and makes a spot on the opposite side if the elevator is at rest (Fig. 6–18a). If the elevator is accelerating upward as in Fig. 6–18b, the light beam still travels straight as observed in the original reference frame at rest. In the upwardly accelerating elevator, however, the beam is observed to curve downward. Why? Because during the time the light travels from one side of the elevator to the other, the elevator is moving upward at ever-increasing speed.

Now, according to the equivalence principle, an upwardly accelerating reference frame is equivalent to a downward gravitational field. Hence, we can picture the curved light path in Fig. 6–18b as being the effect of a gravitational field. Thus we expect gravity to exert a force on a beam of light and to bend it out of a straight-line path!

Einstein, in his general theory of relativity, predicted just this, that light would be affected by gravity. It was calculated that light from a distant star would be deflected by 1.75″ of arc (tiny but detectable) as it passed near the Sun, as shown in Fig. 6–19. Such a deflection was measured in 1919 during an eclipse of the Sun. (The eclipse reduced the brightness of the Sun so that the stars in line with its edge at that moment would be visible.)

A light beam travels by the shortest, most direct, path between two points. If it didn't, some other object could travel between the two points in a shorter time and thus have a greater speed than the speed of light, which would contradict the special theory of relativity. If a light beam can follow a curved path (as discussed above), then this curved path must be the shortest distance between the two points. This suggests that *space itself is curved* and that it is the gravitational field that causes the curvature. The curvature is greatest near very massive bodies. To visualize this curvature of space, we might think of space as being like a thin rubber sheet; if a heavy weight is hung from it, it curves as shown in Fig. 6–20. The weight corresponds to a huge mass that causes space (space itself!) to curve.

The extreme curvature of space-time shown in Fig. 6–20, could be produced by a **black hole**. A black hole is a hugely massive star that is so dense and so massive that gravity would be so strong that even light could not escape it. Light would be pulled back in by the force of gravity. Since no light could escape from such a massive star, we could not see it—it would be black. A body might pass by it and be deflected by its gravitational field, but if the body came too close, it would be swallowed up, never to escape. Hence the name for these hypothesized black holes. Experimentally there is good evidence for their existence, although some scientists remain cautious. One possibility is that there may be a giant black hole at the center of our Galaxy and perhaps at the center of other galaxies.

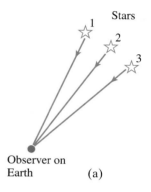

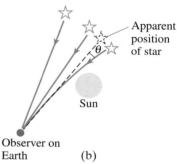

FIGURE 6–19 (a) Three stars in the sky. (b) If the light from one of these stars passes very near the Sun, whose gravity bends the light beam, the star will appear higher than it actually is.

FIGURE 6–20 Rubber sheet analogy for space (technically space-time) curved by matter.

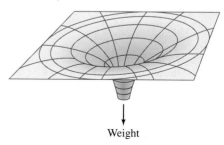

Summary

Newton's **law of universal gravitation** states that every particle in the universe attracts every other particle with a force proportional to the product of their masses and inversely proportional to the square of the distance between them:

$$F = G \frac{m_1 m_2}{r^2}.$$

The direction of this force is along the line joining the two particles. It is this gravitational force that keeps the Moon revolving around the Earth and the planets revolving around the Sun.

The total gravitational force on any body is the vector sum of the forces exerted by all other bodies; frequently the effects of all but one or two bodies can be ignored.

Satellites revolving around the Earth are acted on by gravity, but "stay up" because of their high tangential speed.

Newton's three laws of motion, plus his law of universal gravitation, constituted a wide-ranging theory of the universe. With them, motion of objects on Earth and in the heavens could be accurately described. And they provided a theoretical base for **Kepler's laws** of planetary motion.

According to the **field** concept, a **gravitational field** surrounds every body that has mass, and it permeates all of space. The gravitational field at any point in space is the vector sum of the fields due to all massive bodies and can be defined as

$$\mathbf{g} = \frac{\mathbf{F}}{m}$$

where $\mathbf{F}$ is the force acting on a small "test" mass m placed at that point.

Questions

1. Does an apple exert a gravitational force on the Earth? If so, how large a force? Consider an apple (a) attached to a tree and (b) falling.

2. The Sun's gravitational pull on the Earth is much larger than the Moon's. Yet the Moon's is mainly responsible for the tides. Explain. [*Hint:* Consider the difference in gravitational pull from one side of the Earth to the other.]

3. Will an object weigh more at the equator or at the poles? What two effects are at work? Do they oppose each other?

4. Why is more fuel required for a spacecraft to travel from the Earth to the Moon than it does to return from the Moon to the Earth?

5. The gravitational force on the Moon due to the Earth is only about half the force on the Moon due to the Sun (see Example 6–3). Why isn't the Moon pulled away from the Earth?

6. If you were in a satellite orbiting the Earth, how might you cope with walking, drinking, or putting a pair of scissors on a table?

7. An antenna loosens and becomes detached from a satellite in a circular orbit around the Earth. Describe the antenna's motion subsequently. If it will land on the Earth, describe where; if not, describe how it could be made to land on the Earth.

8. Describe how careful measurements of the variation in g in the vicinity of an ore deposit might be used to estimate the amount of ore present.

9. The Sun is directly below us at midnight, in line with the Earth's center. Are we then heavier at midnight, due to the Sun's gravitational force on us, than we are at noon? Explain.

10. When will your apparent weight be the greatest, as measured by a scale in a moving elevator: when the elevator (a) accelerates downward, (b) when it accelerates upward, (c) when it is in free fall, or (d) when it moves upward at constant speed? In which case would your weight be the least? When would it be the same as when you are on the ground?

11. If the Earth's mass were double what it is, in what ways would the Moon's orbit be different?

12. The source of the Mississippi River is closer to the center of the Earth than is its outlet in Louisiana (since the Earth is fatter at the equator than at the poles). Explain how the Mississippi can flow "uphill."

13. People sometimes ask, "What keeps a satellite up in its orbit around the Earth?" How would you respond?

14. Explain how a runner experiences free fall or apparent weightlessness between steps.

15. The Earth moves faster in its orbit around the Sun in December than in June. Is it closer to the Sun in June or in December? Does this affect the seasons? Explain.

16. The mass of the planet Pluto was not known until it was discovered to have a moon. Explain how this enabled an estimate of Pluto's mass.

17. Does your body directly sense a gravitational field? (Compare to what you would feel in free fall.)

18. Discuss the conceptual differences between $\mathbf{g}$ as acceleration due to gravity and $\mathbf{g}$ as gravitational field.

19. Describe a general procedure to determine the mass of a planet from observations on the orbit of one of its satellites.

20. Which pulls harder gravitationally, the Earth on the Moon, or the Moon on the Earth? Which accelerates more?

Problems

Sections 6–1 to 6–3

1. (I) Calculate the force of gravity on a spacecraft 12,800 km (2 earth radii) above the Earth's surface if its mass is 1400 kg.

2. (I) Calculate the acceleration due to gravity on the Moon. The Moon's radius is about 1.74×10^6 m and its mass is 7.35×10^{22} kg.

3. (I) A hypothetical planet has a radius 2.5 times that of Earth, but has the same mass. What is the acceleration due to gravity near its surface?

4. (I) A hypothetical planet has a mass 3.0 times that of Earth, but the same radius. What is g near its surface?

5. (II) You are explaining to friends why astronauts feel weightless orbiting in the space shuttle, and they respond that they thought gravity was just a lot weaker up there. Convince them and yourself that it isn't so by calculating how much weaker gravity is 300 km above the Earth's surface.

6. (II) Calculate the effective value of g, the acceleration of gravity, at (a) 3200 m, and (b) 3200 km, above the Earth's surface.

7. (II) Four 8.5-kg spheres are located at the corners of a square of side 0.70 m. Calculate the magnitude and direction of the gravitational force on one sphere due to the other three.

8. (II) Every few hundred years most of the planets line up on the same side of the Sun. Calculate the total net force on the Earth due to Venus, Jupiter, and Saturn, assuming all four planets are in a line, Fig. 6–21. The masses are respectively $M_V = 0.815\,M_E$, $M_J = 318\,M_E$, $M_{Sa} = 95.1\,M_E$, and their mean distances from the Sun are 108, 150, 778, and 1430 million km.

FIGURE 6–21 Problem 8 (not to scale).

9. (II) Four masses are arranged as shown in Fig. 6–22. Determine the x and y components of the gravitational force on the mass at the origin (m). Write the force in vector notation $(\mathbf{i}, \mathbf{j})$.

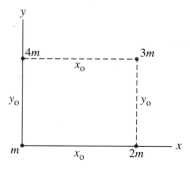

FIGURE 6–22 Problem 9.

10. (II) Suppose a spacecraft travels directly from Earth to Mars (where $g = 3.7$ m/s² at the surface) at constant velocity. Graph the weight of (net gravitational force on) a 70-kg passenger as a function of distance between the two planets. Assume any acceleration is negligible.

11. (II) Suppose the mass of the Earth were doubled, but it kept the same density and spherical shape. How would the weight of objects at the Earth's surface change?

12. (II) Equation 6–4 can be applied to other bodies, such as other planets, to obtain useful information, if appropriate variables are used. For example, suppose a spacecraft to Mars has measured g at the surface to be $g_{Mars} = 3.7$ m/s², and astronomical observations show the radius of Mars to be $r_{Mars} = 3.4 \times 10^6$ m. Using this information, determine the mass of Mars.

13. (II) At what distance from the Earth will a spacecraft on the way to the Moon experience zero net force due to these two bodies because the Earth and Moon pull with equal and opposite forces?

14. (II) Determine the mass of the Sun using the known value for the period of the Earth and its distance from the Sun. [Note: Compare your answer to that obtained using Kepler's laws, Example 6–9.]

15. (III) (a) Use the binomial expansion

$$(1 \pm x)^n = 1 \pm nx + \frac{n(n-1)}{2}x^2 \pm \cdots$$

to show that the value of g is altered by approximately

$$\Delta g \approx -2g\frac{\Delta r}{r_E}$$

at a height Δr above the Earth's surface, where r_E is the radius of the Earth, as long as $\Delta r \ll r_E$. (b) What is the meaning of the minus sign in this relation? (c) Use this result to compute the effective value of g at 100 km above the Earth's surface. Compare to a direct use of Eq. 6–1.

16. (III) Determine the magnitude and direction of the effective value of $\mathbf{g}$ at a latitude of 45° on the Earth. Assume the Earth is a rotating sphere.

17. (III) A ship steams at speed v along the equator. Show that the apparent weight w of an object as weighed on the ship is given approximately by $w = w_0(1 \pm 4\pi f v/g)$, where f is the frequency of rotation (revolutions/second) of the Earth. Why is there a $\pm$ sign? Let w_0 be the measured weight of the object when the ship is at rest relative to the Earth.

Section 6–4

18. (I) Calculate the velocity of a satellite moving in a stable circular orbit about the Earth at a height of 5200 km.

19. (I) The space shuttle releases a satellite into a circular orbit 600 km above the Earth. How fast must the shuttle be moving (relative to Earth) when the release occurs?

20. (II) A 16.0-kg monkey hangs from a cord suspended from the ceiling of an elevator. The cord can withstand a tension of 200 N and breaks as the elevator accelerates. What was the elevator's minimum acceleration (magnitude and direction)?

21. (II) Calculate the period of a satellite orbiting the Moon, 100 km above the Moon's surface. Ignore effects of the Earth. The radius of the Moon is 1740 km.

22. (II) Determine the time it takes for a satellite to orbit the Earth in a circular "near-Earth" orbit. The definition of "near-Earth" orbit is one which is at a height above the surface of the Earth that is very small compared to the radius of the Earth, and you may take the acceleration due to gravity as essentially the same as that on the surface. Does your result depend on the mass of the satellite?

23. (II) What will a spring scale read for the weight of a 56-kg woman in an elevator that moves (a) with constant speed upward of 3.0 m/s, (b) with constant speed downward of 3.0 m/s, (c) with upward acceleration of 0.33 g, (d) with downward acceleration 0.33 g, and (e) in free fall?

24. (II) A ferris wheel 27.5 m in diameter rotates once every 10.5 s. What is the fractional change in a person's apparent weight (a) at the top and (b) at the bottom, as compared to her weight at rest?

25. (II) What is the apparent weight of a 75-kg astronaut 4100 km from the center of the Moon in a space vehicle (a) moving at constant velocity and (b) accelerating toward the Moon at 2.6 m/s²? State "direction" in each case.

26. (II) How long would a day be if the Earth were rotating so fast that objects at the equator were weightless?

27. (II) Two equal-mass stars maintain a constant distance apart of 8.0×10^{10} m and rotate about a point midway between them at a rate of one revolution every 12.6 yr. (a) Why don't the two stars crash into one another due to the gravitational force between them? (b) What must be the mass of each star?

28. (II) (a) Show that if a satellite orbits very near the surface of a planet with period T, the density (= mass per unit volume) of the planet is $\rho = m/V = 3\pi/GT^2$. (b) Estimate the density of the Earth, given that a satellite near the surface orbits with a period of about 90 minutes.

29. (II) At what horizontal velocity would a satellite have to be launched from the top of Mt. Everest to be placed in a circular orbit around the Earth?

30. (II) You are an astronaut in the space shuttle pursuing a satellite in need of repair. You find yourself in a circular orbit of the same radius as the satellite, but 30 km behind it. (a) How long will it take to reach the satellite if you reduce your orbital radius by 1.0 km? (b) By how much must you reduce your orbital radius to catch up in 8.0 hours?

31. (III) Three bodies of identical mass M form the vertices of an equilateral triangle of side L and rotate in circular orbits about the center of the triangle. They are held in place by their mutual gravitation. What is the speed of each?

32. (III) An inclined plane, fixed to the inside of an elevator, makes a 30° angle with the floor. A mass m slides on the plane without friction. What is its acceleration relative to the plane if the elevator (a) accelerates upward at 0.50 g, (b) accelerates downward at 0.50 g, (c) falls freely, or (d) moves upward at constant speed?

Section 6–5

33. (I) Use Kepler's laws and the period of the Moon (27.4 d) to determine the period of an artificial satellite orbiting very near the Earth's surface.

34. (I) The asteroid Icarus, though only a few hundred meters across, orbits the Sun like the other planets. Its period is about 410 d. What is its mean distance from the Sun?

35. (I) Neptune is an average distance of 4.5×10^9 km from the Sun. Estimate the length of the Neptunian year using the fact that the Earth is 1.50×10^8 km from the Sun on the average.

36. (I) Determine the mass of the Earth from the known period and distance of the Moon.

37. (II) Our Sun rotates about the center of the Galaxy $(M \approx 4 \times 10^{41} \text{ kg})$ at a distance of about 3×10^4 light years $(1 \text{ ly} = 3 \times 10^8 \text{ m/s} \times 3.16 \times 10^7 \text{ s/y})$. What is the period of our orbital motion about the center of the Galaxy?

TABLE 6–3
Principal Moons of Jupiter (Problems 38 and 39)

Moon	Mass (kg)	Period (Earth days)	Mean distance from Jupiter (km)
Io	8.9×10^{22}	1.77	422×10^3
Europa	4.9×10^{22}	3.55	671×10^3
Ganymede	15×10^{22}	7.16	1070×10^3
Callisto	11×10^{22}	16.7	1883×10^3

38. (II) Table 6–3 gives the mean distance, period, and mass for the four largest moons of Jupiter (those discovered by Galileo in 1609). (a) Determine the mass of Jupiter using the data for Io. (b) Determine the mass of Jupiter using data for each of the other three moons. Are the results consistent?

39. (II) Determine the mean distance from Jupiter for each of Jupiter's moons, using the distance of Io and the periods given in Table 6–3. Compare to the values in the Table.

40. (II) The asteroid belt between Mars and Jupiter consists of many fragments, once hypothesized to have been a planet. (a) If the center of mass of the asteroid belt is about 3 times farther from the Sun than the Earth is, how long would it have taken this hypothetical planet to orbit the Sun? (b) Can we use these data to deduce the mass of that planet?

41. (III) A science fiction tale describes an artificial "planet" in the form of a band encircling a sun as shown: Fig. 6–23. The inhabitants live on the inside surface (where it is always noon). Imagine the sun is exactly like our own, that the distance to the band is the same as the Earth–Sun distance (so the climate is temperate), and that the ring rotates quickly enough to produce an apparent gravity of one g as on Earth. What will be the band's period of revolution, this planet's year, in Earth days?

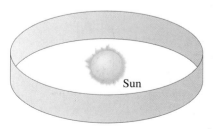

FIGURE 6–23
Problem 41.

42. (III) (a) Use Kepler's second law to show that the ratio of the speeds of a planet at its nearest and farthest points from the Sun is equal to the inverse ratio of the near and far distances: $v_N/v_F = d_F/d_N$. (b) Given that the Earth's distance from the Sun varies from 1.47 to 1.52×10^{11} m, determine the minimum and maximum velocities of the Earth in its orbit around the Sun.

43. (III) The recently discovered comet Hale–Bopp has a period of 3000 years. (a) What is its mean distance from the Sun? (b) At its closest approach, the comet is about 1 A.U. from the Sun (1 A.U. = distance from Earth to the Sun). What is the farthest distance? (c) What is the ratio of the speed at the closest point to the speed at the farthest point?

Section 6–6

44. (II) What is the magnitude and direction of the gravitational field midway between the Earth and Moon? Ignore effects of the Sun.

45. (II) (a) What is the gravitational field at the surface of the Earth due to the Sun? (b) Will this affect your weight significantly?

46. (III) Two identical particles, each of mass m, are located on the x axis at $x = +x_0$ and $x = -x_0$. (a) Determine a formula for the gravitational field due to these two particles for points on the y axis; that is, write **g** as a function of y, m, x_0, and so on. (b) At what point (or points) on the y axis is the magnitude of **g** a maximum value, and what is its value there? [Hint: Take the derivative $d\mathbf{g}/dy$.]

General Problems

47. How far above the Earth's surface will the acceleration of gravity be half what it is at the surface?

48. At the surface of a certain planet, the gravitational acceleration g has a magnitude of $12.0 \, \text{m/s}^2$. A 3.0-kg brass ball is transported to this planet. What is (a) the mass of the brass ball on the Earth and on the planet, and (b) the weight of the brass ball on the Earth and on the planet?

49. An exotic finish to massive stars is that of a neutron star, which might have as much as five times the mass of our Sun packed into a sphere about 10 km in radius! Estimate the surface gravity on this monster.

50. What is the distance from the Earth's center to a point outside the Earth where the gravitational acceleration due to the Earth is $\frac{1}{10}$ of its value at the Earth's surface?

51. A typical white dwarf star, which once was an average star like our Sun but is now in the last stage of its evolution, is the size of our Moon but has the mass of our Sun. What is the surface gravity on this star?

52. During an *Apollo* lunar landing mission, the command module continued to orbit the Moon at an altitude of about 100 km. How long did it take to go around the Moon once?

53. Estimate what the value of G would need to be if you could actually "feel" yourself gravitationally attracted to someone else near you. Make reasonable assumptions.

54. (a) Calculate the gravitational force of the Sun on the Earth, and the gravitational force of the Moon on the Earth. (b) Why is the gravitational attraction of the Moon rather than the Sun the dominant mechanism for the occurrence of ocean tides?

55. The rings of Saturn are composed of chunks of ice that orbit the planet. The inner radius of the rings is 73,000 km, while the outer radius is 170,000 km. Find the period of an orbiting chunk of ice at the inner radius and the period of a chunk at the outer radius. Compare your numbers with Saturn's mean rotation period of 10 hours and 39 minutes. The mass of Saturn is 5.69×10^{26} kg.

56. The Navstar Global Positioning System (GPS) utilizes a group of 24 satellites orbiting the Earth. Using "triangulation" and signals transmitted by these satellites, the position of a receiver on the Earth can be determined to within an accuracy of a few centimeters. The satellite orbits are distributed evenly around the Earth, with four satellites in each of six orbits, allowing continuous navigational "fixes." The satellites orbit at an altitude of approximately 11,000 nautical miles [1 nautical mile = 1.852 km = 6076 ft]. (a) Determine the speed of each satellite. (b) Determine the period of each satellite.

57. NASA launched the Near Earth Asteroid Rendevous (NEAR), which, after traveling 1.3 billion miles, is meant to orbit the asteroid Eros at a height of about 15 km. Eros is potato-shaped: 40 km × 6 km × 6 km. Assume Eros has a density (mass/volume) of about $2.3 \times 10^3 \, \text{kg/m}^3$. (a) What will be the period of NEAR as it orbits Eros? (b) Suppose Eros to be a sphere with the same mass and density. What would its radius be? (c) What would g be at the surface of this spherical Eros?

58. Suppose that a binary star system consists of two stars of equal mass. They are observed to be separated by 360 million km and to take 5.0 Earth years to orbit about a point midway between them. What is the mass of each?

59. Halley's comet orbits the Sun roughly once every 76 years. It comes very close to the surface of the Sun on its closest approach (Fig. 6–24). About how far out from the Sun is it at its farthest? Is it still "in" the Solar System? What planet's orbit is nearest when it is out there?

Halley's comet

Sun

FIGURE 6–24
Problem 59.

60. The Sun rotates about the center of the Milky Way Galaxy (Fig. 6–25) at a distance of about 30,000 light years from the center $(1\,\text{ly} = 9.5 \times 10^{15}\,\text{m})$. If it takes about 200 million years to make one rotation, estimate the mass of our Galaxy. Assume that the mass distribution of our galaxy is concentrated mostly in a central uniform sphere. If all the stars had about the mass of our Sun $(2 \times 10^{30}\,\text{kg})$, how many stars would there be in our Galaxy?

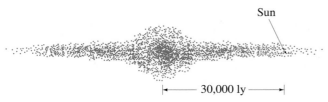

FIGURE 6–25 Edge-on view of our galaxy. Problem 60.

61. The planet Jupiter is about 320 times as massive as the Earth. Thus, it has been claimed that a person would be crushed by the force of gravity on Jupiter since people can't survive more than a few g's. Calculate the number of g's a person would experience if she could stand on the equator of Jupiter. Use the following astronomical data for Jupiter: mass $= 1.9 \times 10^{27}\,\text{kg}$, equatorial radius $= 7.1 \times 10^4\,\text{km}$, rotation period 9 hr 55 min. Take the centripetal acceleration into account.

62. Astronomers using the Hubble Space Telescope have recently deduced the presence of an extremely massive core in the distant galaxy M87, so dense that it could well be a black hole (from which no light escapes). They measured the speed of gas clouds orbiting the core to be 780 km/s at a distance of 60 light-years $(5.7 \times 10^{17}\,\text{m})$ from the core. Deduce the mass of the core, and compare it to the mass of our Sun.

63. Derive a formula for the mass of a planet in terms of its radius, r, the acceleration due to gravity at its surface, g_P, and the gravitational constant, G.

64. A plumb bob is deflected from the vertical by an angle θ due to a massive mountain nearby (Fig. 6–26). (a) Find an approximate formula for θ in terms of the mass of the mountain, M_M, the distance to its center, D_M, and the radius and mass of the Earth. (b) Make a rough estimate of the mass of Mt. Everest, assuming it has the shape, say, of an equilateral pyramid (or cone) 4000 m high above its base, and then (c) estimate the angle θ of the pendulum bob if it is 5 km from the center of Mt. Everest.

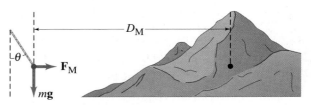

FIGURE 6–26 Problem 64.

65. Show that the rate of change of your weight is

$$-2G\frac{M_E m}{r^3}v$$

if you are traveling directly away from Earth at constant speed v. Your mass is m, and r is your distance from the center of the Earth at any moment.

66. A geologist searching for oil finds that the gravity at a certain location is 2 parts in 10^7 smaller than average. Assume that the location contains oil and the oil is located 2000 m below the Earth's crust. Estimate the size of the deposit. Take the density (mass per unit volume) of rock to be $3000\,\text{kg/m}^3$ and that of oil to be $1000\,\text{kg/m}^3$.

* 67. A particle is released at a height r_E (radius of Earth) above the Earth's surface. Determine its velocity when it hits the Earth. Ignore air resistance. [*Hint:* Use Newton's second law, the law of universal gravitation, the chain rule, and integrate.]

This baseball pitcher is about to accelerate the baseball to a high velocity by exerting a force on it. He will be doing work on the ball as he exerts the force over a displacement of perhaps several meters, from behind his head until he releases the ball with arm outstretched in front of him. The work he does on the ball will be equal to the kinetic energy $\left(\frac{1}{2}mv^2\right)$ he gives to the ball, a result known as the work-energy principle.

CHAPTER 7

Work and Energy

Until now we have been studying the translational motion of an object in terms of Newton's three laws of motion. In this analysis, *force* has played a central role as the quantity determining the motion. In this chapter and the two that follow, we discuss an alternative analysis of the motion of an object in terms of the quantities *energy* and *momentum*. The significance of these quantities is that they are *conserved*. That is, in quite general circumstances they remain constant. That conserved quantities exist not only gives us a deeper insight into the nature of the world, but also gives us another way to approach practical problems.

The conservation laws of energy and momentum are especially valuable in dealing with systems of many objects, in which a detailed consideration of the forces involved would be difficult or impossible. These laws are applicable to a wide range of phenomena, including the atomic and subatomic world, where Newton's laws do not apply.

This chapter is devoted to the very important concept of *energy* and the closely related concept of *work*. These two quantities are scalars and so have no direction associated with them, which often makes them easier to work with than vector quantities such as acceleration and force. Energy has great importance for two reasons. First it is conserved. Second, energy is a concept that is useful not only in the study of motion, but in all areas of physics and in other sciences as well. But before discussing energy itself, we first examine the notion of work.

We consider only translational motion for now and, unless otherwise explained, objects are assumed to be rigid and therefore particle-like with no complicating internal motion.

7–1 | Work Done by a Constant Force

The word *work* has a variety of meanings in everyday language. But in physics, work is given a very specific meaning to describe what is accomplished by the action of a force when it acts on an object over some distance. Specifically, the **work** done on an object by a constant force (constant in both magnitude and direction) is defined to be *the product of the magnitude of the displacement times the component of the force parallel to the displacement*. In equation form, we can write

$$W = F_{\parallel} d$$

where $F_{\parallel}$ is the component of the constant force **F** parallel to the displacement **d**. We can also write

Work defined (for constant force)

$$W = Fd \cos\theta, \qquad\qquad (7–1)$$

where F is the magnitude of the constant force, d is the magnitude of the displacement of the object, and θ is the angle between the directions of the force and the displacement. The $\cos\theta$ factor appears in Eq. 7–1 because $F\cos\theta \, (= F_{\parallel})$ is the component of **F** parallel to **d**. See Fig. 7–1. Work is a scalar quantity—it has only magnitude.

Let's first consider the case in which the motion and the force are in the same direction, so $\theta = 0$ and $\cos\theta = 1$, and then $W = Fd$. For example, if you push a loaded grocery cart a distance of 50 m by exerting a horizontal force of 30 N on the cart, you do $30\,\text{N} \times 50\,\text{m} = 1500\,\text{N}\cdot\text{m}$ of work on the cart.

Units for work: the joule

As this example shows, in SI units, work is measured in newton-meters. A special name is given to this unit, the **joule** (J): $1\,\text{J} = 1\,\text{N}\cdot\text{m}$. In the cgs system, the unit of work is called the *erg* and is defined as $1\,\text{erg} = 1\,\text{dyne}\cdot\text{cm}$. In British units, work is measured in foot-pounds. It is easy to show that $1\,\text{J} = 10^7\,\text{erg} = 0.7376\,\text{ft}\cdot\text{lb}$.

Force without work

A force can be exerted on an object and yet do no work. For example, if you hold a heavy bag of groceries in your hands at rest, you do no work on it. A force is exerted, but the displacement is zero, so the work $W = 0$. You also do no work

FIGURE 7–1 A person pulling a crate along the floor. The work done by the force **F** is $W = Fd\cos\theta$, where **d** is the displacement.

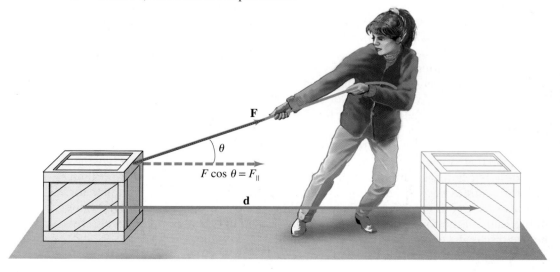

on the bag of groceries if you carry it as you walk horizontally across the floor at constant velocity, as shown in Fig. 7–2. No horizontal force is required to move the package at a constant velocity. However, the person shown in Fig. 7–2 does exert an upward force $\mathbf{F}_P$ on the package equal to its weight. But this upward force is perpendicular to the horizontal motion of the package and thus has nothing to do with that motion. Hence, the upward force is doing no work. This conclusion comes from our definition of work, Eq. 7–1: $W = 0$, because $\theta = 90°$ and $\cos 90° = 0$. Thus, when a particular force is perpendicular to the motion, no work is done by that force. (When you start or stop walking, there is a horizontal acceleration and you do briefly exert a horizontal force, and thus do work.)

When dealing with work, as with force, it is necessary to specify whether you are talking about work done *by* a specific object or done *on* a specific object. It is also important to specify whether the work done is due to one particular force (and which one), or the total (net) work done by the *net force* on the object.

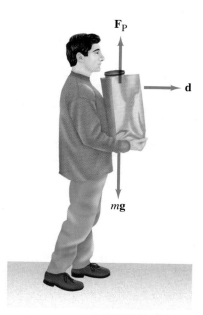

FIGURE 7–2 Work done on the bag of groceries by the person in this case is zero since $\mathbf{F}_P$ is perpendicular to the displacement $\mathbf{d}$.

EXAMPLE 7–1 **Work done on a crate.** A 50-kg crate is pulled 40 m along a horizontal floor by a constant force exerted by a person, $F_P = 100$ N, which acts at a 37° angle as shown in Fig. 7–3. The floor is rough and exerts a friction force $F_{fr} = 50$ N. Determine the work done by each force acting on the crate, and the net work done on the crate.

SOLUTION We choose our coordinate system so that $\mathbf{x}$ can be the vector that represents the 40-m displacement (that is, along the x axis). There are four forces acting on the crate, as shown in Fig. 7–3 which is the free-body diagram: the force exerted by the person $\mathbf{F}_P$; the friction force $\mathbf{F}_{fr}$; the crate's weight $m\mathbf{g}$; and the normal force $\mathbf{F}_N$ exerted upward by the floor. The work done by the gravitational and normal forces is zero, since they are perpendicular to the displacement $\mathbf{x}$ ($\theta = 90°$ in Eq. 7–1):

$$W_G = mgx \cos 90° = 0$$

$$W_N = F_N x \cos 90° = 0.$$

The work done by $\mathbf{F}_P$ is

$$W_P = F_P x \cos \theta = (100 \text{ N})(40 \text{ m}) \cos 37° = 3200 \text{ J}.$$

The work done by the friction force is

$$W_{fr} = F_{fr} x \cos 180°$$
$$= (50 \text{ N})(40 \text{ m})(-1) = -2000 \text{ J}.$$

The angle between the displacement $\mathbf{x}$ and $\mathbf{F}_{fr}$ is 180° because they point in opposite directions. Since the force of friction is opposing the motion (and $\cos 180° = -1$), it does *negative* work on the crate.

FIGURE 7–3 Example 7–1: a 50-kg crate being pulled along a floor.

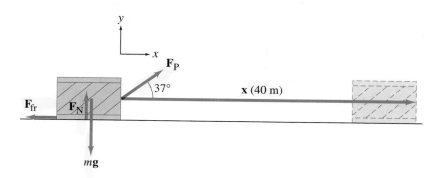

Finally, the net work can be calculated in two equivalent ways.

(1) The net work done on an object is the algebraic sum of the work done by each force, since work is a scalar:

$$W_{net} = W_G + W_N + W_P + W_{fr}$$
$$= 0 + 0 + 3200\,J - 2000\,J = 1200\,J.$$

(2) The net work can also be calculated by first determining the net force on the object and then taking its component along the displacement: $(F_{net})_x = F_P \cos\theta - F_{fr}$. Then the net work is

$$W_{net} = (F_{net})_x\, x = (F_P \cos\theta - F_{fr})x$$
$$= (100\,N \cos 37° - 50\,N)(40\,m) = 1200\,J.$$

In the vertical (y) direction, there is no displacement and no work done.

Negative work

In Example 7–1 we saw that friction did negative work. In general, the work done by a force is negative whenever the force (or the component of the force, $F_{\parallel}$) acts in the direction opposite to the direction of motion.

FIGURE 7–4 Example 7–2.

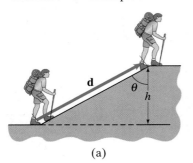

(a)

(b)

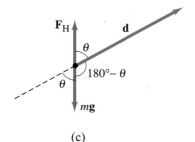

(c)

EXAMPLE 7–2 **Work on a carried backpack.** (*a*) Determine the work a hiker must do on a 15.0-kg backpack to carry it up a hill of height $h = 10.0\,m$, as shown in Fig. 7–4a. Determine also (*b*) the work done by gravity on the backpack, and (*c*) the net work done on the backpack. For simplicity, assume the motion is smooth and at constant velocity (i.e., there is negligible acceleration).

SOLUTION (*a*) The forces on the backpack are shown in Fig. 7–4b: the force of gravity, $m\mathbf{g}$, acting downward; and $\mathbf{F}_H$, the force the hiker must exert upward to support the pack. Since we assume there is negligible acceleration, horizontal forces are negligible. In the vertical (y) direction, we choose up as positive. Newton's second law applied to the backpack gives

$$\Sigma F_y = ma_y$$
$$F_H - mg = 0.$$

Hence,

$$F_H = mg = (15.0\,kg)(9.80\,m/s^2) = 147\,N.$$

To calculate the work done by the hiker on the backpack, Eq. 7–1 can be written

$$W_H = F_H(d \cos\theta),$$

and we note from Fig. 7–4a that $d \cos\theta = h$. So the work done by the hiker can be written:

$$W_H = F_H(d \cos\theta) = F_H h = mgh$$
$$= (147\,N)(10.0\,m) = 1470\,J.$$

Note that the work done depends only on the change in elevation and not on the angle of the hill, θ. The same work would be done to lift the pack vertically the same height h.

(*b*) The work done by gravity is (from Eq. 7–1 and Fig. 7–4c):

$$W_G = (F_G)(d) \cos(180° - \theta).$$

Since $\cos(180° - \theta) = -\cos\theta$, we have

$$W_G = (F_G)(d)(-\cos\theta) = mg(-d \cos\theta)$$
$$= -mgh$$
$$= -(15.0\,kg)(9.80\,m/s^2)(10.0\,m) = -1470\,J.$$

Note that the work done by gravity doesn't depend on the angle of the incline but only on the vertical height h of the hill. This is because gravity does work only in the vertical direction. We will make use of this important result later.

(c) The *net* work done on the backpack is $W_{net} = 0$, since the net force on the backpack is zero (it is assumed not to accelerate significantly). We can also determine the net work done by writing

$$W_{net} = W_G + W_H = -1470\,J + 1470\,J = 0$$

which is, as it should be, the same result.

Note in this example that even though the *net* work on the backpack is zero, the hiker nonetheless does do work on the backpack equal to 1470 J.

Work done by gravity depends on the height of the hill and not on the angle of incline

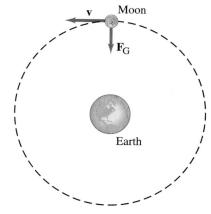

FIGURE 7–5
Conceptual Example 7–3.

CONCEPTUAL EXAMPLE 7–3 **Does the Earth do work on the Moon?**
The Moon revolves around the Earth in a nearly circular orbit, kept there by the gravitational force exerted by the Earth. Does gravity do (a) positive work, (b) negative work, or (c) no work at all on the Moon?

RESPONSE The gravitational force on the Moon (Fig. 7–5) acts toward the Earth (as a centripetal force), inward along the radius of the Moon's orbit. The Moon's displacement at any moment is along the circle, in the direction of its velocity, perpendicular to the radius and perpendicular to the force of gravity. Hence the angle θ between the force $\mathbf{F}_G$ and the instantaneous displacement of the Moon is 90°, and the work done by gravity is therefore zero ($\cos 90° = 0$). This is why the Moon, as well as artificial satellites, can stay in orbit without expenditure of fuel, since no work needs to be done against the force of gravity.

PROBLEM	Work
SOLVING	

1. Choose an xy coordinate system. If the body is in motion, it may be convenient to choose the direction of motion as one of the coordinate directions. [Thus, for an object on an incline, you might choose one coordinate axis to be parallel to the incline.]

2. Draw a free-body diagram showing all the forces acting on the body.

3. Determine any unknown forces using Newton's laws.

4. Find the work done *by* a specific force *on* the body by using $W = Fd\cos\theta$. Note that the work done is negative when a force tends to oppose the displacement.

5. To find the *net* work done on the body, either (a) find the work done by each force and add the results algebraically; or (b) find the net force on the object, F_{net}, and then use it to find the net work done:

$$W_{net} = F_{net}\,d\cos\theta.$$

7–2 Scalar Product of Two Vectors

Although work is a scalar, it involves the product of two quantities, force and displacement, each of which are vectors. Therefore, we now investigate the multiplication of vectors, which will be useful throughout the book.

Because vectors have direction as well as magnitude, they cannot be multiplied in the same way that scalars are. Instead we must *define* what the operation of vector multiplication means. Among the many possible ways to define how to multiply vectors, there are three ways that we will find useful in physics: (1) multiplication of a vector by a scalar, which was already discussed in Section 3–3; (2) multiplication of one vector by a second vector so as to produce a scalar; (3) multiplication of one vector by a second vector so as to produce another vector. The third type, called the *vector product*, will be discussed later, in Section 11–1.

We now discuss the second type, called the *scalar product*, or *dot product* (because a dot is used to indicate the multiplication). If we have two vectors, **A** and **B**, then their **scalar** (or **dot**) **product** is defined to be

*Scalar product
(dot product)*

$$\mathbf{A} \cdot \mathbf{B} = AB \cos \theta, \tag{7-2}$$

where A and B are the magnitudes of the vectors and θ is the angle ($< 180°$) between them when their tails touch, Fig. 7–6. Since A, B, and $\cos \theta$ are scalars, then so is the scalar product $\mathbf{A} \cdot \mathbf{B}$ (read "A dot B"). This definition, Eq. 7–2, fits perfectly with our definition of the work done by a constant force, Eq. 7–1. That is, we can write the work done by a constant force as the scalar product of force and displacement:

$$W = \mathbf{F} \cdot \mathbf{d} = Fd \cos \theta. \tag{7-3}$$

FIGURE 7–6 The scalar product, or dot product, of two vectors **A** and **B**, is $\mathbf{A} \cdot \mathbf{B} = AB \cos \theta$.

Indeed, the definition of scalar product, Eq. 7–2, is so chosen because many physically important quantities, such as work (and others we will meet later), can be described as the scalar product of two vectors.

An equivalent definition of the scalar product is that it is the product of the magnitude of one vector (say A) and the component of the other vector along the direction of the first ($B \cos \theta$).

Since A, B, and $\cos \theta$ are scalars, it doesn't matter in what order they are multiplied. Hence the scalar product is **commutative**:

Commutative property

$$\mathbf{A} \cdot \mathbf{B} = \mathbf{B} \cdot \mathbf{A}.$$

It is also easy to show that it is **distributive** (see Problem 29 for the proof):

Distributive property

$$\mathbf{A} \cdot (\mathbf{B} + \mathbf{C}) = \mathbf{A} \cdot \mathbf{B} + \mathbf{A} \cdot \mathbf{C}.$$

Let us use these properties and write each vector in terms of its rectangular components using unit vectors (Section 3–5, Eq. 3–5) as

$$\mathbf{A} = A_x \mathbf{i} + A_y \mathbf{j} + A_z \mathbf{k} \quad \text{and} \quad \mathbf{B} = B_x \mathbf{i} + B_y \mathbf{j} + B_z \mathbf{k};$$

then we can see that

*Scalar product
(in terms of components)*

$$\mathbf{A} \cdot \mathbf{B} = A_x B_x + A_y B_y + A_z B_z, \tag{7-4}$$

because the unit vectors, **i**, **j**, and **k** are perpendicular to each other so that

$$\mathbf{i} \cdot \mathbf{i} = \mathbf{j} \cdot \mathbf{j} = \mathbf{k} \cdot \mathbf{k} = 1$$

$$\mathbf{i} \cdot \mathbf{j} = \mathbf{i} \cdot \mathbf{k} = \mathbf{j} \cdot \mathbf{k} = 0.$$

Equation 7–4 is particularly useful.

If **A** is perpendicular to **B**, then Eq. 7–2 tells us $\mathbf{A} \cdot \mathbf{B} = 0$. But the converse, given that $\mathbf{A} \cdot \mathbf{B} = 0$, can come about in three different ways: $\mathbf{A} = 0$, $\mathbf{B} = 0$, or $\mathbf{A} \perp \mathbf{B}$.

EXAMPLE 7–4 **Using the dot product.** The force shown in Fig. 7–7 has magnitude $F_P = 20 \text{ N}$ and makes an angle of $30°$ to the ground. Calculate the work done by this force using Eq. 7–4 when the wagon is dragged 100 m along the ground.

FIGURE 7–7 Work done by a force $\mathbf{F}_P$ acting at an angle θ to the ground is $W = \mathbf{F}_P \cdot \mathbf{d}$.

SOLUTION We choose the x axis horizontal to the right and the y axis vertically upward. Then

$$\mathbf{F}_P = F_x\mathbf{i} + F_y\mathbf{j} = (F_P\cos 30°)\mathbf{i} + (F_P\sin 30°)\mathbf{j} = (17\,\text{N})\mathbf{i} + (10\,\text{N})\mathbf{j},$$

whereas $\mathbf{d} = (100\,\text{m})\mathbf{i}$. Then, using Eq. 7–4,

$$W = \mathbf{F}_P \cdot \mathbf{d} = (17\,\text{N})(100\,\text{m}) + (10\,\text{N})(0) + (0)(0) = 1700\,\text{J}.$$

Note that by choosing the x axis along $\mathbf{d}$ we simplified the calculation because $\mathbf{d}$ then has only one component.

7–3 Work Done by a Varying Force

If the force acting on an object is constant, the work done by that force can be calculated using Eq. 7–1. But in many cases, the force varies in magnitude or direction during a process. For example, as a rocket moves away from Earth, work is done to overcome the force of gravity, which varies as the inverse square of the distance from the Earth's center. Other examples are the force exerted by a spring, which increases with the amount of stretch, or the work done by a varying force in pulling a box or cart up an uneven hill.

Figure 7–8 shows the path of an object in the xy plane as it moves from point a to point b. The path has been divided into short intervals each of length Δl_1, $\Delta l_2, \ldots \Delta l_7$. A force $\mathbf{F}$ acts at each point on the path, and is indicated at two points as $\mathbf{F}_1$ and $\mathbf{F}_5$. During each small interval Δl, the force is approximately constant. So, for the first interval, the force does work ΔW of approximately (see Eq. 7–1)

$$\Delta W \approx F_1\cos\theta_1\,\Delta l_1.$$

In the second interval the work done is approximately $F_2\cos\theta_2\,\Delta l_2$, and so on. The total work done in moving the particle the total distance $l = \Delta l_1 + \Delta l_2 + \ldots + \Delta l_7$ is the sum of all these terms:

$$W \approx \sum_{i=1}^{7} F_i\cos\theta_i\,\Delta l_i. \tag{7-5}$$

We can examine this graphically by plotting $F\cos\theta$ versus l as shown in Fig. 7–9a. The distance l has been subdivided into the same seven intervals indicated by the vertical dashed lines. The value of $F\cos\theta$ at the center of each interval is indicated by the horizontal dashed lines. Each of the shaded rectangles has an area $(F_i\cos\theta)(\Delta l_i)$, which is the work done during the interval. Thus the estimate of the work done given by Eq. 7–5 equals the sum of the areas of all the rectangles. If we subdivide the distance into a greater number of intervals so that each Δl_i is smaller, the estimate of the work done given by Eq. 7–5 becomes more accurate (the assumption that F is constant over each interval is more accurate). If we let each Δl_i approach zero (so we approach an infinite number of intervals), then we obtain an exact result for the work done:

$$W = \lim_{\Delta l_i \to 0} \Sigma F_i\cos\theta_i\,\Delta l_i = \int_a^b F\cos\theta\,dl. \tag{7-6}$$

This limit as $\Delta l_i \to 0$ is the *integral* of $(F\cos\theta\,dl)$ from point a to point b. The symbol for the integral, $\int$, is an elongated S to indicate an infinite sum; and Δl has been replaced by dl, meaning an infinitesimal distance.

In this limit as Δl approaches zero, the total area of the rectangles (Fig. 7–9a) approaches the area between the $(F\cos\theta)$ curve and the l axis from a to b as shown shaded in Fig. 7–9b. That is, *the work done by a variable force in moving an object between two points is equal to the area under the $(F\cos\theta)$ versus (l) curve between those two points.*

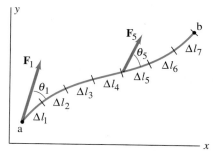

FIGURE 7–8 A particle acted on by a variable force, **F**, moves along the path shown from point a to point b.

FIGURE 7–9 Work done by a force F is (a) approximately equal to the sum of the areas of the rectangles, (b) exactly equal to the area under the curve of $F\cos\theta$ vs. l.

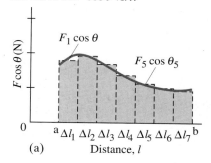

(a)

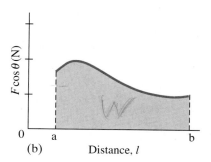

(b)

$W = $ area under $F\cos\theta$ vs. l curve

In the limit as Δl approaches zero, the infinitesimal distance dl equals[†] the magnitude of the infinitesimal displacement vector $d\mathbf{l}$. The direction of the vector $d\mathbf{l}$ is along the tangent to the curve at that point, so θ is the angle between $\mathbf{F}$ and $d\mathbf{l}$ at any point. Thus we can rewrite Eq. 7–6, using dot-product notation:

Work defined in general

$$W = \int_a^b F \cos\theta \, dl = \int_a^b \mathbf{F} \cdot d\mathbf{l}. \tag{7–7}$$

This is the *most general definition of work*. The integral in Eq. 7–7 is called a *line integral* since it is the integral of $F \cos\theta$ along the line that represents the path of the object. (Equation 7–1 for a constant force is a special case of Eq. 7–7.) In rectangular coordinates, the force can be written

$$\mathbf{F} = F_x\mathbf{i} + F_y\mathbf{j} + F_z\mathbf{k}$$

and the displacement $d\mathbf{l}$ is

$$d\mathbf{l} = dx\,\mathbf{i} + dy\,\mathbf{j} + dz\,\mathbf{k}.$$

Then the work done can be written

$$W = \int_a^b F_x\,dx + \int_a^b F_y\,dy + \int_a^b F_z\,dz.$$

To actually use Eq. 7–6 or 7–7 to calculate the work, there are several options. (1) If $F \cos\theta$ is known as a function of position, a graph like that of Fig. 7–9b can be made and the area determined graphically. (2) Another possibility is to use numerical integration (numerical summing), perhaps with the aid of a computer or calculator. (3) A third possibility is to use the analytical methods of integral calculus. To do this, we must be able to write $\mathbf{F}$ as a function of position, $F(x, y, z)$, and we must know the path.

For example, let us consider a one-dimensional situation to determine analytically the work done by a coiled spring, such as that shown in Fig. 7–11. For a person to hold a spring either stretched or compressed an amount x from its normal (unstretched) length requires a force F_P that is directly proportional to x. That is,

Person
$$F_P = kx,$$

where k is a constant, called the *spring constant*, and is a measure of the stiffness of the particular spring. The spring itself exerts a force in the opposite direction (Fig. 7–11):

Spring force

Spring
$$F_S = -kx. \tag{7–8}$$

This force is sometimes called a "restoring force" because the spring exerts its force in the direction opposite the displacement (hence the minus sign), acting to return it to its normal length. Equation 7–8 is known as the **spring equation** and also as **Hooke's law** (see Chapter 12), and is accurate for springs as long as x is not too great.

[†]The distance along a curve, Δl, is not generally equal to the magnitude of the displacement, $\Delta\mathbf{r}$, as shown in Fig. 7–10 below. But in the limit of infinitesimals they are equal: $dl = dr$, and in this limit the vector $d\mathbf{l} = d\mathbf{r}$. Note that we cannot define a vector $\Delta\mathbf{l}$ since we cannot assign it a unique direction when the path curves; we can define a direction for $d\mathbf{l}$—namely, the direction of the tangent to the curve at that point, so $d\mathbf{l} = d\mathbf{r}$ and $dl = dr$.

FIGURE 7–10 The displacement vector, $\Delta\mathbf{r}$, has a magnitude that is not necessarily equal to the distance traveled, Δl.

Let us calculate the work a person does to stretch (or compress) a spring from its normal (unstretched) length, $x_a = 0$, to an extra length, $x_b = x$. We assume the stretching is done slowly, so that the acceleration is essentially zero. The force $\mathbf{F}_P$ is exerted parallel to the axis of the spring, along the x axis, so $\mathbf{F}_P$ and $d\mathbf{l}$ are parallel. Hence, since $d\mathbf{l} = dx\,\mathbf{i}$ in this case, the work done by the person is[†]

$$W_P = \int_{x_a=0}^{x_b=x} [F_P(x)\mathbf{i}] \cdot [dx\,\mathbf{i}] = \int_0^x F_P(x)\,dx = \int_0^x kx\,dx = \tfrac{1}{2}kx^2 \Big|_0^x = \tfrac{1}{2}kx^2.$$

(As is frequently done, we have used x to represent both the variable of integration, and the particular value of x at the end of the interval $x_a = 0$ to $x_b = x$.) Thus we see that the work needed is proportional to the square of the distance stretched (or compressed), x. This same result can be obtained by computing the area under the F-versus-x graph (with $\cos\theta = 1$ in this case) as shown in Fig. 7–12. Since the area is a triangle of altitude kx and base x, the work a person does to stretch or compress a spring an amount x, equal to the area, is

$$W = \tfrac{1}{2}(x)(kx) = \tfrac{1}{2}kx^2,$$

which is the same result as before.

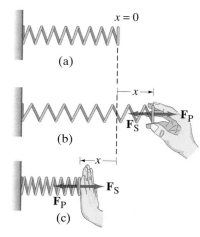

FIGURE 7–11 (a) Spring in normal (unstretched) position. (b) Spring is stretched by a person exerting a force $\mathbf{F}_P$ to the right (positive direction). The spring pulls back with a force $\mathbf{F}_S$ where $F_S = -kx$. (c) Person compresses the spring ($x < 0$), and the spring pushes back with a force $F_S = -kx$ where $F_S > 0$ because $x < 0$.

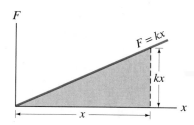

FIGURE 7–12 Work done to stretch a spring a distance x equals the triangular area under the curve $F = kx$. The area of a triangle is $\tfrac{1}{2} \times$ base $\times$ altitude, so $W = \tfrac{1}{2}(x)(kx) = \tfrac{1}{2}kx^2$.

EXAMPLE 7–5 **Work done on a spring.** (a) A person pulls on the spring in Fig. 7–11, stretching it 3.0 cm, which requires a maximum force of 75 N. How much work does the person do? (b) If, instead, the person compresses the spring 3.0 cm, how much work does the person do?

SOLUTION (a) First we need to calculate the spring constant k:

$$k = \frac{F_{max}}{x_{max}} = \frac{75\ \text{N}}{0.030\ \text{m}} = 2.5 \times 10^3\ \text{N/m}.$$

Then the work done by the person on the spring is

$$W = \tfrac{1}{2}kx_{max}^2 = \tfrac{1}{2}(2.5 \times 10^3\ \text{N/m})(0.030\ \text{m})^2 = 1.1\ \text{J},$$

(b) The force that the person exerts is still $F_P = kx$, though now both x and F_P are negative (x is positive to the right). The work done is

$$W_P = \int_{x=0}^{x=-0.030\,\text{m}} F_P(x)\,dx = \int_0^{x=-0.030\,\text{m}} kx\,dx = \tfrac{1}{2}kx^2 \Big|_0^{-0.030\,\text{m}}$$

$$= \tfrac{1}{2}(2.5 \times 10^3\ \text{N/m})(-0.030\ \text{m})^2 = 1.1\ \text{J},$$

which is the same as for stretching it.

Note that we cannot use $W = Fd$ (Eq. 7–1) for a spring because the force is not constant.

[†] See the Table of Integrals, Appendix B.

EXAMPLE 7-6 **Force as function of x.** A robot arm that controls the position of a video camera (Fig. 7–13) in an automated surveillance system is manipulated by a servo motor that exerts a force on a push-rod. The force is given by

$$F(x) = F_0\left(1 + \frac{1}{6}\frac{x^2}{x_0^2}\right)$$

where $F_0 = 2.0\,\text{N}$, $x_0 = 0.0070\,\text{m}$, and x is the position of the end of the push-rod. If the push-rod moves from $x_1 = 0.010\,\text{m}$ to $x_2 = 0.050\,\text{m}$, how much work did the servo motor do?

SOLUTION The force applied by the motor is not a linear function of x. We can determine the integral $\int F(x)\,dx$, or the area under the $F(x)$ curve (shown in Fig. 7–14). We integrate to find the work done by the motor:

$$W_M = F_0\int_{x_1}^{x_2}\left(1 + \frac{x^2}{6x_0^2}\right)dx = F_0\int_{x_1}^{x_2}dx + F_0\int_{x_1}^{x_2}\frac{x^2\,dx}{6x_0^2}$$

$$= F_0\left(x + \frac{1}{3}\frac{x^3}{6x_0^2}\right)\Big|_{x_1}^{x_2}.$$

We put in the values given and obtain

$$W_M = 2.0\,\text{N}\left[(0.050\,\text{m} - 0.010\,\text{m}) + \frac{(0.050\,\text{m})^3 - (0.010\,\text{m})^3}{(3)(6)(0.0070\,\text{m})^2}\right] = 0.361\,\text{J}.$$

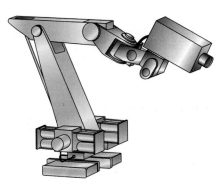

FIGURE 7–13 Robot arm positions a video camera.

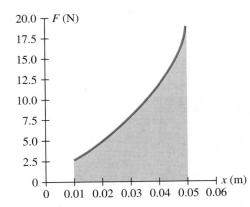

FIGURE 7–14 Example 7–6.

7-4 | Kinetic Energy and the Work-Energy Principle

Energy is one of the most important concepts in science. Yet we cannot give a simple general definition of energy in only a few words. Nonetheless, each specific type of energy can be defined fairly simply. In this chapter we define translational kinetic energy; in the next chapter, we take up potential energy. In later chapters we will define other types of energy, such as that related to heat (Chapters 19 and 20). The crucial aspect of all the types of energy is that they can be defined consistently in such a way that the sum of all types, the *total energy*, is the same after any process occurs as it was before: that is, the quantity "energy" can be defined so that it is a conserved quantity. But more on this later.

For the purposes of this chapter, we could define energy in the usual way as "the ability to do work." This simple definition is not very precise, nor is it really valid for all types of energy.[†] However, for mechanical energy which we discuss in

[†] Energy associated with heat is often not available to do work, as we will discuss in detail in Chapter 20.

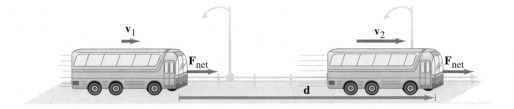

this chapter and the next, it serves to underscore the fundamental connection between work and energy. We now define and discuss one of the basic types of energy, kinetic energy.

A moving object can do work on another object it strikes. A flying cannonball does work on a brick wall it knocks down; a moving hammer does work on a nail it strikes. In either case, a moving object exerts a force on a second object and moves it through a distance. An object in motion has the ability to do work and thus can be said to have energy. The energy of motion is called **kinetic energy**, from the Greek word *kinetikos*, meaning "motion."

To obtain a quantitative definition for kinetic energy, let us consider a rigid object of mass m that is moving in a straight line with an initial speed v_1. To accelerate it uniformly to a speed v_2, a constant net force F_{net} is exerted on it parallel to its motion over a distance d, Fig. 7–15. Then the net work done on the object is $W_{net} = F_{net} d$. We apply Newton's second law, $F_{net} = ma$, and use Eq. 2–12c, which we now write as $v_2^2 = v_1^2 + 2ad$, with v_1 as the initial speed and v_2 the final speed, and we find

$$W_{net} = F_{net} d = mad = m\left(\frac{v_2^2 - v_1^2}{2d}\right)d$$

or

$$W_{net} = \tfrac{1}{2}mv_2^2 - \tfrac{1}{2}mv_1^2. \tag{7–9}$$

We *define* the quantity $\tfrac{1}{2}mv^2$ to be the **translational kinetic energy**, K, of the object:

$$K = \tfrac{1}{2}mv^2. \tag{7–10}$$

Translational kinetic energy defined

(We call this "translational" kinetic energy to distinguish it from rotational kinetic energy, which we will discuss later, in Chapter 10.)

We can rewrite Eq. 7–9 as:

$$W_{net} = K_2 - K_1$$

or

$$W_{net} = \Delta K. \tag{7–11}$$

WORK-ENERGY PRINCIPLE

WORK-ENERGY PRINCIPLE

Equation 7–11 (or Eq. 7–9) is an important result. It can be stated in words:

The net work done on an object is equal to the change in its kinetic energy.

This is known as the **work-energy principle**. Notice, however, that we made use of Newton's second law, $F_{net} = ma$, where F_{net} is the *net* force—the sum of all forces acting on the object. Thus, the work-energy principle is valid only if W is the *net work* done on the object—that is, the work done by all forces acting on the object.

The work-energy principle tells us that if (positive) net work W is done on a body, its kinetic energy increases by an amount W. The principle also holds true for the reverse situation: if negative net work W is done on the body, the body's kinetic energy decreases by an amount W. That is, a net force exerted on a body opposite to the body's direction of motion reduces its speed and its kinetic energy.

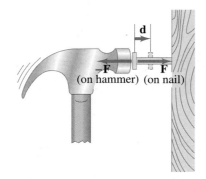

FIGURE 7–16 A moving hammer strikes a nail and comes to rest. The hammer exerts a force F on the nail; the nail exerts a force $-F$ on the hammer (Newton's third law). The work done on the nail by the hammer is positive $\left(W_n = Fd > 0\right)$. The work done on the hammer is negative $\left(W_h = -Fd\right)$.

General derivation of the work-energy principle

An example is a moving hammer (Fig. 7–16) striking a nail. The net force on the hammer ($-\mathbf{F}$ in the figure, where $\mathbf{F}$ is assumed constant for simplicity) acts toward the left, whereas the displacement $\mathbf{d}$ is toward the right. So the net work done on the hammer, $W_h = (F)(d)(\cos 180°) = -Fd$, is negative and the hammer's kinetic energy decreases (usually to zero).

Figure 7–16 also illustrates how energy can be considered the ability to do work. The hammer, as it slows down, does positive work on the nail: if the nail exerts a force $-\mathbf{F}$ on the hammer to slow it down, the hammer exerts a force $+\mathbf{F}$ on the nail (Newton's third law) through the distance d. Hence the work done on the nail by the hammer is $W_n = (+F)(+d) = Fd = -W_h$, and W_n is positive. Thus the decrease in kinetic energy of the hammer is equal to the work the hammer can do on another object—which is consistent with energy being the ability to do work.

Note that whereas the translational kinetic energy $\left(= \frac{1}{2}mv^2\right)$ is directly proportional to the mass of the object, it is proportional to the *square* of the speed. Thus, if the mass is doubled, the kinetic energy is doubled. But if the speed is doubled, the object has four times as much kinetic energy and is therefore capable of doing four times as much work.

To summarize, the connection between work and kinetic energy (Eq. 7–11) operates both ways. If the net work W done on an object is positive, then the object's kinetic energy increases. If the net work W done on an object is negative, its kinetic energy decreases. If the net work done on the object is zero, its kinetic energy remains constant (which also means its speed is constant).

We derived the work-energy principle, Eq. 7–11, for motion in one dimension with a constant force. It is valid even if the force is variable and the motion is in two or three dimensions, as we now show. Suppose the net force $\mathbf{F}_{net}$ on an object varies in both magnitude and direction, and the path of the object is a curve as in Fig. 7–8. The net force may be considered to be a function of l, the distance along the curve. The net work done is (Eq. 7–6):

$$W_{net} = \int F_{net} \cos\theta \, dl = \int F_{\parallel} \, dl,$$

where $F_{\parallel}$ represents the component of the net force parallel to the curve at any point. By Newton's second law,

$$F_{\parallel} = ma_{\parallel} = m\frac{dv}{dt},$$

where $a_{\parallel}$, the component of a parallel to the curve at any point, is equal to the rate of change of speed, dv/dt. We can think of v as a function of l, and using the chain rule for derivatives, we have

$$\frac{dv}{dt} = \frac{dv}{dl}\frac{dl}{dt} = v\frac{dv}{dl},$$

since dl/dt is the speed v. Thus (letting 1 and 2 refer to the initial and final quantities, respectively):

$$W_{net} = \int_1^2 F_{\parallel} \, dl = \int_1^2 m\frac{dv}{dt} \, dl = \int_1^2 mv\frac{dv}{dl} \, dl = \int_1^2 mv \, dv,$$

which integrates to

$$W_{net} = \frac{1}{2}mv_2^2 - \frac{1}{2}mv_1^2 = \Delta K.$$

This is again the work-energy principle, which we have now derived for motion in three dimensions with a variable net force. Note, incidentally, that the work-energy principle is not a new independent law. Rather, it has been derived from the definitions of work and kinetic energy using Newton's second law.

Notice in this derivation that only the component of $\mathbf{F}_{net}$ parallel to the motion, $F_{\parallel}$, contributes to the work. Indeed, a force (or component of a force) acting perpendicular to the velocity vector does no work. Such a force changes only the direction of the velocity. It does not affect the magnitude of the velocity. One example of this is uniform circular motion in which an object moving with constant speed in a circle has a ("centripetal") force acting on it toward the center of the circle. This force does no work on the object, because (as we saw in Example 7–3) it is always perpendicular to the object's displacement $d\mathbf{l}$.

Centripetal force does no work

Because of the direct connection between work and kinetic energy, Eq. 7–11, energy is measured in the same units as work: joules in SI units, ergs in the cgs, and foot-pounds in the British system. Like work, kinetic energy is a scalar quantity. The kinetic energy of a set of objects is the (scalar) sum of the kinetic energies of the individual objects.

Energy unit: the joule

EXAMPLE 7–7 **A baseball.** A 145-g baseball is thrown with a speed of 25 m/s. (*a*) What is its kinetic energy? (*b*) How much work was done to reach this speed, starting from rest?

SOLUTION (*a*) The kinetic energy is

$$K = \tfrac{1}{2}mv^2 = \tfrac{1}{2}(0.145 \text{ kg})(25 \text{ m/s})^2 = 45 \text{ J}.$$

(*b*) Since the initial kinetic energy was zero, the net work done is just equal to the final kinetic energy, 45 J.

CONCEPTUAL EXAMPLE 7–8 **Work to stop a car.** An automobile traveling 60 km/h can brake to a stop within a distance of 20 m (Fig. 7–17a). If the car is going twice as fast, 120 km/h, what is its stopping distance (Fig. 7–17b)? The maximum braking force is approximately independent of speed.

RESPONSE We treat the car as if it were a particle or simple rigid body. Since the stopping force F is approximately constant, the work needed to stop the car, Fd, is proportional to the distance traveled. We apply the work-energy principle, noting that $\mathbf{F}$ and $\mathbf{d}$ are in opposite directions and that the final speed of the car is zero:

$$W_{\text{net}} = Fd \cos 180° = -Fd$$
$$= \Delta K = 0 - \tfrac{1}{2}mv^2.$$

Thus, since the force and mass are constant, we can see that the stopping distance, d, increases with the square of the speed:

$$d \propto v^2.$$

Stopping distance $\propto$ initial speed squared

If the car's initial speed is doubled, the stopping distance is $(2)^2 = 4$ times as great, or 80 m.

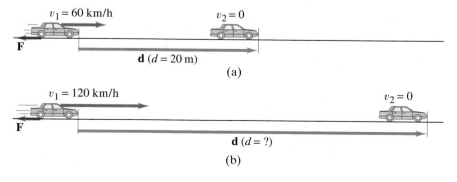

$v_1 = 60$ km/h $\qquad$ $v_2 = 0$

$\mathbf{F}$

$\mathbf{d}$ ($d = 20$ m)

(a)

$v_1 = 120$ km/h $\qquad\qquad$ $v_2 = 0$

$\mathbf{F}$

$\mathbf{d}$ ($d = ?$)

(b)

FIGURE 7–17
Example 7–8.

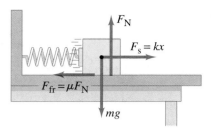

FIGURE 7–18 Example 7–9.

EXAMPLE 7–9 **A compressed spring.** A horizontal spring has spring constant $k = 360\,\text{N/m}$. (a) How much work is required to compress it from its uncompressed length ($x = 0$) to $x = 11.0\,\text{cm}$? (b) If a 1.85-kg block is placed against the spring and the spring is released, what will be the speed of the block when it separates from the spring at $x = 0$? Ignore friction. (c) Repeat part (b) but assume that the block is moving on a table as in Fig. 7–18 and that the coefficient of kinetic friction is $\mu_k = 0.38$.

SOLUTION (a) We saw in Section 7–3 that the net work, W, needed to stretch or compress a spring by a distance x is $W = \frac{1}{2}kx^2$. Therefore the work required to compress the spring a distance $x = 0.110\,\text{m}$ is

$$W = \tfrac{1}{2}(360\,\text{N/m})(0.110\,\text{m})^2 = 2.18\,\text{J},$$

where we have converted all units to SI.

(b) In returning to its uncompressed length, the spring does 2.18 J of work on the block (same calculation as in part (a), only in reverse). According to the work-energy principle, the block acquires kinetic energy of 2.18 J. Since $K = \frac{1}{2}mv^2$, the block's speed must be

$$v = \sqrt{\frac{2K}{m}} = \sqrt{\frac{2(2.18\,\text{J})}{1.85\,\text{kg}}} = 1.54\,\text{m/s}.$$

(c) There are two forces on the block: that exerted by the spring and that exerted by friction. The spring does 2.18 J of work on the block. Since the normal force F_N equals the weight mg (there is no vertical motion) the work done by the friction force on the block, $\mu_k F_N = \mu_k mg$, is

$$W_{\text{fr}} = (-\mu_k mg)(x) = -(0.38)(1.85\,\text{kg})(9.8\,\text{m/s}^2)(0.110\,\text{m}) = -0.76\,\text{J}.$$

This work is negative because the force of friction is in the direction opposite to the displacement x. The net work done on the block is $W_{\text{net}} = 2.18\,\text{J} - 0.76\,\text{J} = 1.42\,\text{J}$. From the work-energy principle, Eq. 7–9 (with $v_2 = v$ and $v_1 = 0$), we have

$$v = \sqrt{\frac{2W_{\text{net}}}{m}} = \sqrt{\frac{2(1.42\,\text{J})}{1.85\,\text{kg}}} = 1.24\,\text{m/s}.$$

FIGURE 7–19 Example 7–10.

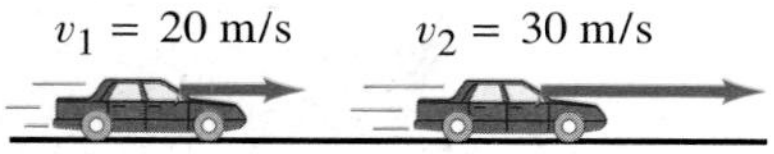

$v_1 = 20\,\text{m/s}$ $v_2 = 30\,\text{m/s}$

EXAMPLE 7–10 **Work to accelerate a car.** How much work is required to accelerate a 1000-kg car from 20 m/s to 30 m/s? See Fig. 7–19.

SOLUTION We need to be careful, for this is (if looked at in detail) a complicated situation. However, if we treat the car as a particle, then we can write that the net work needed is equal to the increase in its kinetic energy:

$$\begin{aligned} W_{\text{net}} &= \tfrac{1}{2}mv_2^2 - \tfrac{1}{2}mv_1^2 \\ &= \tfrac{1}{2}(1000\,\text{kg})\big[(30\,\text{m/s})^2 - (20\,\text{m/s})^2\big] = 2.5 \times 10^5\,\text{J}. \end{aligned}$$

This conclusion is useful. The difficulty comes if we look deeper. First, we note that the net force that accelerates the car is the friction force that the road exerts on the tires (the reaction to the tires pushing against the road). In reality, this force does no work because the tires are not sliding and have no component of motion parallel to the road or to the friction force. In principle, this Example might be approached by examining the work done by the engine, transmission to the wheels, and so on—sometimes referred to as "internal work"—but it becomes very complicated. What counts here, is that we do get a useful result from our calculation above in which we treated the car as a particle (or simple rigid body) undergoing translational motion.

This last Example, though useful, also shows us that the concept of work is not always helpful. The problem is that we are trying to examine a deformable (non-rigid) object that can have internal motions. To delve deeper into the car example, to deal with the engine, transmission, and connections to the wheels, we need to use the powerful concept of energy. With energy in its various forms, we can talk about the energy stored in the gasoline being used to do work on the piston, and the energy eventually getting transferred to the wheels. In the next chapter we will extend our discussion of energy to other forms of energy, particularly potential energy. Energy is a central topic in physics and it will come up a lot in later chapters as well. An important result discussed in the next chapter is that energy can be transformed from one type to another, but the total energy never increases or decreases. This is the law of conservation of energy, one of the most important laws in physics.

* 7-5 | Kinetic Energy at Very High Speed

Einstein's theory of special relativity, which we will discuss in Chapter 37, points out some difficulties with classical Newtonian mechanics. First published in 1905, Einstein's great theory has caused us to reevaluate some aspects of mechanics. For now we need only point out that, according to relativity theory, the kinetic energy of a particle of mass m moving with velocity v is given by

$$K = mc^2 \left(\frac{1}{\sqrt{1 - \dfrac{v^2}{c^2}}} - 1 \right) \qquad \text{[Relativistic kinetic energy]} \quad \textbf{(7-12)}$$

FIGURE 7–20 (a) Interior photo of the Stanford linear collider. (b) Photo in Brookhaven bubble chamber that reveals tracks of tiny elementary particles.

where c is the speed of light, $c = 3.00 \times 10^8$ m/s. Einstein's formula, Eq. 7–12, gives a significantly different result from the simpler classical form, $K = \frac{1}{2}mv^2$, only when the object's speed v is extremely high, greater than about $\frac{1}{10}c$. Indeed, Eq. 7–12 reduces to the classical equation when $v \ll c$, which is easy to show using the binomial expansion,

$$(1 + x)^n = 1 + nx + n(n - 1)x^2/2! + \cdots .$$

Thus, setting $n = -\frac{1}{2}$ and $x = -v^2/c^2$, then for $v \ll c$, Eq. 7–12 becomes

$$K = mc^2 \left(\frac{1}{\sqrt{1 - \dfrac{v^2}{c^2}}} - 1 \right) = mc^2 \left(1 + \frac{1}{2}\frac{v^2}{c^2} + \cdots - 1 \right) \approx \frac{1}{2}mv^2$$

where we ignore higher terms in the expansion because they are very small given that $v/c \ll 1$.

Experiments with subatomic particles (Fig. 7–20), such as electrons and protons, confirm Eq. 7–12. An interesting aspect of Eq. 7–12 is that velocities equal to or greater than the speed of light c are not possible. Why? If v were equal to c, then the denominator in the first term, $\sqrt{1 - v^2/c^2}$, would equal zero; so the kinetic energy would be infinite. That is, it would take an infinite amount of work to accelerate an object of non-zero mass to $v = c$. And if v were greater than c, the factor $\sqrt{1 - v^2/c^2}$ would be the square root of a negative number, which is imaginary and, we believe, non-physical. Indeed, experiment confirms that no particle with mass reaches (or exceeds) the speed of light.

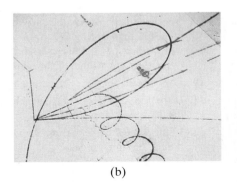

(a)

(b)

Summary

Work is done on a body by a force when that force acts on the body as it moves the object through some distance. The **work** W done by a constant force $\mathbf{F}$ on an object whose position changes by a displacement $\mathbf{d}$ is given by

$$W = Fd\cos\theta = \mathbf{F}\cdot\mathbf{d},$$

where θ is the angle between $\mathbf{F}$ and $\mathbf{d}$.

The last expression is called the scalar product of $\mathbf{F}$ and $\mathbf{d}$. In general, the **scalar product** of any two vectors $\mathbf{A}$ and $\mathbf{B}$ is defined as

$$\mathbf{A}\cdot\mathbf{B} = AB\cos\theta$$

where θ is the angle between $\mathbf{A}$ and $\mathbf{B}$. In rectangular coordinates we can also write $\mathbf{A}\cdot\mathbf{B} = A_x B_x + A_y B_y + A_z B_z$.

The work W done by a variable force $\mathbf{F}$ on an object that moves from point a to point b is

$$W = \int_a^b \mathbf{F}\cdot d\mathbf{l} = \int_a^b F\cos\theta\, dl,$$

where $d\mathbf{l}$ represents an infinitesimal displacement along the path of the object and θ is the angle between $d\mathbf{l}$ and $\mathbf{F}$ at each point of the object's path.

The translational **kinetic energy** K of an object of mass m moving with speed v is defined to be

$$K = \tfrac{1}{2}mv^2.$$

The **work-energy principle** states that the net work done on a body by the net resultant force is equal to the change in kinetic energy of the body:

$$W_{\text{net}} = \tfrac{1}{2}mv_2^2 - \tfrac{1}{2}mv_1^2.$$

Questions

1. In what ways is the word "work" as used in everyday language the same as defined in physics? In what ways is it different?

2. A woman swimming upstream in a fast river is not moving with respect to the shore. Is she doing any work? If she stops swimming and merely floats, is work done on her?

3. Is the work done by a kinetic friction force always negative? [*Hint*: Consider what happens to the dishes when pulling a tablecloth from under your mom's best china.]

4. Does the scalar product of two vectors depend on the choice of coordinate system?

5. Can a dot product ever be negative? If yes, under what conditions?

6. If $\mathbf{A}\cdot\mathbf{C} = \mathbf{B}\cdot\mathbf{C}$, is it necessarily true that $\mathbf{A} = \mathbf{B}$?

7. Does the dot product of two vectors have direction as well as magnitude?

8. Can the normal force on an object ever do work? Explain.

9. You have two springs that are identical except that spring 1 is stiffer than spring 2 $(k_1 > k_2)$. On which spring is more work done: (*a*) if they are stretched using the same force; (*b*) if they are stretched the same distance?

10. Can kinetic energy ever be negative? Explain.

11. If the kinetic energy of a particle is doubled, by what factor has its speed increased?

12. If the speed of a particle triples, by what factor does its kinetic energy increase?

13. In Example 7–9, it was stated that the block separates from the compressed spring when the spring reached its equilibrium length $(x = 0)$. Explain why separation doesn't take place before (or after) this point.

14. Two bullets are fired at the same time with the same kinetic energy. If one bullet has twice the mass of the other, which has the greater speed and by what factor? Which can do the most work?

15. Does the net work done on a particle depend on the choice of reference frame? How does this affect the work-energy principle?

16. A hand exerts a constant horizontal force on a block that is free to slide on a frictionless surface, as shown below (Fig. 7–21). The block starts from rest at point A, and by the time it has traveled a distance d to point B it is traveling with speed v_B. When the block has traveled another distance d to point C, will its speed be greater than, less than, or equal to $2v_B$? Explain your reasoning.

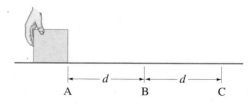

FIGURE 7–21 Question 16.

Problems

Section 7–1

1. (I) How much work is done by the gravitational force when a 250-kg pile driver falls 2.80 m?

2. (I) If the average retarding force on a car is 535 N, how much work is done by this force as the car moves 1.25 km?

3. (I) A 65.0-kg firefighter climbs a flight of stairs 20.0 m high. How much work is required?

4. (I) A 1200-N crate rests on the floor. How much work is required to move it at constant speed (a) 4.0 m along the floor against a friction force of 230 N, and (b) 4.0 m vertically?

5. (I) How much work did the movers do (horizontally) pushing a 160-kg crate 10.3 m across a rough floor without acceleration, if the effective coefficient of friction was 0.50?

6. (I) How high will a 1.85-kg rock go if thrown straight up by someone who does 80.0 J of work on it? Neglect air resistance.

7. (I) Determine the conversion factor between joules and ergs and between joules and foot-pounds.

8. (I) A hammerhead with a mass of 2.0 kg is allowed to fall onto a nail from a height of 0.50 m. What is the maximum amount of work it could do on the nail? Why do people not just "let it fall" but add their own force to the hammer as it falls?

9. (II) Estimate the work you do to mow a lawn 10 m by 20 m. Assume you push with a force of about 15 N.

10. (II) What is the minimum work needed to push a 950-kg car 310 m up along a 9.0° incline? (a) Ignore friction. (b) Assume the effective coefficient of friction retarding the car is 0.25.

11. (II) A lever such as that shown in Fig. 7–22 can be used to lift objects we might not otherwise be able to lift. Show that the ratio of output force, F_O, to input force, F_I, is related to the lengths l_I and l_O from the pivot by $F_O/F_I = l_I/l_O$ (ignoring friction and the mass of the lever), given that the work output equals work input.

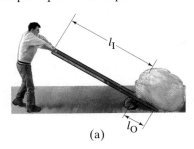

(a)

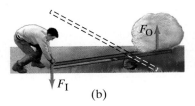

(b)

FIGURE 7–22
A simple lever.
Problem 11.

12. (II) Eight books, each 4.3 cm thick with mass 1.7 kg, lie flat on a table. How much work is required to stack them one on top of another?

13. (II) A 380-kg piano slides 3.5 m down a 27° incline and is kept from accelerating by a man who is pushing back on it *parallel to the incline* (Fig. 7–23). The effective coefficient of kinetic friction is 0.40. Calculate: (a) the force exerted by the man, (b) the work done by the man on the piano, (c) the work done by the friction force, (d) the work done by the force of gravity, and (e) the net work done on the piano.

FIGURE 7–23 Problem 13.

14. (II) A 20,000-kg jet takes off from an aircraft carrier via a catapult (Fig. 7–24a). The jet's engines exert a constant force of 130 kN on the jet; the force exerted on the jet by the catapult is plotted in Fig. 7–24b. Determine: (a) the work done on the jet by its engines during launch of the jet; and (b) the work done on the jet by the catapult during launch of the jet.

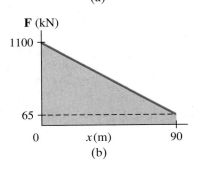

(a)

(b)

FIGURE 7–24 Problem 14.

15. (II) A grocery cart with mass of 18 kg is being pushed at constant speed along an aisle by a force $F_P = 14$ N which acts at an angle of 20° to the horizontal. Find the work done by each of the forces $(m\mathbf{g}, \mathbf{F}_N, \mathbf{F}_P, \mathbf{F}_{fr})$ on the cart if the aisle is 15 m long.

16. II (a) Find the force required to give a helicopter of mass M an acceleration of 0.10 g upward. (b) Find the work done by this force as the helicopter moves a distance h upward.

Section 7–2

17. (I) For any vector $\mathbf{V} = V_x\mathbf{i} + V_y\mathbf{j} + V_z\mathbf{k}$ show that
$$V_x = \mathbf{i} \cdot \mathbf{V}, \quad V_y = \mathbf{j} \cdot \mathbf{V}, \quad V_z = \mathbf{k} \cdot \mathbf{V}.$$

18. (I) Calculate the angle between the vectors:
$$\mathbf{A} = 6.8\mathbf{i} + 4.6\mathbf{j} + 6.2\mathbf{k} \quad \text{and} \quad \mathbf{B} = 8.2\mathbf{i} + 2.3\mathbf{j} - 7.0\mathbf{k}.$$

19. (I) Show that $\mathbf{A} \cdot (-\mathbf{B}) = -\mathbf{A} \cdot \mathbf{B}$.

20. (I) Vector $\mathbf{V}_1$ points along the z axis and has magnitude $V_1 = 75$. Vector $\mathbf{V}_2$ lies in the xz plane, has magnitude $V_2 = 50$, and makes a $-48°$ angle with the x axis (points below x axis). What is the scalar product $\mathbf{V}_1 \cdot \mathbf{V}_2$?

21. (II) If $\mathbf{A} = 7.0\mathbf{i} - 8.5\mathbf{j}$, $\mathbf{B} = -8.0\mathbf{i} + 8.1\mathbf{j} + 4.2\mathbf{k}$, and $\mathbf{C} = 6.8\mathbf{i} - 7.2\mathbf{j}$, determine (a) $\mathbf{A} \cdot (\mathbf{B} + \mathbf{C})$; (b) $(\mathbf{A} + \mathbf{C}) \cdot \mathbf{B}$; (c) $(\mathbf{B} + \mathbf{A}) \cdot \mathbf{C}$.

22. (II) Prove that $\mathbf{A} \cdot \mathbf{B} = A_x B_x + A_y B_y + A_z B_z$, starting from Eq. 7–2 and using the distributive law (proved in Problem 29).

23. (II) Given vectors $\mathbf{A} = -4.8\mathbf{i} + 7.8\mathbf{j}$ and $\mathbf{B} = 9.6\mathbf{i} + 6.7\mathbf{j}$, determine the vector $\mathbf{C}$ that lies in the xy plane perpendicular to $\mathbf{B}$ and whose dot product with $\mathbf{A}$ is 20.0.

24. (II) Show that if two non-parallel vectors have the same magnitude, their sum must be perpendicular to their difference.

25. (II) Let $\mathbf{V} = 20.0\mathbf{i} + 12.0\mathbf{j} - 14.0\mathbf{k}$. What angles does this vector make with the x, y, and z axes?

26. (II) Use the scalar product to prove the law of cosines for a triangle:
$$c^2 = a^2 + b^2 - 2ab \cos\theta,$$
where a, b, and c are the lengths of the sides of a triangle and θ is the angle opposite side c.

27. (II) Vectors $\mathbf{A}$ and $\mathbf{B}$ are in the xy plane and their scalar product is 20.0 units. If $\mathbf{A}$ makes a 30° angle with the x axis and has magnitude $A = 12.0$ units, and $\mathbf{B}$ has magnitude $B = 4.0$ units, what can you say about the direction of $\mathbf{B}$?

28. (II) $\mathbf{A}$ and $\mathbf{B}$ are two vectors in the xy plane that make angles α and β with the x axis respectively. Evaluate the scalar product of $\mathbf{A}$ and $\mathbf{B}$ and deduce the following trigonometric identity: $\cos(\alpha - \beta) = \cos\alpha \cos\beta + \sin\alpha \sin\beta$.

29. (III) Show that the scalar product of two vectors is distributive: $\mathbf{A} \cdot (\mathbf{B} + \mathbf{C}) = \mathbf{A} \cdot \mathbf{B} + \mathbf{A} \cdot \mathbf{C}$. [Hint: Use a diagram showing all three vectors in a plane and indicate dot products on the diagram.]

Section 7–3

30. (I) In pedaling a bicycle uphill, a cyclist exerts a downward force of 470 N during each stroke. If the diameter of the circle traced by each pedal is 36 cm, calculate how much work is done in each stroke.

31. (I) A spring has $k = 84$ N/m. Draw a graph like that in Fig. 7–12 and use it to determine the work needed to stretch the spring from $x = 3.0$ cm to $x = 5.5$ cm, where $x = 0$ refers to the spring's unstretched length.

32. (II) If the hill in Example 7–2 (Fig. 7–4) were not an even slope but rather an irregular curve as in Fig. 7–25, show that the same result would be obtained as in Example 7–2: namely, that the work done by gravity depends only on the height of the hill and not on its shape or the path taken.

FIGURE 7–25 Problem 32.

33. (II) The net force exerted on a particle acts in the positive x direction. Its magnitude increases linearly from zero at $x = 0$, to 300 N at $x = 3.0$ m. It remains constant at 300 N from $x = 3.0$ m to $x = 7.0$ m and then decreases linearly to zero at $x = 11.0$ m. Determine the work done to move the particle from $x = 0$ to $x = 11.0$ m by graphically determining the area under the F versus x graph.

34. (II) The force on a particle, acting along the x axis, varies as shown in Fig. 7–26. Determine the work done by this force to move the particle along the x axis: (a) from $x = 0.0$ to $x = 10.0$ m; (b) from $x = 0.0$ to $x = 15.0$ m.

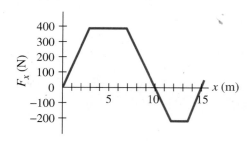

FIGURE 7–26 Problem 34.

35. (II) In Fig. 7–9 assume the distance axis is linear and that a = 10.0 m and b = 30.0 m. Estimate the work done by this force in moving a 2.50-kg object from a to b.

36. (II) The resistance of a packing material to a sharp object penetrating it is a force proportional to the fourth power of the penetration depth, x: $\mathbf{F} = kx^4\mathbf{i}$. Calculate the work done to force the sharp object a distance d.

37. (II) The force needed to hold a particular spring compressed an amount x from its normal length is given by $F = kx + ax^3 + bx^4$. How much work must be done to compress it by an amount X, starting from $x = 0$?

38. (III) A 3.0-m-long steel chain is stretched out along the top level of a horizontal scaffold at a construction site, in such a way that 2.0 m of the chain remains on the top level and 1.0 m hangs vertically, Fig. 7–27. At this point, the force on the hanging segment is sufficient to pull the entire chain over the edge. Once the chain is moving, the kinetic friction is so small that it can be neglected. How much work is performed on the chain by the force of gravity as the chain falls from the point where 2.0 m remains on the scaffold to the point where the entire chain has left the scaffold? (Assume that the chain has a linear weight density of 20 N/m.)

FIGURE 7–27 Problem 38.

39. (III) A 2500-kg space vehicle, initially at rest, falls vertically from a height of 3000 km above the Earth's surface. (a) Determine how much work is done by the force of gravity in bringing the vehicle to the Earth's surface by first constructing an F versus r graph (using Eq. 7–1), where r is the distance from the Earth's center. Then determine the work graphically to an accuracy of 3 percent. (b) Repeat using integration.

Section 7–4

40. (I) At room temperature, an oxygen molecule, with mass of 5.31×10^{-26} kg, typically has a kinetic energy of about 6.21×10^{-21} J. How fast is it moving?

41. (I) (a) If the kinetic energy of a particle is tripled, by what factor has its speed increased? (b) If the speed of a particle is halved, by what factor does its kinetic energy change?

42. (I) How much work is required to stop an electron $(m = 9.11 \times 10^{-31}$ kg$)$ which is moving with a speed of 1.70×10^6 m/s?

43. (I) How much work must be done to stop a 1300-kg car traveling at 100 km/h?

44. (II) An 85-g arrow is fired from a bow whose string exerts an average force of 105 N on the arrow over a distance of 80 cm. What is the speed of the arrow as it leaves the bow?

45. (II) A baseball $(m = 145$ g$)$ traveling 32 m/s moves a fielder's glove backward 25 cm when the ball is caught. What was the average force exerted by the ball on the glove?

46. (II) If the speed of a car is increased by 50%, by what factor will its minimum braking distance be increased, assuming all else is the same? Ignore the driver's reaction time.

47. (II) At an accident scene on a level road, investigators measure a car's skid mark to be 78 m long. It was a rainy day and the coefficient of friction was estimated to be 0.38. Use these data to determine the speed of the car when the driver slammed on (and locked) the brakes. (Why does the car's mass not matter?)

48. (II) One car has twice the mass of a second car, but only half as much kinetic energy. When both cars increase their speed by 7.0 m/s, they then have the same kinetic energy. What were the original speeds of the two cars?

49. (II) A force of 6.0 N is used to accelerate a mass of 1.0 kg from rest for a distance of 12 m. The force is applied along the direction of travel. The coefficient of kinetic friction is 0.30. What is the work done (a) by the applied force? (b) by friction? (c) What is the kinetic energy at the 12-m mark?

50. (II) A 1200-kg car rolling on a horizontal surface has speed $v = 60$ km/h when it strikes a horizontal coiled spring and is brought to rest in a distance of 2.2 m. What is the spring constant of the spring?

51. (II) At 66.0-kg crate, starting from rest, is pulled across a floor with a constant horizontal force of 225 N. For the first 11.0 m the floor is frictionless, and for the next 10.0 m the coefficient of friction is 0.20. What is the final speed of the crate after being pulled these 21.0 m?

52. (II) A mass m is attached to a spring which is held stretched a distance x by a force F (Fig. 7–28), and then released. The spring compresses, pulling the mass. Assuming there is no friction, determine the speed of the mass m when the spring returns: (a) to its normal length ($x = 0$); (b) to half its original extension ($x/2$).

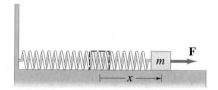

FIGURE 7–28 Problems 52 and 53.

53. (II) Suppose that there is friction in the preceding problem and the mass on the end of the stretched spring in Fig. 7–28, after being released, comes to rest just when it reaches the spring's equilibrium position. Determine the coefficient of friction, μ_k, in terms of F, x, g, and m.

54. (II) A 355-kg load is lifted 33.0 m vertically by a single cable with an acceleration $a = 0.15$ g. Determine (a) the tension in the cable; (b) the net work done on the load; (c) the work done by the cable on the load; (d) the work done by gravity on the load; (e) the final speed of the load assuming it starts from rest.

55. (II) (*a*) How much work is done by the horizontal force $F_P = 150$ N on the 20-kg block of Fig. 7–29 when the force pushes the block 5.0 m up the 30° frictionless incline? (*b*) How much work is done by the gravitational force on the block during this displacement? (*c*) How much work is done by the normal force? (*d*) What is the speed of the block (assume that it is zero initially) after this displacement?

56. (II) Repeat Problem 55 assuming a coefficient of friction $\mu_k = 0.10$.

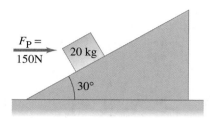

FIGURE 7–29 Problems 55 and 56.

57. (III) An elevator cable breaks when a 755-kg elevator is 22.5 m above the top of a huge spring $(k = 8.00 \times 10^4 \, \text{N/m})$ at the bottom of the shaft. Calculate (*a*) the work done by gravity on the elevator before it hits the spring; (*b*) the speed of the elevator just before striking the spring; (*c*) the amount the spring compresses (note that here work is done by both the spring and gravity).

58. (III) We usually neglect the mass of a spring if it is small compared to the mass attached to the spring. But in some applications, the mass of the spring must be taken into account. Consider a spring of unstretched length L and mass M_S. This mass is uniformly distributed along the length of the spring. A mass m is attached to the end of the spring. One

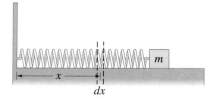

FIGURE 7–30 Problem 58.

end of the spring is fixed and the mass m is allowed to vibrate horizontally without friction (see Fig. 7–30). Each point on the spring moves with a velocity proportional to the distance from that point to the fixed end. For example, if the mass on the end moves with speed v, the midpoint of the spring moves with speed $v/2$. Show that the kinetic energy of the mass plus spring when the mass is moving with velocity v is

$$K = \tfrac{1}{2} M v^2$$

where $M = m + \tfrac{1}{3} M_S$ is the "effective mass" of the system.

* **Section 7–5**

* 59. (II) In two separate experiments, a proton (with mass of 1.67×10^{-27} kg) and an electron (with mass of 9.1×10^{-31} kg) are each accelerated by a device that performs 3.2×10^{-13} J of work on each particle. Using the relativistic formula for kinetic energy, determine the resulting speed of each particle. Compare your calculations with the results that would have been achieved using the classical formula.

* 60. (II) Calculate, to three significant figures, the kinetic energy of a particle of mass m moving with a speed of (*a*) 3.00×10^4 m/s, (*b*) 3.00×10^6 m/s (1% of the speed of light), and (*c*) 3.00×10^7 m/s (= $0.1c$). Compare each result to a classical calculation.

▨ General Problems

61. In a certain library the first shelf is 12.0 cm off the ground, and the remaining 4 shelves are each spaced 33.0 cm above the previous one. If the average book has a mass of 1.60 kg with a height of 22.0 cm, and an average shelf holds 25 books (standing vertically), how much work is required to fill this bookshelf from scratch, assuming the books are all laying flat on the floor to start?

62. (*a*) A 3.0-g locust reaches a speed of 3.0 m/s during its jump. What is its kinetic energy at this speed? (*b*) If the locust transforms energy with 40 percent efficiency, how much energy is required for the jump?

63. A 6.0-kg block is pushed 7.0 m up a rough 37° inclined plane by a horizontal force of 75 N. If the initial speed of the block is 2.2 m/s up the plane and a constant kinetic friction force of 25 N opposes the motion, calculate (*a*) the initial kinetic energy of the block; (*b*) the work done by the 75-N force; (*c*) the work done by the friction force; (*d*) the work done by gravity; (*e*) the work done by the normal force; (*f*) the final kinetic energy of the block.

64. A mass m slides on a circular horizontal track in a circle of radius R, Fig. 7–31. Its initial speed is v_0, but after one revolution the speed has dropped to $0.75v_0$ because of friction.

Determine (*a*) the work done by friction during one revolution; (*b*) the coefficient of friction; (*c*) the number of revolutions the mass will make before coming to rest.

FIGURE 7–31 Problem 64.

65. Two forces, $\mathbf{F}_1 = (1.50\mathbf{i} - 0.80\mathbf{j} + 0.70\mathbf{k})$ N and $\mathbf{F}_2 = (-0.70\mathbf{i} + 1.20\mathbf{j})$ N, are applied on a moving object of mass 0.20 kg. The displacement vector produced by the two forces is $\mathbf{d} = (8.0\mathbf{i} + 6.0\mathbf{j} + 5.0\mathbf{k})$ m. (*a*) What is the work done by the two forces? (*b*) If there is a frictional force given by $\mathbf{F}_{fr} = -0.20\mathbf{F}_1$, what is the net work done? (*c*) What is the work done by the frictional force in (*b*)?

66. The barrels of the 16-inch guns (bore diameter = 16 in. = 40 cm) on the World-War II battleship *U.S.S. Massachusetts* were each 15 m long. The shells each had a mass of 1250 kg and were fired with sufficient explosive to provide them with a muzzle velocity of 750 m/s. Use the work-energy principle to determine the explosive force (assumed to be a constant) that was applied to the shell within the barrel of the gun. Express your answer in both newtons and in pounds.

67. A force $\mathbf{F} = (10.0\mathbf{i} + 9.0\mathbf{j} + 12.0\mathbf{k})$kN acts on a small object of mass 100 g. If the displacement of the object is $\mathbf{d} = (5.0\mathbf{i} + 4.0\mathbf{j})$m, find the work done by the force. What is the angle between $\mathbf{F}$ and $\mathbf{d}$?

68. The arrangement of atoms in zinc is an example of "hexagonal close-packed" structure. Three of the nearest neighbors are found at the following (x, y, z) coordinates, given in nanometers $(10^{-9}$ m$)$: atom 1 is at $(0, 0, 0)$; atom 2 is at $(0.230, 0.133, 0)$; atom 3 is at $(0.077, 0.133, 0.247)$. Find the angle between two vectors: one that connects atom 1 with atom 2 and another that connects atom 1 with atom 3.

69. A varying force is given by $F = Ae^{-kx}$, where x is the position; A and k are constants that have units of N and m^{-1}, respectively. What is the work done when x goes from 0.10 m to infinity?

70. The force required to compress an imperfect horizontal spring an amount x is given by $F = 150x + 12x^3$, where x is in meters and F in newtons. If the spring is compressed 2.0 m, what speed will it give to a. 3.0-kg ball held against it and then released?

71. A softball having a mass of 0.25 kg is pitched at 110 km/h. By the time it reaches the plate, it may have slowed by 10 percent. Neglecting gravity, estimate the average force of air resistance during a pitch, if the distance between the plate and the pitcher is about 15 m.

72. Today's cars have "5 mi/h (8 km/h) bumpers" that are designed to elastically compress and rebound without any physical damage at speeds below 8 km/h. If the material of the bumpers permanently deforms after a compression of 1.5 cm, but remains like an elastic spring up to that point, what must the effective spring constant of the bumper material be, assuming the car has a mass of 1150 kg and is tested by ramming into a solid wall?

73. What should be the spring constant k of a spring designed to bring a 1300-kg car to rest from a speed of 90 km/h so that the occupants undergo a maximum acceleration of $5.0 g$?

74. An airplane pilot fell 370 m after jumping without his parachute opening. He landed in a snowbank, creating a crater 1.1 m deep, but survived with only minor injuries. Assuming the pilot's mass was 80 kg and his terminal velocity was 50 m/s, estimate: (a) the work done by the snow in bringing him to rest; (b) the average force exerted on him by the snow to stop him; and (c) the work done on him by air resistance as he fell.

75. Assume a cyclist of weight mg can exert a force on the pedals equal to 0.90 mg on the average. If the pedals rotate in a circle of radius 18 cm, the wheels have a radius of 34 cm, and the front and back sprockets on which the chain runs have 42 and 19 teeth respectively (Fig. 7–32), determine the maximum steepness of hill the cyclist can climb at constant speed. Assume the mass of the bike is 12 kg and that of the rider is 60 kg. Ignore friction. Assume the cyclist's average force is always: (a) downward; (b) tangential to pedal motion.

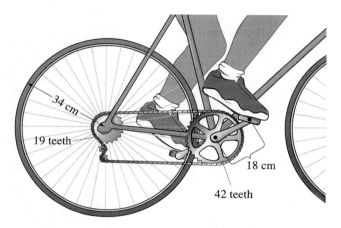

FIGURE 7–32 Problem 75.

76. A simple pendulum consists of a small object of mass m (the "bob") suspended by a cord of length L (Fig. 7–33) of negligible mass. A force $\mathbf{F}$ is applied in the horizontal direction (so $\mathbf{F} = F\mathbf{i}$), moving the bob very slowly so the acceleration is essentially zero. (Note that the magnitude of $\mathbf{F}$ will need to vary with the angle θ that the cord makes with the vertical at any moment.) (a) Determine the work done by this force, $\mathbf{F}$, to move the pendulum from $\theta = 0$ to $\theta = \theta_0$. (b) Determine the work done by the gravitational force on the bob, $\mathbf{F}_G = m\mathbf{g}$, and the work done by the force $\mathbf{F}_T$ that the cord exerts on the bob.

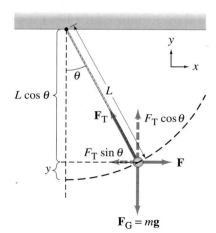

FIGURE 7–33
Problem 76.

A polevaulter running toward the high bar has kinetic energy. When he plants the pole and he puts his weight on it, his kinetic energy gets transformed: first into elastic potential energy of the bent pole and then into gravitational potential energy as his body rises. As he crosses the bar, the pole is straight and has given up all its elastic potential energy to the athlete's gravitational potential energy. Nearly all his kinetic energy has disappeared, also becoming gravitational potential energy of his body at the great height of the bar (world record over 6 m), which is exactly what he wants. In these, and all other energy transformations that continually take place in the world, the total energy is always conserved. Indeed, the conservation of energy is one of the greatest laws of physics, and finds applications in a wide range of other fields.

CHAPTER 8

Conservation of Energy

This chapter continues the discussion of the concepts of work and energy begun in the preceding chapter and introduces additional types of energy, in particular potential energy. Now we will see why the concept of energy is so important. The reason, ultimately, is that energy is conserved—the total energy remains constant in any process. That such a quantity can be defined which remains constant, as far as our best experiments can tell, is a remarkable statement about nature. The law of conservation of energy is, in fact, one of the great unifying principles of science.

The law of conservation of energy also gives us another tool, another approach, to solving problems. There are many situations for which an analysis based on Newton's laws would be difficult or impossible—the forces may not be known or accessible to measurement. But often these situations can be dealt with using the law of conservation of energy and in some cases other conservation laws (such as the conservation of momentum).

In this chapter we will mainly treat objects as if they were particles, or rigid objects, that undergo only translational motion, with no internal or rotational motion.

8-1 Conservative and Nonconservative Forces

We will find it important to categorize forces into two types: conservative and non-conservative. By definition, we call any force a **conservative force** if

> *the work done by the force on an object moving from one point to another depends only on the initial and final positions and is independent of the particular path taken.*

Conservative force defined

We can readily show that the force of gravity is a conservative force. The gravitational force on an object of mass m near the Earth's surface is $\mathbf{F} = m\mathbf{g}$, where $\mathbf{g}$ is a constant. We saw in Chapter 7 (see Example 7–2) that the work done by the gravitational force near the Earth's surface is $W_G = Fd = mgh$, where h is the vertical height through which an object of mass m falls (see Fig. 8–1a). Now suppose instead of moving vertically downward or upward, an object follows some arbitrary path in the xy plane, as shown in Fig. 8–1b. The object starts at a vertical height y_1 and reaches a height y_2, where $y_2 - y_1 = h$. To calculate the work done by gravity, W_G, we use Eq. 7–7:

$$W_G = \int_1^2 \mathbf{F}_G \cdot d\mathbf{l} = \int_1^2 mg \cos \theta \, dl.$$

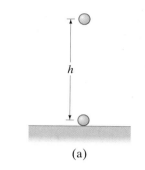

(a)

We now let $\phi = 180° - \theta$ be the angle between $d\mathbf{l}$ and its vertical component dy, as shown in Fig. 8–1b. Then, since $\cos \theta = -\cos \phi$ and $dy = dl \cos \phi$, we have

$$W_G = -\int_{y_1}^{y_2} mg \, dy$$
$$= -mg(y_2 - y_1). \tag{8-1}$$

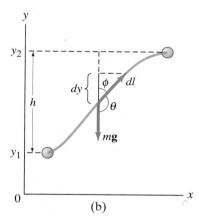

(b)

Since $(y_2 - y_1)$ is the vertical height h, we see that the work done depends only on the vertical height and does *not* depend on the particular path taken! Hence, by definition, gravity is a conservative force. (Note that in the case shown in Fig. 8–1b, $y_2 > y_1$ and therefore the work done by gravity is negative. If on the other hand $y_2 < y_1$, so that the object is falling, then W_G is positive.)

We can give the definition of a conservative force in another, completely equivalent way: a force is conservative if

> *the net work done by the force on an object moving around any closed path is zero.*

To see why this is equivalent to our earlier definition, consider a small object that moves from point 1 to point 2 via either of two paths labeled A and B in Fig. 8–2a. If we assume a conservative force acts on the object, the work done by this force is the same whether the object takes path A or path B, by our first definition. This work to get from point 1 to point 2 we will call W. Now consider the round trip shown in Fig. 8–2b. The object moves from 1 to 2 via path A and our force does work W. Our object then returns to point 1 via path B. How much work is done during the return? In going from 1 to 2 via path B the work done is W, which by definition equals $\int_1^2 \mathbf{F} \cdot d\mathbf{l}$. In doing the reverse, going from 2 to 1, the force $\mathbf{F}$ at each point is the same, but $d\mathbf{l}$ is directed in precisely the opposite direction. Consequently $\mathbf{F} \cdot d\mathbf{l}$ has the opposite sign at each point so the total work done in making the return trip from 2 to 1 must be $-W$. Hence the total work done in going from 1 to 2 and back to 1 is $W + (-W) = 0$, which proves the equivalence of the two above definitions for conservative force.

FIGURE 8–1 Object of mass m: (a) falls a height h vertically; (b) is raised along an arbitrary two-dimensional path.

FIGURE 8–2 (a) A tiny object moves between points 1 and 2 via two different paths, A and B. (b) The object makes a round trip, via path A from point 1 to point 2 and via path B back to point 1.

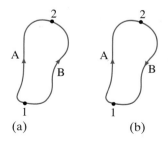

(a)   (b)

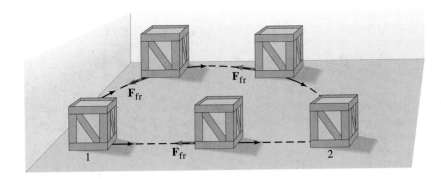

FIGURE 8–3 A crate is pulled across the floor from position 1 to position 2 via two paths, one straight and one curved. The friction force is always in the direction exactly opposed to the direction of motion. Hence for a constant magnitude friction force, $W_{fr} = -F_{fr}d$, so if d is greater (as for the curved path), then W is greater.

The second definition of a conservative force illuminates an important aspect of such a force: the *work done by a conservative force is recoverable* in the sense that if positive work is done *by* an object (on something else) on one part of a closed path, an equivalent amount of negative work will be done by our object on its return.

As we saw above, the force of gravity is conservative, and it is easy to show that the elastic force ($F = -kx$) is also conservative.

But not all forces are conservative. The force of friction, for example, is a **nonconservative** force. The work done in moving a heavy crate across a level floor is equal to the product of the (constant) friction force and the total distance traveled, since the friction force is directed precisely opposite to the direction of motion. Hence the work done to move the object between two points depends on the path length; the work done along a straight line is less than if the path between the two points is curved, as in Fig. 8–3.

Note also in this example of kinetic friction that since the force of friction always opposes the motion, the work done on a body by friction is negative. Thus when an object is moved in a round trip, say from some point 1 to point 2 and back to 1, the total work done by friction is never zero—it is always negative. Thus the work done by a nonconservative force is not recoverable, as it is for a conservative force.

8–2 | Potential Energy

In Chapter 7 we discussed the energy associated with a moving object, which is its kinetic energy $K = \frac{1}{2}mv^2$. Now we introduce **potential energy**, which is the energy associated with the position or configuration of an object (or objects). Various types of potential energy can be defined and each type is associated with a particular conservative force.

A wound-up clock spring is an example of potential energy. The clock spring acquired its potential energy because work was done *on* it by the person winding the clock. As the spring unwinds, it exerts a force and does work to move the clock hands around.

Gravitational Potential Energy

FIGURE 8–4 A person exerts an upward force $F_{ext} = mg$ to lift a brick from y_1 to y_2.

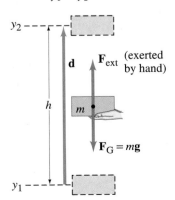

Perhaps the most common example of potential energy is *gravitational potential energy*. A heavy brick held high in the air has potential energy because of its position relative to the Earth. It has the ability to do work, for if it is released, it will fall to the ground due to the gravitational force, and can do work on, say, a stake, driving it into the ground. Let us determine quantitatively the gravitational potential energy of an object near the surface of the Earth. In order to lift an object of mass m vertically, an upward force at least equal to its weight, mg, must be exerted on it, say, by a person's hand. In order to lift it without acceleration to a height h, from position y_1 to y_2 in Fig. 8–4 (upward direction chosen positive), a person must do work equal to the product of the needed external force, $F_{ext} = mg$ upward, and the vertical distance h. That is,

$$W_{ext} = \mathbf{F}_{ext} \cdot \mathbf{d} = mgh \cos 0° = mgh = mg(y_2 - y_1)$$

where both $\mathbf{F}_{ext}$ and $\mathbf{d}$ point upward. Gravity is also acting on the object as it moves from y_1 to y_2, and does work on it equal to

$$W_G = \mathbf{F}_G \cdot \mathbf{d} = mgh \cos 180° = -mgh = -mg(y_2 - y_1).$$

Since $\mathbf{F}_G$ is downward and $\mathbf{d}$ is upward, W_G is negative. If the object follows an arbitrary path, as in Fig. 8–1b, the work done by gravity still depends only on the change in vertical height (see Eq. 8–1):

$$W_G = -mg(y_2 - y_1) = -mgh.$$

Next, if we allow the object to start from rest and fall freely under the action of gravity, it acquires a velocity given by $v^2 = 2gh$ (Eq. 2–12c) after falling a height h. It then has kinetic energy $\frac{1}{2}mv^2 = \frac{1}{2}m(2gh) = mgh$. If it then strikes a stake, the falling object can do work on the stake equal to mgh (work-energy principle, Section 7–4). Thus, to raise an object of mass m to a height h *requires* an amount of work equal to mgh. And once at height h, the object has the *ability* to do an amount of work equal to mgh. We can say that the work done in lifting the object has been stored as gravitational potential energy.

Indeed, we can define the *change in gravitational potential energy U* when an object moves from a height y_1 to a height y_2 as equal to the work done by an external force to accomplish this at constant acceleration:

$$\Delta U = U_2 - U_1 = W_{ext} = mg(y_2 - y_1).$$

Equivalently, we can define the change in gravitational potential energy as equal to the negative of the work done by gravity itself in the process:

$$\Delta U = U_2 - U_1 = -W_G = mg(y_2 - y_1). \qquad \textbf{(8–2)}$$

Change in gravitational potential energy

Equation 8–2 defines the change in gravitational potential energy between two points. The gravitational potential energy, U, at any point a vertical height y above some reference point (the origin of the coordinate system) can be defined as

$$U = mgy. \qquad \text{[gravity only]} \quad \textbf{(8–3)}$$

Gravitational potential energy

Note that the potential energy is associated with the force of gravity between the Earth and the mass m. Hence U represents the gravitational potential energy, not simply of the mass m alone, but of the mass-Earth system.

We could also define the gravitational potential energy at a point to be

$$U = mgy + C,$$

where C is a constant. This is consistent with Eq. 8–2 (the constants C cancel each other when we subtract U_1 from U_2). We usually choose C to be zero, for convenience, since U depends on the choice of coordinate system anyway (that is, where we choose y to be zero). The gravitational potential energy of a book held high above a table, for example, depends on whether we measure y from the top of the table, from the floor, or from some other reference point. What is physically important in any situation is the *change* in potential energy because that is what is related to the work done. We can thus choose the potential energy to be zero at any point that is convenient, but we must be consistent throughout any given problem. The *change* in potential energy between any two points does not depend on this choice.

Change in potential energy is what is physically meaningful

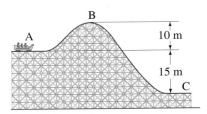

FIGURE 8–5 Example 8–1.

Potential energy changes for a roller coaster. A 1000-kg roller-coaster car moves from point A, Fig. 8–5, to point B and then to point C. (*a*) What is its gravitational potential energy at B and C relative to point A? That is, take $y = 0$ at point A. (*b*) What is the change in potential energy when it goes from B to C? (*c*) Repeat parts (*a*) and (*b*), but take the reference point ($y = 0$) to be at point C.

SOLUTION (*a*) We take upward as the positive direction, and measure the heights from point A, which means initially that the potential energy is zero. At point B, where $y_B = 10\,\mathrm{m}$,

$$U_B = mgy_B = (1000\,\mathrm{kg})(9.8\,\mathrm{m/s^2})(10\,\mathrm{m}) = 9.8 \times 10^4\,\mathrm{J}.$$

At point C, $y_C = -15\,\mathrm{m}$, since C is below A. Therefore,

$$U_C = mgy_C = (1000\,\mathrm{kg})(9.8\,\mathrm{m/s^2})(-15\,\mathrm{m}) = -1.5 \times 10^5\,\mathrm{J}.$$

(*b*) In going from B to C, the potential energy change is

$$U_C - U_B = (-1.5 \times 10^5\,\mathrm{J}) - (9.8 \times 10^4\,\mathrm{J})$$
$$= -2.5 \times 10^5\,\mathrm{J}.$$

The gravitational potential energy decreases by $2.5 \times 10^5\,\mathrm{J}$.
(*c*) In this instance, $y_A = +15\,\mathrm{m}$ at point A, so the potential energy initially (at A) is equal to

$$U_A = (1000\,\mathrm{kg})(9.8\,\mathrm{m/s^2})(15\,\mathrm{m}) = 1.5 \times 10^5\,\mathrm{J}.$$

At B, $y_B = 25\,\mathrm{m}$, so the potential energy is

$$U_B = 2.5 \times 10^5\,\mathrm{J}.$$

At C, $y_C = 0$, so $U_C = 0$. The change in potential energy going from B to C is

$$U_C - U_B = 0 - 2.5 \times 10^5\,\mathrm{J} = -2.5 \times 10^5\,\mathrm{J},$$

which is the same as in part (*b*).

Potential Energy in General

We have defined the change in gravitational potential energy (Eq. 8–2) to be equal to the negative of the work done by gravity[†] when the object moves from height y_1 to y_2:

$$\Delta U = -W_G = -\int_1^2 \mathbf{F}_G \cdot d\mathbf{l}.$$

There are other types of potential energy besides gravitational. In general, we define the *change in potential energy associated with a particular conservative force* **F** *as the negative of the work done by that force*:

General definition of potential energy

$$\Delta U = U_2 - U_1 = -\int_1^2 \mathbf{F} \cdot d\mathbf{l} = -W. \tag{8–4}$$

However, we cannot use this definition to define a potential energy for all possible forces. It makes sense only for conservative forces such as gravity, for which the

[†] This is equivalent to saying that the change in potential energy, ΔU, equals the work done (not its negative) by a second force (of equal magnitude)—say, exerted by a person—to raise the object against the force of gravity.

integral depends only on the end points and not on the path taken. It does not apply to nonconservative forces like friction, because the integral in Eq. 8–4 would not have a unique value depending on the end points 1 and 2. That is, ΔU would depend on path and we could not say that the U had a particular value at each point in space. Thus the concept of potential energy is meaningless for a nonconservative force.

Elastic Potential Energy

We consider now another type of potential energy, that associated with elastic materials. This includes a great variety of practical applications.

Consider a spring such as the coil spring shown in Fig. 8–6. The spring has potential energy when compressed (or stretched) for when released, it can do work on a ball as shown. Like other elastic materials, a spring is described by Hooke's law (as already discussed in Section 7–3) as long as the displacement x is not too great. Let us take our coordinate system so the end of the uncompressed spring is at $x = 0$ (Fig. 8–6a) and x is positive to the right. To hold the spring compressed (or stretched) a distance x, a person must exert a force $F_P = kx$ (recall Fig. 7–11), and the spring pushes back with a force (Newton's third law),

$$F_S = -kx.$$

The negative sign appears because the force F_S is in the direction opposite to the displacement x (see Fig. 8–6b). From Eq. 8–4, the change in potential energy of the spring between $x_1 = 0$ (its uncompressed position) and $x_2 = x$ is

$$\Delta U = U(x) - U(0) = -\int_1^2 \mathbf{F} \cdot d\mathbf{l} = -\int_0^x (-kx)\, dx = \tfrac{1}{2}kx^2.$$

Here, $U(x)$ means the potential energy at x, and $U(0)$ means U at $x = 0$. It is usually convenient to choose the potential energy at $x = 0$ to be zero: $U(0) = 0$, so the potential energy of a spring compressed or stretched an amount x from equilibrium is

$$U(x) = \tfrac{1}{2}kx^2. \qquad \text{[elastic]} \quad \textbf{(8–5)}$$

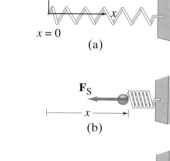

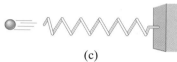

FIGURE 8–6 (a) A spring can (b) store energy (elastic potential energy) when compressed, which (c) can be used to do work when released.

Elastic potential energy

Potential Energy Summarized

In each of the preceding examples of potential energy—gravitational or elastic—an object has the capacity or *potential* to do work even though it is not yet actually doing it. That is why we use the term "potential" energy. From these examples we can also see that energy can be stored, for later use, in the form of potential energy. Note that the mathematical form of each type of potential energy depends on the force involved.

Let us summarize here the important aspects of potential energy:

1. A potential energy is always associated with a conservative force, and the difference in potential energy between two points is defined as the negative of the work done by that force, Eq. 8–4.
2. The choice of where $U = 0$ is arbitrary and can be chosen wherever it is most convenient.
3. Since a force is always exerted on one body by another body (the Earth exerts a gravitational force on a falling stone; a compressed spring exerts a force on a ball; and so on), potential energy is not something a body "has" by itself, but rather is associated with the interaction of two (or more) bodies.

In the one-dimensional case, where a conservative force can be written as a function, say, of x, the potential energy can be written

$$U(x) = -\int F(x)\,dx. \tag{8–6}$$

This relation tells us how to obtain $U(x)$ when given $F(x)$. If, instead, we are given $U(x)$, we can obtain $F(x)$ by inverting the above equation: that is, we take the derivative of both sides, remembering that integration and differentiation are inverse operations:

$$\frac{d}{dx}\int F(x)\,dx = F(x).$$

Thus

$$F(x) = -\frac{dU(x)}{dx}. \tag{8–7}$$

[In three dimensions, we can write the relation between $\mathbf{F}(x, y, z)$ and U as:

$$F_x = -\frac{\partial U}{\partial x}, \quad F_y = -\frac{\partial U}{\partial y}, \quad F_z = -\frac{\partial U}{\partial z},$$

or

$$\mathbf{F}(x, y, z) = -\mathbf{i}\frac{\partial U}{\partial x} - \mathbf{j}\frac{\partial U}{\partial y} - \mathbf{k}\frac{\partial U}{\partial z}.$$

Here, $\partial/\partial x$, and so on, are called partial derivatives; $\partial/\partial x$, for example, means that although U may be a function of x, y, and z, written $U(x, y, z)$, we take the derivative only with respect to x with the other variables held constant.]

EXAMPLE 8–2 **Determine F from U.** Suppose $U(x) = -ax/(b^2 + x^2)$, where a and b are constants. What is F as a function of x?

SOLUTION Since $U(x)$ depends only on x, this is a one-dimensional problem and we don't need to use partial derivatives, so

$$F(x) = -\frac{dU}{dx} = -\frac{d}{dx}\left[-\frac{ax}{b^2 + x^2}\right] = \frac{a}{b^2 + x^2} - \frac{ax}{(b^2 + x^2)^2}2x = \frac{a(b^2 - x^2)}{(b^2 + x^2)^2}.$$

8–3 | Mechanical Energy and Its Conservation

Let us consider a conservative system (meaning only conservative forces do work) in which energy is transformed from kinetic to potential or vice versa. Again, we must consider a system because potential energy does not exist for an isolated body. Our system might be a mass m oscillating on the end of a spring or moving in the Earth's gravitational field.

According to the work-energy principle (Eq. 7–11), the net work W_{net} done on an object is equal to its change in kinetic energy:

$$W_{net} = \Delta K.$$

(If more than one object of our system has work done on it, then W_{net} and ΔK can represent the sum for all of them.) Since we assume a conservative system, we can write the net work done in terms of the total potential energy (see Eqs. 7–7 and 8–4):

$$\Delta U_{total} = -\int_1^2 \mathbf{F}_{net} \cdot d\mathbf{l} = -W_{net}. \qquad (8\text{–}8)$$

We combine the previous two equations, letting U be the total potential energy:

$$\Delta K + \Delta U = 0 \qquad \text{[conservative forces only]} \qquad (8\text{–}9a)$$

or

$$(K_2 - K_1) + (U_2 - U_1) = 0. \qquad (8\text{–}9b)$$

We now define a quantity E, called the **total mechanical energy** of our system, as the sum of the kinetic energy plus the potential energy of the system at any moment

$$E = K + U,$$

Total mechanical energy defined

Now we can rewrite Eq. 8–9b as

$$K_2 + U_2 = K_1 + U_1 \qquad \text{[conservative forces only]} \qquad (8\text{–}10a)$$

or

$$E_2 = E_1 = \text{constant}. \qquad \text{[conservative forces only]} \qquad (8\text{–}10b)$$

CONSERVATION OF
MECHANICAL ENERGY

Equations 8–10 express a useful and profound principle regarding the total mechanical energy—namely, that it is a **conserved quantity**. The total mechanical energy E remains constant as long as no nonconservative forces do work: $(K + U)$ at some initial point 1 is equal to the $(K + U)$ at any later point 2. To say it another way, consider Eq. 8–9 which tells us $\Delta U = -\Delta K$; that is, if the kinetic energy K increases, then the potential energy U must decrease by an equivalent amount to compensate. Thus the total, $K + U$, remains constant. This is called the **principle of conservation of mechanical energy** for conservative forces:

> **If only conservative forces are doing work, the total mechanical energy of a system neither increases nor decreases in any process. It stays constant—it is conserved.**

CONSERVATION OF
MECHANICAL ENERGY

We now see the reason for the term "conservative force"—because for such forces, mechanical energy is conserved.

If only one object of a system[†] has significant kinetic energy, then Eqs. 8–10 become

$$E = \tfrac{1}{2}mv^2 + U = \text{constant}. \qquad \text{[conservative forces only]} \qquad (8\text{–}11a)$$

If we let v_1 and U_1 represent the velocity and potential energy at one instant, and v_2 and U_2 represent them at a second instant, then we can rewrite this as

$$\tfrac{1}{2}mv_1^2 + U_1 = \tfrac{1}{2}mv_2^2 + U_2. \qquad \text{[conservative system]} \qquad (8\text{–}11b)$$

From this equation we can see again that it doesn't make any difference where we choose the potential energy to be zero: adding a constant to U (as discussed in Section 8–2) merely adds a constant to both sides of the above equation, and these cancel. A constant also doesn't affect the force obtained from Eq. 8–7, $F = -dU/dx$, since the derivative of a constant is zero. Because we deal only with changes in the potential energy, the absolute value of U doesn't matter.

[†] For an object moving under the influence of Earth's gravity, the kinetic energy of the Earth can usually be ignored, as long as the object's mass is small compared to the Earth's mass. For a mass oscillating at the end of a spring, for example, the mass of the spring, and hence its kinetic energy, can often be ignored.

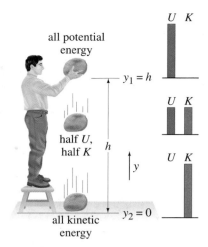

all potential energy

$y_1 = h$

half U, half K

h

y

all kinetic energy

$y_2 = 0$

U K

U K

U K

FIGURE 8–7 The rock's potential energy changes to kinetic energy as it falls. Note bar graphs representing potential energy U and kinetic energy K for the three different positions.

FIGURE 8–8 A roller-coaster car moving without friction illustrates the conservation of mechanical energy.

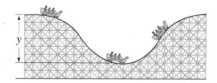

y

8–4 Problem Solving Using Conservation of Mechanical Energy

A simple example of the conservation of mechanical energy is a rock allowed to fall from a height h under gravity (neglecting air resistance), as shown in Fig. 8–7. At the instant it is dropped, a rock, starting at rest, initially has only potential energy. As it falls, its potential energy decreases (because y decreases), but its kinetic energy increases to compensate, so that the sum of the two remains constant. At any point along the path, the total mechanical energy is given by

$$E = K + U = \tfrac{1}{2}mv^2 + mgy$$

where y is the rock's height above the ground at a given instant and v is its speed at that point. If we let the subscript 1 represent the rock at one point along its path (for example, the initial point), and 2 represent it at some other point, then we can write

$$\tfrac{1}{2}mv_1^2 + mgy_1 = \tfrac{1}{2}mv_2^2 + mgy_2. \qquad \text{[gravity only]} \quad \textbf{(8–12)}$$

Just before the rock hits the ground ($y_2 = 0$), the potential energy will be zero: all of the initial potential energy will have been transformed into kinetic energy.

EXAMPLE 8–3 **Falling rock.** If the original height of the stone in Fig. 8–7 is $y_1 = h = 3.0$ m, calculate the stone's speed when it has fallen to 1.0 m above the ground.

SOLUTION Since $v_1 = 0$ (the moment of release), $y_1 = 3.0$ m, $y_2 = 1.0$ m, and $g = 9.8$ m/s^2, Eq. 8–12 gives

$$\tfrac{1}{2}mv_1^2 + mgy_1 = \tfrac{1}{2}mv_2^2 + mgy_2$$

$$0 + (m)(9.8 \text{ m/s}^2)(3.0 \text{ m}) = \tfrac{1}{2}mv_2^2 + (m)(9.8 \text{ m/s}^2)(1.0 \text{ m}).$$

The m's cancel out, and solving for v_2^2 (which we see doesn't depend on m), we find

$$v_2^2 = 2\big[(9.8 \text{ m/s}^2)(3.0 \text{ m}) - (9.8 \text{ m/s}^2)(1.0 \text{ m})\big] = 39.2 \text{ m}^2/\text{s}^2,$$

and

$$v_2 = \sqrt{39.2} \text{ m/s} = 6.3 \text{ m/s}.$$

Equation 8–12 can be applied to any object moving without friction under the action of gravity. For example, Fig. 8–8 shows a roller-coaster car starting from rest at the top of a hill, and coasting without friction to the bottom and up the hill on the other side. True, there is another force besides gravity acting on the car, the normal force exerted by the tracks. But this "constraint" force acts perpendicular to the direction of motion at each point and so does zero work. We ignore rotational motion of the car's wheels and treat the car as a particle undergoing simple translation. Initially, the car has only potential energy. As it coasts down the hill, it loses potential energy and gains in kinetic energy, but the sum of the two remains constant. At the bottom of the hill it has its maximum kinetic energy, and as it climbs up the other side the kinetic energy changes back to potential energy. When the car comes to rest again, all of its energy will be potential energy. Given that the potential energy is proportional to the vertical height, energy conservation tells us that (in the absence of friction) the car comes to rest at a height equal to its original height. If the two hills are the same height, the car will just barely reach the top of the second hill when it stops. If the second hill is lower than the first, not all of the car's kinetic energy will be transformed to potential energy and the car can continue over the top and down the other side. If the second hill is higher, the car will only reach a height on it equal to its original height on the first hill. This is true (in the absence of friction) no matter how steep the hill is, since potential energy depends only on the vertical height.

EXAMPLE 8–4 **Roller-coaster speed using energy conservation.** Assuming the height of the hill in Fig. 8–8 is 40 m, and the roller-coaster car starts from rest at the top, calculate (*a*) the speed of the roller-coaster car at the bottom of the hill, and (*b*) at what height it will have half this speed. Take $y = 0$ (and $U = 0$) at the bottom of the hill.

SOLUTION (*a*) We use Eq. 8–12 with $v_1 = 0$, $y_1 = 40$ m, and $y_2 = 0$. Then

$$\tfrac{1}{2}mv_1^2 + mgy_1 = \tfrac{1}{2}mv_2^2 + mgy_2$$
$$0 + (m)(9.8 \text{ m/s}^2)(40 \text{ m}) = \tfrac{1}{2}mv_2^2 + 0.$$

The m's cancel out and we find $v_2 = \sqrt{2(9.8 \text{ m/s}^2)(40 \text{ m})} = 28$ m/s.
(*b*) We use the same equation, but now $v_2 = 14$ m/s (half of 28 m/s) and y_2 is unknown:

$$\tfrac{1}{2}mv_1^2 + mgy_1 = \tfrac{1}{2}mv_2^2 + mgy_2$$
$$0 + (m)(9.8 \text{ m/s}^2)(40 \text{ m}) = \tfrac{1}{2}(m)(14 \text{ m/s})^2 + (m)(9.8 \text{ m/s}^2)(y_2).$$

We cancel the m's and solve for y_2 and find $y_2 = 30$ m. That is, the car has a speed of 14 m/s when it is 30 *vertical* meters above the lowest point, both when descending the left-hand hill and when ascending the right-hand hill in Fig. 8–8.

The mathematics of this Example is almost the same as that in Example 8–3. But there is an important difference between them. Example 8–3 could have been solved using force and acceleration. But here, where the motion is not vertical, using $F = ma$ would have been very difficult, whereas energy conservation readily gives us the answer.

CONCEPTUAL EXAMPLE 8–5 **Speeds on two water slides.** Two water slides at a pool are shaped differently but start at the same height h (Fig. 8–9). Two riders, Paul and Kathleen, start from rest at the same time on different slides. (*a*) Which rider, Paul or Kathleen, is traveling faster at the bottom? (*b*) Which rider makes it to the bottom first? Ignore friction.

RESPONSE (*a*) Each rider's initial potential energy mgh gets transformed to kinetic energy, so the speed v at the bottom is obtained from $\tfrac{1}{2}mv^2 = mgh$. The mass cancels in this equation and so the speed will be the same, regardless of the mass of the rider. Since they descend the same vertical height, they will finish with the same speed.
(*b*) Note that Kathleen is consistently at a lower elevation than Paul for the entire trip. This means she has converted her potential energy to kinetic energy earlier. Consequently, she is traveling faster than Paul for the whole trip, except toward the end where Paul finally gets up to the same speed. Since she was going faster for almost the whole trip, and the distance is roughly the same, Kathleen gets to the bottom first.

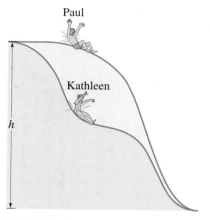

FIGURE 8–9 Example 8–5.

FIGURE 8–10 Transformation of energy during a pole vault.

There are many interesting examples of the conservation of energy in sports, one of which is the pole vault illustrated in Fig. 8–10. We often have to make approximations, but the sequence of events in broad outline for this case is as follows. The kinetic energy of the running athlete is transformed into elastic potential energy of the bending pole and, as the athlete leaves the ground, into gravitational potential energy. When the vaulter reaches the top and the pole has straightened out again, the energy has all been transformed into gravitational potential energy (if we ignore the vaulter's low horizontal speed over the bar). The pole does not supply any energy, but it acts as a device to *store* energy and thus aid in the transformation of kinetic energy into gravitational potential energy, which is the net result. The energy required to pass over the bar depends on how high the center of mass[†] (CM) of the vaulter must be raised. By bending their bodies, pole vaulters

[†] The center of mass (CM) of a body is that point at which the entire mass of the body can be considered as concentrated, for purposes of describing its translational motion. (This is discussed in Chapter 9.) In Eq. 8–12, y is the position of the CM.

keep their CM so low that it can actually pass slightly beneath the bar (Fig. 8–11), thus enabling them to cross over a higher bar than would otherwise be possible.

➥ **PHYSICS APPLIED**

Sports

FIGURE 8–11 By bending their bodies, pole vaulters can keep their center of mass so low that it may even pass below the bar. By changing their kinetic energy (of running) into gravitational potential energy (= mgy) in this way, vaulters can cross over a higher bar than if the change in potential energy were accomplished without carefully bending the body.

EXAMPLE 8–6 **ESTIMATE** **Pole vault.** Estimate the kinetic energy and the speed required for a 70-kg pole vaulter to just pass over a bar 5.0 m high. Assume the vaulter's center of mass is initially 0.90 m off the ground and reaches its maximum height at the level of the bar itself.

SOLUTION We equate the total energy just before the vaulter places the end of the pole onto the ground (and the pole begins to bend and store potential energy) with the vaulter's total energy when passing over the bar (we ignore the small amount of kinetic energy at this point). We choose the initial position of the vaulter's center of mass to be $y_1 = 0$. The vaulter's body must then be raised to a height $y_2 = 5.0\,\text{m} - 0.9\,\text{m} = 4.1\,\text{m}$. Thus, using Eq. 8–12,

$$\tfrac{1}{2}mv_1^2 + 0 = 0 + mgy_2$$

and

$$K_1 = \tfrac{1}{2}mv_1^2 = mgy_2 = (70\,\text{kg})(9.8\,\text{m/s}^2)(4.1\,\text{m}) = 2.8 \times 10^3\,\text{J}.$$

The speed is

$$v_1 = \sqrt{\frac{2K_1}{m}} = \sqrt{\frac{2(2800\,\text{J})}{70\,\text{kg}}} = 8.9\,\text{m/s}.$$

This is an approximation because we have ignored such things as the vaulter's speed while crossing over the bar, mechanical energy transformed when the pole is planted in the ground, and work done by the vaulter on the pole.

As another example of the conservation of mechanical energy, let us consider a mass m connected to a horizontal spring whose own mass can be neglected and whose stiffness constant is k. The mass m has speed v at any moment and the potential energy of the system is $\tfrac{1}{2}kx^2$, where x is the displacement of the spring from its unstretched length. If neither friction nor any other force is acting, the conservation-of-energy principle tells us that

Conservation of mechanical energy (elastic force only)

$$\tfrac{1}{2}mv_1^2 + \tfrac{1}{2}kx_1^2 = \tfrac{1}{2}mv_2^2 + \tfrac{1}{2}kx_2^2 \qquad \text{[elastic potential energy only]} \qquad \textbf{(8–13)}$$

where the subscripts 1 and 2 refer to the velocity and displacement at two different points.

FIGURE 8–12 Example 8–7. (a) A dart is pushed against a spring, compressing it 6.0 cm. The dart is then released, and in (b) it leaves the spring at high velocity (v_2).

(a)

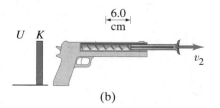

(b)

EXAMPLE 8–7 **Toy dart gun.** A dart of mass 0.100 kg is pressed against the spring of a toy dart gun as shown in Fig. 8–12. The spring (with spring constant $k = 250\,\text{N/m}$) is compressed 6.0 cm and released. If the dart detaches from the spring when the latter reaches its normal length ($x = 0$), what speed does the dart acquire?

SOLUTION In the horizontal direction, the only force on the dart (neglecting friction) is the force exerted by the spring. Vertically, gravity is counterbalanced by the normal force exerted on the dart by the gun barrel. (After the dart leaves the barrel it will follow a projectile's path under gravity.) We use Eq. 8–13 with point 1 being at the maximum compression of the spring, so $v_1 = 0$ (dart not yet released) and $x_1 = -0.060\,\text{m}$. Point 2 we choose to be the instant the dart flies off the end of the spring (Fig. 8–12b), so $x_2 = 0$ and we want to find v_2. Thus Eq. 8–13 can be written

$$0 + \tfrac{1}{2}kx_1^2 = \tfrac{1}{2}mv_2^2 + 0,$$

so

$$v_2 = \sqrt{\frac{kx_1^2}{m}} = \sqrt{\frac{(250\,\text{N/m})(0.060\,\text{m})^2}{(0.100\,\text{kg})}} = 3.0\,\text{m/s}.$$

EXAMPLE 8-8 **Two kinds of Potential Energy.** A ball of mass $m = 2.60 \text{ kg}$, starting from rest, falls a vertical distance $h = 55.0 \text{ cm}$ before striking a vertical coiled spring, which it compresses (see Fig. 8–13) an amount $Y = 15.0 \text{ cm}$. Determine the spring constant of the spring. Assume the spring has negligible mass. Measure all distances from the point where the ball first touches the uncompressed spring ($y = 0$ at this point).

SOLUTION Since the motion is vertical, we use y instead of x (y positive upward). We divide this solution into two parts. (See also alternate solution below.) *Part 1*: Let us first consider the energy changes of the ball as it falls from a height $y_1 = h = 0.55 \text{ m}$, Fig. 8–13a, to $y_2 = 0$, just as it touches the spring, Fig. 8–13b. Our system is the ball acted on by gravity (so far, the spring does nothing) so

$$\tfrac{1}{2}mv_1^2 + mgy_1 = \tfrac{1}{2}mv_2^2 + mgy_2$$

$$0 + mgh = \tfrac{1}{2}mv_2^2 + 0$$

and

$$v_2 = \sqrt{2gh} = \sqrt{2(9.80 \text{ m/s}^2)(0.550 \text{ m})} = 3.28 \text{ m/s}.$$

Part 2: As the ball compresses the spring, Fig. 8–13b to c, there are two conservative forces on the ball—gravity and the spring force. So our energy equation becomes

$$E \text{ (ball touches spring)} = E \text{ (spring compressed)}$$

$$\tfrac{1}{2}mv_2^2 + mgy_2 + \tfrac{1}{2}ky_2^2 = \tfrac{1}{2}mv_3^2 + mgy_3 + \tfrac{1}{2}ky_3^2.$$

We take point 2 to be the instant when the ball just touches the spring, so $y_2 = 0$ and $v_2 = 3.28 \text{ m/s}$. Point 3 we take to be when the ball comes to rest and the spring is fully compressed, so $v_3 = 0$ and $y_3 = -Y = -0.150 \text{ m}$ (given). Putting these into the above energy equation we get

$$\tfrac{1}{2}mv_2^2 + 0 + 0 = 0 - mgY + \tfrac{1}{2}kY^2.$$

We know m, v_2, and Y, so we can solve for k:

$$k = \frac{2}{Y^2}\left[\tfrac{1}{2}mv_2^2 + mgY\right]$$

$$= \frac{m}{Y^2}\left[v_2^2 + 2gY\right]$$

$$= \frac{2.60 \text{ kg}}{(0.150 \text{ m})^2}\left[(3.28 \text{ m/s})^2 + 2(9.80 \text{ m/s}^2)(0.150 \text{ m})\right] = 1580 \text{ N/m}.$$

Alternate Solution: Instead of dividing the solution into two parts, we can do it all at once. After all, we get to choose what two points are used on the left and right of the energy equation. Let us write the energy equation for points 1 and 3 (Fig. 8–13). Point 1 is the initial point just before the ball starts to fall (Fig. 8–13a), so $v_1 = 0$, $y_1 = h = 0.550 \text{ m}$; and point 3 is when the spring is fully compressed (Fig. 8–13c), so $v_3 = 0$, $y_3 = -Y = -0.150 \text{ m}$. The forces on the ball in this process are gravity and (at least part of the time) the spring. So conservation of energy tells us

$$\tfrac{1}{2}mv_1^2 + mgy_1 + \tfrac{1}{2}k(0)^2 = \tfrac{1}{2}mv_3^2 + mgy_3 + \tfrac{1}{2}ky_3^2$$
$$0 + mgh + 0 = 0 - mgY + \tfrac{1}{2}kY^2$$

where we have set $y = 0$ for the spring at point 1 because it is not acting and is not compressed or stretched at this point. We solve for k:

$$k = \frac{2mg(h + Y)}{Y^2} = \frac{2(2.60 \text{ kg})(9.80 \text{ m/s}^2)(0.550 \text{ m} + 0.150 \text{ m})}{(0.150 \text{ m})^2} = 1580 \text{ N/m}$$

just as in our first method of solution.

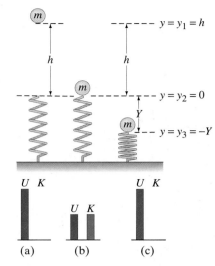

FIGURE 8-13 Example 8-8. (not to scale)

Conservation of energy: gravity and elastic potential energy

➡ **PROBLEM SOLVING**

Alternate Solution

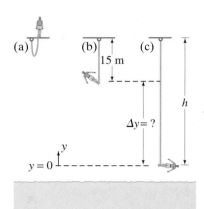

FIGURE 8–14 Example 8–9.
(a) Bungee jumper about to jump.
(b) Bungee cord at its unstretched
length. (c) Maximum stretch of cord.

FIGURE 8–15 Example 8–10: a
simple pendulum; y is measured
positive upwards.

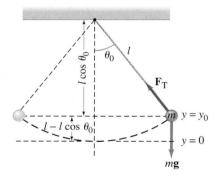

EXAMPLE 8–9 **Bungee jump.** Dave jumps off a bridge with a bungee cord (a heavy stretchable cord) tied around his ankle, Fig. 8–14. He falls for 15 meters before the bungee cord begins to stretch. Dave's mass is 75 kg and we assume the cord obeys Hooke's law, $F = -kx$, with $k = 50$ N/m. If we neglect air resistance, estimate how far below the bridge Dave will fall before coming to a stop. Ignore the mass of the cord (not realistic, however).

SOLUTION Dave starts out with gravitational potential energy, which during his fall is tranformed into both kinetic energy and elastic potential energy. Assuming no frictional forces act on our system, the total energy at the start must be the same as the total energy at the end. If we define our coordinate system such that $y = 0$ at the lowest point in Dave's jump, and let the stretch of the cord at this point be represented by Δy, then the total fall is (see Fig. 8–14)

$$h = 15 \text{ m} + \Delta y.$$

Conservation of energy then gives:

$$K_1 + U_1 = K_2 + U_2$$
$$0 + mg(15 \text{ m} + \Delta y) = 0 + \tfrac{1}{2}k(\Delta y)^2.$$

To solve for Δy, we use the quadratic formula and get two solutions:

$$\Delta y = 40 \text{ m} \quad \text{and} \quad \Delta y = -11 \text{ m}.$$

The negative solution is nonphysical, so the distance that Dave drops in his fall is:

$$h = 15 \text{ m} + 40 \text{ m} = 55 \text{ m}.$$

EXAMPLE 8–10 **A swinging pendulum.** The simple pendulum shown in Fig. 8–15 consists of a small bob of mass m suspended by a massless cord of length l. The bob is released (without a push) at $t = 0$, where the cord makes an angle $\theta = \theta_0$ to the vertical. (a) Describe the motion of the bob in terms of kinetic energy and potential energy. Then determine the speed of the bob: (b) as a function of position θ as it swings back and forth, and (c) at the lowest point of the swing. (d) Find the tension in the cord, $\mathbf{F}_T$. Ignore friction and air resistance.

SOLUTION (a) At the moment of release, the bob is at rest, so $K = 0$. As the bob falls, it loses potential energy and gains kinetic energy. At the lowest point its kinetic energy is a maximum and the potential energy is a minimum. The bob continues its swing until it reaches an equal height and angle (θ_0) on the opposite side, at which point the potential energy is a maximum and $K = 0$. It continues the swinging motion as $U \to K \to U$ and so on, but it can never go higher than $\theta = \pm\theta_0$ (conservation of mechanical energy).
(b) The cord is assumed to be massless, so we don't need to be concerned with the energy of the cord but only with the bob's kinetic energy, and the potential energy. The bob has two forces acting on it at any moment: gravity, mg, and the force the cord exerts on it, $\mathbf{F}_T$. The latter (a constraint force) always acts perpendicular to the motion, so it does no work. We need be concerned only with gravity, for which we can write the potential energy. The mechanical energy of the system is

$$E = \tfrac{1}{2}mv^2 + mgy,$$

where y is the vertical height of the bob at any moment. We take $y = 0$ at the lowest point of the bob's swing. Hence at $t = 0$,

$$y = y_0 = l - l \cos\theta_0 = l(1 - \cos\theta_0)$$

as can be seen from the diagram. At the moment of release

$$E = mgy_0,$$

since $v = v_0 = 0$. At any other point along the swing

$$E = \tfrac{1}{2}mv^2 + mgy = mgy_0.$$

We solve this for v:

$$v = \sqrt{2g(y_0 - y)}.$$

In terms of the angle θ of the cord, we can write

$$v = \sqrt{2gl(\cos\theta - \cos\theta_0)},$$

since $y = l - l\cos\theta$ and $y_0 = l - l\cos\theta_0$.
(c) At the lowest point, $y = 0$ so

$$v = \sqrt{2gy_0} \quad \text{or} \quad v = \sqrt{2gl(1 - \cos\theta_0)}.$$

(d) The tension in the cord is the force $\mathbf{F}_T$ that the cord exerts on the bob. As we've seen, there is no work done by this force, but we can calculate the force simply by using Newton's second law $\Sigma\mathbf{F} = m\mathbf{a}$ and by noting that at any point the acceleration of the bob in the inward radial direction is v^2/l, since the bob is constrained to move in an arc of a circle. In the radial direction, $\mathbf{F}_T$ acts inward, and a component of gravity equal to $mg\cos\theta$ acts outward. Hence

$$m\frac{v^2}{l} = F_T - mg\cos\theta.$$

We solve for F_T and use the result of part (b) for v^2:

$$F_T = m\left(\frac{v^2}{l} + g\cos\theta\right) = 2mg(\cos\theta - \cos\theta_0) + mg\cos\theta$$
$$= (3\cos\theta - 2\cos\theta_0)mg.$$

8–5 | The Law of Conservation of Energy

We now take into account nonconservative forces such as friction, since they are important in real situations. For example, consider again the roller-coaster car in Fig. 8–8, but this time let us include friction. The car will not in this case reach the same height on the second hill as it had on the first hill because of friction.

In this, and in other natural processes, the mechanical energy (sum of the kinetic and potential energies) does not remain constant but decreases. Because frictional forces reduce the total mechanical energy, they are called **dissipative forces**. Historically, the presence of dissipative forces hindered the formulation of a comprehensive conservation of energy law until well into the nineteenth century. It was not until then that heat, which is always produced when there is friction (try rubbing your hands together), was interpreted as a form of energy. Quantitative studies by nineteenth-century scientists (Chapter 19) demonstrated that if heat is considered as energy, now called **thermal energy**, then the total energy is conserved in any process. For example, if the roller-coaster car in Fig. 8–8 is subject to frictional forces, then the initial total energy of the car will be equal to the car's kinetic energy plus the potential energy at any subsequent point along its path plus the amount of thermal energy produced in the process. The thermal energy produced by a constant friction force F_{fr} is equal to the work done by this force. A block sliding freely across a table, for example, comes to rest because of friction. Its initial kinetic energy is all transformed into thermal energy. The block and table are a little warmer as a result of this process. A more obvious example of the transformation of kinetic energy into thermal energy can be observed by vigorously striking a nail several times with a hammer and then gently touching the nail with your finger.

According to the atomic theory, thermal energy represents kinetic energy of rapidly moving molecules. We shall see in Chapter 18 that a rise in temperature corresponds to an increase in the average kinetic energy of the molecules. Because

Dissipative forces

FIGURE 8–16 The burning of fuel (a chemical reaction) releases energy to boil water in this steam engine. The steam produced expands against a piston to do work in turning the wheels.

LAW OF CONSERVATION OF ENERGY

thermal energy represents the energy of atoms and molecules that make up a body, it is often called *internal energy*. Internal energy,[†] from the atomic point of view, can include not only kinetic energy of molecules but also potential energy (electrical in nature) because of the relative positions of atoms within molecules. On a macroscopic level, internal energy corresponds to nonconservative forces such as friction. But at the atomic level, the energy is kinetic and potential and the corresponding forces are mainly conservative. For example, the energy stored in food or in a fuel such as gasoline can be regarded as potential energy stored by virtue of the relative positions of the atoms within a molecule. For this energy to be used to do work, it must be released, usually through a chemical reaction (Fig. 8–16). This is something like a compressed spring which, when released, can do work.

To establish the more general law of conservation of energy, it required nineteenth-century physicists to recognize electrical, chemical, and other forms of energy in addition to heat and to explore if in fact they could fit into a conservation law. For each type of force, conservative or nonconservative, it has always been found possible to define a type of energy that corresponds to the work done by such a force. And it has been found experimentally that the total energy E always remains constant. That is, the change in the total energy, kinetic plus potential plus all other forms of energy, equals zero:

$$\Delta K + \Delta U + [\text{change in all other forms of energy}] = 0. \qquad \textbf{(8–14)}$$

This is one of the most important principles in physics. It is called the **law of conservation of energy** and can be stated as follows:

> **The total energy is neither increased nor decreased in any process. Energy can be transformed from one form to another, and transferred from one body to another, but the total amount remains constant.**

For conservative mechanical systems, this law can be derived from Newton's laws (Section 8–3) and thus is equivalent to them. But in its full generality, the validity of the law of conservation of energy rests on experimental observation. And even though Newton's laws have been found to fail in the submicroscopic world of the atom, the law of conservation of energy has been found to hold there and in every experimental situation so far tested.

8–6 Energy Conservation with Dissipative Forces: Solving Problems

In Section 8–4 we discussed several Examples of the law of conservation of energy for conservative systems. Now let us consider in detail some examples that involve nonconservative forces.

Suppose, for example, that the roller-coaster car rolling on the hills of Fig. 8–8 is subject to frictional forces. In going from some point 1 to a second point 2, the work done by the friction force $\mathbf{F}_{fr}$ acting on the car is $W_{fr} = \int_1^2 \mathbf{F}_{fr} \cdot d\boldsymbol{l}$. If $\mathbf{F}_{fr}$ is constant in magnitude, $W_{fr} = -F_{fr}l$, where l is the actual distance along the path traveled by the object from point 1 to point 2. (The minus sign appears because $\mathbf{F}_{fr}$ opposes the motion and is thus in the opposite direction to $d\boldsymbol{l}$.) From the work-energy principle (Eq. 7–11), the net work W_{net} done on a body is equal to the change in its kinetic energy:

$$\Delta K = W_{net}.$$

The forces that do work on the car in the present case are gravity and friction (the

---

[†] The term *internal energy* can also be used to refer to kinetic and potential energy of the internal parts of a body, such as vibration, when we are primarily interested in the body's motion as a whole.

normal force exerted by the roadbed or track on the car does no work because it acts perpendicular to the motion). Hence we can write

$$W_{\text{net}} = W_{\text{C}} + W_{\text{NC}}$$

where in general W_{C} represents the work done by conservative forces (gravity for our car) and W_{NC} is the work done by nonconservative forces (friction). We saw in Section 8–2 (Eq. 8–4) that the work done by a conservative force, such as gravity, can be written in terms of potential energy:

$$W_{\text{C}} = \int_1^2 \mathbf{F} \cdot d\mathbf{l} = -\Delta U.$$

Hence we can now write

$$\Delta K = -\Delta U + W_{\text{NC}}$$

or

$$\Delta K + \Delta U = W_{\text{NC}}. \qquad (8\text{–}15)$$

Conservation of energy (work-energy principle: general form)

This equation represents the general form of the work-energy principle. It also represents the conservation of energy. For our car, W_{NC} is the work done by the friction force and represents thermal energy. Equation 8–15 tells us that the change in mechanical energy, $\Delta(K + U)$, which is a decrease here since $W_{\text{NC}} < 0$ ($\mathbf{F}_{\text{fr}}$ and $d\mathbf{l}$ are in opposite directions), goes into thermal energy. But Eq. 8–15 is valid in general. Based on our derivation from the work-energy principle, W_{NC} on the right side of Eq. 8–15 must be the total work done by all forces that are not included in the potential energy term, ΔU, on the left side.[†] The potential energy term, U, should include all conservative forces acting.

Let us rewrite Eq. 8–15 for our roller-coaster car example of Fig. 8–8, shown here in Fig. 8–17, setting $W_{\text{NC}} = -F_{\text{fr}}l$ as we discussed above:

$$W_{\text{fr}} = -F_{\text{fr}}l = \Delta K + \Delta U = \left(\tfrac{1}{2}mv_2^2 - \tfrac{1}{2}mv_1^2\right) + (mgy_2 - mgy_1)$$

or

$$\tfrac{1}{2}mv_1^2 + mgy_1 = \tfrac{1}{2}mv_2^2 + mgy_2 + F_{\text{fr}}d. \qquad \begin{bmatrix} \text{gravity and} \\ \text{friction acting} \end{bmatrix} \quad (8\text{–}16)$$

Conservation of energy with gravity and friction

We can write this last equation as

initial energy = final energy (including thermal energy).

On the left we have the mechanical energy of the system initially. It equals the mechanical energy at any subsequent point along the path plus the amount of thermal (or internal) energy produced in the process.

EXAMPLE 8–11 **Friction on the roller coaster.** The roller-coaster car in Example 8–4, which starts at a height $y_1 = 40$ m, is found to reach a vertical height of only 25 m on the second hill before coming to a stop (Fig. 8–17). It traveled a total distance of 400 m. Estimate the average friction force (assume constant) on the car, whose mass is 1000 kg.

SOLUTION We use conservation of energy, here in the form of Eq. 8–16, taking point 1 to be the instant when the car started coasting and point 2 to be the instant it stopped. Then $v_1 = 0$, $y_1 = 40$ m, $v_2 = 0$, $y_2 = 25$ m, and $d = 400$ m. Thus

$$0 + (1000\,\text{kg})(9.8\,\text{m/s}^2)(40\,\text{m}) = 0 + (1000\,\text{kg})(9.8\,\text{m/s}^2)(25\,\text{m}) + F_{\text{fr}}(400\,\text{m}).$$

We solve this for F_{fr} to find $F_{\text{fr}} = 370$ N.

FIGURE 8–17 Roller-coaster car coasting, as in Fig. 8–8, but now with friction. Example 8–11.

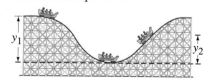

[†] A conservative force, if desired, could be considered as doing work (and therefore included in W_{NC} on the right in Eq. 8–15) rather than as a change in potential energy, whereas nonconservative forces (such as friction) *must* be included in the work term, W_{NC}. It must be emphasized that all forces acting on a body must be included in one term or the other. But avoid the error of including the same force twice, once in the potential energy term, U, and again in the work term, W.

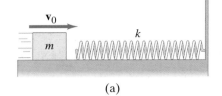

(a)

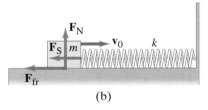

(b)

FIGURE 8–18 Example 8–12.

EXAMPLE 8–12 **Friction with a spring.** A block of mass m sliding along a rough horizontal surface is traveling at a speed v_0 when it strikes a massless spring head-on (see Fig. 8–18) and compresses the spring a maximum distance X. If the spring has stiffness constant k, determine the coefficient of kinetic friction between block and surface.

SOLUTION At the moment of collision, the block has $K = \frac{1}{2}mv_0^2$ and the spring is presumably uncompressed, so $U = 0$. Initially the mechanical energy of the system is $\frac{1}{2}mv_0^2$. By the time the spring reaches maximum compression, $K = 0$ and $U = \frac{1}{2}kX^2$. In the meantime, the friction force $(= \mu_k F_N = \mu_k mg)$ has done work $W = -\mu_k mgX$ which goes into thermal energy. From conservation of energy we can write

$$\text{energy (initial)} = \text{energy (final)}$$

$$\tfrac{1}{2}mv_0^2 = \tfrac{1}{2}kX^2 + \mu_k mgX.$$

We solve for μ_k and find

$$\mu_k = \frac{v_0^2}{2gX} - \frac{kX}{2mg}.$$

Problem solving is not a process that can be done by simply following a set of rules. The following Problem Solving box, like all others, is thus *not* a prescription, but is a summary to help you get started solving problems related to energy.

PROBLEM SOLVING **Conservation of Energy**

1. Draw a picture.
2. Determine the system for which energy will be conserved: the object or objects and the forces acting. Identify all forces that do work.
3. Ask yourself what quantity you are looking for, and decide what the initial (point 1) and final (point 2) locations are.
4. If the body under investigation changes its height during the problem, then choose a $y = 0$ level for gravitational potential energy. This may be chosen for convenience; the lowest point in the problem is often a good choice.
5. If springs are involved, choose the unstretched spring position to be x (or y) = 0.

6. If no friction or other nonconservative forces act, then apply conservation of mechanical energy:

$$K_1 + U_1 = K_2 + U_2.$$

7. Solve for the unknown quantity.
8. If friction or other nonconservative forces are present and significant, then an additional term, W_{NC}, will be needed:

$$K_1 + U_1 = K_2 + U_2 + W_{NC}.$$

To be sure which sign to give W_{NC}, or on which side of the equation to put it, use your intuition: is the total mechanical energy E increased or decreased in the process?

8–7 Gravitational Potential Energy and Escape Velocity

We have been dealing with gravitational potential energy so far in this chapter assuming the force of gravity is constant, $\mathbf{F} = m\mathbf{g}$. This is an accurate assumption for ordinary objects near the Earth's surface. But to deal with gravity more generally, for points far from the Earth's surface, we must consider that the gravitational force exerted by the Earth on a particle of mass m decreases inversely as the square of the distance r from the Earth's center. The precise relationship is given by Newton's law of universal gravitation (Sections 6–1 and 6–2):

$$\mathbf{F} = -G\frac{mM_E}{r^2}\hat{\mathbf{r}} \qquad\qquad [r > r_E]$$

where M_E is the mass of the Earth and $\hat{\mathbf{r}}$ is a unit vector (at the position of m) directed radially away from the Earth's center. The minus sign indicates that the force on m is directed toward the Earth's center, in the direction opposite to $\hat{\mathbf{r}}$. This equation can also be used to describe the gravitational force on a mass m in the vicinity of other heavenly bodies, such as the Moon, planets, or Sun, in which case M_E must be replaced by that body's mass.

Suppose an object of mass m moves from one position to another along an arbitrary path (Fig. 8–19) so that its distance from the Earth's center changes from r_1 to r_2. The work done by the gravitational force is

$$W = \int_1^2 \mathbf{F} \cdot d\mathbf{l} = -GmM_E\int_1^2\frac{\hat{\mathbf{r}} \cdot d\mathbf{l}}{r^2},$$

where $d\mathbf{l}$ represents an infinitesimal displacement. Since $\hat{\mathbf{r}} \cdot d\mathbf{l} = dr$ is the component of $d\mathbf{l}$ along $\hat{\mathbf{r}}$ (see Fig. 8–19), then

$$W = -GmM_E\int_{r_1}^{r_2}\frac{dr}{r^2} = GmM_E\left(\frac{1}{r_2} - \frac{1}{r_1}\right)$$

or

$$W = \frac{GmM_E}{r_2} - \frac{GmM_E}{r_1}.$$

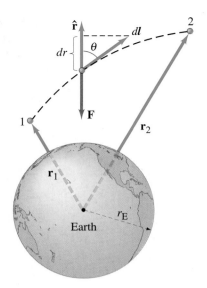

FIGURE 8–19 Arbitrary path of particle of mass m moving from point 1 to point 2.

Work done by gravity

Since the value of the integral depends only on the position of the end points $(r_1$ and $r_2)$ and not on the path taken, the gravitational force is a conservative force. We can therefore use the concept of potential energy for the gravitational force. Since the change in potential energy is always defined (Section 8–2) as the negative of the work done by the force, we have

$$\Delta U = U_2 - U_1 = -\frac{GmM_E}{r_2} + \frac{GmM_E}{r_1}. \qquad (8\text{–}17)$$

From Eq. 8–17 the potential energy at any distance r from the Earth's center can be written:

$$U(r) = -\frac{GmM_E}{r} + C,$$

where C is a constant. It is usual to choose $C = 0$, so that

$$U(r) = -\frac{GmM_E}{r}. \qquad \begin{bmatrix}\text{gravity} \\ (r > r_E)\end{bmatrix} \quad (8\text{–}18)$$

FIGURE 8–20 Gravitational potential energy plotted as a function of r, the distance from Earth's center. Valid only for points $r > r_E$, the radius of the Earth.

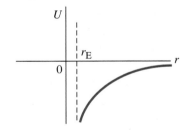

Gravitational potential energy

With this choice for C, $U = 0$ at $r = \infty$. As an object approaches the Earth, its potential energy decreases and is always negative (Fig. 8–20).

Equation 8–17 reduces to Eq. 8–2, $\Delta U = mg(y_2 - y_1)$, for objects near the surface of the Earth (see Problem 40).

The total energy of a particle of mass m, which feels only the force of the Earth's gravity, is conserved since gravity is a conservative force. Therefore we can write

$$\tfrac{1}{2}mv_1^2 - G\frac{mM_E}{r_1} = \tfrac{1}{2}mv_2^2 - G\frac{mM_E}{r_2} = \text{constant}. \qquad \begin{bmatrix}\text{gravity} \\ \text{only}\end{bmatrix} \quad (8\text{–}19)$$

EXAMPLE 8–13 **Package dropped from high speed rocket.** A box of empty film canisters is dumped from a rocket traveling outward from Earth at a speed of 1800 m/s when 1600 km above the Earth's surface. The package eventually falls to the Earth. Estimate its speed just before impact. Ignore air resistance.

SOLUTION The package initially has a speed relative to Earth equal to the speed of the rocket from which it falls. We use conservation of energy,

$$\tfrac{1}{2}mv_1^2 - G\frac{mM_E}{r_1} = \tfrac{1}{2}mv_2^2 - G\frac{mM_E}{r_2}$$

where $v_1 = 1.80 \times 10^3$ m/s, $r_1 = 1.60 \times 10^6$ m $+ 6.38 \times 10^6$ m $= 7.98 \times 10^6$ m, and $r_2 = 6.38 \times 10^6$ m (the radius of the Earth). We solve for v_2:

$$v_2 = \sqrt{v_1^2 - 2GM_E\left(\frac{1}{r_1} - \frac{1}{r_2}\right)}$$

$$= \sqrt{\begin{array}{c}(1.80 \times 10^3 \text{ m/s})^2 - 2(6.67 \times 10^{-11}\, \text{N} \cdot \text{m}^2/\text{kg}^2)(5.98 \times 10^{24}\, \text{kg}) \\ \times \left(\dfrac{1}{7.98 \times 10^6 \text{ m}} - \dfrac{1}{6.38 \times 10^6 \text{ m}}\right)\end{array}}$$

$$= 5320 \text{ m/s}.$$

In reality, the speed will be somewhat less than this because of air resistance. Note, incidentally, that the direction of the velocity never entered into the problem and this is one of the advantages of the energy method. The rocket could have been heading away from the Earth, or toward it, or at some other angle, and the result would be the same.

When a body is projected into the air from the Earth, it will return to Earth unless its speed is very high. But if the speed *is* high enough, it will continue out into space never to return to Earth (barring other forces or collisions). The minimum initial *Escape velocity* velocity needed to prevent an object from returning to the Earth is called the **escape velocity** from Earth, v_{esc}. To determine v_{esc} from the Earth's surface (ignoring air resistance), we use Eq. 8–19 with $v_1 = v_{esc}$ and $r_1 = r_E = 6.38 \times 10^6$ m, the radius of the Earth. Since we want the minimum speed for escape, we need the object to reach $r_2 = \infty$ with merely zero speed, $v_2 = 0$. Applying Eq. 8–19 we have

$$\tfrac{1}{2}mv_{esc}^2 - G\frac{mM_E}{r_E} = 0 + 0$$

or

$$v_{esc} = \sqrt{2GM_E/r_E} = 1.12 \times 10^4 \text{ m/s} \qquad (8\text{–}20)$$

or 11.2 km/s. It is important to note that although a mass can escape from the Earth (or solar system) never to return, the force on it due to the Earth's gravitational field is never actually zero for a finite value of r. However, the force does become very small and usually ignorable at great distances.

EXAMPLE 8–14 **Escaping the Earth or the Moon.** (*a*) Compare the escape velocities of a rocket from the Earth and from the Moon. (*b*) Compare the energies required to launch the rockets. For the Moon, $M_M = 7.35 \times 10^{22}$ kg and $r_M = 1.74 \times 10^6$ m, and for the Earth, $M_E = 5.97 \times 10^{24}$ kg and $r_E = 6.38 \times 10^6$ m.

SOLUTION (*a*) Using Eq. 8–20, the ratio of the escape velocities is

$$\frac{v_{esc}(\text{Earth})}{v_{esc}(\text{Moon})} = \sqrt{\frac{M_E}{M_M}\frac{r_M}{r_E}} = 4.7.$$

To escape Earth requires a speed 4.7 times that required to escape the Moon. (*b*) The fuel that must be burned provides energy proportional to v^2 $\left(K = \tfrac{1}{2}mv^2\right)$; so to launch a rocket to escape Earth requires $(4.7)^2 = 22$ times as much energy as to escape from the Moon.

8–8 | Power

Power is defined as the rate at which work is done. The *average power*, $\overline{P}$, when an amount of work W is done in a time t is

$$\overline{P} = \frac{W}{t}.$$ **(8–21a)**

Power defined

The *instantaneous power, P,* is

$$P = \frac{dW}{dt}.$$ **(8–21b)**

The work done in a process is equal to the energy transformed from one form or from one body to another. For example, as the potential energy stored in the spring of Fig. 8–6b is transformed to kinetic energy of the ball, the spring is doing work on the ball. Similarly, when you throw a ball or push a grocery cart, *whenever work is done, energy is being transformed or transferred from one body to another.* Hence we can also say that power is the *rate at which energy is transformed*:

Energy is transferred whenever work is done

$$P = \frac{dE}{dt}.$$ **(8–21c)**

The power of a horse refers to how much work it can do per unit of time. The power rating of an engine refers to how much chemical or electrical energy can be transformed into mechanical energy per unit of time. In SI units, power is measured in joules per second and this is given a special name, the **watt** (W): $1\,\text{W} = 1\,\text{J/s}$. We are most familiar with the watt for measuring the rate at which an electric light bulb or heater changes electric energy into light or thermal energy, but it is used for other types of energy transformation as well. In the British system, the unit of power is the foot-pound per second ($\text{ft} \cdot \text{lb/s}$). For practical purposes a larger unit is often used, the **horsepower**. One horsepower[†] (hp) is defined as $550\,\text{ft} \cdot \text{lb/s}$, which equals 746 watts.

Units: the watt ($1\,\text{W} = 1\,\text{J/s}$)

The horsepower ($1\,\text{hp} = 746\,\text{W}$)

To see the distinction between energy and power, consider the following example. A person is limited in the work he or she can do, not only by the total energy required, but also by the rate this energy is used; that is, by power. For example, a person may be able to walk a long distance or climb many flights of stairs before having to stop because so much energy has been used up. On the other hand, a person who runs very quickly up stairs, may feel exhausted after only a flight or two. He or she is limited in this case by power, the rate at which his or her body can transform chemical energy into mechanical energy.

Energy and power distinguished

EXAMPLE 8–15 **Stair-climbing power.** A 70-kg jogger runs up a long flight of stairs in 4.0 s. The vertical height of the stairs is 4.5 m. (*a*) Estimate the jogger's power output in watts and horsepower. (*b*) How much energy did this require?

SOLUTION (*a*) The work done is against gravity, and equals $W = mgy$. Then the average power output was

$$\overline{P} = \frac{W}{t} = \frac{mgy}{t} = \frac{(70\,\text{kg})(9.8\,\text{m/s}^2)(4.5\,\text{m})}{4.0\,\text{s}} = 770\,\text{W}.$$

Since there are 746 W in 1 hp, the jogger is doing work at a rate of just over 1 hp. It is worth noting that a human cannot do work at this rate for very long.
(*b*) The energy required is $E = \overline{P}t = (770\,\text{J/s})(4.0\,\text{s}) = 3100\,\text{J}$. [Note that the person had to transform more energy than this. The total energy transformed by a person or an engine always includes some thermal energy (recall how hot you get running up stairs).]

[†]The unit was first chosen by James Watt (1736–1819), who needed a way to specify the power of his newly developed steam engines. He found by experiment that a good horse can work all day at an average rate of about $360\,\text{ft} \cdot \text{lb/s}$. So as not to be accused of exaggeration in the sale of his steam engines, he multiplied this by roughly $1\frac{1}{2}$ when he defined the hp.

Automobiles do work to overcome the force of friction (and air resistance), to climb hills, and to accelerate. A car is limited by the rate it can do work, which is why automobile engines are rated in horsepower. A car needs power most when it is climbing hills and when accelerating. In the next Example, we will calculate how much power is needed in these situations for a car of reasonable size. Even when a car travels on a level road at constant speed, it needs some power just to do work to overcome the retarding forces of internal friction and air resistance. These forces depend on the conditions and speed of the car, but are typically in the range 400 N to 1000 N.

It is often convenient to write the power in terms of the net force **F** applied to an object and its velocity **v**. Since $P = dW/dt$ and $dW = \mathbf{F} \cdot d\mathbf{l}$ (Eq. 7–7), then

$$P = \frac{dW}{dt} = \mathbf{F} \cdot \frac{d\mathbf{l}}{dt} = \mathbf{F} \cdot \mathbf{v}. \qquad (8\text{–}22)$$

EXAMPLE 8–16 Power needs of a car. Calculate the power required of a 1400-kg car under the following circumstances: (a) the car climbs a 10° hill (a fairly steep hill) at a steady 80 km/h; and (b) the car accelerates along a level road from 90 to 110 km/h in 6.0 s to pass another car. Assume the retarding force on the car is $F_R = 700$ N throughout. See Fig. 8–21. (Be careful not to confuse $\mathbf{F_R}$, which is due to air resistance and friction that retard the motion, with the force **F** needed to accelerate the car, which is the frictional force exerted by the road on the tires—the reaction to the motor-driven tires pushing against the road.)

FIGURE 8–21 Example 8–16: calculation of power needed for a car (a) to climb a hill, (b) to pass another car.

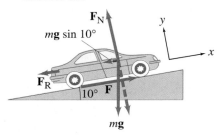

SOLUTION (a) To move at a steady speed up the hill, the car must exert a force equal to the sum of the retarding force, 700 N, and the component of gravity parallel to the hill, $mg \sin 10° = (1400\,\text{kg})(9.80\,\text{m/s}^2)(0.174) = 2400$ N. Since $\bar{v} = 80$ km/h $= 22$ m/s and is parallel to **F**, then (Eq. 8–22):

$$P = Fv$$
$$= (2400\,\text{N} + 700\,\text{N})(22\,\text{m/s}) = 6.80 \times 10^4\,\text{W} = 91\,\text{hp}.$$

(b) The car accelerates from 25.0 m/s to 30.6 m/s (90 to 110 km/h). Thus the car must exert a force that overcomes the 700 N retarding force plus that required to give it the acceleration $\bar{a}_x = (30.6\,\text{m/s} - 25.0\,\text{m/s})/6.0\,\text{s} = 0.93\,\text{m/s}^2$. We apply Newton's second law with x being the direction of motion:

$$ma_x = \Sigma F_x = F - F_R.$$

Then the force required, F, is

$$F = ma_x + F_R$$
$$= (1400\,\text{kg})(0.93\,\text{m/s}^2) + 700\,\text{N} = 2000\,\text{N}.$$

Since $P = \mathbf{F} \cdot \mathbf{v}$, the required power increases with speed and the motor must be able to provide a maximum power output of

$$\bar{P} = (2000\,\text{N})(30.6\,\text{m/s}) = 6.12 \times 10^4\,\text{W} = 82\,\text{hp}.$$

Even taking into account the fact that only 60 to 80 percent of the engine's power output reaches the wheels, it is clear from these calculations that an engine of 100 to 150 hp is more than adequate from a practical point of view.

We mentioned in the Example above that only part of the energy output of a car engine reaches the wheels. Not only is some energy wasted in getting from the engine to the wheels, in the engine itself much of the input energy (from the gasoline) does not end up doing useful work. An important characteristic of all engines is their overall efficiency e, defined as the ratio of the useful power output of the engine, P_{out}, to the power input, P_{in}:

$$e = \frac{P_{out}}{P_{in}}.$$

Efficiency

The efficiency is always less than 1.0 because no engine can create energy, and in fact, cannot even transform energy from one form to another without some going to friction, thermal energy and other nonuseful forms of energy. For example, an automobile engine converts chemical energy released in the burning of gasoline into mechanical energy that moves the pistons and eventually the wheels. But nearly 85% of the input energy is "wasted" as thermal energy that goes out the exhaust pipe, plus friction in the moving parts. Thus car engines are roughly only about 15% efficient. We will discuss efficiency in detail in Chapter 20.

* 8–9 Potential Energy Diagrams; Stable and Unstable Equilibrium

For an object acted on by only a conservative force, we can learn a great deal about its motion simply by examining a potential energy diagram—the graph of $U(x)$ versus x. An example of a potential energy diagram is shown in Fig. 8–22. The rather complex curve represents some complicated potential energy $U(x)$. The total energy $E = K + U$ is constant and can be represented as a horizontal line on this graph. Four different possible values for E are shown, labeled E_0, E_1, E_2, and E_3. What the actual value of E will be for a given system depends on the initial conditions. (For example, the total energy E of a mass oscillating on the end of a spring depends on the amount the spring is initially compressed or stretched.) Because $E = U + K = $ constant, then $U(x)$ must be less than or equal to E for all situations: $U(x) \leq E$. Thus the minimum value which the total energy can take for the potential energy shown in Fig. 8–22 is that labeled E_0. For this value of E, the mass can only be at rest at $x = x_0$. It has potential energy but no kinetic energy.

If the object's total energy E is greater than E_0, say it is E_1 on our plot, the object can have both kinetic and potential energy. Since energy is conserved,

$$K = E - U(x).$$

Since the curve represents $U(x)$ at each x, the kinetic energy at any value of x is represented by the distance between the E line and the curve $U(x)$ at that value of x. In the diagram, the kinetic energy for an object at x_1, when its total energy is E_1, is indicated by the notation K_1.

An object with energy E_1 can oscillate only between the points x_2 and x_3. This is because if $x > x_2$ or $x < x_3$, the potential energy would be greater than E, meaning $K = \frac{1}{2}mv^2 < 0$ and v would be imaginary, and so impossible. At x_2 and x_3 the velocity is zero, since $E = U$ at these points. Hence x_2 and x_3 are called the **turning points** of the motion. If the object is at x_0, say, moving to the right, its kinetic energy (and speed) decreases until it reaches zero at $x = x_2$. The object then reverses direction, proceeding to the left and increasing in speed until it passes x_0 again. It continues to move, decreasing in speed until it reaches $x = x_3$, where again $v = 0$, and the object again reverses direction.

Turning points

If the object has energy $E = E_2$ in Fig. 8–22, there are four turning points. The object can move in only one of the two potential energy "valleys," depending on where it is initially. It cannot get from one valley to the other because of the barrier between them—for example at a point such as x_4, $U > E_2$, which means v would be imaginary.[†] For energy E_3, there is only one turning point since $U(x) < E_3$ for all $x > x_5$. Thus our object, if moving initially to the left, varies in

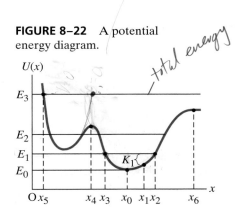

FIGURE 8–22 A potential energy diagram.

[†] Although this is true according to Newtonian physics, modern quantum mechanics predicts that objects can "tunnel" through such a barrier, and such processes have been observed at the atomic and subatomic level.

speed as it passes the potential valleys but eventually stops and turns around at $x = x_5$. It then proceeds to the right indefinitely, never to return.

How do we know the object reverses direction at the turning points? Because of the force exerted on it. The force F is related to the potential energy U by Eq. 8–7, $F = -dU/dx$. The force F is equal to the negative of the slope of the U-versus-x curve at any point x. At $x = x_2$, for example, the slope is positive so the force is negative, which means it acts to the left (toward decreasing values of x) acting in the direction opposite to the object's motion.

At $x = x_0$ the slope is zero, so $F = 0$. At such a point the particle is said to be in **equilibrium**. This term means simply that the net force on the object is zero. Hence, its acceleration is zero, and so if it is initially at rest, it remains at rest. If the object at rest at $x = x_0$ were moved slightly to the left or right, a nonzero force would act on it in the direction to move it back toward x_0. An object that returns toward its equilib-

Stable equilibrium   rium point when displaced slightly is said to be at a point of **stable equilibrium**. Any *minimum* in the potential energy curve represents a point of stable equilibrium.

An object at $x = x_4$ would also be in equilibrium, since $F = -dU/dx = 0$. If the object were displaced a bit to either side of x_4, a force would act to pull the object *away* from the equilibrium point. Points like x_4, where the potential energy

Unstable equilibrium   curve has a maximum, are points of **unstable equilibrium**. The object will *not* return to equilibrium if displaced slightly, but instead will move farther away.

When an object is in a region over which U is constant, such as at $x = x_6$ in Fig. 8–22, the force is zero over some distance. The object is in equilibrium and if displaced slightly to one side the force is still zero. The object is said to be in

Neutral equilibrium   **neutral equilibrium** in this region.

☐ Summary

A **conservative force** is one for which the work done by the force in moving an object from one position to another depends only on the two positions and not on the path taken. The work done by a conservative force is recoverable, which is not true for nonconservative forces, such as friction.

Potential energy is energy associated with the position or configuration of bodies. Examples are: gravitational potential energy

$$U = mgy$$

where the mass m is near the Earth's surface, a height y above some reference point; elastic potential energy

$$U = \tfrac{1}{2}kx^2$$

such as a spring with stiffness constant k stretched or compressed a distance x from equilibrium; and chemical, electrical, and nuclear energy. Potential energy is always associated with a conservative force, and the change in potential energy, ΔU, between two points under the action of a conservative force $\mathbf{F}$ is defined as the negative of the work done by the force:

$$\Delta U = U_2 - U_1 = -\int_1^2 \mathbf{F} \cdot d\boldsymbol{l}.$$

Inversely, we can write, for the one-dimensional case,

$$F = -\frac{dU(x)}{dx}.$$

Only *changes* in potential energy are physically meaning-

ful, so the choice of where $U = 0$ can be chosen for convenience. Potential energy is not a property of a body but is associated with the interaction of two or more bodies.

When only conservative forces act, the total **mechanical energy**, E, defined as the sum of kinetic and potential energies, is conserved:

$$E = K + U = \text{constant}.$$

If nonconservative forces also act, additional types of energy are involved, such as thermal energy. It has been found experimentally that, when all forms of energy are included, the total energy is conserved. This is the **law of conservation of energy**:

$$\Delta K + \Delta U = W_{\text{NC}}.$$

The gravitational force as described by Newton's law of universal gravitation is a conservative force. The potential energy of an object of mass m due to the gravitational force exerted on it by the Earth is given by

$$U(r) = -GmM_{\text{E}}/r,$$

where M_{E} is the mass of the Earth and r is the distance of the object from the Earth's center ($r \geq$ radius of Earth).

Power is defined as the rate at which work is done or the rate at which energy is transformed from one form to another: $P = dW/dt = dE/dt$, or $P = \mathbf{F} \cdot \mathbf{v}$.

Questions

1. List some everyday forces that are not conservative, and explain why they aren't.

2. You lift a heavy book from a table to a high shelf. List the forces on the book during this process, and state whether each is conservative or nonconservative.

3. The net force acting on a particle is conservative and increases the kinetic energy by 300 J. What is the change in (a) the potential energy, and (b) the total energy, of the particle?

4. When a "superball" is dropped, can it rebound to a greater height than its original height?

5. A hill has a height h. A child on a sled (total mass m) slides down starting from rest at the top. Does the velocity at the bottom depend on the angle of the hill if (a) it is icy and there is no friction, and (b) there is friction (deep snow)?

6. Why is it tiring to push hard against a solid wall even though no work is done?

7. Analyze the motion of a simple swinging pendulum in terms of energy, (a) ignoring friction; and (b) taking friction into account. Explain why a grandfather clock has to be wound up.

8. Describe precisely what is "wrong" physically in the famous Escher drawing shown in Fig. 8–23.

FIGURE 8–23 Question 8.

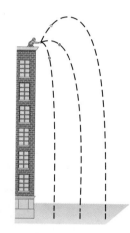

FIGURE 8–24 Question 9.

9. In Fig. 8–24, water balloons are tossed from the roof of a building, all with the same speed but with different launch angles. Which one has the highest speed on impact? Ignore air resistance.

10. Suppose you lift a suitcase from the floor to a table. Does the work you do on the suitcase depend on (a) whether you lift it straight up or along a more complicated path, (b) the time the lifting takes, (c) the height of the table, and/or (d) the weight of the suitcase?

11. A coil spring of mass m rests upright on a table. If you compress the spring by pressing down with your hand and then release it, can the spring actually leave the table? Explain using the law of conservation of energy.

12. What happens to the gravitational potential energy when water at the top of a waterfall falls to the pool below?

13. By approximately how much does your gravitational potential energy change when you jump as high as you can?

14. Experienced hikers prefer to step over a fallen log in their path rather than stepping on top and jumping down on the other side. Explain.

15. Consider two observers who are in different inertial reference frames that move with velocity v with respect to each other. They both observe an object being pulled on a rough horizontal surface. Do they agree as to the value of (a) the object's kinetic energy; (b) the total work done on the object; (c) the amount of energy transformed from mechanical to thermal energy due to friction? Does your answer to (c) contradict (a) and (b)? Explain.

16. (a) Where does the kinetic energy come from when a car accelerates uniformly starting from rest? (b) How is the increase in kinetic energy related to the friction force the road exerts on the tires?

17. The Earth is closest to the Sun in winter (Northern Hemisphere). When is the gravitational potential energy the greatest?

18. Can the total mechanical energy $E = K + U$ ever be negative? Explain.

19. Suppose that you wish to launch a rocket from the surface of the Earth so that it escapes the Earth's gravitational field. You wish to use minimum fuel in doing this. From what point on the surface of the Earth should you make the launch and in what direction? Do the launch location and direction matter? Explain.

20. Recall from Chapter 4, Example 4–14, that you can use a pulley and ropes to decrease the force needed to raise a heavy load (see Fig. 8–25). But for every meter the load is raised, how much rope must be pulled up? Account for this, using energy concepts.

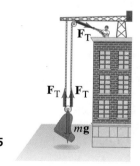

FIGURE 8–25 Question 20.

21. Two identical arrows, one with twice the speed of the other, are fired into a bale of hay. Assuming the hay exerts a constant "frictional" force on the arrows, the faster arrow will penetrate how much farther than the slower arrow? Explain.

22. Why is it easier to climb a mountain via a zigzag trail rather than to climb straight up?

*23. Give several examples of stable, unstable, and neutral equilibrium.

*24. In what state of equilibrium is a cube (a) when resting on its face; (b) when on its edge?

*25. (a) Describe in detail the velocity changes of a particle that has energy E_3 in Fig. 8–22 as it moves from x_6 to x_5 and back to x_6. (b) Where is its kinetic energy the greatest and least?

*26. Name the type of equilibrium for each position of the balls in Fig. 8–26.

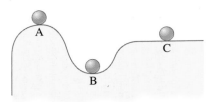

FIGURE 8–26   Question 26.

*27. Figure 8–27 shows a potential energy curve, $U(x)$. (a) At which point does the force have greatest magnitude? (b) For each labeled point, state whether the force acts to the left or to the right, or is zero. (c) Where is there equilibrium and of what type is it?

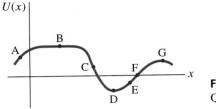

FIGURE 8–27
Question 27.

Problems

Sections 8–1 and 8–2

1. (I) A spring has a spring constant k of 82.0 N/m. How far must this spring be compressed to store 35.0 J of potential energy?

2. (I) A 5.0-kg monkey swings from one branch to another 1.5 m higher. What is the change in gravitational potential energy?

3. (I) By how much does the gravitational potential energy of a 58–kg pole vaulter change if her center of mass rises about 3.8 m during the jump?

4. (I) A 66.5-kg hiker starts at an elevation of 1500 m and climbs to the top of a 2660-m peak. (a) What is the hiker's change in potential energy? (b) What is the minimum work required of the hiker? (c) Can the actual work done be greater than this? Explain.

5. (I) In starting an exercise, a 1.70-m tall person lifts a 2.20-kg book off the ground so it is 2.40 m above the ground. What is the potential energy of the book relative to (a) the ground, and (b) the top of the person's head? (c) How is the work done by the person related to the answers in parts (a) and (b)?

6. (II) If $U = 3x^2 + 2xy + 4y^2 z$, what is the force, $\mathbf{F}$?

7. (II) A particular spring obeys the force law $\mathbf{F} = (-kx + ax^3 + bx^4)\mathbf{i}$. (a) Is this force conservative? Explain why or why not. (b) If it is conservative, determine the form of the potential energy function.

8. (II) Air resistance can be represented by a force proportional to the velocity $\mathbf{v}$ of an object: $\mathbf{F} = -k\mathbf{v}$. Is this force conservative? Explain.

9. (II) (a) A spring of spring constant k is initially compressed a distance x_0 from its unstretched length. What is the change in potential energy if it is then compressed to an amount x from its unstretched length? (b) Suppose the spring is then *stretched* a distance x_0 from the unstretched length. What is the change in potential energy as compared to when it is compressed by an amount x_0?

Sections 8–3 and 8–4

10. (I) Jane, looking for Tarzan, is running at top speed (5.0 m/s) and grabs a vine hanging 4.0 m vertically from a tall tree in the jungle. How high can she swing upward? Does the length of the vine (or rope) affect your answer?

11. (I) A novice skier, starting from rest, slides down a frictionless 32.0° incline whose vertical height is 105 m. How fast is she going when she reaches the bottom?

12. (I) A sled is initially given a shove up a frictionless 25.0° incline. It reaches a maximum vertical height 1.22 m higher than where it started. What was its initial speed?

13. (II) In the high jump, the kinetic energy of an athlete is transformed into gravitational potential energy without the aid of a pole. With what minimum speed must the athlete leave the ground in order to lift his center of mass 2.10 m and cross the bar with a speed of 0.70 m/s?

14. (II) A 75-kg trampoline artist jumps vertically upward from the top of a platform with a speed of 5.0 m/s. (a) How fast is he going as he lands on the trampoline, 2.0 m below (Fig. 8–28)? (b) If the trampoline behaves like a spring of spring constant 5.2×10^4 N/m, how far does he depress it?

FIGURE 8–28
Problem 14.

15. (II) A 60-kg bungee jumper jumps from a bridge. She is tied to a bungee cord that is 12 m long when unstretched, and falls a total of 31 m. (a) Calculate the spring constant k of the bungee cord assuming Hooke's law applies. (b) Calculate the maximum acceleration experienced by the jumper.

16. (II) A roller coaster, shown in Fig. 8–29, is pulled up to point A where it and its screaming occupants are released from rest. Assuming no friction, calculate the speed at points B, C, D.

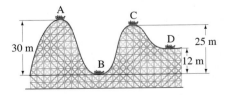

FIGURE 8–29
Problems 16 and 30.

17. (II) A vertical spring (ignore its mass), whose spring constant is 900 N/m, is attached to a table and is compressed down by 0.150 m. (a) What upward speed can it give to a 0.300-kg ball when released? (b) How high above its original position (spring compressed) will the ball fly?

18. (II) A 0.40-kg ball is thrown with a speed of 10 m/s at an angle of 30°. (a) What is its speed at its highest point, and (b) how high does it go? (Use conservation of energy.)

19. (II) A mass m is attached to the end of a spring (constant k) as shown in Fig. 8–30. The mass is given an initial displacement x_0 from equilibrium, and an initial speed v_0. Ignoring friction and the mass of the spring, use energy methods to find (a) its maximum speed, and (b) its maximum stretch from equilibrium, in terms of the given quantities.

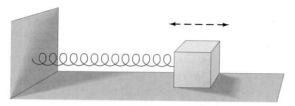

FIGURE 8–30   Problems 19, 33 and 34.

20. (II) A cyclist intends to cycle up a 9.50° hill whose vertical height is 92.0 m. The pedals turn in a circle of diameter 36.0 cm. Assuming the mass of bicycle plus person is 75.0 kg, (a) calculate how much work must be done against gravity. (b) If each complete revolution of the pedals moves the bike 5.10 m along its path, calculate the average force that must be exerted on the pedals tangent to their circular path. Neglect work done by friction and other losses.

21. (II) A pendulum 2.00 m long is released (from rest) at an angle $\theta_0 = 30.0°$ (Fig. 8–15). Determine the speed of the 70.0-g bob (a) at the lowest point ($\theta = 0$); (b) at $\theta = 15.0°$; (c) at $\theta = -15.0°$ (i.e., on the opposite side). (d) Determine the tension in the cord at each of these three points. (e) If the bob is given an initial speed $v_0 = 1.20$ m/s when released at $\theta = 30.0°$, recalculate the speeds for parts (a), (b), and (c).

22. (II) What should be the spring constant k of a spring designed to bring a 1200-kg car to rest from a speed of 100 km/h so that the occupants undergo a maximum acceleration of 5.0 g?

23. (III) An engineer is designing a spring to be placed at the bottom of an elevator shaft. If the elevator cable should happen to break when the elevator is at a height h above the top of the spring, calculate the value that the spring constant k should have so that passengers undergo an acceleration of no more than 5.0 g when brought to rest. Let M be the total mass of the elevator and passengers.

24. (III) A skier of mass m starts from rest at the top of a solid sphere of radius r and slides down its frictionless surface. (a) At what angle θ (Fig. 8–31) will the skier leave the sphere? (b) If friction were present, would the skier fly off at a greater or lesser angle?

FIGURE 8–31
Problem 24.

Sections 8–5 and 8–6

25. (I) Two railroad cars, each of mass 6500 kg and traveling 95 km/h, collide head-on and come to rest. How much thermal energy is produced in this collision?

26. (I) A 16.0-kg child descends a slide 2.50 m high and reaches the bottom with a speed of 2.25 m/s. How much thermal energy due to friction was generated in this process?

27. (II) A ski starts from rest and slides down a 20° incline 100 m long. (a) If the coefficient of friction is 0.090, what is the ski's speed at the base of the incline? (b) If the snow is level at the foot of the incline and has the same coefficient of friction, how far will the ski travel along the level? Use energy methods.

28. (II) A 145-g baseball is dropped from a tree 12.0 m above the ground. (a) With what speed would it hit the ground if air resistance could be ignored? (b) If it actually hits the ground with a speed of 8.00 m/s, what is the average force of air resistance exerted on it?

29. (II) A 90-kg crate, starting from rest, is pulled across a floor with a constant horizontal force of 350 N. For the first 15 m the floor is frictionless, and for the next 15 m the coefficient of friction is 0.25. What is the final speed of the crate?

30. (II) Suppose the roller coaster in Fig. 8–29 passes point A with a speed of 1.70 m/s. If the average force of friction is equal to one fifth of its weight, with what speed will it reach point B? The distance traveled is 45.0 m.

31. (II) A skier traveling 11 m/s reaches the foot of a steady upward 17° incline and glides 12 m up along this slope before coming to rest. What was the average coefficient of friction?

32. (II) Consider the track shown in Figure 8–32. The section AB is one quadrant of a circle of radius 2.0 m and is frictionless. B to C is a horizontal span 3.0 m long with a coefficient of kinetic friction $\mu_k = 0.25$. The section CD under the spring is frictionless. A block of mass 1.0 kg is released from rest at A. After sliding on the track, it is observed to compress the spring by 0.20 m. Determine: (a) the velocity of the block at point B; (b) the work done by friction as the block slides from B to C; (c) the velocity of the block at point C; (d) the stiffness constant k for the spring.

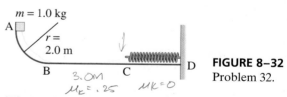

FIGURE 8–32
Problem 32.

33. (II) A 0.620-kg wood block is firmly attached to a very light horizontal spring ($k = 180$ N/m) as shown in Fig. 8–30. It is noted that the block-spring system, when compressed 5.0 cm and released, stretches out 2.3 cm beyond the equilibrium position before stopping and turning back. What is the coefficient of kinetic friction between the block and the table?

34. (II) A 180-g wood block is firmly attached to a very light horizontal spring, Fig. 8–30. The block can slide along a table where the coefficient of friction is 0.30. A force of 22 N compresses the spring 18 cm. If the spring is released from this position, how far beyond its equilibrium position will it stretch on its first swing?

35. (III) A 2.0-kg block slides along a horizontal surface with a coefficient of kinetic friction $\mu_k = 0.30$. The block has a speed $v = 1.3$ m/s when it strikes a massless spring head-on (as in Fig. 8–18). (a) If the spring has force constant $k = 120$ N/m, how far is the spring compressed? (b) What minimum value of the coefficient of static friction, μ_s, will assure that the spring remains compressed at the maximum compressed position? (c) If μ_s is less than this, what is the speed of the block when it detaches from the decompressing spring? [Hint: Detachment occurs when the spring reaches its natural length ($x = 0$); explain why.]

36. (III) In early test flights for the space shuttle using a "glider" (mass of 1000 kg including pilot), it was noted that after a horizontal launch at 500 km/h at a height of 3500 m, the glider eventually landed at a speed of 200 km/h. (a) What would its landing speed have been in the absence of air resistance? (b) What was the average force of air resistance exerted on it if it came in at a constant glide angle of 10° to the Earth?

Section 8–7

37. (I) For a satellite of mass m_S in a circular orbit of radius r_S, determine (a) its kinetic energy K, (b) its potential energy U ($U = 0$ at infinity), and (c) the ratio K/U.

38. (I) Jill and her friends have built a small rocket that soon after lift-off reaches a speed of 850 m/s. How high above the Earth can it rise? Ignore air friction.

39. (II) Determine the escape velocity from the Sun for an object (a) at the Sun's surface ($r = 7.0 \times 10^5$ km, $M = 2.0 \times 10^{30}$ kg), and (b) at the average distance of the Earth (1.50×10^8 km). Compare to the speed of the Earth in its orbit.

40. (II) Show that Eq. 8–17 for gravitational potential energy reduces to Eq. 8–2, $\Delta U = mg(y_2 - y_1)$, for objects near the surface of the Earth.

41. (II) Show that the change in potential energy of an object between the Earth's surface and a height h above is

$$\Delta U = \frac{mgh}{1 + h/r_E},$$

where r_E is the radius of the Earth and h is not necessarily small.

42. (II) (a) Show that the total mechanical energy of a satellite (mass m) orbiting at a distance r from the center of the Earth (mass M_E) is

$$E = -\frac{1}{2}\frac{GmM_E}{r},$$

if $U = 0$ at $r = \infty$. (b) Show that although friction causes the value of E to decrease slowly, kinetic energy must actually increase if the orbit remains a circle.

43. (II) Show that the escape velocity for any satellite in a circular orbit is $\sqrt{2}$ times its velocity.

44. (II) The Earth's distance from the sun varies from 1.471×10^8 km to 1.521×10^8 km during the year. Determine the difference in (a) the potential energy, (b) the Earth's kinetic energy, and (c) the total energy between these extreme points. Take the Sun to be at rest.

45. (II) Take into account the Earth's rotational speed (1 rev/day) and determine the necessary speed, with respect to Earth, for a rocket to escape if fired from the Earth at the equator in a direction (a) eastward; (b) westward; (c) vertically upward.

46. (II) (a) Determine a formula for the maximum height h that a rocket will reach if launched vertically from the Earth's surface with speed $v_0 (< v_{esc})$. Express in terms of v_0, r_E, M_E, and G. (b) How high does a rocket go if $v_0 = 8.2$ km/s? Ignore air resistance and the Earth's rotation.

47. (II) (a) Determine the rate at which the escape velocity from the Earth changes with height above the Earth's surface, dv_{esc}/dr. (b) Use the approximation $\Delta v \approx (dv/dr)\Delta r$ to determine the escape velocity for a spacecraft orbiting the Earth at a height of 300 km.

48. (II) A meteor has a speed of 90.0 m/s when 800 km above the Earth. It is falling vertically (ignore air resistance) and strikes a bed of sand in which it is brought to rest in 3.25 m. (a) What is its speed just before striking the sand? (b) How much work does the sand do to stop the meteor (mass = 575 kg)? (c) What is the average force exerted by the sand on the meteor? (d) How much thermal energy is produced?

49. (II) How much work would be required to move a satellite of mass m from a circular orbit of radius $r_1 = 2r_E$ about the Earth to another circular orbit of radius $r_2 = 3r_E$? (r_E is the radius of the Earth.)

50. (II) (a) Suppose we have three masses, m_1, m_2, and m_3, that initially are infinitely far apart from each other. Show that the work needed to bring them to the positions shown in Fig. 8–33 is

$$W = -G\left(\frac{m_1 m_2}{r_{12}} + \frac{m_1 m_3}{r_{13}} + \frac{m_2 m_3}{r_{23}}\right).$$

(b) Can we say that this formula also gives the potential energy of the system, or the potential energy of one or two of the bodies? (c) Is W equal to the binding energy of the system—that is, equal to the energy required to separate the components by an infinite distance? Explain.

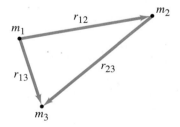

FIGURE 8–33 Problem 50.

51. (II) A NASA satellite has just observed an asteroid that is on a collision course with the Earth. The asteroid has an estimated mass, based on its size, of 5×10^9 kg. It is approaching the Earth on a head-on course with a velocity of 600 m/s relative to the Earth and is now 5.0×10^6 km away. With what speed will it hit the Earth's surface, neglecting friction with the atmosphere?

52. (II) A sphere of radius r_1 has a concentric spherical cavity of radius r_2 (Fig. 8–34). Assume this spherical shell of thickness $r_1 - r_2$ is uniform and has a total mass M. Show that the gravitational potential energy of a mass m at a distance r from the center of the shell $(r > r_1)$ is given by

$$U = -\frac{GmM}{r}.$$

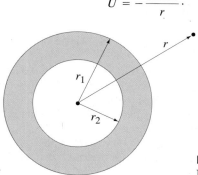

FIGURE 8–34
Problem 52.

53. (III) To escape the solar system, an interstellar spacecraft must overcome the gravitational attraction of both the Earth and Sun. Ignore the effects of other bodies in the solar system. (a) Show that the escape velocity is

$$v = \sqrt{v_E^2 + (v_S - v_0)^2} = 16.7 \text{ km/s},$$

where: v_E is the escape velocity from the Earth (Eq. 8–20); $v_S = \sqrt{2GM_S/r_{SE}}$ is the escape velocity from the gravitational field of the Sun at the orbit of the Earth but far from the Earth's influence (r_{SE} is Sun–Earth distance); and v_0 is the Earth's orbital velocity about the Sun. (b) Show that the energy required is 1.40×10^8 J per kilogram of spacecraft mass. [*Hint*: Write the energy equation for escape from Earth with v' as the velocity, relative to Earth, but far from Earth; then let $v' + v_0$ be the escape velocity from the Sun.]

Section 8–8

54. (I) How long will it take a 1750-W motor to lift a 285-kg piano to a sixth-story window 16.0 m above?

55. (I) If a car generates 18 hp when traveling at a steady 90 km/h, what must be the average retarding force exerted on the car?

56. (I) (a) Show that a British horsepower (550 ft·lb/s) is equal to 746 W. (b) What is the horsepower rating of a 100-W lightbulb?

57. (I) An 80-kg football player traveling 5.0 m/s is stopped in 1.0 s by a tackler. (a) What is the original kinetic energy of the player? (b) What average power is required to stop him?

58. (II) What minimum horsepower must a motor have to be able to drag a 300-kg box along a level floor at a speed of 1.20 m/s if the coefficient of friction is 0.45?

59. (II) A driver notices that her 1000-kg car slows down from 90 km/h to 70 km/h in about 6.0 s on the level when it is in neutral. Approximately what power (watts and hp) is needed to keep the car traveling at a constant 80 km/h?

60. (II) How much work can a 3.0-hp motor do in 1.0 h?

61. (II) A shot-putter accelerates a 7.3-kg shot from rest to 14 m/s. If this motion takes 1.5 s, what average power was developed?

62. (II) A pump is to lift 18.0 kg of water per minute through a height of 3.50 m. What output rating (watts) should the pump motor have?

63. (II) During a workout, the football players at State U ran up the stadium stairs in 61 s. The stairs are 140 m long and inclined at an angle of 30°. If a typical player has a mass of 105 kg, estimate the average power output on the way up. Ignore friction and air resistance.

64. (II) A 1000-kg car has a maximum power output of 120 hp. How steep a hill can it climb at a constant speed of 70 km/h if the frictional forces add up to 600 N?

65. (II) Squaw Valley ski area in California claims that its lifts can move 47,000 people per hour. If the average lift carries people about 200 m (vertically) higher, estimate the maximum total power needed.

66. (III) A bicyclist coasts down a 7.0° hill at a steady speed of 5.0 m/s. Assuming a total mass of 75 kg (bicycle plus rider), what must be the cyclist's power output to climb the same hill at the same speed?

67. (III) The position of a 280-g object is given (in meters) by $x = 5.0t^3 - 8.0t^2 - 30t$, where t is in seconds. Determine the net rate of work done on this object (a) at $t = 2.0$ s and (b) at $t = 4.0$ s. (c) What is the average net power input during the interval from $t = 0$ s to $t = 2.0$ s, and in the interval from $t = 2.0$ s to 4.0 s?

*Section 8–9

*68. (II) Draw a potential energy diagram and analyze the motion of a mass m resting on a frictionless horizontal table and connected to a horizontal spring with stiffness constant k. The mass is pulled a distance to the right so that the spring is stretched a distance x_0 initially, and then the mass is released from rest.

*69. (II) The spring of Problem 68 has a stiffness constant $k = 160$ N/m. The mass $m = 5.0$ kg is released from rest when the spring is stretched $x_0 = 1.0$ m from equilibrium. Determine (a) the total energy of the system; (b) the kinetic energy when $x = \frac{1}{2}x_0$; (c) the maximum kinetic energy; (d) the maximum speed and at what positions it occurs; (e) the maximum acceleration and where it occurs.

*70. (III) The potential energy of the two atoms in a diatomic (two-atom) molecule can be written

$$U(r) = -\frac{a}{r^6} + \frac{b}{r^{12}},$$

where r is the distance between the two atoms and a and b are positive constants. (a) At what values of r is $U(r)$ a minimum? A maximum? (b) At what values of r is $U(r) = 0$? (c) Plot $U(r)$ as a function of r from $r = 0$ to r at a value large enough for all the features in (a) and (b) to show. (d) Describe the motion of one atom with respect to the second atom when $E < 0$, and for $E > 0$. (e) Let F be the force one atom exerts on the other. For what values of r is $F > 0$, $F < 0$, $F = 0$? (f) Determine F as a function of r.

*71. (III) The *binding energy* of a two-particle system is defined as the energy required to separate the two particles from their state of lowest energy to $r = \infty$. Determine the binding energy for the molecule discussed in Problem 70.

72. A projectile is fired at an upward angle of 45.0° from the top of a 165-m cliff with a speed of 180 m/s. What will be its speed when it strikes the ground below? (Use conservation of energy.)

73. In a film of Jesse Owens's famous long jump in the 1936 Olympics, it is observed that his center of mass rose 1.1 m from launch point to the top of the arc. What minimum speed did he need at launch if he was also noted to be traveling at 6.5 m/s at the top of the arc?

74. How fast must a cyclist climb a 12° hill to maintain a power output of 0.20 hp? Ignore friction and assume the mass of cyclist plus bicycle is 85 kg.

75. What is the average power output of an elevator that lifts 850 kg a vertical height of 32.0 m in 11.0 s?

76. A 0.20-kg pinecone falls from a branch 18 m above the ground. (a) With what speed would it hit the ground if air resistance could be ignored? (b) If it actually hits the ground with a speed of 10.0 m/s, what was the average force of air resistance exerted on it?

77. A 60-kg skier starts from rest at the top of a ski jump, point A in Fig. 8–35, and travels down the ramp. If friction and air resistance can be neglected, (a) determine her speed v_B when she reaches the horizontal end of the ramp at B. (b) Determine the distance s to where she strikes the ground at C.

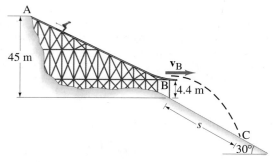

FIGURE 8–35 Problems 77 and 78.

78. Repeat Problem 77, but now assume the skier jumps upward upon reaching point B and acquires a vertical component of velocity (at B) of 3.0 m/s.

79. A ball is attached to a horizontal cord of length L whose other end is fixed, Fig. 8–36. (a) If the ball is released, what will be its speed at the lowest point of its path? (b) A peg is located a distance h directly below the point of attachment of the cord. If $h = 0.80\,L$, what will be the speed of the ball when it reaches the top of its circular path about the peg?

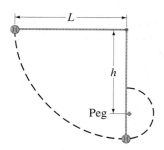

FIGURE 8–36
Problems 79 and 80.

80. Show that only if $h \geq 0.60\,L$ can the ball in Fig. 8–36 make a complete circle about the peg.

81. A 65-kg hiker climbs to the top of a 3900-m-high mountain. The climb is made in 5.0 h starting at an elevation of 2200 m. Calculate (a) the work done by the hiker against gravity, (b) the average power output in watts and in horsepower, and (c) assuming the body is 15% efficient, what rate of energy input was required.

82. The small mass m sliding without friction along the looped track shown in Fig. 8–37 is to remain on the track at all times, even at the very top of the loop of radius r. (a) Calculate, in terms of the given quantities, the minimum release height h. Next, if the actual release height is $2h$, calculate (b) the normal force exerted by the track at the bottom of the loop, (c) the normal force exerted by the track at the top of the loop, and (d) the normal force exerted by the track after the block exits the loop onto the flat section.

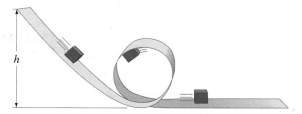

FIGURE 8–37 Problem 82.

83. Water flows over a dam at the rate of 550 kg/s and falls vertically 80 m before striking the turbine blades. Calculate (a) the speed of the water just before striking the turbine blades (neglect air resistance), and (b) the rate at which mechanical energy is transferred to the turbine blades, assuming 60% efficiency.

84. A bicyclist of mass 75 kg (including the bicycle) can coast down a 4.0° hill at a steady speed of 10 km/h. Pumping hard, the cyclist can descend the hill at a speed of 30 km/h. Using the same power, at what speed can the cyclist climb the same hill? Assume the force of friction is proportional to the square of the speed v; that is, $F_{fr} = bv^2$, where b is a constant.

85. Show that on a roller coaster with a circular vertical loop (Fig. 8–38), the difference in your apparent weight at the top of the loop and the bottom of the loop is 6 g's—that is, six times your weight. Ignore friction. Show also that as long as your speed is above the minimum needed, this answer doesn't depend on the size of the loop or how fast you go through it.

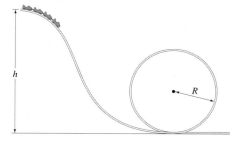

FIGURE 8–38
Problem 85.

86. If you stand on a bathroom scale, the spring inside the scale compresses 0.50 mm, and it tells you your weight is 700 N. Now if you jump on the scale from a height of 1.0 m, what does the scale read at its peak?

87. A 75-kg student runs at 5.0 m/s, grabs a hanging rope, and swings out over a lake (Fig. 8–39). He releases the rope when his velocity is zero. (a) What is the angle θ when he releases the rope? (b) What is the tension in the rope just before he releases it? (c) What is the maximum tension in the rope?

θ

10.0 m

FIGURE 8–39
Problem 87.

88. In the rope climb, a 70-kg athlete climbs a vertical distance of 5.0 m in 9.0 s. What minimum power output was used to accomplish this feat?

89. The nuclear force between two neutrons in a nucleus is described roughly by the Yukawa potential

$$U(r) = -U_0 \frac{r_0}{r} e^{-r/r_0},$$

where r is the distance between the neutrons and U_0 and $r_0 (\approx 10^{-15} \text{ m})$ are constants. (a) Determine the force $F(r)$. (b) What is the ratio $F(3r_0)/F(r_0)$? (c) Calculate this same ratio for the force between two electrically charged particles where $U(r) = -C/r$, with C a constant. Why is the Yukawa force referred to as a "short-range" force?

90. A 20-kg sled starts up a 30° incline with a speed of 2.4 m/s. The coefficient of kinetic friction is $\mu_k = 0.25$. (a) How far up the incline does the sled travel? (b) What condition must you put on the coefficient of static friction if the sled is not to get stuck at the point determined in part (a)? (c) If the sled slides back down, what is its speed when it returns to its starting point?

91. A fire hose for use in urban areas must be able to shoot a stream of water to a maximum height of 30 meters. The water leaves the hose at ground level in a circular stream 3.0 cm in diameter. What minimum power is required to create such a stream of water? Every cubic meter of water has a mass of 1000 kg.

92. Proper design of automobile braking systems must account for heat buildup under heavy braking. Calculate the thermal energy dissipated from brakes in a 1500-kg car that descends a 20° hill. The car begins braking when its speed is 90 km/h and slows to a speed of 30 km/h in a distance of 0.30 km measured along the road.

93. The Lunar Module could make a safe landing if its vertical velocity at impact is 3.0 m/s or less. Suppose that you want to determine the greatest height h at which the pilot could shut off the engine if the velocity of the lander relative to the surface is (a) zero; (b) 2.0 m/s downward; (c) 2.0 m/s upward. Use conservation of energy to determine h in each case. The acceleration due to gravity at the surface of the Moon is 1.62 m/s².

94. Some electric power companies use water to store energy. Water is pumped by reversible turbine pumps from a low to a high reservoir. If we wish to store the energy produced in 1.0 hour by a 100-MW (electrical) power plant, how many cubic meters of water will have to be pumped from the lower to the upper pool? Assume the upper pool is 500 m above the lower and we can neglect the small change in depths of each pool. Water has a mass of 1000 kg for every 1.0 m³.

95. As a city engineer, you need to estimate the power required to pump water from a new well. The well is 400 m deep and the estimated demand is 1,000,000 kg per day. The pump motor is about 80% efficient in converting electrical energy to mechanical energy.

96. Estimate the energy required from fuel to launch a 12,000-kg satellite into orbit 1000 km above the Earth's surface. Consider two cases: (a) the satellite is launched into an equatorial orbit from a point on the Earth's equator, and (b) it is launched from the north pole into a polar orbit.

97. A satellite is in an elliptic orbit around the Earth (Fig. 8–40). Its velocity at the perigee A is 8650 m/s. (a) Use conservation of energy to determine its velocity at B. The radius of the Earth is 6380 km. (b) Use conservation of energy to determine the velocity at the apogee C.

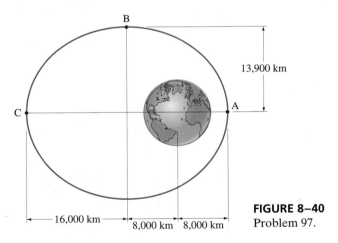

B

13,900 km

C

A

16,000 km

8,000 km 8,000 km

FIGURE 8–40
Problem 97.

*98. A particle moves where its potential energy is given by $U(r) = U_0[(2/r^2) - (1/r)]$. (a) Sketch the approximate shape of $U(r)$ versus r. Where does the curve cross the $U(r) = 0$ axis? At what value of r does the minimum value of $U(r)$ occur? (b) Suppose that the particle has an energy of $E = -0.050U_0$. Sketch in the approximate "turning points" of the motion of the particle on your diagram. What is the maximum kinetic energy of the particle, and for what value of r does this occur?

Conservation of linear momentum is another of the great conservation laws of physics. Collisions, as between billiard or pool balls, illustrate this vector law very nicely: the total vector momentum $(\Sigma m_i \mathbf{v}_i)$ before the collision equals the total vector momentum just after the collision. In this photo, the moving cue ball strikes the 11 ball at rest. Both balls move after the collision, at angles, but the sum of their vector momenta equals the initial momentum of the incoming cue ball. We will consider both elastic collisions (where kinetic energy is also conserved) and inelastic collisions. We also will examine the concept of center of mass, and how it can make the study of complex motion more readily analyzed and understood.

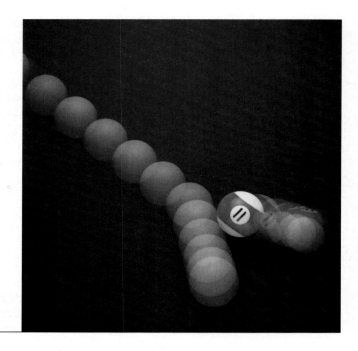

CHAPTER 9

Linear Momentum and Collisions

The law of conservation of energy, which we discussed in the previous chapter, is one of several great conservation laws in physics. Among the other quantities found to be conserved are linear momentum, angular momentum, and electric charge. We will eventually discuss all of these because the conservation laws are among the most important in all of science. In this chapter, we discuss linear momentum and its conservation. We then make use of the laws of conservation of linear momentum and of energy to analyze collisions. Indeed, the law of conservation of momentum is particularly useful when dealing with two or more bodies that interact with each other, as in collisions.

Our focus up to now has been mainly on the motion of a single object. In this chapter we will deal with systems of two or more objects. An important concept for this study is that of center of mass, which we discuss later in the chapter.

9–1 Momentum and Its Relation to Force

The **linear momentum** (or "momentum" for short) of an object is defined as the product of its mass and its velocity. Momentum (plural is momenta) is usually represented by the symbol $\mathbf{p}$. If we let m represent the mass of an object and $\mathbf{v}$ represent its velocity, then its momentum $\mathbf{p}$ is

Linear momentum

$$\mathbf{p} = m\mathbf{v}. \tag{9–1}$$

Since velocity is a vector, momentum is a vector. The direction of the momentum is the direction of the velocity, and the magnitude of the momentum is $p = mv$. Since $\mathbf{v}$ depends on the reference frame, this frame must be specified. The unit of momentum is simply that of mass $\times$ velocity, which in SI units is $kg \cdot m/s$. There is no special name for this unit.

Everyday usage of the term *momentum* is in accord with the definition above. For according to Eq. 9–1, a fast-moving car has more momentum than a slow-moving car of the same mass, and a heavy truck has more momentum than a small car moving with the same speed. The more momentum an object has, the harder it is to stop it, and the greater effect it will have if it is brought to rest by impact or collision. A football player is more likely to be stunned if tackled by a heavy opponent running at top speed than by a lighter or slower-moving tackler. A heavy, fast-moving truck can do more damage than a slow-moving motorcycle.

A force is required to change the momentum of an object, whether it is to increase the momentum, to decrease it (such as to bring a moving object to rest), or to change its direction. Newton originally stated his second law in terms of momentum (although he called the product mv the "quantity of motion"). Newton's statement of the **second law of motion**, translated into modern language, is as follows:

> **The rate of change of momentum of an object is equal to the net force applied to it.**

We can write this as an equation,

$$\Sigma \mathbf{F} = \frac{d\mathbf{p}}{dt},$$

(9–2)

NEWTON'S SECOND LAW

where $\Sigma \mathbf{F}$ is the net force applied to the object (the vector sum of all forces acting on it). We can readily derive the familiar form of the second law, $\Sigma \mathbf{F} = m\mathbf{a}$, from Eq. 9–2 for the case of constant mass. If $\mathbf{v}_0$ is the initial velocity of an object and $\mathbf{v}$ is its velocity after a time dt has elapsed, then

$$\Sigma \mathbf{F} = \frac{d\mathbf{p}}{dt} = \frac{d(m\mathbf{v})}{dt} = m\frac{d\mathbf{v}}{dt} = m\mathbf{a},$$ [constant mass]

because, by definition, $\mathbf{a} = d\mathbf{v}/dt$. Newton's statement, Eq. 9–2, is actually more general than the more familiar one because it includes the situation in which the mass may change. This is important in certain circumstances, such as for rockets which lose mass as they burn fuel (Section 9–10) and in relativity theory (Chapter 37).

EXAMPLE 9–1 **Washing a car: momentum change and force.** Water leaves a hose at a rate of 1.5 kg/s with a speed of 20 m/s and is aimed at the side of a car, which stops it, Fig. 9–1. (That is, we ignore any splashing back.) What is the force exerted by the water on the car?

SOLUTION We take the x direction positive to the right. In each second, water with a momentum of $p_x = mv_x = (1.5\,\text{kg})(20\,\text{m/s}) = 30\,\text{kg·m/s}$ is brought to rest at the moment it hits the car. The magnitude of the force (assumed constant) that the car must exert to change the momentum of the water by this amount is

$$F = \frac{\Delta p}{\Delta t} = \frac{p_{\text{final}} - p_{\text{initial}}}{\Delta t} = \frac{0 - 30\,\text{kg·m/s}}{1.0\,\text{s}} = -30\,\text{N}.$$

The minus sign indicates that the force on the water is opposite to the water's original velocity. The car exerts a force of 30 N to the left to stop the water, so by Newton's third law, the water exerts a force of 30 N on the car.

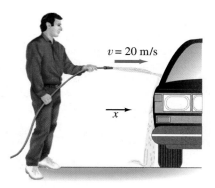

FIGURE 9–1 Example 9–1.

CONCEPTUAL EXAMPLE 9–2 **Water splashes back.** What if the water splashes back from the car in Example 9–1? Would the force on the car be greater or less?

RESPONSE If the water splashes back toward the hose, the change in momentum will be greater in magnitude, and so the force on the car will be greater in magnitude. Note that $\mathbf{p}_{\text{final}}$ will now point in the negative x direction, as shown in Fig. 9–2 (instead of being zero as in Example 9–1). So the result for F (see displayed equation in Example 9–1), will be minus something more than -30 N (i.e., -35 to -40 N, depending on the water's rebound speed). To put it simply, the car exerts not only a force to stop the water, but also an additional force to give it momentum in the opposite direction.

FIGURE 9–2 Example 9–2. Momentum of water before and after splashing back, and $\Delta\mathbf{p}$.

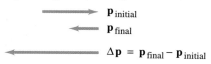

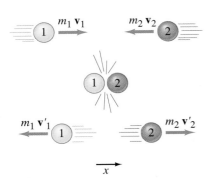

FIGURE 9–3 Momentum is conserved in a collision of two balls.

FIGURE 9–4 Collision of two objects. (a) Their momenta before collision are $\mathbf{p}_1$ and $\mathbf{p}_2$, and after collision are $\mathbf{p}'_1$ and $\mathbf{p}'_2$. At any moment during the collision each exerts a force on the other of equal magnitude but opposite direction.

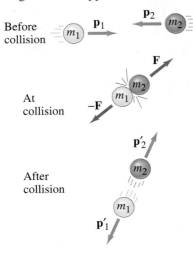

9–2 | Conservation of Momentum

The concept of momentum is particularly important because, under certain circumstances, momentum is a conserved quantity. In the mid-seventeenth century, shortly before Newton's time, it had been observed that the vector sum of the momenta of two colliding objects remains constant. Consider, for example, the head-on collision of two billiard balls, as shown in Fig. 9–3. We assume the net external force on this system of two balls is zero—that is, the only significant forces are those that each ball exerts on the other during the collision. Although the momentum of each of the two balls changes as a result of the collision, the *sum* of their momenta is found to be the same before as after the collision. If $m_1 \mathbf{v}_1$ is the momentum of ball number 1 and $m_2 \mathbf{v}_2$ the momentum of ball 2, both measured before the collision, then the total momentum of the two balls before the collision is $m_1 \mathbf{v}_1 + m_2 \mathbf{v}_2$. After the collision, the balls each have a different velocity and momentum, which we will designate by a "prime" on the velocity: $m_1 \mathbf{v}'_1$ and $m_2 \mathbf{v}'_2$. The total momentum after the collision is $m_1 \mathbf{v}'_1 + m_2 \mathbf{v}'_2$. No matter what the velocities and masses involved are, it is found that the total momentum before the collision is the same as afterwards, whether the collision is head-on or not, as long as no net external force acts:

$$\text{momentum before} = \text{momentum after}$$

or

$$m_1 \mathbf{v}_1 + m_2 \mathbf{v}_2 = m_1 \mathbf{v}'_1 + m_2 \mathbf{v}'_2$$
$$\mathbf{p}_1 + \mathbf{p}_2 = \mathbf{p}'_1 + \mathbf{p}'_2. \tag{9–3}$$

That is, the total vector momentum of the system of two balls is conserved: it stays constant.

Although the law of conservation of momentum was discovered experimentally, it is closely connected to Newton's laws of motion and they can be shown to be equivalent, which we now do.

Let us consider two bodies of mass m_1 and m_2 that have momenta $\mathbf{p}_1$ and $\mathbf{p}_2$ before they collide and $\mathbf{p}'_1$ and $\mathbf{p}'_2$ after they collide, as in Fig. 9–4. During the collision, suppose the force exerted by body 1 on body 2 at any instant is $\mathbf{F}$. Then, by Newton's third law, the force exerted by body 2 on body 1 is $-\mathbf{F}$. During the brief collision time, we assume no other (external) forces are acting (or that $\mathbf{F}$ is much greater than any other external forces acting). We rewrite Eq. 9–2 as $d\mathbf{p} = \mathbf{F}\, dt$ and integrate both sides of this equation over the brief time interval of the collision, from t_i to t_f:

$$\int_{t_i}^{t_f} d\mathbf{p} = \int_{t_i}^{t_f} \mathbf{F}\, dt.$$

We apply this first to body 2, on which the force $\mathbf{F}$ acts (Fig. 9–4),

$$\Delta \mathbf{p}_2 = \mathbf{p}'_2 - \mathbf{p}_2 = \int_{t_i}^{t_f} \mathbf{F}\, dt. \tag{9–4a}$$

On body 1, the force is $-\mathbf{F}$, so

$$\Delta \mathbf{p}_1 = \mathbf{p}'_1 - \mathbf{p}_1 = -\int_{t_i}^{t_f} \mathbf{F}\, dt. \tag{9–4b}$$

We compare these two equations and see that

$$\Delta \mathbf{p}_1 = -\Delta \mathbf{p}_2$$
$$\mathbf{p}'_1 - \mathbf{p}_1 = -(\mathbf{p}'_2 - \mathbf{p}_2).$$

Rearranging, we have

$$\mathbf{p}_1 + \mathbf{p}_2 = \mathbf{p}'_1 + \mathbf{p}'_2,$$

which is Eq. 9–3, the law of conservation of momentum.

We have put this derivation in the context of a collision. As long as no external forces act, Eqs. 9–4 are valid over any time interval, and conservation of momentum is always valid as long as no external forces act. In the real world, external forces do act: friction on billiard balls, gravity acting on a baseball, and so on. So it may seem that conservation of momentum cannot be applied. Or can it? In a collision, the force each body exerts on the other acts only over a very brief time interval, and is very strong. During the brief time, $\mathbf{F}$ in Eqs. 9–4 is very much larger than the other forces acting (gravity, friction). If we measure the momenta just before, and just after, the collision, momentum will be very nearly conserved. We cannot wait for the external forces to produce their effect before measuring $\mathbf{p}'_1$ and $\mathbf{p}'_2$.

For example, when a racket hits a tennis ball or a bat hits a baseball, both before and after the "collision" the ball moves as a projectile under the action of gravity and air resistance. However, when the bat or racket hits the ball, during this brief time of the collision, these external forces are insignificant compared to the collision force the bat exerts on the ball (and vice versa). From Eqs. 9–4 we see then that momentum is conserved (or very nearly so) as long as we measure $\mathbf{p}_1$ and $\mathbf{p}_2$ just before the collision and $\mathbf{p}'_1$ and $\mathbf{p}'_2$ immediately after the collision.

Our derivation of the conservation of momentum via Eqs. 9–4 can be extended to include any number of interacting bodies. Let $\mathbf{P}$ represent the total momentum of a system of n interacting objects:

$$\mathbf{P} = m_1 \mathbf{v}_1 + m_2 \mathbf{v}_2 + \cdots + m_n \mathbf{v}_n = \Sigma \mathbf{p}_i.$$

We differentiate with respect to time:

$$\frac{d\mathbf{P}}{dt} = \Sigma \frac{d\mathbf{p}_i}{dt} = \Sigma \mathbf{F}_i \tag{9–5}$$

where $\mathbf{F}_i$ represents the *net* force on the ith object. The forces can be of two types: (1) *external forces* on objects of the system, exerted by objects outside the system, and (2) *internal forces* that objects within the system exert on other objects in the system. By Newton's third law, the internal forces occur in pairs: if one object exerts a force on a second object, the second exerts an equal and opposite force on the first object. Thus, in the sum over all the forces in Eq. 9–5, all the internal forces cancel each other in pairs. Thus we have

$$\frac{d\mathbf{P}}{dt} = \Sigma \mathbf{F}_{\text{ext}}, \tag{9–6}$$

NEWTON'S SECOND LAW
(for a system of objects)

where $\Sigma \mathbf{F}_{\text{ext}}$ is the sum of all external forces acting on our system. If the net external force is zero, then $d\mathbf{P}/dt = 0$, so $\Delta \mathbf{P} = 0$ or $\mathbf{P}$ = constant. Thus we have

when the net external force on a system is zero, the total momentum remains constant.

This is the **law of conservation of momentum**. It can also be stated as

LAW OF

CONSERVATION

OF LINEAR

MOMENTUM

the total momentum of an isolated system of bodies remains constant.

By an **isolated system**, we mean one on which no external forces act—the only forces acting are those between objects of the system.

If a net external force acts on a system, then the law of conservation of momentum will not apply. However, if the "system" can be redefined so as to include the other objects exerting these forces, then the conservation of momentum principle can apply. For example, if we take as our system a falling rock, it does not conserve momentum since an external force, the force of gravity exerted by the Earth, is acting on it and its momentum changes. However, if we include the Earth in the system, the total momentum of rock plus Earth is conserved. (This of course means that the Earth comes up to meet the ball. Since the Earth's mass is so great, its upward velocity is very tiny.)

As with energy, the importance of the momentum concept is that momentum is conserved under quite general conditions. Although the law of conservation of momentum follows from Newton's second law, as we have seen, it is in fact more general than Newton's laws. In the tiny world of the atom, Newton's laws fail, but the great conservation laws—those of energy, momentum, angular momentum, and electric charge—have been found to hold in every experimental situation tested. It is for this reason that the conservation laws are considered more basic than Newton's laws.

EXAMPLE 9–3 **Railroad cars collide: momentum conserved.** A 10,000-kg railroad car traveling at a speed of 24.0 m/s strikes an identical car at rest. If the cars lock together as a result of the collision, what is their common speed afterward? See Fig. 9–5.

SOLUTION The initial total momentum is simply $m_1 v_1$ (since $v_2 = 0$), to the right in the $+x$ direction. After the collision, the total momentum will be the same, and it will be shared by both cars. Since the two cars become attached, they will have the same speed, call it v'. Then:

$$(m_1 + m_2)v' = m_1 v_1$$

$$v' = \frac{m_1}{m_1 + m_2} v_1 = \tfrac{1}{2} v_1$$

$$= 12.0 \text{ m/s}^2$$

to the right. Their mutual speed after collision is half the initial speed of car 1.

FIGURE 9–5 Example 9–3.

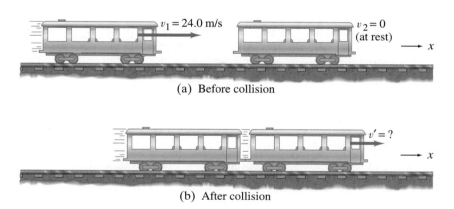

(a) Before collision

(b) After collision

FIGURE 9–6 (a) A rocket, containing fuel, at rest in some reference frame. (b) In the same reference frame, the rocket fires and gases are expelled at high speed out the rear. The total vector momentum, $\mathbf{p}_{gas} + \mathbf{p}_{rocket}$, remains zero.

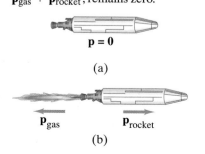

The law of conservation of momentum is particularly useful when we are dealing with fairly simple systems such as collisions and certain types of explosions. For example, *rocket propulsion*, which we saw in Chapter 4 can be understood on the basis of action and reaction, can also be explained on the basis of the conservation of momentum. Before a rocket is fired, the total momentum of rocket plus fuel is zero. As the fuel burns, the total momentum remains unchanged: the backward momentum of the expelled gases is just balanced by the forward momentum gained by the rocket itself (see Fig. 9–6). Thus, a rocket can accelerate in empty space. There is no need for the expelled gases to push against the Earth or the air (as is sometimes erroneously thought), as we already discussed in Chapter 4. Similar examples are the recoil of a gun and the throwing of a package from a boat.

EXAMPLE 9–4 Rifle recoil. Calculate the recoil velocity of a 5.0-kg rifle that shoots a 0.050-kg bullet at a speed of 120 m/s, Fig. 9–7.

SOLUTION The total momentum of the system is conserved. We let the subscripts B represent the bullet and R the rifle; the final velocities are indicated by primes. Then conservation of momentum in the x direction gives

$$m_B v_B + m_R v_R = m_B v_B' + m_R v_R'$$
$$0 + 0 = (0.050\,\text{kg})(120\,\text{m/s}) + (5.0\,\text{kg})(v_R')$$
$$v_R' = -1.2\,\text{m/s}.$$

Since the rifle has a much larger mass, its (recoil) velocity is much less than that of the bullet. The minus sign indicates that the velocity (and momentum) of the rifle is in the negative x direction, opposite to that of the bullet. Notice that it is the *vector sum* of the momenta that is conserved.

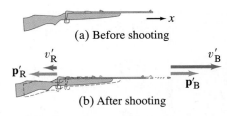

(a) Before shooting

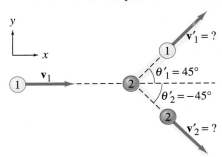

(b) After shooting

FIGURE 9–7 Example 9–4.

EXAMPLE 9–5 Billiard ball collision in 2-D. A billiard ball moving with speed $v_1 = 3.0$ m/s in the $+x$ direction (Fig. 9–8) strikes an equal-mass ball initially at rest. The two balls are observed to move off at 45°, ball 1 above the x axis and ball 2 below. That is, $\theta_1' = 45°$ and $\theta_2' = -45°$ in Fig. 9–8. What are the speeds of the two balls after the collision?

FIGURE 9–8 Example 9–5.

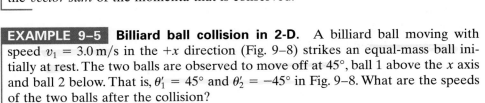

SOLUTION We set up an xy coordinate system as shown in Fig. 9–8. From symmetry, we might guess that the two balls have the same speed. But let us not assume that now. Instead, let us apply conservation of momentum. We are given $m_1 = m_2(= m)$ so

$$\mathbf{P}_{\text{initial}} = \mathbf{P}_{\text{final}}$$
$$m\mathbf{v}_1 = m\mathbf{v}_1' + m\mathbf{v}_2'.$$

Vector momentum is conserved, which means that each component is conserved. We can write the x and y components of this vector equation as

$$mv_1 = mv_1' \cos(45°) + mv_2' \cos(-45°)$$

and

$$0 = mv_1' \sin(45°) + mv_2' \sin(-45°).$$

The m's cancel out in both equations. The second equation yields [recall $\sin(-\theta) = -\sin\theta$]:

$$v_2' = -v_1' \frac{\sin(45°)}{\sin(-45°)} = -v_1'\left(\frac{\sin 45°}{-\sin 45°}\right) = v_1',$$

so they do have equal speeds as we guessed at first. The x component equation gives [recall $\cos(-\theta) = \cos\theta$]:

$$v_1 = v_1' \cos(45°) + v_2' \cos(45°) = 2v_1' \cos(45°)$$

so

$$v_1' = v_2' = \frac{v_1}{2\cos(45°)} = \frac{3.0\,\text{m/s}}{2(0.707)} = 2.1\,\text{m/s}.$$

9–3 Collisions and Impulse

Conservation of momentum is a very useful tool for dealing with collision processes, as we already saw in the Examples of the previous Section. Collisions are a common occurrence in everyday life: a tennis racket or a baseball bat striking a ball, two billiard balls colliding, one railroad car striking another, a hammer hitting a nail. At the subatomic level, scientists learn about the structure of nuclei and their constituents, and about the nature of the forces involved, by careful study of collisions between nuclei and/or elementary particles.

FIGURE 9–9 Tennis racket striking a ball. Note the deformation of both ball and racket due to the large force each exerts on the other.

Impulse

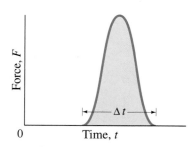

FIGURE 9–10 Force as a function of time during a typical collision.

FIGURE 9–11 The average force $\overline{F}$ over Δt gives the same impulse $(\overline{F} \Delta t)$ as the actual force.

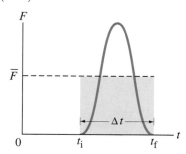

In a collision of two ordinary objects, both objects are deformed, often considerably, because of the large forces involved (Fig. 9–9). When the collision occurs, the force usually jumps from zero at the moment of contact to a very large value within a very short time, and then abruptly returns to zero again. A graph of the magnitude of the force one object exerts on the other during a collision, as a function of time, is something like that shown by the red curve in Fig. 9–10. The time interval Δt is usually very distinct and usually very small.

From Newton's second law, the *net* force on an object is equal to the rate of change of its momentum:

$$\mathbf{F} = \frac{d\mathbf{p}}{dt}.$$

(We have written $\mathbf{F}$ instead of $\Sigma\mathbf{F}$ for the net force, which we assume is entirely due to the brief but large force that acts during the collision.) This equation applies, of course, to *each* of the objects in a collision. During the infinitesimal time interval dt, the momentum changes by

$$d\mathbf{p} = \mathbf{F}\, dt.$$

If we integrate this over the duration of a collision, we have

$$\int_i^f d\mathbf{p} = \mathbf{p}_f - \mathbf{p}_i = \int_{t_i}^{t_f} \mathbf{F}\, dt,$$

where $\mathbf{p}_i$ and $\mathbf{p}_f$ are the momenta of the object just before and just after the collision. The integral of the force over the time interval during which it acts is called the **impulse**, $\mathbf{J}$:

$$\mathbf{J} = \int_{t_i}^{t_f} \mathbf{F}\, dt.$$

Thus the change in momentum of an object, $\Delta\mathbf{p} = \mathbf{p}_f - \mathbf{p}_i$, is equal to the impulse acting on it:

$$\Delta\mathbf{p} = \mathbf{p}_f - \mathbf{p}_i = \int_{t_i}^{t_f} \mathbf{F}\, dt = \mathbf{J}. \tag{9–7}$$

The units for impulse are the same as for momentum, $\text{kg} \cdot \text{m/s}$ (or $\text{N} \cdot \text{s}$) in SI. Since $\mathbf{J} = \int \mathbf{F}\, dt$, we can state that the impulse $\mathbf{J}$ of a force is equal to the area under the F versus t curve, as indicated by the shading in Fig. 9–10.

Equation 9–7 is true only if $\mathbf{F}$ is the *net* force on the object. It is valid for any net force $\mathbf{F}$ where $\mathbf{p}_i$ and $\mathbf{p}_f$ correspond precisely to the times t_i and t_f. But the impulse concept is really most useful for so-called *impulsive forces*—that is, for a force like that shown in Fig. 9–10, which has a very large magnitude over a very short time interval and is essentially zero outside this time interval. For most collision processes, the impulsive force is much larger than any other force acting, and the others can be neglected. Then the impulsive force is essentially the net force, and the change in momentum of an object during a collision is due almost entirely to the impulsive force. For such an impulsive force, the time interval over which we take the integral in Eq. 9–7 is not critical as long as we start before t_i and end after t_f, since $\mathbf{F}$ is essentially zero outside this time interval $\Delta t = t_f - t_i$. (Of course, if the chosen time interval is too large, the effect of the other forces does become significant—such as the flight of a tennis ball which, after the impulsive force administered by the racket, begins to slowly fall under gravity.)

It is sometimes useful to speak of the average force, $\overline{F}$, during a collision. It is defined as that constant force which, if acting over the same time interval $\Delta t = t_f - t_i$ as the actual force, would produce the same impulse and change in momentum. Thus

$$\overline{\mathbf{F}} \Delta t = \int_{t_i}^{t_f} \mathbf{F}\, dt.$$

Figure 9–11 shows the magnitude of the average force, $\overline{F}$, for the impulsive force of Fig. 9–10. The rectangular area $\overline{F} \Delta t$ equals the area under the impulsive force curve.

EXAMPLE 9–6 **Bend your knees when landing.** (*a*) Calculate the impulse experienced when a 70-kg person lands on firm ground after jumping from a height of 3.0 m. Then estimate the average force exerted on the person's feet by the ground, if the landing is (*b*) stiff-legged, and (*c*) with bent legs. In the former case, assume the body moves 1.0 cm during impact, and in the second case, when the legs are bent, about 50 cm.

SOLUTION (*a*) We don't know F and can't calculate the impulse $\mathbf{J} = \int \mathbf{F}\,dt$ directly, but we can use the fact that the impulse equals the change in momentum of the object. We need to determine the velocity of the person just before striking the ground, which we can do using conservation of energy (Eq. 8–9), $\Delta K = -\Delta U$:

$$\tfrac{1}{2}mv^2 - 0 = -mg(y - y_0),$$

where we assume he started from rest $(v_0 = 0)$, and $y_0 = 3.0$ m and $y = 0$. Thus, after falling 3.0 m, the person's velocity just before hitting the ground will be

$$v = \sqrt{2g(y_0 - y)} = \sqrt{2(9.8 \text{ m/s}^2)(3.0 \text{ m})} = 7.7 \text{ m/s}.$$

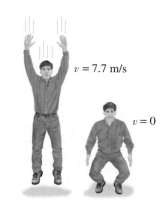

FIGURE 9–12 Period during which impulse acts (Example 9–6).

As the person strikes the ground, the momentum is quickly brought to zero, Fig. 9–12. The impulse on the person is

$$J = \bar{F}\,\Delta t = \Delta p = p_f - p_i$$
$$= 0 - (70 \text{ kg})(7.7 \text{ m/s}) = -540 \text{ N} \cdot \text{s}.$$

The negative sign tells us that the force is opposed to the original momentum—that is, the force acts upward.

(*b*) In coming to rest, the body decelerates from 7.7 m/s to zero in a distance $d = 1.0$ cm $= 1.0 \times 10^{-2}$ m. The average speed during this brief period is

$$\bar{v} = (7.7 \text{ m/s} + 0 \text{ m/s})/2 = 3.8 \text{ m/s},$$

so the collision lasts a time

$$\Delta t = \frac{d}{\bar{v}} = \frac{(1.0 \times 10^{-2} \text{ m})}{(3.8 \text{ m/s})} = 2.6 \times 10^{-3} \text{ s}.$$

Since the magnitude of the impulse is $\bar{F}\,\Delta t = 540 \text{ N} \cdot \text{s}$, and $\Delta t = 2.6 \times 10^{-3}$ s, the average net force $\bar{F}$ has magnitude

$$\bar{F} = \frac{J}{\Delta t} = \frac{540 \text{ N} \cdot \text{s}}{2.6 \times 10^{-3} \text{ s}} = 2.1 \times 10^5 \text{ N}.$$

The force $\bar{F}$ is the *net* force upward on the person (we calculated it from Newton's second law). $\bar{F}$ is the sum of the average force upward on the legs exerted by the ground, F_{grd}, which we take as positive, plus the downward force of gravity, $-mg$ (see Fig. 9–13):

$$\bar{F} = F_{grd} - mg.$$

Since $mg = (70 \text{ kg})(9.8 \text{ m/s}^2) = 690 \text{ N}$, then

$$F_{grd} = \bar{F} + mg = 2.1 \times 10^5 \text{ N} + 0.690 \times 10^3 \text{ N} \approx 2.1 \times 10^5 \text{ N}.$$

FIGURE 9–13 When the person lands on the ground, the average net force during impact is $\bar{F} = F_{grd} - mg$, where F_{grd} is the force the ground exerts upward on the person.

(*c*) This is like part (*b*), except $d = 0.50$ m, so $\Delta t = (0.50 \text{ m})/(3.8 \text{ m/s}) = 0.13$ s, and

$$\bar{F} = \frac{540 \text{ N} \cdot \text{s}}{0.13 \text{ s}} = 4.2 \times 10^3 \text{ N}.$$

The upward force exerted on the person's feet by the ground is, as in part (*b*):

$$F_{grd} = \bar{F} + mg = 4.2 \times 10^3 \text{ N} + 0.69 \times 10^3 \text{ N} = 4.9 \times 10^3 \text{ N}.$$

Clearly, the force on the feet and legs is much less when the knees are bent. In fact, the ultimate strength of the leg bone (see Chapter 12, Table 12–2) is not great enough to support the force calculated in part (*b*), so the leg would likely break in such a stiff landing, whereas it probably wouldn't in part (*c*).

9–4 Conservation of Energy and Momentum in Collisions

During most collisions, we usually don't know how the collision force varies over time, and so analysis using Newton's second law becomes difficult or impossible. But we can still determine a lot about the motion after a collision, given the initial motion, by making use of the conservation laws for momentum and energy. We saw in Section 9–2 that in the collision of two objects such as billiard balls, the total momentum is conserved, as long as any other "external" forces can be ignored. If the two objects are very hard and no heat is produced in the collision, then kinetic energy is conserved as well. By this we mean that the sum of the kinetic energies of the two objects is the same after the collision as before. Of course, for the brief moment during which the two objects are in contact, some (or all) of the energy is stored momentarily in the form of elastic potential energy. But if we compare the total kinetic energy before the collision with the total after the collision, they are found to be the same. Such a collision, in which the total kinetic energy is conserved, is called an **elastic collision**. If we use the subscripts 1 and 2 to represent the two objects, we can write the equation for conservation of total kinetic energy as

$$\frac{1}{2} m_1 v_1^2 + \frac{1}{2} m_2 v_2^2 = \frac{1}{2} m_1 v_1'^2 + \frac{1}{2} m_2 v_2'^2. \qquad \text{[elastic collision]} \quad \textbf{(9–8)}$$

Here, primed quantities (′) mean after the collision and unprimed mean before the collision, just as in Eq. 9–3 for conservation of momentum.

At the atomic level the collisions of atoms and molecules are often elastic. But in the "macroscopic" world of ordinary objects, an elastic collision is an ideal that is never quite reached, since at least a little thermal energy (and perhaps sound and other forms of energy) is always produced during a collision. Collisions of two air table pucks or two hard elastic balls, such as billiard balls, are very close to perfectly elastic, and we often treat them as such. Even when the kinetic energy is not conserved, the *total* energy is, of course, always conserved.

Collisions in which kinetic energy is not conserved are said to be **inelastic collisions**. The kinetic energy that is lost is changed into other forms of energy, often thermal energy, so that the total energy (as always) is conserved. In this case, we can write that

$$K_1 + K_2 = K_1' + K_2' + \text{thermal and other forms of energy.}$$

See Fig. 9–14.

Elastic collision

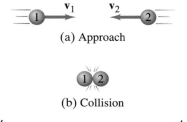

(a) Approach

(b) Collision

(c) If elastic

(d) If inelastic

FIGURE 9–14 Two equal mass objects (a) approach each other with equal speeds, (b) collide, and then (c) bounce off with equal speeds in the opposite directions if the collision is elastic, or (d) bounce back much less or not at all, if the collision is inelastic.

9–5 Elastic Collisions in One Dimension

We now apply the conservation laws for momentum and kinetic energy to an elastic collision between two small objects that collide head-on, so all the motion is along a line. We deal here only with translational motion, but our analysis works well even for hard billiard balls because at the instant of collision the rolling has little effect. Let us assume that both objects are initially moving with velocities v_1 and v_2 along the x axis, Fig. 9–15a. After the collision, their velocities are v_1' and v_2', Fig. 9–15b. For any $v > 0$, the object is moving to the right (increasing x), whereas for $v < 0$, the object is moving to the left (toward decreasing values of x).

From conservation of momentum, we have

Momentum conservation

$$m_1 v_1 + m_2 v_2 = m_1 v_1' + m_2 v_2'.$$

Because the collision is assumed to be elastic, kinetic energy is also conserved:

Kinetic energy conservation

$$\frac{1}{2} m_1 v_1^2 + \frac{1}{2} m_2 v_2^2 = \frac{1}{2} m_1 v_1'^2 + \frac{1}{2} m_2 v_2'^2.$$

We have two equations, so we can solve for two unknowns. If we know the masses and initial velocities, then we can solve these two equations for the velocities after the collision, v_1' and v_2'. We will do this in a moment in some Examples, but first we derive a useful result. To do so we rewrite the momentum equation as

$$m_1(v_1 - v_1') = m_2(v_2' - v_2), \qquad \text{(i)}$$

and we rewrite the kinetic energy equation as

$$m_1(v_1^2 - v_1'^2) = m_2(v_2'^2 - v_2^2)$$

or [noting that $(a - b)(a + b) = a^2 - b^2$] we write this as

$$m_1(v_1 - v_1')(v_1 + v_1') = m_2(v_2' - v_2)(v_2' + v_2). \qquad \text{(ii)}$$

We divide Eq. (ii) by Eq. (i), and (assuming $v_1 \neq v_1'$ and $v_2 \neq v_2'$)[†] obtain

$$v_1 + v_1' = v_2' + v_2.$$

We can rewrite this equation as

$$
\begin{aligned}
v_1 - v_2 &= v_2' - v_1' \\
&= -(v_1' - v_2'). \qquad \text{[head-on elastic collision]} \quad \text{(9–9)}
\end{aligned}
$$

This is an interesting result: it tells us that for any elastic head-on collision, the relative speed of the two objects after the collision has the same magnitude as before (but opposite direction), no matter what the masses are.

Now let us look at some special cases of elastic head-on collisions. We assume v_1, v_2, m_1, and m_2 are known, and we wish to solve for v_1' and v_2', the velocities of the two objects after the collision.

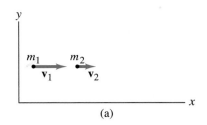

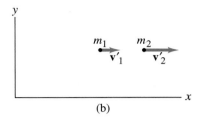

FIGURE 9–15 Two small objects, of masses m_1 and m_2, (a) before the collision, and (b) after the collision.

EXAMPLE 9–7 **Equal masses.** A billiard ball of mass m moving with speed v collides head-on with a second ball of equal mass. What are the speeds of the two balls after the collision, assuming it is elastic? Assume (a) both balls are moving, (b) ball 2 is initially at rest ($v_2 = 0$).

SOLUTION (a) Since $m_1 = m_2 = m$, then conservation of momentum gives

$$v_1 + v_2 = v_1' + v_2'.$$

We need a second equation, since there are two unknowns. We could use the conservation of kinetic energy equation, or the simpler Eq. 9–9 derived above:

$$v_1 - v_2 = v_2' - v_1'.$$

Now we add these two equations and get

$$v_2' = v_1$$

and then subtract the two equations to get

$$v_1' = v_2.$$

That is, the balls exchange velocities as a result of the collision: ball 2 acquires the velocity that ball 1 had before the collision, and vice versa.

(b) If ball 2 is at rest initially, so that $v_2 = 0$, we have

$$v_2' = v_1 \quad \text{and} \quad v_1' = 0.$$

That is, ball 1 is brought to rest by the collision, whereas ball 2 acquires the original velocity of ball 1. This result is often observed by billiard and pool players, and is valid only if the two balls have equal masses (and no spin is given to the balls). See Fig. 9–16.

FIGURE 9–16 In this multi-flash photo of a head-on collision between two balls of equal mass, the white cue ball is accelerated from rest by the cue stick and then strikes the red ball, initially at rest. The white ball stops in its tracks and the (equal mass) red ball moves off with the same speed as the white ball had before the collision. See Example 9–7.

[†] Note that Eqs. (i) and (ii), which are the conservation laws for momentum and kinetic energy, are both satisfied by the solution $v_1' = v_1$ and $v_2' = v_2$. This is a valid solution, but not very interesting. It corresponds to no collision at all—when the two objects miss each other.

EXAMPLE 9–8 **Unequal masses, target at rest.** A very common practical situation is for a moving object (m_1) to strike a second object $(m_2$, the "target") at rest $(v_2 = 0)$. Assume the objects have unequal masses and that the collision occurs along a line (head-on). (a) Derive equations for v_2' and v_1' in terms of the initial velocity v_1 of mass m_1 and the masses m_1 and m_2. (b) Determine the final velocities if the moving object is much more massive than the target $(m_1 \gg m_2)$. (c) Determine the final velocities if the moving object is much less massive than the target $(m_1 \ll m_2)$.

SOLUTION (a) We combine the momentum equation, which (with $v_2 = 0$) is

$$m_2 v_2' = m_1(v_1 - v_1'),$$

with Eq. 9–9, rewritten as $v_1 + v_1' = v_2'$, to obtain

$$v_2' = v_1\left(\frac{2m_1}{m_1 + m_2}\right)$$

$$v_1' = v_1\left(\frac{m_1 - m_2}{m_1 + m_2}\right).$$

As a check, we let $m_1 = m_2$, and we obtain

$$v_2' = v_1 \quad \text{and} \quad v_1' = 0.$$

This is the same case treated in Example 9–7, and we get the same result: for objects of equal mass, one of which is initially at rest, the velocity of the one moving initially is completely transferred to the object originally at rest.

(b) Let $v_2 = 0$ and $m_1 \gg m_2$. A very heavy moving object strikes a light object at rest, and we have, using the relations for v_2' and v_1' above,

$$v_2' \approx 2v_1$$

$$v_1' \approx v_1.$$

Thus the velocity of the heavy incoming object is practically unchanged, whereas the light object, originally at rest, takes off with twice the velocity of the heavy one. The velocity of a heavy bowling ball, for example, is hardly affected by striking the much lighter bowling pins.

(c) Finally, we let $v_2 = 0$ and $m_1 \ll m_2$. A moving light object strikes a very massive object at rest. In this case, using the equations in part (a)

$$v_2' \approx 0$$

$$v_1' \approx -v_1.$$

The massive object remains essentially at rest and the very light incoming object rebounds with essentially its same speed but in the opposite direction. For example, a tennis ball colliding head-on with a stationary bowling ball will hardly affect the bowling ball, but will rebound with nearly the same speed it had initially, just as if it had struck a hard wall.

It can readily be shown (it is given as Problem 40) for any elastic head-on collision that

$$v_2' = v_1\left(\frac{2m_1}{m_1 + m_2}\right) + v_2\left(\frac{m_2 - m_1}{m_1 + m_2}\right)$$

and

$$v_1' = v_1\left(\frac{m_1 - m_2}{m_1 + m_2}\right) + v_2\left(\frac{2m_2}{m_1 + m_2}\right).$$

These general equations, however, should not be memorized. They can always be derived quickly from the conservation laws. For many problems, it is simplest just to start from scratch, as we did in the special cases above and as shown in the next Example.

PHYSICS APPLIED
Nuclear collision

EXAMPLE 9–9 **A nuclear collision.** A proton of mass 1.01 u (unified atomic mass units) traveling with a speed of 3.60×10^4 m/s has an elastic head-on collision with a helium (He) nucleus ($m_{He} = 4.00$ u) initially at rest. What are the velocities of the proton and helium nucleus after the collision? (As mentioned in Chapter 1, 1 u $= 1.66 \times 10^{-27}$ kg, but we won't need this fact.)

SOLUTION Call the initial direction of motion the $+x$ direction. We have $v_2 = v_{He} = 0$ and $v_1 = v_p = 3.60 \times 10^4$ m/s. We want to find the velocities v'_p and v'_{He} after the collision. From conservation of momentum we have

$$m_p v_p + 0 = m_p v'_p + m_{He} v'_{He}.$$

Because the collision is elastic, kinetic energy is conserved and we can use Eq. 9–9, which becomes

$$v_p - 0 = v'_{He} - v'_p.$$

Thus

$$v'_p = v'_{He} - v_p,$$

and substituting this into the momentum equation we get

$$m_p v_p = m_p v'_{He} - m_p v_p + m_{He} v'_{He}.$$

Solving for v'_{He}, we obtain

$$v'_{He} = \frac{2m_p v_p}{m_p + m_{He}} = \frac{2(1.01 \text{ u})(3.60 \times 10^4 \text{ m/s})}{5.01 \text{ u}} = 1.45 \times 10^4 \text{ m/s}.$$

The other unknown is v'_p, which we can now obtain from

$$v'_p = v'_{He} - v_p$$
$$= 1.45 \times 10^4 \text{ m/s} - 3.60 \times 10^4 \text{ m/s} = -2.15 \times 10^4 \text{ m/s}.$$

The minus sign tells us that the proton reverses direction upon collision, and we see that its speed is less than its initial speed (see Fig. 9–17). This makes sense from ordinary experience: the lighter proton would be expected to "bounce back" somewhat from the more massive helium nucleus, but not with its full original velocity as it would from a rigid wall (which would correspond to extremely large, or infinite, mass).

(a)

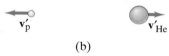

(b)

FIGURE 9–17 Example 9–9: (a) before collision, (b) after collision.

9–6 | Inelastic Collisions

Collisions in which kinetic energy is not conserved are called inelastic collisions. Some of the initial kinetic energy in such collisions is transformed into other types of energy, such as thermal or potential energy, so the total final kinetic energy is less than the total initial kinetic energy. The inverse can also happen when potential energy (such as chemical or nuclear) is released, in which case the total final kinetic energy can be greater than the initial kinetic energy. Explosives are examples of this type. Typical macroscopic collisions are inelastic, at least to some extent, and often to a large extent. If two objects stick together as a result of a collision, the collision is said to be **completely inelastic**. Two colliding balls of putty

Completely inelastic collision

that stick together or two railroad cars that couple together when they collide are examples of completely inelastic collisions. The kinetic energy in some cases is all transformed to other forms of energy in an inelastic collision, but in other cases only part of it is. In Example 9–3, for instance, we saw that when a traveling railroad car collided with a stationary one, the coupled cars traveled off with some kinetic energy.

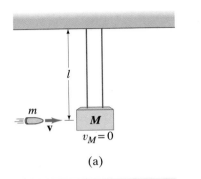

Ballistic pendulum

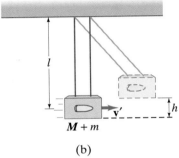

FIGURE 9–18 Ballistic pendulum (Example 9–10).

EXAMPLE 9–10 **Ballistic pendulum.** The *ballistic pendulum* is a device used to measure the speed of a projectile, such as a bullet. The projectile, of mass m, is fired into a large block (of wood or other material) of mass M, which is suspended like a pendulum. (Usually, M is somewhat greater than m.) As a result of the collision, the pendulum-projectile system swings up to a maximum height h, Fig. 9–18. Determine the relationship between the initial horizontal speed of the projectile, v, and the height h.

SOLUTION We analyze this process by dividing it into two parts: (1) the collision itself, and (2) the subsequent motion of the pendulum from the vertical hanging position to height h. In part (1), Fig. 9–18a, we assume the collision time is very short, and so the projectile comes to rest in the block before the block has moved significantly from its position directly below its support. Thus there is no net external force and momentum is conserved:

$$mv = (m + M)v', \tag{i}$$

where v' is the speed of the block and embedded projectile just after the collision, before they have moved significantly. Once the pendulum begins to move (part 2, Fig. 9–18b), there will be a net external force (gravity, tending to pull it back to the vertical position). So, for part (2), we cannot use conservation of momentum. But we can use conservation of mechanical energy since the kinetic energy immediately after the collision is changed entirely to gravitational potential energy when the pendulum reaches its maximum height, h. Therefore (letting $y = 0$ for the pendulum in the vertical position):

$$K_1 + U_1 = K_2 + U_2$$

or

$$\tfrac{1}{2}(m + M)v'^2 + 0 = 0 + (m + M)gh, \tag{ii}$$

so $v' = \sqrt{2gh}$. We combine equations (i) and (ii) to obtain

$$v = \frac{m + M}{m} v' = \frac{m + M}{m} \sqrt{2gh},$$

which is the final result. To obtain this result, we had to be opportunistic, in that we used whichever conservation laws we could: in (1) we could use only conservation of momentum, since the collision is inelastic and conservation of mechanical energy is not valid[†]; and in (2), conservation of mechanical energy is valid, but not conservation of momentum. In part (1), if there were significant motion of the pendulum during the deceleration of the projectile in the block, then there *would* be an external force during the collision—so conservation of momentum would not be valid, and this would have to be taken into account.

[†] Total energy is conserved, of course.

Collisions in Two or Three Dimensions

Conservation of momentum and energy can also be applied to collisions in two or three dimensions, and the vector nature of momentum is especially important. One common type of non-head-on collision is that in which a moving object (called the "projectile") strikes a second object initially at rest (the "target") object. This is the common situation in games such as billiards, and for experiments in atomic and nuclear physics (the projectiles, from radioactive decay or a high-energy accelerator, strike a stationary target nucleus).

Figure 9–19 shows the incoming projectile, m_1 heading along the x axis toward the target, m_2, which is initially at rest. If these are billiard balls, m_1 strikes m_2 and they go off at the angles θ_1' and θ_2', respectively, which are measured relative to m_1's initial direction (the x axis). The objects may begin to deflect even before they touch if electric, magnetic, or nuclear forces act between them.

Let us apply the law of conservation of momentum to a collision like that of Fig. 9–19. We choose the xy plane to be the plane in which the initial and final momenta lie. Because momentum is a vector, and is conserved, its components in the x and y directions remain constant. In the x direction, we have $p_{1x} + p_{2x} = p_{1x}' + p_{2x}'$, or:

$$m_1 v_1 = m_1 v_1' \cos \theta_1' + m_2 v_2' \cos \theta_2'.$$ **(9–10a)**   *p_x conserved*

Because there is no motion in the y direction initially, the y component of the total momentum is zero:

$$0 = m_1 v_1' \sin \theta_1' + m_2 v_2' \sin \theta_2'.$$ **(9–10b)**   *p_y conserved*

When we have two independent equations, we can solve for, at most, two unknowns.

If we know that a collision is elastic, we can then apply conservation of kinetic energy and obtain a third equation:

$$K_1 + K_2 = K_1' + K_2'$$

or, for the collision shown in Fig. 9–19,

$$\tfrac{1}{2} m v_1^2 = \tfrac{1}{2} m_1 v_1'^2 + \tfrac{1}{2} m v_2'^2.$$ [elastic collision] **(9–10c)**   *Kinetic energy conserved*

If the collision is elastic, we have three independent equations and we can solve for three unknowns. If we are given m_1, m_2, v_1 (and v_2, if it is not zero), we cannot, for example, predict the final variables, v_1', v_2', θ_1', and θ_2', because there are four of them. However, if we measure one of these variables, say θ_1', then the other three variables $\left(v_1', v_2', \text{ and } \theta_2'\right)$ are uniquely determined, and we can determine them using Eqs. 9–10a, b, c.

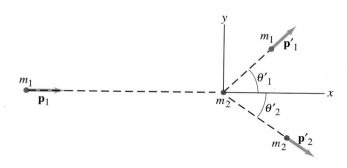

FIGURE 9–19 Object 1, the projectile, collides with object 2, the target. They move off, after the collision, with momenta $\mathbf{p}_1'$ and $\mathbf{p}_2'$ at angles θ_1' and θ_2'.

EXAMPLE 9–11 Proton–proton collision. A proton traveling with speed 8.2×10^5 m/s collides elastically with a stationary proton in a hydrogen target as in Fig. 9–19. One of the protons is observed to be scattered at a 60° angle. At what angle will the second proton be observed, and what will be the velocities of the two protons after the collision?

SOLUTION We saw a two-dimensional collision in Example 9–5, where we needed to use only conservation of momentum. Now we are given less information: we have three unknowns instead of two, so we need the kinetic energy equation as well as the two momentum equations. Since $m_1 = m_2$, Eqs. 9–10a, b, and c become

$$v_1^2 = v_1'^2 + v_2'^2 \tag{i}$$

$$v_1 = v_1' \cos\theta_1' + v_2' \cos\theta_2' \tag{ii}$$

$$0 = v_1' \sin\theta_1' + v_2' \sin\theta_2', \tag{iii}$$

where $v_1 = 8.2 \times 10^5$ m/s and $\theta_1' = 60°$ are given. In the second and third equations, we move the v_1' terms to the left side and square both sides of the equations:

$$v_1^2 - 2v_1 v_1' \cos\theta_1' + v_1'^2 \cos^2\theta_1' = v_2'^2 \cos^2\theta_2'$$

$$v_1'^2 \sin^2\theta_1' = v_2'^2 \sin^2\theta_2'.$$

We add these two equations and use $\sin^2\theta + \cos^2\theta = 1$ to get:

$$v_1^2 - 2v_1 v_1' \cos\theta_1' + v_1'^2 = v_2'^2.$$

Into this equation we substitute $v_2'^2 = v_1^2 - v_1'^2$, from equation (i) above, and get

$$2v_1'^2 = 2v_1 v_1' \cos\theta_1'$$

or

$$v_1' = v_1 \cos\theta_1' = (8.2 \times 10^5 \text{ m/s})(\cos 60°) = 4.1 \times 10^5 \text{ m/s}.$$

To obtain v_2', we use equation (i) above (conservation of kinetic energy):

$$v_2' = \sqrt{v_1^2 - v_1'^2} = 7.1 \times 10^5 \text{ m/s}.$$

Finally, from equation (iii), we have

$$\sin\theta_2' = -\frac{v_1'}{v_2'} \sin\theta_1' = -\left(\frac{4.1 \times 10^5 \text{ m/s}}{7.1 \times 10^5 \text{ m/s}}\right)(0.866) = -0.50,$$

so $\theta_2' = -30°$. (The minus sign means particle 2 moves at an angle below the x axis if particle 1 is above the axis, as in Fig. 9–19.) An example of such a collision is shown in the bubble chamber photo of Fig. 9–20. Notice that the two trajectories are at right angles to each other after the collision. This can be shown to be true in general for non-head-on elastic collisions of two particles of equal mass, one of which was at rest initially (see Problem 57).

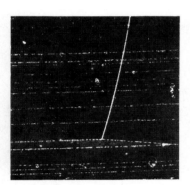

FIGURE 9–20 Photo of a proton–proton collision in a hydrogen bubble chamber (a device that makes visible the paths of elementary particles). The many lines represent incoming protons which can strike the protons of the hydrogen in the chamber.

PROBLEM SOLVING	Momentum Conservation and Collisions

1. Be sure no significant external force acts on your chosen system. That is, the forces that act between the interacting bodies must be the only significant ones if momentum conservation is to be used. [If this is valid for a portion of the problem, you can use momentum conservation for that portion only.]

2. Draw a diagram of the initial situation, just before the interaction (collision, explosion) takes place, and represent the momentum of each object with an arrow and label. Do the same for the final situation, just after the interaction.

3. Choose a coordinate system and "+" and "−" directions. (For a head-on collision, you will need only an

x axis.) It is often convenient to choose the $+x$ axis in the direction of one object's initial velocity.

4. Write momentum conservation equation(s):

total initial momentum = total final momentum.

You have one equation for each component (x, y, z); only one equation for a head-on collision.

5. If the collision is elastic, you can also write down a conservation of kinetic energy equation:

total initial kinetic energy = total final kinetic energy.

[Alternately, you could use Eq. 9–9: $v_1 - v_2 = v_2' - v_1'$, if the collision is one dimensional (head-on).]

6. Solve algebraically for the unknown(s).

9-8 Center of Mass (CM)

Until now, when we have dealt with an extended body (that is, a body that has size), we have assumed that it could be approximated as a point particle or that it undergoes only translational motion. Real extended bodies, however, can undergo rotational and other types of motion as well. For example, the diver in Fig. 9–21a undergoes only translational motion (all parts of the body follow the same path), whereas the diver in Fig. 9–21b undergoes both translational and rotational motion. We will refer to motion that is not pure translation as *general motion*.

Observations indicate that even if a body rotates, or there are several bodies that move relative to one another, there is one point that moves in the same path that a particle would if subjected to the same net force. This point is called the **center of mass** (abbreviated CM). The general motion of an extended body (or system of bodies) can be considered as *the sum of the translational motion of the* CM, *plus rotational, vibrational, or other types of motion about the* CM.

As an example, consider the motion of the center of mass of the diver in Fig. 9–21; the CM follows a parabolic path even when the diver rotates, as shown in Fig. 9–21b. This is the same parabolic path that a projected particle follows when acted on only by the force of gravity (that is, projectile motion). Other points in the rotating diver's body follow more complicated paths.

Figure 9–22 shows a wrench translating and rotating along a horizontal surface—note that its CM, marked by a red +, moves in a straight line, as shown by the dashed white line.

We show below that the important properties of the CM follow from Newton's laws if the CM is defined in the following way. We can consider any extended body as being made up of many tiny particles. But first we consider a system made up of only two particles (or small objects), of mass m_1 and m_2. We choose a coordinate system so that both particles lie on the x axis at positions x_1 and x_2, Fig. 9–23. The center of mass of this system is defined to be at the position, x_{CM}, given by

$$x_{CM} = \frac{m_1 x_1 + m_2 x_2}{m_1 + m_2} = \frac{m_1 x_1 + m_2 x_2}{M},$$

where $M = m_1 + m_2$ is the total mass of the system. The center of mass lies on the line joining m_1 and m_2. If the two masses are equal $(m_1 = m_2 = m)$, x_{CM} is midway between them, since in this case $x_{CM} = m(x_1 + x_2)/2m = (x_1 + x_2)/2$. If one mass is greater than the other, say, $m_1 > m_2$, then the CM is closer to the larger mass. If all the mass is concentrated at x_2, say, so $m_1 = 0$, then $x_{CM} = (0x_1 + m_2 x_2)/(0 + m_2) = x_2$, as we would expect.

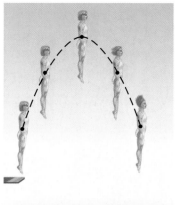
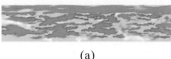

(a)

(b)

FIGURE 9–21 The motion of the diver is pure translation in (a), but is translation plus rotation in (b).

FIGURE 9–22 Translation plus rotation: a wrench moving over a horizontal surface. The CM, marked with a red **+**, moves in a straight line.

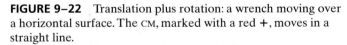

FIGURE 9–23 The center of mass of a two-particle system lies on the line joining the two masses. Here $m_1 > m_2$ so the CM is closer to m_1 than to m_2.

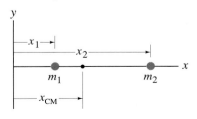

Now let us consider a system consisting of n particles, where n could be very large. This system could be an extended body which we consider as being made up of n tiny particles. If these n particles are all along a straight line (call it the x axis), we define the CM of the system to be located at

$$x_{CM} = \frac{m_1 x_1 + m_2 x_2 + \cdots + m_n x_n}{m_1 + m_2 + \cdots + m_n} = \frac{\sum_{i=1}^{n} m_i x_i}{M},$$

where $m_1, m_2, \ldots m_n$ are the masses of each particle and $x_1, x_2, \ldots x_n$ are their positions. The symbol $\sum_{i=1}^{n}$ is the summation sign meaning to sum over all the particles, where i takes on integer values from 1 to n. (Often we simply write $\Sigma m_i x_i$, leaving out the $i = 1$ to n.) The total mass of the system is $M = \Sigma m_i$.

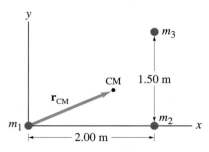

FIGURE 9–24 Example 9–12.

EXAMPLE 9–12 **CM of three guys on a raft.** Three people of roughly equivalent mass m on a lightweight (air-filled) banana boat sit along the x axis at positions $x_1 = 1.0$ m, $x_2 = 5.0$ m, and $x_3 = 6.0$ m (Fig. 9–24). Find the position of the CM.

SOLUTION We assume each x is the horizontal location of each person's CM. Treating them as particles, the overall CM is:

$$x_{CM} = \frac{mx_1 + mx_2 + mx_3}{m + m + m} = \frac{m(x_1 + x_2 + x_3)}{3m}$$

$$= \frac{(1.0\,\text{m} + 5.0\,\text{m} + 6.0\,\text{m})}{3} = \frac{12.0\,\text{m}}{3} = 4.0\,\text{m}.$$

If the particles are spread out in two or three dimensions, as for a typical extended body, then we define the coordinates of the CM as

Center of mass

$$x_{CM} = \frac{\Sigma m_i x_i}{M}, \quad y_{CM} = \frac{\Sigma m_i y_i}{M}, \quad z_{CM} = \frac{\Sigma m_i z_i}{M}, \quad \textbf{(9–11)}$$

where x_i, y_i, z_i are the coordinates of the particle of mass m_i and again $M = \Sigma m_i$ is the total mass.

Although from a practical point of view we usually calculate the components of the CM (Eq. 9–11), it is sometimes convenient (for example, for derivations) to write Eq. 9–11 in vector form. If $\mathbf{r}_i = x_i \mathbf{i} + y_i \mathbf{j} + z_i \mathbf{k}$ is the position vector of the ith particle, and $\mathbf{r}_{CM} = x_{CM}\mathbf{i} + y_{CM}\mathbf{j} + z_{CM}\mathbf{k}$ is the position vector of the center of mass, then

$$\mathbf{r}_{CM} = \frac{\Sigma m_i \mathbf{r}_i}{M}. \quad \textbf{(9–12)}$$

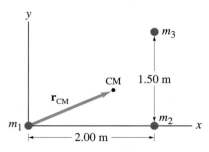

FIGURE 9–25 Example 9–13.

EXAMPLE 9–13 **Three particles in 2-D.** Three particles, each of mass 2.50 kg, are located at the corners of a right triangle whose sides are 2.00 m and 1.50 m long, as shown in Fig. 9–25. Locate the center of mass.

SOLUTION We choose our coordinate system as shown (to simplify calculations) with m_1 at the origin and m_2 on the x axis. Then m_1 has coordinates $x_1 = y_1 = 0$; m_2 has coordinates $x_2 = 2.0$ m, $y_2 = 0$; and m_3 has coordinates $x_3 = 2.0$ m, $y_3 = 1.5$ m. Then, from Eq. 9–11,

$$x_{CM} = \frac{(2.50\,\text{kg})(0) + (2.50\,\text{kg})(2.00\,\text{m}) + (2.50\,\text{kg})(2.00\,\text{m})}{3(2.50\,\text{kg})} = 1.33\,\text{m}$$

$$y_{CM} = \frac{(2.50\,\text{kg})(0) + (2.50\,\text{kg})(0) + (2.50\,\text{kg})(1.50\,\text{m})}{7.50\,\text{kg}} = 0.50\,\text{m}.$$

The CM and the position vector $\mathbf{r}_{CM}$ are shown in Fig. 9–25.

Note that the numerical values of x_{CM} and y_{CM} depend on the coordinate system chosen, but the physical position of the CM (relative to the three particles) does not depend on the reference frame.

It is often convenient to think of an extended body as made up of a continuous distribution of matter. In other words, we consider the body to be made up of n particles, each of mass Δm_i in a tiny volume around a point x_i, y_i, z_i, and we take the limit of n approaching infinity (Fig. 9–26). Then Δm_i becomes the infinitesimal mass dm at points x, y, z. The summations in Eqs. 9–11 and 9–12 become integrals:

Continuous distribution of mass

$$x_{CM} = \frac{1}{M} \int x \, dm, \quad y_{CM} = \frac{1}{M} \int y \, dm, \quad z_{CM} = \frac{1}{M} \int z \, dm, \qquad (9\text{--}13)$$

where the sum over all the mass elements is $\int dm = M$, the total mass of the object. In vector notation, this becomes

$$\mathbf{r}_{CM} = \frac{1}{M} \int \mathbf{r} \, dm. \qquad (9\text{--}14)$$

A concept similar to *center of mass* is **center of gravity** (CG). The CG of a body is that point at which the force of gravity can be considered to act. Of course, the force of gravity actually acts on all the different parts or particles of a body, but for purposes of determining the translational motion of a body as a whole, we can assume that the entire weight of the body (which is the sum of the weights of all its parts) acts at the CG. Strictly speaking, there is a conceptual difference between the center of gravity and the center of mass, but for practical purposes, they are generally the same point. [There would be a difference between the two only if a body were large enough so that the acceleration due to gravity was different at different parts of the body.]

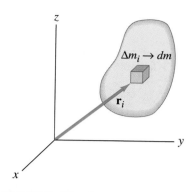

FIGURE 9–26 An extended body, here shown in only two dimensions, can be considered to be made up of many tiny particles (n), each having a mass Δm_i. One such particle is shown located at a point $\mathbf{r}_i = x_i\mathbf{i} + y_i\mathbf{j} + z_i\mathbf{k}$. We take the limit of $n \to \infty$ so Δm_i becomes the infinitesimal dm.

EXAMPLE 9–14 **CM of a thin rod.** (*a*) Show that the CM of a uniform thin rod of length L and mass M is at its center. (*b*) Determine the CM of the rod assuming its linear mass density λ (its mass per unit length) varies linearly from $\lambda = \lambda_0$ at the left end to double that value, $\lambda = 2\lambda_0$, at the right end.

SOLUTION We choose a coordinate system so that the rod lies on the x axis with the left end at $x = 0$, Fig. 9–27. Then $y_{CM} = 0$ and $z_{CM} = 0$.
(*a*) The rod is uniform, so its mass per unit length (linear mass density λ) is constant and we write it as $\lambda = M/L$. We now imagine the rod as divided into infinitesimal elements of length dx, each of which has mass $dm = \lambda \, dx$. We use Eq. 9–13:

$$x_{CM} = \frac{1}{M} \int_{x=0}^{L} x \, dm = \frac{1}{M} \int_0^L \lambda x \, dx = \frac{\lambda}{M} \left. \frac{x^2}{2} \right|_0^L = \frac{\lambda L^2}{2M} = \frac{L}{2}$$

where we used $\lambda = M/L$. This result, x_{CM} at the center, is what we expected.
(*b*) Now we have $\lambda = \lambda_0$ at $x = 0$ and we are told that λ increases linearly to $\lambda = 2\lambda_0$ at $x = L$. So we write

$$\lambda = \lambda_0(1 + \alpha x)$$

which satisfies $\lambda = \lambda_0$ at $x = 0$ and gives $\lambda = 2\lambda_0$ at $x = L$ if $(1 + \alpha L) = 2$. In other words, $\alpha = 1/L$. Again we use Eq. 9–13:

$$x_{CM} = \frac{1}{M} \int_{x=0}^{L} \lambda x \, dx = \frac{1}{M} \lambda_0 \int_0^L \left(1 + \frac{x}{L}\right) x \, dx = \frac{\lambda_0}{M} \left. \left(\frac{x^2}{2} + \frac{x^3}{3L}\right) \right|_0^L = \frac{5}{6} \frac{\lambda_0}{M} L^2.$$

But what is M in terms of λ_0 and L? We can write

$$M = \int_{x=0}^{L} dm = \int_0^L \lambda \, dx = \lambda_0 \int_0^L \left(1 + \frac{x}{L}\right) dx = \lambda_0 \left. \left(x + \frac{x^2}{2L}\right) \right|_0^L = \frac{3}{2} \lambda_0 L.$$

Then

$$x_{CM} = \frac{5}{6} \frac{\lambda_0}{M} L^2 = \frac{5}{9} L,$$

which is more than half way along the rod, as we would expect since there is more mass to the right.

FIGURE 9–27 Example 9–14.

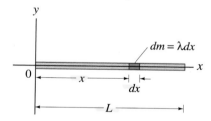

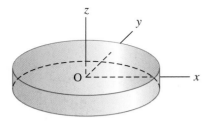

FIGURE 9–28 Cylindrical disk with origin of coordinates at geometric center.

For symmetrically shaped bodies of uniform composition, such as spheres, cylinders, and rectangular solids, the CM is located at the geometric center of the body. Consider a uniform circular cylinder, such as a solid circular disk. We expect the CM to be at the center of the circle. To show that it is, we first choose a coordinate system whose origin is at the center of the circle with the z axis perpendicular to the disk (Fig. 9–28). When we take the sum $\Sigma m_i x_i$ in Eq. 9–11, there is as much mass at any $+x_i$ as there is at $-x_i$. So all terms cancel out in pairs and $x_{CM} = 0$. The same is true for y_{CM}. In the vertical (z) direction, the CM must lie halfway between the circular faces: if we choose our origin of coordinates at that point, there is as much mass at any $+z_i$ as at $-z_i$, so $z_{CM} = 0$. For other uniform, symmetrically shaped bodies, we can make similar arguments to show that the CM must lie on a line of symmetry. If a symmetric body is *not* uniform, then these arguments do not hold. For example, the CM of a wheel or disk weighted on one side is not at the geometric center but closer to the weighted side.

To locate the center of mass of a group of extended bodies, we can use Eq. 9–11, where the m_i are the masses of these bodies and x_i, y_i, and z_i are the coordinates of the CM of each of the bodies.

FIGURE 9–29 Example 9–15. This L-shaped object has thickness t (not shown on diagram).

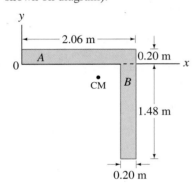

EXAMPLE 9–15 **CM of L-shaped flat object.** Determine the CM of the uniform thin L-shaped construction brace shown in Fig. 9–29.

SOLUTION Consider the object as two rectangles: rectangle A, which is 2.06 m × 0.20 m, and rectangle B, which is 1.48 m × 0.20 m. With the origin at 0 as shown, the CM of A is at

$$x_A = 1.03 \text{ m}, \quad y_A = 0.10 \text{ m}.$$

The CM of B is at

$$x_B = 1.96 \text{ m}, \quad y_B = -0.74 \text{ m}.$$

The mass of A, whose thickness is t, is

$$M_A = (2.06 \text{ m})(0.20 \text{ m})(t)(\rho) = (0.412 \text{ m}^2)(\rho t),$$

where ρ is the density. The mass of B is

$$M_B = (1.48 \text{ m})(0.20 \text{ m})(\rho t) = (0.296 \text{ m}^2)(\rho t),$$

and the total mass is $M = (0.708 \text{ m}^2)(\rho t)$. Thus

$$x_{CM} = \frac{M_A x_A + M_B x_B}{M} = \frac{(0.412 \text{ m}^2)(1.03 \text{ m}) + (0.296 \text{ m}^2)(1.96 \text{ m})}{(0.708 \text{ m}^2)} = 1.42 \text{ m},$$

where ρt was canceled out in numerator and denominator. Similarly,

$$y_{CM} = \frac{(0.412 \text{ m}^2)(0.10 \text{ m}) + (0.296 \text{ m}^2)(-0.74 \text{ m})}{(0.708 \text{ m}^2)} = -0.25 \text{ m},$$

which puts the CM approximately at the point so labeled in Fig. 9–29. For an object not perfectly thin, $z_{CM} = t/2$, since the object is assumed to be uniform.

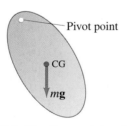

FIGURE 9–30 Determining the CM of a flat uniform body.

FIGURE 9–31 Finding the CG.

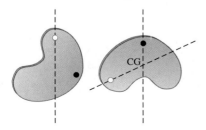

Note in this last Example that the CM can actually lie *outside* the body. Another example is a doughnut whose CM is at the center of the hole.

It is often easier to determine the CM or CG of an extended body experimentally rather than analytically. If a body is suspended from any point, it will swing (Fig. 9–30) unless it is placed so its CG lies on a vertical line directly below the point from which it is suspended. If the object is two-dimensional, or has a plane of symmetry, it need only be hung from two different pivot points and the respective vertical (plumb) lines drawn. Then the CG will be at the intersection of the two lines, as in Fig. 9–31. If the object doesn't have a plane of symmetry, the CG with respect to the third dimension is found by suspending the object from at least three points whose plumb lines do not lie in the same plane.

9–9 Center of Mass and Translational Motion

As mentioned in Section 9–8, a major reason for the importance of the concept of center of mass is that the translational motion of the CM for a system of particles (or extended bodies) is directly related to the net force on the system as a whole. We now show this, by examining the motion of a system of n particles of total mass M, which we assume remains constant. We begin by rewriting Eq. 9–12 as

$$M\mathbf{r}_{CM} = \Sigma m_i \mathbf{r}_i .$$

We differentiate this equation with respect to time:

$$M \frac{d\mathbf{r}_{CM}}{dt} = \Sigma m_i \frac{d\mathbf{r}_i}{dt}$$

or

$$M\mathbf{v}_{CM} = \Sigma m_i \mathbf{v}_i , \qquad (9\text{–}15)$$

where $\mathbf{v}_i = d\mathbf{r}_i/dt$ is the velocity of the ith particle of mass m_i and $\mathbf{v}_{CM}$ is the velocity of the CM. We take the derivative with respect to time again and obtain

$$M \frac{d\mathbf{v}_{CM}}{dt} = \Sigma m_i \mathbf{a}_i ,$$

where $\mathbf{a}_i = d\mathbf{v}_i/dt$ is the acceleration of the ith particle. Now $d\mathbf{v}_{CM}/dt$ is the acceleration of the CM, $\mathbf{a}_{CM}$. By Newton's second law, $m_i\mathbf{a}_i = \mathbf{F}_i$ where $\mathbf{F}_i$ is the net force on the ith particle. Therefore

$$M\mathbf{a}_{CM} = \mathbf{F}_1 + \mathbf{F}_2 + \cdots + \mathbf{F}_n = \Sigma \mathbf{F}_i . \qquad (9\text{–}16)$$

That is, the vector sum of all the forces acting on the system is equal to the total mass of the system times the acceleration of its center of mass. Note that our system of n particles could be the n particles that make up one or more extended bodies.

The forces $\mathbf{F}_i$ exerted on the particles of the system can be divided into two types: (1) *external forces* exerted by objects outside the system and (2) *internal forces* that particles within the system exert on one another. By Newton's third law, the internal forces occur in pairs: if one particle exerts a force on a second particle in our system, the second must exert an equal and opposite force on the first. Thus, in the sum over all the forces in Eq. 9–16, these internal forces cancel each other in pairs. We are left, then, with only the external forces on the right side of Eq. 9–16:

$$M\mathbf{a}_{CM} = \Sigma \mathbf{F}_{ext} , \qquad \text{[constant } M] \quad (9\text{–}17)$$

where $\Sigma \mathbf{F}_{ext}$ is the sum of all the external forces acting on our system, which is the *net force* acting on the system. Thus

the sum of all the forces acting on the system is equal to the total mass of the system times the acceleration of its center of mass.

This is **Newton's second law** for a system of particles. It also applies to an extended body (which can be thought of as a collection of particles), and to a system of bodies. Thus we conclude that

the center of mass of a system of particles (or bodies) with total mass M moves like a single particle of mass M acted upon by the same net external force.

That is, the system moves as if all its mass were concentrated at the CM and all the external forces acted at that point. We can thus treat the *translational motion* of any body or system of bodies as the motion of a particle (see Figs. 9–21 and 9–22).

This theorem clearly simplifies our analysis of the motion of complex systems and extended bodies. Although the motion of various parts of the system may be complicated, we may often be satisfied with knowing the motion of the CM. This theorem also allows us to solve certain types of problems very easily, as illustrated by the following Example.

CONCEPTUAL EXAMPLE 9–16 **A two-stage rocket.** A rocket is shot into the air as shown in Fig. 9–32. At the moment it reaches its highest point, a horizontal distance d from its starting point, a prearranged explosion separates it into two parts of equal mass. Part I is stopped in midair by the explosion and falls vertically to Earth. Where does part II land? Assume $\mathbf{g}$ = constant.

SOLUTION After the rocket is fired, the path of the CM of the system continues to follow the parabolic trajectory of a projectile acted on only by a constant gravitational force. The CM will thus arrive at a point $2d$ from the starting point. Since the masses of I and II are equal, the CM must be midway between them. Therefore, II lands a distance $3d$ from the starting point. (If part I had been given a kick up or down, instead of merely falling, the solution would have been somewhat more complicated.)

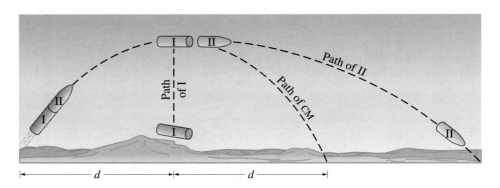

FIGURE 9–32 Example 9–16.

We can write Eq. 9–17, $M\mathbf{a}_{CM} = \Sigma\mathbf{F}_{ext}$, in terms of the total momentum $\mathbf{P}$ of a system of particles. $\mathbf{P}$ is defined, as we saw in Section 9–2 as

$$\mathbf{P} = m_1\mathbf{v}_1 + m_2\mathbf{v}_2 + \cdots + m_n\mathbf{v}_n = \Sigma\mathbf{p}_i.$$

From Eq. 9–15 $\left(M\mathbf{v}_{CM} = \Sigma m_i\mathbf{v}_i\right)$ we have

Total momentum

$$\mathbf{P} = M\mathbf{v}_{CM}. \tag{9–18}$$

Thus, *the total linear momentum of a system of particles is equal to the product of the total mass M and the velocity of the center of mass of the system.* Or, *the linear momentum of an extended body is the product of the body's mass and the velocity of its CM.*

If we differentiate Eq. 9–18 with respect to time, we obtain (assuming the total mass M is constant)

$$\frac{d\mathbf{P}}{dt} = M\frac{d\mathbf{v}_{CM}}{dt} = M\mathbf{a}_{CM}.$$

From Eq. 9–17, we see that

NEWTON'S SECOND LAW (for a system)

$$\frac{d\mathbf{P}}{dt} = \Sigma\mathbf{F}_{ext}, \tag{9–6}$$

where $\Sigma\mathbf{F}_{ext}$ is the net external force on the system. This is just Eq. 9–6 obtained earlier: **Newton's second law for a system of objects**. It is valid for any definite fixed system of particles or objects. If we know $\Sigma\mathbf{F}_{ext}$, we can determine how the total momentum changes.

An interesting example is the discovery since 1996 of nearby stars that seem to "wobble." What could cause such a wobble? Very likely a planet that orbits about the star, each exerting a gravitational force on the other. The planets are too small and too far away to have been observed directly by existing telescopes. But the slight wobble in the motion of the star suggests that both the planet and the star (its sun) orbit about their mutual center of mass, and hence the star appears to have a wobble. Irregularities in the star's motion can be obtained to an accuracy of 3 m/s, and from the data, the size of the planets' orbits can be obtained, as well as their masses (if as large as Jupiter's). See Fig. 6–17 in Chapter 6.

➡ PHYSICS APPLIED

Distant planets discovered

* 9–10 Systems of Variable Mass; Rocket Propulsion

We now treat systems whose mass varies. Such systems could be treated as a type of inelastic collision, but it is simpler to use Eq. 9–6, $d\mathbf{P}/dt = \Sigma\mathbf{F}_{ext}$, where $\mathbf{P}$ is the total momentum of the system and $\Sigma\mathbf{F}_{ext}$ is the net external force exerted on it. Great care must be taken to define the system, and to include all changes in momentum. An important application is to rockets, which propel themselves forward by the ejection of burnt gases: the force exerted by the gases on the rocket accelerates the rocket. The mass M of the rocket decreases during this process, so $dM/dt < 0$. Another application is the dropping of material (gravel, packaged goods) onto a conveyor belt. In this situation, the mass M of the loaded conveyor belt increases, so $dM/dt > 0$.

To treat the general case of a system of variable mass, let us consider the system shown in Fig. 9–33. At some time t, we have a system of mass M and momentum $M\mathbf{v}$; we also have a tiny (infinitesimal) mass dM traveling with velocity $\mathbf{u}$ which is about to enter our system. An infinitesimal time dt later, the mass dM combines with the system. For simplicity we will refer to this as a "collision." So our system has changed in mass from M to $M + dM$ in the time dt. Note that dM can be less than zero, as for a rocket propelled by ejected gases.

In order to apply Eq. 9–6, $d\mathbf{P}/dt = \Sigma\mathbf{F}_{ext}$, we must consider a definite fixed system of particles. That is, in considering the change in momentum, $d\mathbf{P}$, we must consider the momentum of the same particles initially and finally. We will define our *total system* as including M plus dM. Then initially, at time t, the total momentum is $M\mathbf{v} + \mathbf{u}\,dM$ (Fig. 9–33). At time $t + dt$, after dM has combined with M, the velocity of the whole is now $\mathbf{v} + d\mathbf{v}$ and the total momentum is $(M + dM)(\mathbf{v} + d\mathbf{v})$. So the change in momentum $d\mathbf{P}$ is

$$d\mathbf{P} = (M + dM)(\mathbf{v} + d\mathbf{v}) - (M\mathbf{v} + \mathbf{u}\,dM)$$

$$= M\,d\mathbf{v} + \mathbf{v}\,dM + dM\,d\mathbf{v} - \mathbf{u}\,dM.$$

Then, by Eq. 9–6, we have

$$\Sigma\mathbf{F}_{ext} = \frac{d\mathbf{P}}{dt} = \frac{M\,d\mathbf{v} + \mathbf{v}\,dM - \mathbf{u}\,dM}{dt},$$

where we have dropped the term $dM\,d\mathbf{v}/dt$, since in the limit of infinitesimals it is zero ($d\mathbf{v}/dt$ may not approach zero but dM surely does). Thus we get

$$\Sigma\mathbf{F}_{ext} = M\frac{d\mathbf{v}}{dt} - (\mathbf{u} - \mathbf{v})\frac{dM}{dt}. \qquad \textbf{(9–19a)}$$

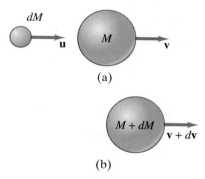

FIGURE 9–33 (a) At time t, a mass dM is about to be added to our system M. (b) At time $t + dt$, the mass dM has been added to our system.

Note that the quantity $(\mathbf{u} - \mathbf{v})$ is the relative velocity, $\mathbf{v}_{rel}$, of dM with respect to M. That is,

$$\mathbf{v}_{rel} = \mathbf{u} - \mathbf{v}$$

is the velocity of the entering mass dM as seen by an observer on M. So we can rearrange Eq. 9–19a:

$$M \frac{d\mathbf{v}}{dt} = \Sigma \mathbf{F}_{ext} + \mathbf{v}_{rel} \frac{dM}{dt}. \qquad \textbf{(9–19b)}$$

We can interpret this equation as follows. $M dv/dt$ is the mass times the acceleration of M. The first term on the right, $\Sigma \mathbf{F}_{ext}$, refers to the external force on the mass M (for a rocket, it would include the force of gravity and air resistance). It does *not* include the force that dM exerts on M as a result of their collision. This is taken care of by the second term on the right, $\mathbf{v}_{rel}(dM/dt)$, which represents the rate at which momentum is being transferred to (or from) the mass M because of the mass that is added to (or leaves) it. It can thus be interpreted as the force exerted on the mass M due to the addition (or ejection) of mass. For a rocket this term is called the *thrust*, since it represents the force exerted on the rocket by the expelled gases.

Thrust on rocket

⇒ **PHYSICS APPLIED**

Moving conveyor belt

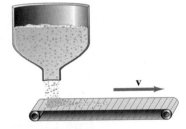

FIGURE 9–34 Gravel dropped from hopper onto conveyor belt.

EXAMPLE 9–17 Conveyor belt. You are designing a conveyor system for a gravel yard. A hopper drops gravel at a rate of 75.0 kg/s onto a conveyor belt that moves at a constant speed $v = 2.20$ m/s (Fig. 9–34). (a) Determine the force needed to keep the conveyor belt moving. (b) What power output must the motor have that drives the conveyor belt.

SOLUTION We assume that the hopper is at rest so $u = 0$ and that the hopper has just begun dropping gravel, so $dM/dt = 75.0$ kg/s. Since the belt moves at a constant speed ($dv/dt = 0$), we have from Eq. 9–19a:

$$F_{ext} = M \frac{dv}{dt} - (u - v) \frac{dM}{dt}$$

$$= 0 - (0 - v) \frac{dM}{dt}$$

$$= v \frac{dM}{dt} = (2.20 \text{ m/s})(75.0 \text{ kg/s}) = 165 \text{ N}.$$

(b) From Eq. 8–22, this force does work at the rate

$$\frac{dW}{dt} = \mathbf{F}_{ext} \cdot \mathbf{v} = v^2 \frac{dM}{dt}$$

$$= 363 \text{ W},$$

which is the power output required of the motor.
[But note that this work does not all go into kinetic energy of the gravel, since

$$\frac{dK}{dt} = \frac{d}{dt}\left(\frac{1}{2}Mv^2\right) = \frac{1}{2}\frac{dM}{dt}v^2,$$

which is only half the work done by $\mathbf{F}_{ext}$. The other half of the external work done goes into thermal energy produced by friction between the gravel and the belt (the same friction force that accelerates the gravel).]

EXAMPLE 9–18 **Rocket propulsion.** A fully fueled rocket has a mass of 21,000 kg, of which 15,000 kg is fuel. The burned fuel is spewed out the rear at a rate of 190 kg/s with a speed of 2800 m/s relative to the rocket. If the rocket is fired vertically upward (Fig. 9–35) calculate: (a) the thrust of the rocket; (b) the net force on the rocket at blastoff, and just before burnout (when all the fuel has been used up); (c) the rocket's velocity as a function of time, and (d) its final velocity at burnout. Ignore air resistance and assume the acceleration due to gravity is constant at $g = 9.80\ \text{m/s}^2$.

➡ PHYSICS APPLIED

Rocket propulsion

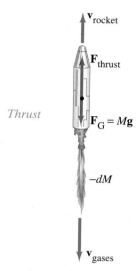

Thrust

FIGURE 9–35 Example 9–18; $\mathbf{v}_{\text{rel}} = \mathbf{v}_{\text{gases}} - \mathbf{v}_{\text{rocket}}$. M is the mass of the rocket at any instant and is decreasing until burnout.

SOLUTION (a) The thrust (see discussion after Eq. 9–19b) is:

$$F_{\text{thrust}} = v_{\text{rel}}\frac{dM}{dt} = (-2800\ \text{m/s})(-190\ \text{kg/s}) = 5.3 \times 10^5\ \text{N}.$$

where we have taken upward as positive so v_{rel} is negative because it is downward, and dM/dt is negative because the rocket's mass is diminishing.
(b) $F_{\text{ext}} = Mg = (2.1 \times 10^4\ \text{kg})(9.80\ \text{m/s}^2) = 2.1 \times 10^5\ \text{N}$ initially, and $(6.0 \times 10^3\ \text{kg})(9.80\ \text{m/s}^2) = 5.9 \times 10^4\ \text{N}$ at burnout. Hence, the net force on the rocket at blastoff is

$$F_{\text{net}} = 5.3 \times 10^5\ \text{N} - 2.1 \times 10^5\ \text{N} = 3.2 \times 10^5\ \text{N}, \qquad [\text{blastoff}]$$

and just before burnout it is

$$F_{\text{net}} = 5.3 \times 10^5\ \text{N} - 5.9 \times 10^4\ \text{N} = 4.7 \times 10^5\ \text{N}. \qquad [\text{burnout}]$$

After burnout, of course, the net force is that of gravity, $-5.9 \times 10^4\ \text{N}$.
(c) From Eq. 9–19b we have

$$dv = \frac{F_{\text{ext}}}{M}dt + v_{\text{rel}}\frac{dM}{M},$$

where $F_{\text{ext}} = -Mg$, and M is the mass of the rocket and is a function of time. Since v_{rel} is constant, we can integrate this easily:

$$\int_{v_0}^{v} dv = -\int_0^t g\,dt + v_{\text{rel}}\int_{M_0}^{M}\frac{dM}{M}$$

or

$$v(t) = v_0 - gt + v_{\text{rel}}\ln\frac{M}{M_0}, \qquad\qquad\qquad v\ as\ function\ of\ t$$

where $v(t)$ is the rocket's velocity and M its mass at any time t. Note that v_{rel} is negative ($-2800\ \text{m/s}$ in our case) because it is opposite to the motion, and that $\ln(M/M_0)$ is also negative because $M_0 > M$. Hence, the last term—which represents the thrust—is positive and acts to increase the velocity.
(d) The time required to reach burnout is the time needed to use up all the fuel (15,000 kg) at a rate of 190 kg/s; so at burnout,

$$t = \frac{1.50 \times 10^4\ \text{kg}}{190\ \text{kg/s}} = 79.0\ \text{s}.$$

If we take $v_0 = 0$, then using the result of part (c):

$$v = -(9.80\ \text{m/s}^2)(79\ \text{s}) + (-2800\ \text{m/s})\left(\ln\frac{6000\ \text{kg}}{21,000\ \text{kg}}\right) = 2730\ \text{m/s}. \qquad v\ at\ burnout$$

Summary

The **linear momentum**, **p**, of an object is defined as the product of its mass times its velocity,

$$\mathbf{p} = m\mathbf{v}.$$

In terms of momentum, **Newton's second law** can be written as

$$\Sigma\mathbf{F} = \frac{d\mathbf{p}}{dt}.$$

That is, the rate of change of momentum of an object equals the net force exerted on it.

When the net external force on a system of objects is zero, the total momentum remains constant. This is the **law of conservation of momentum**. Stated another way, the total momentum of an isolated system of objects remains constant.

The law of conservation of momentum is very useful in dealing with the class of events known as **collisions**. In a collision two (or more) bodies interact with each other for a very short time and the force between them during this time is very large compared to any other forces acting. The **impulse** of such a force on a body is defined as

$$\mathbf{J} = \int \mathbf{F}\, dt$$

and is equal to the change in momentum of the body as long as **F** is the net force on the body:

$$\Delta\mathbf{p} = \mathbf{p}_f - \mathbf{p}_i = \int_{t_i}^{t_f} \mathbf{F}\, dt = \mathbf{J}.$$

Total momentum is conserved in any collision:

$$\mathbf{p}_1 + \mathbf{p}_2 = \mathbf{p}_1' + \mathbf{p}_2'.$$

The total energy is also conserved, but this may not be useful unless the only type of energy transformation involves kinetic energy. In this case, kinetic energy is conserved and the collision is called an **elastic collision**:

$$\tfrac{1}{2}mv_1^2 + \tfrac{1}{2}mv_2^2 = \tfrac{1}{2}m_1 v_1'^2 + \tfrac{1}{2}m_2 v_2'^2.$$

If kinetic energy is not conserved, the collision is called **inelastic**. If two colliding objects stick together as the result of a collision, the collision is said to be **completely inelastic**.

For a system of particles, or for an extended body that can be considered as having a continuous distribution of matter, the **center of mass** (CM) is defined as

$$x_{CM} = \frac{\Sigma m_i x_i}{M}, \qquad y_{CM} = \frac{\Sigma m_i y_i}{M}, \qquad z_{CM} = \frac{\Sigma m_i z_i}{M}$$

or

$$x_{CM} = \frac{1}{M}\int x\, dm, \quad y_{CM} = \frac{1}{M}\int y\, dm, \quad z_{CM} = \frac{1}{M}\int z\, dm,$$

where M is the total mass of the system.

The center of mass (CM) of a system is important because this point moves like a single particle of mass M acted on by the same net external force, $\Sigma\mathbf{F}_{ext}$. In equation form, this is just Newton's second law for a system of particles (or extended bodies):

$$M\mathbf{a}_{CM} = \Sigma\mathbf{F}_{ext},$$

where M is the total mass of the system, $\mathbf{a}_{CM}$ is the acceleration of the CM of the system, and $\Sigma\mathbf{F}_{ext}$ is the total (net) external force acting on all parts of the system.

For a system of particles of total linear momentum $\mathbf{P} = \Sigma m_i \mathbf{v}_i = M\mathbf{v}_{CM}$, Newton's second law is

$$\frac{d\mathbf{P}}{dt} = \Sigma\mathbf{F}_{ext}.$$

Questions

1. We claim that momentum is conserved. Yet most moving objects eventually slow down and stop. Explain.
2. Two blocks of mass m_1 and m_2 rest on a frictionless table and are connected by a spring. The blocks are pulled apart, stretching the spring, and then released. Describe the subsequent motion of the two blocks.
3. A light body and a heavy body have the same kinetic energy. Which has the greater momentum?
4. When a person jumps from a tree to the ground, what happens to the momentum of the person upon landing?
5. Explain, on the basis of conservation of momentum, how a fish propels itself forward by swishing its tail back and forth.
6. Why, when you release an inflated, untied balloon, does it fly across the room?
7. It is said that in ancient times a rich man with a bag of gold coins was stranded on the surface of a frozen lake and froze to death. Because the ice was frictionless, he could not push himself to shore. What could he have done to save himself had he not been so miserly?
8. If a falling ball were to make a perfectly elastic collision with the floor, would it rebound to its original height? Explain.
9. (a) An empty sled is sliding on frictionless ice when Susan drops vertically from a tree above onto the sled. When she lands, does the sled speed up, slow down, or keep the same speed? (b) Later, Susan falls off the sled. When she drops off, does the sled speed up, slow down, or keep the same speed?
10. Is it easier to hit a home run from a pitched ball than from one tossed in the air by the batter? Explain.

11. The speed of a tennis ball on the return of a serve can be just as fast as the serve, even though the racket isn't swung very fast. How can this be?

12. Is it possible for a body to receive a larger impulse from a small force than from a large force?

13. How could a force give zero impulse over a nonzero time interval even though the force is not zero for at least a part of that time interval?

14. In a collision between two cars, which would you expect to be more damaging to the occupants: if the cars collide and remain together, or if the two cars collide and rebound backward? Explain.

15. A superball is dropped from a height h onto a hard steel plate (fixed to the Earth), from which it rebounds at very nearly its original speed. (a) Is the momentum of the ball conserved during any part of this process? (b) If we consider the ball and the Earth as our system, during what parts of the process is momentum conserved? (c) Answer part (b) for a piece of putty that falls and sticks to the steel plate.

16. It used to be common wisdom to build cars to be as rigid as possible to withstand collisions. Today, though, cars are designed to have "crumple zones" that collapse upon impact. What advantage does this have?

17. A rocket following a parabolic path through the air suddenly explodes into many pieces. What can you say about the motion of this system of pieces?

18. Why is the CM of a 1-m length of pipe at the pipe's midpoint, whereas this is not true for your arm or leg?

19. Show on a diagram how your CM shifts when you move from a lying position to a sitting position.

20. Describe an analytic way of determining the CM of any thin, triangular-shaped, uniform plate.

21. Place yourself facing the edge of an open door. Position your feet astride the door with your nose and abdomen touching the door's edge. Try to rise on your tiptoes. Why can't this be done?

22. If only an external force can change the momentum of the center of mass of an object, how can the internal force of the engine accelerate a car?

23. How can a rocket change direction when it is far out in space and essentially in a vacuum?

24. In observations of nuclear β-decay, the electron and recoil nucleus often do not separate along the same line. Use conservation of momentum in two dimensions to explain why this implies the emission of at least one other particle in the disintegration.

25. Bob and Jim decide to play tug-of-war on a frictionless lake. Jim is considerably stronger than Bob, but Bob weighs 160 lbs while Jim weighs 145 lbs. Who loses by crossing over the midline first?

26. At a carnival game you try to knock over a heavy cylinder by throwing a small ball at it. You have a choice of throwing either a ball that will stick to the cylinder, or a second ball of equal mass and speed that will bounce backward off the cylinder. Which ball is more likely to make the cylinder move?

Problems

Section 9–1

1. (I) Calculate the force exerted on a rocket at takeoff, given that the propelling gases are expelled at a rate of 1200 kg/s with a speed of 50,000 m/s.

2. (I) The momentum of a particle, in SI units, is given by $\mathbf{p} = 4.8t^2\mathbf{i} - 8.0\mathbf{j} - 8.9t\mathbf{k}$. What is the force as a function of time?

3. (II) (a) What is the momentum of a 30.0-g sparrow with a speed of 12 m/s? (b) What will be its momentum 12 s later if a constant 2.0×10^{-2} N force due to air resistance acts on it?

4. (II) A 145-g baseball, moving along the x axis with speed 30.0 m/s, strikes a fence at a 45° angle and rebounds along the y axis with unchanged speed. Give its change in momentum using unit vector notation.

5. (II) The force on a particle of mass m is given by $\mathbf{F} = 26\mathbf{i} - 12t^2\mathbf{j}$ where F is in N and t in s. What will be the change in the particle's momentum between $t = 1.0$ s and $t = 2.0$ s?

6. (II) A 4200-kg rocket is traveling in outer space with a velocity of 120 m/s toward the Sun. It needs to alter its course by 23.0°, which can be done by shooting its rockets briefly in a direction perpendicular to its original motion. If the rocket gases are expelled at a speed of 2200 m/s relative to the rocket, what mass of gas must be expelled?

7. (II) A ball of mass m falls freely a distance h from rest. It hits the floor and rebounds to its starting height. (a) What total time is required for the motion? (Ignore the time of contact with the floor.) (b) What is the speed of the ball as it strikes the floor? (c) What is the change in momentum of the ball as it strikes the floor? (d) Calculate the average force the ball exerts on the floor averaged over the time interval of part (a). Is this result surprising?

8. (III) Air in a 100-km/h wind strikes head-on the face of a building 40 m wide by 60 m high and is brought to rest. If air has a mass of 1.3 kg per cubic meter, determine the average force of the wind on the building.

Section 9–2

9. (I) A 9700-kg boxcar traveling 18 m/s strikes a second car. The two stick together and move off with a speed of 4.0 m/s. What is the mass of the second car?

10. (I) A 130-kg tackler moving at 2.5 m/s meets head-on (and tackles) a 90-kg halfback moving at 5.0 m/s. What will be their mutual speed immediately after the collision?

11. (I) An atomic nucleus at rest decays radioactively into an alpha particle and a smaller nucleus. What will be the speed of this recoiling nucleus if the speed of the alpha particle is 2.5×10^5 m/s? Assume the nucleus has a mass 57 times greater than that of the alpha particle.

12. (I) A 10,500-kg railroad car travels alone on a level frictionless track with a constant speed of 15.0 m/s. An additional 6350 kg load is dropped from a tower onto the car. What then will be its speed?

13. (II) A child in a boat throws a 5.40-kg package out horizontally with a speed of 10.0 m/s, Fig. 9–36. Calculate the velocity of the boat immediately after, assuming it was initially at rest. The mass of the child is 26.0 kg and that of the boat is 55.0 kg.

$v = 10$ m/s

FIGURE 9–36 Problem 13.

14. (II) A 12-gram bullet traveling 190 m/s penetrates a 2.0-kg block of wood and emerges going 150 m/s. If the block is stationary on a frictionless surface when hit, how fast does it move after the bullet emerges?

15. (II) An explosion breaks an object, originally at rest, into two fragments. One fragment acquires twice the kinetic energy of the other. What is the ratio of their masses? Which is greater?

16. (II) A ball moving with a speed of 17 m/s strikes an identical ball that is initially at rest. After the collision, the incoming ball has been deviated by 45° from its original direction, and the struck ball moves off at 30° from the original direction (Fig. 9–37). What are the speeds of the two balls after the collision?

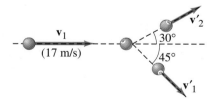

$\mathbf{v}_1$
(17 m/s)

30°

45°

$\mathbf{v}'_2$

$\mathbf{v}'_1$

FIGURE 9–37 Problem 16.

17. (II) A rocket of mass m traveling with speed v_0 along the x axis suddenly shoots out one-third its mass parallel to the y axis (as seen by an observer at rest) with speed $2v_0$. Express the velocity of the remainder of the object in $\mathbf{i}, \mathbf{j}, \mathbf{k}$ notation.

18. (II) The decay of a neutron into a proton, an electron, and a neutrino is an example of a three-particle decay process. Use the vector nature of momentum to show that if the neutron is initially at rest, the velocity vectors of the three must be coplanar (that is, all in the same plane). The result is not true for numbers greater than three.

19. (II) A radioactive nucleus at rest decays into a second nucleus, an electron, and a neutrino. The electron and neutrino are emitted at right angles and have momenta of 8.6×10^{-23} kg·m/s and 6.2×10^{-23} kg·m/s, respectively. What is the magnitude and the direction of the momentum of the recoiling nucleus?

20. (II) A 900-kg two-stage rocket is traveling at a speed of 6.50×10^3 m/s away from Earth when a predesigned explosion separates the rocket into two sections of equal mass that then move with a relative speed (relative to each other) of 2.80×10^3 m/s along the original line of motion. (a) What is the speed and direction of each section (relative to Earth) after the explosion? (b) How much energy was supplied by the explosion? [Hint: What is the change in kinetic energy as a result of the explosion?]

21. (III) A 200-kg projectile, fired with a speed of 100 m/s at a 60° angle, breaks into three pieces of equal mass at the highest point of its arc. Two of the fragments move with the same equal speed right after the explosion as the entire projectile had just before the explosion; one of these moves vertically downward and the other horizontally. Determine (a) the velocity of the third fragment immediately after the explosion and (b) the energy released in the explosion.

Section 9–3

22. (I) A 0.145-kg baseball pitched at 35.0 m/s is hit on a horizontal line drive straight back at the pitcher at 56.0 m/s. If the contact time between bat and ball is 5.00×10^{-3} s, calculate the force (assumed to be constant) between the ball and bat.

23. (I) The tennis ball may leave the racket of a top player on the serve with a speed of 65.0 m/s. If the ball's mass is 0.0600 kg and is in contact with the racket for 0.0300 s, what is the average force on the ball? Would this force be large enough to lift a 60-kg person?

24. (II) A tennis ball of mass $m = 0.060$ kg and speed $v = 28$ m/s strikes a wall at a 45° angle and rebounds with the same speed at 45° (Fig. 9–38). What is the impulse given the wall?

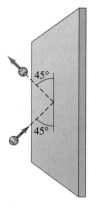

45°

45°

FIGURE 9–38
Problem 24.

25. (II) Water strikes the turbine blades of a generator so that its rebounding velocity is 0.75 of its original magnitude and reversed in direction. If the flow rate is 60 kg/s and the original water speed is 10 m/s, what is the average force on the blades?

26. (II) A 140-kg astronaut (including space suit) acquires a speed of 2.50 m/s by pushing off with his legs from a 1800-kg space capsule. (a) What is the change in speed of the space capsule? (b) If the push lasts 0.500 s, what is the average force exerted by each on the other? As the reference frame, use the position of the capsule before the push. (c) What is the kinetic energy of each after the push?

27. (II) (a) A molecule of mass m and speed v strikes a wall at right angles and rebounds with the same speed. If the collision time is Δt, what is the average force on the wall during the collision? (b) If molecules, all of this type, strike the wall at intervals a time t apart (on the average) what is the average force on the wall averaged over a long time?

28. (II) Suppose the force acting on a tennis ball (mass 0.060 kg) as a function of time is given by the graph of Fig. 9–39. Use the graph to estimate (a) the total impulse given the ball, and (b) the speed of the ball after being struck, assuming the ball is being served so it is nearly at rest initially.

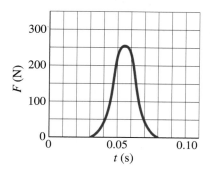

FIGURE 9–39 Problem 28.

29. (II) The force on a bullet is given by the formula $F = (780 - 2.6 \times 10^5 t)\,\text{N}$ over the time interval $t = 0$ to $t = 3.0 \times 10^{-3}\,\text{s}$. In this formula t is in seconds and F is in newtons. (a) Plot a graph of F versus t for $t = 0$ to $t = 3.0 \times 10^{-3}\,\text{s}$. (b) Estimate, using graphs, the impulse given the bullet. (c) Determine the impulse by integration. (d) If the bullet achieves a speed of 300 m/s as a result of this impulse, given to it in the barrel of a gun, what must the bullet's mass be?

30. (III) From what maximum height can a 75-kg person jump without breaking the lower leg bone on either leg? Ignore air resistance and assume the body of the person moves a distance of 0.60 m from the standing to the seated position (that is, in breaking the fall). Assume the breaking strength (force per unit area) of bone is $170 \times 10^6\,\text{N/m}^2$, and its smallest cross-sectional area is $2.5 \times 10^{-4}\,\text{m}^2$.

31. (III) A scale is adjusted so that when a large, flat pan is placed on it, it reads zero. A water faucet at height $h = 2.5\,\text{m}$ above is turned on and water falls into the pan at a rate $R = 0.12\,\text{kg/s}$. Determine (a) a formula for the scale reading as a function of time t and (b) the reading for $t = 15\,\text{s}$. (c) Repeat (a) and (b), but replace the flat pan with a tall, narrow cylindrical container of area $A = 20\,\text{cm}^2$ (the level rises in this case).

Sections 9–4 and 9–5

32. (II) A ball of mass 0.540 kg moving east (+x direction) with a speed of 3.90 m/s collides head-on with a 0.320-kg ball at rest. If the collision is perfectly elastic, what will be the speed and direction of each ball after the collision?

33. (II) A 0.450-kg hockey puck, moving east with a speed of 4.20 m/s, has a head-on collision with a 0.900-kg puck initially at rest. Assuming a perfectly elastic collision, what will be the speed and direction of each object after the collision?

34. (II) A 0.060-kg tennis ball, moving with a speed of 7.50 m/s, has a head-on collision with a 0.090-kg ball initially moving away from it at a speed of 3.00 m/s. Assuming a perfectly elastic collision, what is the speed and direction of each ball after the collision?

35. (II) A softball of mass 0.220 kg that is moving with a speed of 6.5 m/s collides head-on and elastically with another ball initially at rest. Afterward it is found that the incoming ball has bounced backward with a speed of 3.8 m/s. Calculate (a) the velocity of the target ball after the collision, and (b) the mass of the target ball.

36. (II) A 0.280-kg croquet ball makes an elastic head-on collision with a second ball initially at rest. The second ball moves off with half the original speed of the first ball. (a) What is the mass of the second ball? (b) What fraction of the original kinetic energy gets transferred to the second ball?

37. (II) An atomic nucleus initially moving at 500 m/s emits an alpha particle in the direction of its velocity, and the new nucleus slows to 450 m/s. If the alpha particle has a mass of 4.0 u and the original nucleus has a mass of 222 u, what speed does the alpha particle have when it is emitted?

38. (II) A ball of mass m makes a head-on-elastic collision with a second ball (at rest) and rebounds with a speed equal to one-fourth its original speed. What is the mass of the second ball?

39. (II) Determine the fraction of kinetic energy lost by a neutron ($m_1 = 1.01\,\text{u}$) when it collides head-on and elastically with (a) ^1_1H ($m = 1.01\,\text{u}$); (b) ^2_1H (heavy hydrogen, $m = 2.01\,\text{u}$); (c) $^{12}_6\text{C}$ ($m = 12.00\,\text{u}$); (d) $^{208}_{82}\text{Pb}$ (lead, $m = 208\,\text{u}$).

40. (II) Show that, in general, for any head-on one-dimensional elastic collision, the speeds after collision are

$$v_2' = v_1 \left(\frac{2m_1}{m_1 + m_2} \right) + v_2 \left(\frac{m_2 - m_1}{m_1 + m_2} \right)$$

and

$$v_1' = v_1 \left(\frac{m_1 - m_2}{m_1 + m_2} \right) + v_2 \left(\frac{2m_2}{m_1 + m_2} \right),$$

where v_1 and v_2 are the initial speeds of the two objects of mass m_1 and m_2.

41. (III) A 2.0-kg block slides along a frictionless tabletop at 8.0 m/s toward a second block (at rest) of mass 4.5 kg. A coil spring, which obeys Hooke's law and has spring constant $k = 850\,\text{N/m}$, is attached to the second block in such a way that it will be compressed when struck by the moving block, Fig. 9–40. (a) What will be the maximum compression of the spring? (b) What will be the final velocities of the blocks after the collision? (c) Is the collision elastic?

FIGURE 9–40 Problem 41.

42. (II) An 18-g rifle bullet traveling 180 m/s buries itself in a 3.6-kg pendulum hanging on a 2.8-m-long string, which makes the pendulum swing upward in an arc. Determine the horizontal component of the pendulum's displacement.

43. (II) (a) Derive a formula for the fraction of kinetic energy lost, $\Delta K/K$, for the ballistic pendulum collision of Example 9–10. (b) Evaluate for $m = 14.0$ g and $M = 380$ g.

44. (II) An explosion breaks an object into two pieces, one of which has 1.5 times the mass of the other. If 17,500 J were released in the explosion, how much kinetic energy did each piece acquire?

45. (II) After a completely inelastic collision between two objects of equal mass, each having initial speed, v, the two move off with speed $v/3$. What was the angle between their initial directions?

46. (II) A 0.95×10^3-kg Toyota collides into the rear end of a 2.2×10^3-kg Cadillac stopped at a red light. The bumpers lock, the brakes are locked, and the two cars skid forward 4.8 m before stopping. The police officer, knowing that the coefficient of kinetic friction between tires and road is 0.40, calculates the speed of the Toyota at impact. What was that speed?

47. (II) A measure of inelasticity in a head-on collision of two bodies is the *coefficient of restitution*, e, defined as

$$e = \frac{v_1' - v_2'}{v_2 - v_1},$$

where $v_1' - v_2'$ is the relative velocity of the two bodies after the collision and $v_2 - v_1$ is their relative velocity before. (a) Show that for a perfectly elastic collision, $e = 1$, and for a completely inelastic collision, $e = 0$. (b) A simple method for measuring the coefficient of restitution for a body colliding with a very hard surface like steel is to drop the body onto a heavy steel plate, as shown in Fig. 9–41. Determine a formula for e in terms of the original height h and the maximum height reached after collision h'.

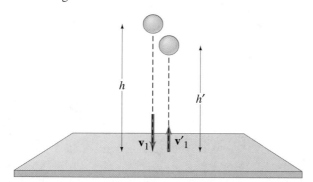

FIGURE 9–41 Problem 47. Measurement of coefficient of restitution.

48. (II) A wooden block is cut into two pieces, one with three times the mass of the other. A depression is made in both faces of the cut, so that a firecracker can be placed in it with the block reassembled. The reassembled block is set on a rough-surfaced table and the fuse is lit. When the firecracker explodes, the two blocks separate and slide apart. What is the ratio of distances each block travels?

49. (III) A 5.0-kg body moving in the $+x$ direction at 5.5 m/s collides head-on with a 3.0-kg body moving in the $-x$ direction at 4.0 m/s. Find the final velocity of each mass if: (a) the bodies stick together; (b) the collision is elastic; (c) the 5.0-kg body is at rest after the collision; (d) the 3.0-kg body is at rest after the collision; (e) the 5.0-kg body has a velocity of 4.0 m/s in the $-x$ direction after the collision. Are the results in (c), (d), and (e) "reasonable"? Explain.

50. (III) For the ballistic pendulum, let us estimate the accuracy of our approximation that the block (mass M) doesn't move during the collision. Use the symbols of Example 9–10 and Fig. 9–18, and assume the projectile is decelerated uniformly within the block in a distance d. Estimate: (a) the collision time Δt; (b) by how much is momentum not conserved, $\Delta \mathbf{p}$, because there is a net external force acting during this time; and (c) by what fraction the calculated speed of the bullet will be in error if we use the equation given in Example 9–10.

Section 9–7

51. (II) An eagle $(m_1 = 3.3$ kg$)$ moving with speed $v_1 = 7.8$ m/s is on a collision course with a second eagle $(m_2 = 4.6$ kg$)$ moving at $v_2 = 10.2$ m/s in a direction at right angles to the first. After they collide, they hold onto one another. In what direction, and with what speed, are they moving after the collision?

52. (II) A billiard ball of mass $m_A = 0.400$ kg moving with speed $v_A = 1.80$ m/s strikes a second ball, initially at rest, of mass $m_B = 0.500$ kg. As a result of the collision, the first ball is deflected off at an angle of 30.0° with a speed $v_A' = 1.10$ m/s. (a) Taking the x axis to be the original direction of motion of ball A, write down the equations expressing the conservation of momentum for the components in the x and y directions separately. (b) Solve these equations for the speed, v_B', and angle, θ', of ball B. Do not assume the collision is elastic.

53. (II) An atomic nucleus of mass m traveling with speed v collides elastically with a target particle of mass $2m$ (initially at rest) and is scattered at 90°. (a) At what angle does the target particle move after the collision? (b) What are the final speeds of the two particles? (c) What fraction of the initial kinetic energy is transferred to the target particle?

54. (II) Two billiard balls of equal mass move at right angles and meet at the origin of an xy coordinate system. One is moving upward along the y axis at 2.0 m/s, and the other is moving to the right along the x axis with speed 3.7 m/s. After the collision (assumed elastic), the second ball is moving along the positive y axis (Fig. 9–42). What is the final direction of the first ball, and what are their two speeds?

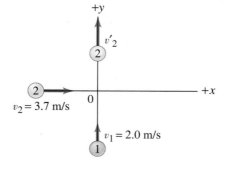

FIGURE 9–42 Problem 54. (Ball 1 after the collision is not shown.)

55. (II) A neutron collides elastically with a helium nucleus (at rest initially) whose mass is four times that of the neutron. The helium nucleus is observed to rebound at an angle $\theta_2' = 45°$. Determine the angle of the neutron, θ_1', and the speeds of the two particles, v_n' and v_{He}', after the collision. The neutron's initial speed is 6.2×10^5 m/s.

56. (III) A neon atom ($m = 20$ u) makes a perfectly elastic collision with another atom. After the impact, the neon atom travels away at a $55.6°$ angle from its original direction and the unknown atom travels away at a $50.0°$ angle. What is the mass (in u) of the unknown atom?

57. (III) Prove that in the elastic collision of two objects of identical mass, with one being a target initially at rest, the angle between their final velocity vectors is always $90°$.

58. (III) For an elastic collision between a projectile particle of mass m_1 and a target particle (at rest) of mass m_2, show that the scattering angle, θ_1', of the projectile (a) can take any value, 0 to $180°$, for $m_1 < m_2$, but (b) has a maximum angle ϕ given by $\cos^2\phi = 1 - (m_2/m_1)^2$ for $m_1 > m_2$.

Section 9–8

59. (I) The distance between a carbon atom ($m = 12$ u) and an oxygen atom ($m = 16$ u) in the CO molecule is 1.13×10^{-10} m. How far from the carbon atom is the center of mass of the molecule?

60. (I) An empty 1150-kg car has its CM 2.50 m behind the front of the car. How far from the front of the car will the CM be when two people sit in the front seat 2.80 m from the front of the car, and three people sit in the back seat 3.90 m from the front? Assume that each person has a mass of 70.0 kg.

61. (II) Find the center of mass of the ammonia molecule. The chemical formula is NH_3. The hydrogens are at the corners of an equilateral triangle (with sides 0.16 nm) that forms the base of a pyramid, with nitrogen at the apex (0.037 nm vertically above the plane of the triangle).

62. (II) Three cubes, of side l_0, $2l_0$, and $3l_0$, are placed next to one another (in contact) with their centers along a straight line and the $l = 2l_0$ cube in the center (Fig. 9–43). What is the position, along this line, of the CM of this system? Assume the cubes are made of the same uniform material.

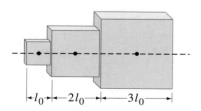

FIGURE 9–43 Problem 62.

63. (II) A square uniform raft, 18 m by 18 m, of mass 6200 kg, is used as a ferryboat. If three cars, each of mass 1200 kg, occupy the NE, SE, and SW corners, determine the CM of the loaded ferryboat.

64. (II) A uniform thin machine part is a flat circular plate of radius $2R$ that has a circular hole of radius R cut out of it. The center of the hole is a distance $0.80R$ from the center of the plate, Fig. 9–44. What is the position of the center of mass of the plate? [*Hint*: Think of a solid plate as being the sum of the given plate plus a small one of radius R.]

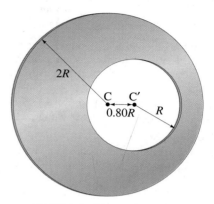

FIGURE 9–44 Problem 64.

65. (II) A uniform thin wire is bent into a semicircle of radius r. Determine the coordinates of its center of mass with respect to an origin of coordinates at the center of the "full" circle.

66. (III) Determine the CM of a thin, uniform, semicircular plate.

67. (III) Determine the CM of a machine part that is a uniform cone of height h and radius R, Fig. 9–45. [*Hint*: Divide the cone into an infinite number of disks of thickness dz, one of which is shown.]

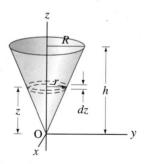

FIGURE 9–45 Problem 67.

68. (III) Determine the CM of a uniform pyramid that has four triangular faces and a square base with equal sides all of length s. [*Hint*: See Problem 67.]

Section 9–9

69. (I) The masses of the Earth and Moon are 5.98×10^{24} kg and 7.35×10^{22} kg, respectively, and their centers are separated by 3.84×10^8 m. (a) Where is the CM of this system located? (b) What can you say about the motion of the Earth-Moon system around the Sun, and of the Earth and Moon separately around the Sun?

70. (II) Two 35-kg masses have velocities (in m/s) of $\mathbf{v}_1 = 12\mathbf{i} - 16\mathbf{j}$ and $\mathbf{v}_2 = -20\mathbf{i} + 14\mathbf{j}$. Determine the velocity of the center of mass of the system.

71. (II) A 50-kg woman and a 70-kg man stand 11.0 m apart on frictionless ice. (a) How far from the man is their CM? (b) If they hold onto the two ends of a rope, and the man pulls on the rope so that he moves 2.8 m, how far from the woman will he now be? (c) How far will the man have moved when he collides with the woman?

72. (II) Suppose that in Example 9–16 (Fig. 9–32) $m_{II} = 3m_I$. (a) Where then would m_{II} land? (b) What if $m_I = 3m_{II}$?

73. (II) A helium balloon and its gondola, of mass M, are in the air and stationary with respect to the ground. A passenger, of mass m, then climbs out and slides down a rope with speed v, measured with respect to the balloon. With what speed and direction (relative to Earth) does the balloon then move? What happens if the passenger stops?

74. (II) Two people, one of mass 75 kg and the other of mass 55 kg, sit in a rowboat of mass 80 kg. With the boat initially at rest, the two people, who have been sitting at opposite ends of the boat, 3.0 m apart from each other, exchange seats. How far and in what direction will the boat move?

75. (III) A 200-kg flatcar 25 m long is moving with a speed of 5.0 m/s along horizontal frictionless rails. A 90-kg man then starts walking from one end of the car to the other in the direction of motion, with speed 2.0 m/s with respect to the car. In the time it takes for him to reach the other end, how far has the flatcar moved?

* Section 9–10

* 76. (II) A 2500-kg rocket is to be accelerated at 3.0 g at take-off from the Earth. If the gases can be ejected at a rate of 30 kg/s, what must be their exhaust speed?

* 77. (II) Suppose the conveyor belt of Example 9–17 is retarded by a friction force of 140 N. Determine the required output power (hp) of the motor as a function of time from the moment gravel first starts falling ($t = 0$) until 3.0 s after the gravel begins to be dumped off the end of the 20-m-long conveyor belt.

* 78. (II) A rocket traveling 1850 m/s away from the Earth at an altitude of 6400 km fires its rockets, which eject gas at a speed of 1200 m/s (relative to the rocket). If the mass of the rocket at this moment is 25,000 kg and an acceleration of 1.7 m/s^2 is desired, at what rate must the gases be ejected?

* 79. (II) The jet engine of an airplane takes in 100 kg of air per second, which is burned with 4.2 kg of fuel per second. The burned gases leave the plane at a speed of 550 m/s (relative to the plane). If the plane is traveling 270 m/s (600 mi/h), determine (a) the thrust due to ejected fuel; (b) the thrust due to accelerated air passing through the engine; and (c) the power (hp) delivered.

* 80. (III) A sled filled with sand slides without friction down a 30° slope. Sand leaks out a hole in the sled at a rate of 2.0 kg/s. If the sled starts from rest with an initial total mass of 40.0 kg, how long does it take the sled to travel 120 m along the slope?

General Problems

81. A novice pool player is faced with the corner pocket shot shown in Fig. 9–46. The relative size of some of the dimensions (the units aren't important, only their ratios) are also shown. Should the player be worried about this being a "scratch shot," one where the cue ball will also fall into a pocket? Give details.

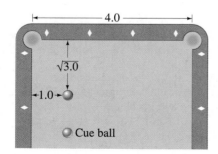

FIGURE 9–46 Problem 81.

82. During a Chicago storm, winds can whip horizontally at speeds of 100 km/h. If the air strikes a person at the rate of 40 kg/s per square meter and is brought to rest, calculate the force of the wind on a person. Assume the person's area to be 1.50 m high and 0.50 m wide. Compare to the typical maximum force of friction ($\mu \approx 1.0$) between the person and the ground, if the person has a mass of 70 kg.

83. A 0.145-kg pitched baseball moving horizontally at 35.0 m/s strikes a bat and is popped straight up to a height of 55.6 m before turning around. If the contact time is 0.50 ms, calculate the average force on the ball during the contact.

84. In order to convert a tough split in bowling, it is necessary to strike the pin a glancing blow as shown in Fig. 9–47. Assume that the bowling ball, initially traveling at 12.0 m/s, has five times the mass of a pin and that the pin goes off at 80° from the original direction of the ball. Calculate the speed (a) of the pin and (b) of the ball just after collision, and (c) calculate the angle through which the ball was deflected. Assume the collision is elastic and ignore any spin of the ball.

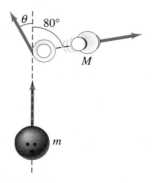

FIGURE 9–47 Problem 84.

85. A 15-g bullet strikes and becomes embedded in a 1.10-kg block of wood placed on a horizontal surface just in front of the gun. If the coefficient of kinetic friction between the block and the surface is 0.25, and the impact drives the block a distance of 9.5 m before it comes to rest, what was the muzzle speed of the bullet?

86. A gun is fired vertically into a 1.40-kg block of wood at rest on a thin horizontal sheet directly above it, Fig. 9–48. If the bullet has a mass of 21.0 g and a speed of 310 m/s, how high will the block rise into the air after the bullet becomes embedded in it?

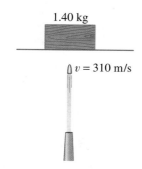

FIGURE 9–48 Problem 86.

87. A ball of mass m makes a head-on elastic collision with a second ball (at rest) and rebounds with a speed equal to 0.600 of its original speed. What is the mass of the second ball?

88. You have been hired as an expert witness in a court case involving an automobile accident. The accident involved a car of mass 2000 kg (car A) which approached a stationary car of mass 1000 kg (car B). The driver of car A applied his brakes 15 m before he crashed into car B. After the collision, car A slid 15 m while car B slid 30 m. The coefficient of kinetic friction between the locked wheels and the road was measured to be 0.60. Prove to the court that the driver of car A was exceeding the 55-mph speed limit before applying the brakes.

89. For the completely inelastic collision of two railroad cars that we considered in Example 9–3, calculate how much of the initial kinetic energy is transformed to thermal or other forms of energy.

90. A hockey puck of mass $4m$ has been rigged to explode, as part of a practical joke. Initially the puck is at rest on a frictionless ice rink. Then it bursts into three pieces. One chunk, of mass m, slides across the ice at speed v. Another chunk, of mass $2m$, slides across the ice at speed $2v$, in a direction at right angles to the direction of the first piece. From this information, find the velocity of the final chunk.

91. Two blocks of mass m_1 and m_2, resting on a frictionless table, are connected by a stretched spring and then released. (a) Is there a net external force on the system? (b) Determine the ratio of their velocities, v_1/v_2. (c) What is the ratio of their kinetic energies? (d) Describe the motion of the CM of this system. (e) How would the presence of friction alter the above results?

92. A (light) pallet has a load of identical cases of tomato paste (see Fig. 9–49), each of which is a cube of length l. Find the center of gravity in the horizontal plane, so that a crane operator can pick up the load without tipping it.

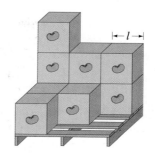

FIGURE 9–49 Problem 92.

93. A 5800-kg open railroad car coasts along with a constant speed of 8.60 m/s on a level track. Snow begins to fall vertically and fills the car at a rate of 3.50 kg/min. Ignoring friction with the tracks, determine the speed of the car after 60.0 min using momentum conservation.

*94. Consider the railroad car of Problem 93, which is slowly filling with snow. (a) Determine the speed of the car as a function of time using Eq. 9–19. (b) What is the speed of the car after 60.0 min. Does this agree with the simpler calculation (Problem 93) based on Section 9–2?

95. A meteor whose mass was about 10^8 kg struck the Earth $(m = 6.0 \times 10^{24}$ kg$)$ with a speed of about 15 km/s and came to rest in the Earth. (a) What was the Earth's recoil speed? (b) What fraction of the meteor's kinetic energy was transformed to kinetic energy of the Earth? (c) By how much did the Earth's kinetic energy change as a result of this collision?

96. A block of mass $m = 2.20$ kg slides down a 30.0° incline which is 3.60 m high. At the bottom, it strikes a block of mass $M = 7.00$ kg which is at rest on a horizontal surface, Fig. 9–50. (Assume a smooth transition at the bottom of the incline.) If the collision is elastic, and friction can be ignored, determine (a) the speeds of the two blocks after the collision, and (b) how far back up the incline the smaller mass will go.

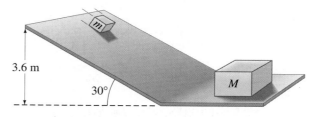

FIGURE 9–50 Problems 96 and 97.

97. In Problem 96 (Fig. 9–50), what is the upper limit on the mass m if it is to rebound from M, slide up the incline, stop, slide down the incline, and collide with M again?

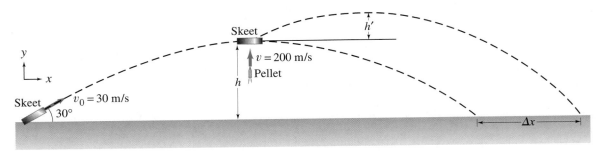

FIGURE 9–51   Problem 98.

98. A 0.25-kg skeet (clay target) is fired at an angle of 30° to the horizon with a speed of 30 m/s (Fig. 9–51). When it reaches the maximum height, it is hit from below by a 15-g pellet traveling vertically upward at a speed of 200 m/s. The pellet is embedded in the skeet. (*a*) How much higher did the skeet go up? (*b*) How much extra distance, Δx, does the skeet travel because of the collision?

99. The gravitational slingshot effect in Fig. 9–52, shows the planet Saturn moving in the negative x direction at its orbital speed (with respect to the Sun) of 9.6 km/s. The mass of Saturn is 5.69×10^{26} kg. A spacecraft with mass 825 kg approaches Saturn, moving initially in the $+x$ direction at 10.4 km/s. The gravitational attraction of Saturn (a conservative force) causes the spacecraft to swing around it (orbit shown as dashed line) and head off in the opposite direction. Estimate the final speed of the spacecraft after it is far enough away to be nearly free of Saturn's gravitational pull.

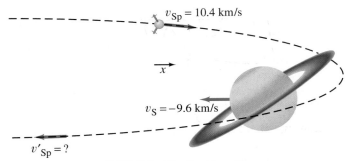

FIGURE 9–52   Problem 99.

100. A mallet consists of a uniform cylindrical head of mass 2.00 kg and a diameter 0.0800 m mounted on a uniform cylindrical handle of mass 0.500 kg and length 0.240 m, as shown in Fig. 9–53. If this mallet is tossed, spinning, into the air, how far above the bottom of the handle is the point that will follow a parabolic trajectory?

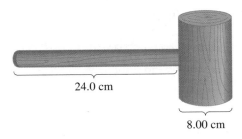

24.0 cm

8.00 cm

FIGURE 9–53   Problem 100.

101. A golf ball of mass 0.045 kg is hit off the tee at a speed of 50 m/s. The golf club was in contact with the ball for 5.0×10^{-3} s. Find (*a*) the impulse imparted to the golf ball, and (*b*) the average force exerted on the ball by the golf club.

102. The space shuttle launches an 800-kg satellite by ejecting it from the cargo bay. The ejection mechanism is activated and is in contact with the satellite for 4.0 s to give it a velocity of 0.30 m/s in the z-direction relative to the shuttle. The mass of the shuttle is 90,000 kg. (*a*) Determine the component of velocity v_f of the shuttle in the minus z-direction resulting from the ejection. (*b*) Find the average force that the shuttle exerts on the satellite during the ejection.

103. Car manufacturers conduct crash tests on new vehicles to ensure the safety of the passengers. In one such test a car of mass 1000 kg traveling at 50 km/h strikes a brick wall. (*a*) Is this an elastic or inelastic collision? (*b*) Assume that the front of the vehicle collapsed by 0.70 m. What is the impact time? (*c*) What is the average impulsive force?

104. A massless spring with spring constant k is placed between a block of mass m and a block of mass $3m$. Initially the blocks are at rest on a frictionless surface and they are held together so that the spring between them is compressed by an amount D from its equilibrium length. The blocks are then released and the spring pushes them off in opposite directions. Find the final speeds of the two blocks.

105. In a physics lab, a small cube slides down a frictionless incline as shown in Fig. 9–54, and elastically strikes a cube that is only one-half its mass. If the incline is 20 cm high and the table is 90 cm off the floor, where does each cube land?

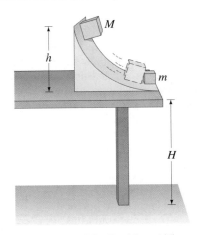

FIGURE 9–54   Problem 105.

When you want to go faster on a bicycle, as Lance Armstrong does in this 1999 Tour de France photo, you exert a strong force on the pedals. The speed of a bike depends on the rotation speed of the wheels and the pedals. The angular velocity of the rotating pedals increases not simply in proportion to the force applied, but rather in proportion to the net torque. Torque is the product of the force times the lever arm (= perpendicular distance from the axis of rotation to the line of action of the force). That is: the angular acceleration of a rotating body is proportional to the net applied torque. When an object is rotating, it has rotational kinetic energy and angular momentum. Angular momentum, and its conservation, play an interesting and often unexpected role in the real world, as we will see.

CHAPTER 10

Rotational Motion About a Fixed Axis

Until now, we have been concerned with translational motion. In this chapter and the next, we will deal with rotational motion. The material has been divided so that this Chapter 10 covers all the basic aspects of a rigid body rotating about a fixed axis. Chapter 11, on the other hand, presents a more general approach and some slightly more advanced (optional) topics.

Although we will sometimes deal with systems of many particles, we will mainly be concerned with rigid bodies. By a **rigid body** we mean a body with a definite shape that doesn't change, so that the particles that compose it stay in fixed positions relative to one another. Of course, any real body is capable of vibrating or deforming when a force is exerted on it. But these effects are often small, so the concept of an ideal rigid body is very useful as a good approximation.

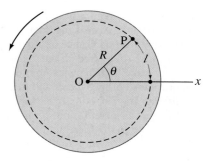

FIGURE 10–1 Looking down on a wheel that is rotating counterclockwise about an axis through the wheel's center at O (perpendicular to the page).

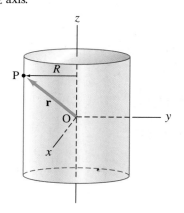

FIGURE 10–2 Showing the distinction between **r** (the position vector) and R (the distance from the rotation axis) for a point P on the edge of a cylinder rotating about the z axis.

The motion of a rigid body can be analyzed as the translational motion of its center of mass plus rotational motion about its center of mass. We have already discussed translational motion in detail, so now we focus our attention on purely rotational motion. By *purely rotational motion* we mean that all points in the body move in circles, such as the point P in the rotating wheel of Fig. 10–1, and that the centers of these circles all lie on a line called the **axis of rotation** (which in Fig. 10–1 is perpendicular to the page and passes through point O). We assume the axis is fixed in an inertial reference frame, but we will not necessarily insist that the axis pass through the center of mass.

10–1 Angular Quantities

For three-dimensional rigid bodies rotating about a fixed axis, we will find it useful to use the symbol R to represent the perpendicular distance of a point or particle from the axis of rotation. We do this to distinguish R from r, which will continue to represent the position of a particle with reference to the origin (a point) of some coordinate system.[†] This distinction is illustrated in Fig. 10–2. For a flat, very thin body, like a wheel, with the origin in the plane of the body (at the center of the wheel, for example), R and r will be the same.

Every point in a body rotating about a fixed axis moves in a circle (shown dashed in Fig. 10–1 for point P) whose center is on the axis and whose radius is R, the distance of that point from the axis of rotation. A straight line drawn from the axis to any point sweeps out the same angle θ in the same time.

To indicate the angular position of the body, or how far it has rotated, we specify the angle θ of some particular line in the body with respect to some reference line, such as the x axis (see Fig. 10–1). A point in the body (such as P in Fig. 10–1) moves through an angle θ when it travels the distance l measured along the circumference of its circular path. Angles are commonly stated in degrees, but the mathematics of circular motion is much simpler if we use the *radian* for angular measure. One **radian** (rad) is defined as the angle subtended by an arc whose length is equal to the radius. For example, in Fig. 10–1, point P is a distance R from the axis of rotation, and it has moved a distance l along the arc of a circle. The arc length l is said to "subtend" the angle θ. If $l = R$, then θ is exactly equal to 1 rad. In general, any angle θ is given by

$$\theta = \frac{l}{R}, \tag{10–1}$$

where R is the radius of the circle and l is the arc length subtended by the angle θ, which is specified in radians. Note that the radian is dimensionless (has no units) since it is the ratio of two lengths.

Radians can be related to degrees in the following way. In a complete circle there are 360°, which of course must correspond to an arc length equal to the circumference of the circle, $l = 2\pi r$. Thus $\theta = l/R = 2\pi R/R = 2\pi$ rad in a complete circle, so

$$360° = 2\pi \text{ rad}.$$

1 rad ≈ 57.3°   One radian is therefore $360°/2\pi \approx 360°/6.28 \approx 57.3°$.

[†] Some books use the symbol ρ instead of R. Since we use ρ to represent density (mass per unit volume), we avoid confusion by using R rather than ρ to represent the perpendicular distance of a point from an axis.

EXAMPLE 10–1 **Birds of prey—in radians.** A particular bird's eye can just distinguish objects that subtend an angle no smaller than about 3×10^{-4} rad. (a) How many degrees is this? (b) How small an object can the bird just distinguish when flying at a height of 100 m (Fig. 10–3a)?

SOLUTION (a) One radian is $360°/2\pi$, so 3×10^{-4} rad is

$$\left(3 \times 10^{-4}\,\text{rad}\right)\left(\frac{360°}{2\pi\,\text{rad}}\right) = 0.017°.$$

(b) From Eq. 10–1, $l = R\theta$. For small angles, the arc length and the chord length are approximately the same (Fig. 10–3b). Since $R = 100$ m and $\theta = 3 \times 10^{-4}$ rad, we find

$$l = (100\,\text{m})(3 \times 10^{-4}\,\text{rad}) = 3 \times 10^{-2}\,\text{m} = 3\,\text{cm}.$$

Had the angle been given in degrees, we would first have had to change it to radians to make this calculation.

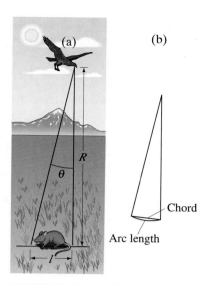

FIGURE 10–3 (a) Example 10–1. (b) For small angles, arc length and the chord length (straight line) are nearly equal. For an angle as large as 15°, the error in making this estimate is only 1 percent. For larger angles the error increases rapidly.

When an object, such as the bicycle wheel in Fig. 10–4, rotates from some initial position, specified by θ_1, to some final position, θ_2, its angular displacement is $\Delta\theta = \theta_2 - \theta_1$. The *angular velocity* (denoted by ω, the Greek lowercase letter omega) is defined in analogy with ordinary linear velocity. Instead of linear displacement, we use the angular displacement. Then the magnitude of the **average angular velocity** is defined as

$$\overline{\omega} = \frac{\Delta\theta}{\Delta t}, \tag{10–2a}$$

where $\Delta\theta$ is the angle through which the body has rotated in the time Δt. The magnitude of the **instantaneous angular velocity** is the limit of this ratio as Δt approaches zero:

$$\omega = \lim_{\Delta t \to 0} \frac{\Delta\theta}{\Delta t} = \frac{d\theta}{dt}. \tag{10–2b}$$

Angular velocity is generally specified in radians per second (rad/s). Note that *all points in a rigid body rotate with the same angular velocity*, since every position in the body moves through the same angle in the same time interval.

Angular acceleration (denoted by α, the Greek lowercase letter alpha), in analogy to ordinary linear acceleration, is defined as the change in angular velocity divided by the time required to make this change. The **average angular acceleration** is defined as

$$\overline{\alpha} = \frac{\omega_2 - \omega_1}{\Delta t} = \frac{\Delta\omega}{\Delta t}, \tag{10–3a}$$

where ω_1 is the angular velocity initially, and ω_2 is the angular velocity after a time Δt. **Instantaneous angular acceleration** is defined as the limit of this ratio as Δt approaches zero:

$$\alpha = \lim_{\Delta t \to 0} \frac{\Delta\omega}{\Delta t} = \frac{d\omega}{dt}. \tag{10–3b}$$

Since ω is the same for all points of a rotating body, Eqs. 10–3 tell us that α also will be the same for all points. Thus, ω and α are properties of the rotating body as a whole. With ω measured in radians per second and t in seconds, α will be expressed as radians per second squared (rad/s^2).

FIGURE 10–4 A wheel rotates from (a) initial position θ_1, to (b) final position θ_2. The angular displacement is $\Delta\theta = \theta_2 - \theta_1$.

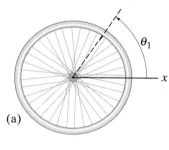

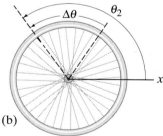

Angular acceleration (rad/s^2)

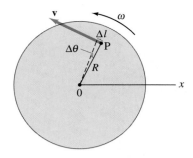

FIGURE 10–5 A particle P on a rotating wheel has a linear velocity **v** at any moment.

Linear and angular velocity related

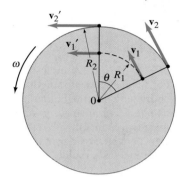

FIGURE 10–6 A wheel rotating uniformly counterclockwise. Two points on the wheel, at distances R_1 and R_2 from the center, have different linear velocities because they travel different distances in the same time interval. Since $R_2 > R_1$, then $v_2 > v_1$ ($v = R\omega$). But the two points have the same angular velocity ω because they travel through the same angle θ in the same time interval.

Centripetal (or radial) acceleration

FIGURE 10–7 On a rotating wheel whose rotation speed is increasing, a point P has both tangential and radial (centripetal) components of acceleration (See also Chapter 5.)

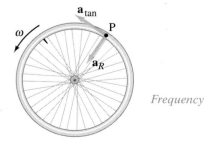

Frequency

Each particle or point of a rotating rigid body has, at any moment, a linear velocity v and a linear acceleration a. We can relate these linear quantities, v and a, of each particle, to the angular quantities, ω and α, of the rotating body as a whole. Consider a particle located a distance R from the axis of rotation, as in Fig. 10–5. If the body rotates with angular velocity ω, any particle will have a linear velocity whose direction is tangent to its circular path. The magnitude of its linear velocity, v, is $v = dl/dt$. From Eq. 10–1, a change in rotation angle $d\theta$ is related to the linear distance traveled by $dl = R\,d\theta$. Hence

$$v = \frac{dl}{dt} = R\frac{d\theta}{dt}$$

or

$$v = R\omega. \tag{10–4}$$

Thus, although ω is the same for every point in a rotating body at any instant, the linear velocity v is greater for points farther from the axis. See Fig. 10–6. Note that Eq. 10–4 is valid both instantaneously and on the average.

If a rotating body speeds up or slows down, so $\alpha \neq 0$, then a point in the object will have not only a (tangential) linear velocity but a tangential acceleration as well. We can use Eq. 10–4 to show that the angular acceleration α is related to the tangential linear acceleration a_{tan}, of a particle in the rotating body by

$$a_{\text{tan}} = \frac{\Delta v}{\Delta t} = R\frac{\Delta \omega}{\Delta t}$$

or

$$a_{\text{tan}} = R\alpha. \tag{10–5}$$

In this equation, R is the radius of the circle in which the particle is moving, and the subscript tan in a_{tan} stands for "tangential" since the acceleration considered here is along the circle (that is, tangent to it).

We saw in Chapter 3 that a particle moving in a circle of radius R with linear speed v has a radial or "centripetal" acceleration given by Eq. 3–14:

$$a_R = \frac{v^2}{R},$$

and this acceleration points toward the center of the circular path (Fig. 10–7). Using Eq. 10–4, we can write this radial acceleration as

$$a_R = \frac{v^2}{R} = \frac{(\omega R)^2}{R} = \omega^2 R. \tag{10–6}$$

Equation 10–6 applies to any particle of a rotating object. Thus the centripetal acceleration is greater the farther you are from the axis of rotation: the children farthest out on a merry-go-round feel the greatest acceleration. The total linear acceleration of a particle in a rotating object is the sum of the two components:

$$\mathbf{a} = \mathbf{a}_{\text{tan}} + \mathbf{a}_R.$$

We can relate the angular velocity ω to the frequency of rotation, f, where by **frequency** we mean the number of complete revolutions (rev) per second. One revolution (of a wheel, say) corresponds to an angle of 2π radians, and thus $1\,\text{rev/s} = 2\pi\,\text{rad/s}$. Hence, in general, the frequency f is related to the angular velocity ω by

$$f = \frac{\omega}{2\pi}$$

or

$$\omega = 2\pi f. \tag{10–7}$$

The unit for frequency, revolutions per second (rev/s), is given the special name, the hertz (Hz). That is

$$1\,\text{Hz} = 1\,\text{rev/s} = 1\,\text{s}^{-1}.$$

The hertz (Hz)

Note that "revolution" is not really a unit, so we can also write $1\,\text{Hz} = 1\,\text{s}^{-1}$.

The time required for one complete revolution is the **period**, T, and it is related to the frequency by

$$T = \frac{1}{f}. \qquad\qquad (10\text{--}8)$$

Period

For example, if a particle rotates at a frequency of three revolutions per second, then each revolution takes $\frac{1}{3}$ s.

EXAMPLE 10–2 **Hard drive.** The platter of the hard disk of a computer rotates at 5400 rpm (revolutions per minute). (*a*) What is the angular velocity of the disk? (*b*) If the reading head of the drive is located 3.0 cm from the rotation axis, what is the speed of the disk below it? (*c*) What is the linear acceleration of this point? (*d*) If a single bit requires 5.0 μm of length along the motion direction, how many bits per second can the writing head write when it is 3.0 cm from the axis? (*e*) If the disk took 3.6 s to spin up to 5400 rpm from rest, what was the average acceleration?

➡ **PHYSICS APPLIED**

*Hard drive
and bit speed*

SOLUTION (*a*) First we find the frequency:

$$f = \frac{(5400\,\text{rev/min})}{(60\,\text{s/min})} = 90\,\text{rev/s} = 90\,\text{Hz}.$$

Then the angular velocity is

$$\omega = 2\pi f = 570\,\text{rad/s}.$$

(*b*) The speed of a point 3.0 cm out from the axis is

$$v = R\omega = (3.0 \times 10^{-2}\,\text{m})(570\,\text{rad/s}) = 17\,\text{m/s}.$$

(*c*) The linear acceleration has two components, tangential and radial. Since $\omega = $ constant, then $\alpha = 0$, so $a_{\text{tan}} = r\alpha = 0$. The radial acceleration is

$$a_{\text{R}} = \omega^2 R = (570\,\text{rad/s})^2(0.030\,\text{m}) = 9700\,\text{m/s}^2$$

toward the axis.

(*d*) Each bit requires 5.0×10^{-6} m, so at a speed of 17 m/s, the number of bits passing the head per second is

$$\frac{17\,\text{m/s}}{5.0 \times 10^{-6}\,\text{m}} = 3.4 \times 10^6 \text{ bits per second.}$$

(*e*) The average angular acceleration was

$$\overline{\alpha} = \frac{\Delta\omega}{\Delta t} = \frac{570\,\text{rad/s} - 0}{3.6\,\text{s}} = 160\,\text{rad/s}^2.$$

Thus, on average, during each second the disk increased its angular speed by 160 rad/s or by $(160/2\pi) = 25$ rev/s.

10–2 | Kinematic Equations for Uniformly Accelerated Rotational Motion

In Chapter 2, we derived the important equations (2–12) that relate acceleration, velocity, and distance for the situation of uniform linear acceleration. Those equations were derived from the definitions of linear velocity and acceleration, assuming constant acceleration. The definitions of angular velocity and angular acceleration are the same as for their linear counterparts, except that θ has replaced the linear displacement x, ω has replaced v, and α has replaced a. Therefore, the angular equations for **constant angular acceleration** will be analogous to Eqs. 2–12 with x replaced by θ, v by ω, and a by α, and they can be derived in exactly the same way. We summarize them here, opposite their linear equivalents (we've chosen $x_0 = 0$, and $\theta_0 = 0$, for $t = 0$):

	Angular	Linear		
Uniformly	$\omega = \omega_0 + \alpha t$	$v = v_0 + at$	[constant α, a]	**(10–9a)**
accelerated	$\theta = \omega_0 t + \frac{1}{2}\alpha t^2$	$x = v_0 t + \frac{1}{2}at^2$	[constant α, a]	**(10–9b)**
rotational	$\omega^2 = \omega_0^2 + 2\alpha\theta$	$v^2 = v_0^2 + 2ax$	[constant α, a]	**(10–9c)**
motion	$\overline{\omega} = \dfrac{\omega + \omega_0}{2}$	$\overline{v} = \dfrac{v + v_0}{2}$	[constant α, a]	**(10–9d)**

Note that ω_0 represents the angular velocity at $t = 0$, whereas θ and ω represent the angular position and velocity, respectively, at time t. Since the angular acceleration is constant, $\alpha = \overline{\alpha}$. These equations are of course also valid for constant angular velocity, for which case $\alpha = 0$ and we have $\omega = \omega_0$, $\theta = \omega_0 t$ and $\overline{\omega} = \omega$.

EXAMPLE 10–3 **Hard drive again.** Through how many revolutions did the hard drive of Example 10–2 turn to reach 5400 rpm during its acceleration period? Assume constant angular acceleration.

SOLUTION We are given that $\omega_0 = 0$, $\omega = 570\ \text{rad/s}$, $\alpha = \overline{\alpha} = 160\ \text{rad/s}^2$, and $t = 3.6\ \text{s}$. We could use either Eq. 10–9b or 10–9c to find θ. The former gives

$$\theta = 0 + \tfrac{1}{2}(160\ \text{rad/s}^2)(3.6\ \text{s})^2 = 1.04 \times 10^3\ \text{rad}.$$

To find the total number of revolutions, we divide by 2π and obtain

$$\frac{1.04 \times 10^3\ \text{rad}}{2\pi\ \text{rad/rev}} = 165\ \text{rev}.$$

10–3 Rolling Motion (without slipping)

The rolling motion of a ball or wheel is familiar in everyday life: a ball rolling across the floor, or the wheels and tires of a car or bicycle rolling along the pavement. Rolling *without slipping* is readily analyzed and depends on static friction between the rolling object and the ground. The friction is static because the rolling object's point of contact with the ground is at rest at each moment. (Kinetic friction comes in if, for example, you brake too hard so the tires skid, or you accelerate so fast that you "burn rubber"—but these are more complicated situations.)

Rolling without slipping involves both rotation and translation. There is then a simple relation between the linear speed v of the axle and the angular velocity ω of the rotating wheel or sphere, namely $v = R\omega$ where R is the radius, as we now show. Figure 10–8a shows a wheel rolling to the right without slipping. At the moment shown, point P on the wheel is in contact with the ground and is momentarily at rest. The velocity of the axle at the wheel's center C is **v**. In Fig. 10–8b we have put ourselves in the reference frame of the wheel—that is, we are moving to the right with velocity **v** relative to the ground. In this reference frame the axle C is at rest, whereas the ground and point P are moving to the left with velocity $-\mathbf{v}$ as shown. Here we are seeing pure rotation. So we can use Eq. 10–4 to obtain $v = R\omega$ where R is the radius of the wheel. This is the same v as in Fig. 10–8a, so we see that the linear speed v of the axle relative to the ground is related to the angular velocity ω by $v = R\omega$. Note that this is valid only if there is no slipping.

FIGURE 10–8 (a) A wheel rolling to the right. Its center C moves with velocity **v**. (b) The same wheel as seen from a reference frame in which the axle of the wheel C is at rest—that is, we are moving to the right with velocity **v** relative to part (a). Point P, which was at rest in (a), here in (b) moves to the left with velocity $-\mathbf{v}$ as shown.

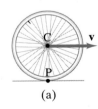

(a)

(b)

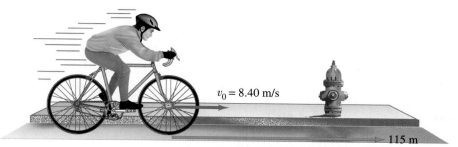

(a) Bike as seen from the ground ($t = 0$).

(b) From rider's reference frame, the ground is moving (to our left) at an initial speed of 8.40 m/s ($t = 0$).

FIGURE 10–9 Example 10–4.

EXAMPLE 10–4 **Bicycle.** A bicycle slows down uniformly from $v_0 = 8.40$ m/s to rest over a distance of 115 m, Fig. 10–9a. Each wheel and tire has an overall diameter of 68.0 cm. Determine (a) the angular velocity of the wheels at the initial instant ($t = 0$), (b) the total number of revolutions each wheel rotates before coming to rest, (c) the angular acceleration of the wheel, and (d) the time it took to come to a stop.

SOLUTION (a) Let us put ourselves in the reference frame of the bike—that is, as if we were riding the bike. Then the ground is going past us, initially, at a speed of 8.40 m/s, Fig. 10–9b. Since the tire is in contact with the ground at any moment, then a point on the rim of the tire (such as that touching the ground) moves at a speed of $v = 8.40$ m/s in this reference frame. Hence the angular velocity of the wheel is

$$\omega_0 = \frac{v_0}{R} = \frac{8.40 \text{ m/s}}{0.340 \text{ m}} = 24.7 \text{ rad/s}.$$

(b) In coming to a stop, 115 m of ground passes beneath the tire. Because the tire is in firm contact with the ground, any point on the edge of the rotating tire travels 115 m total. Each revolution corresponds to a distance of $2\pi R$, so the number of revolutions the wheel makes in coming to a stop is

$$\frac{115 \text{ m}}{2\pi R} = \frac{115 \text{ m}}{(2\pi)(0.340 \text{ m})} = 53.8 \text{ rev}.$$

(c) The angular acceleration of the wheel can be obtained from Eq. 10–9c:

$$\alpha = \frac{\omega - \omega_0^2}{2\theta} = \frac{0 - (24.7 \text{ rad/s})^2}{2(2\pi)(53.8 \text{ rev})} = -0.902 \text{ rad/s}^2,$$

where we have set $\theta = 2\pi$ rad/rev $\times$ 53.8 rev ($= 338$ rad) because each revolution corresponds to 2π radians. [Alternatively, we could have used Eq. 10–1 to get the total θ: $\theta = l/R = 115$ m/0.340 m $= 338$ rad.]

(d) Eq. 10–9a or b allows us to solve for the time. The first is easier:

$$t = \frac{\omega - \omega_0}{\alpha} = \frac{0 - 24.7 \text{ rad/s}}{-0.902 \text{ rad/s}^2} = 27.4 \text{ s}.$$

When a wheel rolls without slipping, the point of contact of the wheel with the ground is instantaneously at rest. It is sometimes useful to think of the motion of the wheel as pure rotation about this "instantaneous axis" passing through that point P (Fig. 10–10a). Points close to the ground have a small linear speed, as they are close to this instantaneous axis, whereas points farther away have a greater linear speed. This can be seen in a photograph of a real rolling wheel (Fig. 10–10b): spokes near the top of the wheel appear more blurry because they are moving faster than those near the bottom of the wheel.

FIGURE 10–10 (a) A rolling wheel rotates about the instantaneous axis (perpendicular to the page) passing through the point of contact with the ground, P. The arrows represent the instantaneous velocity of each point. (b) Photograph of a rolling wheel. The spokes are more blurred where the speed is greater.

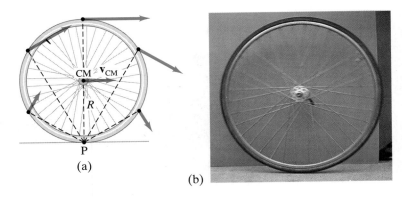

(a)

(b)

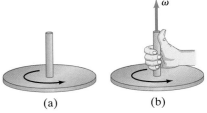

(a)   (b)

FIGURE 10–11 (a) Rotating wheel. (b) Right-hand rule for obtaining direction of **ω**.

Right-hand rule

10–4 | Vector Nature of Angular Quantities

Both **ω** and **α** can be treated as vectors, and we define their directions in the following way. Consider the rotating wheel shown in Fig. 10–11a. The linear velocities of different particles of the wheel point in all different directions. The only unique direction in space associated with the rotation is along the axis of rotation, perpendicular to the actual motion. We therefore choose the axis of rotation to be the direction of the angular velocity vector, **ω**. Actually, there is still an ambiguity since **ω** could point in either direction along the axis of rotation (up or down in Fig. 10–11a). The convention we use, called the **right-hand rule**, is the following: when the fingers of the right hand are curled around the rotation axis and point in the direction of the rotation, then the thumb points in the direction of **ω**. This is shown in Fig. 10–11b. Note that **ω** points in the direction a right-handed screw would move when turned in the direction of rotation. Thus, if the rotation of the wheel in Fig. 10–11a is counterclockwise, the direction of **ω** is upward as shown in Fig. 10–11b. If the wheel rotates clockwise, then **ω** points in the opposite direction, downward.[†] Note that no part of the rotating body moves in the direction of **ω**.

FIGURE 10–12 (a) Velocity is a true vector. The reflection of **v** points in the same direction. (b) Angular velocity is a pseudovector since it does not follow this rule. As can be seen, the reflection of the wheel rotates in the opposite direction, so the direction of *ω* is opposite for the reflection.

If the axis of rotation is fixed in direction, then **ω** can change only in magnitude. Thus **α** $= d\omega/dt$ must also point along the axis of rotation. If the rotation is counterclockwise as in Fig. 10–11a and the magnitude ω is increasing, then **α** points upward; but if ω is decreasing (the wheel is slowing down), **α** points downward.

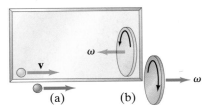

(a)   (b)

[†] Strictly speaking, **ω** and **α** are not quite vectors. The problem is that they do not behave like vectors under reflection. Suppose, as we are looking directly into a mirror, a particle moving with velocity **v** to the right passes in front of and parallel to the mirror. In the reflection of the mirror, **v** still points to the right, Fig. 10–12a. Thus a true vector, like velocity, when pointing parallel to the face of mirror has the same direction in the reflection as in actuality. Now consider a wheel rotating in front of the mirror, so **ω** points to the right. (We will be looking at the edge of the wheel.) As viewed in the mirror, Fig. 10–12b, the wheel will be rotating in the opposite direction. So **ω** will point in the opposite direction (to the left) in the mirror. Because of this difference under reflection between **ω** and true vectors, **ω** is called a *pseudovector* or *axial vector*. The angular acceleration **α** is also a pseudovector, as are all cross products of true vectors (Section 11–1). The difference between true vectors and pseudovectors will not concern us in this book.

If the rotation is clockwise, $\boldsymbol{\alpha}$ will point downward if ω is increasing, and point upward if ω is decreasing. In other words, if ω is increasing, $\boldsymbol{\alpha}$ points in the same direction as $\boldsymbol{\omega}$, whereas if ω is decreasing, $\boldsymbol{\alpha}$ points in the opposite direction.

Since $\boldsymbol{\omega}$ always points along the axis of rotation, if the axis of rotation changes direction, $\boldsymbol{\omega}$ changes direction. In this case $\boldsymbol{\alpha}$ will not point along the axis of rotation. We will see examples of this in Chapter 11, but in this chapter we consider only motion about a fixed axis, so $\boldsymbol{\omega}$ and $\boldsymbol{\alpha}$ are both along the rotation axis.

10–5 | Torque

We have so far discussed rotational kinematics—the description of rotational motion in terms of angle, angular velocity, and angular acceleration. Now we discuss the dynamics, or causes, of rotational motion. Just as we found analogies between linear and rotational motion for the description of motion, so rotational equivalents for dynamics exist as well.

To make an object start rotating about an axis clearly requires a force. But the direction of this force, and where it is applied, are also important. Take, for example, an ordinary situation such as the door in Fig. 10–13. If you apply a force $\mathbf{F}_1$ to the door as shown, you will find that the greater the magnitude, F_1, the more quickly the door opens. But now if you apply the same magnitude force at a point closer to the hinge, say $\mathbf{F}_2$ in Fig. 10–13, you will find that the door will not open so quickly. The effect of the force is less. Indeed, if only this one force acts, it is found that the angular acceleration of the door is proportional not only to the magnitude of the force, but is also directly proportional to *the perpendicular distance from the axis of rotation to the line along which the force acts*. This distance is called the **lever arm**, or **moment arm**, of the force, and is labeled R_1 and R_2 for the two forces in Fig. 10–13. Thus, if R_1 in Fig. 10–13 is three times larger than R_2, then the angular acceleration of the door will be three times as great, assuming that the magnitudes of the forces are the same. To say it another way, if $R_1 = 3R_2$, then F_2 must be three times as large as F_1 to give the same angular acceleration. (Figure 10–14 shows two examples of tools whose long lever arms are very effective.)

The angular acceleration, then, is proportional to the product of the *force times the lever arm*. This product is called the *moment of the force* about the axis, or, more commonly, it is called the **torque**, and is abbreviated τ (Greek lowercase letter tau). Thus, the angular acceleration α of an object is directly proportional to the net applied torque, τ:

$$\alpha \propto \tau,$$

and we see that it is torque that gives rise to angular acceleration. This is the rotational analog of Newton's second law for linear motion, $a \propto F$. (In Section 10–6, we will see what factor is needed to make this proportionality an equation.)

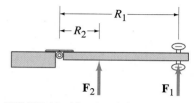

FIGURE 10–13 Applying the same force with different lever arms, R_1 and R_2.

Lever arm

Torque defined

(a)   (b)

FIGURE 10–14 (a) A plumber can exert greater torque using a wrench with a long lever arm. (b) A tire iron too can have a long lever arm.

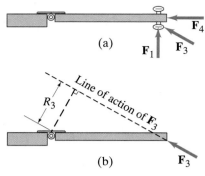

(a)

(b)

FIGURE 10–15 (a) Forces acting at different angles at the doorknob. (b) The lever arm is defined as the perpendicular distance from the axis of rotation (the hinge) to the line of action of the force.

FIGURE 10–16
Torque $= R_\perp F = RF_\perp$.

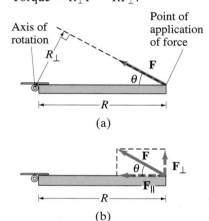

(a)

(b)

Magnitude of a torque

Units for torque: m · N

We defined the lever arm as the *perpendicular* distance of the axis of rotation from the line of action of the force—that is, the distance which is perpendicular both to the axis of rotation and to an imaginary line drawn along the direction of the force. We do this to take into account the effect of forces acting at an angle. It is clear that a force applied at an angle, such as $\mathbf{F}_3$ in Fig. 10–15a, will be less effective than the same magnitude force applied straight on, such as $\mathbf{F}_1$ (Fig. 10–15a). And if you push on the end of the door so that the force is directed at the hinge (the axis of rotation), as indicated by $\mathbf{F}_4$, the door will not rotate at all.

The lever arm for a force such as $\mathbf{F}_3$ is found by drawing a line along the direction of $\mathbf{F}_3$ (this is the "line of action" of $\mathbf{F}_3$), and then drawing another line (perpendicular to the rotation axis) from the axis of rotation perpendicular to this "line of action". The length of this second line is the lever arm for $\mathbf{F}_3$ and is labeled R_3 in Fig. 10–15b.

The torque associated with $\mathbf{F}_3$ is then $R_3 F_3$. This short lever arm and the corresponding smaller torque associated with $\mathbf{F}_3$ is consistent with the observation that $\mathbf{F}_3$ is less effective in accelerating the door than is $\mathbf{F}_1$. When the lever arm is defined in this way, experiment shows that the relation $\alpha \propto \tau$ is valid in general. Notice in Fig. 10–15 that the line of action of the force $\mathbf{F}_4$ passes through the hinge and hence its lever arm is zero. Consequently, zero torque is associated with $\mathbf{F}_4$ and it gives rise to no angular acceleration, in accord with everyday experience.

In general, then, we can write the torque about a given axis as

$$\tau = R_\perp F, \tag{10–10a}$$

where $R_\perp$ is the lever arm, and the perpendicular symbol ($\perp$) reminds us that we must use the distance from the axis of rotation that is perpendicular to the line of action of the force (Fig. 10–16a). An alternate but equivalent way of determining the torque associated with a force is to resolve the force into components parallel and perpendicular to the line joining the point of application of the force to the axis, as shown in Fig. 10–16b. Then the torque will be equal to $F_\perp$ times the distance R from the axis to the point of application of the force:

$$\tau = RF_\perp. \tag{10–10b}$$

This gives the same result as Eq. 10–10a since $F_\perp = F \sin \theta$ and $R_\perp = R \sin \theta$. So

$$\tau = RF \sin \theta \tag{10–10c}$$

in either case. We can use any of Eqs. 10–10 to calculate the torque, which ever is easiest.[†]

Since torque is a distance times a force, it is measured in units of m · N in SI units,[‡] cm · dyne in the cgs system, and ft · lb in the English system.

When more than one torque acts on a body, the acceleration α is found to be proportional to the *net* torque. If all the torques acting on a body tend to rotate it in the same direction, the net torque is the sum of the torques. But if, say, one torque acts to rotate a body in one direction, and a second torque acts to rotate the body in the opposite direction (as in Fig. 10–17), the net torque is the difference of the two torques. We assign a positive sign to torques that act to rotate the body in one direction (say counterclockwise) and a negative sign to torques that act to rotate the body in the opposite direction (clockwise).

[†] We will see in Chapter 11 that Eq. 10–10c gives the magnitude of the torque vector, $\boldsymbol{\tau} = \mathbf{r} \times \mathbf{F}$.

[‡] Note that the units for torque are the same as those for energy. We write the unit for torque here as m · N (in SI) to help distinguish it from energy (N · m) because the two quantities are very different. An obvious difference is that energy is a scalar, whereas torque has a direction and is a vector. The special name *joule* (1 J = 1 N · m) is used only for energy (and for work), *never* for torque.

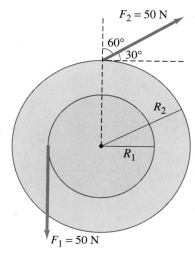

EXAMPLE 10–5 **Torque on a compound wheel.** Two thin cylindrical wheels, of radii $R_1 = 30$ cm and $R_2 = 50$ cm, are attached to each other on an axle that passes through the center of each, as shown in Fig. 10–17. Calculate the net torque on the two-wheel system due to the two forces shown, each of magnitude 50 N.

SOLUTION The force $\mathbf{F}_1$ acts to rotate the system counterclockwise, whereas $\mathbf{F}_2$ acts to rotate it clockwise. So the two forces act in opposition to each other. We must choose one direction of rotation to be positive—say, counterclockwise. Then $\mathbf{F}_1$ exerts a positive torque, $\tau_1 = R_1 F_1$, since the lever arm is R_1. On the other hand, $\mathbf{F}_2$ produces a negative (clockwise) torque and does not act perpendicular to R_2, so we must use its perpendicular component to calculate the torque it produces: $\tau_2 = -R_2 F_{2\perp} = -R_2 F_2 \sin\theta$, where $\theta = 60°$. (Note that θ must be the angle between $\mathbf{F}_2$ and a radial line from the rotation axis.) Hence the net torque is

$$\tau = R_1 F_1 - R_2 F_2 \sin 60°$$
$$= (0.30\,\text{m})(50\,\text{N}) - (0.50\,\text{m})(50\,\text{N})(0.866) = -6.7\,\text{m}\cdot\text{N}.$$

This net torque acts to accelerate the rotation of the wheel in the clockwise direction. Note that the two forces have equal magnitude, yet produce a net torque because their lever arms are different.

FIGURE 10–17 Example 10–5. The torque due to $\mathbf{F}_1$ tends to accelerate the wheel counterclockwise, whereas the torque due to $\mathbf{F}_2$ tends to accelerate the wheel clockwise.

10–6 Rotational Dynamics; Torque and Rotational Inertia

We have discussed that the angular acceleration α of a rotating body is proportional to the net torque τ applied to it:

$$\alpha \propto \Sigma\tau$$

where we write $\Sigma\tau$ to remind us that it is the net torque (sum of all torques acting on the body) that is proportional to α. This corresponds to Newton's second law for translational motion, $a \propto \Sigma F$, where torque has taken the place of force, and, correspondingly, the angular acceleration α takes the place of the linear acceleration a. In the linear case, the acceleration is not only proportional to the net force, but it is also inversely proportional to the inertia of the body, which we call its mass, m. Thus we could write $a = \Sigma F/m$. But what plays the role of mass for the rotational case? That is what we now set out to determine. At the same time, we will see that the relation $\alpha \propto \Sigma\tau$ follows directly from Newton's second law, $\Sigma F = ma$.

We first consider a very simple case: a particle of mass m rotating in a circle of radius r at the end of a string or rod whose mass we can ignore (Fig. 10–18), and we assume a single force F acts on it as shown. The torque that gives rise to the angular acceleration is $\tau = RF$. If we make use of Newton's second law for linear quantities, $\Sigma F = ma$, and Eq. 10–5 relating the angular acceleration to the tangential linear acceleration, $a_{\text{tan}} = R\alpha$, we have

$$F = ma$$
$$= mR\alpha.$$

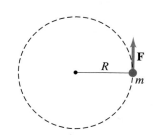

FIGURE 10–18 A mass m rotating in a circle of radius R about a fixed point.

When we multiply both sides by R, the torque $\tau = RF$ is given by

$$\tau = mR^2\alpha. \qquad\qquad \text{[single particle]} \quad (10\text{–}11)$$

Here we have the direct relation between the angular acceleration and the applied torque τ. The quantity mR^2 represents the *rotational inertia* of the particle and is called its *moment of inertia*.

Now let us consider a rotating rigid body, such as a wheel rotating about an axis through its center, such as an axle. We can think of the wheel as consisting of many particles located at various distances from the axis of rotation. We can apply Eq. 10–11 to each particle of the body; that is, we write $\tau_i = m_i R_i^2 \alpha$ for the ith particle of the body. Then we sum over all the particles. The sum of the various torques is just the total torque, $\Sigma\tau$, so we obtain:

$$\Sigma\tau_i = \left(\Sigma m_i R_i^2\right)\alpha \qquad \text{[axis fixed]} \quad \textbf{(10–12)}$$

where we factored out the α since it is the same for all the particles of the body. The resultant torque, $\Sigma\tau$, represents the sum of all internal torques that each particle exerts on another, plus all external torques applied from the outside: $\Sigma\tau = \Sigma\tau_{\text{ext}} + \Sigma\tau_{\text{int}}$. Now the sum of the internal torques is zero from Newton's third law.[†] Hence $\Sigma\tau$ represents the resultant *external* torque.

The sum $\Sigma m_i R_i^2$ in Eq. 10–12 represents the sum of the masses of each particle in the body multiplied by the square of the distance of that particle from the axis of rotation. If we give each particle a number (1, 2, 3, ···), then $\Sigma m_i R_i^2 = m_1 R_1^2 + m_2 R_2^2 + m_3 R_3^2 + \cdots$. This quantity is called the **moment of inertia** (or *rotational inertia*) of the body, I:

$$I = \Sigma m_i R_i^2 = m_1 R_1^2 + m_2 R_2^2 + \cdots. \qquad \textbf{(10–13)}$$

Combining Eqs. 10–12 and 10–13, we can write

$$\Sigma\tau = I\alpha. \qquad \begin{bmatrix} \text{axis fixed in} \\ \text{inertial reference frame} \end{bmatrix} \quad \textbf{(10–14)}$$

This is the rotational equivalent of Newton's second law. It is valid for the rotation of a rigid body about a fixed axis.[‡] It can be shown (see Chapter 11) that Eq. 10–14 is also valid when the body is translating with acceleration, as long as I and α are calculated about the center of mass of the body, and the rotation axis through the CM doesn't change direction. (A ball rolling straight down a ramp is an example.) Then

$$(\Sigma\tau)_{\text{CM}} = I_{\text{CM}}\alpha_{\text{CM}}, \qquad \begin{bmatrix} \text{axis fixed in direction,} \\ \text{but may accelerate} \end{bmatrix} \quad \textbf{(10–15)}$$

where the subscript CM means "calculated about the center of mass".

We see that the moment of inertia, I, which is a measure of the rotational inertia of a body, plays the same role for rotational motion that mass does for translational motion. As can be seen from Eq. 10–13, the rotational inertia of an object depends not only on its mass, but also on how that mass is distributed with respect to the axis. For example, a large-diameter cylinder will have greater rotational inertia than one of equal mass but smaller diameter (and therefore greater length), Fig. 10–19. The former will be harder to start rotating, and harder to stop. When the mass is concentrated farther from the axis of rotation, the rotational inertia is greater. For rotational motion, the mass of a body *cannot* be considered as concentrated at its center of mass.

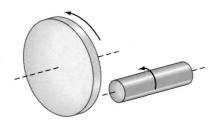

FIGURE 10–19 A large-diameter wheel has greater rotational inertia than one of smaller diameter but equal mass.

Careful:
Mass can not be considered concentrated at CM for rotational motion

10–7 Solving Problems in Rotational Dynamics

Whenever dealing with torque and angular acceleration (Eq. 10–14), it is important to use a consistent set of units, which in SI is: α in rad/s²; τ in m·N; and the moment of inertia, I, in kg·m².

[†] This depends on the so-called "strong" form of Newton's third law in which not only is the force one particle exerts on a second equal and opposite to the force the second exerts on the first, but these two forces act along the same line.

[‡] That is, the axis is fixed relative to the body and is fixed in an inertial reference frame. This includes an axis moving at uniform velocity in an inertial frame, since the axis can be considered fixed in a second inertial frame that moves with respect to the first.

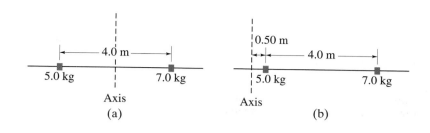

FIGURE 10–20 Example 10–6: calculating the moment of inertia.

EXAMPLE 10–6 **Two weights on a bar: different axis, different I.** Two small "weights" of mass 5.0 kg and 7.0 kg are mounted 4.0 m apart on a light rod (whose mass can be ignored), as shown in Fig. 10–20. Calculate the moment of inertia of the system (a) when rotated about an axis halfway between the weights, Fig. 10–20a, and (b) when the system rotates about an axis 0.50 m to the left of the 5.0-kg mass (Fig. 10–20b).

SOLUTION (a) Both weights are the same distance, 2.0 m, from the axis of rotation. Thus

$$I = \Sigma mR^2 = (5.0\,\text{kg})(2.0\,\text{m})^2 + (7.0\,\text{kg})(2.0\,\text{m})^2$$
$$= 20\,\text{kg}\cdot\text{m}^2 + 28\,\text{kg}\cdot\text{m}^2 = 48\,\text{kg}\cdot\text{m}^2.$$

(b) The 5.0-kg mass is now 0.50 m from the axis and the 7.0-kg mass is 4.50 m from the axis. Then

$$I = \Sigma mR^2 = (5.0\,\text{kg})(0.50\,\text{m})^2 + (7.0\,\text{kg})(4.5\,\text{m})^2$$
$$= 1.3\,\text{kg}\cdot\text{m}^2 + 142\,\text{kg}\cdot\text{m}^2 = 143\,\text{kg}\cdot\text{m}^2.$$

The above Example illustrates two important points. First, the moment of inertia of a given system is different for different axes of rotation. Second, we see in part (b) that mass close to the axis of rotation contributes little to the total moment of inertia; the 5.0-kg object contributed less than 1 percent to the total in this case.

I depends on axis of rotation and on distribution of mass

For most ordinary bodies, the mass is distributed continuously, and the calculation of the moment of inertia, ΣmR^2, can be difficult. Expressions for the moments of inertia of regularly shaped bodies in terms of their dimensions can, however, be

---

PROBLEM SOLVING   **Rotational Motion (Axis fixed)**

1. As always, draw a clear and complete diagram.

2. Draw a *free-body diagram* for the body under consideration (or for each body if more than one), showing only (and all) the forces acting on that body and exactly where they act, so you can determine the torque due to each. Gravity acts at the CG of the body.

3. Identify the axis of rotation and calculate the torques about it. Choose positive and negative directions of rotation (counter-clockwise and clockwise), and assign the correct sign to each torque.

4. Apply Newton's second law for rotation, $\Sigma\tau = I\alpha$. If the moment of inertia is not given, and it is not the unknown sought, you need to determine it first. Use consistent units, which in SI are: α in rad/s^2; τ in m·N; and I in kg·m^2.

5. Also apply Newton's second law for translation, $\Sigma F = m\mathbf{a}$, if needed.

6. Solve the resulting equation(s) for the unknown(s).

7. As always, do a rough estimate to determine if your answer is reasonable: does it make sense?

Object	Location of axis	Moment of inertia
(a) Thin hoop of radius R_0	Through center	MR_0^2
(b) Thin hoop of radius R_0 and width w	Through central diameter	$\frac{1}{2}MR_0^2 + \frac{1}{12}Mw^2$
(c) Solid cylinder of radius R_0	Through center	$\frac{1}{2}MR_0^2$
(d) Hollow cylinder of inner radius R_1 and outer radius R_2	Through center	$\frac{1}{2}M(R_1^2 + R_2^2)$
(e) Uniform sphere of radius r_0	Through center	$\frac{2}{5}Mr_0^2$
(f) Long uniform rod of length l	Through center	$\frac{1}{12}Ml^2$
(g) Long uniform rod of length l	Through end	$\frac{1}{3}Ml^2$
(h) Rectangular thin plate, of length l and width w	Through center	$\frac{1}{12}M(l^2 + w^2)$

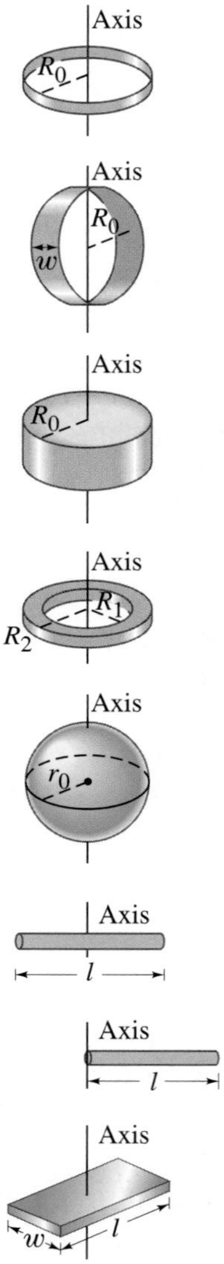

FIGURE 10–21
Moments of inertia for various objects of uniform composition.

worked out (using calculus) as discussed in the next Section. Figure 10–21 gives these expressions for a number of solids rotated about the axes specified. The only one for which the result is obvious is that for the thin hoop or ring of radius R_0 rotated about an axis passing through its center perpendicular to the plane of the hoop (Fig. 10–21a). For this object, all the mass is concentrated at the same distance from the axis, R_0. Thus $\Sigma mR^2 = (\Sigma m)R_0^2 = MR_0^2$, where M is the total mass of the hoop. When calculation is difficult, I can be determined experimentally by measuring the angular acceleration α about a fixed axis due to a known net torque, $\Sigma\tau$, and applying Eq. 10–14, $I = \Sigma\tau/\alpha$.

EXAMPLE 10–7 **A heavy pulley.** A 15.0-N force (represented by $\mathbf{F}_T$) is applied to a cord wrapped around a pulley of mass $M = 4.00\,kg$ and radius $R_0 = 33.0\,cm$, Fig. 10–22. The pulley is observed to accelerate uniformly from rest to reach an angular speed of 30.0 rad/s in 3.00 s. If there is a frictional torque (at the axle), $\tau_{fr} = 1.10\,m \cdot N$, determine the moment of inertia of the pulley. The pulley is assumed to rotate about its center.

SOLUTION The free-body diagram for the pulley is shown in Fig. 10–22, although the friction force is not shown since we are given only its torque. We can calculate the moment of inertia from Eq. 10–14, $\Sigma\tau = I\alpha$, since from the mea-

FIGURE 10–22 Example 10–7.

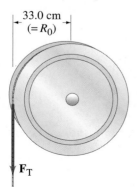

33.0 cm
$(= R_0)$

$\mathbf{F}_T$

surements given we can determine $\Sigma\tau$ and α. The cord leaves the wheel's edge perpendicular to the radius, so the angle between the force $\mathbf{F}_T$ and its lever arm is $90°$. The net torque is the applied torque due to $\mathbf{F}_T$ minus the frictional torque; we take positive to be counterclockwise:

$$\Sigma\tau = (0.330\,\text{m})(15.0\,\text{N}) - 1.10\,\text{m}\cdot\text{N} = 3.85\,\text{m}\cdot\text{N}.$$

The angular acceleration is

$$\alpha = \frac{\Delta\omega}{\Delta t} = \frac{30.0\,\text{rad/s} - 0}{3.00\,\text{s}} = 10.0\,\text{rad/s}^2.$$

Hence

$$I = \Sigma\tau/\alpha = (3.85\,\text{m}\cdot\text{N})/(10.0\,\text{rads/s}^2) = 0.385\,\text{kg}\cdot\text{m}^2.$$

EXAMPLE 10–8 **Pulley and bucket.** Consider again the pulley in Fig. 10–22. But this time, suppose that, instead of a constant 15.0-N force being exerted on the cord, we now have a bucket of weight 15.0 N (mass $m = 1.53\,\text{kg}$) hanging from the cord, which we assume not to stretch or slip on the pulley. See Fig. 10–23a. (*a*) Calculate the angular acceleration α of the pulley and the linear acceleration a of the bucket. (*b*) Determine the angular velocity ω of the pulley and the linear velocity v of the bucket at $t = 3.00\,\text{s}$ if the pulley (and bucket) start from rest at $t = 0$.

SOLUTION (*a*) Let F_T be the tension in the cord. Then a force F_T acts at the edge of the pulley, and we use Eq. 10–14 for the rotation of the pulley:

$$I\alpha = \Sigma\tau = F_T R_0 - \tau_{\text{fr}}. \qquad\text{[pulley]}$$

Next we look at the (linear) motion of the bucket of mass m. Figure 10–23b shows a free-body diagram for the bucket. Two forces act on the bucket: the force of gravity mg acts downward, and the tension of the cord F_T pulls upward. So by $\Sigma F = ma$ for the bucket we have (taking downward as positive):

$$mg - F_T = ma. \qquad\text{[bucket]}$$

Note that the tension F_T, which is the force exerted on the edge of the pulley, is *not* equal to the weight of the bucket ($= mg = 15.0\,\text{N}$). There must be a net force (hence $F_T < mg$) if the bucket is accelerating. Indeed, by the last equation above, $F_T = mg - ma$. To obtain α, we note that the tangential acceleration of a point on the edge of the pulley is the same as the acceleration of the bucket if the cord doesn't stretch or slip. Hence we can use Eq. 10–5, $a_{\text{tan}} = a = R_0\alpha$. Substituting $F_T = mg - ma$ into the first equation above, we obtain

$$I\alpha = \Sigma\tau = F_T R_0 - \tau_{\text{fr}} = (mg - mR_0\alpha)R_0 - \tau_{\text{fr}} = mgR_0 - mR_0^2\alpha - \tau_{\text{fr}}.$$

α appears in the middle term on the right, so we bring that term to the left side and solve for α:

$$\alpha = \frac{mgR_0 - \tau_{\text{fr}}}{I + mR_0^2}.$$

Then, since $I = 0.385\,\text{kg}\cdot\text{m}^2$ and $\tau_{\text{fr}} = 1.10\,\text{m}\cdot\text{N}$ (Example 10–7),

$$\alpha = \frac{(15.0\,\text{N})(0.330\,\text{m}) - 1.10\,\text{m}\cdot\text{N}}{0.385\,\text{kg}\cdot\text{m}^2 + (1.53\,\text{kg})(0.330\,\text{m})^2} = 6.98\,\text{rad/s}^2.$$

The angular acceleration is somewhat less in this case than the 10.0 rad/s² of Example 10–7. Why? Because $F_T(= mg - ma)$ is somewhat less than the weight of the bucket, mg. The linear acceleration of the bucket is

$$a = R_0\alpha = (0.330\,\text{m})(6.98\,\text{rad/s}^2) = 2.30\,\text{m/s}^2.$$

(*b*) Since the angular acceleration is constant, after 3.00 s

$$\omega = \omega_0 + \alpha t = 0 + (6.98\,\text{rad/s}^2)(3.00\,\text{s}) = 20.9\,\text{rad/s}.$$

The velocity of the bucket is the same as that of a point on the wheel's edge:

$$v = R_0\omega = (0.330\,\text{m})(20.9\,\text{rad/s}) = 6.91\,\text{m/s}.$$

The same result can also be obtained by using the linear equation $v = v_0 + at = 0 + (2.30\,\text{m/s}^2)(3.00\,\text{s}) = 6.90\,\text{m/s}$. (The difference is due to rounding off.)

FIGURE 10–23 Example 10–8. With free-body diagram for the falling bucket of mass m shown in (b).

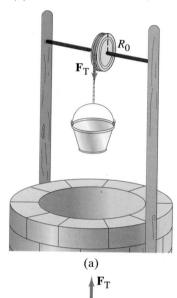

(a)

(b)

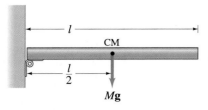

FIGURE 10–24 Example 10–9.

EXAMPLE 10–9 Rotating rod. A uniform rod of mass M and length l can pivot freely (i.e., we ignore friction) about a hinge or pin attached to the case of a large machine, as in Fig. 10–24. The rod is held horizontally and then released. At the moment of release, determine (a) the angular acceleration of the rod and (b) the linear acceleration of the tip of the rod. Assume the force of gravity acts at the center of mass of the rod, as shown.

SOLUTION (a) The only torque on the rod is that due to gravity, which acts with a force $F = Mg$ downward with a lever arm (at the moment of release) of $l/2$, since the CM is at the center of a uniform rod. (There is also a force on the rod at the pin, but with the pin as axis of rotation, the lever arm of this force is zero.) The moment of inertia of a uniform rod pivoted about its end is (Fig. 10–21g) $I = \frac{1}{3}Ml^2$. Thus, from Eq. 10–14,

$$\alpha = \frac{\tau}{I} = \frac{Mg\dfrac{l}{2}}{\frac{1}{3}Ml^2} = \frac{3}{2}\frac{g}{l}$$

is the initial angular acceleration of the rod. As the rod descends, the force of gravity on it is constant but the torque due to this force is not constant; hence the rod's angular acceleration is not constant.

(b) The linear acceleration of the tip of the rod is found from the relation $a_{\text{tan}} = R\alpha$ (Eq. 10–5) with $R = l$:

$$a_{\text{tan}} = l\alpha = \tfrac{3}{2}g.$$

Thus the tip of the rod falls with an acceleration greater than g! A small object balanced on the tip of the rod would be left behind when the rod is released. In contrast, the CM of the rod, at a distance $l/2$ from the pivot, has acceleration $a_{\text{tan}} = (l/2)\alpha = \tfrac{3}{4}g$.

10–8 Determining Moments of Inertia

By Experiment

The moments of inertia of any body about any axis can be determined experimentally by measuring the net torque $\Sigma\tau$ required to give the body an angular acceleration α. Then, from Eq. 10–14, $I = \Sigma\tau/\alpha$. See Example 10–7.

Using Calculus

For many bodies, or systems of particles, the moment of inertia can be calculated directly, as in Example 10–6. Many bodies can be considered as a continuous distribution of mass. In this case, Eq. 10–13 defining moment of inertia becomes

$$I = \int R^2 \, dm, \tag{10–16}$$

where dm represents the mass of any infinitesimal particle of the body and R is the perpendicular distance of this particle from the axis of rotation. The integral is taken over the whole body. This is easily done only for bodies of simple geometric shape.

EXAMPLE 10–10 **Hollow cylinder.** Show that the moment of inertia of a uniform hollow cylinder of inner radius R_1, outer radius R_2, and mass M, is $I = \frac{1}{2}M(R_1^2 + R_2^2)$, as stated in Fig. 10–21d, if the rotation axis is through the center along the axis of symmetry.

SOLUTION We know that the moment of inertia of a thin ring of radius R is mR^2. So we divide the cylinder into thin concentric cylindrical rings or hoops of thickness dR, one of which is indicated in Fig. 10–25. If the density (mass per unit volume) is ρ, then

$$dm = \rho\, dV,$$

where dV is the volume of the thin ring of radius R, thickness dR, and height h. Since $dV = (2\pi R)(dR)(h)$, we have

$$dm = 2\pi\rho h R\, dR.$$

Then the moment of inertia is obtained by integrating (summing) over all these hoops:

$$I = \int R^2\, dm$$

$$= \int_{R_1}^{R_2} 2\pi\rho h R^3\, dR = 2\pi\rho h \left[\frac{R_2^4 - R_1^4}{4}\right],$$

where we are given that the cylinder has uniform density, $\rho = $ constant. (If this were not so, we would have to know ρ as a function of R before the integration could be carried out.) The volume V of this hollow cylinder is $V = (\pi R_2^2 - \pi R_1^2)h$, so its mass M is

$$M = \rho V = \rho\pi(R_2^2 - R_1^2)h.$$

Since $(R_2^4 - R_1^4) = (R_2^2 - R_1^2)(R_2^2 + R_1^2)$, we have

$$I = \frac{\pi\rho h}{2}(R_2^2 - R_1^2)(R_2^2 + R_1^2)$$

$$= \frac{1}{2}M(R_1^2 + R_2^2),$$

as stated in Fig. 10–21d. As a check, note that for a solid cylinder, $R_1 = 0$ and we obtain, with $R_2 = R_0$:

$$I = \frac{1}{2}MR_0^2,$$

which is that given in Fig. 10–21c for a solid cylinder of mass M and radius R_0.

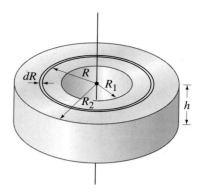

FIGURE 10–25 Determining the moment of inertia of a hollow cylinder (Example 10–10).

The Parallel-axis and Perpendicular-axis Theorems

There are two simple theorems that are helpful in obtaining moments of inertia. The first is called the **parallel-axis theorem**. It states that if I is the moment of inertia of a body of total mass M about any axis, and I_{CM} is the moment of inertia about an axis passing through the center of mass and parallel to the first axis but a distance h away, then

$$I = I_{CM} + Mh^2. \qquad \text{[parallel axis]} \quad \textbf{(10–17)}$$

Thus, for example, if the moment of inertia about an axis through the CM is known, the moment of inertia about any axis parallel to this axis is easily obtained.

Parallel-axis theorem

FIGURE 10–26 Example 10–11.

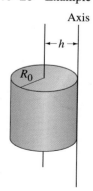

EXAMPLE 10–11 **Parallel axis.** Determine the moment of inertia of a solid cylinder of radius R_0, and mass M about an axis tangent to its edge and parallel to its symmetry axis, Fig. 10–26.

SOLUTION We use the parallel axis theorem with $I_{CM} = \frac{1}{2}MR_0$ (Fig. 10–21c). Since $h = R_0$, we have

$$I = I_{CM} + Mh^2 = \frac{3}{2}MR_0^2.$$

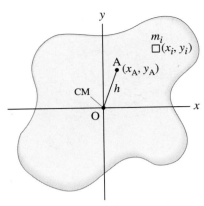

FIGURE 10–27 Derivation of the parallel-axis theorem.

FIGURE 10–28 The perpendicular axis theorem, and Example 10–12.

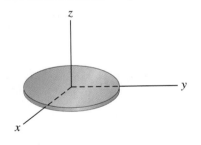

The proof of the parallel-axis theorem is as follows. We choose our coordinate system so the origin is at the CM, and I_{CM} is the moment of inertia about the z axis. Figure 10–27 shows a cross section of a body of arbitrary shape in the xy plane. We let I represent the moment of inertia of the body about an axis parallel to the z axis that passes through the point A in Fig. 10–27 where the point A has coordinates x_A and y_A. Let x_i, y_i, and m_i represent the coordinates and mass of an arbitrary particle of the body. The square of the distance from this point to A is $\left[(x_i - x_A)^2 + (y_i - y_A)^2\right]$. So the moment of inertia, I, about the axis through A is

$$I = \Sigma m_i\left[(x_i - x_A)^2 + (y_i - y_A)^2\right]$$
$$= \Sigma m_i(x_i^2 + y_i^2) - 2x_A \Sigma m_i x_i - 2y_A \Sigma m_i y_i + (\Sigma m_i)(x_A^2 + y_A^2).$$

The first term on the right is just $I_{CM} = \Sigma m_i(x_i^2 + y_i^2)$ since the CM is at the origin. The second and third terms are zero since, by definition of the CM, $\Sigma m_i x_i = \Sigma m_i y_i = 0$ because $x_{CM} = y_{CM} = 0$. The last term is Mh^2 since $\Sigma m_i = M$ and $(x_A^2 + y_A^2) = h^2$ where h is the distance of A from the CM. Thus we have proved $I = I_{CM} + Mh^2$, which is Eq. 10–17.

The parallel-axis theorem can be applied to any body. The second theorem, the **perpendicular-axis theorem**, can be applied only to plane figures—that is, to two-dimensional bodies, or bodies of uniform thickness whose thickness can be neglected compared to the other dimensions. This theorem states that the sum of the moments of inertia of a plane body about any two perpendicular axes in the plane of the body is equal to the moment of inertia about an axis through their point of intersection perpendicular to the plane of the object. That is, if the figure is in the xy plane (Fig. 10–28),

$$I_z = I_x + I_y. \qquad \text{[object in } xy \text{ plane]} \quad \textbf{(10–18)}$$

Here I_z, I_x, I_y are moments of inertia about the z, x, and y axes. The proof is simple: since $I_x = \Sigma m_i y_i^2$, $I_y = \Sigma m_i x_i^2$, and $I_z = \Sigma m_i(x_i^2 + y_i^2)$, Eq. 10–18 follows directly.

EXAMPLE 10–12 Perpendicular axis. Determine the moment of inertia of a thin circular coin (a cylinder) about an axis through its center in the plane of the coin (Fig. 10–28).

SOLUTION We want to calculate the moment of inertia about the x axis in Fig. 10–28, and we can use the perpendicular-axis theorem. From Fig. 10–21c we have $I_z = \frac{1}{2}MR_0^2$, and, from symmetry, $I_x = I_y$. Hence $2I_x = I_z$, or $I_x = \frac{1}{2}I_z = \frac{1}{4}MR_0^2$.

10–9 Angular Momentum and Its Conservation

Throughout this chapter we have seen that if we use the appropriate angular variables, the kinematic and dynamic equations for rotational motion are analogous to those for ordinary linear motion. In like manner, the linear momentum, $p = mv$, has a rotational analog. It is called **angular momentum**, L, and for a body rotating about a fixed axis with angular velocity ω, it is defined as[†]

$$L = I\omega, \qquad \textbf{(10–19)}$$

where I is the moment of inertia. The SI units for L are $\text{kg} \cdot \text{m}^2/\text{s}$.

We saw in Chapter 9 (Section 9–1) that Newton's second law can be written not only as $\Sigma F = ma$, but also more generally in terms of momentum (Eq. 9–2), $\Sigma F = dp/dt$. In a similar way, the rotational equivalent of Newton's second law, which we saw in Eqs. 10–14 and 10–15 can be written as $\Sigma \tau = I\alpha$, can also be written in terms of angular momentum: since the angular acceleration $\alpha = d\omega/dt$ (Eq. 10–3), then $I\alpha = I(d\omega/dt) = d(I\omega)/dt = dL/dt$, so

$$\Sigma \tau = I\alpha = \frac{dL}{dt}. \qquad \textbf{(10–20)}$$

This simple derivation assumes that the moment of inertia, I, remains constant.

[†] Angular momentum is a vector, and we will see in Chapter 11 that the angular momentum discussed here, $L = I\omega$, is the component along the rotation axis. There may or may not be other components.

However, it is valid even if the moment of inertia changes (see Chapter 11). Equation 10–20 is Newton's second law for rotational motion of a rigid body about a fixed axis, and is also valid for a moving body if its rotation is about an axis passing through its CM (as for Eq. 10–15).

Conservation of Angular Momentum

Angular momentum is an important concept in physics because, under certain conditions, it is a conserved quantity. What are the conditions for which it is conserved? From Eq. 10–20 we see immediately that if the net external torque $\Sigma\tau$ on a body is zero, then dL/dt equals zero. That is, L does not change. This, then, is the **law of conservation of angular momentum** for a rotating body:

> **The total angular momentum of a rotating body remains constant if the net external torque acting on it is zero.**

CONSERVATION OF ANGULAR MOMENTUM

The law of conservation of angular momentum is one of the great laws of physics.

When there is zero net torque acting on a body, and the body is rotating about a fixed axis or about an axis through its CM such that its direction doesn't change, we can write

$$I\omega = I_0\omega_0 = \text{constant}.$$

I_0 and ω_0 are the moment of inertia and angular velocity, respectively, about the axis at some initial time ($t = 0$), and I and ω are their values at some other time. The parts of the body may alter their positions relative to one another, so that I changes. But then ω changes as well and the product $I\omega$ remains constant.

Many interesting phenomena can be understood on the basis of conservation of angular momentum. Consider a skater doing a spin on the tips of her skates, Fig. 10–29. She rotates at a relatively low speed when her arms are outstretched, but when she brings her arms in close to her body, she suddenly spins much faster. By remembering the definition of moment of inertia as $I = \Sigma mR^2$, it is clear that when she pulls her arms in closer to the axis of rotation, R is reduced for the arms so her moment of inertia is reduced. Since the angular momentum $I\omega$ remains constant (we ignore the small torque due to friction), if I decreases, then the angular velocity ω must increase. If the skater reduces her moment of inertia by a factor of 2, she will then rotate with twice the angular velocity.

A similar example is the diver shown in Fig. 10–30. The push as she leaves the board gives her an initial angular momentum about her CM. When she curls herself into the tuck position, she rotates quickly one or more times. She then stretches out again, increasing her moment of inertia, which reduces the angular velocity to a small value, and then she enters the water. The change in moment of inertia from the straight position to the tuck position can be a factor of as much as $3\frac{1}{2}$.

Note that for angular momentum to be conserved, the net torque must be zero, but the net force does not necessarily have to be zero. The net force on the diver in Fig. 10–30, for example, is not zero (gravity is acting), but the net torque on her is zero.

➡ **PHYSICS APPLIED**

Spinning skaters and divers

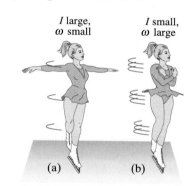

FIGURE 10–29 A skater doing a spin on ice, illustrating conservation of angular momentum: in (a), I is large and ω is small; in (b), I is smaller so ω is larger.

FIGURE 10–30 A diver rotates faster when arms and legs are tucked in than when they are outstretched. Angular momentum is conserved.

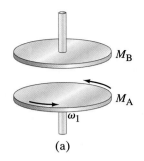

(a)

FIGURE 10–31 Example 10–13.

EXAMPLE 10–13 **Clutch design.** You are designing a clutch assembly for a piece of machinery. The clutch assembly consists of two cylindrical plates, of mass $M_A = 6.0$ kg and $M_B = 9.0$ kg, with equal radii $R_0 = 0.60$ m. They are initially separated (Fig. 10–31). Plate M_A is accelerated from rest to an angular velocity $\omega_1 = 7.2$ rad/s in time $\Delta t = 2.0$ s. Calculate (a) the angular momentum of M_A, and (b) the torque required to have accelerated M_A from rest to ω_1. Next, plate M_B, initially at rest but free to rotate without friction, is allowed to fall vertically (or could be pushed by a spring, as in a car) so it is in firm contact with plate M_A (their contact surfaces are made of high-friction material). Before contact, M_A was rotating at constant ω_1, with no friction or other torque exerted on it. Upon contact, the two plates both rotate at a constant angular velocity ω_2, which is considerably less than ω_1. (c) Why does this happen, and what is ω_2?

SOLUTION (a) The angular momentum of M_A will be

$$L_A = I_A \omega_1 = \tfrac{1}{2} M_A R_0^2 \omega_1 = \tfrac{1}{2}(6.0 \text{ kg})(0.60 \text{ m})^2(7.2 \text{ rad/s}) = 7.8 \text{ kg·m}^2/\text{s}.$$

(b) The plate started from rest so the torque, assumed constant, was

$$\tau \approx \frac{\Delta L}{\Delta t} = \frac{7.8 \text{ kg·m}^2/\text{s} - 0}{2.0 \text{ s}} = 3.9 \text{ m·N}.$$

(c) Initially, M_A is rotating at constant ω_1 since we assume no torques (such as friction) are acting. When plate B comes in contact, it too is free to rotate, and they rotate together because their surfaces stick together. But why does their rotation speed decrease? You might think in terms of the torque each exerts on the other upon contact. But quantitatively, it's easier to use conservation of angular momentum (which we can, since no external torques are assumed to act). Thus

$$\text{angular momentum before} = \text{angular momentum after}$$
$$I_A \omega_1 = (I_A + I_B)\omega_2.$$

Solving for ω_2 we find

$$\omega_2 = \left(\frac{I_A}{I_A + I_B}\right)\omega_1 = \left(\frac{M_A}{M_A + M_B}\right)\omega_1 = \left(\frac{6.0 \text{ kg}}{15.0 \text{ kg}}\right)(7.2 \text{ rad/s}) = 2.9 \text{ rad/s}.$$

➡ **PHYSICS APPLIED**

Neutron star

EXAMPLE 10–14 **ESTIMATE** **Star collapse.** Astronomers often detect stars that are rotating extremely rapidly, known as neutron stars. They are believed to have been formed from the inner core of a larger star that collapsed, due to its own gravitation, to a star of very small radius and very high density. Suppose the core of such a star is the size of our Sun. $(R \approx 7 \times 10^5 \text{ km})$ before collapse, but of mass 2.0 times as great, and is rotating at a speed of 1.0 revolution every 10 days. If it were to undergo gravitational collapse to a neutron star of radius 10 km, what would its rotation speed be? Assume the star is a uniform sphere at all times.

SOLUTION The star is isolated (no external forces) so we can use conservation of angular momentum for this process:

$$I_i \omega_i = I_f \omega_f$$

where the subscript i and f refer to initial (normal star) and final (neutron star). Then

$$\omega_f = \left(\frac{I_i}{I_f}\right)\omega_i = \left(\frac{\tfrac{2}{5} M_i R_i^2}{\tfrac{2}{5} M_f R_f^2}\right)\omega_i = \frac{R_i^2}{R_f^2}\omega_i,$$

where we assume no mass is lost in the process. The frequency $f = \omega/2\pi$ so

$$f_f = \left(\frac{7 \times 10^5 \text{ km}}{10 \text{ km}}\right)^2\left(\frac{1 \text{ rev}}{10 \text{ d}(24 \text{ h/d})(3600 \text{ s/h})}\right) \approx 6 \times 10^3 \text{ rev/s}.$$

Vector Nature of Angular Momentum

Although we will not go into the vector nature of angular momentum in detail until the next chapter, we can deal with some simple cases involving conservation of angular momentum here. For a body rotating about a fixed axis, the direction of the angular momentum can be taken as the direction of the angular velocity, $\boldsymbol{\omega}$. That is,

$$\mathbf{L} = I\boldsymbol{\omega}.$$

However, this is strictly true[†] only if the rotation axis is an axis of symmetry of the body or if the body is thin and flat and rotates about an axis perpendicular to the plane of the body (such as a wheel rotating on an axle).

As a simple example, consider a person standing at rest on a circular platform capable of rotating friction-free about an axis through its center (that is, a simplified merry-go-round, Fig. 10–32). If the person now starts to walk along the edge of the platform, the platform starts rotating in the opposite direction. Why? One explanation is that the person's foot exerts a force on it. Another explanation (and this is the most useful analysis here) is that this is an example of the conservation of angular momentum. If the person starts walking counterclockwise, the person's angular momentum will be pointed upward along the axis of rotation (remember how we defined the direction of $\boldsymbol{\omega}$ using the right-hand rule in Section 10–4). The magnitude of the person's angular momentum will be $L = I\omega = (mR^2)(v/R)$, where v is the person's speed (relative to the Earth, not the platform), R is his distance from the rotation axis, m is his mass, and mR^2 is his moment of inertia since all of his mass is roughly the same distance R from the rotation axis. The platform rotates in the opposite direction, so its angular momentum points downward. If the initial total angular momentum was zero (person and platform at rest), it will remain zero after the person starts walking—that is, the upward angular momentum of the person just balances the oppositely directed angular momentum of the platform, Fig. 10–32b, so the total vector angular momentum remains zero. Even though the person exerts a force (and torque) on the platform, the platform exerts an equal and opposite torque on the person. So the net torque on the *system* of person plus platform is zero (ignoring friction) and the total angular momentum remains constant.

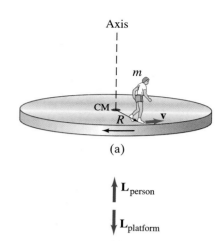

FIGURE 10–32 (a) A person standing on a circular platform, both initially at rest, begins walking along the edge at speed v. The platform, assumed to be mounted on friction-free bearings, begins rotating in the opposite direction, so that the total angular momentum remains zero, as shown in (b).

EXAMPLE 10–15 **Running on a circular platform.** Suppose a 60-kg person stands at the edge of a 6.0-m-diameter circular platform, which is mounted on frictionless bearings and has a moment of inertia of $1800\ \text{kg}\cdot\text{m}^2$. The platform is at rest initially, but when the person begins running at a speed of 4.2 m/s (with respect to the ground) around its edge, the platform begins to rotate in the opposite direction as in Fig. 10–32. Calculate the angular velocity of the platform.

SOLUTION The total angular momentum is zero initially. Since there is no net torque, $\mathbf{L}$ is conserved and will remain zero, as in Fig. 10–32. The person's angular momentum is $L_{\text{per}} = (mR^2)(v/R)$, and we take this as positive. The angular momentum of the platform is $L_{\text{plat}} = -I\omega$. Thus

$$L = L_{\text{per}} + L_{\text{plat}}$$
$$0 = mR^2\left(\frac{v}{R}\right) - I\omega.$$

So

$$\omega = \frac{mRv}{I} = \frac{(60\ \text{kg})(3.0\ \text{m})(4.2\ \text{m/s})}{1800\ \text{kg}\cdot\text{m}^2} = 0.42\ \text{rad/s}.$$

The frequency of rotation is $f = \omega/2\pi = 0.067$ rev/s and the period $T = 1/f = 15$ s per revolution.

[†] For more complicated situations of bodies rotating about a fixed axis, there will be a component of $\mathbf{L}$ along the direction of $\boldsymbol{\omega}$ and its magnitude will be equal to $I\omega$. (There could be other components as well.) If the total angular momentum is conserved, then the component $I\omega$ will also be conserved. So our results here can be applied to any rotation about a fixed axis.

FIGURE 10–33 Example 10–16.

CONCEPTUAL EXAMPLE 10–16 **Spinning bicycle wheel.** Your physics teacher is holding a spinning bicycle wheel while standing on a stationary frictionless turntable (Fig. 10–33). What will happen if the teacher suddenly flips the bicycle wheel over so that it is spinning in the opposite direction?

RESPONSE The total angular momentum initially is **L** vertically upward, and that is what the system's angular momentum must be afterward since **L** is conserved. Thus, if the wheel's angular momentum afterward is −**L** downward, then the angular momentum of teacher plus turntable will have to be +2**L** upward. We can safely predict that the teacher will begin spinning around in the same direction the wheel was spinning originally.

10–10 Rotational Kinetic Energy

The quantity $\frac{1}{2}mv^2$ is the kinetic energy of a body undergoing translational motion. A body rotating about an axis is said to have **rotational kinetic energy**. By analogy with translational kinetic energy, we would expect this to be given by the expression $\frac{1}{2}I\omega^2$ where I is the moment of inertia of the body and ω is its angular velocity. We can indeed show that this is true. Consider any rigid rotating object as made up of many tiny particles, each of mass m_i. If we let R_i represent the distance of any one particle from the axis of rotation, then its linear velocity is $v_i = R_i\omega$. The total kinetic energy of the whole body will be the sum of the kinetic energies of all its particles:

$$K = \Sigma\left(\tfrac{1}{2}m_i v_i^2\right) = \Sigma\left(\tfrac{1}{2}m_i R_i^2 \omega^2\right)$$
$$= \tfrac{1}{2}\Sigma\left(m_i R_i^2\right)\omega^2.$$

We have factored out the $\frac{1}{2}$ and the ω^2 since they are the same for every particle of a rigid body.

Since $\Sigma m_i R_i^2 = I$, the moment of inertia, we see that the kinetic energy, K, of an object rotating about a fixed axis is, as expected,

$$K = \tfrac{1}{2}I\omega^2. \qquad\qquad \text{[fixed axis]} \quad \textbf{(10–21)}$$

If the axis is not fixed in space, the rotational kinetic energy can take on a more complicated form.

The work done on a body rotating about a fixed axis can be written in terms of angular quantities. Suppose a force **F** is exerted at a point whose distance from the axis of rotation is R, as in Fig. 10–34. The work done by this force is

$$W = \int \mathbf{F} \cdot d\boldsymbol{l} = \int F_\perp R\, d\theta,$$

Rotational kinetic energy

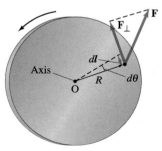

FIGURE 10–34 Calculating the work done by a torque acting on a rigid body rotating about a fixed axis.

Work done by a torque

where $d\boldsymbol{l}$ is an infinitesimal distance perpendicular to R with magnitude $dl = R\, d\theta$, and $F_\perp$ is the component of **F** perpendicular to R and parallel to $d\boldsymbol{l}$ (Fig. 10–34). But $F_\perp R$ is the torque about the axis, so

$$W = \int \tau\, d\theta. \qquad\qquad\qquad \textbf{(10–22)}$$

The rate of work done, or power P, is

$$P = \frac{dW}{dt} = \tau\frac{d\theta}{dt} = \tau\omega. \qquad\qquad \textbf{(10–23)}$$

The work-energy principle holds for rotation of a rigid body about a fixed axis. From Eq. 10–14 we have

$$\tau = I\alpha = I\frac{d\omega}{dt} = I\frac{d\omega}{d\theta}\frac{d\theta}{dt} = I\omega\frac{d\omega}{d\theta},$$

where we used the chain rule and $\omega = d\theta/dt$. Then $\tau\,d\theta = I\omega\,d\omega$ and

$$W = \int_{\theta_1}^{\theta_2} \tau\,d\theta = \int_{\omega_1}^{\omega_2} I\omega\,d\omega = \tfrac{1}{2}I\omega_2^2 - \tfrac{1}{2}I\omega_1^2. \qquad \textbf{(10–24)}$$

Work-energy principle for rotation

This is the work-energy principle for a body rotating about a fixed axis. It states that the work done in rotating a body through an angle $\theta_2 - \theta_1$ is equal to the change in rotational kinetic energy of the body.

EXAMPLE 10–17 **ESTIMATE** **Flywheel.** Flywheels, which are simply large rotating disks, have been suggested as a means of storing energy for solar-powered generating systems. Estimate the kinetic energy that can be stored in a 20,000-kg (10-ton) flywheel with a diameter of 20 m (a six-story building). Assume it could hold together (without flying apart due to internal stresses) at 100 rpm.

➡ **PHYSICS APPLIED**

Energy from a flywheel

SOLUTION We are given

$$\omega = 100\ \text{rpm} = \left(100\ \frac{\text{rev}}{\text{min}}\right)\left(\frac{1\ \text{min}}{60\ \text{sec}}\right)\left(\frac{2\pi\ \text{rad}}{\text{rev}}\right) = 10.5\ \text{rad/s}.$$

The kinetic energy stored in the disk $\left(I = \tfrac{1}{2}MR_0^2\right)$ is, from Eq. 10–21,

$$K = \tfrac{1}{2}\left(\tfrac{1}{2}MR_0^2\right)\omega^2 = \tfrac{1}{4}\left(2.0 \times 10^4\ \text{kg}\right)(10\ \text{m})^2(10.5\ \text{rad/s})^2 = 5.2 \times 10^6\ \text{J}.$$

In terms of kilowatt-hours $\left[1\ \text{kWh} = (1000\ \text{J}\cdot\text{s})(3600\ \text{s/h})(1\ \text{h}) = 3.6 \times 10^6\ \text{J}\right]$, this energy is only about 1.5 kWh, which is not a lot of energy $\left(\text{one 3-kW oven would use it all in }\tfrac{1}{2}\ \text{h}\right)$. Thus flywheels seem unlikely for this application.

EXAMPLE 10–18 **Rotating rod.** A rod of mass M is pivoted on a frictionless hinge at one end, as shown in Fig. 10–35. The rod is held at rest horizontally and then released. Determine the angular velocity of the rod when it reaches the vertical position, and the speed of the rod's tip at this moment.

SOLUTION We can use the work-energy principle here. The work done is due to gravity, and is equal to the change in gravitational potential energy of the rod. Since the CM of the rod drops a vertical distance $l/2$, the work done by gravity is

$$W = Mg\frac{l}{2}.$$

The initial kinetic energy is zero. Hence, from the work-energy principle,

$$\tfrac{1}{2}I\omega^2 = Mg\frac{l}{2}.$$

Since $I = \tfrac{1}{3}Ml^2$ for a rod pivoted about its end (Fig. 10–21), we can solve for ω:

$$\omega = \sqrt{\frac{3g}{l}}.$$

The tip of the rod will have a linear speed (see Eq. 10–4)

$$v = l\omega = \sqrt{3gl}.$$

By comparison, an object that falls vertically a height l has a speed $v = \sqrt{2gl}$.

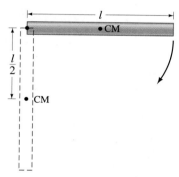

FIGURE 10–35 Example 10–18.

Kinetic Energy = $K_{CM} + K_{rot}$

An object that rotates while its center of mass (CM) undergoes translational motion will have both translational and rotational kinetic energy. Equation 10–21, $K = \frac{1}{2}I\omega^2$, gives the rotational kinetic energy if the rotation axis is fixed. If the object is moving (such as a wheel rolling along the ground, Fig. 10–36), this equation is still valid as long as the rotation axis is fixed in direction. Then the total kinetic energy is

$$K = \tfrac{1}{2}Mv_{CM}^2 + \tfrac{1}{2}I_{CM}\omega^2, \qquad\qquad (10\text{–}25)$$

where v_{CM} is the linear velocity of the CM, I_{CM} is the moment of inertia about an axis through the CM, ω is the angular velocity about this axis, and M is the total mass of the body.

Before we prove this important theorem, Eq. 10–25, which is done in the following optional subsection, we first do some Examples to get a good feel for it.

EXAMPLE 10–19 **Sphere rolling down an incline.** What will be the speed of a solid sphere of mass M and radius R_0 when it reaches the bottom of an incline if it starts from rest at a vertical height H and rolls without slipping? See Fig. 10–37. Ignore losses due to dissipative forces, and compare your result to that for an object *sliding* down a frictionless incline.

SOLUTION We use the law of conservation of energy, and we must now include rotational kinetic energy. The total energy at any point a vertical distance y above the base of the incline is

$$\tfrac{1}{2}Mv^2 + \tfrac{1}{2}I_{CM}\omega^2 + Mgy,$$

where v is the speed of the CM. We equate the total energy at the top ($y = H$ and $v = \omega = 0$) to the total energy at the bottom ($y = 0$):

$$0 + 0 + MgH = \tfrac{1}{2}Mv^2 + \tfrac{1}{2}I_{CM}\omega^2 + 0.$$

From Fig. 10–21e, the moment of inertia of a solid sphere about an axis through its CM is $I_{CM} = \frac{2}{5}MR_0^2$. Since the sphere rolls without slipping, the speed, v, of the center of mass with respect to the point of contact (which is momentarily at rest at any instant) is equal to the speed of a point on the edge relative to the center, as we saw in Section 10–3 (Fig. 10–8). We therefore have $\omega = v/R$. Hence

$$\tfrac{1}{2}Mv^2 + \tfrac{1}{2}\left(\tfrac{2}{5}MR_0^2\right)\left(\frac{v^2}{R_0^2}\right) = MgH.$$

Dividing out the M's and R's, we obtain

$$\left(\tfrac{1}{2} + \tfrac{1}{5}\right)v^2 = gH$$

or

$$v = \sqrt{\tfrac{10}{7}gH}.$$

Note first that v is independent of both the mass M and the radius R of the sphere. Also, we can compare this result for the speed of a rolling sphere to that for an object sliding down a plane without rotating and without friction (see Chapter 8, $\frac{1}{2}mv^2 = mgH$), in which case $v = \sqrt{2gH}$, which is greater. An object sliding without friction transforms its initial potential energy entirely into translational kinetic energy (none into rotational kinetic energy), so its speed is greater.

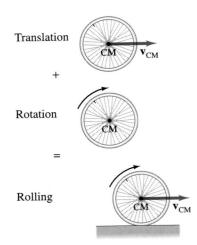

Translation
+
Rotation
=
Rolling

FIGURE 10–36 A wheel rolling without slipping can be considered as translation of the wheel as a whole with velocity $\mathbf{v}_{CM}$ plus rotation about the CM.

➡ **P R O B L E M   S O L V I N G**

Rotational energy adds to other forms of energy to get the total energy which is conserved

FIGURE 10–37 A sphere rolling down a hill has both translational and rotational kinetic energy. Example 10–19.

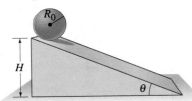

CONCEPTUAL EXAMPLE 10–20 | **Who's fastest?** Several objects roll without slipping down an incline of vertical height H, all starting from rest at the same moment. The objects are a thin hoop (or a plain wedding band), a marble, a solid cylindrical battery (D-cell), an empty soup can, and an unopened soup can. In addition a greased box slides down without friction. In what order do they reach the bottom of the incline?

RESPONSE The sliding box wins every time! As we saw in Example 10–19, the speed of a rolling sphere at the bottom of the incline is less than that for a sliding box (without friction) because the potential energy loss (MgH) is transformed completely into translational kinetic energy for the box, whereas for rolling objects, the initial potential energy is shared between translational and rotational kinetic energy. For each of the rolling objects we can state that the loss in potential energy equals the increase in kinetic energy:

$$MgH = \tfrac{1}{2}Mv^2 + \tfrac{1}{2}I_{CM}\omega^2.$$

First we note that for all our rolling objects, the moment of inertia I_{CM} is a numerical factor times the mass M and the radius R^2 (Fig. 10–21). The mass M is in each term, so the translational speed v doesn't depend on M, nor does it depend on the radius R since $\omega = v/R$ so R^2 cancels out for all the rolling objects, just as in Example 10–19. Thus the speed v at the bottom depends only on that numerical factor in I_{CM} which expresses how the mass is distributed. Consequently, the hoop, with all its mass concentrated at radius R ($I_{CM} = MR^2$), will have the lowest speed and will arrive at the bottom behind the D-cell ($I_{CM} = \tfrac{1}{2}MR^2$) which in turn will be behind the marble ($I_{CM} = \tfrac{2}{5}MR^2$). The empty can, which is mainly a hoop plus a small disk, has its mass concentrated almost all at R; so it will be a bit faster than the pure hoop but slower than the D-cell. See Fig. 10–38. The unopened soup can is more complicated. It cannot be considered a solid cylinder because the soup can move about inside, and that will dissipate some energy; so we expect it to be slower than the D-cell, but that's about all we can safely say. For all the other objects, note that the speed at the bottom does not depend on the object's mass M or radius R, but only on its shape (and the height of the hill H).

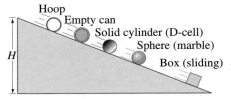

FIGURE 10–38 Example 10–20.

If there had been no friction between the sphere (and other rolling objects) and the plane in these Examples, the sphere would have slid rather than rolled. Friction must be present to make a round object roll. We did not need to take friction into account in the energy equation because it is *static* friction and does no work. If we assume the sphere is perfectly rigid, it is in contact with the surface at a point, and the force of friction acts parallel to the plane. But the point of contact of the sphere at each instant does not slide—it moves perpendicular to the plane (first down and then up) as the sphere rolls (Fig. 10–39). Thus, no work is done by the friction force because the force and the motion are perpendicular. The reason the rolling objects in Examples 10–19 and 10–20 move down the slope more slowly than if they were sliding is *not* because friction is doing work. Rather it is because some of the gravitional potential energy is converted to rotational kinetic energy, leaving less for the translational kinetic energy.

FIGURE 10–39 A sphere rolling without slipping to the right on a plane surface. The point in contact with the ground at any moment, point P, is momentarily at rest. Point A on the left of P is moving nearly vertically upward at the instant shown, and point B on the right is moving nearly vertically downward. (An instant later, point B will touch the plane and be at rest momentarily.)

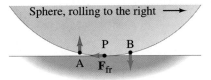

Using $\Sigma\tau_{CM} = I_{CM}\alpha_{CM}$

We can examine objects rolling down a plane not only from the point of view of kinetic energy, as we did in Examples 10–19 and 10–20, but also in terms of forces and torques. If we calculate torques about an axis fixed in direction (even though accelerating) which passes through the center of mass of the rolling sphere, then

$$\Sigma\tau_{CM} = I_{CM}\alpha_{CM}$$

is valid, as we discussed in Section 10–6. See Eq. 10–15, whose validity we will show in Chapter 11. Be careful however: Do not assume $\Sigma\tau = I\alpha$ is always valid. You cannot just calculate τ, I, and α about any axis unless the axis is (1) fixed in an inertial reference frame or (2) fixed in direction but passes through the CM of the body.

Careful:
When is $\Sigma\tau = I\alpha$ valid?

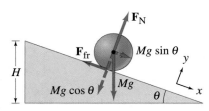

FIGURE 10–40 Example 10–21.

What must μ_s be if the sphere is to roll without slipping?

EXAMPLE 10–21 **Analysis of a sphere on an incline using forces.** Analyze the rolling sphere of Example 10–19, Fig. 10–37, in terms of forces and torques. In particular, find the velocity v and the magnitude of the friction force, F_{fr}, Fig. 10–40. [F_{fr} is due to static friction and we cannot assume $F_{fr} = \mu_s F_N$, only $F_{fr} \leq \mu_s F_N$.]

SOLUTION We analyze the motion as translation of the CM plus rotation about the CM. For translation in the x direction we have from $\Sigma F = ma$,

$$Mg \sin \theta - F_{fr} = Ma,$$

and in the y direction

$$F_N - Mg \cos \theta = 0$$

since there is no acceleration perpendicular to the plane. This last equation merely tells us the magnitude of the normal force,

$$F_N = Mg \cos \theta.$$

For the rotational motion about the CM, we use Newton's second law for rotation $\Sigma \tau_{CM} = I_{CM} \alpha_{CM}$ (Eq. 10–15), calculating about an axis passing through the CM but fixed in direction:

$$F_{fr} R_0 = \left(\tfrac{2}{5} M R_0^2\right)\alpha.$$

Since the other forces, $\mathbf{F}_N$ and $M\mathbf{g}$, have lever arms equal to zero, they do not appear here. As we saw in Example 10–19, $\omega = v/R_0$ where v is the speed of the CM. Taking derivatives with respect to time we have $\alpha = a/R_0$ and substituting this into the last equation we find

$$F_{fr} = \tfrac{2}{5} Ma.$$

When we substitute this into the top equation, we get

$$Mg \sin \theta - \tfrac{2}{5} Ma = Ma,$$

or

$$a = \tfrac{5}{7} g \sin \theta.$$

We thus see that the acceleration of the CM of a rolling sphere is less than that for an object sliding without friction ($a = g \sin \theta$). To find the speed v at the bottom we use Eq. 2–12c where the total distance traveled along the plane is $x = H/\sin \theta$, and H is the height of the plane (see Fig. 10–40). Thus

$$v = \sqrt{2ax} = \sqrt{2\left(\frac{5}{7} g \sin \theta\right)\left(\frac{H}{\sin \theta}\right)} = \sqrt{\frac{10}{7} gH}.$$

This is the same result obtained in Example 10–19 although less effort was needed there. To get the magnitude of the force of friction, we use the equations obtained above:

$$F_{fr} = \tfrac{2}{5} Ma = \tfrac{2}{5} M\left(\tfrac{5}{7} g \sin \theta\right) = \tfrac{2}{7} Mg \sin \theta.$$

If the coefficient of static friction is sufficiently small, or θ sufficiently large so that $F_{fr} > \mu_s F_N$ (that is, $\tan \theta > \tfrac{7}{2} \mu_s$), the sphere will not simply roll but will slip as it moves down the plane.

* Derivation of $K = K_{CM} + K_{rot}$

Now we derive a general theorem, that the *total kinetic energy of a moving body will be equal to the translational kinetic energy of its CM plus the kinetic energy of motion relative to the center of mass.* This is a general theorem, and to prove it we proceed as follows. Let $\mathbf{r}_{CM} = x_{CM}\mathbf{i} + y_{CM}\mathbf{j} + z_{CM}\mathbf{k}$ represent the position of the CM

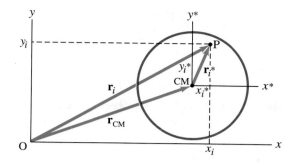

FIGURE 10–41 Point P has coordinates x_i, y_i, z_i, $(\mathbf{r}_i)$ relative to origin O of an inertial frame, and coordinates x_i^*, y_i^*, z_i^*, $(\mathbf{r}_i^*)$ relative to the center of mass (CM).

at any moment in some inertial reference frame. Let $\mathbf{r}_i = x_i\mathbf{i} + y_i\mathbf{j} + z_i\mathbf{k}$ be the position vector of the ith particle of mass m_i in this inertial reference frame, and let $\mathbf{r}_i^* = x_i^*\mathbf{i} + y_i^*\mathbf{j} + z_i^*\mathbf{k}$ be the position vector of this particle with reference to the CM (not necessarily an inertial reference frame). Then (see Fig. 10–41):

$$x_i = x_{\text{CM}} + x_i^*, \qquad y_i = y_{\text{CM}} + y_i^*, \quad \text{and} \quad z_i = z_{\text{CM}} + z_i^*.$$

The velocity of the ith particle in the inertial reference frame is

$$\mathbf{v}_i = \mathbf{v}_{\text{CM}} + \mathbf{v}_i^*,$$

where $\mathbf{v}_{\text{CM}}$ is the velocity of the CM in this reference frame and $\mathbf{v}_i^*$ is the velocity of the ith particle relative to the CM. We can use the vector dot product and write $v^2 = \mathbf{v} \cdot \mathbf{v}$, so that the total kinetic energy K is

$$
\begin{aligned}
K &= \tfrac{1}{2}\Sigma m_i v_i^2 = \tfrac{1}{2}\Sigma m_i(\mathbf{v}_i \cdot \mathbf{v}_i) \\
&= \tfrac{1}{2}\Sigma m_i(\mathbf{v}_{\text{CM}} + \mathbf{v}_i^*) \cdot (\mathbf{v}_{\text{CM}} + \mathbf{v}_i^*) \\
&= \tfrac{1}{2}\Sigma m_i v_{\text{CM}}^2 + \mathbf{v}_{\text{CM}} \cdot (\Sigma m_i \mathbf{v}_i^*) + \tfrac{1}{2}\Sigma m_i v_i^{*2}.
\end{aligned}
$$

Now $\Sigma m_i = M$, the total mass of the body. Also, $\Sigma m_i\mathbf{v}_i^* = 0$, as can be seen by taking the derivative of the definition of the center of mass when the CM is at the origin of coordinates $\left[\mathbf{r}_{\text{CM}} = (1/M)\Sigma m_i\mathbf{r}_i^* = 0\right]$. Thus

$$K = \tfrac{1}{2}Mv_{\text{CM}}^2 + \tfrac{1}{2}\Sigma m_i v_i^{*2} \tag{10–26}$$

Total kinetic energy

and we have proved that the total kinetic energy is the sum of the translational kinetic energy of the center of mass, $\tfrac{1}{2}Mv_{\text{CM}}^2$, plus the kinetic energy of motion relative to the CM.

Our proof of Eq. 10–26 is general. We can apply this equation to a rigid body moving in a plane, such as a wheel rolling down a hill. In this case the axis of rotation is fixed in direction (perpendicular to the plane in which the body moves), although it moves along with the center of mass and so is not fixed in position. For each particle, $v_i^* = \omega R_i^*$, where R_i^* is the perpendicular distance of the ith particle from a line passing through the CM and perpendicular to the plane of motion. Then

Rolling wheel

$$\tfrac{1}{2}\Sigma m_i v_i^{*2} = \tfrac{1}{2}\left(\Sigma m_i R_i^{*2}\right)\omega^2 = \tfrac{1}{2}I_{\text{CM}}\omega^2$$

and

$$K = \tfrac{1}{2}Mv_{\text{CM}}^2 + \tfrac{1}{2}I_{\text{CM}}\omega^2. \qquad \text{[rigid body, axis fixed in direction]}$$

This is just Eq. 10–25. In this equation, I_{CM} is the moment of inertia of the body about an axis through its center of mass and perpendicular to the plane of motion. Thus the total kinetic energy of a body moving in a plane with both translational and rotational motion, such that the rotation axis doesn't change direction, is the sum of the translational kinetic energy of the CM plus the rotational kinetic energy about the CM, $\tfrac{1}{2}I_{\text{CM}}\omega^2$.

*More Advanced Examples

Here we do three more Examples, all of them fun and interesting. They each make use of $\Sigma\tau = I\alpha$, and again we must remember that this equation is valid only if τ, α and I are calculated about an axis that is either (1) fixed in an inertial reference frame, or (2) passes through the CM of the object and remains fixed in direction.

PHYSICS APPLIED

Braking distribution of a car

EXAMPLE 10–22 **Braking a car.** When the brakes of a car are applied, the force on the front tires is much greater than that on the rear tires. To see why, calculate the magnitude of the friction forces, F_1 and F_2, on the front and rear tires of the car shown in Fig. 10–42 when the car decelerates at a rate $a = 0.50\, g$. The car has mass $M = 1200$ kg, the distance between the front and rear axles is 3.0 m, and its CM (where the force of gravity acts) is midway between the axles 75 cm above the ground.

Forces on the tires for a braking car

SOLUTION Figure 10–42 is the free-body diagram showing all the forces on the car. F_1 and F_2 are the frictional forces that decelerate the car. We let F_1 be the sum of the forces on both front tires, and F_2 likewise for the two rear tires. Thus

$$F_1 + F_2 = Ma$$
$$= (1200\,\text{kg})(0.50)(9.8\,\text{m/s}^2) = 5900\,\text{N}. \qquad \textbf{(i)}$$

F_{N1} and F_{N2} are the normal forces the road exerts on the tires as shown and, for simplicity, we assume the static friction force acts the same for all the tires, so that F_{N1} and F_{N2} are proportional respectively to F_1 and F_2 $(F_{N1} = F_1/\mu, F_{N2} = F_2/\mu)$.[†] While the car is braking it does not rotate, so the net torque on it is zero. If we calculate the torques about the CM as axis, we see that whereas F_1, F_2, and F_{N2} all act to rotate the car clockwise, only F_{N1} acts to rotate it counterclockwise and so must balance the other three. Hence, F_{N1} must be significantly greater than F_{N2}, and hence F_1 must be significantly greater than F_2. Mathematically, we have for the torques calculated about the CM:

$$(1.5\,\text{m})F_{N1} - (1.5\,\text{m})F_{N2} - (0.75\,\text{m})F_1 - (0.75\,\text{m})F_2 = 0.$$

Since F_{N1} and F_{N2} are proportional to F_1 and F_2, we can write this as

$$(1.5\,\text{m})\frac{(F_1 - F_2)}{\mu} - (0.75\,\text{m})(F_1 + F_2) = 0. \qquad \textbf{(ii)}$$

Also, since the car does not accelerate vertically, we have

$$Mg = F_{N1} + F_{N2} = \frac{F_1 + F_2}{\mu}. \qquad \textbf{(iii)}$$

Comparing (iii) to (i), we see that $\mu = a/g = 0.50$. We solve (ii) for F_1:

$$F_1 = \frac{1.5\,\text{m}/\mu + 0.75\,\text{m}}{1.5\,\text{m}/\mu - 0.75\,\text{m}} F_2 = \tfrac{5}{3}F_2.$$

Thus F_1 is $1\tfrac{2}{3}$ times greater than F_2. Actual magnitudes, from (i), are $F_1 = 3700$ N and $F_2 = 2200$ N. Because the force on the front tires is generally greater than on the rear tires, cars are often designed with larger brake pads on the front wheels than on the rear. Or, to say it another way, if the brake pads are equal, the front ones wear out a lot faster.

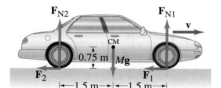

F_{N2} F_{N1}

v

CM

0.75 m   Mg

F_2 F_1

|←1.5 m→|←1.5 m→|

FIGURE 10–42 Forces on a braking car (Example 10–22).

EXAMPLE 10–23 **A falling yo-yo.** String is wrapped around a uniform solid cylinder (something like a yo-yo) of mass M and radius R, and the cylinder starts falling from rest, Fig. 10–43a. As the cylinder falls, find (*a*) its acceleration and (*b*) the tension in the string.

[†] Our proportionality constant μ is not equal to μ_s, the static coefficient of friction $(F_{fr} \le \mu_s F_N)$, unless the car is just about to skid.

SOLUTION As always we begin with a free–body diagram, Fig. 10–43b, which shows the weight of the cylinder acting at the CM and the tension of the string $\mathbf{F}_T$ acting at the edge of the cylinder. We write Newton's second law for the linear motion (down is positive)

$$Ma = \Sigma F$$
$$= Mg - F_T.$$

Since we do not know the tension in the string, we cannot immediately solve for a. We write Newton's second law for the rotational motion, calculated about the center of mass:

$$\Sigma \tau_{CM} = I_{CM}\alpha_{CM}$$
$$F_T R = \tfrac{1}{2}MR^2\alpha.$$

Because the cylinder "rolls without slipping" down the string, we have the additional relation that $a = \alpha R$ (Eq. 10–5). Then the torque equation becomes

$$F_T R = \tfrac{1}{2}MR^2\left(\frac{a}{R}\right) = \tfrac{1}{2}MRa$$

so

$$F_T = \tfrac{1}{2}Ma.$$

Substituting this into the force equation, we obtain

$$Ma = Mg - F_T$$
$$= Mg - \tfrac{1}{2}Ma.$$

Solving for a, we find that $a = \tfrac{2}{3}g$. That is, the linear acceleration is less than what it would be if the cylinder were simply dropped. This makes sense since gravity is not the only vertical force acting; the tension in the string is acting as well.
(b) Since $a = \tfrac{2}{3}g$, $F_T = \tfrac{1}{2}Ma = \tfrac{1}{3}Mg$.

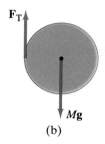

(a)

(b)

FIGURE 10–43 Example 10–23.

EXAMPLE 10–24 What if a rolling ball slips? A bowling ball of mass M and radius R is thrown along a level surface so that initially ($t = 0$) it slides with a linear speed v_0 but does not rotate. As it slides, it begins to spin, and eventually rolls without slipping. How long does it take to begin rolling without slipping?

Rolling with slipping

SOLUTION The free-body diagram is shown in Fig. 10–44, with the ball moving to the right. Newton's second law for translation gives

$$Ma_x = \Sigma F_x = -\mu_k F_N = -\mu_k Mg.$$

where μ_k is the coefficient of kinetic friction because the ball is sliding. The friction force does two things: it acts to slow down the translational motion of the CM; and it immediately acts to start the ball rotating clockwise. The velocity of the CM is

$$v_{CM} = v_0 + a_x t = v_0 - \mu_k gt.$$

Next we apply Newton's second law for rotation about the CM, $I_{CM}\alpha_{CM} = \Sigma \tau_{CM}$:

$$\tfrac{2}{5}MR^2\alpha_{CM} = \mu_k MgR.$$

The angular acceleration is thus $\alpha_{CM} = 5\mu_k g/2R$, which is constant. Then the angular velocity of the ball is (Eq. 10–9a)

$$\omega_{CM} = \omega_0 + \alpha_{CM}t = 0 + \frac{5\mu_k gt}{2R}.$$

The ball starts rolling immediately after it touches the ground, but it rolls and slips at the same time to begin with. It eventually stops slipping, and then rolls without slipping. The condition for rolling without slipping is that

$$v_{CM} = \omega_{CM}R,$$

Eq. 10–4, which is *not* valid if there is slipping. This rolling without slipping begins at a time $t = t_1$ given by (see equations for v_{CM} and ω_{CM} above)

$$v_0 - \mu_k gt_1 = \frac{5\mu_k gt_1}{2R}R \qquad \text{so} \qquad t_1 = \frac{2v_0}{7\mu_k g}.$$

FIGURE 10–44 Example 10–24.

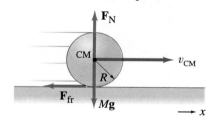

Why Does a Rolling Sphere Slow Down?

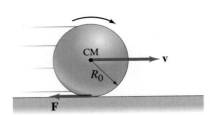

FIGURE 10–45 Sphere rolling to the right.

A sphere of mass M and radius R_0 rolling on a horizontal flat surface eventually comes to rest. What force causes it to come to rest? You might think it is friction, but when you examine the problem from a simple, sraightforward point of view, a paradox seems to arise.

Suppose a sphere is rolling to the right as shown in Fig. 10–45, and is slowing down. By Newton's second law, $\Sigma \mathbf{F} = M\mathbf{a}$, there must be a force $\mathbf{F}$ (presumably frictional) acting to the left as shown, so that the acceleration $\mathbf{a}$ will also point to the left and v will decrease. Curiously enough, though, if we now look at the torque equation (calculated about the center of mass), $\Sigma \tau_{CM} = I_{CM}\alpha$, we see that the force $\mathbf{F}$ acts to increase the angular acceleration α, and thus to *increase* the velocity of the sphere. Thus the paradox! The force $\mathbf{F}$ acts to decelerate the sphere if we look at the translational motion, but speeds it up if we look at the rotational motion.

The resolution of this apparent paradox is that some other force must be acting. The only other forces acting are gravity, $M\mathbf{g}$, and the normal force $\mathbf{F}_N (= -M\mathbf{g})$. These act vertically and hence do not affect the horizontal translational motion. If we assume the sphere and plane are rigid, so the sphere is in contact at only one point, these forces give rise to no torques about the CM either, since they act through the CM.

The only recourse we have to resolve the paradox is to give up our idealization that the bodies are rigid. In fact, all bodies are deformable to some extent. Our sphere flattens slightly and the level surface also acquires a slight depression where the two are in contact. There is an *area* of contact, not a point. Hence there can be a torque at this area of contact which acts in the opposite direction to the torque associated with $\mathbf{F}$, and thus acts to slow down the rotation of the sphere. This torque is associated with the normal force $\mathbf{F}_N$ that the table exerts on the sphere over the whole area of contact. The net effect is that we can consider $\mathbf{F}_N$ acting vertically a distance l in front of the CM as shown in Fig. 10–46 (where the deformation is greatly exaggerated).

FIGURE 10–46 The normal force, $\mathbf{F}_N$, exerts a torque that slows down the sphere. The deformation of the sphere and the surface it moves on has been exaggerated for detail.

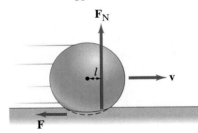

Is it reasonable that the normal force $\mathbf{F}_N$ should effectively act in *front* of the CM as shown in Fig. 10–46? Yes. The sphere is rolling, and the leading edge strikes the surface with a slight impulse. The table therefore pushes upward a bit more strongly on the front part of the sphere than it would if the sphere were at rest. At the back part of the area of contact, the sphere is starting to move upward and so the table pushes upward on it less strongly than when the sphere is at rest. The table pushing up more strongly on the front part of the area of contact gives rise to the necessary torque and justifies the effective acting point of $\mathbf{F}_N$ being in front of the CM.

When other forces are present, the tiny torque τ_N due to $\mathbf{F}_N$ can usually be ignored. For example, when a sphere or cylinder rolls down an incline, the force of gravity has far more influence than τ_N, so the latter can be ignored. For many purposes (but not all), we can assume a hard sphere is in contact with a hard surface at essentially one point.

Summary

When a rigid body rotates about a fixed axis, each point of the body moves in a circular path. Lines drawn perpendicularly from the rotation axis to different points in the body all sweep out the same angle θ in any given time interval.

Angles are conveniently measured in **radians**. One radian is the angle subtended by an arc whose length is equal to the radius, or

$$2\pi \text{ rad} = 360° \qquad \text{so} \qquad 1 \text{ rad} \approx 57.3°.$$

All parts of a rigid body rotating about a fixed axis have the same **angular velocity** ω and the same **angular**

acceleration α at any instant, where

$$\omega = \frac{d\theta}{dt} \qquad \text{and} \qquad \alpha = \frac{d\omega}{dt}.$$

The linear velocity and acceleration of any point in a body rotating about a fixed axis are related to the angular quantities by

$$v = R\omega, \qquad a_{\tan} = R\alpha, \qquad a_R = \omega^2 R$$

where R is the perpendicular distance of the point from the rotation axis, and $a_{\tan}$ and a_R are the tangential and radial components of the linear acceleration. The frequency f and period T are related to ω by

$$f = \frac{1}{T} = \frac{\omega}{2\pi}.$$

If a rigid body undergoes uniformly accelerated rotational motion ($\alpha = $ constant), equations analogous to those for linear motion are valid:

$$\omega = \omega_0 + \alpha t; \qquad \theta = \omega_0 t + \tfrac{1}{2}\alpha t^2;$$

$$\omega^2 = \omega_0^2 + 2\alpha\theta; \qquad \overline{\omega} = \frac{\omega + \omega_0}{2}.$$

Angular velocity and angular acceleration are vectors. For a rigid body rotating about a fixed axis, both $\boldsymbol{\omega}$ and $\boldsymbol{\alpha}$ point along the rotation axis. The direction of $\boldsymbol{\omega}$ is given by the **right-hand rule**.

The **torque** due to a force $\mathbf{F}$ exerted on a rigid body is equal to

$$\tau = R_\perp F = RF_\perp = RF \sin\theta,$$

where $R_\perp$, called the **lever arm**, is the perpendicular distance from the axis of rotation to the line along which the force acts, and θ is the angle between $\mathbf{F}$ and $\mathbf{R}$.

The rotational equivalent of Newton's second law is

$$\Sigma\tau = I\alpha,$$

where $I = \Sigma m_i R_i^2$ is the **moment of inertia** of the body

about the axis of rotation. This relation is valid for a rigid body rotating about an axis fixed in an inertial reference frame, or when τ, I, and α are calculated about the center of mass of a body even if the CM is moving.

The **angular momentum**, L, of a body about a fixed rotation axis is given by

$$L = I\omega.$$

Newton's second law, in terms of angular momentum, becomes

$$\Sigma\tau = \frac{dL}{dt}.$$

If the net torque on the body is zero, $dL/dt = 0$, so $L = $ constant. This is the **law of conservation of angular momentum** for a rotating body.

The **rotational kinetic energy** of a body rotating about a fixed axis with angular velocity ω is

$$K = \tfrac{1}{2}I\omega^2.$$

For a body undergoing both translational and rotational motion, the total kinetic energy is the sum of the translational kinetic energy of the body's CM plus the rotational kinetic energy of the body about its CM:

$$K = \tfrac{1}{2}Mv_{\text{CM}}^2 + \tfrac{1}{2}I_{\text{CM}}\omega^2$$

as long as the rotation axis is fixed in direction.

Questions

1. A bicycle odometer (which measures distance traveled) is attached near the wheel hub and is designed for 27-in. wheels. What happens if you use it on a bicycle with 24-in. wheels?

2. Suppose a circular platform rotates at constant angular velocity. Does a point on the rim have radial and/or tangential acceleration? If the platform accelerates uniformly, does the point have radial and/or tangential acceleration? What would cause the magnitude of either (linear) acceleration to change?

3. A rotating carousel has one child sitting on a horse near the outer edge and another child seated on a lion halfway out from the center. (a) Which child has the greater linear speed? (b) Which child has the greater angular speed?

4. In what direction is the Earth's angular velocity for its daily rotation on its axis?

5. The angular velocity of a wheel rotating on a horizontal axle points west. In what direction is the linear velocity of a point on the top of the wheel? If the angular acceleration points east, describe the tangential linear acceleration of this point. Is the angular speed increasing or decreasing?

6. If the angular quantities θ, ω and α were specified in terms of degrees rather than radians, how would Eqs. 10–9 for uniformly accelerated rotational motion have to be altered?

7. Can a small force exert a greater torque than a larger force? Explain.

8. If a force $\mathbf{F}$ acts on a body such that its lever arm is zero, does it have any effect on the body's motion?

9. If the net force on a system is zero, is the net torque also zero? If the net torque on a system is zero, is the net force zero?

10. The moment of inertia of this textbook would be the least about what axis?

11. Why is it more difficult to do a situp with your hands behind your head than when they are outstretched in front of you? A diagram may help you to answer this.

12. Expert bicyclists use very lightweight tubular ("sew-up") tires. They claim that reducing the mass of the tires is far more significant than an equal reduction in mass elsewhere on the bicycle. Explain.

13. Why do tightrope walkers carry a long, narrow beam (Fig. 10–47)?

FIGURE 10–47 Question 13.

14. A quarterback leaps into the air to throw a forward pass. As he throws the ball, the upper part of his body rotates. If you look quickly you will notice that his hips and legs rotate in the opposite direction (Fig. 10–48). Explain.

FIGURE 10–48
Quarterback in the air, throwing a pass. Question 14.

15. We claim that momentum and angular momentum are conserved. Yet most moving or rotating bodies eventually slow down and stop. Explain.

16. If there were a great migration of people toward the equator, how would this affect the length of the day?

17. Can the diver of Fig. 10–30 do a somersault without having any initial rotation when she leaves the board?

18. Suppose you are standing on the edge of a large rotating turntable. What happens if you walk toward the center?

19. When a motorcyclist leaves the ground on a jump and leaves the throttle on (so the rear wheel spins), why does the front of the cycle rise up?

20. A stick stands vertically on its end on a frictionless surface. Describe the motion of its CM, and of each end, when it is tipped slightly to one side and falls.

21. Two inclines have the same height but make different angles with the horizontal. The same steel ball is rolled down each incline. On which incline will the speed of the ball at the bottom be greatest? Explain.

22. Two solid spheres simultaneously start rolling (from rest) down an incline. One sphere has twice the radius and twice the mass of the other. Which reaches the bottom of the incline first? Which has the greater speed there? Which has the greater total kinetic energy at the bottom?

23. A sphere and a cylinder have the same radius and the same mass. They start from rest at the top of an incline. Which reaches the bottom first? Which has the greater speed at the bottom? Which has the greater total kinetic energy at the bottom? Which has the greater rotational kinetic energy?

24. A cyclist rides over the top of a hill. Is the bicycle's motion rotational, translational, or a combination of both?

* **25.** The total kinetic energy of a system of particles is equal to the translational kinetic energy of the center of mass of the system plus the kinetic energy of motion relative to the CM (Eq. 10–26). Is it possible or useful to make a similar statement about the total linear momentum?

Problems

Sections 10–1 to 10–3

1. (I) What are the following angles expressed in radians: (a) 30°, (b) 57°, (c) 90°, (d) 360°, and (e) 420°? Give as numerical values and as fractions of π.

2. (I) The Sun subtends an angle of about 0.5° to us on Earth, 150 million km away. Estimate the radius of the Sun.

3. (I) The Eiffel Tower is 300 m tall. When you are standing at a certain place in Paris, it subtends an angle of 7.5°. How far are you, then, from the Eiffel Tower?

4. (I) A centrifuge rotor is accelerated from rest to 20,000 rpm in 5.0 min. What is its average angular acceleration?

5. (II) Calculate the angular velocity of (a) the second hand, (b) the minute hand, and (c) the hour hand, of a clock. State in rad/s. (d) What is the angular acceleration in each case?

6. (II) Calculate the angular velocity of the Earth (a) in its orbit around the Sun and (b) about its axis.

7. (II) What is the linear speed of a point (a) on the Equator, (b) on the Arctic Circle (latitude 66.5° N), (c) at a latitude of 40.0° N, due to the Earth's rotation?

8. (II) Estimate the angle subtended by the Moon using a ruler and your finger or other object to just blot out the Moon. Describe your measurement and the result obtained and then use it to estimate the diameter of the Moon. The Moon is about 380,000 km from the Earth.

9. (II) (a) A 0.35-m diameter grinding wheel rotates at 2500 rpm. Calculate its angular velocity in rad/s. (b) What is the linear speed and acceleration of a point on the edge of the grinding wheel?

10. (II) A rotating merry-go-round makes one complete revolution in 4.0 s (Fig. 10–49). (a) What is the linear speed of a child seated 1.2 m from the center? (b) What is her acceleration?

FIGURE 10–49 Problem 10.

11. (II) A child rolls a ball on a level floor 3.5 m to another child. If the ball makes 15.0 revolutions, what is its diameter?

12. (II) A 60.0-cm-diameter wheel accelerates uniformly from 210 rpm to 380 rpm in 6.5 s. How far will a point on the edge of the wheel have traveled in this time?

13. (II) In traveling to the Moon, astronauts aboard the Apollo spacecraft put it into a slow rotation in order to distribute the Sun's energy evenly. At the start of their trip, they accelerated uniformly during a 10-min time interval from no rotation to one revolution every minute. The spacecraft can be thought of as a cylinder with a diameter of 8.5 m. Determine (a) the angular acceleration, and (b) the radial and tangential components of the linear acceleration of a point on the surface of the spacecraft 5.0 min after it started this acceleration.

14. (II) Using calculus, derive the angular kinematic equations 10–9a and 10–9b for constant angular acceleration. Start with $\alpha = d\omega/dt$.

15. (II) A small rubber wheel is used to drive a large pottery wheel, and they are mounted so that their circular edges touch. If the small wheel has a radius of 2.0 cm and accelerates at the rate of 7.2 rad/s², and it is in contact with the pottery wheel (radius 25.0 cm) without slipping, calculate (a) the angular acceleration of the pottery wheel, and (b) the time it takes the pottery wheel to reach its required speed of 65 rpm.

16. (II) The angle through which a rotating wheel has turned in time t is given by $\theta = 6.0t - 8.0t^2 + 4.5t^4$, where θ is in radians and t in seconds. Determine an expression (a) for the instantaneous angular velocity ω and (b) for the instantaneous angular acceleration α. (c) Evaluate ω and α at $t = 3.0$ s. (d) What is the average angular velocity, and (e) the average angular acceleration between $t = 2.0$ s and $t = 3.0$ s?

17. (II) The angular acceleration of a wheel, as a function of time, is $\alpha = 5.0t^2 - 3.5t$, where α is in rad/s² and t in seconds. If the wheel starts from rest ($\theta = \omega = 0$ at $t = 0$), determine a formula for (a) the angular velocity ω and (b) the angular position θ, both as a function of time. (c) Evaluate ω and θ at $t = 2.0$ s.

18. (II) The tires of a car make 85 revolutions as the car reduces its speed uniformly from 90.0 km/h to 60.0 km/h. The tires have a diameter of 0.90 m. (a) What was the angular acceleration? (b) If the car continues to decelerate at this rate, how much more time is required for it to stop?

Section 10–4

19. (II) The axle of a wheel is mounted on supports that rest on a rotating turntable as shown in Fig. 10–50. The wheel has angular velocity $\omega_1 = 50.0$ rad/s about its axle, and the turntable has angular velocity $\omega_2 = 35.0$ rad/s about a vertical axis. (Note arrows showing these motions in the figure.) (a) What are the directions of ω_1 and ω_2? (b) What is the resultant angular velocity of the wheel, as seen by an outside observer, at the instant shown? Give the magnitude and direction. (c) What is the magnitude and direction of the angular acceleration of the wheel at the instant shown? Take the z axis vertically upward and the direction of the axle at the moment shown to be the x axis pointing to the right.

FIGURE 10–50
Problem 19.

Section 10–5

20. (I) A person exerts a force of 38 N on the end of a door 96 cm wide. What is the magnitude of the torque if the force is exerted (a) perpendicular to the door and (b) at a 60.0° angle to the face of the door?

21. (I) The biceps muscle exerts a vertical force on the lower arm as shown in Figs. 10–51a and b. For each case, calculate the torque about the axis of rotation through the elbow joint, assuming the muscle is attached 5.0 cm from the elbow as shown.

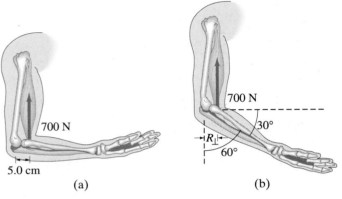

FIGURE 10–51   Problem 21.

22. (II) A wheel of diameter 27.0 cm is constrained to rotate in the xy plane, about the z axis, which passes through its center. A force $\mathbf{F} = (-31.0\mathbf{i} + 38.6\mathbf{j})$ N acts at a point on the edge of the wheel that lies exactly on the x axis at a particular instant. What is the torque about the rotation axis at this instant?

23. (II) Calculate the net torque about the axle of the wheel shown in Fig. 10–52. Assume that a friction torque of 0.30 m·N opposes the motion.

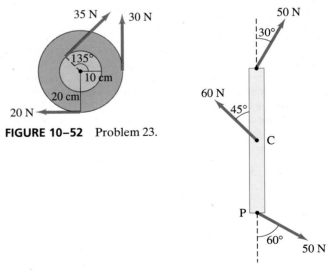

FIGURE 10–52   Problem 23.

FIGURE 10–53   Problem 24.

24. (II) Determine the net torque on the 2.0-m-long beam shown in Fig. 10–53. Calculate about (a) point C, the CM, and (b) point P at one end.

25. (II) The bolts on the cylinder head of certain engines require tightening to a torque of 90 m·N. If a wrench is 26 cm long, what force perpendicular to the wrench must the mechanic exert at its end? If the six-sided bolt is 15 mm in diameter, estimate the force applied near each of the six points by a socket wrench (Fig. 10–54).

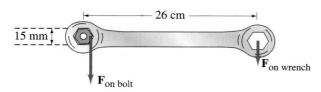

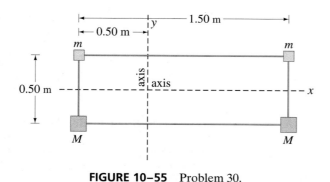

FIGURE 10–54 Problem 25.

Sections 10–6 to 10–7

26. (I) Determine the moment of inertia of a 12.0-kg solid sphere of radius 0.80 m when the axis of rotation is through its center.

27. (I) A 1.4-kg grindstone in the shape of a uniform cylinder of radius 0.20 m acquires a rotational rate of 1800 rev/s from rest over a 6.0-s interval at constant angular acceleration. Calculate the torque delivered by the motor.

28. (I) An oxygen molecule consists of two oxygen atoms whose total mass is 5.3×10^{-26} kg and whose moment of inertia about an axis perpendicular to the line joining the two atoms, midway between them, is 1.9×10^{-46} kg·m². Estimate, from these data, the effective distance between the atoms.

29. (II) A 2.4-kg ball on the end of a thin, light rod is rotated in a horizontal circle of radius 1.2 m. Calculate (a) the moment of inertia of the ball about the center of the circle, and (b) the torque needed to keep the ball rotating at constant angular velocity if air resistance exerts a force of 0.020 N on the ball. Ignore the rod's moment of inertia and air resistance.

30. (II) Calculate the moment of inertia of the array of point objects shown in Fig.10–55 about (a) the vertical axis, and (b) the horizontal axis. Assume the objects are connected by very light rigid wires. About which axis would it be harder to accelerate this array? In Fig. 10–55, $m = 1.8$ kg and $M = 3.1$ kg. The array is rectangular and it is split through the middle by the horizontal axis.

FIGURE 10–55 Problem 30.

31. (II) Consider a helicopter rotor blade as a long thin rod, as shown in Fig. 10–56. If each of the three helicopter blades is 3.75 m long and has a mass of 160 kg, calculate the moment of inertia of the three rotor blades about the axis of rotation. How much torque must the motor apply to bring the blades up to a speed of 5.0 rev/s in 8.0 s?

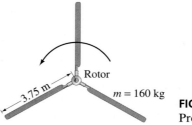

FIGURE 10–56
Problem 31.

32. (II) A 0.84-m-diameter solid sphere can be rotated about an axis through its center by a torque of 10.8 m·N which accelerates it uniformly from rest through a total of 180 revolutions in 15.0 s. What is the mass of the sphere?

33. (II) A merry-go-round accelerates from rest to 3.0 rad/s in 24 s. Assuming the merry-go-round is a uniform disk of radius 7.0 m and mass 31,000 kg, calculate the net torque required to accelerate it.

34. (II) Four equal masses M are spaced at equal intervals, l, along a horizontal straight rod whose mass can be ignored. The system is to be rotated about a vertical axis passing through the mass at the left end of the rod and perpendicular to it. (a) What is the moment of inertia of the system about this axis? (b) What minimum force, applied to the farthest mass, will impart an angular acceleration α? (c) What is the direction of this force?

35. (II) Suppose the force F_T in the cord hanging from the wheel of Example 10–7, Fig. 10–22, is given by the relation $F_T = 3.00t - 0.20t^2$ (newtons) where t is in seconds. If the wheel starts from rest, what is the linear speed of a point on its rim 8.0 s later?

36. (II) A centrifuge rotor rotating at 10,000 rpm is shut off and is eventually brought to rest by a frictional torque of 1.00 m·N. If the mass of the rotor is 4.70 kg and it can be considered a solid cylinder of radius 0.0780 m, through how many revolutions will the rotor turn before coming to rest, and how much time will this take?

37. (II) Two blocks are connected by a light string passing over a pulley of radius 0.25 m and moment of inertia I. The blocks move to the right with an acceleration of 1.00 m/s² on inclines with frictionless surfaces (see Fig. 10–57). (a) Draw free-body diagrams for each of the two blocks and the pulley. (b) Determine F_{T1} and F_{T2}, the tensions in the two parts of the string. (c) Find the net torque acting on the pulley, and determine its moment of inertia, I.

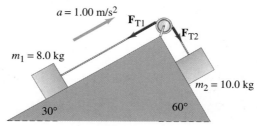

FIGURE 10–57
Problem 37.

38. (II) A string passing over a pulley has a 3.80-kg mass hanging from one end and a 3.40-kg mass hanging from the other end. The pulley is a uniform solid cylinder of radius 3.0 cm and mass 0.80 kg. (*a*) If the bearings of the pulley were frictionless, what would be the acceleration of the two masses? (*b*) In fact, it is found that if the heavier mass is given a downward speed of 0.20 m/s, it comes to rest in 6.2 s. What is the average frictional torque acting on the pulley?

39. (II) When discussing moments of inertia, especially for unusual or irregularly shaped objects, it is sometimes convenient to work with the **radius of gyration**, k. This radius is defined so that if all the mass of the object were concentrated at this distance from the axis, the moment of inertia would be the same as that of the original object. Thus, the moment of inertia of any object can be written in terms of its mass M and the radius of gyration as $I = Mk^2$. Determine the radius of gyration for each of the objects (hoop, cylinder, sphere, etc.) shown in Fig. 10–21.

40. (III) A thin rod of length l stands vertically on a table. The rod begins to fall, but its lower end does not slide. (*a*) Determine the angular velocity of the rod as a function of the angle ϕ it makes with the tabletop. (*b*) What is the speed of the tip of the rod just before it strikes the table?

41. (III) A hammer thrower accelerates the hammer (mass = 7.30 kg) from rest within four full turns (revolutions) and releases it at a speed of 29.0 m/s. Assuming a uniform rate of increase in angular velocity and a radius of 2.00 m, calculate (*a*) the angular acceleration, (*b*) the (linear) tangential acceleration, (*c*) the centripetal acceleration just before release, (*d*) the net force exerted on the hammer by the athlete just before release, and (*e*) the angle of this force with respect to the radius of the circular motion. Neglect the effect of gravity.

Section 10–8

42. (I) Use the parallel-axis theorem to show that the moment of inertia of a thin rod about an axis perpendicular to the rod at one end is $I = \frac{1}{3}Ml^2$, given that if the axis passes through the center, $I = \frac{1}{12}Ml^2$ (Fig. 10–21f and g).

43. (II) Use the perpendicular-axis theorem and Fig. 10–21h to determine a formula for the moment of inertia of a thin, square plate of side s about an axis (*a*) through its center and along a diagonal of the plate, (*b*) through the center and parallel to a side.

44. (II) Determine the moment of inertia of a 25-kg door that is 2.5 m high and 1.0 m wide and is hinged along one side. Ignore the thickness of the door.

45. (II) Two uniform solid spheres of mass M and radius R_0 are connected by a thin (massless) rod of length R_0 so that the centers are $3R_0$ apart. (*a*) Determine the moment of inertia of this system about an axis perpendicular to the rod at its center. (*b*) What would be the percentage error if the masses of each sphere were assumed to be concentrated at their centers and a very simple calculation made?

46. (II) A ball of mass M and radius R_1 on the end of a thin massless rod is rotated in a horizontal circle of radius R_0 about an axis of rotation AB, as shown in Fig. 10–58. (*a*) Considering the mass of the ball to be concentrated at

its center of mass, calculate its moment of inertia about AB. (*b*) Using the parallel-axis theorem and considering the finite radius of the ball, calculate the moment of inertia of the ball about AB. (*c*) Calculate the percentage error introduced by the point mass approximation for $R_1 = 10$ cm and $R_0 = 1.0$ m.

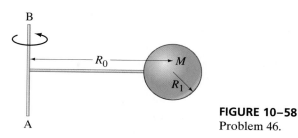

FIGURE 10–58
Problem 46.

47. (II) What is the moment of inertia of a uniform horizontal disk of radius R_0 and mass M about (*a*) a vertical axis a distance $0.25R_0$, from the center, (*b*) a horizontal axis through the center, and (*c*) a horizontal axis tangent to the edge.

48. (II) A thin 7.0-kg wheel of radius 32 cm is weighted to one side by a 1.50 kg weight, small in size, placed 22 cm from the center of the wheel. Calculate (*a*) the position of the center of mass of the weighted wheel and (*b*) the moment of inertia about an axis through its center of mass, perpendicular to its face.

49. (III) Derive the formula for the moment of inertia of a uniform sphere of radius r_0 and mass M about an axis through its center (Fig. 10–21e). [*Hint*: Divide the sphere into infinitesimally thin disks (cylinders) of thickness dy. Then use the result of Example 10–10 or Fig. 10–21c, and integrate over these cylindrical disks.]

50. (III) Derive the formula for the moment of inertia of a uniform thin rod of length l about an axis through its center, perpendicular to the rod (see Fig. 10–21f).

51. (III) (*a*) Derive the formula given in Fig. 10–21h for the moment of inertia of a uniform, flat, rectangular plate of dimensions $l \times w$ about an axis through its center, perpendicular to the plate. (*b*) What is the moment of inertia about each of the axes through the center that are parallel to the edges of the plate?

Section 10–9

52. (I) (*a*) What is the angular momentum of a 2.8-kg uniform cylindrical grinding wheel of radius 18 cm when rotating at 1500 rpm? (*b*) How much torque is required to stop it in 7.0 s?

53. (I) A diver (such as the one in Fig. 10–30) can reduce her moment of inertia by a factor of about 3.5 when changing from the straight position to the tuck position. If she makes two rotations in 1.5 s when in the tuck position, what is her angular speed (rev/s) when in the straight position?

54. (I) A figure skater during her finale can increase her rotation rate from an initial rate of 1.0 rev every 2.0 s to a final rate of 3.0 rev/s. If her initial moment of inertia was 4.6 kg·m², what is her final moment of inertia? How does she physically accomplish this change?

55. (II) A person stands, hands at the side, on a platform that is rotating at a rate of 1.30 rev/s. If the person now raises his arms to a horizontal position, Fig. 10–59, the speed of rotation decreases to 0.80 rev/s. (a) Why does this occur? (b) By what factor has the moment of inertia of the person changed?

FIGURE 10–59 Problem 55.

56. (II) (a) What is the angular momentum of a figure skater spinning (with arms in close to her body) at 3.5 rev/s, assuming her to be a uniform cylinder with a height of 1.5 m, a radius of 15 cm, and a mass of 55 kg. (b) How much torque is required to slow her to a stop in 5.0 s, assuming she does *not* move her arms?

57. (II) Determine the angular momentum of the Earth (a) about its rotation axis (assume the Earth is a uniform sphere), and (b) in its orbit around the Sun (treat the Earth as a particle orbiting the Sun). The Earth has mass = 6.0×10^{24} kg, radius = 6.4×10^6 m, and is 1.5×10^8 km from the Sun.

58. (II) A uniform horizontal rod of mass M and length l rotates with angular velocity ω about a vertical axis through its center. Attached to each end of the rod is a small mass m. Determine the angular momentum of the system about the axis.

59. (II) A 4.8-m-diameter merry-go-round is rotating freely with an angular velocity of 0.80 rad/s. Its total moment of inertia is 1950 kg·m². Four people standing on the ground, each of 65-kg mass, suddenly step onto the edge of the merry-go-round. What is the angular velocity of the merry-go-round now? What if the people were on it initially and then jumped off in a radial direction (relative to the merry-go-round)?

60. (II) A woman of mass m stands at the edge of a solid cylindrical platform of mass M and radius R. At $t = 0$, the platform is rotating with negligible friction at angular velocity ω_0 about a vertical axis through its center, and the person begins walking with speed v (relative to the platform) toward the center of the platform. (a) Determine the angular velocity of the system as a function of time. (b) What will be the angular velocity when the woman reaches the center?

Section 10–10

61. (I) A centrifuge rotor has a moment of inertia of 4.25×10^{-2} kg·m². How much energy is required to bring it from rest to 10,000 rpm?

62. (I) An automobile engine develops a torque of 280 m·N at 4000 rpm. What is the horsepower of the engine?

63. (II) For the star in Example 10–14 undergoing gravitational collapse, by what factor did the original kinetic energy change? Where did this energy come from?

64. (II) A rotating uniform cylindrical platform of mass 280 kg and radius 5.5 m slows down from 3.8 rev/s to rest in 18 s when the driving motor is disconnected. Estimate the power output of the motor (hp) required to maintain a steady speed of 3.8 rev/s.

65. (II) Two masses, $m_1 = 35.0$ kg and $m_2 = 38.0$ kg, are connected by a rope that hangs over a pulley (as in Fig. 10–60). The pulley is a uniform cylinder of radius 0.30 m and mass 4.8 kg. Initially m_1 is on the ground and m_2 rests 2.5 m above the ground. If the system is released, use conservation of energy to determine the speed of m_2 just before it strikes the ground. Assume the pulley bearing is frictionless.

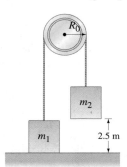

FIGURE 10–60
Problem 65.

66. (II) A uniform thin rod of length l and mass M is suspended freely from one end. It is pulled to the side an angle θ and released. If friction can be ignored, what is its angular velocity, and the speed of its free end, at the lowest point?

67. (II) A 70-kg person stands on a tiny rotating platform with arms outstretched. (a) Estimate the moment of inertia of the person using the following approximations: the body (including head and legs) is a 60-kg cylinder, 12 cm in radius and 1.70 m high; and each arm is a 5.0-kg thin rod, 60 cm long, attached to the cylinder. (b) Using the same approximations, estimate the moment of inertia when the arms are at the person's sides. (c) If one rotation takes 1.5 s when the person's arms are outstretched, what is the time for each rotation with arms at the sides? Ignore the moment of inertia of the lightweight platform. (d) Determine the change in kinetic energy when the arms are lifted from the sides to the horizontal position. (e) From your answer to part (d), would you expect it to be harder or easier to lift your arms when rotating or when at rest?

68. (III) A 4.00-kg mass and a 3.00-kg mass are attached to opposite ends of a thin 50.0-cm-long horizontal rod (Fig. 10–61). The system is rotating at angular speed $\omega = 8.00$ rad/s about a vertical axle at the center of the rod. Determine (a) the kinetic energy K of the system, and (b) the net force on each mass. (c) Repeat parts (a) and (b) assuming that the axle passes through the center of mass of the system.

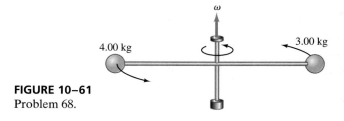

FIGURE 10–61
Problem 68.

69. (I) Calculate the translational speed of a cylinder when it reaches the foot of an incline 11.8 m high. Assume it starts from rest and rolls without slipping.

70. (I) Estimate the kinetic energy of the Earth with respect to the Sun as the sum of two terms, (a) that due to its daily rotation about its axis and (b) that due to its yearly revolution about the Sun (assume the Earth is a uniform sphere, mass = 6.0×10^{24} kg, radius = 6.4×10^6 m, and is 1.5×10^8 km from the Sun).

71. (I) A bowling ball of mass 7.3 kg and radius 9.0 cm rolls without slipping down a lane at 5.3 m/s. Calculate its total kinetic energy.

72. (II) A narrow but solid spool of thread has radius R and mass M. If you pull up on the thread so that the CM of the spool remains suspended in the air at the same place, (a) what force must you exert on the thread? (b) How much work have you done by the time the spool turns with angular velocity ω?

73. (II) A thin, hollow 60.0-g section of pipe of radius 10.0 cm starts rolling (from rest) down a 21.5° incline 5.60 m long. (a) If the pipe rolls without slipping, what will be its speed at the base of the incline? (b) What will be its total kinetic energy at the base of the incline? (c) What minimum value must the coefficient of static friction have if the pipe is not to slip?

74. (II) A solid rubber ball rests on the floor of a railroad car when the car begins moving with acceleration a. Assuming the ball rolls without slipping, what is its acceleration relative to (a) the car and (b) the ground?

75. (II) A ball of radius r_0 rolls on the inside of a track of radius R_0 (see Fig. 10–62). If the ball starts from rest at the vertical edge of the track, what will be its speed when it reaches the lowest point of the track, rolling without slipping?

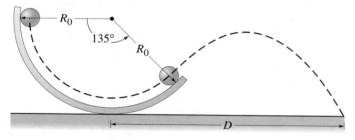

FIGURE 10–62 Problems 75 and 76.

76. (III) A small sphere of radius $r_0 = 1.5$ cm rolls without slipping on the track shown in Fig. 10–62 whose radius is $R_0 = 26.0$ cm. The sphere starts rolling at a height R_0 above the bottom of the track. When it leaves the track after passing through an angle of 135° as shown, (a) what will be its speed, and (b) at what distance D from the base of the track will the sphere hit the ground?

77. (III) The 1100-kg mass of a car includes four tires, each of mass (including wheels) 35 kg and diameter 0.80 m. Assume each tire and wheel combination acts as a solid cylinder. Determine (a) the total kinetic energy of the car when traveling 100 km/h and (b) the fraction of the kinetic energy in the tires and wheels. (c) If the car is initially at rest and is then pulled by a tow truck with a force of 2000 N, what is the acceleration of the car? Ignore frictional losses. (d) What percent error would you make in part (c) if you ignored the rotational inertia of the tires and wheels?

*__78.__ (III) Determine the speed of the falling cylinder in Example 10–23 in terms of the distance fallen h. Choose $h = 0$ and $v = 0$ at $t = 0$.

*__79.__ (III) In Example 10–24, (a) how far has the ball moved down the lane when it starts rolling without slipping? (b) What are its final linear and rotational speeds?

*__80.__ (III) A wheel with rotational inertia $I = \frac{1}{2}MR^2$ about its central axle is set spinning with initial angular speed ω_0 and is then lowered onto the ground so that it touches the ground with no horizontal speed. Initially it slips, but then begins to move forward and eventually rolls without slipping. (a) In what direction does friction act on the slipping wheel? (b) How long does the wheel slip before it begins to roll without slipping? (c) What is the wheel's final translational speed? [*Hint:* Use $\Sigma F = ma$, $\Sigma \tau_{CM} = I_{CM}\alpha_{CM}$, and recall that only when there is rolling without slipping is $v_{CM} = \omega R$.]

*__81.__ (III) In Examples 10–19 and 10–21 suppose the sphere does not simply roll, but slips as well. To be concrete, suppose $\theta = 33.0°$ and $\mu = \mu_s = \mu_k = 0.10$ with the incline having a height $H = 2.0$ m and the sphere a radius $R_0 = 11.0$ cm with mass $M = 850$ g. The sphere starts from rest at the top of the incline. Determine: (a) the acceleration of the sphere; (b) the speed of the CM of the sphere when it reaches the foot of the incline; (c) the total kinetic energy of the sphere at the foot; and (d) the loss in mechanical energy. (e) Determine these same quantities for $\mu = 0.30$, for which there is no slipping, and compare. (f) Compare your answers in (a), (b), and (c) to a box of the same mass sliding down the incline. [*Hint:* Use $\Sigma F = ma$, $\Sigma \tau_{CM} = I_{CM}\alpha_{CM}$, and recall that $v_{CM} = \omega R$ is valid only when there is no slipping.]

*## Section 10–12

*__82.__ (I) A rolling ball slows down because the normal force does not pass exactly through the CM of the ball, but passes in front of the CM. Using Fig. 10–46, show that the torque resulting from the normal force ($\tau_N = lF_N$ in Fig. 10–46) is 7/5 of that due to the frictional force, $\tau_{fr} = R_0 F$; that is, show that $\tau_N = \frac{7}{5}\tau_{fr}$.

83. Eclipses happen on Earth because of an amazing coincidence. Calculate, using the information inside the front cover, the angular diameter (in radians) of the Sun and the angular diameter of the Moon, as seen on Earth.

84. A hollow cylinder (hoop) is rolling on a horizontal surface at speed $v = 2.10$ m/s when it starts up a $21.5°$ incline. (a) How far along the incline will it go? (b) How long will it be on the incline before it arrives back at the bottom?

85. A solid wheel of radius R_1 is turned by a circular rubber roller of radius R_2 in contact with it at their outer edges. What is the ratio of their angular velocities, ω_1/ω_2?

86. The Moon orbits the Earth so that the same side always faces the Earth. Determine the ratio of its spin angular momentum (about its own axis) to its orbital angular momentum. (In the latter case, treat the Moon as a particle orbiting the Earth.)

87. A large spool of rope stands on the ground with the end of the rope lying on the top edge of the spool. A person grabs the end of the rope and walks a distance l, holding onto it, Fig. 10–63. The spool rolls behind the person without slipping. What length of rope unwinds from the spool? How far does the spool's CM move?

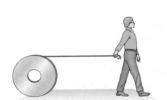

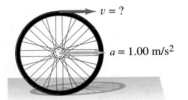

FIGURE 10–63   Problem 87.   FIGURE 10–64   Problem 88.

88. A cyclist accelerates from rest at a rate of 1.00 m/s². How fast will a point on the rim of the tire (diameter $= 68$ cm) at the top be moving after 3.0 s? See Fig. 10–64.

89. A person stands on a platform, initially at rest, that can rotate freely without friction. The moment of inertia of the person plus the platform is I_p. The person holds a spinning bicycle wheel with axis horizontal. The wheel has moment of inertia I_w and angular velocity ω_w. What will be the angular velocity ω_p of the platform if the person moves the axis of the wheel so that it points (a) vertically upward, (b) at a $60°$ angle to the vertical, (c) vertically downward? (d) What will ω_p be if the person reaches up and stops the wheel in part (a)?

90. The density (mass per unit length) of a thin rod of length l increases uniformly from λ_0 at one end to $2\lambda_0$ at the other end. Determine the moment of inertia about an axis perpendicular to the rod through its geometric center.

91. (a) For a bicycle (see Fig. 10–65), how is the angular speed of the rear wheel (ω_R) related to that of the pedals and front sprocket (ω_F)? That is, derive a formula for ω_R/ω_F. Let N_F and N_R be the number of teeth on the front and rear sprockets, respectively. The teeth are spaced equally on all sprockets so that the chain meshes properly. (b) Then evaluate the ratio ω_R/ω_F when the front and rear sprockets have 52 and 13 teeth, respectively, and (c) when they have 42 and 28 teeth.

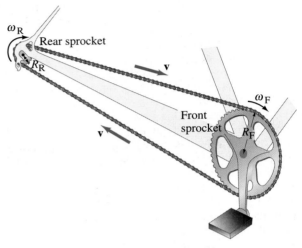

FIGURE 10–65   Problem 91.

92. The forearm in Fig. 10–66 accelerates a 1.00-kg ball at 7.0 m/s² by means of the triceps muscle as shown. Calculate (a) the torque needed and (b) the force that must be exerted by the triceps muscle. Ignore the mass of the arm.

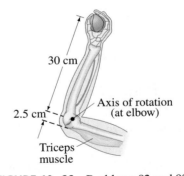

FIGURE 10–66   Problems 92 and 93.

93. Assume that a 1.00-kg ball is thrown solely by the action of the forearm (Fig. 10–66), which rotates about the elbow joint under the action of the triceps muscle. The ball is accelerated from rest to 10.0 m/s in 0.22 s, at which point it is released. Calculate (a) the angular acceleration of the arm and (b) the force required of the triceps muscle. Assume the forearm has a mass of 3.4 kg and rotates like a uniform rod about an axis at its end.

94. A softball player swings a bat, accelerating it from rest to 3.0 rev/s in a time of 0.20 s. Approximate the bat as a 2.2-kg uniform rod of length 0.95 m, and compute the torque the player applies to one end of it.

95. Pilots can be tested for the stresses of flying highspeed jets in a whirling "human centrifuge" which takes 1.0 min to turn through 20 complete revolutions before reaching its final speed. (*a*) What was its angular acceleration, and (*b*) what was its final speed in rpm?

96. A uniform disk turns at 7.0 rev/s around a frictionless spindle. A nonrotating rod, of the same mass as the disk and length equal to the disk's diameter, is dropped onto the freely spinning disk. They then both turn around the spindle with their centers superposed, Fig. 10–67. What is the angular velocity in rev/s of the combination?

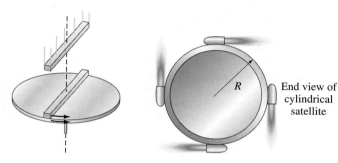

FIGURE 10–67
Problem 96.

FIGURE 10–68 Problem 97.

97. In order to get a flat uniform cylindrical satellite spinning at the correct rate, engineers fire four tangential rockets as shown in Fig. 10–68. If the satellite has a mass of 2000 kg and a radius R of 3.0 m, what is the required steady force of each rocket if the satellite is to reach 30 rpm in 5.0 min?

98. (*a*) Calculate the translational and rotational speeds of a sphere (radius 20.0 cm and mass 2.20 kg), that rolls without slipping down a 30.0° incline that is 10.0 m long, when it reaches the bottom. Assume it started from rest. (*b*) What is its ratio of translational to rotational kinetic energy at the bottom? Avoid putting in numbers until the end so you can answer: (*c*) do your answers in (*a*) and (*b*) depend on the radius of the sphere or its mass?

99. One possibility for a low-pollution automobile is for it to use energy stored in a heavy rotating flywheel. Suppose such a car has a total mass of 1400 kg, uses a 1.50-m diameter uniform cylindrical flywheel of mass 240 kg, and should be able to travel 300 km without needing a flywheel "spinup." (*a*) Make reasonable assumptions (average frictional retarding force = 500 N, twenty acceleration periods from rest to 90 km/h, equal uphill and downhill—assuming during downhill, energy can be put back into the flywheel), and show that the total energy needed to be stored in the flywheel is about 1.6×10^8 J. (*b*) What is the angular velocity of the flywheel when it has a full "energy charge"? (*c*) About how long would it take a 150-hp motor to give the flywheel a full energy charge before a trip?

100. If the coefficient of static friction between tires and pavement is 0.75, calculate the minimum torque that must be applied to the 66-cm-diameter tire of a 1250-kg automobile in order to "lay rubber" (make the wheels spin, slipping as the car accelerates). Assume each wheel supports an equal share of the weight.

101. The radius of the roll of paper shown in Fig. 10–69 is 7.6 cm and its moment of inertia is $I = 2.9 \times 10^{-3}$ kg·m². A force of 3.2 N is exerted on the end of the roll for 1.3 s, but the paper does not tear so it begins to unroll. A constant friction torque of 0.11 m·N is exerted on the roll which gradually brings it to a stop. Assuming that the paper's thickness is negligible, calculate (*a*) the length of paper that unrolls during the time that the force is applied (1.3 s) and (*b*) the length of paper that unrolls from the time the force ends to the time when the roll has stopped moving.

FIGURE 10–69
$\mathbf{F}$ Problem 101.

102. An *Atwood's machine* consists of two masses, m_1 and m_2, which are connected by a massless inelastic cord that passes over a pulley, Fig. 10–70. If the pulley has radius R and moment of inertia I about its axle, determine the acceleration of the masses m_1 and m_2, and compare to the situation in which the moment of inertia of the pulley is ignored. [*Hint*: The tensions F_{T1} and F_{T2} are not necessarily equal.]

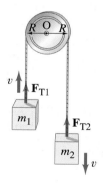

FIGURE 10–70
Atwood's machine. Problem 102.

103. A person of mass 75 kg stands at the center of a rotating merry-go-round platform of radius 3.0 m and moment of inertia 1000 kg·m². The platform rotates without friction with angular velocity 2.0 rad/s. The person walks radially to the edge of the platform. (*a*) Calculate the angular velocity when the person reaches the edge. (*b*) Compare the rotational kinetic energies of the system of platform plus person before and after the person's walk.

104. A wheel of mass M has radius R. It is standing vertically on the floor, and we want to exert a horizontal force F at its axle so that it will climb a step against which it rests (Fig. 10–71). The step has height h, where $h < R$. What minimum force F is needed?

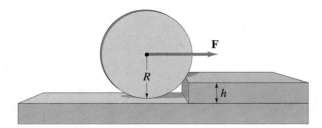

FIGURE 10–71 Problem 104.

105. A cord connected at one end to a block which can slide on an inclined plane has its other end wrapped around a cylinder resting in a depression at the top of the plane as shown in Fig. 10–72. Determine the speed of the block after it has traveled 1.80 m along the plane, starting from rest. Assume (a) there is no friction, (b) the coefficient of friction between all surfaces is $\mu = 0.035$. [Hint: In part (b) first determine the normal force on the cylinder, and make any reasonable assumptions needed.]

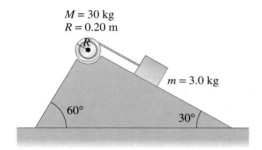

FIGURE 10–72 Problem 105.

106. A crucial part of a piece of machinery starts as a flat uniform cylindrical disk of radius R_0 and mass M. It then has a circular hole of radius R_1 drilled into it (Fig. 10–73). The hole's center is a distance h from the center of the disk. Find the moment of inertia of this disk (with off-center hole) when rotated about its center, C. [Hint: Consider a solid disk and "subtract" the hole; use the parallel-axis theorem.]

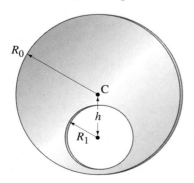

FIGURE 10–73 Problem 106.

107. A marble of mass m and radius r_0 rolls along the looped rough track of Fig. 10–74. What is the minimum value of h if the marble is to reach the highest point of the loop without leaving the track? (a) Assume $r_0 \ll R_0$, (b) do not make this assumption. Ignore frictional losses.

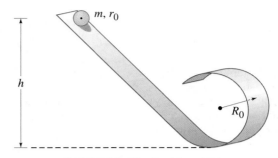

FIGURE 10–74 Problem 107.

108. A solid uniform disk of mass 21.0 kg and radius 85.0 cm is at rest flat on a frictionless surface. Figure 10–75 shows a view from above. A string is wrapped around the rim of the disk and a constant force of 30 N is applied to the string. The string does not slip on the rim. (a) In what direction does the CM move? When the disk has moved a distance of 9.0 m, (b) how fast is it moving? (c) How fast is it spinning (in radians per second)? (d) How much string has unwrapped from around the rim?

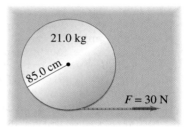

FIGURE 10–75 Problem 108, looking down on the disk.

109. (a) A yo-yo is made of two solid cylindrical disks, each of mass 0.050 kg and diameter 0.075 m, joined by a (concentric) thin solid cylindrical hub of mass 0.0050 kg and diameter 0.010 m. Use conservation of energy to calculate the linear speed of the yo-yo when it reaches the end of its 1.0-m-long string, if it is released from rest. (b) What fraction of its kinetic energy is rotational?

The child is seeing the world from a rotating frame of reference. Newton's laws of motion do not hold in such a non-inertial reference frame.

To people on the ground, she and her merry-go-round have angular momentum. The rate of change of angular momentum is proportional to the net applied torque—which, if zero, means the angular momentum is conserved. In this chapter we develop further many of the themes already discussed in Chapter 10.

CHAPTER 11

General Rotation

In Chapter 10 we dealt with the kinematics and dynamics of the rotation of a rigid body about an axis whose direction is fixed in an inertial reference frame. We analyzed the motion in terms of the rotational equivalent of Newton's laws (where torque plays the role that force does for translational motion) as well as in terms of angular momentum and rotational kinetic energy.

To keep the axis of a rotating body fixed, the body must usually be constrained by external supports (such as bearings at the end of an axle). The motion of bodies that are not constrained to move about a fixed axis, which we introduced in Section 10–11, is more difficult to describe and analyze. Indeed, the complete analysis of the general rotational motion of a body (or system of bodies) in space is very complicated, and we will only look at some aspects of general rotational motion in this chapter. We will prove some general theorems and apply them to some interesting types of motion. In particular, we will deal with the vector nature of torque and angular momentum.

Parts of this chapter may seem like a repeat of material in Chapter 10. But as a careful reading will reveal, this chapter contains a more general and useful treatment of rotational motion, making use of the vector nature of rotational quantities and deriving some of the assertions we made in Chapter 10.

11–1 | Vector Cross Product

To deal with the vector nature of angular momentum and torque in general, we will need the concept of the *vector cross product* (often called simply the *vector product* or *cross product*).

In general, the **vector** or **cross product** of two vectors **A** and **B** is defined as another vector $C = A \times B$ *whose magnitude is*

$$C = |A \times B| = AB \sin\theta, \qquad \textbf{(11–1a)}$$

where θ is the angle ($< 180°$) between **A** *and* **B**, *and whose direction is perpendicular to both* **A** *and* **B** *in the sense of the right-hand rule*, Fig. 11–1. The angle θ is measured between **A** and **B** when their tails are together at the same point. According to the right-hand rule, as shown in the figure, you orient your right hand so your fingers point along **A**, and when you bend your fingers they point along **B**. When your hand is correctly oriented in this way, your thumb will point along the direction of $C = A \times B$.

The cross product of two vectors, $A = A_x\mathbf{i} + A_y\mathbf{j} + A_z\mathbf{k}$, and $B = B_x\mathbf{i} + B_y\mathbf{j} + B_z\mathbf{k}$, can be written in component form (see Problem 6) as

$$A \times B = \begin{vmatrix} \mathbf{i} & \mathbf{j} & \mathbf{k} \\ A_x & A_y & A_z \\ B_x & B_y & B_z \end{vmatrix} \qquad \textbf{(11–1b)}$$

$$= (A_y B_z - A_z B_y)\mathbf{i} + (A_z B_x - A_x B_z)\mathbf{j} + (A_x B_y - A_y B_x)\mathbf{k}. \qquad \textbf{(11–1c)}$$

Equation 11–1b is not actually a determinant, but rather a memory aid—we evaluate Eq. 11–1b using the rules of determinants (and obtain Eq. 11–1c).

Some properties of the cross product are the following:

$$A \times A = 0 \qquad \textbf{(11–2a)}$$

$$A \times B = -B \times A \qquad \textbf{(11–2b)}$$

$$A \times (B + C) = (A \times B) + (A \times C) \qquad \text{[distributive law]} \quad \textbf{(11–2c)}$$

$$\frac{d}{dt}(A \times B) = \frac{dA}{dt} \times B + A \times \frac{dB}{dt}. \qquad \textbf{(11–2d)}$$

Equation 11–2a follows from Eqs. 11–1 (since $\theta = 0$). So does Eq. 11–2b, since the magnitude of $B \times A$ is the same as that for $A \times B$, but by the right-hand rule the direction is opposite. Thus the order of the two vectors is crucial. If you change the order, you change the result (see Fig. 11–2). That is, the commutative law does *not* hold for the cross product, although it does for the dot product of two vectors and for the product of scalars. The proofs for Eqs. 11–2c and d are given as Problems 4 and 5. Note in Eq. 11–2d that the order of quantities in the two products on the right must not be changed (because of Eq. 11–2b).

11–2 The Torque Vector

Torque is an example of a quantity that can be expressed as a cross product. To see this, let us take a simple example: the thin wheel shown in Fig. 11–3 which is free to rotate about an axis through its center at point O. A force **F** acts at the edge of the wheel, at a point whose position relative to the center O is given by the position vector **r** as shown. The force **F** tends to rotate the wheel (assumed initially at rest) counterclockwise, so the angular velocity $\boldsymbol{\omega}$ will point out of the page toward the viewer (remember the right-hand rule from Section 10–4). The torque due to **F** will tend to increase $\boldsymbol{\omega}$ so $\boldsymbol{\alpha}$ also points outward along the rotation axis. The relation between angular acceleration and torque that we developed in Chapter 10 for a body rotating about a fixed axis is

$$\Sigma\tau = I\alpha,$$

(Eq. 10–14) where I is the moment of inertia. This scalar equation is the rotational equivalent of $\Sigma F = ma$, and we would like to make it a vector equation just

Vector cross product

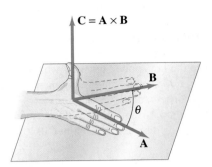

FIGURE 11–1 The vector $C = A \times B$ is perpendicular to the plane containing **A** and **B**; its direction is given by the right-hand rule.

Properties of the cross product

FIGURE 11–2 The vector $B \times A$ equals $-A \times B$; compare to Fig. 11–1.

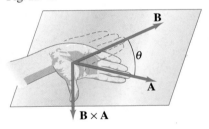

FIGURE 11–3 The torque due to the force **F** starts the wheel rotating counterclockwise so $\boldsymbol{\omega}$ and $\boldsymbol{\alpha}$ point out of the page.

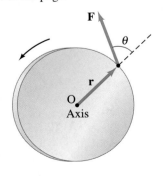

as $\Sigma\mathbf{F} = m\mathbf{a}$ is a vector equation. To do so in the case of Fig. 11–3 we must have the direction of $\boldsymbol{\tau}$ point outward along the rotation axis, since $\boldsymbol{\alpha}$ $(= d\boldsymbol{\omega}/dt)$ does; and the magnitude of the torque must be (see Eqs. 10–10 and Fig. 11–3) $\tau = rF_\perp = rF\sin\theta$. We can achieve this by defining the torque vector to be the cross product of $\mathbf{r}$ and $\mathbf{F}$:

$$\boldsymbol{\tau} = \mathbf{r} \times \mathbf{F}. \tag{11–3}$$

Vector torque (defined)

From the definition of the cross product above (Eq. 11–1) the magnitude of $\boldsymbol{\tau}$ will be $rF\sin\theta$ and the direction will be along the axis, as required for this special case.

We will see in Sections 11–3 through 11–5 that if we take Eq. 11–3 as the *general definition of torque*, then the vector relation $\Sigma\boldsymbol{\tau} = I\boldsymbol{\alpha}$ will hold in general. Thus we state now that Eq. 11–3 is the general definition of torque. It contains both magnitude and direction information. Note that this definition involves the position vector $\mathbf{r}$ and thus the torque is being calculated about a point. We can choose that point O as we wish.

For a particle of mass m on which a force $\mathbf{F}$ is applied, we define the torque about a point O as

$$\boldsymbol{\tau} = \mathbf{r} \times \mathbf{F}$$

where $\mathbf{r}$ is the position vector of the particle relative to O (Fig. 11–4). If we have a system of particles (which could be the particles making up a rigid body) the total torque $\boldsymbol{\tau}$ on the system will be the sum of the torques on the individual particles:

$$\boldsymbol{\tau} = \Sigma\mathbf{r}_i \times \mathbf{F}_i,$$

where $\mathbf{r}_i$ is the position vector of the ith particle and $\mathbf{F}_i$ is the net force on the ith particle.

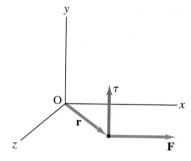

FIGURE 11–4 $\boldsymbol{\tau} = \mathbf{r} \times \mathbf{F}$, where $\mathbf{r}$ is the position vector.

EXAMPLE 11–1 **Torque Vector.** Suppose the vector $\mathbf{r}$ is in the xz plane, as in Fig. 11–4, and is given by $\mathbf{r} = (1.2\,\text{m})\mathbf{i} + (1.2\,\text{m})\mathbf{k}$. Calculate the torque vector $\boldsymbol{\tau}$ if $\mathbf{F} = (150\,\text{N})\mathbf{i}$.

SOLUTION We use the determinant form, Eq. 11–1b:

$$\boldsymbol{\tau} = \mathbf{r} \times \mathbf{F} = \begin{vmatrix} \mathbf{i} & \mathbf{j} & \mathbf{k} \\ 1.2\,\text{m} & 0 & 1.2\,\text{m} \\ 150\,\text{N} & 0 & 0 \end{vmatrix} = 0\mathbf{i} + (180\,\text{m}\cdot\text{N})\mathbf{j} + 0\mathbf{k}.$$

So τ has magnitude $180\,\text{m}\cdot\text{N}$ and points along the positive y axis.

11–3 | Angular Momentum of a Particle

The most general way of writing Newton's second law for the translational motion of a particle (or system of particles) is in terms of the linear momentum $\mathbf{p} = m\mathbf{v}$ as given by Eq. 9–2 (or 9–6):

$$\Sigma\mathbf{F} = \frac{d\mathbf{p}}{dt}.$$

The rotational analog of linear momentum is *angular momentum*. Just as the rate of change of $\mathbf{p}$ is related to the net force $\Sigma\mathbf{F}$, so we might expect the rate of change of angular momentum to be related to the net torque. Indeed, we saw this was true in Chapter 10 for the special case of a rigid body rotating about a fixed axis. Now we will see it is true in general. We first treat a single particle.

Suppose a particle of mass m has momentum $\mathbf{p}$ and position vector $\mathbf{r}$ with respect to the origin O in some chosen inertial reference frame. Then the general

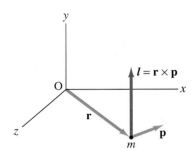

FIGURE 11–5 The angular momentum of a particle of mass m is given by $\boldsymbol{l} = \mathbf{r} \times \mathbf{p} = \mathbf{r} \times m\mathbf{v}$.

definition of the angular momentum $\boldsymbol{l}$ of the particle about point O is the vector cross product of $\mathbf{r}$ and $\mathbf{p}$:

$$\boldsymbol{l} = \mathbf{r} \times \mathbf{p}. \qquad \text{[particle]} \quad \textbf{(11–4)}$$

Angular momentum is a vector.[†] Its direction is perpendicular to both $\mathbf{r}$ and $\mathbf{p}$ as given by the right-hand rule (Fig. 11–5). Its magnitude is given by

$$l = rp \sin \theta$$

or

$$l = rp_\perp = r_\perp p$$

where θ is the angle between $\mathbf{r}$ and $\mathbf{p}$ and $p_\perp (= p \sin \theta)$ and $r_\perp (= r \sin \theta)$ are the components of $\mathbf{p}$ and $\mathbf{r}$ perpendicular to $\mathbf{r}$ and $\mathbf{p}$ respectively.

Now let us find the relation between angular momentum and torque for a particle. If we take the derivative of $\boldsymbol{l}$ with respect to time we have

$$\frac{d\boldsymbol{l}}{dt} = \frac{d}{dt}(\mathbf{r} \times \mathbf{p}) = \frac{d\mathbf{r}}{dt} \times \mathbf{p} + \mathbf{r} \times \frac{d\mathbf{p}}{dt}.$$

But

$$\frac{d\mathbf{r}}{dt} \times \mathbf{p} = \mathbf{v} \times m\mathbf{v} = m(\mathbf{v} \times \mathbf{v}) = 0,$$

since $\sin \theta = 0$ for this case. Thus

$$\frac{d\boldsymbol{l}}{dt} = \mathbf{r} \times \frac{d\mathbf{p}}{dt}.$$

If we let $\Sigma \mathbf{F}$ represent the resultant force on the particle, then in an inertial reference frame, $\Sigma \mathbf{F} = d\mathbf{p}/dt$ and

$$\mathbf{r} \times \Sigma \mathbf{F} = \mathbf{r} \times \frac{d\mathbf{p}}{dt} = \frac{d\boldsymbol{l}}{dt}.$$

But $\mathbf{r} \times \Sigma \mathbf{F} = \Sigma \boldsymbol{\tau}$ is the net torque on our particle. Hence

$$\Sigma \boldsymbol{\tau} = \frac{d\boldsymbol{l}}{dt}. \qquad \text{[particle, inertial frame]} \quad \textbf{(11–5)}$$

The time rate of change of angular momentum of a particle is equal to the net torque applied to it. Equation 11–5 is the rotational equivalent of Newton's second law for a particle, written in its most general form. Equation 11–5 is valid only in an inertial frame since only then is it true that $\Sigma \mathbf{F} = d\mathbf{p}/dt$, which was used in the proof.

FIGURE 11–6 The angular momentum of a particle of mass m rotating in a circle of radius $\mathbf{r}$ with velocity $\mathbf{v}$ is $\boldsymbol{l} = \mathbf{r} \times m\mathbf{v}$ (Example 11–2).

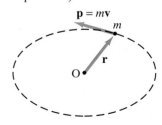

EXAMPLE 11–2 **A particle's angular momentum.** Determine the angular momentum of a particle of mass m moving with speed v in a circle of radius r in a counterclockwise direction.

SOLUTION The value of the angular momentum depends on the choice of the point O. Let us calculate $\boldsymbol{l}$ with respect to the center of the circle, Fig. 11–6. Then $\mathbf{r}$ is perpendicular to $\mathbf{p}$ so $l = |\mathbf{r} \times \mathbf{p}| = rmv$. By the right-hand rule, the direction of $\boldsymbol{l}$ is perpendicular to the plane of the circle, upward and outward toward the viewer. Since $v = \omega r$ and $I = mr^2$ for a single particle rotating about an axis a distance r away, we can write

$$l = mvr = mr^2 \omega = I\omega.$$

[†] Actually a pseudovector; see footnote in Section 10–4. Also note that we are using small $\boldsymbol{l}$ for the angular momentum of a particle and capital $\mathbf{L}$ for a collection of particles or an extended body (as in the previous chapter and later in this chapter).

11–4 Angular Momentum and Torque for a System of Particles; General Motion

Relation between Angular Momentum and Torque

Consider a system of n particles which have angular momenta $l_1, l_2, \ldots l_n$. The system could be anything from a rigid body to a loose assembly of particles whose positions are not fixed relative to each other. The total angular momentum $\mathbf{L}$ of the system is defined as the vector sum of the angular momenta of all the particles in the system:

$$\mathbf{L} = \sum_{i=1}^{n} l_i. \tag{11–6}$$

The resultant torque acting on the system is the sum of the net torques acting on all the particles:

$$\boldsymbol{\tau}_{\text{net}} = \sum \boldsymbol{\tau}_i.$$

This sum includes (1) internal torques due to internal forces that particles of the system exert on other particles of the system, and (2) external torques due to forces exerted by objects outside our system. By Newton's third law, the force each particle exerts on another is equal and opposite (and acts along the same line) as the force that the second particle exerts on the first. Hence the sum of all internal torques adds to zero, and

$$\boldsymbol{\tau}_{\text{net}} = \sum_i \boldsymbol{\tau}_i = \sum \boldsymbol{\tau}_{\text{ext}}.$$

Now we take the time derivative of Eq. 11–6 and use Eq. 11–5 for each particle to obtain

$$\frac{d\mathbf{L}}{dt} = \sum_i \frac{dl_i}{dt} = \sum \boldsymbol{\tau}_{\text{ext}}$$

or

$$\frac{d\mathbf{L}}{dt} = \sum \boldsymbol{\tau}. \qquad \text{[inertial reference frame]} \tag{11–7a}$$

NEWTON'S SECOND LAW
(rotation, system of particles)

(We have dropped the subscript on τ for convenience.) This fundamental result states that the time rate of change of the total angular momentum of a system of particles (or a rigid body) equals the resultant external torque on the system. It is the rotational equivalent of Eq. 9–6, $d\mathbf{P}/dt = \Sigma\mathbf{F}_{\text{ext}}$ for translational motion. Note that $\mathbf{L}$ and $\Sigma\boldsymbol{\tau}$ must be calculated about the same origin O.

Equation 11–7a is valid when $\mathbf{L}$ and $\boldsymbol{\tau}$ are calculated with reference to a point fixed in an inertial reference frame. (In the derivation, we used Eq. 11–5 which is valid only in this case.) It is also valid when $\boldsymbol{\tau}$ and $\mathbf{L}$ are calculated about a point which is moving uniformly in an inertial reference frame since such a point can be considered the origin of a second inertial reference frame. It is *not* valid in general when $\boldsymbol{\tau}$ and $\mathbf{L}$ are calculated about a point that is *accelerating*, except for one special (and very important) case—when that point is the center of mass of the system:

Accelerating reference frame

$$\frac{d\mathbf{L}_{\text{CM}}}{dt} = \sum \boldsymbol{\tau}_{\text{CM}}. \qquad \text{[even if accelerating]} \tag{11–7b}$$

NEWTON'S SECOND LAW
(for CM, even if accelerating)

Equation 11–7b is valid no matter how the CM moves, and $\Sigma\boldsymbol{\tau}_{\text{CM}}$ is the net external torque calculated about the center of mass.

It is because of the validity of Eq. 11–7b that we are justified in describing the general motion of a system of particles, as we did in Chapter 10, as the *translational* motion of the CM plus rotation about the CM. [Eq. 10–25 or 10–26, which tell us the total $K = K_{\text{CM}} + K_{\text{rot}}$, applies to the energy but not to the other aspects of the motion. Equations 11–7b plus 9–6 $(d\mathbf{P}_{\text{CM}}/dt = \Sigma\mathbf{F}_{\text{ext}})$ provide the more general statement of this principle. See also Section 9–9.]

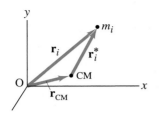

FIGURE 11–7 The position of m_i in the inertial frame is $\mathbf{r}_i$; with regard to the CM (which could be accelerating) it is $\mathbf{r}_i^*$, where $\mathbf{r}_i = \mathbf{r}_i^* + \mathbf{r}_{CM}$ and $\mathbf{r}_{CM}$ is the position of the CM in the inertial frame.

*Derivation of $d\mathbf{L}_{CM}/dt = \Sigma\boldsymbol{\tau}_{CM}$

The proof of Eq. 11–7b is as follows. Let $\mathbf{r}_i$ be the position vector of the ith particle in an inertial reference frame, and $\mathbf{r}_{CM}$ be the position vector of the center of mass of the system in this reference frame. The position of the ith particle with respect to the CM is $\mathbf{r}_i^*$ where (see Fig. 11–7)

$$\mathbf{r}_i = \mathbf{r}_{CM} + \mathbf{r}_i^*.$$

When we take the derivative of this equation, we can write

$$\mathbf{p}_i = m_i \frac{d\mathbf{r}_i}{dt} = m_i \frac{d}{dt}(\mathbf{r}_i^* + \mathbf{r}_{CM}) = \mathbf{p}_i^* + m_i\mathbf{v}_{CM}.$$

The angular momentum with respect to the CM is

$$\mathbf{L}_{CM} = \sum_i (\mathbf{r}_i^* \times \mathbf{p}_i^*) = \sum_i \mathbf{r}_i^* \times (\mathbf{p}_i - m_i\mathbf{v}_{CM}).$$

Then, taking the time derivative, we have

$$\frac{d\mathbf{L}_{CM}}{dt} = \sum_i \left(\frac{d\mathbf{r}_i^*}{dt} \times \mathbf{p}_i^*\right) + \sum_i \left(\mathbf{r}_i^* \times \frac{d\mathbf{p}_i^*}{dt}\right).$$

The first term is $\mathbf{v}_i^* \times m\mathbf{v}_i^* = 0$, since $\mathbf{v}_i^*$ is parallel to itself. Thus

$$\frac{d\mathbf{L}_{CM}}{dt} = \sum_i \mathbf{r}_i^* \times \frac{d}{dt}(\mathbf{p}_i - m_i\mathbf{v}_{CM})$$

$$= \sum_i \mathbf{r}_i^* \times \frac{d\mathbf{p}_i}{dt} - \left(\sum_i m_i\mathbf{r}_i^*\right) \times \frac{d\mathbf{v}_{CM}}{dt}.$$

The second term on the right is zero since, by Eq. 9–12, $\Sigma m_i\mathbf{r}_i^* = M\mathbf{r}_{CM}^*$, and $\mathbf{r}_{CM}^* = 0$ by definition (the position of the CM is at the origin of the CM reference frame). Furthermore, by Newton's second law we have

$$\frac{d\mathbf{p}_i}{dt} = \mathbf{F}_i,$$

where $\mathbf{F}_i$ is the net force on m_i. (Note that $d\mathbf{p}_i^*/dt \neq \mathbf{F}_i$ because the CM may be accelerating and Newton's second law does not hold in a noninertial reference frame.) Consequently

$$\frac{d\mathbf{L}_{CM}}{dt} = \sum_i \mathbf{r}_i^* \times \mathbf{F}_i = \sum_i (\boldsymbol{\tau}_i)_{CM} = \Sigma\boldsymbol{\tau}_{CM},$$

where $\Sigma\boldsymbol{\tau}_{CM}$ is the resultant external torque on the entire system calculated about the CM. (By Newton's third law, the sum over all the $\boldsymbol{\tau}_i$ eliminates the net torque due to internal forces, as we saw earlier.) This last equation is Eq. 11–7b, and this concludes its proof.

Summary

To summarize, the relation

$$\sum\boldsymbol{\tau} = \frac{d\mathbf{L}}{dt}$$

is *not* valid in general. It is true only when $\boldsymbol{\tau}$ and $\mathbf{L}$ are calculated with respect to either (1) the origin of an inertial reference frame or (2) the center of mass of a system of particles (or of a rigid body).

11–5 Angular Momentum and Torque for a Rigid Body

Let us now consider the rotation of a rigid body about an axis that has a fixed direction in space. We will use the general principles developed earlier in this chapter and will arrive at the conclusions already used in Chapter 10.

Let us calculate the component of angular momentum along the rotation axis of the rotating body. We will call this component L_ω since the angular velocity $\boldsymbol{\omega}$ points along the rotation axis. For each particle of the body,

$$\boldsymbol{l}_i = \mathbf{r}_i \times \mathbf{p}_i.$$

Let θ be the angle between $\boldsymbol{l}_i$ and the rotation axis. (See Fig. 11–8; θ is *not* the angle between $\mathbf{r}_i$ and $\mathbf{p}_i$, which is 90°). Then the component of $\boldsymbol{l}_i$ along the rotation axis is

$$l_{i\omega} = r_i p_i \cos\theta = m_i v_i r_i \cos\theta,$$

where m_i is the mass and v_i the velocity of the ith particle. Now $v_i = R_i \omega$ where ω is the angular velocity of the body and R_i is the perpendicular distance of m_i from the axis of rotation. Furthermore, $R_i = r_i \cos\theta$, as can be seen in Fig. 11–8, so

$$l_{i\omega} = m_i v_i (r_i \cos\theta) = m_i R_i^2 \omega.$$

We sum over all the particles to obtain

$$L_\omega = \sum_i l_{i\omega} = \left(\sum_i m_i R_i^2 \right)\omega.$$

But $\Sigma m_i R_i^2$ is the moment of inertia I of the body about the axis of rotation. Therefore the component of the total angular momentum along the rotation axis is given by

$$L_\omega = I\omega. \tag{11–8}$$

Note that we would obtain Eq. 11–8 no matter where we choose the point O (for measuring $\mathbf{r}_i$) as long as it is on the axis of rotation. Equation 11–8 is just Eq. 10–19 of Chapter 10, which we have now proved from the general definition of angular momentum.

If the body rotates about a symmetry axis through the CM, then L_ω is the only component of $\mathbf{L}$, as we now show. For each point on one side of the axis there will be a corresponding point on the opposite side. We can see from Fig. 11–8 that each $\boldsymbol{l}_i$ has a component parallel to the axis $(l_{i\omega})$ and a component perpendicular to the axis. The components parallel to the axis add together for each pair of opposite points, but the components perpendicular to the axis for opposite points will have the same magnitude but opposite direction and so will cancel. Hence, for a body rotating about a symmetry axis, the angular momentum vector is parallel to the axis and we can write

$$\mathbf{L} = I\boldsymbol{\omega}, \qquad \text{[rotation axis = symmetry axis, through CM]} \tag{11–9}$$

where $\mathbf{L}$ is measured relative to the CM.

The general relation between angular momentum and torque is Eq. 11–7:

$$\sum \boldsymbol{\tau} = \frac{d\mathbf{L}}{dt}$$

where $\Sigma\boldsymbol{\tau}$ and $\mathbf{L}$ are calculated either about (1) the origin of an inertial reference frame, or (2) the CM of the system. This is a vector relation, and must therefore be valid for each component. Hence, for a rigid body, the component along the rotation axis is

$$\sum \tau_{\text{axis}} = \frac{dL_\omega}{dt} = \frac{d}{dt}(I\omega) = I\frac{d\omega}{dt} = I\alpha,$$

which is valid for a rigid body rotating about an axis fixed relative to the body and also either (1) fixed in an inertial system or (2) passing through the CM of the body. This is equivalent to Eqs. 10–14 and 10–15, which we now see are special cases of Eq. 11–7, $\Sigma\boldsymbol{\tau} = d\mathbf{L}/dt$.

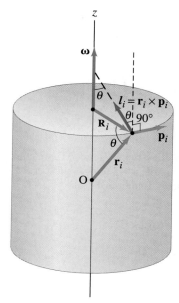

FIGURE 11–8
Calculating $L_z = \Sigma l_{iz}$.

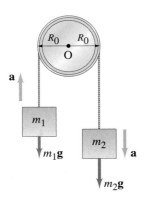

FIGURE 11–9 Atwood's machine, Example 11–3. We also discussed this in Example 4–13.

EXAMPLE 11–3 **Atwood's Machine.** An *Atwood Machine* consists of two masses, m_1 and m_2, which are connected by a massless inelastic cord that passes over a pulley, Fig. 11–9. If the pulley has radius R_0 and moment of inertia I about its axle, determine the acceleration of the masses m_1 and m_2, and compare to the situation where the moment of inertia of the pulley is ignored.

SOLUTION The angular momentum is calculated about an axis along the axle through the center O of the pulley. The pulley has angular momentum $I\omega$, where $\omega = v/R_0$ and v is the velocity of m_1 and m_2 at any instant. The angular momentum of m_1 is $R_0 m_1 v$ and that of m_2 is $R_0 m_2 v$. The total angular momentum is

$$L = (m_1 + m_2)vR_0 + I\frac{v}{R_0}.$$

The external torque on the system is

$$\tau = m_2 g R_0 - m_1 g R_0$$

(the force on the pulley exerted by the support on its axle gives rise to no torque because the lever arm is zero). We apply Eq. 11–7a:

$$\tau = \frac{dL}{dt}$$

$$(m_2 - m_1)g R_0 = (m_1 + m_2)R_0\frac{dv}{dt} + \frac{I}{R_0}\frac{dv}{dt}.$$

Solving for $a = dv/dt$, we get

$$a = \frac{dv}{dt} = \frac{(m_2 - m_1)g}{(m_1 + m_2) + I/R_0^2}.$$

If we were to ignore I, $a = (m_2 - m_1)g/(m_2 + m_1)$ and we see that the effect of the moment of inertia of the pulley is to slow down the system. This is just what we would expect.

FIGURE 11–10 When you try to tilt a rotating bicycle wheel vertically upward, it swerves to the side.

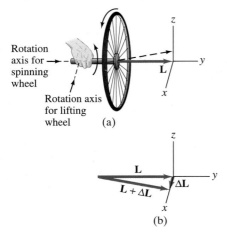

CONCEPTUAL EXAMPLE 11–4 **Bicycle wheel.** Suppose you are holding a bicycle wheel by a handle connected to its axle as in Fig. 11–10a. The wheel is spinning rapidly so its angular momentum **L** points horizontally as shown. Now you suddenly try to tilt the axle upward as shown by the dashed line in Fig. 11–10a (so the CM moves vertically). You expect the wheel to go up (and it would if it weren't rotating), but it unexpectedly swerves to the right! Explain.

SOLUTION To explain this seemingly odd behavior—you may need to do it to believe it—we only need to use the relation $\boldsymbol{\tau}_{\text{net}} = d\mathbf{L}/dt$. In the short time Δt, you exert a net torque (about an axis through your wrist) that points along the x axis perpendicular to **L**. Thus the change in **L** is

$$\Delta\mathbf{L} \approx \boldsymbol{\tau}_{\text{net}}\,\Delta t;$$

so $\Delta\mathbf{L}$ must also point (approximately) along the x axis, since $\boldsymbol{\tau}_{\text{net}}$ does (Fig. 11–10b). Thus the new angular momentum, $\mathbf{L} + \Delta\mathbf{L}$, points to the right, looking along the axis of the wheel, as shown in Fig. 11–10b. Since the angular momentum is directed along the axle of the wheel, we see that the axle, which now is along $\mathbf{L} + \Delta\mathbf{L}$, must move sideways to the right, which is what we observe.

$\mathbf{L} = I\boldsymbol{\omega}$ is not always valid

Although Eq. 11–9, $\mathbf{L} = I\boldsymbol{\omega}$, is often very useful, it is not valid in general if the rotation axis is not along a symmetry axis through the center of mass. Nonetheless, it can be shown that every rigid body, no matter what its shape, has three "principal axes" about which Eq. 11–9 is valid (we won't go into the details here).

As an example of a case where Eq. 11–9 is not valid, consider the nonsymmetrical body shown in Fig. 11–11. It consists of two equal masses, m_1 and m_2, attached to the ends of a rigid (massless) rod which makes an angle ϕ with the axis of rotation. We calculate the angular momentum about the CM at point O. At the moment shown, m_1 is coming toward the viewer, and m_2 is moving away, so $\mathbf{L}_1 = \mathbf{r}_1 \times \mathbf{p}_1$ and $\mathbf{L}_2 = \mathbf{r}_2 \times \mathbf{p}_2$ are as shown. The total angular momentum is $\mathbf{L} = \mathbf{L}_1 + \mathbf{L}_2$, which is clearly *not* along $\boldsymbol{\omega}$.

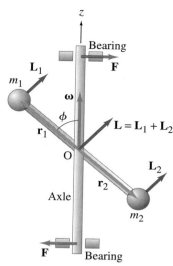

FIGURE 11–11 In this system $\mathbf{L}$ and $\boldsymbol{\omega}$ are not parallel. This is an example of rotational imbalance.

*11–6 Rotational Imbalance

Let us go one step further with the system shown in Fig. 11–11, since it is a fine illustration of $\Sigma\boldsymbol{\tau} = d\mathbf{L}/dt$. If the system rotates with constant angular velocity, ω, the magnitude of $\mathbf{L}$ will not change, but its direction will. As the rod and two masses rotate about the z axis, $\mathbf{L}$ also rotates about the axis. At the moment shown in Fig. 11–11, $\mathbf{L}$ is in the plane of the paper. A time dt later, when the rod has rotated through an angle $d\theta = \omega\, dt$, $\mathbf{L}$ will also have rotated through an angle $d\theta$ (it remains perpendicular to the rod). $\mathbf{L}$ will then have a component pointing into the page. Thus $d\mathbf{L}$ points into the page and so must $d\mathbf{L}/dt$. Since

$$\sum \boldsymbol{\tau} = \frac{d\mathbf{L}}{dt}$$

we see that a net torque, directed into the page at the moment shown, must be applied to the axle on which the rod is mounted. The torque is supplied by bearings (or other constraint) at the end of the axle. The forces $\mathbf{F}$ exerted by the bearings on the axle are shown in Fig. 11–11. The direction of each force $\mathbf{F}$ rotates as the system does, always being in the plane of $\mathbf{L}$ and $\boldsymbol{\omega}$ for this system. If the torque due to these forces were not present, the system would not rotate about the fixed axis as desired.

The axle itself, by Newton's third law, must exert forces of $-\mathbf{F}$ on the bearings. Hence the axle tends to move in the direction of $-\mathbf{F}$ and thus tends to wobble as it rotates. This has many practical applications, such as the vibrations felt in a car whose wheels are not balanced. Consider an automobile wheel that is symmetrical except for an extra mass m_1 on one rim and an equal mass m_2 opposite it on the other rim, as shown in Fig. 11–12. Because of the nonsymmetry of m_1 and m_2, the wheel bearings would have to exert a force perpendicular to the axle at all times simply to keep the wheel rotating, just as in Fig. 11–11. The bearings would wear excessively and the wobble of the wheel would be felt by occupants of the car. When the wheels are balanced, they rotate smoothly without wobble. This is why "dynamic balancing" of automobile wheels and tires is important. The wheel of Fig. 11–12 would balance *statically* just fine. If equal masses m_3 and m_4 are added symmetrically, below m_1 and above m_2, the wheel will be balanced dynamically as well $\big(\mathbf{L}$ will be parallel to $\boldsymbol{\omega}$, and $\boldsymbol{\tau}_{\text{ext}} = 0\big)$.

➡ **PHYSICS APPLIED**

Automobile wheel balancing

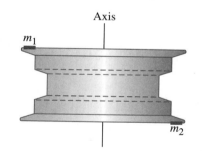

FIGURE 11–12
Unbalanced automobile wheel.

FIGURE 11–13 Angular momentum vector looking down along the rotation axis of the system of Fig. 11–11 as it rotates during a time dt.

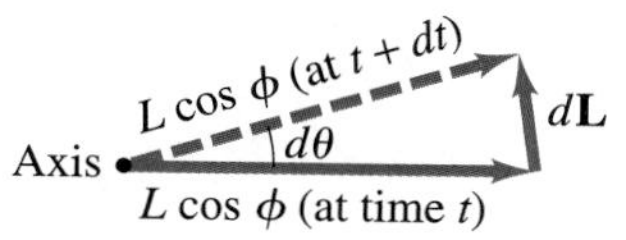

EXAMPLE 11–5 Torque on imbalanced system. Determine the magnitude of the net torque τ_{net} needed to keep the system turning in Fig. 11–11.

SOLUTION Figure 11–13 is a view of the angular momentum vector, looking down the rotation axis (z axis) of the object depicted in Fig. 11–11, as it rotates. $L\cos\phi$ is the component of $\mathbf{L}$ perpendicular to the axle. In a time dt, $\mathbf{L}$ changes by an amount

$$dL = (L\cos\phi)d\theta = L\cos\phi\,\omega\,dt.$$

Hence

$$\tau_{\text{net}} = \frac{dL}{dt} = \omega L \cos\phi.$$

Now $L = L_1 + L_2 = r_1 m_1 v_1 + r_2 m_2 v_2 = r_1 m_1(\omega r_1 \sin\phi) + r_2 m_2(\omega r_2 \sin\phi) = (m_1 r_1^2 + m_2 r_2^2)\omega \sin\phi$. Since $I = (m_1 r_1^2 + m_2 r_2^2)\sin^2\phi$ is the moment of inertia about the axis of rotation, then $L = I\omega/\sin\phi$. So

$$\tau_{\text{net}} = (m_1 r_1^2 + m_2 r_2^2)\omega^2 \sin\phi \cos\phi = I\omega^2/\tan\phi.$$

The situation of Fig. 11–11 illustrates the usefulness of the vector nature of torque and angular momentum. If we had considered only the components of angular momentum and torque along the rotation axis, we could not have calculated the torque due to the bearings (since the forces **F** act at the axle and hence produce no torque along that axis). By using the concept of vector angular momentum we have a far more powerful technique for understanding and for attacking problems.

11–7 Conservation of Angular Momentum

In Chapter 9 we saw that the most general form of Newton's second law for the translational motion of a particle or system of particles is

$$\sum \mathbf{F}_{\text{ext}} = \frac{d\mathbf{P}}{dt},$$

where **P** is the (linear) momentum, defined as $m\mathbf{v}$ for a particle, or $M\mathbf{v}_{\text{CM}}$ for a system of particles of total mass M whose CM moves with velocity $\mathbf{v}_{\text{CM}}$, and $\Sigma\mathbf{F}_{\text{ext}}$ is the net external force acting on the particle or system. This relation is valid only in an inertial reference frame.

In this chapter, we have found a similar relation to describe the general rotation of a system of particles (including rigid bodies):

$$\sum \boldsymbol{\tau} = \frac{d\mathbf{L}}{dt},$$

where $\Sigma\boldsymbol{\tau}$ is the net external torque acting on the system, and **L** is the total angular momentum. This relation is valid when $\Sigma\boldsymbol{\tau}$ and **L** are calculated about a point fixed in an inertial reference frame, or about the CM of the system. For a rigid body rotating with angular velocity ω about an axis fixed in direction, the component of angular momentum about this axis is (Eq. 11–8):

$$L_\omega = I\omega,$$

where I is the moment of inertia of the body about this axis.

For translational motion, if the net force on the system is zero, $d\mathbf{P}/dt = 0$, so the total linear momentum of the system remains constant. This is the law of conservation of momentum. For rotational motion, if the net torque on the system is zero, then

$$\frac{d\mathbf{L}}{dt} = 0 \quad \text{and} \quad \mathbf{L} = \text{constant}. \qquad\qquad \left[\Sigma\boldsymbol{\tau} = 0\right] \quad \textbf{(11–10)}$$

In words:

CONSERVATION OF
ANGULAR MOMENTUM

The total angular momentum of a system remains constant if the net external torque acting on the system is zero.

This is the **law of conservation of angular momentum**, and ranks with the laws of conservation of energy and linear momentum (and others to be discussed later) as one of the great laws of physics. In Chapter 10 we already saw some examples of this important law applied to the special case of a rigid body rotating about a fixed axis. Here we have it in general form. We use it now in interesting Examples.

EXAMPLE 11–6 **Derivation of Kepler's second law.** *Kepler's second law* states that each planet moves so that a line from the Sun to the planet sweeps out equal areas in equal periods of time (see Section 6–5). Use the law of conservation of angular momentum to show this.

SOLUTION The planet moves in an ellipse as shown in Fig. 11–14. In a time dt, the planet moves a distance $v\,dt$ and sweeps out an area dA equal to the area of a triangle of base r and height $v\,dt\sin\theta$ (shown exaggerated in Fig. 11–14). Hence

$$dA = \tfrac{1}{2}(r)(v\,dt\sin\theta)$$

and

$$\frac{dA}{dt} = \tfrac{1}{2}rv\sin\theta.$$

The magnitude of the angular momentum $\mathbf{L}$ about the Sun is

$$L = |\mathbf{r} \times m\mathbf{v}| = mrv\sin\theta,$$

so

$$\frac{dA}{dt} = \frac{1}{2m}L.$$

But $L =$ constant, since the gravitational force $\mathbf{F}$ is directed toward the Sun so the torque it produces is zero (we ignore the pull of the other planets). Hence $dA/dt =$ constant, which is what we set out to prove.

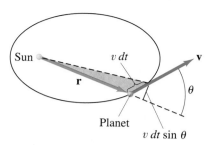

FIGURE 11–14 Kepler's second law of planetary motion (Example 11–6).

EXAMPLE 11–7 **Bullet strikes cylinder edge.** A bullet of mass m moving with velocity v strikes and becomes embedded at the edge of a cylinder of mass M and radius R_0, as shown in Fig. 11–15. The cylinder, initially at rest, begins to rotate about its symmetry axis, which remains fixed in position. Assuming no frictional torque, what is the angular velocity of the cylinder after this collision? Is kinetic energy conserved?

FIGURE 11–15 Bullet strikes and becomes embedded in cylinder at its edge (Example 11–7).

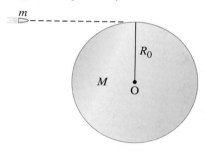

SOLUTION We take as our system the bullet and cylinder, on which there is no net external torque. Thus we can use conservation of angular momentum, and we calculate all angular momenta about the center O of the cylinder. Initially, because the cylinder is at rest, the total angular momentum is solely that of the bullet:

$$L = |\mathbf{r} \times \mathbf{p}| = R_0 mv,$$

since R_0 is the perpendicular distance of $\mathbf{p}$ from O. After the collision, the cylinder $\left(I_{\text{cyl}} = \tfrac{1}{2}MR_0^2\right)$ rotates with the bullet $\left(I = mR_0^2\right)$ embedded in it at angular velocity ω:

$$L = \left(I_{\text{cyl}} + mR_0^2\right)\omega = \left(\tfrac{1}{2}M + m\right)R_0^2\omega.$$

Hence, since angular momentum is conserved,

$$\omega = \frac{mv}{\left(\tfrac{1}{2}M + m\right)R_0}.$$

Angular momentum is conserved in this collision, but kinetic energy is not:

$$K_{\text{f}} - K_{\text{i}} = \tfrac{1}{2}I_{\text{cyl}}\omega^2 + \tfrac{1}{2}\left(mR_0^2\right)\omega^2 - \tfrac{1}{2}mv^2$$

$$= -\frac{mM}{2M + 4m}v^2,$$

which is less than zero. Hence $K_{\text{f}} < K_{\text{i}}$. This energy is transformed to thermal energy as a result of the inelastic collision.

Another example of the conservation of angular momentum is the **gyroscope**, used by mariners and as the basis for guidance systems of aircraft. The rapidly spinning wheel is mounted on a complicated set of bearings, so that even when the mount moves, no net torque acts to change the direction of the angular momentum. Thus the axis of the rotating wheel remains pointed in the same direction in space.

Gyroscope

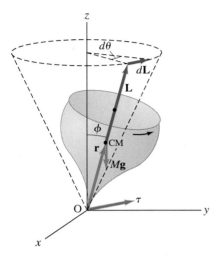

FIGURE 11–16 Spinning top.

The motion of a rapidly spinning top, or a gyroscope, is an interesting example of rotational motion and of the use of the vector equation

$$\sum \boldsymbol{\tau} = \frac{d\mathbf{L}}{dt}.$$

Consider a symmetrical top of mass M spinning rapidly about its symmetry axis, as in Fig. 11–16. The top is balanced on its tip at point O in an inertial reference frame. If the axis of the top makes an angle ϕ to the vertical (z axis), when the top is carefully released its axis will move, sweeping out a cone about the vertical as shown by the dashed lines in Fig. 11–16. This type of motion, in which a torque produces a change in the direction of the rotation axis, is called **precession**. The rate at which the rotation axis moves about the vertical (z) axis is called the angular velocity of precession, Ω (capital Greek omega). Let us now try to understand the reasons for this motion, and calculate Ω.

If the top were not spinning, it would immediately fall to the ground when released due to the pull of gravity. The apparent mystery of a top is that when it is spinning, it does not immediately fall to the ground but instead precesses—it moves sideways. But this is not really so mysterious if we examine it from the point of view of angular momentum and torque, which we calculate about the point O. When the top is spinning with angular velocity ω about its symmetry axis, it has an angular momentum $\mathbf{L}$ directed along its axis, as shown in Fig. 11–16. (There is also angular momentum due to the precessional motion, so that the total $\mathbf{L}$ is not exactly along the axis of the top; but if $\Omega \ll \omega$, which is usually the case, we can ignore this.) Now to change the angular momentum, a torque is required. If no torque were applied to the top, $\mathbf{L}$ would remain constant in magnitude and direction; the top would neither fall nor precess. But the slightest tip to the side results in a net torque about O, equal to $\boldsymbol{\tau}_{\text{net}} = \mathbf{r} \times M\mathbf{g}$, where $\mathbf{r}$ is the position vector of the top's center of mass with respect to O and M is the mass of the top. The direction of $\boldsymbol{\tau}_{\text{net}}$ is perpendicular to both $\mathbf{r}$ and $M\mathbf{g}$ and by the right-hand rule is, as shown in Fig. 11–16, in the horizontal (xy) plane. The change in $\mathbf{L}$ in a time dt is

$$d\mathbf{L} = \boldsymbol{\tau}_{\text{net}}\, dt,$$

which is perpendicular to $\mathbf{L}$ and horizontal (parallel to $\boldsymbol{\tau}_{\text{net}}$), as shown in Fig. 11–16. Since $d\mathbf{L}$ is perpendicular to $\mathbf{L}$, the magnitude of $\mathbf{L}$ does not change. Only the direction of $\mathbf{L}$ changes. Since $\mathbf{L}$ points along the axis of the top, we see that this axis moves to the right in the figure. That is, the upper end of the top's axis moves in a horizontal direction perpendicular to $\mathbf{L}$. This explains why the top precesses rather than falls. The vector $\mathbf{L}$ and the top's axis move together in a horizontal circle. As they do so, $\boldsymbol{\tau}_{\text{net}}$ and $d\mathbf{L}$ rotate as well so as to be horizontal and perpendicular to $\mathbf{L}$.

To determine Ω, we see from Fig. 11–16 that the angle $d\theta$ (which is in a horizontal plane) is related to dL by

$$dL = L \sin \phi \, d\theta,$$

since $\mathbf{L}$ makes an angle ϕ to the z axis. The angular velocity of precession is $\Omega = d\theta/dt$, which becomes (since $d\theta = dL/L \sin \phi$)

$$\Omega = \frac{1}{L \sin \phi} \frac{dL}{dt} = \frac{\tau}{L \sin \phi}. \qquad \text{[spinning top]} \quad \textbf{(11–11a)}$$

But $\tau_{\text{net}} = |\mathbf{r} \times M\mathbf{g}| = rMg \sin \phi$ [because $\sin (\pi - \phi) = \sin \phi$] so we can also write

$$\Omega = \frac{Mgr}{L}. \qquad \text{[spinning top]} \quad \textbf{(11–11b)}$$

Thus the rate of precession does not depend on the angle ϕ; but it is inversely proportional to the top's angular momentum. The faster the top spins, the greater L is and the slower the top precesses.

Rotating Frames of Reference; Inertial Forces

Inertial and Noninertial Reference Frames

Up to now, we have examined the motion of bodies, including circular and rotational motion, from the outside, as observers fixed on the Earth. Sometimes it is convenient to place ourselves (in theory, if not physically) into a reference frame that is rotating. Let us examine the motion of objects from the point of view, or frame of reference, of persons seated on a rotating platform such as a merry-go-round. It looks to them as if the rest of the world is going around *them*—see chapter opening photo. But let us focus attention on what they observe when they place a tennis ball on the floor of the rotating platform, which we assume is frictionless. If they put the ball down gently, without giving it any push, they will observe that it accelerates from rest and moves outward as shown in Fig. 11–17a. According to Newton's first law, an object initially at rest should stay at rest if no force acts on it. But, according to the observers on the rotating platform, the ball starts moving even though there is no force applied to it. To observers in an inertial reference frame, this is all very clear: the ball has an initial velocity when it is released (because the platform is moving), and it simply continues moving in a straight-line path as shown in Fig. 11–17b, in accordance with Newton's first law.

But what shall we do about the frame of reference of the observers on the rotating platform? Clearly, Newton's first law, the law of inertia, does not hold in this rotating frame of reference. For this reason, such a frame is called a **noninertial reference frame**. An **inertial reference frame** is one in which the law of inertia—Newton's first law—does hold, and so do Newton's second and third laws. In a noninertial reference frame, such as our rotating platform, Newton's second law also does not hold. For instance in the situation described above, there is no net force on the ball; yet, with respect to the rotating platform, the ball accelerates.

Fictitious (Inertial) Forces

Because Newton's laws do not hold when observations are made with respect to a rotating frame of reference, calculation of motion can be complicated. However, we can still make use of Newton's laws in such a reference frame if we make use of a trick. If the ball is kept at rest on the merry-go-round, then to keep it rotating in a circle (as seen from an inertial reference frame) an inward force equal to mv^2/r would be needed, as we saw in Chapter 5 (Eq. 5–1). In the rotating reference frame of the ball, this inward force still acts, and if we want to somehow use Newton's laws, we need to figure out some way to balance it since the ball is at rest in this reference frame. So the trick we use is to write down the equation $\Sigma F = ma$ as if a force equal to mv^2/r (or $m\omega^2 r$) were acting radially outward on the object in addition to any other forces that may be acting. This extra force, which might be designated as "centrifugal force" since it *seems* to act outward, is called a **fictitious force** or **pseudoforce**. It is a pseudoforce ("pseudo" means "false") because there is no object that exerts this force. Furthermore, when viewed from an inertial reference frame, the effect doesn't exist at all. We have made up this pseudoforce so that we can make calculations in a noninertial frame using Newton's second law, $\Sigma F = ma$. Thus the observer in the non-inertial frame of Fig. 11–17a uses Newton's second law for the ball's outward motion by assuming that a pseudoforce equal to mv^2/r acts on it. Such pseudoforces are also called **inertial forces** since they arise only because the reference frame is not an inertial one.

The Earth itself is rotating on its axis. Thus, strictly speaking, Newton's laws are not valid on the Earth. However, the effect of the Earth's rotation is usually so small that it can be ignored, although it does influence the movement of large air masses and ocean currents. Because of the Earth's rotation, the material of the Earth is concentrated slightly more at the equator. The Earth is thus not a perfect sphere but is slightly fatter at the equator than it is at the poles.

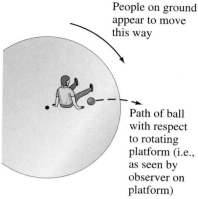

People on ground appear to move this way

(a) Rotating reference frame

Path of ball with respect to rotating platform (i.e., as seen by observer on platform)

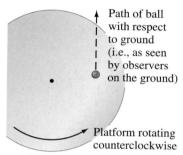

Path of ball with respect to ground (i.e., as seen by observers on the ground)

Platform rotating counterclockwise

(b) Inertial reference frame

FIGURE 11–17 Path of a ball released on a rotating merry-go-round (a) in the reference frame of the merry-go-round, and (b) in an inertial reference frame.

Fictitious force (pseudoforce)

Inertial force

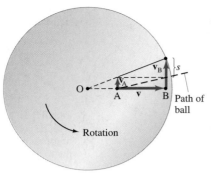

(a) Inertial reference frame

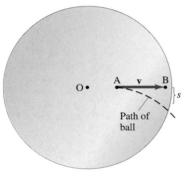

(b) Rotating reference frame

FIGURE 11–18 The origin of the Coriolis effect. Looking down on a rotating platform, (a) as seen from an inertial reference frame, and (b) as seen from the rotating platform as frame of reference.

In a reference frame that rotates at a constant angular speed ω (relative to an inertial system), there exists another pseudoforce known as the *Coriolis force*. It appears to act on a body in a rotating reference frame only if the body is moving relative to that reference frame, and it acts to deflect the body sideways. It, too, is an effect of the reference frame being noninertial and hence is referred to as an *inertial force*. To see how the Coriolis force arises, consider two people, A and B, at rest on a platform rotating with angular speed ω, as shown in Fig. 11–18a. They are situated at distances r_A and r_B from the axis of rotation (at O). The woman at A throws a ball with a horizontal velocity $\mathbf{v}$ (in her reference frame) radially outward toward the man at B on the outer edge of the platform. In Fig. 11–18a, we view the situation from an inertial reference frame. The ball initially has not only the velocity $\mathbf{v}$ radially outward, but also a tangential velocity $\mathbf{v}_A$ due to the rotation of the platform. Now Eq. 10–4 tells us that $v_A = r_A\omega$, where r_A is the woman's radial distance from the axis of rotation at O. If the man at B had this same velocity v_A, the ball would reach him perfectly. But his speed is $v_B = r_B\omega$, which is greater than v_A because $r_B > r_A$. Thus, when the ball reaches the outer edge of the platform, it passes a point that the man at B has already passed because his speed in that direction is greater than the ball's. So the ball passes behind him.

Figure 11–18b shows the situation as seen from the rotating platform as frame of reference. Both A and B are at rest, and the ball is thrown with velocity $\mathbf{v}$ toward B, but the ball deflects to the right as shown and passes behind B as previously described. This is not a centrifugal-force effect, for the latter acts radially outward. Instead, this effect acts sideways, perpendicular to $\mathbf{v}$, and is called a **Coriolis acceleration**; it is said to be due to the Coriolis force, which is a fictitious, inertial force. Its explanation as seen from an inertial system was given above: it is an effect of being in a rotating system, wherein points that are farther from the rotation axis have higher linear speeds. On the other hand, when viewed from the rotating system, we can describe the motion using Newton's second law, $\Sigma\mathbf{F} = m\mathbf{a}$, if we add a "pseudoforce" term corresponding to this Coriolis effect.

Let us determine the magnitude of the Coriolis acceleration for the simple case described above. (We assume v is large and distances short, so we can ignore gravity.) We do the calculation from the inertial reference frame (Fig. 11–18a). The ball moves radially outward a distance $r_B - r_A$ at speed v in a time t given by

$$r_B - r_A = vt.$$

During this time, the ball moves to the side a distance s_A given by

$$s_A = v_A t.$$

The man at B, in this time t, moves a distance

$$s_B = v_B t.$$

The ball therefore passes behind him a distance s (Fig. 11–18a) given by

$$s = s_B - s_A = (v_B - v_A)t.$$

We saw earlier that $v_A = r_A\omega$ and $v_B = r_B\omega$, so

$$s = (r_B - r_A)\omega t.$$

We substitute $r_B - r_A = vt$ (see above) and get

$$s = \omega v t^2. \qquad \textbf{(11–12)}$$

This same s equals the sideways displacement as seen from the noninertial rotating system (Fig. 11–18b).

We see immediately that Eq. 11–12 corresponds to motion at constant acceleration. For as we saw in Chapter 2 (Eq. 2–12b), $y = \frac{1}{2}at^2$ for a constant acceleration (with zero initial velocity in the y direction). Thus, if we write Eq. 11–12 in the form $s = \frac{1}{2} a_{\text{cor}} t^2$, we see that the Coriolis acceleration a_{cor} is

$$a_{\text{cor}} = 2\omega v. \qquad \textbf{(11–13)}$$

This relation is valid for any velocity in the plane of rotation—that is, in the plane perpendicular to the axis of rotation[†] (in Fig. 11–18, the axis through point O perpendicular to the page).

Because the Earth rotates, the Coriolis effect has some interesting manifestations on the Earth. It affects the movement of air masses and thus has an influence on weather. In the absence of the Coriolis effect, air would rush directly into a region of low pressure, as shown in Fig. 11–19a. But because of the Coriolis effect, the winds are deflected to the right in the Northern Hemisphere (Fig. 11–19b), since the Earth rotates from west to east. So there tends to be a counterclockwise wind pattern around a low-pressure area. The reverse is true in the Southern Hemisphere. Thus cyclones rotate counterclockwise in the Northern Hemisphere and clockwise in the Southern Hemisphere. The same effect explains the easterly trade winds near the equator: any winds heading south toward the equator will be deflected toward the west (that is, as if coming from the east).

The Coriolis effect also acts on a falling body. A body released from the top of a high tower will not hit the ground directly below the release point, but will be deflected slightly to the east. Equations 11–12 and 11–13 do not apply directly on Earth except at the poles, because they are valid only for velocity in the plane perpendicular to the axis of rotation. A velocity, or component of velocity, parallel to the axis of rotation will not experience a Coriolis acceleration since r (distance from axis) doesn't change. We can alter Eqs. 11–12 and 11–13 for use on Earth by considering only the component of $\mathbf{v}$ perpendicular to the axis of rotation. From Fig. 11–20 we see that this is $v \cos \lambda$ for a vertically falling body, where λ is the latitude of the place on the Earth.

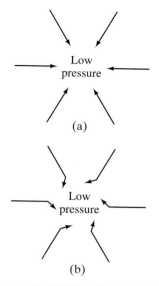

(a)

(b)

(c)

(d)

EXAMPLE 11–8 **A ball does not fall straight.** A lead ball is dropped vertically from a 110-m-high tower in Florence, Italy (latitude = 44°). How far from the base of the tower is it deflected by the Coriolis force?

SOLUTION The Coriolis acceleration is $a_{\text{cor}} = 2\omega v \cos \lambda$, where ω is the angular velocity of rotation of the Earth. Now v is the vertical velocity of free fall, $v = gt$, so

$$a_{\text{cor}} = (2\omega g \cos \lambda)t$$

and acts sideways. To obtain the horizontal displacement, which we take to be in the x direction, perpendicular to the fall, we must integrate twice:

$$\frac{dv_x}{dt} = a_{\text{cor}} = (2\omega g \cos \lambda)t$$

so

$$\int_0^{v_x} dv_x = \int_0^t (2\omega g \cos \lambda)t \, dt$$

$$v_x = (\omega g \cos \lambda)t^2$$

since $v_{x0} = 0$. Then, since $v_x = dx/dt$, we have

$$x = \int_0^t v_x \, dt = \omega g \cos \lambda \int_0^t t^2 \, dt = \tfrac{1}{3} \omega g \cos \lambda t^3.$$

The value of t is easily obtained, since it is the time for the ball to fall a

FIGURE 11–19 (a) Winds (moving air masses) would flow directly toward a low-pressure area if the Earth did not rotate. (b) and (c) Because of the Earth's rotation, the winds are deflected to the right in the Northern Hemisphere (as in Fig. 11–18) as if a fictitious (Coriolis) force were acting. (d) The reverse occurs in the Southern Hemisphere.

[†]The Coriolis acceleration can be written in general in terms of the vector cross product as $\mathbf{a}_C = -2\boldsymbol{\omega} \times \mathbf{v}$ where $\boldsymbol{\omega}$ has direction along the rotation axis; its magnitude is $a_C = 2wv_\perp$ where $v_\perp$ is the component of velocity perpendicular to the rotation axis.

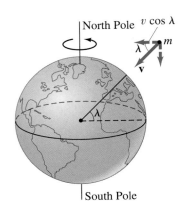

North Pole $v \cos \lambda$

m

λ

v

South Pole

FIGURE 11–20 Object of mass m falling vertically to Earth at a latitude λ.

distance $h = 110\,\text{m}$ from the top of the tower—namely, $h = \frac{1}{2}gt^2$—so $t = (2h/g)^{\frac{1}{2}}$. Thus the deflection is

$$x = \frac{1}{3}\omega g \cos \lambda \left(\frac{2h}{g}\right)^{\frac{3}{2}} = \frac{1}{3}\omega \cos \lambda \sqrt{\frac{8h^3}{g}}.$$

We set $h = 110\,\text{m}$ and $\lambda = 44°$, and write ω for the rotation of the Earth (1 revolution in 24 h) as

$$\omega = 2\pi f = 2\pi(1\,\text{rev})/(24\,\text{h})(3600\,\text{s/h}) = 7.27 \times 10^{-5}\,\text{rad/s}.$$

Then

$$x = \frac{1}{3}(7.27 \times 10^{-5}\,\text{rad/s})(\cos 44°)\sqrt{\frac{8(110\,\text{m})^3}{9.80\,\text{m/s}^2}} = 0.018\,\text{m}$$

or about 2 cm. This distance is small and difficult to measure because of wind currents, and so on. Finally, we determine the direction of this deflection. From Fig. 11–20 we see that the ball starts at a point farther from the axis and moves closer to the Earth's axis as it falls, which is the reverse direction of the motion shown in Fig. 11–18. That is, due to the Earth's rotation, the top of the tower is moving slightly faster than the bottom is. Hence the ball deflects to the east (into the page in Fig. 11–20).

Summary

The **vector product** or **cross product** of two vectors $\mathbf{A}$ and $\mathbf{B}$ is another vector $\mathbf{C} = \mathbf{A} \times \mathbf{B}$ whose magnitude is $AB \sin \theta$ and whose direction is perpendicular to both $\mathbf{A}$ and $\mathbf{B}$ in the sense of the right-hand rule.

The **torque** $\boldsymbol{\tau}$ due to a force $\mathbf{F}$ is a vector quantity and is always calculated about some point O (the origin of a coordinate system) as follows:

$$\boldsymbol{\tau} = \mathbf{r} \times \mathbf{F},$$

where $\mathbf{r}$ is the position vector of the point at which the force $\mathbf{F}$ acts.

Angular momentum is also a vector. For a particle having momentum $\mathbf{p} = m\mathbf{v}$, the angular momentum $\boldsymbol{l}$ about some point O is

$$\boldsymbol{l} = \mathbf{r} \times \mathbf{p},$$

where $\mathbf{r}$ is the position vector of the particle relative to the point O at any instant. The net torque $\Sigma\boldsymbol{\tau}$ on a particle is related to its angular momentum by

$$\Sigma\boldsymbol{\tau} = \frac{d\boldsymbol{l}}{dt}.$$

For a system of particles, the total angular momentum $\mathbf{L} = \Sigma\boldsymbol{l}_i$. The total angular momentum of the system is related to the total net torque $\Sigma\boldsymbol{\tau}$ on the system by

$$\Sigma\boldsymbol{\tau} = \frac{d\mathbf{L}}{dt}.$$

This last relation is the vector rotational equivalent of Newton's second law. It is valid when $\mathbf{L}$ and $\Sigma\boldsymbol{\tau}$ are calculated about an origin (1) fixed in an inertial reference system or (2) situated at the CM of the system. For a rigid body rotating about a fixed axis, the component of angular momentum about the rotation axis is given by $L_\omega = I\omega$. If a body rotates about an axis of symmetry, then the vector relation $\mathbf{L} = I\boldsymbol{\omega}$ holds, but this is not true in general.

If the total net torque on a system is zero, then the total vector angular momentum $\mathbf{L}$ remains constant. This is the important **law of conservation of angular momentum**. It applies to the vector $\mathbf{L}$ and therefore also to each of its components.

Questions

1. If all the components of the vectors $\mathbf{V}_1$ and $\mathbf{V}_2$ were reversed, how would this alter $\mathbf{V}_1 \times \mathbf{V}_2$?

2. Name the three different conditions that could make $\mathbf{V}_1 \times \mathbf{V}_2 = 0$.

3. A force $\mathbf{F} = \mathbf{j}F$ is applied to a body at a position $\mathbf{r} = x\mathbf{i} + y\mathbf{j} + z\mathbf{k}$ where the origin is at the CM. Does the torque about the CM depend on x? On y? On z?

4. A particle moves with constant speed along a straight line. How does its angular momentum, calculated about any point not on its path, change in time?

5. If the net force on a system is zero, is the net torque also zero? If the net torque on a system is zero, is the net force zero?

6. Explain how a child "pumps" on a swing to make it go higher.

* 7. For the nonsymmetrical system of Fig. 11–11, is there a point about which the angular momentum ($\mathbf{L}$) could be calculated so that $\mathbf{L}$ would be in the same direction as the angular velocity $\boldsymbol{\omega}$? What if there were only one mass (say, $m_2 = 0$)? If your answer to either question is yes, how do you explain the need for forces at the bearings for both cases and the requirement $\boldsymbol{\tau} = d\mathbf{L}/dt$?

8. An astronaut floats freely in a weightless environment. Describe how the astronaut can move her limbs so as to (a) turn her body upside down and (b) turn her body about face.

9. On the basis of the law of conservation of angular momentum, discuss why a helicopter must have more than one rotor (or propeller). Discuss one or more ways the second propeller can operate in order to keep the body stable.

10. A wheel is rotating freely about a vertical axis with constant angular velocity. Small parts of the wheel come loose and fly off. How does this affect the rotational speed of the wheel? Is angular momentum conserved? Is kinetic energy conserved? Explain.

11. Consider the following vector quantities: displacement, velocity, acceleration, momentum, angular momentum, torque. (a) Which of these are independent of the choice of origin of coordinates? (Consider different points as origin which are at rest with respect to each other.) (b) Which are independent of the velocity of the coordinate system?

12. Describe the torque needed if the person in Fig. 11–10 is to tilt the axle of the rotating wheel directly upward with no swerving to the side.

13. How does a car make a right turn? Where does the torque come from that is needed to change the angular momentum?

14. Suppose you are standing on a turntable that can freely rotate. When you hold a rotating bicycle wheel over your head, with its axis vertical, you are at rest. If you now move the wheel so its axis is horizontal, what happens to you? What happens if you then point the axis of the wheel downward?

* 15. The axis of the Earth precesses with a period of about 25,000 years. This is much like the precession of a top. Explain how the Earth's equatorial bulge gives rise to a torque exerted by the Sun and Moon on the Earth; see Fig. 11–21, which is drawn for the winter solstice (December 21). About what axis would you expect the Earth's rotation axis to precess as a result of the torque due to the Sun? Does the torque exist 3 months later? Explain.

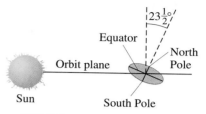

FIGURE 11–21 Question 15.

* 16. Why is it that at most locations on the Earth, a plumb bob does not hang precisely in the direction of the Earth's center?

* 17. In the battle of the Falkland Islands in 1914, the shots of British gunners initially fell wide of their marks because their calculations were based on naval battles fought in the Northern Hemisphere. The Falklands are in the Southern Hemisphere. Explain the origin of their problem.

Problems

Section 11–1

1. (I) Show that (a) $\mathbf{i} \times \mathbf{i} = \mathbf{j} \times \mathbf{j} = \mathbf{k} \times \mathbf{k} = 0$, (b) $\mathbf{i} \times \mathbf{j} = \mathbf{k}$, $\mathbf{i} \times \mathbf{k} = -\mathbf{j}$, and $\mathbf{j} \times \mathbf{k} = \mathbf{i}$.

2. (I) If vector $\mathbf{A}$ points along the negative x axis and vector $\mathbf{B}$ along the positive z axis, what is the direction of (a) $\mathbf{A} \times \mathbf{B}$ and (b) $\mathbf{B} \times \mathbf{A}$? (c) What is the magnitude of $\mathbf{A} \times \mathbf{B}$ and $\mathbf{B} \times \mathbf{A}$?

3. (II) Consider a particle of a rigid body rotating about a fixed axis. Show that the tangential and radial vector components of the linear acceleration are:

$$\mathbf{a}_{tan} = \boldsymbol{\alpha} \times \mathbf{r} \quad \text{and} \quad \mathbf{a}_R = \boldsymbol{\omega} \times \mathbf{v}.$$

4. (II) Prove the distributive law for the cross product, Eq. 11–2c.

5. (II) Use the limiting process to obtain Eq. 11–2d. [Hint: Let $\Delta \mathbf{A}$, $\Delta \mathbf{B}$, and Δt approach zero.]

6. (II) (a) Show that the cross product of two vectors, $\mathbf{A} = A_x\mathbf{i} + A_y\mathbf{j} + A_z\mathbf{k}$, and $\mathbf{B} = B_x\mathbf{i} + B_y\mathbf{j} + B_z\mathbf{k}$ is

$$\mathbf{A} \times \mathbf{B} = (A_yB_z - A_zB_y)\mathbf{i} + (A_zB_x - A_xB_z)\mathbf{j}$$
$$+ (A_xB_y - A_yB_x)\mathbf{k}.$$

(b) Then show that the cross product can be written

$$\mathbf{A} \times \mathbf{B} = \begin{vmatrix} \mathbf{i} & \mathbf{j} & \mathbf{k} \\ A_x & A_y & A_z \\ B_x & B_y & B_z \end{vmatrix},$$

where we use the rules for evaluating a determinant. (Note, however, that this is not actually a determinant, but a memory aid.)

7. (II) Use the result of Problem 6 to determine (a) the vector product $\mathbf{A} \times \mathbf{B}$ and (b) the angle between $\mathbf{A}$ and $\mathbf{B}$ if $\mathbf{A} = 7.0\mathbf{i} - 3.5\mathbf{j}$ and $\mathbf{B} = -8.5\mathbf{i} + 7.0\mathbf{j} + 2.0\mathbf{k}$.

8. (III) Show that the velocity $\mathbf{v}$ of any point in a body rotating with angular velocity $\boldsymbol{\omega}$ about a fixed axis can be written

$$\mathbf{v} = \boldsymbol{\omega} \times \mathbf{r}$$

where $\mathbf{r}$ is the position vector of the point relative to an origin O located on the axis of rotation. Can O be anywhere on the rotation axis? Will $\mathbf{v} = \boldsymbol{\omega} \times \mathbf{r}$ if O is located at a point not on the axis of rotation?

9. (III) Let $\mathbf{A}, \mathbf{B}, \mathbf{C}$ be three vectors that do not all lie in the same plane. Show that $\mathbf{A} \cdot (\mathbf{B} \times \mathbf{C}) = \mathbf{B} \cdot (\mathbf{C} \times \mathbf{A}) = \mathbf{C} \cdot (\mathbf{A} \times \mathbf{B})$.

Section 11–2

10. (II) A particle is located at $\mathbf{r} = (4.0\mathbf{i} + 8.0\mathbf{j} + 6.0\mathbf{k})$ m. A force $\mathbf{F} = (16.0\mathbf{j} - 4.0\mathbf{k})$N acts on it. What is the torque, calculated about the origin?

11. (II) The origin of a coordinate system is at the center of a wheel. The wheel can rotate in the xy plane about an axis through the origin. A force $F = 188$ N, which acts at a $+33.0°$ angle to the x axis, is applied to the wheel at the point $x = 22.0$ cm, $y = 33.5$ cm. What are the magnitude and direction of the torque produced by this force about the axis?

12. (II) A thin 43.0-cm-diameter wheel is constrained to rotate about an axis through its center, which we choose to be the z axis (the wheel rotates in the xy plane). A force $\mathbf{F} = 22.8\mathbf{j} - 21.6\mathbf{k}$ (newtons) is exerted at a point on the edge of the wheel which is exactly on the x axis. (a) Determine the torque (magnitude and direction), produced by this force, calculated about the wheel's center. (b) Is the direction of this torque parallel to the direction of $\boldsymbol{\alpha}$? If not, explain how $\boldsymbol{\tau}$ can be proportional to $\boldsymbol{\alpha}$.

13. (II) An engineer estimates that under the most adverse expected weather conditions, the total force on the highway sign in Fig. 11–22 will be $\mathbf{F} = \pm 2.4\mathbf{i} - 3.0\mathbf{j}$ kN, acting at the CM. What torque does this force exert about the base O?

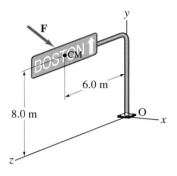

FIGURE 11–22
Problem 13.

Section 11–3

14. (I) What are the x, y, and z components of the angular momentum of a particle located at $\mathbf{r} = x\mathbf{i} + y\mathbf{j} + z\mathbf{k}$ which has momentum $\mathbf{p} = p_x\mathbf{i} + p_y\mathbf{j} + p_z\mathbf{k}$?

15. (I) Show that the kinetic energy K of a particle of mass m, moving in a circular path, is $K = l^2/2I$, where l is its angular momentum and I is its moment of inertia about the center of the circle.

16. (I) Calculate the angular momentum of a particle of mass m moving with constant velocity v for two cases (see Fig. 11–23): (a) about origin O, and (b) about O′.

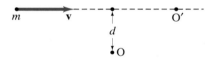

FIGURE 11–23
Problem 16.

17. (II) Two identical particles have equal but opposite momenta, $\mathbf{p}$ and $-\mathbf{p}$, but they are not traveling along the same line. Show that the total angular momentum of this system does not depend on the choice of origin.

18. (II) Determine the angular momentum of a 60-g particle about the origin of coordinates when the particle is at $x = 7.0$ m, $y = -6.0$ m, and it has velocity $v = (2.0\mathbf{i} - 8.0\mathbf{k})$ m/s.

19. (II) A particle is at the position $(x, y, z) = (1.0, 2.0, 3.0)$ m. It is traveling with a vector velocity $(-5.0, -4.5, -3.1)$ m/s. Its mass is 7.6 kg. What is its vector angular momentum about the origin?

20. (II) Two particles rest a distance d apart on the edge of a table. One of the particles, of mass m, falls off the edge and falls vertically. Using the other particle (still at rest) as the origin, calculate (a) the torque τ on m as a function of time (before hitting the ground below) and (b) the angular momentum l of m as a function of time. (c) Show that $\tau = dl/dt$.

Sections 11–4 to 11–5

21. (II) Four identical particles of mass m are mounted at equal intervals on a thin rod of length l and mass M, with one mass at each end of the rod. If the system is rotated with angular velocity ω about an axis perpendicular to the rod through one of the end masses, determine (a) the kinetic energy and (b) the angular momentum of the system.

22. (II) An Atwood machine (Fig. 11–9) consists of two masses, $m_1 = 7.0$ kg and $m_2 = 8.8$ kg, connected by a cord that passes over a pulley free to rotate about a fixed axis. The pulley is a solid cylinder of radius $R_0 = 0.50$ m and mass 0.80 kg. (a) Determine the acceleration a of each mass. (b) What percentage of error in a would be made if the moment of inertia of the pulley were ignored? Ignore friction in the pulley bearings.

23. (II) Use $\tau = dL/dt$ to determine the acceleration of the bucket in Example 10–8.

24. (II) Two lightweight rods 20 cm in length are mounted perpendicular to an axle and at 180° to each other (Fig. 11–24). At the end of each rod is a 600-g mass. The rods are spaced 40 cm apart along the axle. The axle rotates at 30 rad/s. (a) What is the component of the total angular momentum along the axle? (b) What angle does the vector angular momentum make with the axle? [*Hint:* Remember that the vector angular momentum must be calculated about the *same point* for *both* masses.]

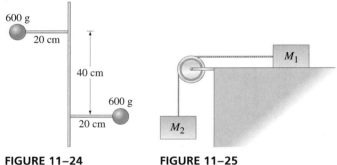

FIGURE 11–24
Problem 24.

FIGURE 11–25
Problems 25 and 26.

25. (II) Figure 11–25 shows two masses connected by a cord passing over a pulley of radius R_0 and moment of inertia I. Mass M_1 slides on a frictionless surface, and M_2 hangs freely. Determine a formula for (a) the angular momentum of the system about the pulley axis, as a function of the speed v of mass M_1 or M_2, and (b) the acceleration of the system.

26. (III) Repeat Problem 25, but this time suppose that the block M_1 in Fig. 11–25 slides on the horizontal surface with coefficient of friction μ_k. [*Hint:* The cord attaches to M_1 a distance R_0 above the surface. Assume M_1 does not tip.]

27. (III) A uniform thin rod, 7.0 cm long with a mass of 40.0 g, lies on a frictionless horizontal table. It is struck with a horizontal impulse, at right angles to its length, at a point 2.0 cm from one end. If the impulse is 8.5 mN · s, describe the resulting motion of the stick.

28. (III) A thin rod of length L and mass M rotates about a vertical axis through its center with angular velocity ω. The rod makes an angle ϕ with the rotation axis. Determine the magnitude and direction of $\mathbf{L}$.

29. (III) Show that the total angular momentum $\mathbf{L} = \Sigma \mathbf{r}_i \times \mathbf{p}_i$ of a system of particles about the origin of an inertial reference frame can be written as the sum of the angular momentum about the CM, $\mathbf{L}^*$ (spin angular momentum), plus the angular momentum of the CM about the origin (orbital angular momentum): $\mathbf{L} = \mathbf{L}^* + \mathbf{r}_{\text{CM}} \times M\mathbf{v}_{\text{CM}}$. [*Hint:* See the derivation of Eq. 10–26.]

*Section 11–6

*30. (II) What is the magnitude of the force **F** exerted by each bearing in Fig. 11–11 (Example 11–5)? The bearings are a distance d from point O. Ignore the effects of gravity.

*31. (II) Suppose in Fig. 11–11 that $m_2 = 0$; that is, only one mass, m_1, is actually present. If the bearings are each a distance d from O, determine the forces F_1 and F_2 at the upper and lower bearings respectively. [*Hint*: Choose an origin—different than O in Fig. 11–11—such that **L** is parallel to **ω**. Ignore effects of gravity.]

*32. (II) For the system shown in Fig. 11–11, suppose that $m_1 = m_2 = 0.60$ kg, $r_1 = r_2 = 0.30$ m, and the distance between the bearings is 0.20 m. What will be the force that each bearing must exert on the axle if $\phi = 23.0°$ and $\omega = 11.0$ rad/s? [Ignore effects of gravity.]

*33. (II) What would your answer be to Problem 32 if $m_1 = 0.60$ kg and $m_2 = 0$? [Ignore effects of gravity.]

*34. (II) A uniform cylindrical wheel 70.0 cm in diameter and of mass 11.8 kg rotates about an axle displaced 1.00 cm from its true center. (*a*) If the bearings are mounted 9.50 cm apart, what is the force exerted on them when the wheel rotates at 11.2 rev/s? (*b*) To balance this wheel, where should a 1.00 kg mass be placed? [Ignore effects of gravity.]

Section 11–7

35. (II) A thin rod of mass M and length l is suspended vertically from a frictionless pivot at its upper end. A mass m of putty traveling horizontally with a speed v strikes the rod at its CM and sticks there. How high does the bottom of the rod swing?

36. (II) A uniform stick 1.0 m long with a total mass of 300 g is pivoted at its center. A 3.0-g bullet is shot through the stick midway between the pivot and one end (Fig. 11–26). The bullet approaches at 250 m/s and leaves at 160 m/s. With what angular speed is the stick spinning after the collision?

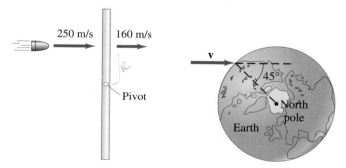

FIGURE 11–26 Problem 36.   **FIGURE 11–27** Problem 37.

37. (II) Suppose a 7.0×10^{10} kg meteor struck the Earth at the equator with a speed $v = 1.0 \times 10^4$ m/s, as shown in Fig. 11–27 and remained stuck. By what factor would this affect the rotational frequency of the Earth (1 rev/day)?

38. (II) A person of mass 55 kg stands at the center of a rotating merry-go-round platform of radius 2.5 m and moment of inertia 670 kg·m². The platform rotates without friction with an angular velocity of 2.0 rad/s. The person walks radially to the edge of the platform. (*a*) Calculate the angular velocity when the person reaches the edge. (*b*) Compare the rotational kinetic energies of the system of platform plus person before and after the person's walk.

39. (III) A 200-kg beam 2.0 m in length slides broadside down the ice with a speed of 18 m/s (Fig. 11–28). A 50-kg man at rest grabs one end as it goes past and hangs on as both he and the beam go spinning down the ice. Assume frictionless motion. (*a*) How fast does the center of mass of the system move after the collision? (*b*) With what angular velocity does the system rotate about its CM?

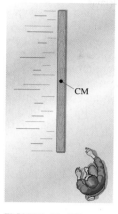

FIGURE 11–28
Problem 39.

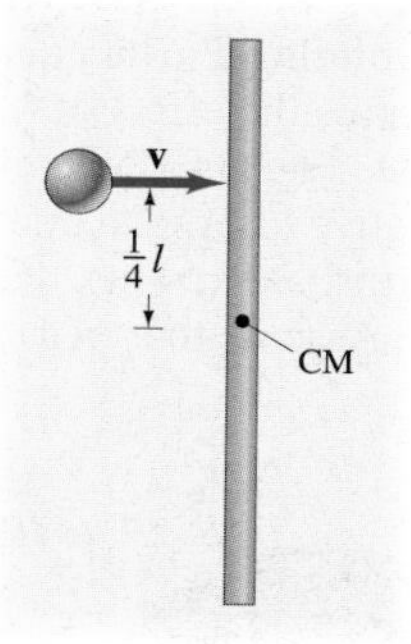

FIGURE 11–29
Problem 40.

40. (III) A thin rod of mass M and length l rests on a frictionless table and is struck at a point $l/4$ from its CM by a clay ball of mass m moving at speed v (Fig. 11–29). The ball sticks to the rod. Determine the translational and rotational motion of the rod after the collision.

*Section 11–8

*41. (II) A 220-g top spinning at 15 rev/s makes an angle of 30° to the vertical and precesses at a rate of 1 rev per 8.0 s. If its CM is 3.5 cm from its tip along its symmetry axis, what is the moment of inertia of the top?

*42. (II) A toy gyroscope consists of a 150-g disk with a radius of 5.5 cm mounted at the center of an axle 17 cm long (Fig. 11–30). The gyroscope spins at 70 rev/s. One end of its axle rests on a stand and the other end precesses horizontally about the stand as shown. (*a*) How long does it take the gyroscope to precess once around? (*b*) If all the dimensions of the gyroscope were doubled (radius = 11 cm, axle = 34 cm), how long would it take to precess once?

FIGURE 11–30
A wheel, rotating about a horizontal axle supported at one end, precesses. Problems 42, 43, and 44.

*43. (II) Suppose the solid wheel of Fig. 11–30 has a mass of 300 g and rotates at 250 rad/s; it has radius 6.0 cm and is mounted at the center of a horizontal thin axle 20 cm long. At what rate does the axle precess?

*44. (II) If a mass equal to half the mass of the wheel in Problem 43 is placed at the free end of the axle, what will be the precession rate now?

* 45. (II) If a plant is allowed to grow from seed on a rotating platform, it will grow at an angle, pointing inward. Calculate what this angle will be (put yourself in the rotating frame) in terms of g, r, and ω. Why does it grow inward rather than outward?

* 46. (II) In a rotating frame of reference, Newton's first and second laws remain useful if we assume that a pseudoforce equal to $m\omega^2 r$ is acting. What effect does this assumption have on the validity of Newton's third law?

* 47. (III) Let $\mathbf{g}'$ be the effective acceleration of gravity at a point on the rotating Earth, equal to the vector sum of the "true" value $\mathbf{g}$ plus the effect of the rotating reference frame ($m\omega^2 r$ term). See Fig. 11–31. Determine the magnitude and direction of $\mathbf{g}'$ relative to a radial line from the center of the Earth (a) at the North Pole, (b) at a latitude of 45° north, and (c) at the equator. Assume that g (if ω were zero) is a constant 9.80 m/s².

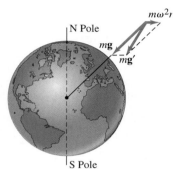

FIGURE 11–31
Problem 47.

* 48. (II) Suppose the man at B in Fig. 11–18 throws the ball toward the woman at A. (a) In what direction is the ball deflected as seen in the noninertial system? (b) Determine a formula for the amount of deflection and for the (Coriolis) acceleration in this case.

* 49. (II) For what directions of velocity would the Coriolis effect on a body moving at the Earth's equator be zero?

* 50. (II) Determine the Coriolis "force" (magnitude and direction) acting on a 1200-kg race car traveling due north at 500 km/h in Utah (latitude 40°). [*Hint:* Use a component of velocity perpendicular to the Earth's axis.]

* 51. (III) A projectile is fired nearly horizontally at high velocity v_0 toward the east. (a) In what direction is it deflected by the Coriolis effect? (b) Determine a formula for the deflection in terms of v_0, ω, λ, and the distance traveled D. (c) Evaluate for $v_0 = 1000$ m/s, $\lambda = 45°$, and $D = 3.0$ km.

* 52. (III) An ant crawls with constant speed along a radial spoke of a wheel rotating at constant angular velocity ω about a vertical axis. Write a vector equation for all the forces (including inertial forces) acting on the ant. Take the x axis along the spoke, y perpendicular to the spoke pointing to the ant's left, and the z axis vertically upward. The wheel rotates counterclockwise as seen from above.

General Problems

53. A particle of mass 0.50 kg is moving with velocity $\mathbf{v} = (8.0\mathbf{i} + 6.0\mathbf{j})$ m/s. (a) Find the angular momentum $\mathbf{L}$ relative to the origin when the particle is at $\mathbf{r} = (2.0\mathbf{j} + 3.0\mathbf{k})$ m. (b) At position $\mathbf{r}$ a force of $\mathbf{F} = 3.0\,\mathrm{N}\,\mathbf{i}$ is applied to the particle. Find the torque relative to the origin.

54. An asteroid of mass 10^5 kg traveling at a speed of 30 km/s relative to the Earth hits the Earth at the equator. It hits the Earth tangentially and in the direction of the Earth's rotation. Use angular momentum to estimate the fractional change in the angular speed of the Earth as a result of the collision.

55. A boy rolls a tire along a straight level street. The tire has mass 9.0 kg, radius 0.32 m and moment of inertia about its central axis of symmetry of 0.83 kg·m². The boy shoves the tire away from him at a speed of 2.1 m/s and sees that the tire leans 10° to the right (Fig. 11–32). (a) How will the gravitational torque affect the subsequent motion of the tire? (b) Compare the change in angular momentum caused by this torque in 0.20 s to the original magnitude of angular momentum.

56. In the Bohr model of the hydrogen atom, the electron (mass m) is held in a circular orbit about the nucleus (a proton) by the electric force on it, $F = ke^2/r^2$ (where e is the electric charge on the electron and on the proton, and k is a constant), and only certain orbits are allowed: those for which the angular momentum l of the electron about the nucleus is an integer multiple n of $h/2\pi$, where h is a constant called Planck's constant. That is, $l = nh/2\pi$ where n is an integer. Show that the possible radii for electron orbits are given by

$$r = \frac{n^2 h^2}{4\pi^2 kme^2} \qquad n = 1, 2, 3, \cdots.$$

To do this, first show that the radius of an orbit is $r = ke^2/mv^2$ and then apply $l = nh/2\pi$.

57. Water drives a waterwheel (or turbine) of radius $R = 3.0$ m as shown in Fig. 11–33. The water enters at a speed $v_1 = 7.0$ m/s and exits from the waterwheel at a speed $v_2 = 3.0$ m/s. (a) If 150 kg of water passes through per second, what is the rate at which the water delivers angular

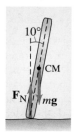

FIGURE 11–32
Problem 55.

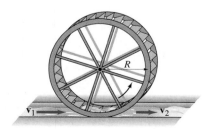

FIGURE 11–33
Problem 57.

momentum to the waterwheel? (b) What is the torque the water applies to the waterwheel? (c) If the water causes the waterwheel to make one revolution every 5.5 s, how much power is delivered to the wheel?

58. A bicyclist traveling with speed $v = 3.2$ m/s on a flat road is making a turn with a radius $r = 2.8$ m. The forces acting on the cyclist and cycle are the normal force $(\mathbf{F}_N)$ and friction force $(\mathbf{F}_{fr})$ exerted by the road on the tires and $m\mathbf{g}$, the total weight of the cyclist and cycle. (a) Explain carefully why the angle θ the bicycle makes with the vertical (Fig. 11–34) must be given by $\tan\theta = F_{fr}/F_N$ if the cyclist is to maintain balance. (b) Calculate θ for the values given. [Hint: Consider the "circular" translational motion of the bicycle and rider.] (c) If the coefficient of static friction between tires and road is $\mu_s = 0.65$, what is the minimum turning radius?

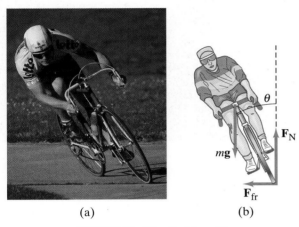

FIGURE 11–34 Problem 58.

59. A thin uniform stick of mass M and length l is positioned vertically, with its tip on a frictionless table. It is released and allowed to slip and fall (Fig. 11–35). Determine the speed of its center of mass just before it hits the table.

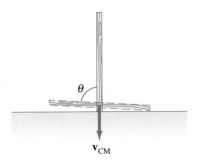

FIGURE 11–35 Problem 59.

60. A thin string is wrapped around a cylindrical hoop of radius R and mass M. One end of the string is fixed, and the hoop is allowed to fall vertically, starting from rest, as the string unwinds. (a) Determine the angular momentum of the hoop about its CM as a function of time. (b) What is the tension in the string as function of time?

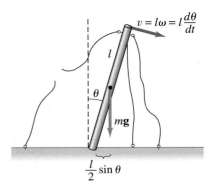

FIGURE 11–36 Problem 61.

61. A radio transmission tower has a mass of 80 kg and is 12 m high. The tower is anchored to the ground by a flexible joint at its base, but it is secured by three cables 120° apart (Fig. 11–36). In an analysis of a potential failure, a mechanical engineer needs to determine the behavior of the tower if one of the cables broke. The tower would fall away from the broken cable, rotating about its base. Determine the speed of the top of the tower as a function of the rotation angle θ. Start your analysis with the rotational dynamics equation of motion $d\mathbf{L}/dt = \boldsymbol{\tau}_{net}$. Approximate the tower as a tall thin rod.

62. A baseball bat has a "sweet spot" where a ball can be hit with almost effortless transmission of energy. A careful analysis of baseball dynamics shows that this special spot is located at the point where an applied force would result in pure rotation of the bat about the handle. Determine the location of the sweet spot of the bat shown in Fig. 11–37. The linear mass density of the bat is given roughly by $(0.56 + 3.5x^2)$ kg/m, where x is in meters measured from the end of the handle. The entire bat is 0.84 m long. The desired rotation point should be 5.0 cm from the end where the bat is held. [Hint: Where is the CM of the bat?]

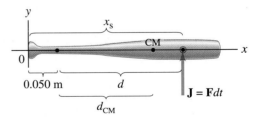

FIGURE 11–37 Problem 62.

63. Suppose a star the size of our Sun, but of mass 8.0 times as great, were rotating at a speed of 1.0 revolution every 10 days. If it were to undergo gravitational collapse to a neutron star of radius 10 km, losing $\frac{3}{4}$ of its mass in the process, what would its rotation speed be? Assume the star is a uniform sphere at all times. Assume also that the thrown-off mass carries off (a) no angular momentum, (b) its proportional share $(\frac{3}{4})$ of the initial angular momentum.

Our whole built environment, from modern bridges to skyscrapers and even 200 year old forts (on the right, beneath the Golden Gate bridge in San Francisco) has required engineers and architects to determine the forces and stresses within these structures. The object is to keep these structures static—that is, not in motion, especially not falling down.

Static Equilibrium; Elasticity and Fracture

This chapter deals with forces within objects at rest

I n this chapter, we will study a special case of motion—when the net force and the net torque on an object, or system of objects, are both zero. In this case the object or system is not accelerating; it is either at rest, or its center of mass is moving at constant velocity. We will be concerned mainly with the first case, when the object or objects are all at rest. Now you may think that the study of objects at rest is not very interesting since the objects will have neither velocity nor acceleration; and the net force and the net torque will be zero. But this does not imply that no forces at all act on the objects. In fact it is virtually impossible to find a body on which no forces act at all. Sometimes the forces may be so great that the object is seriously deformed, or it may even fracture (break)—and avoiding such problems gives this field of **statics** great importance.

FIGURE 12–1 The book is in equilibrium; the net force on it is zero.

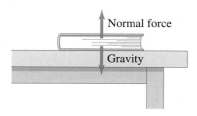

12–1 | Statics—The Study of Forces in Equilibrium

Objects within our experience have at least one force acting on them (gravity), and if they are at rest then there must be other forces acting on them as well so that the net force is zero. An object at rest on a table, for example, has two forces acting on it, the downward force of gravity and the normal force the table exerts upward on it (Fig. 12–1). Since the net force is zero, the upward force exerted by the table must be equal in magnitude to the force of gravity acting downward.

Such a body is said to be in **equilibrium** (Latin for "equal forces" or "balance") under the action of these two forces. [Do not confuse the two forces in Fig. 12–1 with the equal and opposite forces of Newton's third law which act on different bodies; here both forces act on the same body.]

The subject of statics is concerned with the calculation of the forces acting on and within structures that are in equilibrium. Determination of these forces, which occupies us in the first part of this chapter, then allows a determination of whether the structures can sustain the forces without significant deformation or fracture, subjects we discuss later in this chapter. These techniques can be applied in a wide range of fields. Architects and engineers must be able to calculate the forces on the structural components of buildings, bridges, machines, vehicles, and other structures, since any material will break or buckle if too much force is applied (Fig. 12–2). In the human body, a knowledge of the forces in muscles and joints is of great value for medicine and physical therapy, and is also valuable for the study of athletic activity.

FIGURE 12–2   Elevated walkway collapse in a Kansas City hotel in 1981. How a simple physics calculation could have prevented the tragic loss of 100 lives is considered in Example 12–14.

12–2 | The Conditions for Equilibrium

For an object to be in equilibrium it must not be accelerating, so the *vector sum of all external forces acting on the body must be zero*:

$$\Sigma \mathbf{F} = 0. \tag{12–1a}$$

This is called the **first condition for equilibrium** and is equivalent to three component equations:

$$\Sigma F_x = 0, \qquad \Sigma F_y = 0, \qquad \Sigma F_z = 0. \tag{12–1b}$$

First condition for equilibrium: the sum of all forces is zero

EXAMPLE 12–1 **Pullups on a scale.** A 90-kg weakling cannot do even one pull-up. By standing on a scale (Fig. 12–3), he can determine how close he gets. His best effort results in a scale reading of 23 kg. What force is he exerting?

SOLUTION There are three forces acting on our nonathlete, as shown in Fig. 12–3: gravity, $mg = (90\,\text{kg})(9.8\,\text{m/s}^2)$ downward, and two upwards forces which are (1) the force the bar pulls upward on him, F_B (equal and opposite to the force he exerts on the bar), and (2) the force the scale exerts on his feet, F_S. At best, $F_S = (23\,\text{kg})(g)$. The person doesn't move, so the sum of these forces is zero:

$$F_B + F_S - mg = 0.$$

We solve for F_B:

$$F_B = mg - F_S$$
$$= (90\,\text{kg} - 23\,\text{kg})(g) = (67\,\text{kg})(9.8\,\text{m/s}^2) = 660\,\text{N}.$$

That is, he could lift himself if his mass were only 67 kg.

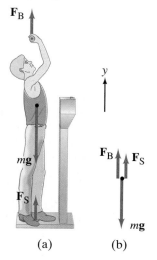

FIGURE 12–3   Example 12–1: (a) A person trying to do a pull-up while standing on a scale. (b) Simple free-body diagram.

(a)   (b)

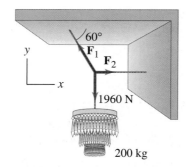

60°

y

F_1

F_2

x

1960 N

200 kg

FIGURE 12–4 Example 12–2.

EXAMPLE 12–2 **Chandelier cord tension.** Calculate the tensions $\mathbf{F}_1$ and $\mathbf{F}_2$ in the two cords that are connected to the cord supporting the 200-kg chandelier in Fig. 12–4.

SOLUTION The three forces, $\mathbf{F}_1$, $\mathbf{F}_2$, and the 200-kg weight of the chandelier, act at the point where the three cords join. We choose this junction point (it could be a knot) as the object for which we write $\Sigma F_x = 0$, $\Sigma F_y = 0$. (We don't bother considering the chandelier itself since only two forces act on it, gravity downward and the equal but opposite force exerted upward by the cord, both of which equal $mg = (200\,\text{kg})(9.8\,\text{m/s}^2) = 1960\,\text{N}$.) There are two unknowns $(F_1$ and $F_2)^\dagger$ and we can solve for them using Eqs. 12–1, $\Sigma F_x = 0$, $\Sigma F_y = 0$.

We first resolve $\mathbf{F}_1$ into its horizontal (x) and vertical (y) components. Although we don't know the value of F_1, we can write $F_{1x} = F_1 \cos 60°$ and $F_{1y} = F_1 \sin 60°$. $\mathbf{F}_2$ has only an x component. In the vertical direction, we have only the weight of the chandelier $= (200\,\text{kg})(g)$ acting downward and the vertical component of $\mathbf{F}_1$ upward. Since $\Sigma F_y = 0$, we have

$$\Sigma F_y = F_1 \sin 60° - (200\,\text{kg})(g) = 0$$

so

$$F_1 = \frac{(200\,\text{kg})g}{\sin 60°} = \frac{(200\,\text{kg})g}{0.866} = (231\,\text{kg})g = 2260\,\text{N}.$$

In the horizontal direction,

$$\Sigma F_x = F_2 - F_1 \cos 60° = 0.$$

Thus

$$F_2 = F_1 \cos 60° = (231\,\text{kg})(g)(0.500) = (115\,\text{kg})g = 1130\,\text{N}.$$

The magnitudes of $\mathbf{F}_1$ and $\mathbf{F}_2$ determine the strength of cord or wire that must be used. In this case, the wire must be able to hold more than 230 kg. Note in this Example that we didn't insert the value of g, the acceleration due to gravity, until the end. In this way we found the magnitude of the force in terms of the number of kilograms (which may be a more familiar quantity than newtons) times g.

Although Eqs. 12–1 are a necessary condition for an object to be in equilibrium, they are not a sufficient condition. Figure 12–5 shows an object on which the net force is zero. Although the two forces labeled $\mathbf{F}$ add up to give zero net force on the object, they do give rise to a net torque that will rotate the object. Thus we need a *second condition for equilibrium*: that the vector sum of all external torques acting on the body must add up to zero:

$$\Sigma \boldsymbol{\tau} = 0. \tag{12–2}$$

This will assure that the angular acceleration, α, about any point will be zero. If the body is not rotating initially ($\omega = 0$) it will not start rotating. Equations 12–1 and 12–2 are the only requirements for a body to be in equilibrium.

Equation 12–2 can also be written in component form

$$\Sigma \tau_x = 0, \quad \Sigma \tau_y = 0, \quad \Sigma \tau_z = 0.$$

The torques are calculated about some chosen point O, and τ_x, τ_y, and τ_z are the components along any three chosen axes. However, we will restrict ourselves in most of this chapter to a common situation that is simpler than the general case: the situation when all the external forces are acting in a plane. If we call this plane the xy plane, then we have only two force equations,

$$\Sigma F_x = 0, \quad \Sigma F_y = 0,$$

and one torque equation,

$$\Sigma \tau_z = 0.$$

The torque is calculated about an axis that is perpendicular to the xy plane. The

Second condition for equilibrium: sum of all torques is zero

FIGURE 12–5 Although the net force on it is zero, the ruler will move (rotate). A pair of equal forces acting in opposite directions but at different points on a body (as shown here) is referred to as a *couple*.

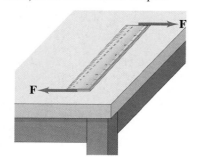

F

F

$\dagger$ The directions of $\mathbf{F}_1$ and $\mathbf{F}_2$ are known, since tension in a rope can only be along the rope—any other direction would cause the rope to bend. Thus, our unknowns are the magnitudes F_1 and F_2.

choice of this axis is arbitrary. Since $\Sigma\tau = I\alpha$ for a rigid body about any fixed axis (Eq. 10–14), $\Sigma\tau = 0$ about any fixed axis if the body is in equilibrium ($\alpha = 0$). Therefore the choice of this axis is arbitrary, so we can choose any axis that makes our calculation easier.

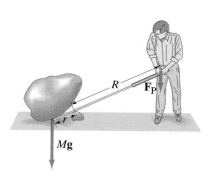

➡ PHYSICS APPLIED
The lever

CONCEPTUAL EXAMPLE 12–3 **A lever.** The bar in Fig. 12–6 is being used as a lever to pry up a large rock. The small rock acts as a fulcrum. The force F_P required at the long end of the bar can be quite a bit smaller than the rock's weight Mg, since it is the *torques* that balance in the rotation about the fulcrum. If, however, the leverage isn't quite good enough, and the rock isn't budged, what are two ways to increase the leverage?

RESPONSE One way is to increase the lever arm of the force F_P by slipping a pipe over the end of the bar and thereby pushing with a longer lever arm. A second, perhaps handier, way is to move the fulcrum closer to the large rock. This may change the long lever arm R only a little, but it changes the short lever arm r by a good fraction and therefore changes the ratio of R/r dramatically. In order to pry the rock, the torque due to F_P must at least balance the torque due to mg, so $mgr = F_P R$ and

$$\frac{r}{R} = \frac{F_P}{Mg}.$$

FIGURE 12–6 Example 12–3.

With r smaller, the weight Mg can be balanced with less force F_P. The ratio R/r is the **mechanical advantage** of the system. A lever is a "simple machine." We discussed another simple machine, the pulley, in Chapter 4, Example 4–14.

12–3 | Solving Statics Problems

This subject of statics is important because it allows us to calculate certain forces on (or within) a structure when some of the forces on it are already known. There is no single technique for attacking such statics problems, but the following procedure, which assumes the force vectors all lie in the xy plane, may be helpful:

PROBLEM SOLVING Statics

1. Choose one body at a time for consideration, and make a careful *free-body diagram* for it by showing all the forces acting on that body and the points at which these forces act.
2. Choose a convenient coordinate system, and resolve the forces into their components.
3. Using letters to represent unknowns, write down equations for $\Sigma F_x = 0$, $\Sigma F_y = 0$, and $\Sigma\tau = 0$.
4. For the $\Sigma\tau = 0$ equation, choose any axis perpendicular to the xy plane that you like. (For example, you can reduce the number of unknowns in the resulting equation by choosing the axis so that one of the unknown forces passes through the axis; then this force will have

zero lever arm and produce zero torque and so won't appear in the equation.) Pay careful attention to determining the lever arm for each force correctly. Give each torque a + or − sign. If torques that would tend to rotate the object counterclockwise are given a + sign, then torques that would tend to rotate it clockwise are negative.

5. Solve these equations for the unknowns. Three equations allow a maximum of three unknowns to be solved for; they can be forces, distances, even angles. (If an unknown force comes out negative in your solution, it means the direction you originally chose for that force is actually the opposite.)

Another way of stating the torque equilibrium condition is that the sum of all clockwise torques is equal to the sum of all counterclockwise torques. And for the forces, the sum of the upward forces is equal to the sum of the downward forces, and the sum of horizontal forces to the left is equal to the sum of horizontal forces to the right.

Alternate form of $\Sigma\tau = 0$, $\Sigma\mathbf{F} = 0$

(a)

Counterweight
M = 9500 kg

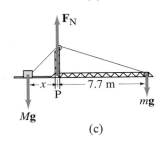

|←3.4 m→|←——7.7 m——→|

m = 2800 kg

(b)

$\mathbf{F_N}$

|←x→|←——7.7 m——→|
P

$M\mathbf{g}$

$m\mathbf{g}$

(c)

FIGURE 12–7 Example 12–4.

FIGURE 12–8 A 1500-kg beam supports 15,000-kg machine. Example 12–5.

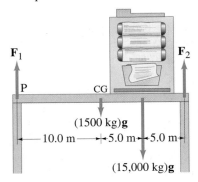

$\mathbf{F_1}$

$\mathbf{F_2}$

P

CG

(1500 kg)$\mathbf{g}$

|←——10.0 m——→|←5.0 m→|←5.0 m→|

(15,000 kg)$\mathbf{g}$

One of the forces that acts on bodies is the force of gravity. Our analysis in this chapter is greatly simplified if we use the concept of center of gravity (CG) or center of mass (CM), which are the same point as long as **g** is the same throughout the body (nearly always valid). As we discussed in Section 9–8, we can consider the force of gravity on the body as a whole as acting at its CG. For uniform symmetrically shaped bodies, the CG is at the geometric center. For more complicated bodies, the CG can be determined as discussed in Section 9–8.

EXAMPLE 12–4 **Tower crane.** A tower crane (Fig. 12–7a) must always be carefully balanced so that there is no net torque tending to tip it. A particular crane at a building site is about to lift a 2800-kg air conditioning unit. The crane's dimensions are shown in Fig. 12–7b. (a) Where must the crane's 9500-kg counterweight be placed when the load is lifted from the ground? (Note—the counterweight is usually moved automatically via sensors and motors to precisely compensate for the load.) (b) Determine the maximum load that can be lifted with this counterweight when it is placed at its full extent. Ignore the mass of the beam.

SOLUTION (a) The free-body diagram, showing the forces on the beam, is drawn in Fig. 12–7c. Let us calculate the torques about the pivot point P. Then the upward force $\mathbf{F_N}$ exerted on the beam by the tower (this force keeps the beam up there) has a lever arm of zero and so doesn't appear in the equation. Thus, setting $\Sigma\tau = 0$,

$$\Sigma\tau = (9500 \text{ kg})(g)(x) - (2800 \text{ kg})(g)(7.7 \text{ m}) = 0$$

where x is the position we wish to determine for the counterweight. Solving for x, we find $x = (2800 \text{ kg})(7.7 \text{ m})/(9500 \text{ kg}) = 2.3 \text{ m}$.

(b) Figure 12–7b suggests that the counterweight can be at most 3.4 m from the tower. Hence the maximum load m that can be lifted is found using the torque equation

$$\Sigma\tau = (9500 \text{ kg})(g)(3.4 \text{ m}) - mg(7.7 \text{ m}) = 0,$$

which gives $m = (9500 \text{ kg})(3.4 \text{ m})/(7.7 \text{ m}) = 4200 \text{ kg}$.

EXAMPLE 12–5 **Forces on a beam and supports.** A uniform 1500-kg beam, 20.0 m long, supports a 15,000-kg printing press 5.0 m from the right support column (Fig. 12–8). Calculate the force on each of the vertical support columns.

SOLUTION We analyze the forces on the beam, since the force the beam exerts on the columns is equal and opposite to the forces exerted by the columns on the beam. We call the latter $\mathbf{F_1}$ and $\mathbf{F_2}$ in Fig. 12–8. The weight of the beam itself acts at its center of gravity, 10.0 m from either end. Since it doesn't matter which point we choose as the axis for writing the torque equation, we can choose one that is convenient. If we calculate the torques about the point of application of $\mathbf{F_1}$, (labelled P), then $\mathbf{F_1}$ will not enter the equation (its lever arm will be zero) and we will have an equation in only one unknown, F_2. We choose the counterclockwise direction as positive, and $\Sigma\tau_P = 0$ gives

$$\Sigma\tau_P = -(10.0 \text{ m})(1500 \text{ kg})g - (15.0 \text{ m})(15,000 \text{ kg})g + (20.0 \text{ m})F_2 = 0.$$

Solving for F_2 we find $F_2 = (12,000 \text{ kg})g = 118,000 \text{ N}$. To find F_1, we use $\Sigma F_y = 0$:

$$\Sigma F_y = F_1 - (1500 \text{ kg})g - (15,000 \text{ kg})g + F_2 = 0.$$

Putting in $F_2 = (12,000 \text{ kg})g$ we find that $F_1 = (4500 \text{ kg})g = 44,100 \text{ N}$.

Figure 12–9 shows a beam that extends beyond its support like a diving board. Such a beam is called a **cantilever**. The forces acting on the beam in this figure are those due to the supports, $\mathbf{F_1}$ and $\mathbf{F_2}$, and the force of gravity, which acts at the CG, 5.0 m to the right of the right-hand support. If you follow the procedure of

Cantilever

the last Example to calculate F_1 and F_2, assuming they point upward as shown in the figure, you will find that F_1 comes out negative. If the beam has a mass of 1200 kg, then $F_2 = 15{,}000$ N and $F_1 = -3000$ N (see Problem 13). Whenever an unknown force comes out negative, it merely means that the force actually points in the opposite direction from what you assumed. Thus in Fig. 12–9, $\mathbf{F}_1$ actually points downward. With a little reflection it should become obvious that the left-hand support must indeed pull downward on the beam (by means of bolts, screws, fasteners and/or glue) if the beam is to be in equilibrium; otherwise the sum of the torques about the CM (or about the point where $\mathbf{F}_2$ acts) could not be zero.

Any building, or part of a building, that overhangs its support is said to be cantilevered (Fig. 12–10). An airplane wing is a cantilever. A cantilevered bridge is one whose span between two supports was designed and analyzed as two arms, each fixed to a separate support, that extend out toward each other to meet in the middle.

Our next Example involves a beam attached to a wall by a hinge or "pin," and is supported by a cable or cord (Fig. 12–11). It is important to remember that a flexible cable can support a force only along its length. (If there were a component of force perpendicular to the cable, it would bend because it is flexible.) But for a rigid device, such as the pin in Fig. 12–11, the force can be in any direction and we can know the direction only after solving the problem. The pin is assumed small and smooth, so it can exert no internal torque (about its center) on the beam.

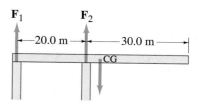

FIGURE 12–9 A cantilever.

FIGURE 12–10 This famous cantilevered house, called "Fallingwater," was designed by the American architect Frank Lloyd Wright.

EXAMPLE 12–6 **Beam supported by pin and cable.** A uniform beam, 2.20 m long with mass $m = 25.0$ kg, is mounted by a pin on a wall as shown in Fig. 12–11. The beam is held in a horizontal position by a cable that makes an angle $\theta = 30.0°$ as shown. The beam supports a sign of mass $M = 280$ kg suspended from its end. Determine the components of the force $\mathbf{F}_P$ that the pin exerts on the beam, and the tension F_T in the supporting cable.

SOLUTION Figure 12–11 is the free-body diagram for the beam, showing all the forces acting on the beam; it also shows the components of $\mathbf{F}_P$ and $\mathbf{F}_T$. We have three unknowns, F_{Px}, F_{Py}, and F_T (we are given θ), so we will need all three equations, $\Sigma F_x = 0$, $\Sigma F_y = 0$, $\Sigma \tau = 0$. The sum of the forces in the vertical (y) direction is

$$\Sigma F_y = F_{Py} + F_{Ty} - mg - Mg = 0. \qquad \text{(i)}$$

In the horizontal (x) direction, the sum of the forces is

$$\Sigma F_x = F_{Px} - F_{Tx} = 0. \qquad \text{(ii)}$$

For the torque equation, we choose the axis at the point where $\mathbf{F}_T$ and $M\mathbf{g}$ act (so our equation then contains only one unknown, F_{Py}, and we can solve it more quickly); we choose torques that tend to rotate the beam counterclockwise as positive. The weight mg of the (uniform) beam acts at its center, so we have:

$$\Sigma \tau = -(F_{Py})(2.20 \text{ m}) + (mg)(1.10 \text{ m}) = 0$$

so

$$F_{Py} = \tfrac{1}{2}mg = \tfrac{1}{2}(25.0 \text{ kg})(9.80 \text{ m/s}) = 123 \text{ N}. \qquad \text{(iii)}$$

Next, since the tension $\mathbf{F}_T$ in the cable acts along the cable ($\theta = 30.0°$),

$$F_{Ty} = F_{Tx} \tan \theta = 0.577 \, F_{Tx}. \qquad \text{(iv)}$$

From Eqs. (i), (ii), and (iv) we get

$$F_{Ty} = (m + M)g - F_{Py} = (305 \text{ kg})(9.80 \text{ m/s}^2) - 123 \text{ N} = 2870 \text{ N}$$

$$F_{Tx} = F_{Ty}/0.577 = 4970 \text{ N}$$

$$F_{Px} = F_{Tx} = 4970 \text{ N}.$$

The components of $\mathbf{F}_P$ are $F_{Py} = 123$ N and $F_{Px} = 4970$ N. The tension in the cable is $F_T = \sqrt{F_{Tx}^2 + F_{Ty}^2} = 5740$ N.

FIGURE 12–11 Beam attached to a wall by a pin. Example 12–6.

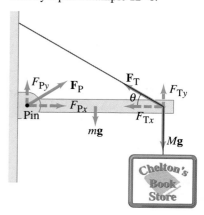

EXAMPLE 12–7 **Force exerted by biceps muscle.** How much force must the biceps muscle exert when a 5.0-kg mass is held in the hand (a) with the arm horizontal as in Fig. 12–12a, and (b) when the arm is at a 30° angle as in Fig. 12–12b? Assume that the mass of forearm and hand together is 2.0 kg and their CG is as shown.

SOLUTION (a) The forces acting on the forearm are shown in Fig. 12–12a and include the upward force F_M exerted by the muscle and a force F_J exerted at the joint by the bone in the upper arm (both assumed to act vertically). We wish to find F_M, which is done most easily by using the torque equation and by choosing our axis through the joint so that F_J does not enter:

$$\Sigma\tau = (0.050\,\text{m})(F_M) - (0.15\,\text{m})(2.0\,\text{kg})(g) - (0.35\,\text{m})(5.0\,\text{kg})(g) = 0.$$

We solve this for F_M and find $F_M = (41\,\text{kg})(g) = 400\,\text{N}$.

(b) The lever arm, as calculated about the joint, is reduced by the factor cos 30° for all three forces. So our torque equation will look like the one just above, except that each term will have a "cos 30°." The latter will cancel out so the same result will be obtained, $F_M = 400\,\text{N}$.

Leverage with muscles

Note that the force required of the muscle (400 N) is quite large compared to the weight of the object lifted (49 N). Indeed, the muscles and joints of the body are generally subjected to quite large forces. The point of insertion of a muscle varies from person to person. A slight increase in the point of insertion of the biceps muscle from 5.0 cm to 5.5 cm can be a considerable advantage for lifting and throwing. Champion athletes are often found to have muscle insertions farther from the joint than the average person.

FIGURE 12–12 Example 12–7.

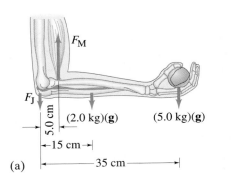
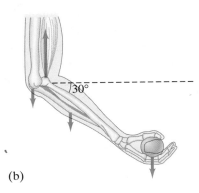

(a)

(b)

FIGURE 12–13 A ladder leaning against a wall. Example 12–8 (a).

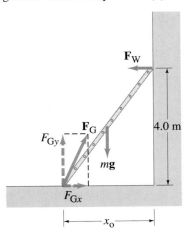

EXAMPLE 12–8 **Ladder.** A 5.0-m-long ladder leans against a wall at a point 4.0 m above the ground as shown in Fig. 12–13. The ladder is uniform and has mass 12.0 kg. Assuming the wall is frictionless (but the ground is not), (a) determine the forces exerted on the ladder by the ground and the wall. (b) If the coefficient of static friction between the ladder and the ground is $\mu_s = 0.40$, how high up the ladder can a 58-kg painter climb (Fig. 12–14) without the ladder slipping?

SOLUTION (a) Figure 12–13 shows the free-body diagram for the ladder, showing all the forces acting on the ladder. The wall, since it is frictionless, can exert a force F_W that must be perpendicular to the wall. The ground, however, can exert both horizontal, F_{Gx}, and vertical, F_{Gy}, force components, the former being frictional and the latter the normal force. Finally, gravity exerts a force $mg = (12.0\,\text{kg})(9.80\,\text{m/s}^2) = 118\,\text{N}$ on the ladder at its midpoint, since the ladder is uniform. The y component of the force equation is

$$\Sigma F_y = F_{Gy} - mg = 0,$$

so immediately we have $F_{Gy} = mg = 118\,\text{N}$. The x component of the force equation is

$$\Sigma F_x = F_{Gx} - F_W = 0.$$

To determine F_{Gx} and F_W, which are both unknowns, we need another equation, namely a torque equation, which we calculate about the point where the ladder touches the ground. This point is a distance $x_0 = \sqrt{(5.0\,\text{m})^2 - (4.0\,\text{m})^2} = 3.0\,\text{m}$ from the wall. The lever arm for mg is half this, or 1.5 m, and the lever arm for F_W is 4.0 m. Since $\mathbf{F}_G$ acts at the axis, its lever arm is zero and so doesn't enter the equation (we planned it like that), and we get

$$\Sigma\tau = (4.0\,\text{m})F_W - (1.5\,\text{m})mg = 0.$$

Thus

$$F_W = \frac{(1.5\,\text{m})(12.0\,\text{kg})(9.8\,\text{m/s}^2)}{4.0\,\text{m}} = 44\,\text{N}.$$

Then, from the x component of the force equation,

$$F_{Gx} = F_W = 44\,\text{N}.$$

Since the components of $\mathbf{F}_G$ are $F_{Gx} = 44\,\text{N}$ and $F_{Gy} = 118\,\text{N}$, then

$$F_G = \sqrt{(44\,\text{N})^2 + (118\,\text{N})^2} = 126\,\text{N} \approx 130\,\text{N}$$

(rounded off to two significant figures) and it acts at an angle

$$\theta = \tan^{-1}(118\,\text{N}/44\,\text{N}) = 70°$$

to the ground. Note the force $\mathbf{F}_G$ does *not* have to act along the ladder's direction because the ladder is rigid and not flexible like a cord or cable.

(*b*) The free-body diagram of the ladder with the person on it is shown in Fig. 12–14, where x is the horizontal distance from the base of the ladder to a point on the ground directly below the painter. The force equations are

⇒ **PHYSICS APPLIED**

Safety on a ladder

$$\Sigma F_x = 0: \qquad F_{Gx} - F_W = 0 \tag{i}$$

$$\Sigma F_y = 0: \qquad F_{Gy} - (m + M)g = 0, \tag{ii}$$

FIGURE 12–14 Example 12–8 (*b*) with a person on the ladder.

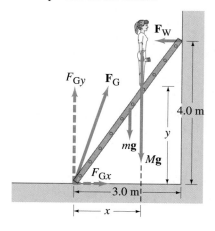

where $m = 12.0\,\text{kg}$ and $M = 58\,\text{kg}$. The torque equation, calculated as before about the base of the ladder, is

$$(4.0\,\text{m})F_W - (1.5\,\text{m})mg - xMg = 0. \tag{iii}$$

We have four unknowns (F_W, F_{Gx}, F_{Gy}, x) but only three equations; but we get a fourth equation when we use the condition imposed on x, that the painter is at the maximum height without the ladder slipping. This condition limits F_{Gx} to be

$$F_{Gx} = \mu_s F_{Gy} = 0.40 F_{Gy}. \tag{iv}$$

Using Eq. (ii) above, this is

$$F_{Gx} = 0.40(12\,\text{kg} + 58\,\text{kg})g = 28\,g = 270\,\text{N}.$$

Then, from Eq. (i),

$$F_W = F_{Gx} = 28\,g$$

and from Eq. (iii)

$$x = \frac{(4.0\,\text{m})F_W - (1.5\,\text{m})mg}{(58\,\text{kg})g} = 1.62\,\text{m}.$$

Since the base of the ladder is 3.0 m from the wall, the painter can stand at a point that is a fraction $(x/3.0\,\text{m})$ of the distance up the ladder, so her feet can be a height $y = (1.62/3.0)4.0\,\text{m} = 2.2\,\text{m}$ above the ground. If the painter climbs any higher, F_{Gx} will exceed $\mu_s F_{Gy}$ (can you see why?) and the ladder will slip.

12–4 Stability and Balance

A body in static equilibrium, if left undisturbed, will undergo no translational or rotational acceleration since the sum of all the forces and the sum of all the torques acting on it are zero. However, if the object is displaced slightly, three different outcomes are possible: (1) the object returns to its original position, in which case it is said to be in **stable equilibrium**; (2) the object moves even farther from its original position, in which case it is said to be in **unstable equilibrium**; or (3) the object remains in its new position, in which case it is said to be in **neutral equilibrium**.

Consider the following examples. A ball suspended freely from a string is in stable equilibrium, for if it is displaced to one side, it will quickly return to its original position (Fig. 12–15a). On the other hand, a pencil standing on its point is in unstable equilibrium. If its CG is directly over its tip (Fig. 12–15b), the net force and net torque on it will be zero. But if it is displaced ever so slightly—say by a slight vibration or tiny air current—there will be a torque on it, and it will continue to fall in the direction of the original displacement. Finally, an example of an object in neutral equilibrium is a sphere resting on a horizontal tabletop. If it is placed slightly to one side, it will remain in its new position.

In most situations, such as in the design of structures and in working with the human body, we are interested in maintaining stable equilibrium or *balance*, as we sometimes say. In general, an object whose CG is below its point of support, such as a ball on a string, will be in stable equilibrium. If the CG is above the base of support, we have a more complicated situation. Consider a standing refrigerator (Fig. 12–16a). If it is tipped slightly, it will return to its original position due to the torque on it as shown in Fig. 12–16b. But if it is tipped too far, Fig. 12–16c, it will fall over. The critical point is reached when the CG is no longer above the base of support. In general, *a body whose CG is above its base of support will be stable if a vertical line projected downward from the CG falls within the base of support.* This is because the normal force upward on the object (which balances out gravity) can be exerted only within the area of contact, so that if the force of gravity acts beyond this area, a net torque will act to topple the object. Stability, then, can be relative. A brick lying on its widest face is more stable than a brick standing on its end, for it will take more of an effort to tip it over. In the extreme case of the pencil in Fig. 12–15b, the base is practically a point and the slightest disturbance will topple it. In general, the larger the base and the lower the CG, the more stable the object.

In this sense, humans are less stable than four-legged mammals, which not only have a larger base of support because of their four legs, but also have a lower center of gravity. Because of their upright position, humans suffer from numerous ailments such as low back pain due to the large forces involved (see Problem 97). When walking and performing other kinds of movement, a person continually shifts the body so that its CG is over the feet, although in the normal adult this requires no conscious thought. Even as simple a movement as bending over requires moving the hips backward so that the CG remains over the feet, and this repositioning is done without thinking about it. To see this, position yourself with your heels and back to a wall and try to touch your toes. You won't be able to do it without falling. Persons carrying heavy loads automatically adjust their posture so that the CG of the total mass is over their feet, Fig. 12–17.

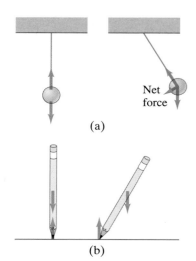

FIGURE 12–15 (a) Stable equilibrium, and (b) unstable equilibrium.

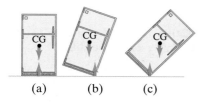

FIGURE 12–16 Equilibrium of a refrigerator resting on a surface.

FIGURE 12–17 Humans adjust their posture to achieve stability when carrying loads.

12–5 Elasticity and Elastic Moduli; Stress and Strain

In the first part of this chapter we studied how to calculate the forces on objects in equilibrium. In this section we study the effects of these forces, for any object changes shape under the action of applied forces. If the forces are great enough, the object will break or *fracture*—as we will discuss in Section 12–6.

If a force is exerted on an object, such as the vertically suspended metal rod shown in Fig. 12–18, the length of the object changes. If the amount of elongation, ΔL, is small compared to the length of the object, experiment shows that ΔL is proportional to the weight or force exerted on the object. This proportionality can be written as an equation and is sometimes called **Hooke's law**[†] (after Robert Hooke (1635–1703) who first noted it):

$$F = k \, \Delta L. \qquad (12\text{–}3)$$

Here F represents the force (or weight) pulling on the object, ΔL is the increase in length, and k is a proportionality constant. Equation 12–3 is found to be valid for almost any solid material from iron to bone, but it is valid only up to a point. For if the force is too great, the object stretches excessively and eventually breaks. Figure 12–19 shows a typical graph of elongation versus applied force. Up to a point called the **proportional limit**, Eq. 12–3 is a good approximation for many common materials and the curve is a straight line. Beyond this point the graph deviates from a straight line and no simple relationship exists between F and ΔL. Nonetheless, up to a point further along the curve called the **elastic limit**, the object will return to its original length if the applied force is removed. The region from the origin to the elastic limit is called the *elastic region*. If the object is stretched beyond the elastic limit, it enters the *plastic region*: it does not return to the original length upon removal of the external force, but remains permanently deformed (like a bent paper clip). The maximum elongation is reached at the *breaking point*. The maximum force that can be applied without breaking is called the **ultimate strength** of the material (actually force per unit area, as discussed in Section 12–6).

The amount of elongation of an object, such as the rod shown in Fig. 12–18, depends not only on the force applied to it, but also on the material from which it is made and on its dimensions. That is, the constant k in Eq. 12–3 can be written in terms of these factors. If we compare rods made of the same material but of different lengths and cross-sectional areas, it is found that for the same applied force the amount of stretch (again assumed small compared to the total length) is proportional to the original length and inversely proportional to the cross-sectional area. That is, the longer the object, the more it elongates for a given force; and the thicker it is, the less it elongates. These findings can be combined with Eq. 12–3 to yield

$$\Delta L = \frac{1}{E} \frac{F}{A} L_0, \qquad (12\text{–}4)$$

where L_0 is the original length of the object, A is the cross-sectional area, and ΔL is the change in length due to the applied force F. E is a constant of proportionality[‡] known as the **elastic modulus**, or **Young's modulus**, and its value depends only on the material. The value of Young's modulus for various materials is given

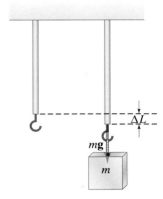

FIGURE 12–18 Hooke's law: $\Delta L \propto$ applied force.

FIGURE 12–19 Applied force vs. elongation for a typical metal under tension.

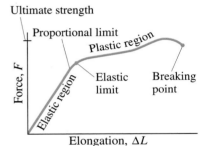

Young's modulus

[†]The term "law" applied to this relation is not really appropriate, since first of all, it is only an approximation, and secondly, it refers only to a limited set of phenomena. Most physicists prefer to reserve the word "law" for those relations that are deeper and more encompassing and precise, such as Newton's laws of motion or the law of conservation of energy.

[‡]That E is in the denominator, so that $1/E$ is the actual proportionality constant, is merely a convention. When we rewrite Eq. 12–4 to get Eq. 12–5, E is in the numerator.

in Table 12–1. Because E is a property only of the material and is independent of the object's size or shape, Eq. 12–4 is far more useful for practical calculation than Eq. 12–3.

TABLE 12–1 Elastic moduli

Material	Young's modulus, E (N/m²)	Shear modulus, G (N/m²)	Bulk modulus, B (N/m²)
Solids			
Iron, cast	100×10^9	40×10^9	90×10^9
Steel	200×10^9	80×10^9	$140 \cdot \times 10^9$
Brass	100×10^9	35×10^9	80×10^9
Aluminum	70×10^9	25×10^9	70×10^9
Concrete	20×10^9		
Brick	14×10^9		
Marble	50×10^9		70×10^9
Granite	45×10^9		45×10^9
Wood (pine) (parallel to grain) (perpendicular to grain)	10×10^9 1×10^9		
Nylon	5×10^9		
Bone (limb)	15×10^9	80×10^9	
Liquids			
Water			2.0×10^9
Alcohol (ethyl)			1.0×10^9
Mercury			2.5×10^9
Gases[†]			
air, H_2, He, CO_2			1.01×10^5

[†] At normal atmospheric pressure, and no temperature variation in the process.

From Eq. 12–4 we see that the change in length of an object is directly proportional to the product of the object's length L_0 and the force per unit area F/A applied to it. It is general practice to define the force per unit area as the **stress**:

Stress (N/m²)

$$\text{stress} = \frac{\text{force}}{\text{area}} = \frac{F}{A},$$

which has units of N/m². Also, the **strain** is defined as the ratio of the change in length to the original length:

Strain

$$\text{strain} = \frac{\text{change in length}}{\text{original length}} = \frac{\Delta L}{L_0},$$

and is dimensionless (no units). Strain is thus the fractional change in length of an object, and is a measure of how much it has been deformed. Stress is applied by outside agents whereas strain is the material's response to the stress. Equation 12–4 can be rewritten as

$$\frac{F}{A} = E \frac{\Delta L}{L_0} \tag{12–5}$$

or

Young's modulus (again)

$$E = \frac{F/A}{\Delta L/L_0} = \frac{\text{stress}}{\text{strain}}.$$

Thus we see that the strain is directly proportional to the stress, in the elastic region.

EXAMPLE 12-9 **Tension in a piano wire.** A 1.60-m-long steel piano wire has a diameter of 0.20 cm. How great is the tension in the wire if it stretches 0.30 cm when tightened?

SOLUTION We solve for F in Eq. 12–5 and note that the cross-sectional area $A = \pi r^2 = (3.14)(0.0010\,\text{m})^2 = 3.1 \times 10^{-6}\,\text{m}^2$. Then

$$F = E \frac{\Delta L}{L_0} A$$

$$= (2.0 \times 10^{11}\,\text{N/m}^2)\left(\frac{0.0030\,\text{m}}{1.60\,\text{m}}\right)(3.1 \times 10^{-6}\,\text{m}^2) = 1200\,\text{N},$$

where the value for E was taken from Table 12–1. The strong tension in all the wires in a piano must be supported by a very strong frame, usually of cast iron.

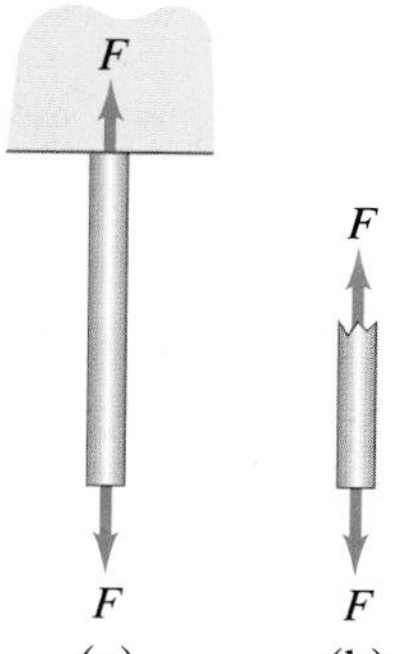

The rod shown in Fig. 12–18 is said to be under **tension** or **tensile stress**. For not only is there a force pulling down on the rod at its lower end, but since the rod is in equilibrium we know that the support at the top is exerting an equal upward force[†] on the rod at its upper end, Fig. 12–20. In fact, this tensile stress exists throughout the material. Consider, for example, the lower half of a suspended rod as shown in Fig. 12–20b. This lower half is in equilibrium so there must be an upward force on it to balance the downward force at its lower end. What exerts this upward force? It must be the upper part of the rod. Thus we see that external forces applied to an object give rise to internal forces, or stress, within the material itself.

Strain or deformation due to tensile stress is but one type of stress to which materials can be subjected. There are two other common types of stress: compressive and shear. **Compressive stress** is the exact opposite of tensile stress. Instead of being stretched, the material is compressed: the forces act inwardly on the body. Columns that support a weight, such as the columns of a Greek temple (Fig. 12–21) are subjected to compressive stress. Equations 12–4 and 12–5 apply equally well to compression and tension, and the values for E are usually the same.

Figure 12–22 compares tensile and compressive stresses as well as the third type, shear stress. An object under **shear stress** has equal and opposite forces applied *across* its opposite faces. An example is a book or brick, firmly attached to a tabletop, on which a force is exerted parallel to the top surface. The table exerts

[†] If we ignore the weight of the bar.

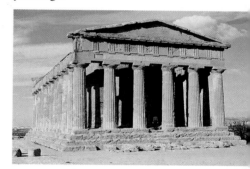

FIGURE 12–20 Stress exists *within* the material.

FIGURE 12–21 This Greek temple (in Agrigento, Sicily) was built 2500 years ago.

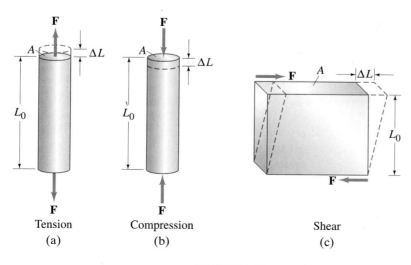

Tension
(a)

Compression
(b)

Shear
(c)

FIGURE 12–22
The three types of stress.

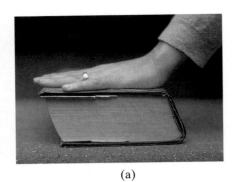

(a)

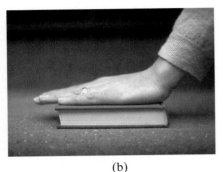

(b)

FIGURE 12–23 The fatter book (a) shifts more than the thinner book (b) with the same applied shear force.

FIGURE 12–24 Balance of forces and torques for shear stress.

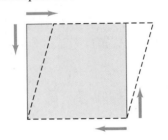

FIGURE 12–25 Fracture as a result of the three types of stress.

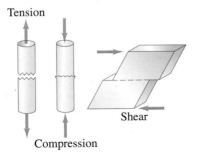

an equal and opposite force along the bottom surface. Although the dimensions of the object do not change significantly, the shape of the object does change as shown in Fig. 12–22c. An equation similar to 12–4 can be used for shear strain:

$$\Delta L = \frac{1}{G}\frac{F}{A}L_0 \qquad (12\text{–}6)$$

but ΔL, L_0, and A must be reinterpreted as indicated in Fig. 12–22c. Note that A is the area of the surface *parallel* to the applied force (and not perpendicular as for tension and compression), and ΔL is *perpendicular* to L_0. The constant of proportionality, G, is called the **shear modulus** and is generally one-half to one-third the value of the elastic modulus, E (see Table 12–1). Figure 12–23 illustrates why $\Delta L \propto L_0$: the fatter book shifts more for the same stress.

The rectangular object undergoing shear in Fig. 12–22c would not actually be in equilibrium under the forces shown, for a net torque would exist. If the object is in fact in equilibrium, there must be two more forces acting on it to balance this torque. One force acts vertically upward on the right, and the other acts vertically downward on the left as shown in Fig. 12–24. This is generally true of shear forces. If the object is a brick or book lying on a table, these two additional forces can be exerted by the table and whatever exerts the other horizontal force (such as your hand pushing across the top of a book).

If an object is subjected to a pressure on all sides, its volume will decrease. A common situation is a body submerged in a fluid; for in this case, the fluid exerts a pressure on the object in all directions, as we shall see in Chapter 13. Pressure is defined as force per unit area and thus is the equivalent of "stress." For this situation the change in volume, ΔV, is found to be proportional to the original volume V_0 and to the increase in the pressure ΔP. We thus obtain a relation in the same form as Eq. 12–4 but with a proportionality constant called the **bulk modulus**, B:

$$\frac{\Delta V}{V_0} = -\frac{1}{B}\Delta P$$

or

$$B = -\frac{\Delta P}{\Delta V/V_0}. \qquad (12\text{–}7a)$$

The minus sign is included to indicate that the volume *decreases* with an increase in pressure. The bulk modulus is sometimes given a more general definition in terms of differentials:

$$B = -V\frac{dP}{dV}. \qquad (12\text{–}7b)$$

Values for the bulk modulus are given in Table 12–1. Since liquids and gases do not have a fixed shape, only the bulk modulus applies to them, and not the shear or Young's modulus.

12–6 Fracture

If the stress on a solid object is too great, the object fractures or breaks, Fig. 12–25. Table 12–2 lists the ultimate tensile strengths, compressive strengths, and shear strengths for a variety of materials. These values give the maximum force per unit area that an object can withstand under each of these three types of stress. They are, however, representative values only, and the actual value for a given specimen can differ considerably. It is therefore necessary to maintain a safety factor of from 3 to perhaps 10 or more—that is, the actual stresses on a structure should not exceed one-tenth to one-third the values given in the Table. One sometimes encounters Tables of "allowable stresses" in which appropriate safety factors have already been included.

TABLE 12–2 Ultimate strengths of materials (force/area)

Material	Tensile strength (N/m^2)	Compressive strength (N/m^2)	Shear strength (N/m^2)
Iron, cast	170×10^6	550×10^6	170×10^6
Steel	500×10^6	500×10^6	250×10^6
Brass	250×10^6	250×10^6	200×10^6
Aluminum	200×10^6	200×10^6	200×10^6
Concrete	2×10^6	20×10^6	2×10^6
Brick		35×10^6	
Marble		80×10^6	
Granite		170×10^6	
Wood (pine) (parallel to grain) (perpendicular to grain)	40×10^6	35×10^6 10×10^6	5×10^6
Nylon	500×10^6		
Bone (limb)	130×10^6	170×10^6	

EXAMPLE 12–10 **Size and compression of support columns.** (a) What minimum cross-sectional area should the two columns have to support the beam of Example 12–5 (Fig. 12–8) assuming the columns are made of concrete and a safety factor of 6 is required? We saw in Example 12–5 that the column on the left supports 4.4×10^4 N and that on the right supports 1.2×10^5 N. (b) How much will the chosen supports compress under the given load?

SOLUTION (a) The right-hand column receives the larger force, $F_2 = 1.2 \times 10^5$ N. It is clearly under compression, and from Table 12–2, we see that the ultimate compressive strength of concrete is 2.0×10^7 N/m². Using a safety factor of 6, the maximum allowable stress is $\frac{1}{6}(2.0 \times 10^7 \text{ N/m}^2) = 3.3 \times 10^6$ N/m², which equals F/A. Since $F = 1.2 \times 10^5$ N, we can solve for A, and we find:

$$A = \frac{1.2 \times 10^5 \text{ N}}{3.3 \times 10^6 \text{ N/m}^2} = 3.6 \times 10^{-2} \text{ m}^2, \text{ or } 360 \text{ cm}^2.$$

A support 18 cm × 20 cm will be adequate.

(b) This column will shorten by a fractional amount

$$\frac{\Delta L}{L_0} = \frac{1}{E}\frac{F}{A} = \left(\frac{1}{2.0 \times 10^{10} \text{ N/m}^2}\right)(3.3 \times 10^6 \text{ N/m}^2) = 1.7 \times 10^{-4}.$$

Thus, if the support has a length $L_0 = 5.0$ m, $\Delta L = 0.85 \times 10^{-3}$ m, or about 1 mm. This calculation was for the right-hand support. If the left-hand support is made of the same cross-sectional area, it will compress less and this should be taken into account.

As can be seen in Table 12–2, concrete (like stone and brick) is reasonably strong under compression but extremely weak under tension. Thus concrete can be used as vertical columns placed under compression but is of little value as a beam since it cannot withstand the tensile forces that arise along the bottom edge when it sags (see Fig. 12–26). *Reinforced concrete*, in which iron rods are embedded in the concrete, is much stronger (Fig. 12–27). But the concrete on the lower edge of a loaded beam still tends to crack because of its weakness under tension. This problem is solved with **prestressed concrete**, which also contains iron rods or a wire mesh, but during the pouring of the concrete, the rods or wire are held under tension. After the concrete dries, the tension on the iron is released, putting the concrete under compression. The amount of compressive stress is carefully predetermined so that when the design loads are applied to the beam, they reduce the compression on the lower edge but never put the concrete into tension.

FIGURE 12–26 A beam sags, at least a little (but is exaggerated here), even under its own weight. The beam thus changes shape so that the upper portion is compressed, and the lower portion is under tension (elongated). Shearing stress also occurs within the beam.

FIGURE 12–27 Steel rods waiting for concrete to be poured around them to form a new highway.

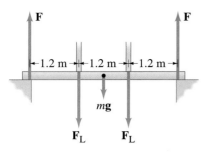

FIGURE 12–28 Example 12–11.

EXAMPLE 12–11 **Shear on a beam.** A uniform pine beam, 3.6 m long and 9.5 cm × 14 cm in cross section, rests on two supports near its ends, as shown in Fig. 12–28. The beam's mass is 25 kg and it is loaded by two vertical roof supports, each 1/3 of the way from each end. What maximum load force F_L can each of the roof supports exert without shearing the pine beam at its supports? Use a safety factor of 5.0.

SOLUTION Each support exerts an upward force F (there is symmetry) that can be at most (see Table 12–2)

$$F = \frac{1}{5} A(5 \times 10^6 \text{ N/m}^2)$$

$$= \frac{1}{5}(0.095 \text{ m})(0.14 \text{ m})(5 \times 10^6 \text{ N/m}^2)$$

$$= 13{,}000 \text{ N}.$$

To determine the maximum load force F_L, we calculate the torque about the left end of the beam (counterclockwise positive):

$$\Sigma\tau = -F_L(1.2 \text{ m}) - (25 \text{ kg})(9.8 \text{ m/s}^2)(1.8 \text{ m}) - F_L(2.4 \text{ m}) + F(3.6 \text{ m}) = 0$$

so

$$F_L = \frac{(13{,}000 \text{ N})(3.6 \text{ m}) - (250 \text{ N})(1.8 \text{ m})}{(1.2 + 2.4)} = 13{,}000 \text{ N}.$$

The total mass of roof the beam can support is $(2)(13{,}000 \text{ N})/(9.8 \text{ m/s}^2) = 2600 \text{ kg}$.

CONCEPTUAL EXAMPLE 12–12 **Maximum stresses.** For the beam in Fig. 12–28, at what points do the shear, compression, and tension stresses reach their maxima?

RESPONSE The shear force is greatest at the ends, given that the beam is uniform and F is slightly greater than F_L (see the last displayed equation in Example 12–11). Compressive stress is greatest at the center on the top side of the beam, and tensile stress is greatest at the center on the bottom side (see Fig. 12–26).

FIGURE 12–29 *Titanic* side wall.

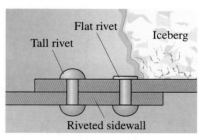

CONCEPTUAL EXAMPLE 12–13 **The *Titanic*.** One hypothesis for why the ship *Titanic* sank when it hit an iceberg on its maiden voyage in 1912 is that the rivet caps were too tall (Fig. 12–29). Explain.

RESPONSE As the ship crashed through an iceberg, the massive iceberg would exert a force on the rivets. With a greater area exposed, more force could be exerted on the rivet heads, causing them to shear off and allowing the side walls to open, letting water in. Note: No rivet heads have ever been found.

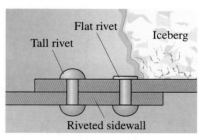

➡ **PHYSICS APPLIED**

A tragic collapse

CONCEPTUAL EXAMPLE 12–14 **A tragic substitution.** Two walkways, one above the other, are suspended from vertical rods attached to the ceiling of a high hotel lobby, as shown in Fig. 12–30a. The original design called for single rods 14 m long, but when such long rods proved to be unwieldy to install, it was decided to replace each long rod with two shorter ones as shown (schematically) in Fig. 12–30b. Determine the net force exerted by the rods on the supporting pin A (assumed the same size) for each design. Assume each vertical rod supports a mass m of each bridge.

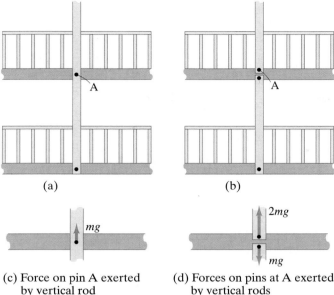

(a)

(b)

(c) Force on pin A exerted by vertical rod

(d) Forces on pins at A exerted by vertical rods

FIGURE 12–30 Example 12–14. Parts (c) and (d) show forces on the pins exerted only by the vertical rods.

RESPONSE The single long vertical rod in Fig. 12–30a exerts an upward force of magnitude mg on pin A to support the mass m of the upper bridge, as shown in Fig. 12–30c. Why? Because the pin is in equilibrium, and the only other force on the pin, and which balances this one, is the downward force mg exerted on it by the upper bridge itself. There is thus a shear stress on the pin. The situation when two shorter rods support the bridges (Fig. 12–30b) is shown in Fig. 12–30d in which only the connections at the upper bridge are shown. The lower rod exerts a force of mg downward on the lower of the two pins because it supports the lower bridge. The upper rod exerts a force of $2mg$ on the upper pin (pin A) because the upper rod supports *both* bridges. Thus we see that when the builders substituted two shorter rods for each single long one, the stress in the supporting pin A was *doubled*. What perhaps seemed like a simple substitution did, in fact, lead to a tragic collapse in 1981 (Fig. 12–2) with a loss of over 100 lives. Having a feel for physics, and being able to make simple calculations based on physics, can have a great effect, literally, on people's lives.

* 12–7 Trusses and Bridges

A beam used to span a wide space, as for a bridge, is subject to strong stresses of all three types as we saw in Fig. 12–26: compression, tension and shear. A basic engineering device to support large spans is the *truss*, an example of which is shown in Fig. 12–31. Wooden truss bridges were first designed by the great architect Andrea Palladio (1518–1580), who is famous for his design of public buildings and villas. With the introduction of steel in the nineteenth century, much stronger steel trusses came into use, although wood trusses are still used to support the roofs of houses and mountain lodges (Fig. 12–32).

Basically, a **truss** is a framework of rods or struts joined together at their ends by pins or rivets, always arranged as triangles. (Triangles are relatively stable, as compared to a rectangle, which easily becomes a parallelogram under sideways forces and then collapses.) The place where the struts are joined by a pin is called a **joint**.

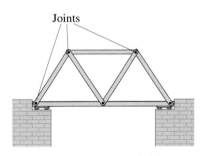

FIGURE 12–31 A truss bridge.

FIGURE 12–32 A roof truss.

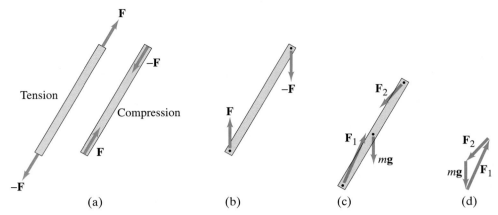

FIGURE 12–33 (a) Each strut (or rod) of a truss is assumed to be under tension or compression. (b) The two equal and opposite forces must be along the same line or a net torque would exist. (c) Real struts have mass, so the forces $\mathbf{F}_1$ and $\mathbf{F}_2$ at the joints do not act precisely along the strut. (d) Vector diagram of part (c).

It is commonly assumed that the struts of a truss are under pure compression or pure tension—that is, the forces act along the length of each strut, Fig. 12–33a. This is an ideal, valid only if a strut has no mass and supports no weight along its length, in which case a strut has only two forces on it, at the ends, as shown in Fig. 12–33a. If the strut is in equilibrium, these two forces must be equal and opposite in direction ($\Sigma\mathbf{F} = 0$). But couldn't they be at an angle, as in Fig. 12–33b? No, because then $\Sigma\tau$ would not be zero. The two forces *must* act along the strut if the strut is in equilibrium. But in a real case of a strut with mass, there are three forces on the strut, as shown in Fig. 12–33c, and $\mathbf{F}_1$ and $\mathbf{F}_2$ do not act along the strut; the vector diagram in Fig. 12–33d shows $\Sigma\mathbf{F} = \mathbf{F}_1 + \mathbf{F}_2 + m\mathbf{g} = 0$. Can you see why $\mathbf{F}_1$ and $\mathbf{F}_2$ both point *above* the strut? (Do $\Sigma\tau$ about each end.)

Consider again the simple beam in Example 12–6, Fig. 12–11. The force $\mathbf{F}_P$ at the pin is *not* along the beam, but acts at an upward angle, albeit only 1.4°. If that beam were massless, we see from Eq. (iii) in Example 12–6 with $m = 0$, that $F_{Py} = 0$, and $\mathbf{F}_P$ would be along the beam.

The assumption that the forces in each strut of a truss are purely along the strut is still very useful whenever the loads act only at the joints and are much greater than the mass of the struts themselves.

PHYSICS APPLIED

A truss bridge

PROBLEM SOLVING

Method of joints

EXAMPLE 12–15 **A truss bridge.** Determine the tension or compression in each of the struts of the truss bridge shown in Fig. 12–34a. The bridge is 64 m long and supports a uniform concrete roadway whose total mass is 1.40×10^6 kg. Use the **method of joints**, which involves (1) drawing a free-body diagram of the truss as a whole, and (2) drawing a free-body diagram for each of the pins (joints), one by one, and setting $\Sigma\mathbf{F} = 0$ for each pin. Ignore the mass of the struts. Assume all triangles are equilateral, and assume that the mass of the roadway acts at the center.

SOLUTION Any bridge has two trusses, one on each side of the roadway. Consider only one truss, Fig. 12–34a, and it will support half the weight of the roadway. That is, our truss supports a total mass $M = 7.0 \times 10^5$ kg. First we draw a free-body diagram for the entire truss as a single unit, which we assume rests on supports at either end that exert upward forces $\mathbf{F}_1$ and $\mathbf{F}_2$, Fig. 12–34b. We assume the mass of the roadway acts entirely at the center, on pin C, as shown. From symmetry [or by doing a torque equation about, say, point A: $(F_2)(l) - Mg(l/2) = 0$] we see that each of the end supports carries half the weight, so

$$F_1 = F_2 = \tfrac{1}{2} Mg.$$

Next we look at pin A and apply $\Sigma \mathbf{F} = 0$ to it. We label the forces on pin A due to each strut with two subscripts: $\mathbf{F}_{AB}$ means the force exerted by the strut AB and $\mathbf{F}_{AC}$ is the force exerted by strut AC. $\mathbf{F}_{AB}$ and $\mathbf{F}_{AC}$ act along their respective struts; but not knowing whether each is compressive or tensile, we could draw four different free-body diagrams, as shown in Fig. 12–34c. Only the one on the left could provide $\Sigma \mathbf{F} = 0$, so we immediately know the directions of $\mathbf{F}_{AB}$ and $\mathbf{F}_{AC}$.[†] These forces act on the pin. The force that pin A exerts on rod AB is opposite in direction to $\mathbf{F}_{AB}$ (Newton's third law), so rod AB is under compression and rod AC is under tension. Now let's calculate the magnitudes of $\mathbf{F}_{AB}$ and $\mathbf{F}_{AC}$. At pin A:

$$\Sigma F_x = F_{AC} - F_{AB} \cos 60° = 0$$
$$\Sigma F_y = F_1 - F_{AB} \sin 60° = 0.$$

Thus

$$F_{AB} = \frac{F_1}{\sin 60°} = \frac{\tfrac{1}{2} Mg}{\tfrac{1}{2}\sqrt{3}} = \frac{1}{\sqrt{3}} Mg,$$

which equals $(7.0 \times 10^5 \text{ kg})(9.8 \text{ m/s}^2)/\sqrt{3} = 4.0 \times 10^6$ N; and

$$F_{AC} = F_{AB} \cos 60° = \frac{1}{2\sqrt{3}} Mg.$$

Next we look at pin B, and Fig. 12–34d is the free-body diagram. [Convince yourself that if $\mathbf{F}_{BD}$ or $\mathbf{F}_{BC}$ were in the opposite direction, $\Sigma \mathbf{F}$ could not be zero; note that $\mathbf{F}_{AB}$ here is in the opposite direction from that in Fig. 12–34c because now we are at the opposite end of strut AB.] We see that BC is under tension and BD compression. (Recall that the forces on the struts are opposite to the forces shown which are on the pin.) We set $\Sigma \mathbf{F} = 0$:

$$\Sigma F_x = F_{AB} \cos 60° + F_{BC} \cos 60° - F_{BD} = 0$$
$$\Sigma F_y = F_{AB} \sin 60° - F_{BC} \sin 60° = 0$$

so

$$F_{BC} = F_{AB} = \frac{1}{\sqrt{3}} Mg,$$

and

$$F_{BD} = F_{AB} \cos 60° + F_{BC} \cos 60° = \frac{1}{\sqrt{3}} Mg \left(\tfrac{1}{2}\right) + \frac{1}{\sqrt{3}} Mg \left(\tfrac{1}{2}\right) = \frac{1}{\sqrt{3}} Mg.$$

The solution is complete. By symmetry, $F_{DE} = F_{AB}$, $F_{CE} = F_{AC}$ and $F_{CD} = F_{BC}$. As a check, calculate ΣF_x and ΣF_y for pin C and see if they equal zero. Figure 12–34e shows the free-body diagram.

[†] If we were to choose the direction of a force on a diagram opposite to what it really is, we would get a minus sign, which tells us so.

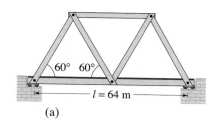

(a)

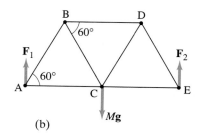

(b)

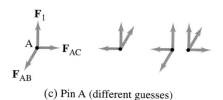

(c) Pin A (different guesses)

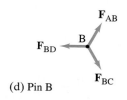

(d) Pin B

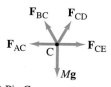

(e) Pin C

FIGURE 12–34
Example 12–15. (a) A truss bridge. Freebody diagrams:
(b) for the entire truss,
(c) for pin A (different guesses),
(d) for pin B and (e) for pin C.

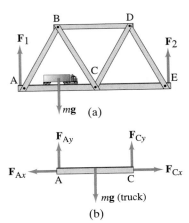

FIGURE 12–35 (a) Truss with truck of mass m at center of strut AC. (b) Forces on strut AC.

Example 12–15 put the roadway load at the center, C. Now consider a heavy load, such as a heavy truck, supported by strut AC at its middle, as shown in Fig. 12–35a. The strut AC sags under this load, telling us there is shear stress in strut AC. Figure 12–35b shows the forces exerted on strut AC: the weight of the truck $m\mathbf{g}$, and the forces $\mathbf{F}_A$ and $\mathbf{F}_C$ that pins A and C exert on the strut. [Note that $\mathbf{F}_1$ does not appear because it is a force (exerted by external supports) that acts on pin A, not on strut AC.] The forces that pins A and C exert on strut AC will be not only along the strut, but will have vertical components too, perpendicular to the strut, creating shear stress to balance the weight of the truck, $m\mathbf{g}$. The other struts, not bearing weight, remain under pure tension or compression. Problems 64 and 65 deal with this situation, and an early step in their solution is to calculate the forces $\mathbf{F}_A$ and $\mathbf{F}_C$ by using torque equations for the strut.

For very large bridges, truss structures are too heavy. One solution is to build suspension bridges, with the load being carried by relatively light suspension cables under tension, supporting the roadway by means of closely spaced vertical wires, as shown in Fig. 12–36, and in the photo on the first page of this chapter.

➡ **PHYSICS APPLIED**

Suspension bridge

FIGURE 12–36
A suspension bridge (Verrazano Narrows, NY).

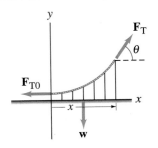

EXAMPLE 12–16 Suspension bridge. Determine the shape of the cable between the two towers in Fig. 12–36, assuming the weight of the roadway is supported uniformly along its length. Ignore the weight of the cable.

SOLUTION We take $x = 0$, $y = 0$ at the center of the span, as shown in Fig. 12–37. Let $\mathbf{F}_{T0}$ be the tension in the cable at $x = 0$; it acts horizontally as shown. Let F_T be the tension in the cable at some other place where the horizontal coordinate is x, as shown. This section of cable supports a portion of the roadway whose weight w is proportional to the distance x, since the roadway is assumed uniform; that is,

$$w = \lambda x$$

where λ is the weight per unit length. We now set $\Sigma\mathbf{F} = 0$:

$$\Sigma F_x = F_T \cos\theta - F_{T0} = 0$$
$$\Sigma F_y = F_T \sin\theta - w = 0.$$

We divide these two equations,

$$\tan\theta = \frac{w}{F_{T0}} = \frac{\lambda x}{F_{T0}}.$$

The slope of our curve (the cable) at any point is

$$\frac{dy}{dx} = \tan\theta$$

$$= \frac{\lambda}{F_{T0}} x.$$

We integrate this:

$$\int dy = \frac{\lambda}{F_{T0}} \int x\, dx$$

$$y = Ax^2 + B$$

where we set $A = \lambda/F_{T0}$ and B is a constant of integration. This is just the equation of a parabola.

Real bridges have cables that do have mass, so the cables hang only approximately as a parabola, although often it is quite close.

FIGURE 12–37 Example 12–16.

There are various methods that engineers and architects can use to span interior spaces as well as exterior spaces (bridges). We have already looked briefly at beams, trusses and suspension bridges. In this Section we discuss arches and domes, beginning with a brief history of how builders throughout the centuries have attacked the problem of spanning a space.

The first important architectural invention was the post-and-beam (or post-and-lintel) construction, in which two upright posts support a horizontal beam. Before steel was introduced in the nineteenth century, the length of a beam was quite limited because the strongest building materials were then stone and brick. Hence the width of a span was limited by the size of available stones. Equally important, stone and brick, though strong under compression—are very weak under tension and shear; all three types of stress occur in a beam, as we saw in Fig. 12–26. The minimal space that could be spanned using stone is shown by the closely spaced columns of the great Greek temples (Fig. 12–21).

The introduction of the semicircular **arch** by the Romans (Figs. 12–38 and 12–39), aside from its aesthetic appeal, was a tremendous technological innovation. The advantage of the "true" or semicircular arch is that, if well designed, its wedge-shaped stones experience stress which is mainly compressive even when supporting a large load such as the wall and roof of a cathedral. Because the stones are forced to squeeze against one another, they are mainly under compression (see Fig. 12–40). Note, however, that the arch transfers horizontal as well as vertical forces to the supports. A round arch consisting of many well-shaped stones could span a very wide space. However, considerable buttressing on the sides was needed to support the horizontal components of the forces.

The pointed arch came into use about A.D. 1100 and became the hallmark of the great Gothic cathedrals. It too was an important technical innovation, and was first used to support heavy loads such as the tower of a cathedral, and as the central arch. Apparently the builders realized that, because of the steepness of the pointed arch, the forces due to the weight above could be brought down more nearly vertically, so less horizontal buttressing would be needed. The pointed arch reduced the load on the walls, so there could be more openness and light. The smaller buttressing needed was provided on the outside by graceful flying buttresses (Fig. 12–41).

➥ PHYSICS APPLIED

Architecture: Beams, arches and domes

FIGURE 12–38 Round arches in the Roman Forum. The one in the background is the Arch of Titus.

FIGURE 12–39 An arch is used here to good effect in spanning a chasm on the California coast.

FIGURE 12–40 Stones in a round arch (see Fig. 12–38) are mainly under compression.

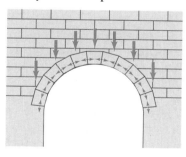

FIGURE 12–41 Flying buttresses (on the cathedral of Notre Dame, in Paris).

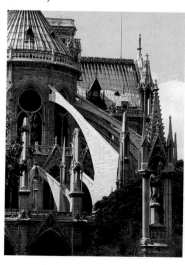

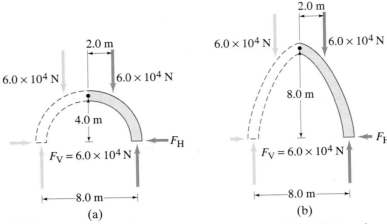

FIGURE 12–42 (a) Forces in a round arch, compared (b) with those in a pointed arch.

FIGURE 12–43 Interior of the Pantheon in Rome, built almost 2,000 years ago. This view, showing the great dome and its central opening for light, was painted about 1740 by Panini. Photographs do not capture its grandeur as well as this painting does.

FIGURE 12–44 The Louisiana Superdome.

The technical innovation of the pointed arch was achieved not through calculation but through experience and intuition, for it was not until much later that detailed calculations, such as those presented earlier in this chapter, came into use. To make an accurate analysis of a stone arch is quite difficult in practice. But if we make some simplifying assumptions, we can show why the horizontal component of the force at the base is less for a pointed arch than for a round one. Figure 12–42 shows a round arch and a pointed arch, each with an 8.0-m span. The height of the round arch is thus 4.0 m, whereas that of the pointed arch is larger and has been chosen to be 8.0 m. Each arch supports a weight of 12.0×10^4 N ($= 12{,}000$ kg $\times g$) which, for simplicity, we have divided into two parts (each 6.0×10^4 N) acting on the two halves of each arch as shown. To be in equilibrium, each of the supports must exert an upward force of 6.0×10^4 N. For rotational equilibrium, each support also exerts a horizontal force, F_H, at the base of the arch, and it is this we want to calculate. We focus only on the right half of each arch. We set equal to zero the total torque calculated about the apex of the arch due to the forces exerted on that half arch. For the round arch, the torque equation is (see Fig. 12–42a)

$$(4.0\,\text{m})(6.0 \times 10^4\,\text{N}) - (2.0\,\text{m})(6.0 \times 10^4\,\text{N}) - (4.0\,\text{m})(F_H) = 0.$$

Thus $F_H = 3.0 \times 10^4$ N. For the pointed arch, the torque equation is (see Fig. 12–42b)

$$(4.0\,\text{m})(6.0 \times 10^4\,\text{N}) - (2.0\,\text{m})(6.0 \times 10^4\,\text{N}) - (8.0\,\text{m})(F_H) = 0.$$

Solving, we find that $F_H = 1.5 \times 10^4$ N—only half as much! From this calculation we can see that the horizontal buttressing force required for a pointed arch is less because the arch is higher, and there is therefore a longer lever arm for this force. Indeed, the steeper the arch, the less the horizontal component of the force needs to be, and hence the more nearly vertical is the force exerted at the base of the arch.

Whereas an arch spans a two-dimensional space, a **dome**—which is basically an arch rotated about a vertical axis—spans a three-dimensional space. The Romans built the first large domes. Their shape was hemispherical and some still stand, such as that of the Pantheon in Rome (Fig. 12–43).

Only in the twentieth century were larger domes built, the largest being that of the Superdome in New Orleans, completed in 1975 (Fig. 12–44). The 5×10^6-kg dome, over 200 m in diameter, is made of steel trusses and concrete. Its outer thrust (see Example 12–17) is supported by a huge steel tension ring, 200 m in diameter, in the form of a truss, which is attached directly to the dome 50 m above the ground.

EXAMPLE 12–17 **A modern dome.** The 1.2×10^6 kg dome of the Small Sports Palace in Rome (Fig. 12–45a) is supported by 36 buttresses positioned at a 38° angle so they connect smoothly with the dome. Calculate the components of the force, F_H and F_V, that each buttress exerts on the dome so that the force acts purely in compression—that is, at a 38° angle (Fig. 12–45b).

SOLUTION The vertical load on *each* buttress is $\frac{1}{36}$ of the total weight. Thus

$$F_V = \frac{(1.2 \times 10^6 \text{ kg})(9.8 \text{ m/s}^2)}{36} = 3.4 \times 10^5 \text{ N}.$$

The force must act at a 38° angle at the base of the dome in order to be purely compressive. Thus

$$F_H = \frac{F_V}{\tan 38°} = \frac{340{,}000 \text{ N}}{\tan 38°} = 430{,}000 \text{ N}.$$

In order that each of the buttresses be able to exert this 430,000-N horizontal force, a prestressed-concrete tension ring surrounds the base of the buttresses beneath the ground (see Problem 68 and Fig. 12–83).

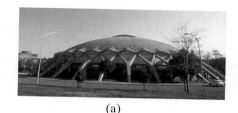

(a)

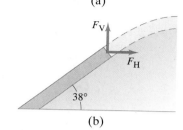

(b)

FIGURE 12–45 (a) The dome of the Small Sports Palace in Rome, built for the 1960 Olympics. (b) Example 12–17.

Summary

A body at rest [or at least not accelerating] is said to be in **equilibrium**. The subject concerned with the determination of the forces within a structure at rest is called **statics**.

The two necessary conditions for a body to be in equilibrium are (1) the vector sum of all the forces on it must be zero, and (2) the sum of all the torques (calculated about any arbitrary axis) must also be zero:

$$\Sigma F_x = 0 \quad \Sigma F_y = 0 \quad \Sigma \tau = 0,$$

when all the forces act in a plane. It is important when doing statics problems to apply the equilibrium conditions to only one body at a time.

A body in static equilibrium is said to be in (*a*) **stable**, (*b*) **unstable**, or (*c*) **neutral equilibrium**, depending on whether a slight displacement leads to (*a*) a return to the original position, (*b*) further movement away from the original position, or (*c*) remaining in the new position. An object in stable equilibrium is also said to be in **balance**.

Hooke's law applies to many elastic solids, and states that the change in length of an object is proportional to the applied force:

$$F = k \, \Delta L.$$

If the force is too great, the object will exceed its **elastic limit**, which means it will no longer return to its original shape when the distorting force is removed. If the force is even greater, the **ultimate strength** of the material can be exceeded and the object fractures.

The force per unit area acting on a body is called the **stress**, and the resulting fractional change in length is called the **strain**.

The stress on a body is present within the body and can be of three types: **compression**, **tension**, and **shear**.

The ratio of stress to strain is called the **elastic modulus** of the material. **Young's modulus** applies for compression and tension, and the **shear modulus** for shear; **bulk modulus** applies to an object whose volume changes as a result of pressure on all sides. All three moduli are constants for a given material when distorted within the elastic region.

Questions

1. Describe several situations where a body is not in equilibrium, even though the net force on it is zero.

2. A bungee jumper momentarily comes to rest at the bottom of the dive before he springs back upward. At that moment, is the bungee jumper in equilibrium? Explain.

3. You can find the center of gravity of a meter stick by resting it horizontally on both your index fingers, and then slowly drawing your fingers together. First the meter stick will slip on one finger, and then on the other, but eventually the fingers meet at the CG. Why does this work?

4. A ladder, leaning against a wall, makes a 60° angle with the ground. When is it more likely to slip: when a person stands near the top or near the bottom? Explain.

5. An earthen retaining wall is shown in Fig. 12–46a. The earth, particularly when wet, can exert a significant force F on the wall. (*a*) What force produces the torque to keep the wall upright? (*b*) Explain why the retaining wall in Fig. 12–46b would be much less likely to overturn.

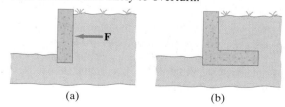

(a)   (b)

FIGURE 12–46 Question 5.

6. Your doctor's scale uses a slidable weight on an arm to balance your weight, Fig. 12–47. These weights are obviously much lighter than you are. How does this work?

FIGURE 12–47
Question 6.

7. Explain why touching the toes while seated on the floor with outstretched legs produces less stress on the lower spinal column than when touching the toes from a standing position. Use a diagram.

8. Explain under what circumstances a moving body can be in equilibrium.

9. If the net torque on a body is zero when calculated about a point O, will it *necessarily* be zero when calculated about a different point O'?

10. A uniform meter stick supported at the 25-cm mark is in equilibrium when a 1-kg rock is suspended at the 0-cm end, as shown in Fig. 12–48. Is the mass of the meter stick greater than, equal to, or less than the mass of the rock? Explain your reasoning.

FIGURE 12–48
Question 10.

11. Place yourself facing the edge of an open door. Position your feet astride the door with your nose and abdomen touching the door's edge. Try to rise on your tiptoes. Why can't this be done?

12. Which of the configurations of brick, (a) or (b) of Fig. 12–49, is the more likely to be stable? Why?

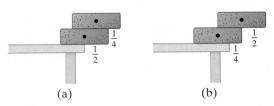

FIGURE 12–49 Question 12. The dots indicate the CG of each brick. The fractions $\frac{1}{4}$ and $\frac{1}{2}$ indicate what portion of each brick is hanging beyond its support.

13. Name the type of equilibrium for each position of the ball in Fig. 12–50.

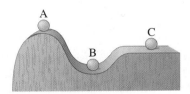

FIGURE 12–50 Question 13.

14. Why is it not possible to sit upright in a chair and rise to your feet without first leaning forward?

15. Why is it more difficult to do sit-ups when your knees are bent than when your legs are stretched out?

16. Examine how a pair of scissors or shears cuts through a piece of cardboard. Is the name "shears" justified?

Problems

Sections 12–1 to 12–3

1. (I) Three forces are applied to a tree sapling, as shown in Fig. 12–51, to stabilize it. If $\mathbf{F}_1 = 380$ N and $\mathbf{F}_2 = 255$ N, find $\mathbf{F}_3$ in magnitude and direction.

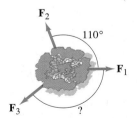

FIGURE 12–51
Problem 1.

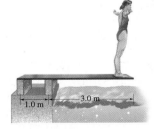

FIGURE 12–52
Problems 3, 4 and 8.

2. (I) Two cords support a chandelier in the manner shown in Fig. 12–4 except that the upper wire makes an angle of 45° with the ceiling. If the cords can sustain a force of 1150 N without breaking, what is the maximum chandelier weight that can be supported?

3. (I) Calculate the torque about the front support of a diving board, Fig. 12–52, exerted by a 56-kg person 3.0 m from that support.

4. (I) How far out on the diving board (Fig. 12–52) would a 56-kg diver have to be to exert a torque of 1000 m · N on the board, relative to the left support post?

5. (I) Calculate the mass m needed in order to suspend the leg shown in Fig. 12–53. Assume the leg (with cast) has a mass of 15.0 kg, and its CG is 35.0 cm from the hip joint; the sling is 80.5 cm from the hip joint.

FIGURE 12–53 Problem 5.

6. (I) A 67.0-kg adult sits at one end of a 10.0 m board, on the other end of which sits his 23.0-kg child. Where should the pivot be placed so the board (ignore its mass) is balanced?

7. (II) Repeat Problem 6 taking into account the board's 12.0-kg mass.

8. (II) Calculate the forces F_1 and F_2 that the supports exert on the diving board of Fig. 12–52 when a 56-kg person stands at its tip. (a) Ignore the weight of the board. (b) Take into account the board's mass of 35 kg. Assume the board's CG is at its center.

9. (II) How close to the edge of the 20.0-kg table shown in Fig. 12–54 can a 66.0-kg person sit without tipping it over?

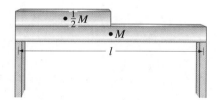

FIGURE 12–54   Problem 9.

10. (II) A uniform steel beam has a mass of 1100 kg. On it is resting half of an identical beam, as shown in Fig. 12–55. What is the vertical support force at each end?

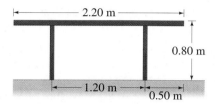

FIGURE 12–55   Problem 10.

11. (II) Find the tension in the two cords shown in Fig. 12–56. Neglect the mass of the cords, and assume that the angle θ is 30° and the mass m is 200 kg.

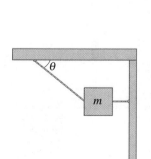

FIGURE 12–56
Problem 11.

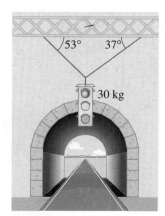

FIGURE 12–57
Problems 12 and 46.

12. (II) Find the tension in the two wires supporting the traffic light shown in Fig. 12–57.

13. (II) Calculate F_1 and F_2 for the uniform cantilever shown in Fig. 12–9 whose mass is 1200 kg.

14. (II) A 0.60-kg sheet hangs from a massless clothesline as shown in Fig. 12–58. The line on either side of the sheet makes an angle of 3.5° with the horizontal. Calculate the tension in the clothesline on either side of the sheet. Why is the tension so much greater than the weight of the sheet?

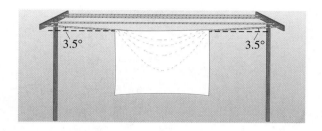

FIGURE 12–58   Problem 14.

15. (II) A door, 2.30 m high and 1.30 m wide, has a mass of 13.0 kg. A hinge 0.40 m from the top and another hinge 0.40 m from the bottom each support half the door's weight (Fig. 12–59). Assume that the center of gravity is at the geometrical center of the door, and determine the horizontal and vertical force components exerted by each hinge on the door.

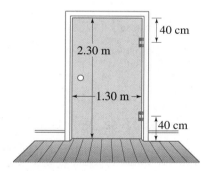

FIGURE 12–59   Problem 15.

16. (II) Figure 12–60 shows a pair of forceps; they are used to firmly hold a thin plastic rod. If each finger squeezes with a force $F_T = F_B = 11.0$ N, what force do the forceps jaws exert on the plastic rod?

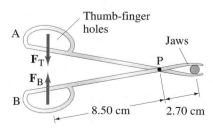

FIGURE 12–60   Problem 16.

17. (II) The mass of the trailer in Fig. 12–61 is 2200 kg. The distances are $a = 2.5$ m and $b = 5.5$ m. The truck is stationary, and the wheels of the trailer can turn freely, meaning the road exerts no horizontal force on them. The hitch at B can be modeled as a pin support. (a) Draw the free-body diagram of the trailer. (b) Determine the total normal force exerted by the road on the rear tires at A and (c) the force exerted on the trailer by the pin support B.

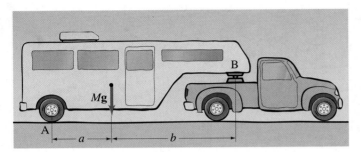

FIGURE 12–61   Problem 17.

18. (II) Approximately what force, F_M, must the extensor muscle in the upper arm exert on the lower arm to hold a 7.3-kg shot put (Fig. 12–62)? Assume the lower arm has a mass of 2.8 kg and its CG is 12 cm from the elbow-joint pivot.

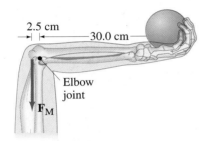

FIGURE 12–62   Problem 18.

19. (II) The Achilles tendon is attached to the rear of the foot as shown in Fig. 12–63. When a person elevates himself just barely off the floor on the "ball of one foot," estimate the tension in the Achilles tendon (pulling upward), and the (downward) force exerted by the lower leg bone on the foot. Assume the person has a mass of 70 kg, and that D is twice as long as d.

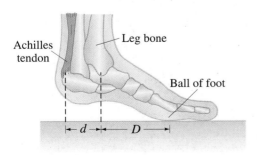

FIGURE 12–63   Problem 19.

20. (II) (a) Calculate the force required of the "deltoid" muscle, F_M, to hold up the outstretched arm shown in Fig. 12–64. The total mass of the arm is 3.3 kg. (b) Calculate the magnitude of the force $\mathbf{F}_J$ exerted by the shoulder joint on the upper arm.

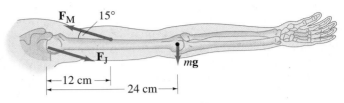

FIGURE 12–64   Problems 20 and 21.

21. (II) Suppose the hand in Problem 20 holds a 15-kg mass. What force, F_M, is required of the deltoid muscle, assuming the mass is 52 cm from the shoulder joint?

22. (II) Three children are trying to balance on a seesaw, which consists of a fulcrum rock, acting as a pivot at the center, and a very light board 3.6 m long (Fig. 12–65). Two boys are already on either end. One has a mass of 50 kg, and the other a mass of 35 kg. Where should the third child, whose mass is 25 kg, place herself so as to balance the seesaw?

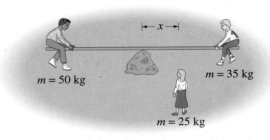

FIGURE 12–65   Problem 22.

23. (II) A 170-cm-tall person lies on a light (massless) board which is supported by two scales, one under the feet and one beneath the top of the head (Fig. 12–66). The two scales read, respectively, 31.6 and 35.1 kg. Where is the center of gravity of this person?

FIGURE 12–66   Problem 23.

24. (II) A 76.0-kg horizontal beam is supported at each end. A 285-kg piano rests a quarter of the way from one end. What is the vertical force on each of the supports?

25. (II) Calculate F_1 and F_2 for the beam shown in Fig. 12–67. Assume it is uniform and has a mass of 250 kg.

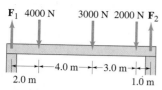

F_1 4000 N   3000 N 2000 N F_2

2.0 m ⟵4.0 m⟶⟵3.0 m⟶ 1.0 m

FIGURE 12–67 Problem 25.

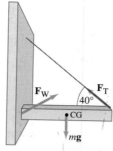

FIGURE 12–68
Problem 26.

26. (II) Calculate the tension F_T in the wire that supports the 30.0-kg beam shown in Fig. 12–68, and the force F_W exerted by the wall on the beam (give magnitude and direction).

27. (II) The two trees in Fig. 12–69 are 7.6 m apart. Calculate the magnitude of the force **F** a backpacker must exert to hold a 16-kg backpack so that the rope sags at its midpoint by (a) 1.5 m, (b) 0.15 m.

FIGURE 12–69
Problem 27 and 83.

28. (II) A large 80.0-kg board is propped at a 45° angle against the edge of a barn door that is 2.6 m wide. How great a horizontal force must a person behind the door exert (at the edge) in order to open it? Assume that there is negligible friction between the door and the board but that the board is firmly set against the ground.

29. (II) Repeat Problem 28 assuming the coefficient of friction between the board and the door is 0.45.

30. (II) A shop sign weighing 215 N is supported by a uniform 135-N beam as shown in Fig. 12–70. Find the tension in the guy wire and the horizontal and vertical forces exerted by the pin on the beam.

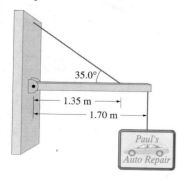

35.0°

1.35 m

1.70 m

Paul's
Auto Repair

FIGURE 12–70
Problem 30.

31. (II) A traffic light hangs from a structure as shown in Fig. 12–71. The uniform aluminum pole AB is 7.5 m long and has a mass of 8.0 kg. The mass of the traffic light is 12.0 kg. Determine the tension in the horizontal massless cable CD, and the vertical and horizontal components of the force exerted by the pivot A on the aluminum pole.

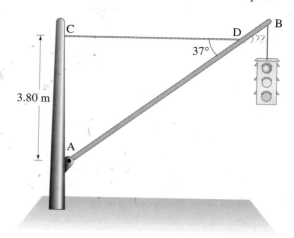

C   D   B

37°

3.80 m

A

FIGURE 12–71 Problem 31.

32. (II) A uniform ladder of mass m and length l leans at an angle θ against a frictionless wall, Fig. 12–72. If the coefficient of static friction between the ladder and the ground is μ_s, what is the minimum angle at which the ladder will not slip?

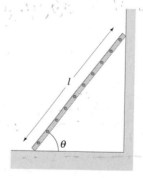

l

θ

FIGURE 12–72
Problem 32 and 82.

33. (II) A meter stick with a mass of 230 g is supported horizontally by two vertical strings, one at the 0-cm mark and the other at the 90-cm mark (Fig. 12–73). (a) What is the tension in the string at 0 cm? (b) What is the tension in the string at 90 cm?

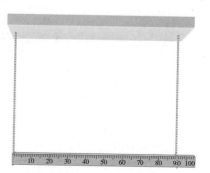

10 20 30 40 50 60 70 80 90 100

FIGURE 12–73 Problem 33.

34. (II) A uniform rod AB of length 5.0 m and mass $M = 3.0$ kg is hinged at A and held in equilibrium by a light rope, as shown in Fig. 12–74. A load $W = 20$ N hangs from the rod at a distance x so that the tension in the rope is 70 N. (a) Draw a free-body diagram for the rod. (b) Determine the vertical and horizontal forces on the rod exerted by the hinge. (c) Determine x from the appropriate torque equation.

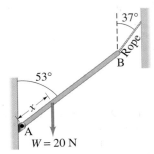

FIGURE 12–74 Problem 34.

35. (III) A cubic crate of side $s = 2.0$ m is top-heavy: its CG is 20 cm above its true center. How steep an incline can the crate rest on without tipping over? What would your answer be if the crate were to slide at constant speed down the plane without tipping over?

36. (III) A refrigerator is approximately a uniform rectangular solid 2.5 m tall, 1.0 m wide, and 1.5 m deep. If it sits upright on a truck with its 1.0-m dimension in the direction of travel, and if the refrigerator cannot slide on the truck, how rapidly can the truck accelerate without tipping the refrigerator over?

37. (III) A 60-kg person stands 2.0 m from the bottom of the stepladder shown in Fig. 12–75. Determine (a) the tension in the horizontal tie rod, which is halfway up the ladder, (b) the normal force the ground exerts on each side of the ladder, and (c) the force (magnitude and direction) that the left side of the ladder exerts on the right side at the hinge on the top. Ignore the mass of the ladder and assume the ground is frictionless. [Hint: Consider free-body diagrams for each section of the ladder.]

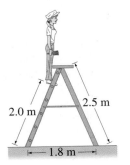

FIGURE 12–75
Problem 37.

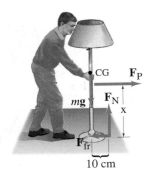

FIGURE 12–76 Problem 38.

38. (III) A person wants to push a lamp (mass 7.2 kg) across the floor, where the coefficient of friction is 0.20 (Fig. 12–76). Calculate the maximum height above the floor at which the person can push the lamp so that it slides rather than tips.

39. (III) Two guy wires run from the top of a pole 2.6 m tall that supports a volleyball net. The two wires are anchored to the ground 2.0 m apart and each is 2.0 m from the pole (Fig. 12–77). The tension in each wire is 95 N. What is the tension in the net, assumed horizontal and attached at the top of the pole?

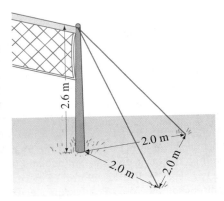

FIGURE 12–77 Problem 39.

Section 12–5

40. (I) A nylon tennis string on a racket is under a tension of 250 N. If its diameter is 1.00 mm, by how much is it lengthened from its untensioned length of 30.0 cm?

41. (I) A marble column of cross-sectional area 2.0 m² supports a mass of 25,000 kg. (a) What is the stress within the column? (b) What is the strain?

42. (I) By how much is the column in the previous problem shortened if it is 12 m high?

43. (I) A vertical steel girder with a cross-sectional area of 0.15 m² has a sign (mass 2000 kg) hanging from its bottom end. (a) What is the stress within the girder? (b) What is the strain on the girder? (c) If the girder is 9.50 m long, how much is it lengthened? (Ignore the mass of the girder itself.)

44. (I) One liter of alcohol $(1000\ \text{cm}^3)$ in a flexible container is carried to the bottom of the sea, where the pressure is $2.6 \times 10^6\ \text{N/m}^2$. What will be its volume there?

45. (I) A 15-cm-long animal tendon was found to stretch 3.7 mm by a force of 13.4 N. The tendon was approximately round with an average diameter of 8.5 mm. Calculate the elastic modulus of this tendon.

46. (II) If the two wires in Fig. 12–57 (Problem 12) are made of steel wire 1.0 mm in diameter, what is the percentage stretch of each because of the load?

47. (II) How much pressure is needed to compress the volume of an iron block by 0.10 percent? Express answer in N/m², and compare it to atmospheric pressure $(1.0 \times 10^5\ \text{N/m}^2)$.

48. (II) At depths of 2000 m in the sea, the pressure is about 200 times atmospheric pressure $(1.0 \times 10^5\ \text{N/m}^2)$. By what percentage does an iron bathysphere's volume change at this depth?

49. (II) Shear forces as shown in Fig. 12–24 are applied to a steel plate that is 0.50 m on a side and 1.1 cm thick. The shear strain produced by these forces is 0.050. Calculate the magnitude of the force.

50. (II) A scallop forces open its shell with an elastic material called abductin, whose elastic modulus is about $2.0 \times 10^6 \, \text{N/m}^2$. If this piece of abductin is 3.0 mm thick and has a cross-sectional area of $0.50 \, \text{cm}^2$, how much potential energy does it store when compressed 1.0 mm?

51. (III) A pole projects horizontally from the front wall of a shop. A 5.1-kg sign hangs from the pole at a point 2.2 m from the wall (Fig. 12–78). (a) What is the torque due to this sign calculated about the point where the pole meets the wall? (b) If the pole is not to fall off, there must be another torque exerted to balance it. What exerts this torque? Use a diagram to show how this torque must act. (c) Discuss whether compression, tension, and/or shear play a role in part (b).

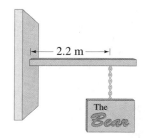

FIGURE 12–78 Problem 51.

Section 12–6

52. (I) The femur bone in the leg has a minimum effective cross section of about $3.0 \, \text{cm}^2 \, (= 3.0 \times 10^{-4} \, \text{m}^2)$. How much compressive force can it withstand before breaking?

53. (II) What is the maximum tension possible in a 1.00-mm-diameter nylon tennis racket string? If you want tighter strings, what do you do to prevent breakage: go to thinner or thicker strings? Why? What causes strings to break when they are hit by the ball?

54. (II) If a compressive force of $3.6 \times 10^4 \, \text{N}$ is exerted on the end of a 20.0-cm-long bone of cross-sectional area $3.6 \, \text{cm}^2$, (a) will the bone break, and (b) if not, by how much does it shorten?

55. (II) (a) What is the minimum cross-sectional area required of a vertical steel cable from which is suspended a 320-kg chandelier? Assume a safety factor of 7.0. (b) If the cable is 7.5 m long, how much does it elongate?

56. (II) Assume the supports of the cantilever shown in Fig. 12–9 (mass = 2600 kg) are made of wood. Calculate the minimum cross-sectional area required of each, assuming a safety factor of 8.5.

57. (II) An iron bolt is used to connect two iron plates together. The bolt must withstand shear forces up to about 3200 N. Calculate the minimum diameter for the bolt, based on a safety factor of 6.0.

58. (III) A steel cable is to support an elevator whose total (loaded) mass is not to exceed 3100 kg. If the maximum acceleration of the elevator is $1.2 \, \text{m/s}^2$, calculate the diameter of cable required. Assume a safety factor of 7.0.

59. (III) Parachutists whose chutes have failed to open have been known to survive if they landed in deep snow. Assume that a 75-kg parachutist hits the ground with an area of impact of $0.30 \, \text{m}^2$ at a velocity of 60 m/s, and that the ultimate strength of body tissue is $5 \times 10^5 \, \text{N/m}^2$. Assume that the person is brought to rest in 1.0 m of snow. Show that the person can escape injury.

* Section 12–7

* 60. (II) Figure 12–79 shows a simple truss that carries a load at the center (C) of $1.25 \times 10^4 \, \text{N}$. (a) Calculate the force on each strut at the pins, A, B, C, D, and (b) determine which struts (ignore their masses) are under tension and which under compression.

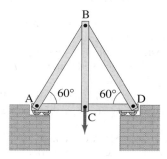

FIGURE 12–79 Problem 60.

* 61. (II) A heavy load Mg hangs at point E of the single cantilever truss shown in Fig. 12–80. (a) Use a torque equation for the truss as a whole to determine the tension F_T in the support cable, and then determine the force F_A at pin A. (b) Determine the force in each member of the truss. Neglect the weight of the trusses, which is small compared to the load.

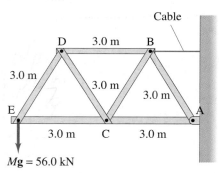

FIGURE 12–80 Problem 61.

* 62. (II) Consider again Example 12–15 but this time assume the roadway is supported uniformly so that $\frac{1}{2}$ its mass M acts at the center and $\frac{1}{4} M$ at each end support (think of the bridge as two spans, AC and CE, so the center pin supports two spans ends). Calculate the magnitude of the force in each truss member and compare to Example 12–15.

* 63. (II) (a) What minimum cross-sectional area must the trusses have in Example 12–15 if they are of steel (and all the same size for looks), using a safety factor of 6.0? (b) If at any time the bridge may carry as many as 50 trucks with an average mass of 1.2×10^4 kg, estimate again the area needed for the truss members.

* 64. (III) The truss shown in Fig. 12–81 supports a railway bridge. Determine the stress in each strut if a 44 ton $(1 \text{ ton} = 10^3 \text{ kg})$ train locomotive is stopped at the midpoint between the center and one end. Ignore the masses of the rails and truss, and use only $\frac{1}{2}$ the mass of train because there are two trusses. Assume all triangles are equilateral (as in Example 12–15). [*Hint*: See Fig. 12–35.]

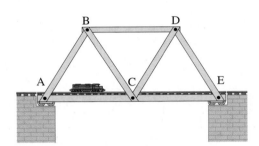

FIGURE 12–81 Problem 64.

* 65. (III) Suppose in Example 12–15, a 28 ton truck $(m = 28 \times 10^3 \text{ kg})$ has its CM located 22 m from the left end of the bridge (point A). Determine the magnitude of the force and type of stress in each strut. [*Hint*: See Fig. 12–35.]

* 66. (III) For the "Pratt truss" shown in Fig. 12–82, determine the force on each member and whether it is tensile or compressive. Assume the truss is loaded as shown, and give results in terms of F. The height is a and each of the four lower horizontal spans is a.

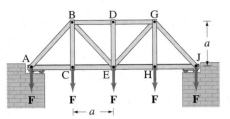

FIGURE 12–82
Problem 66.

* 67. (II) How high must a pointed arch be if it is to span a space 8.0 m wide and exert one third the horizontal force at its base that a round arch would?

* 68. (II) The subterranean tension ring that exerts the balancing horizontal force on the abutments for the dome in Fig. 12–45 is 36-sided, so each segment makes a 10° angle with the adjacent one (Fig. 12–83). Calculate the tension F that must exist in each segment so that the required force of 4.3×10^5 N can be exerted at each corner (Example 12–17).

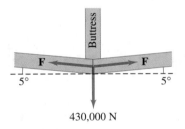

FIGURE 12–83
Problem 68.

General Problems

69. The mobile in Fig. 12–84 is in equilibrium. The object B has mass of 0.735 kg. Determine the masses of the objects A, C, and D. (Neglect the weights of the crossbars.)

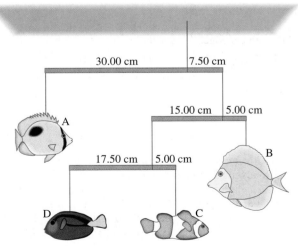

FIGURE 12–84   Problem 69.

70. A tightly stretched "high wire" is 46 m long. It sags 3.4 m when a 60.0-kg tightrope walker stands at its center. What is the tension in the wire? Is it possible to increase the tension in the wire so that there is no sag?

71. What minimum horizontal force F is needed to pull a wheel of radius R and mass M over a step of height h as shown in Fig. 12–85 $(R > h)$? (a) Assume the force is applied at the top edge as shown. (b) Assume the force is applied instead at the wheel's center.

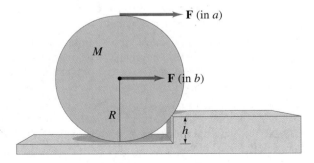

FIGURE 12–85   Problem 71.

72. A 50-story building is being planned. It is to be 200.0 m high with a base 40.0 m by 70.0 m. Its total mass will be about 1.8×10^7 kg and its weight therefore about 1.8×10^8 N. Suppose a 200-km/h wind exerts a force of $950 \, \text{N/m}^2$ over the 70.0-m-wide face (Fig. 12–86). Calculate the torque about the potential pivot point, the rear edge of the building (where $\mathbf{F}_E$ acts in Fig. 12–86), and determine whether the building will topple. Assume the total force of the wind acts at the midpoint of the building's face, and that the building is not anchored in bedrock. [*Hint:* $\mathbf{F}_E$ in Fig. 12–86 represents the force that the Earth exerts on the building in the case where the building is just beginning to tip.]

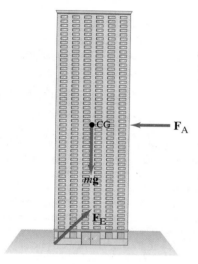

FIGURE 12–86
Force on a building subjected to wind ($\mathbf{F}_A$) and gravity ($m\mathbf{g}$); $\mathbf{F}_E$ is the force on the building due to the Earth in the situation when the building is just about to tip. Problem 72.

73. The center of gravity of a loaded truck depends on how it is packed. If a truck is 4.0 m high and 2.4 m wide, and its CG is 2.2 m above the ground, how steep a slope can the truck be parked on without tipping over sideways (Fig. 12–87)?

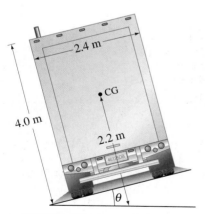

FIGURE 12–87   Problem 73.

74. In Example 9–6 in Chapter 9, we calculated the impulse and average force on the leg of a person who jumps 3.0 m down to the ground. If the legs are not bent upon landing, so that the body moves a distance d of only 1.0 cm during collision, determine (*a*) the stress in the tibia bone (area $= 3.0 \times 10^{-4} \, \text{m}^2$), and (*b*) whether or not the bone will break. (*c*) Repeat for a bent-knees landing ($d = 50.0$ cm).

75. The roof of a 9.0 m $\times$ 10.0 m room in a school has a total mass of 12,600 kg. The roof is to be supported by vertical "2 $\times$ 4s" (actually about 4.0 cm $\times$ 9.0 cm) along the 10.0-m sides. How many supports are required on each side and how far apart must they be? Consider only compression and assume a safety factor of 12.

76. In Fig. 12–88, consider the right-hand (northernmost) section of the Golden Gate Bridge, which has a length $d_1 = 343$ m. Assume the CG of this span is halfway between the tower and anchor. Determine F_{T1} and F_{T2} (which act on the northernmost cable) in terms of mg, the weight of the northernmost span, and calculate the tower height h needed for equilibrium. Assume the roadway is supported only by the suspension cables and neglect the mass of the cables. [*Hint:* F_{T3} does not act on this section.]

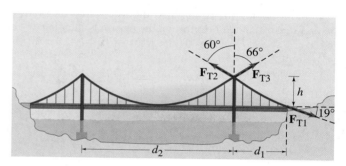

FIGURE 12–88   Problems 76 and 77.

77. Assume that a single-span suspension bridge such as the Golden Gate Bridge has the configuration indicated in Fig. 12–88. Assume that the roadway is uniform over the length of the bridge and that each segment of the suspension cable provides the sole support for the roadway directly below it. The ends of the cable are anchored to the ground only, not to the roadway. What must the ratio of d_2 to d_1 be so that the suspension cable exerts no net horizontal force on the towers? Neglect the mass of the cables and the fact that the roadway isn't precisely horizontal.

78. A 36-kg round table is supported by three legs placed equal distances apart on the edge. What minimum mass, placed on the table's edge, will cause the table to overturn?

79. The forces acting on a 77,000-kg aircraft flying at constant speed are shown in Fig. 12–89. The engine thrust, $F_T = 5.0 \times 10^5$ N, acts on a line 1.6 m below the CM. Determine the drag force F_D and the distance above the CM that it acts.

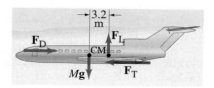

FIGURE 12–89   Problem 79.

80. A uniform flexible steel cable of weight mg is suspended between two equal elevation points as shown in Fig. 12–90, where $\theta = 60°$. Determine the tension in the cable (a) at its lowest point, and (b) at the points of attachment. (c) What is the direction of the tension force in each case?

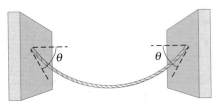

FIGURE 12–90 Problem 80.

81. A 20.0-m-long uniform beam weighing 600 N is supported on walls A and B, as shown in Fig. 12–91. (a) Find the maximum weight a person can be to walk to the extreme end D without tipping the beam. Find the forces that the walls A and B exert on the beam when the person is standing: (b) at D; (c) at a point 2.0 m to the right of B; (d) 2.0 m to the right of A.

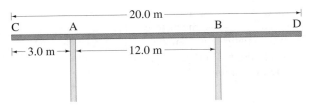

FIGURE 12–91 Problem 81.

82. A uniform 7.0-m-long ladder of mass 15.0 kg leans against a smooth wall (so the force exerted by the wall, $\mathbf{F}_W$, is perpendicular to the wall). The ladder makes an angle $\theta = 20.0°$ with the vertical wall (see Fig. 12–72); and the ground is rough. (a) Calculate the components of the force exerted by the ground on the ladder at its base, and (b) determine what the coefficient of static friction at the base of the ladder must be if the ladder is not to slip when a 70.0-kg person stands three-fourths of the way up the ladder.

83. A 23.0 kg backpack is suspended midway between two trees by a light cord as in Fig. 12–69. A bear grabs the backpack and pulls vertically downward with a constant force, so that each section of cord makes an angle of 30° below the horizontal. Initially, without the bear pulling, the angle was 15°; the tension in the cord with the bear pulling is double what it was when he was not. Calculate the force the bear is exerting on the backpack.

84. There is a maximum height of a uniform vertical column made of any material that can support itself without buckling, and it is independent of the cross-sectional area (why?). Calculate this height for (a) steel (density $7.8 \times 10^3 \, \text{kg/m}^3$), and (b) granite (density $2.7 \times 10^3 \, \text{kg/m}^3$).

85. From what minimum height must a 1.2-kg rectangular brick 15.0 cm × 6.0 cm × 4.0 cm be dropped above a rigid steel floor in order to break the brick? Assume the brick strikes the floor directly on its largest face, and that the compression of the brick is much greater than that of the steel (that is, ignore compression of the steel). State other simplifying assumptions that may be necessary.

86. A cube of side l rests on a rough floor. It is subjected to a steady horizontal pull, F, exerted a distance h above the floor as shown in Fig. 12–92. As F is increased, the block will either begin to slide, or begin to tip over. (a) What must be the coefficient of static faction μ_s so that the block begins to slide rather than tip? (b) What must be the coefficient of static friction so that the block begins to tip? [*Hint*: Where will the normal force on the block act if it tips?]

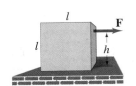

FIGURE 12–92 Problem 86.

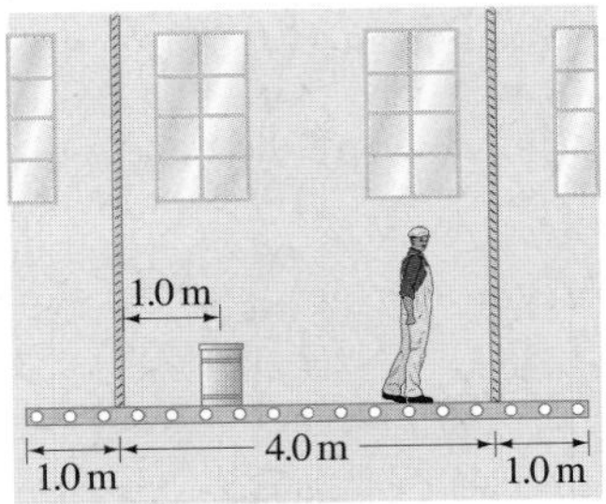

FIGURE 12–93 Problem 87.

87. A 60.0-kg painter is on a scaffold supported from above by ropes (Fig. 12–93). The scaffold has a mass of 25 kg and is uniformly constructed. There is a 4.0-kg pail of paint off to one side, as shown in the figure. Can the painter walk safely to both ends of the scaffold? If not, which end(s) is dangerous and how close to the end can he approach safely?

88. A 95,000-kg train locomotive starts across a 220-m-long bridge at time $t = 0$. The bridge is a uniform beam of mass 23,000 kg and the train travels at 80.0 km/h. What are the magnitudes of the vertical forces, $F_1(t)$ and $F_2(t)$, on the two end supports, written as a function of time during the train's passage?

89. A woman holds a 2.0-m-long uniform 10.0-kg pole as shown in Fig. 12–94. (a) Determine the forces she must exert with each hand (magnitude and direction). To what position should she move her left hand so that neither hand has to exert a force greater than (b) 150 N? (c) 80 N?

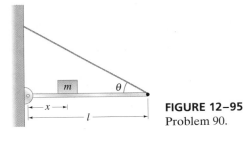

FIGURE 12–94
Problem 89.

90. A uniform beam of mass M and length l is mounted on a hinge at a wall as shown in Fig. 12–95. It is held in a horizontal position by a wire making an angle θ as shown. A mass m is placed on the beam a distance x from the wall, and this distance can be varied. Determine, as a function of x, (a) the tension in the wire and (b) the components of the force exerted by the beam on the hinge.

FIGURE 12–95
Problem 90.

91. Two identical, uniform beams are symmetrically set up against each other (Fig. 12–96) on a floor with which they have a coefficient of friction $\mu_s = 0.60$. What is the minimum angle the beams can make with the floor and still not fall?

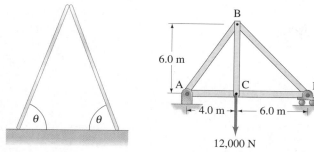

FIGURE 12–96 Problem 91.   **FIGURE 12–97** Problem 92.

*92. Use the method of joints to determine the force in each member of the truss shown in Fig. 12–97. State whether each member is in tension or compression.

93. Four bricks are to be stacked at the edge of a table, each brick overhanging the one below it, so that the top brick extends as far as possible beyond the edge of the table. (*a*) To achieve this, show that successive bricks must extend no more than (starting at the top) $1/2, 1/4, 1/6$, and $1/8$ of their length beyond the one below (Fig. 12–98a). (*b*) Is the top brick completely beyond the base? (*c*) Determine a general formula for the maximum total distance spanned by n bricks if they are to remain stable. (*d*) A builder wants to construct a corbeled arch (Fig. 12–98b) based on the principle of stability discussed in (*a*) and (*c*) above. What minimum number of bricks, each 0.30 m long, is needed if the arch is to span 1.0 m?

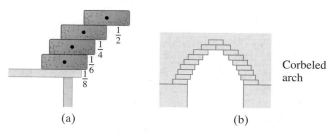

(a)   (b)

FIGURE 12–98   Problem 93.

94. One rod of the square frame shown in Fig. 12–99 contains a turnbuckle which, when turned, can put the rod under tension or compression. If the turnbuckle puts rod AB under a compressive force F, determine the forces produced in the other rods. Ignore the mass of the rods and assume the diagonal rods cross each other freely at the center without friction. [*Hint*: Use the symmetry of the situation.]

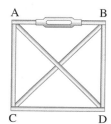

FIGURE 12–99
Problem 94.

95. A 20.0-kg ball is supported from the ceiling by rope A. Rope B pulls downward and to the side on the ball. If the angle of A to the vertical is $20°$ and if B makes an angle of $50°$ to the vertical (Fig. 12–100), find the tensions in ropes A and B.

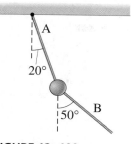

FIGURE 12–100
Problem 95.

FIGURE 12–101   Problem 96.

96. A home mechanic wants to raise the 250-kg engine out of a car. The plan is to stretch a rope vertically from the engine to a branch of a tree 20.0 m above. When the mechanic climbs halfway up the tree and pulls horizontally on the rope at its midpoint, the engine rises out of the car (Fig. 12–101). How much force must the mechanic exert to hold the engine 0.50 m above its normal position?

97. Estimate the magnitude and direction of the force $\mathbf{F_V}$ acting on the fifth lumbar vertebra (exerted by the spine below) in the lower human back. Use the model shown in Fig. 12–102b.

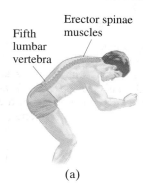

(a)

$w_1 = 0.07w$
(head)

$w_2 = 0.12w$
(arms) $w =$ Total weight

$w_3 = 0.46w$ of person
(trunk)

(b)

FIGURE 12–102   Problem 97.

Under water, sea creatures and scuba divers experience a buoyant force ($\mathbf{F_B}$) that almost exactly balances their weight $m\mathbf{g}$. The buoyant force is equal to the weight of the volume of fluid displaced (Archimedes' principle) and arises because the pressure increases with depth in the fluid. Sea creatures, and even humans, have a density very close to that of water, so their weight very nearly equals the buoyant force. Actually, humans have a density slightly less than water, so they can float. When fluids flow, interesting effects occur because the pressure in the fluid is lower where the fluid velocity is higher (Bernoulli's principle): airplanes can fly, sailboats can sail against the wind, smoke goes up a chimney, and many others.

Fluids

Phases of matter

The three common states, or **phases**, of matter are solid, liquid, and gas. We can distinguish these three phases as follows. A **solid** maintains a fixed shape and a fixed size; even if a large force is applied to a solid, it does not readily change its shape or volume. A **liquid** does not maintain a fixed shape—it takes on the shape of its container—but like a solid it is not readily compressible, and its volume can be changed significantly only by a very large force. A **gas** has neither a fixed shape nor a fixed volume—it will expand to fill its container. For example, when air is pumped into an automobile tire, the air does not all run to the bottom of the tire as a liquid would; it spreads out to fill the whole volume of the tire. Since liquids and gases do not maintain a fixed shape, they both have the ability to flow; they are thus often referred to collectively as **fluids**.

13–1 | Density and Specific Gravity

It is sometimes said that iron is "heavier" than wood. This cannot really be true since a large log clearly weighs more than an iron nail. What we should say is that iron is more *dense* than wood.

The **density**, ρ, of an object (ρ is the lowercase Greek letter "rho") is defined as its mass per unit volume:

Density

$$\rho = \frac{m}{V},$$ (13–1)

where m is the mass of the object and V its volume. Density is a characteristic property of any pure substance. Objects made of a given pure substance, such as pure gold, can have any size or mass, but the density will be the same for each.

(Sometimes we will find Eq. 13–1 useful for writing the mass of an object as $m = \rho V$, and the weight of an object, mg, as $\rho V g$.)

The SI unit for density is kg/m^3. Since $1\, kg/m^3 = 1000\, g/(100\, cm)^3 = 10^{-3}\, g/cm^3$, then a density given in g/cm^3 must be multiplied by 1000 to give the result in kg/m^3. The densities of a variety of substances are given in Table 13–1. The Table specifies temperature and pressure because they affect the density of substances, although the effect is slight for liquids and solids.

EXAMPLE 13–1 **Mass, given volume and density.** What is the mass of a solid iron wrecking ball of radius 18 cm?

SOLUTION The volume of any sphere is $V = \frac{4}{3}\pi r^3$ so we have

$$V = \frac{4}{3}\pi r^3 = \frac{4}{3}(3.14)(0.18\, m)^3 = 0.024\, m^3.$$

From Table 13–1, the density of iron is $\rho = 7800\, kg/m^3$, so we have from Eq. 13–1,

$$m = \rho V = (7800\, kg/m^3)(0.024\, m^3) = 190\, kg.$$

The **specific gravity** of a substance is defined as the ratio of the density of that substance to the density of water at 4.0°C. Specific gravity (abbreviated SG) is a number, without dimensions or units. Since the density of water is $1.00\, g/cm^3 = 1.00 \times 10^3\, kg/m^3$, the specific gravity of any substance will be equal numerically to its density specified in g/cm^3, or 10^{-3} times its density specified in kg/m^3. For example (see Table 13–1), the specific gravity of lead is 11.3, and that of alcohol is 0.79.

13–2 Pressure in Fluids

Pressure is defined as force per unit area, where the force F is understood to be acting perpendicular to the surface area A:

$$\text{pressure} = P = \frac{F}{A}. \qquad \textbf{(13–2)}$$

The SI unit of pressure is N/m^2. This unit has the official name **pascal** (Pa), in honor of Blaise Pascal (see Section 13–4); that is, $1\, Pa = 1\, N/m^2$. However, for simplicity, we will often use N/m^2. Other units sometimes used are $dynes/cm^2$, $lb/in.^2$ (sometimes abbreviated "psi"). We will meet several other units shortly, and will discuss conversions between them in Section 13–5.

As an example of calculating pressure, a 60-kg person whose two feet cover an area of $500\, cm^2$ will exert a pressure of

$$F/A = mg/A = (60\, kg)(9.8\, m/s^2)/(0.050\, m^2) = 12 \times 10^3\, N/m^2$$

on the ground. If the person stands on one foot, the force is the same but the area will be half, so the pressure will be twice as much: $24 \times 10^3\, N/m^2$.

The concept of pressure is particularly useful in dealing with fluids. It is an experimental fact that *a fluid exerts a pressure in all directions*. This is well known to swimmers and divers who feel the water pressure on all parts of their bodies. At any point in a fluid at rest, the pressure is the same in all directions. This is illustrated in Fig. 13–1. Consider a tiny cube of the fluid which is so small that we can ignore the force of gravity on it. The pressure on one side of it must equal the pressure on the opposite side. If this weren't true, there would be a net force on the cube and it would start moving. If the fluid is not flowing, then the pressures must be equal.

TABLE 13–1
Densities of Substances†

Substance	Density, $\rho\,(kg/m^3)$
Solids	
Aluminum	2.70×10^3
Iron and steel	$7.8 \ \times 10^3$
Copper	$8.9 \ \times 10^3$
Lead	$11.3 \ \times 10^3$
Gold	$19.3 \ \times 10^3$
Concrete	$2.3 \ \times 10^3$
Granite	$2.7 \ \times 10^3$
Wood (typical)	$0.3–0.9 \times 10^3$
Glass, common	$2.4–2.8 \times 10^3$
Ice	0.917×10^3
Bone	$1.7–2.0 \times 10^3$
Liquids	
Water (4°C)	$1.00 \ \times 10^3$
Sea water	1.025×10^3
Blood, plasma	$1.03 \ \times 10^3$
Blood, whole	$1.05 \ \times 10^3$
Mercury	$13.6 \ \times 10^3$
Alcohol, ethyl	$0.79 \ \times 10^3$
Gasoline	$0.68 \ \times 10^3$
Gases	
Air	1.29
Helium	0.179
Carbon dioxide	1.98
Water (steam) (100°C)	0.598

† Densities are given at 0°C and 1 atm pressure unless otherwise specified.

FIGURE 13–1 Pressure is the same in every direction in a fluid at a given depth; if it weren't the fluid would start to move.

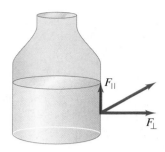

FIGURE 13–2 If there were a component of force parallel to the solid surface, the liquid would move in response to it; for a liquid at rest, $F_\parallel = 0$.

FIGURE 13–3 Calculating the pressure at a depth h in a liquid.

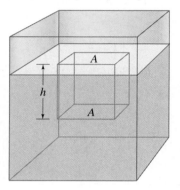

FIGURE 13–4 Forces on a flat, slablike volume of fluid for determining the pressure P at a height y in the fluid.

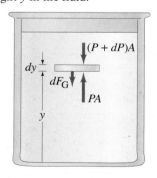

Another important property of a fluid at rest is that the force due to fluid pressure always acts *perpendicular* to any surface it is in contact with. If there were a component of the force parallel to the surface as shown in Fig. 13–2, then according to Newton's third law, the surface would exert a force back on the fluid that also would have a component parallel to the surface. Such a component would cause the fluid to flow, in contradiction to our assumption that the fluid is at rest. Thus the force due to the pressure is perpendicular to the surface.

Let us now calculate quantitatively how the pressure in a liquid of uniform density varies with depth. Consider a point which is at a depth h below the surface of the liquid (that is, the surface is a height h above this point), as shown in Fig. 13–3. The pressure due to the liquid at this depth h is due to the weight of the column of liquid above it. Thus the force due to the weight of liquid acting on the area A is $F = mg = \rho A h g$, where Ah is the volume of the column, ρ is the density of the liquid (assumed to be constant), and g is the acceleration of gravity. The pressure P due to the weight of liquid is then

$$P = \frac{F}{A} = \frac{\rho A h g}{A}$$

$$P = \rho g h. \qquad \text{[liquid]} \quad \textbf{(13–3)}$$

Thus the fluid pressure is directly proportional to the density of the liquid, and to the depth within the liquid. In general, the pressure at equal depths within a uniform liquid is the same.

Equation 13–3 tells us what the pressure is at a depth h in the liquid, due to the liquid itself. But what if there is additional pressure exerted at the surface of the liquid, such as the pressure of the atmosphere or a piston pushing down? And what if the density of the fluid is not constant? Gases are quite compressible and hence their density can vary significantly with depth; and liquids, too, can be compressed, although we can often ignore the variation in density. (One exception is in the depths of the ocean where the great weight of water above significantly compresses the water and increases its density.) To cover these, and other cases, we now treat the general case of determining how the pressure in a fluid varies with depth.

Consider any fluid, and let us determine the pressure at any height y above some reference point (such as the ocean floor or the bottom of a tank or swimming pool), as shown in Fig. 13–4.[†] Within this fluid, at the height y, we consider a tiny, flat, slablike volume of the fluid whose area is A and whose (infinitesimal) thickness is dy, as shown. Let the pressure acting upward on its lower surface (at height y) be P. The pressure acting downward on the top surface of our tiny slab (at height $y + dy$) is designated $P + dP$. The fluid pressure acting on our slab thus exerts a force equal to PA upward on our slab and a force equal to $(P + dP)A$ downward on it. The only other force acting vertically on the slab is the (infinitesimal) force of gravity dF_G, which on our slab of mass dm is

$$dF_G = (dm)g = \rho g\, dV = \rho g A\, dy,$$

where ρ is the density of the fluid at the height y. Since the fluid is assumed to be at rest, our slab is in equilibrium so the net force on it must be zero. Therefore we have

$$PA - (P + dP)A - \rho g A\, dy = 0,$$

which when simplified becomes

$$\frac{dP}{dy} = -\rho g. \qquad \textbf{(13–4)}$$

This relation tells us how the pressure varies with height within the fluid. The minus sign indicates that the pressure decreases with an increase in height; or that the pressure increases with depth (reduced height).

[†] Now we are measuring y positive upwards, the reverse of what we did to get Eq. 13–3 where we measured the depth (i.e. downward as positive).

If the pressure at a height y_1 in the fluid is P_1, and at height y_2 it is P_2, then we can integrate Eq. 13–4 to obtain

$$\int_{P_1}^{P_2} dP = -\int_{y_1}^{y_2} \rho g \, dy$$

$$P_2 - P_1 = -\int_{y_1}^{y_2} \rho g \, dy, \qquad\qquad \textbf{(13–5)}$$

where we assume ρ is a function of height y: $\rho = \rho(y)$. This is a general relation, and we apply it now to two special cases: (1) pressure in liquids of uniform density and (2) pressure variations in the Earth's atmosphere.

For liquids in which any variation in density can be ignored, $\rho = $ constant and Eq. 13–5 is readily integrated:

$$P_2 - P_1 = -\rho g (y_2 - y_1). \qquad\qquad \textbf{(13–6a)}$$

For the everyday situation of a liquid in an open container—such as water in a glass, a swimming pool, a lake, or the ocean—there is a free surface at the top. And it is convenient to measure distances from this top surface. That is, we let h be the *depth* in the liquid where $h = y_2 - y_1$ as shown in Fig. 13–5. If we let y_2 be the position of the top surface, then P_2 represents the atmospheric pressure, P_0, at the top surface. Then, from Eq. 13–6a, the pressure $P(= P_1)$ at a depth h in the fluid is

$$P = P_0 + \rho g h. \qquad\qquad [h \text{ is depth in liquid}] \quad \textbf{(13–6b)}$$

Note that Eq. 13–6b is simply the liquid pressure (Eq. 13–3) plus the pressure P_0 due to the atmosphere above.

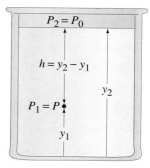

FIGURE 13–5 Pressure at a depth $h = (y_2 - y_1)$ in a liquid of density ρ is $P = P_0 + \rho g h$, where P_0 is the external pressure at the liquid's top surface.

EXAMPLE 13–2 **Pressure at a faucet.** The surface of the water in a storage tank is 30 m above a water faucet in the kitchen of a house, Fig. 13–6. Calculate the water pressure at the faucet.

SOLUTION The same atmospheric pressure acts both at the surface of the water in the storage tank, and on the water leaving the faucet. The pressure difference between the inside and outside of the faucet is

$$\Delta P = \rho g h = (1.0 \times 10^3 \, \text{kg/m}^3)(9.8 \, \text{m/s}^2)(30 \, \text{m}) = 2.9 \times 10^5 \, \text{N/m}^2.$$

The height h is sometimes called the **pressure head**. In this Example, the head of the water is 30 m. Note that the very different diameters of the tank and faucet don't affect the result—only pressure does.

⇒ **PHYSICS APPLIED**
Water supply

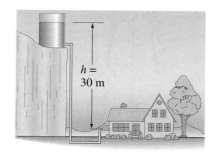

FIGURE 13–6 Example 13–2.

EXAMPLE 13–3 **Force on aquarium window.** Calculate the force due to water pressure exerted on a 1.0 m × 3.0 m aquarium viewing window, Fig. 13–7.

SOLUTION At a depth h, the pressure due to the water is given by Eq. 13–6b. Divide the window up into thin horizontal strips of width $w = 3.0$ m and thickness dy, as shown in Fig. 13–7. We choose a coordinate system with $y = 0$ at the surface of the water and y is positive downward. (With this choice, the minus sign in Eq. 13–6a becomes plus, or we use Eq. 13–6b with $y = h$.) The force due to water pressure on each strip is $dF = P \, dA = \rho g y w \, dy$. The total force on the window is given by the integral:

$$\int_{y_1 = 1.0 \, \text{m}}^{y_2 = 2.0 \, \text{m}} \rho g y w \, dy = \tfrac{1}{2} \rho g w (y_2^2 - y_1^2)$$

$$= \tfrac{1}{2}(1.00 \, \text{kg/m}^3)(9.8 \, \text{m/s}^2)(3.0 \, \text{m})[(2.0 \, \text{m})^2 - (1.0 \, \text{m})^2] = 44 \, \text{N}.$$

FIGURE 13–7 Example 13–3.

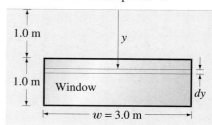

Now let us apply Eq. 13–4 or 13–5 to gases. The density of gases is normally quite small, so the difference in pressure at different heights can usually be ignored if $y_2 - y_1$ is not large (which is why, in Example 13–2, we could ignore the difference in air pressure between the faucet and the top of the storage tank). Indeed, for most ordinary containers of gas, we can assume that the pressure is the same throughout. However, if $y_2 - y_1$, is very large, we cannot make this assumption. An interesting example is the Earth's atmosphere, whose pressure at sea level is about $1.013 \times 10^5 \, \text{N/m}^2$ and decreases slowly with altitude.

Air pressure variation with altitude

EXAMPLE 13–4 **Elevation effect on atmospheric pressure.** (a) Determine the variation in pressure in the Earth's atmosphere as a function of height y above sea level, assuming g is constant and that the density of the air is proportional to the pressure. (This last assumption is not terribly accurate, in part because temperature and other weather effects are important.) (b) At what elevation is the air pressure equal to half the pressure at sea level?

SOLUTION (a) We are assuming that ρ is proportional to P, so we can write

$$\frac{\rho}{\rho_0} = \frac{P}{P_0},$$

where $P_0 = 1.013 \times 10^5 \, \text{N/m}^2$ is atmospheric pressure at sea level and $\rho_0 = 1.29 \, \text{kg/m}^3$ is the density of air at sea level at 0°C (Table 13–1). From the differential change in pressure with height, Eq. 13–4, we have

$$\frac{dP}{dy} = -\rho g = -P\left(\frac{\rho_0}{P_0}\right)g,$$

so

$$\frac{dP}{P} = -\frac{\rho_0}{P_0}g\,dy.$$

We integrate this from $y = 0$ (Earth's surface) and $P = P_0$, to the height y where the pressure is P:

$$\int_{P_0}^{P}\frac{dP}{P} = -\frac{\rho_0}{P_0}g\int_{0}^{y}dy$$

$$\ln\frac{P}{P_0} = -\frac{\rho_0}{P_0}gy$$

since $\ln P - \ln P_0 = \ln(P/P_0)$. Thus

$$P = P_0 e^{-(\rho_0 g/P_0)y}.$$

Exponential decrease in pressure with altitude

So, based on our assumptions, we find that the air pressure in our atmosphere decreases approximately exponentially with height. (Note that the atmosphere does not have a distinct top surface, so there is no natural point from which to measure depth in the atmosphere, as we can do for a liquid.)

(b) The constant $(\rho_0 g/P_0)$ has the value

$$\frac{\rho_0 g}{P_0} = \frac{(1.29 \, \text{kg/m}^3)(9.80 \, \text{m/s}^2)}{(1.013 \times 10^5 \, \text{N/m}^2)}$$

$$= 1.25 \times 10^{-4} \, \text{m}^{-1}.$$

Then, when we set $P = \frac{1}{2}P_0$, we have

$$\tfrac{1}{2} = e^{-(1.25\times 10^{-4}\,\text{m}^{-1})y}$$

or

$$y = (\ln 2.00)/(1.25 \times 10^{-4} \, \text{m}^{-1}) = 5550 \, \text{m},$$

where $\ln 2 = 0.693$. Thus, at an elevation of about 5500 m (about 18,000 ft), atmospheric pressure drops to half what it is at sea level. It is not surprising that mountain climbers often use oxygen tanks at very high altitudes.

13–3 | Atmospheric Pressure and Gauge Pressure

The pressure of the Earth's atmosphere varies with altitude, as we have seen. But even at a given altitude, it varies slightly according to the weather. As already mentioned, at sea level the pressure of the atmosphere on the average is $1.013 \times 10^5 \, \text{N/m}^2$ (or $14.7 \, \text{lb/in}^2$). This value is used to define a commonly used unit of pressure, the **atmosphere** (abbreviated atm):

$$1 \, \text{atm} = 1.013 \times 10^5 \, \text{N/m}^2 = 101.3 \, \text{kPa}.$$

One atmosphere (unit)

Another unit of pressure sometimes used (in meteorology and on weather maps) is the **bar**, which is defined as $1 \, \text{bar} = 1.00 \times 10^5 \, \text{N/m}^2$. Thus standard atmospheric pressure is slightly more than 1 bar.

The bar (unit)

The pressure due to the weight of the atmosphere is exerted on all objects immersed in this great sea of air, including our bodies. How does a human body withstand the enormous pressure on its surface? The answer is that living cells maintain an internal pressure that closely equals the external pressure, just as the pressure inside a balloon only slightly exceeds the outside pressure of the atmosphere. An automobile tire, because of its rigidity, can maintain internal pressures much greater than the external pressure.

It is important to note that tire gauges, and most other pressure gauges, register the pressure over and above atmospheric pressure. This is called **gauge pressure**. Thus, to get the absolute pressure, P, one must add the atmospheric pressure, P_A, to the gauge pressure, P_G:

Gauge pressure

$$P = P_A + P_G.$$

If a tire gauge registers $220 \, \text{kPa}$, the absolute pressure within the tire is $220 \, \text{kPa} + 101 \, \text{kPa} = 321 \, \text{kPa}$. This is equivalent to about 3.2 atm (2.2 atm gauge pressure).

Absolute pressure = atmospheric pressure + gauge pressure

CONCEPTUAL EXAMPLE 13–5 | **Finger holds water in a straw.** You insert a straw of length L into a tall glass of your favorite beverage. You place your finger over the top of the straw so that no air can get in or out, and then lift the straw from the liquid. You find that the straw retains the liquid such that the distance from the bottom of your finger to the top of the liquid is h. (See Fig. 13–8.) Does the air in the space between your finger and the top of the liquid have a pressure P that is greater than, equal to, or less than, the atmospheric pressure P_A outside the straw?

RESPONSE Consider the forces on the column of liquid. Atmospheric pressure on the outside of the straw pushes upward on the liquid at the bottom of the straw, gravity pulls the liquid downward, and the air pressure inside the top of the straw pushes downward on the liquid. Since the liquid is in equilibrium, the upward force due to atmospheric pressure must balance the two downward forces. The only way this is possible is for the air pressure inside the straw to be rather less than the atmosphere pressure outside the straw.

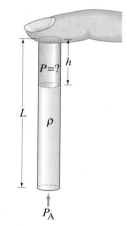

FIGURE 13–8 Example 13–5.

13–4 | Pascal's Principle

The Earth's atmosphere exerts a pressure on all objects with which it is in contact, including other fluids. External pressure acting on a fluid is transmitted throughout that fluid. For instance, according to Eq. 13–3, the pressure due to the water at a depth of $100 \, \text{m}$ below the surface of a lake is $P = \rho g h = (1000 \, \text{kg/m}^3)(9.8 \, \text{m/s}^2)(100 \, \text{m}) = 9.8 \times 10^5 \, \text{N/m}^2$, or 9.7 atm. However, the total pressure at this point is due to the pressure of water plus the pressure of the air above it (Eq. 13–6b). Hence the total pressure (if the lake is near sea level) is $9.7 \, \text{atm} + 1.0 \, \text{atm} = 10.7 \, \text{atm}$. This is just one example of a general principle attributed to the French philosopher and scientist Blaise Pascal (1623–1662). **Pascal's principle** states that *pressure applied to a confined fluid increases the pressure throughout by the same amount.*

Pascal's principle

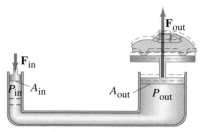

FIGURE 13–9 Application of Pascal's principle: hydraulic lift.

A number of practical devices make use of Pascal's principle. One example is the hydraulic lift, illustrated in Fig. 13–9, in which a small force is used to exert a large force by making the area of the output piston larger than the area of the input piston. To see how this works, we assume the input and output pistons are at the same height (at least approximately). Then the external input force F_{in}, by Pascal's principle, increases the pressure equally throughout so that at the same level (see Fig. 13–9):

$$P_{out} = P_{in}$$

where the input quantities are represented by the subscript "in" and the output by "out." Thus

$$\frac{F_{out}}{A_{out}} = \frac{F_{in}}{A_{in}},$$

or

$$\frac{F_{out}}{F_{in}} = \frac{A_{out}}{A_{in}}.$$

The quantity F_{out}/F_{in} is called the "mechanical advantage" of the hydraulic lift, and is equal to the ratio of the areas. For example, if the area of the output piston is 20 times that of the input cylinder, the force is multiplied by a factor of 20: thus a force of 200 lb could lift a 4000-lb car.

13–5 | Measurement of Pressure; Gauges and the Barometer

Many devices have been invented to measure pressure, some of which are shown in Fig. 13–10. The simplest is the open-tube *manometer* (Fig. 13–10a) which is a U-shaped tube partially filled with a liquid, usually mercury or water. The pressure P being measured is related to the difference in height h of the two levels of the liquid by the relation

$$P = P_0 + \rho g h,$$

where P_0 is atmospheric pressure (acting on the top of the fluid in the left-hand tube), and ρ is the density of the liquid. Note that the quantity $\rho g h$ is the "gauge pressure"—the amount by which P exceeds atmospheric pressure. If the liquid in the left-hand column were lower than that in the right-hand column, this would indicate that P was less than atmospheric pressure (and h would be negative).

FIGURE 13–10 Pressure gauges: (a) open-tube manometer, (b) aneroid gauge, and (c) common tire pressure gauge.

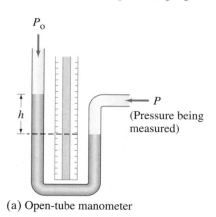

(a) Open-tube manometer

(b) Aneroid gauge (used mainly for air pressure and then called an aneroid barometer)

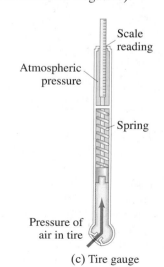

(c) Tire gauge

Instead of calculating the product ρgh, it is common to simply specify the height h. In fact, pressures are sometimes specified as so many "millimeters of mercury" (mm-Hg), and sometimes as so many "mm of water" (mm-H_2O). The unit mm-Hg is equivalent to a pressure of $133\ N/m^2$, since ρgh for $1\ mm = 1.0 \times 10^{-3}\ m$ of mercury gives

$$\rho gh = (13.6 \times 10^3\ kg/m^3)(9.80\ m/s^2)(1.00 \times 10^{-3}\ m) = 1.33 \times 10^2\ N/m^2.$$

The unit mm-Hg is also called the **torr** in honor of Evangelista Torricelli (1608–1647), who invented the barometer (see below). Conversion factors among the various units of pressure (an incredible nuisance!) are given in Table 13–2. It is important that only $N/m^2 = Pa$, the proper SI unit, be used in calculations involving other quantities specified in SI units.

The torr (unit)

➡ **PROBLEM SOLVING**

Use SI unit in calculations:
$1\ Pa = 1\ N/m^2$

TABLE 13–2 Conversion Factors Between Different Units of Pressure

In Terms of $1\ Pa = 1\ N/m^2$	Related to 1 atm
1 atm $= 1.013 \times 10^5\ N/m^2$	1 atm $= 1.013 \times 10^5\ N/m^2$
$= 1.013 \times 10^5\ Pa = 101.3\ kPa$	
1 bar $= 1.000 \times 10^5\ N/m^2$	1 atm $= 1.013\ bar$
1 dyne/$cm^2 = 0.1\ N/m^2$	1 atm $= 1.013 \times 10^6\ dyne/cm^2$
1 lb/$in.^2 = 6.90 \times 10^3\ N/m^2$	1 atm $= 14.7\ lb/in.^2$
1 lb/$ft^2 = 47.9\ N/m^2$	1 atm $= 2.12 \times 10^3\ lb/ft^2$
1 cm-Hg $= 1.33 \times 10^3\ N/m^2$	1 atm $= 76\ cm\text{-}Hg$
1 mm-Hg $= 133\ N/m^2$	1 atm $= 760\ mm\text{-}Hg$
1 torr $= 133\ N/m^2$	1 atm $= 760\ torr$
1 mm-H_2O (4°C) $= 9.81\ N/m^2$	1 atm $= 1.03 \times 10^4\ mm\text{-}H_2O$ (4°C)

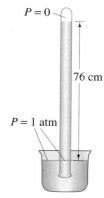

FIGURE 13–11 Diagram of a mercury barometer, when the air pressure is 76 cm-Hg.

FIGURE 13–12 A water barometer: water was added at the top, and the spigot at the top then closed; the water level dropped, leaving a vacuum between its upper surface and the spigot. Why? Because air pressure could not support a column of water more than 10 m high.

Another type of pressure gauge is the aneroid gauge (Fig. 13–10b) in which the pointer is linked to the flexible ends of an evacuated thin metal chamber. In an electronic gauge, the pressure may be applied to a thin metal diaphragm whose resulting distortion is translated into an electrical signal by a transducer. How a common tire gauge is constructed is shown in Fig. 13–10c.

Atmospheric pressure is often measured by a modified kind of mercury manometer with one end closed, called a mercury **barometer** (Fig. 13–11). The glass tube is completely filled with mercury and then inverted into the bowl of mercury. If the tube is long enough, the level of the mercury will drop, leaving a vacuum at the top of the tube, since atmospheric pressure can support a column of mercury only about 76 cm high (exactly 76.0 cm at standard atmospheric pressure). That is, a column of mercury 76 cm high exerts the same pressure as the atmosphere:

$$P = \rho gh = (13.6 \times 10^3\ kg/m^3)(9.80\ m/s^2)(0.760\ m) = 1.013 \times 10^5\ N/m^2 = 1.00\ atm.$$

Household barometers are usually of the aneroid type, either mechanical (Fig. 3–10b), or electronic.

A calculation similar to that above will show that atmospheric pressure can maintain a column of water 10.3 m high in a tube whose top is under vacuum (Fig. 13–12). A few centuries ago, it was a source of wonder and frustration that no matter how good a vacuum pump was, it could not lift water more than about 10 m. The only way to pump water out of deep mine shafts, for example, was to use multiple stages for depths greater than 10 m. Galileo studied this problem, and his student Torricelli was the first to explain it. The point is that a pump does not really suck water up a tube—it merely reduces the pressure at the top of the tube. Atmospheric air pressure *pushes* the water up the tube if the top end is at low pressure (under a vacuum), just as it is air pressure that pushes (or maintains) the mercury 76 cm high in a barometer.

CONCEPTUAL EXAMPLE 13–6 **Suction.** You sit in a meeting where a novice NASA engineer proposes suction cup shoes for Space Shuttle astronauts working on the exterior of the spacecraft. Having just studied this chapter, you gently remind him of the fallacy of this plan. What is it?

RESPONSE Suction cups work by pushing out the air underneath the cup. What holds the cup in place is the air pressure outside the cup. (This can be a substantial force on Earth. For example, a 10 cm diameter cup has an area of $7.8 \times 10^{-3} \, \text{m}^2$. The force of the atmosphere on it is $(7.8 \times 10^{-3} \, \text{m}^2)(1.0 \times 10^5 \, \text{N/m}^2) \approx 800 \, \text{N}$, about 180 lbs!). But in outer space, there is no air pressure to hold the suction cup on the spacecraft.

We sometimes mistakenly think of suction as something we actively do. For example, we intuitively think that we pull the soda up through a straw. Instead, all we do is lower the pressure at the top of the straw, and the atmosphere *pushes* the soda up the straw.

13–6 Buoyancy and Archimedes' Principle

Objects submerged in a fluid appear to weigh less than they do when outside the fluid. For example, a large rock that you would have difficulty lifting off the ground can often be easily lifted from the bottom of a stream. When the rock breaks through the surface of the water, it suddenly seems to be much heavier. Many objects, such as wood, float on the surface of water. These are two examples of *buoyancy*. In each example, the force of gravity is acting downward. But in addition, an upward *buoyant force* is exerted by the liquid.

The buoyant force occurs because the pressure in a fluid increases with depth. Thus the upward pressure on the bottom surface of a submerged object is greater than the downward pressure on its top surface. To see the effect of this, consider a cylinder of height h whose top and bottom ends have an area A and which is completely submerged in a fluid of density ρ_F, as shown in Fig. 13–13. The fluid exerts a pressure $P_1 = \rho_F g h_1$ at the top surface of the cylinder. The force due to this pressure on top of the cylinder is $F_1 = P_1 A = \rho_F g h_1 A$, and it is directed downward. Similarly, the fluid exerts an upward force on the bottom of the cylinder equal to $F_2 = P_2 A = \rho_F g h_2 A$. The net force due to the fluid pressure, which is the **buoyant force**, $\mathbf{F}_B$, acts upward and has the magnitude

$$F_B = F_2 - F_1 = \rho_F g A (h_2 - h_1)$$
$$= \rho_F g A h = \rho_F g V,$$

where $V = Ah$ is the volume of the cylinder. Since ρ_F is the density of the fluid, the product $\rho_F g V = m_F g$ is the weight of fluid which takes up a volume equal to the volume of the cylinder. Thus the buoyant force on the cylinder is equal to the weight of fluid displaced[†] by the cylinder. This result is valid no matter what the shape of the object. Its discovery is credited to Archimedes (287?–212 B.C.), and it is called **Archimedes' principle**: *the buoyant force on a body immersed in a fluid is equal to the weight of the fluid displaced by that object.*

We can derive Archimedes' principle in general by the following simple but elegant argument. The irregularly shaped object D shown in Fig. 13–14a is acted on by the force of gravity (its weight, $m\mathbf{g}$, downward) and the buoyant force, $\mathbf{F}_B$, upward. We wish to determine F_B. To do so, we next consider a body, this time made of our same fluid (D' in Fig. 13–14b) with the same shape and size as the original object, and located at the same depth. You might think of this body of fluid as being separated from the rest of the fluid by an imaginary membrane. The buoyant force F_B on this body of fluid will be exactly the same as that on the

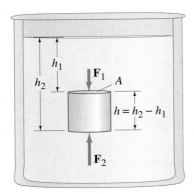

FIGURE 13–13 Determination of the buoyant force.

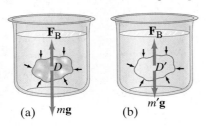

FIGURE 13–14 Archimedes' principle.

[†] By "fluid displaced," we mean a volume of fluid equal to the volume of the object, or that part of the object submerged if it floats or is only partly submerged (the fluid that used to be where the object is). If the object is placed in a glass or tub initially filled to the brim with water, the water that flows over the top represents the water displaced by the object.

original object since the surrounding fluid, which exerts F_B, is in exactly the same configuration. Now the body of fluid D' is in equilibrium (the fluid as a whole is at rest). Therefore, $F_B = m'g$, where $m'g$ is the weight of the body of fluid. Hence the buoyant force F_B is equal to the weight of the body of fluid whose volume equals the volume of the original submerged object, which is Archimedes' principle.

CONCEPTUAL EXAMPLE 13–7 **Two pails of water.** Consider two identical pails of water filled to the brim. One pail contains only water while the other has a piece of wood floating in it. Which one has the greater weight?

RESPONSE Both pails weigh the same. Recall Archimedes' principle: the wood displaces a volume of water with weight equal to the weight of the wood. Some water will overflow the pail, but the spilled water has weight equal to that of the wood; so the pails have the same weight.

EXAMPLE 13–8 **Recovering a submerged statue.** A 70-kg ancient statue lies at the bottom of the sea. Its volume is $3.0 \times 10^4 \, \text{cm}^3$. How much force is needed to lift it?

SOLUTION The buoyant force on the statue due to the water is equal to the weight of $3.0 \times 10^4 \, \text{cm}^3 = 3.0 \times 10^{-2} \, \text{m}^3$ of water (for seawater $\rho = 1.025 \times 10^3 \, \text{kg/m}^3$):

$$F_B = m_{H_2O}\,g = \rho_{H_2O}\,gV$$
$$= (1.025 \times 10^3 \, \text{kg/m}^3)(9.8 \, \text{m/s}^2)(3.0 \times 10^{-2} \, \text{m}^3) = 3.0 \times 10^2 \, \text{N}.$$

The weight of the statue is $mg = (70 \, \text{kg})(9.8 \, \text{m/s}^2) = 6.9 \times 10^2 \, \text{N}$. Hence the force needed to lift it is $690 \, \text{N} - 300 \, \text{N} = 390 \, \text{N}$. It is as if the statue had a mass of only $(390 \, \text{N})/(9.8 \, \text{m/s}^2) = 40 \, \text{kg}$.

Archimedes is said to have discovered his principle in his bath while thinking how he might determine whether the king's new crown was pure gold or a fake. Gold has a specific gravity of 19.3, somewhat higher than that of most metals, but a determination of specific gravity or density is not readily done directly because, even if the mass is easily known, the volume of an irregularly shaped object is not easily calculated. However, if the object is weighed in air $(= w)$ and also "weighed" while it is under water $(= w')$, the density can be determined using Archimedes' principle, as the following Example shows. The quantity w' is called the *apparent weight* in water, and is what a scale reads when the object is submerged in water (see Fig. 13–15); w' equals the true weight $(w = mg)$ minus the buoyant force.

EXAMPLE 13–9 **Archimedes: Is the crown gold?** When a crown of mass 14.7 kg is submerged in water, an accurate scale reads only 13.4 kg. Is the crown made of gold?

SOLUTION See analysis in Fig. 13–15. The apparent weight of the submerged object, w' $(= F'_T$ in Fig. 13–15b), equals its actual weight $w(= mg)$ minus the buoyant force F_B as shown:

$$w' = F'_T = w - F_B = \rho_O gV - \rho_F gV,$$

where V is the volume of the object, ρ_O the object's density, and ρ_F the density of the fluid (water in this case). From this relation, we can see that $F_B = w - w' = \rho_F gV$. Then we can write

$$\frac{w}{w - w'} = \frac{\rho_O gV}{\rho_F gV} = \frac{\rho_O}{\rho_F}.$$

Thus $w/(w - w')$ is equal to the specific gravity of the object if the fluid in which it is submerged is water. For the crown we have

$$\frac{\rho_O}{\rho_{H_2O}} = \frac{w}{w - w'} = \frac{(14.7 \, \text{kg})g}{(14.7 \, \text{kg} - 13.4 \, \text{kg})g} = \frac{14.7 \, \text{kg}}{1.3 \, \text{kg}} = 11.3.$$

This corresponds to a density of $11,300 \, \text{kg/m}^3$. The crown seems to be made of lead (see Table 13–1)!

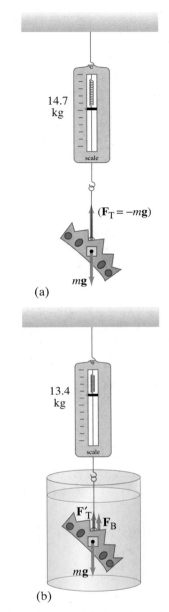

FIGURE 13–15 (a) A scale reads the mass of an object in air—in this case the crown of Example 13–9. All objects are at rest, so the tension F_T in the connecting cord equals the weight w of the object: $F_T = mg$. Note that we show the free-body diagram of the crown, and that F_T is what causes the scale reading (it's equal to the net downward force on the scale). (b) The object submerged has an additional force on it, the buoyant force F_B. The net force is zero, so $F'_T + F_B = mg(= w)$. The scale now reads $m' = 13.4 \, \text{kg}$, where m' is related to the effective weight by $w' = m'g$, where $F'_T = w' = w - F_B$.

FIGURE 13–16 (a) The fully submerged log accelerates upward because $F_B > mg$. It comes to equilibrium (b) when $\Sigma F = 0$, so $F_B = mg = (1200\,\text{kg})g$. Thus 1200 kg, or 1.2 m³, of water is displaced.

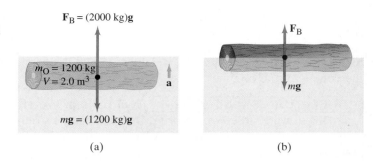

(a)　　　　　　　(b)

Archimedes' principle applies equally well to objects that float, such as wood. In general, *an object floats on a fluid if its density is less than that of the fluid.* This is readily seen from Fig. 13–16a, where a submerged object will experience a net upward force and float to the surface if $F_B > mg$; that is, if $\rho_F V g > \rho_O V g$ or $\rho_F > \rho_O$. At equilibrium—that is, when floating—the buoyant force on an object has magnitude equal to the weight of the object. For example, a log whose specific gravity is 0.60 and whose volume is 2.0 m³ has a mass $m = \rho_O V = (0.60 \times 10^3\,\text{kg/m}^3)(2.0\,\text{m}^3) = 1200\,\text{kg}$. If the log is fully submerged, it will displace a mass of water $m_F = \rho_F V = (1000\,\text{kg/m}^3)(2.0\,\text{m}^3) = 2000\,\text{kg}$. Hence the buoyant force on the log will be greater than its weight, and it will float upward to the surface (Fig. 13–16). It will come to equilibrium when it displaces 1200 kg of water, which means that 1.2 m³ of its volume will be submerged. This 1.2 m³ corresponds to 60 percent of the volume of the log $(1.2/2.0 = 0.60)$, so 60 percent of the log is submerged. In general when an object floats, we have $F_B = mg$, which we can write as (see Fig. 13–17)

$$\rho_F V_{\text{displ}}\, g = \rho_O V_O\, g$$

where V_O is the full volume of the object and V_{displ} is the volume of fluid it displaces (= volume submerged). Thus

$$\frac{V_{\text{displ}}}{V_O} = \frac{\rho_O}{\rho_F}.$$

That is, the fraction of the object submerged is given by the ratio of the object's density to that of the fluid.

Floating

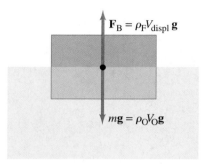

$\mathbf{F}_B = \rho_F V_{\text{displ}}\,\mathbf{g}$

$m\mathbf{g} = \rho_O V_O \mathbf{g}$

FIGURE 13–17 An object floating in equilibrium: $F_B = mg$.

EXAMPLE 13–10 Hydrometer calibration. A **hydrometer** is a simple instrument used to indicate specific gravity of a liquid by measuring how deeply it sinks in the liquid. A particular hydrometer (Fig. 13–18) consists of a glass tube, weighted at the bottom, which is 25.0 cm long, 2.00 cm² in cross-sectional area, and has a mass of 45.0 g. How far from the end should the 1.000 mark be placed?

SOLUTION The hydrometer has an overall density

$$\rho = \frac{m}{V} = \frac{45.0\,\text{g}}{(2.00\,\text{cm}^2)(25.0\,\text{cm})} = 0.900\,\text{g/cm}^3.$$

Thus, when placed in water, it will come to equilibrium when 0.900 of its volume is submerged. Since it is of uniform cross section, $(0.900)(25.0\,\text{cm}) = 22.5\,\text{cm}$ of its length will be submerged. Since the specific gravity of water is defined to be 1.000, the mark should be placed 22.5 cm from the end.

FIGURE 13–18 A hydrometer. Example 13–10.

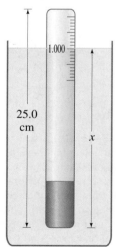

Archimedes' principle is also useful in geology. According to the theory of plate tectonics and continental drift, the continents can be considered to be floating on a fluid "sea" of slightly deformable rock (mantle rock). Some interesting calculations can be done using very simple models, which we consider in the problems at the end of the chapter.

Air is a fluid and it too exerts a buoyant force. Ordinary objects weigh less in air than they do if weighed in a vacuum. Because the density of air is so small, the effect for ordinary solids is slight. There are objects, however, that float in air—helium-filled balloons, for example, because the density of helium is less than the density of air.

➥ **PHYSICS APPLIED**

Continental drift—plate tectonics

Weight affected by buoyancy of air

EXAMPLE 13–11 **Helium balloon.** What volume V of helium is needed if a balloon is to lift a load of 180 kg (including the weight of the empty balloon)?

SOLUTION The buoyant force on the helium balloon, F_B, which is equal to the weight of displaced air, must be at least equal to the weight of the helium plus the load (Fig. 13–19).

$$F_B = (m_{He} + 180 \text{ kg})g.$$

This equation can be written in terms of density:

$$\rho_{air} Vg = (\rho_{He} V + 180 \text{ kg})g.$$

Solving now for V, we find

$$V = \frac{180 \text{ kg}}{\rho_{air} - \rho_{He}} = \frac{180 \text{ kg}}{(1.29 \text{ kg/m}^3 - 0.18 \text{ kg/m}^3)} = 160 \text{ m}^3.$$

This is the volume needed near the Earth's surface, where $\rho_{air} = 1.29 \text{ kg/m}^3$. To reach a high altitude, a greater volume would be needed since the density of air decreases with altitude.

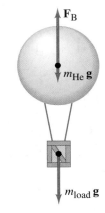

FIGURE 13–19 Example 13–11.

13–7 | Fluids in Motion; Flow Rate and the Equation of Continuity

We now turn from the study of fluids at rest to the more complex subject of fluids in motion, which is called **fluid dynamics**, or (especially if the fluid is water) **hydrodynamics**. Many aspects of fluid motion are still being studied today (for example, turbulence as a manifestation of chaos is a "hot" topic today). Nonetheless, with certain simplifying assumptions, a good understanding of this subject can be obtained.

To begin with, we can distinguish two main types of fluid flow. If the flow is smooth, such that neighboring layers of the fluid slide by each other smoothly, the flow is said to be **streamline** or **laminar flow**.† In this kind of flow, each particle of the fluid follows a smooth path, called a **streamline**, and these paths do not cross over one another (Fig. 13–20a). Above a certain speed, the flow becomes turbulent. **Turbulent flow** is characterized by erratic, small, whirlpool-like circles called eddy currents or eddies (Fig. 13–20b). Eddies absorb a great deal of energy, and

†The word laminar means "in layers."

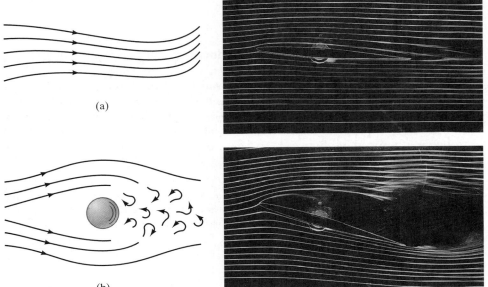

(a)

(b)

FIGURE 13–20 (a) Streamline or laminar flow; (b) turbulent flow.

although a certain amount of internal friction, called **viscosity**, is present even during streamline flow, it is much greater when the flow is turbulent. A few tiny drops of ink or food coloring dropped into a moving liquid can quickly reveal whether the flow is streamline or turbulent.

We will assume in this chapter that the fluid is essentially incompressible (no significant variations in density) and that the flow past any point is steady.

Let us consider the steady laminar flow of a fluid through an enclosed tube or pipe as shown in Fig. 13–21. First we determine how the speed of the fluid changes when the size of the tube changes. The mass **flow rate** is defined as the mass Δm of fluid that passes a given point per unit time Δt: mass flow rate = $\Delta m / \Delta t$. In Fig. 13–21, the volume of fluid passing point 1 (that is, through area A_1) in a time Δt is just $A_1 \Delta l_1$, where Δl_1 is the distance the fluid moves in time Δt. Since the velocity† of fluid passing point 1 is $v_1 = \Delta l_1 / \Delta t$, the mass flow rate $\Delta m_1 / \Delta t$ through area A_1 is

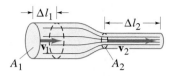

$$\frac{\Delta m_1}{\Delta t} = \frac{\rho_1 \Delta V_1}{\Delta t} = \frac{\rho_1 A_1 \Delta l_1}{\Delta t} = \rho_1 A_1 v_1,$$

FIGURE 13–21 Fluid flow through a pipe of varying diameter.

where $\Delta V_1 = A_1 \Delta l_1$ is the volume of mass Δm_1, and ρ_1 is the fluid density. Similarly, at point 2 (through area A_2), the flow rate is $\rho_2 A_2 v_2$. Since no fluid flows in or out of the sides, the flow rates through A_1 and A_2 must be equal. Thus, since:

$$\frac{\Delta m_1}{\Delta t} = \frac{\Delta m_2}{\Delta t},$$

then

$$\rho_1 A_1 v_1 = \rho_2 A_2 v_2.$$

This is called the **equation of continuity**. If the fluid is incompressible (ρ doesn't change with pressure), which is an excellent approximation for liquids under most circumstances (and sometimes for gases as well), then $\rho_1 = \rho_2$, and the equation of continuity becomes

$$A_1 v_1 = A_2 v_2. \qquad\qquad [\rho = \text{constant}] \quad \textbf{(13–7)}$$

Notice that the product Av represents the *volume rate of flow* (volume of fluid passing a given point per second), since $\Delta V / \Delta t = A \, \Delta l / \Delta t = Av$, which in SI units is m^3/s. Equation 13–7 tells us that where the cross-sectional area is large the velocity is small, and where the area is small the velocity is large. That this is reasonable can be seen by looking at a river. A river flows slowly through a meadow where it is broad, but speeds up to torrential speed when passing through a narrow gorge.

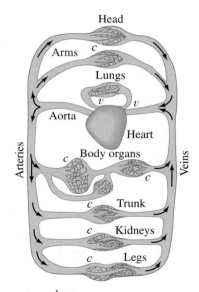

Head
Arms
c
Lungs
v v
Aorta
Heart
Body organs
c
c
c Trunk
c Kidneys
c Legs
Arteries
Veins

v = valves
c = capillaries

FIGURE 13–22
Human circulatory system.

➡ **PHYSICS APPLIED**

Blood flow

EXAMPLE 13–12 ESTIMATE **Blood flow.** In humans, blood flows from the heart into the aorta, from which it passes into the major arteries. These branch into the small arteries (arterioles), which in turn branch into myriads of tiny capillaries, Fig. 13–22. The blood returns to the heart via the veins. The radius of the aorta is about 1.0 cm and the blood passing through it has a speed of about 30 cm/s. A typical capillary has a radius of about 4×10^{-4} cm, and blood flows through it at a speed of about 5×10^{-4} m/s. Estimate how many capillaries there are in the body.

† If there were no viscosity, the velocity would be the same across a cross section of the tube. Real fluids have viscosity, and this internal friction causes different layers of the fluid to flow at different speeds. In this case v_1 and v_2 represent the average speeds at each cross section.

SOLUTION Let A_1 be the area of the aorta, and A_2 be the area of *all* the capillaries through which blood flows. Then $A_2 = N\pi r_{cap}^2$ where N is the number of capillaries and $r_{cap} \approx 4 \times 10^{-4}$ cm is the estimated average radius of one capillary. From the equation of continuity (Eq. 13–7), we have

$$v_2 A_2 = v_1 A_1$$

$$v_2 N\pi r_{cap}^2 = v_1 \pi r_{aorta}^2$$

so

$$N = \frac{v_1}{v_2} \frac{r_{aorta}^2}{r_{cap}^2} = \left(\frac{0.30 \text{ m/s}}{5 \times 10^{-4} \text{ m/s}} \right) \left(\frac{1.0 \times 10^{-2} \text{ m}}{4 \times 10^{-6} \text{ m}} \right)^2 \approx 4 \times 10^9,$$

or about 4 billion capillaries.

Another Example that makes use of the equation of continuity and the argument leading up to it is the following.

EXAMPLE 13–13 **Heating duct to a room.** How large must a heating duct be if air moving 3.0 m/s along it can replenish the air every 15 minutes in a room of 300-m³ volume? Assume the air's density remains constant.

➡ **PHYSICS APPLIED**
Heating duct

SOLUTION We can apply the equation of continuity (Eq. 13–7) if we consider the room (call it point 2) as a large section of the duct, Fig. 13–23. Reasoning in the same way we did to obtain Eq. 13–7 (changing Δt to t), we see that $A_2 v_2 = A_2 l_2/t = V_2/t$ where V_2 is the volume of the room. Then $A_1 v_1 = A_2 v_2 = V_2/t$ and

$$A_1 = \frac{V_2}{v_1 t} = \frac{300 \text{ m}^3}{(3.0 \text{ m/s})(900 \text{ s})} = 0.11 \text{ m}^2.$$

If the duct is square, then each side has length $l = \sqrt{A} = 0.33$ m or 33 cm. A rectangular duct 20 cm × 55 cm will also do.

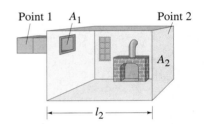

FIGURE 13–23 Example 13–13.

13–8 Bernoulli's Equation

Have you ever wondered why smoke goes up a chimney, why a car's convertible top bulges upward at high speeds, or how a sailboat can move against the wind? These are examples of a principle worked out by Daniel Bernoulli (1700–1782) in the early eighteenth century. In essence, **Bernoulli's principle** states that *where the velocity of a fluid is high, the pressure is low, and where the velocity is low, the pressure is high.* For example, if the pressures at points 1 and 2 in Fig. 13–21 are measured, it will be found that the pressure is lower at point 2, where the velocity is greater, than it is at point 1, where the velocity is smaller. At first glance, this might seem strange; you might expect that the greater speed at point 2 would imply a higher pressure. But this cannot be the case. For if the pressure at point 2 were higher than at 1, this higher pressure would slow the fluid down, whereas in fact it has speeded up in going from point 1 to point 2. Thus the pressure at point 2 must be less than at point 1, to be consistent with the fact that the fluid accelerates.

Bernoulli's principle

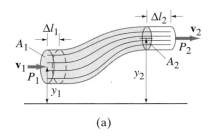

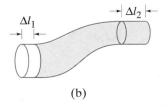

FIGURE 13–24 Fluid flow: for derivation of Bernoulli's equation.

Bernoulli developed an equation that expresses this principle quantitatively. To derive Bernoulli's equation, we assume the flow is steady and laminar, the fluid is incompressible, and the viscosity is small enough to be ignored. To be general, we assume the fluid is flowing in a tube of nonuniform cross section that varies in height above some reference level, Fig. 13–24. We will consider the amount of fluid shown in color and calculate the work done to move it from the position shown in (a) to that shown in (b). In this process, fluid at point 1 flows a distance Δl_1 and forces the fluid at point 2 to move a distance Δl_2. The fluid to the left of point 1 exerts a pressure P_1 on our section of fluid and does an amount of work

$$W_1 = F_1 \Delta l_1 = P_1 A_1 \Delta l_1.$$

At point 2, the work done on our section of fluid is

$$W_2 = -P_2 A_2 \Delta l_2;$$

the negative sign is present because the force exerted on the fluid is opposite to the motion (thus the fluid shown in color does work on the fluid to the right of point 2). Work is also done on the fluid by the force of gravity. Since the net effect of the process shown in Fig. 13–24 is to move a mass m of volume $A_1 \Delta l_1 (= A_2 \Delta l_2$, since the fluid is incompressible) from point 1 to point 2, the work done by gravity is

$$W_3 = -mg(y_2 - y_1),$$

where y_1 and y_2 are heights of the center of the tube above some (arbitrary) reference level. Notice that in the case shown in Fig. 13–24, this term is negative since the motion is uphill against the force of gravity. The net work W done on the fluid is thus:

$$W = W_1 + W_2 + W_3$$

$$W = P_1 A_1 \Delta l_1 - P_2 A_2 \Delta l_2 - mgy_2 + mgy_1.$$

According to the work-energy principle (Section 7–4), the net work done on a system is equal to its change in kinetic energy. Thus

$$\tfrac{1}{2}mv_2^2 - \tfrac{1}{2}mv_1^2 = P_1 A_1 \Delta l_1 - P_2 A_2 \Delta l_2 - mgy_2 + mgy_1.$$

The mass m has volume $A_1 \Delta l_1 = A_2 \Delta l_2$. Thus we can substitute $m = \rho A_1 \Delta l_1 = \rho A_2 \Delta l_2$, and also divide through by $A_1 \Delta l_1 = A_2 \Delta l_2$, to obtain:

$$\tfrac{1}{2}\rho v_2^2 - \tfrac{1}{2}\rho v_1^2 = P_1 - P_2 - \rho g y_2 + \rho g y_1$$

which we rearrange to get

Bernoulli's equation

$$P_1 + \tfrac{1}{2}\rho v_1^2 + \rho g y_1 = P_2 + \tfrac{1}{2}\rho v_2^2 + \rho g y_2. \qquad \textbf{(13–8)}$$

This is **Bernoulli's equation**. Since points 1 and 2 can be any two points along a tube of flow, Bernoulli's equation can be written:

$$P + \tfrac{1}{2}\rho v^2 + \rho g y = \text{constant}$$

at every point in the fluid, where y is the height of the center of the tube above a fixed reference level. [Note that if there is no flow $(v_1 = v_2 = 0)$, then Eq. 13–8 reduces to the hydrostatic equation, Eq. 13–6a: $P_2 - P_1 = -\rho g(y_2 - y_1)$.]

Bernoulli's equation is an expression of the law of energy conservation, since we derived it from the work-energy principle.

EXAMPLE 13–14 Flow and pressure in hot-water heating systems.
Water circulates throughout a house in a hot-water heating system. If the water is pumped at a speed of 0.50 m/s through a 4.0-cm-diameter pipe in the basement under a pressure of 3.0 atm, what will be the flow speed and pressure in a 2.6-cm-diameter pipe on the second floor 5.0 m above? Assume the pipes do not divide into branches.

➡ **PHYSICS APPLIED**

Hot-water heating system

SOLUTION We first calculate the flow speed on the second floor, calling it v_2, using the equation of continuity, Eq. 13–7. Noting that the areas are proportional to the radii squared $(A = \pi r^2)$, we call the basement point 1 and obtain

$$v_2 = \frac{v_1 A_1}{A_2} = \frac{v_1 \pi r_1^2}{\pi r_2^2} = (0.50 \text{ m/s}) \frac{(0.020 \text{ m})^2}{(0.013 \text{ m})^2} = 1.2 \text{ m/s}.$$

To find the pressure, we use Bernoulli's equation:

$$\begin{aligned}
P_2 &= P_1 + \rho g(y_1 - y_2) + \tfrac{1}{2}\rho(v_1^2 - v_2^2) \\
&= (3.0 \times 10^5 \text{ N/m}^2) + (1.0 \times 10^3 \text{ kg/m}^3)(9.8 \text{ m/s}^2)(-5.0 \text{ m}) \\
&\quad + \tfrac{1}{2}(1.0 \times 10^3 \text{ kg/m}^3)\left[(0.50 \text{ m/s})^2 - (1.2 \text{ m/s})^2\right] \\
&= 3.0 \times 10^5 \text{ N/m}^2 - 4.9 \times 10^4 \text{ N/m}^2 - 6.0 \times 10^2 \text{ N/m}^2 \\
&= 2.5 \times 10^5 \text{ N/m}^2,
\end{aligned}$$

or 2.5 atm. Notice that the velocity term contributes very little in this case.

13–9 Applications of Bernoulli's Principle: From Torricelli to Sailboats, Airfoils, and TIA

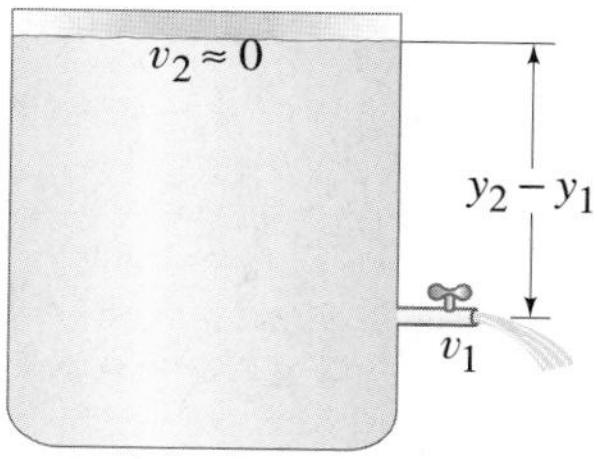

Bernoulli's equation can be applied to many situations. One example is to calculate the velocity, v_1, of a liquid flowing out of a spigot at the bottom of a reservoir, Fig. 13–25. We choose point 2 in Eq. 13–8 to be the top surface of the liquid. Assuming the diameter of the reservoir is large compared to that of the spigot, v_2 will be almost zero. Points 1 (the spigot) and 2 (top surface) are open to the atmosphere so the pressure at both points is equal to atmospheric pressure: $P_1 = P_2$. Then Bernoulli's equation becomes

$$\tfrac{1}{2}\rho v_1^2 + \rho g y_1 = \rho g y_2$$

or

$$v_1 = \sqrt{2g(y_2 - y_1)}. \tag{13–9}$$

FIGURE 13–25 Torricelli's theorem: $v_1 = \sqrt{2g(y_2 - y_1)}$.

Torricelli's theorem

This result is called **Torricelli's theorem**. Although it is seen to be a special case of Bernoulli's equation, it was discovered a century before Bernoulli by Evangelista Torricelli, a student of Galileo, hence its name. Equation 13–9 tells us that the liquid leaves the spigot with the same speed that a freely falling object would attain falling the same height. This should not be too surprising since the derivation of Bernoulli's equation relies on the conservation of energy.

Another special case of Bernoulli's equation arises when a fluid is flowing horizontally with no appreciable change in height; that is, $y_1 = y_2$. Then Eq. 13–8 becomes

$$P_1 + \tfrac{1}{2}\rho v_1^2 = P_2 + \tfrac{1}{2}\rho v_2^2, \tag{13–10}$$

which tells us quantitatively that where the speed is high the pressure is low, and vice versa. It explains many common phenomena, some of which are illustrated in

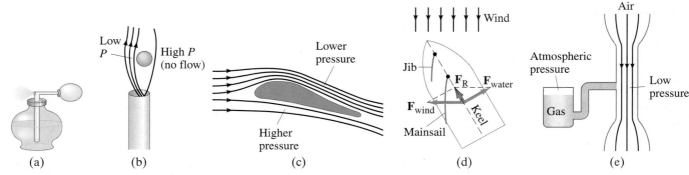

FIGURE 13-26 Examples of Bernoulli's principle: (a) atomizer, (b) Ping-Pong ball in jet of air, (c) airplane wing, (d) sailboat, (e) carburetor barrel.

Fig. 13–26. The pressure in the air blown at high speed across the top of the vertical tube of a perfume atomizer (Fig. 13–26a) is less than the normal air pressure acting on the surface of the liquid in the bowl. Thus perfume is pushed up the tube because of the reduced pressure at the top. A Ping-Pong ball can be made to float above a blowing jet of air (some vacuum cleaners can blow air), Fig. 13–26b; if the ball begins to leave the jet of air, the higher pressure in the still air outside the jet (Bernoulli's principle) pushes the ball back in.

Airplane wings and other airfoils moving rapidly relative to the air are designed to deflect the air so that, although streamline flow is largely maintained, the streamlines are crowded together above the wing, Fig. 13–26c. Just as the flow lines are crowded together in a pipe constriction where the velocity is high (see Fig. 13–21), so the crowded streamlines above the wing indicate that the air speed is greater above the wing than below it. Hence the air pressure above the wing is less than that below and there is thus a net upward force, called **dynamic lift**. Bernoulli's principle is only one aspect of the lift on a wing. Wings are usually tilted slightly upward so that air striking the bottom surface is deflected downward; the change in momentum of the rebounding air molecules results in an additional upward force on the wing. Turbulence also plays an important role.

A sailboat can move against the wind, Fig. 13–26d and the Bernoulli effect aids in this considerably if the sails are arranged so that the air velocity increases in the narrow constriction between the two sails. The normal atmospheric pressure behind the mainsail is larger than the reduced pressure in front of it (due to the fast moving air in the narrow slot between the sails), and this pushes the boat forward. When going against the wind, the mainsail is set at an angle, as shown in Fig. 13–26d, such that the net force on the sail (wind and Bernoulli) acts nearly perpendicular to the sail $\left(\mathbf{F}_{\text{wind}}\right)$. This would tend to make the boat move sideways if it weren't for the keel that extends vertically downward beneath the water—for the water exerts a force $\left(\mathbf{F}_{\text{water}}\right)$ on the keel nearly perpendicular to the keel. The resultant of these two forces $\left(\mathbf{F}_{\text{R}}\right)$ is almost directly forward as shown.

A **venturi tube** is essentially a pipe with a narrow constriction (the throat). One example of a venturi tube is the barrel of a carburetor in a car, Fig. 13–26e. The flowing air speeds up as it passes through this constriction, and so the pressure is lower. Because of the reduced pressure, gasoline under atmospheric pressure in the carburetor reservoir is forced into the air stream in the throat and mixes with the air before entering the cylinders.

FIGURE 13-27 Venturi meter.

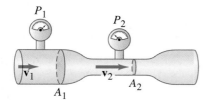

The venturi tube is also the basis of the *venturi meter*, which is used to measure the flow speed of fluids, Fig. 13–27. Venturi meters can be used to measure the flow velocities of gases and liquids, including blood velocity in arteries.

Why does smoke go up a chimney? It's partly because hot air rises (it's less dense and therefore buoyant). But Bernoulli's principle also plays a role. When wind blows across the top of a chimney, the pressure is less there than inside the house. Hence, air and smoke are pushed up the chimney. Even on an apparently still night there is usually enough ambient air flow at the top of a chimney to assist upward flow of smoke.

In medicine, one of many applications of Bernoulli's principle is to explain a TIA, a transient ischemic attack (meaning a temporary lack of blood supply to the brain), caused by the so-called "subclavian steal syndrome." A person suffering a TIA may experience symptoms such as dizziness, double vision, headache, and weakness of the limbs. A TIA can occur as follows. Blood normally flows up to the brain at the back of the head via the two vertebral arteries—one going up each side of the neck—which meet to form the basilar artery just below the brain, as shown in Fig. 13–28. The vertebral arteries issue from the subclavian arteries, as shown, before the latter pass to the arms. When an arm is exercised vigorously, blood flow increases to meet the needs of the arm's muscles. If the subclavian artery on one side of the body is partially blocked, however, say by arteriosclerosis, the blood velocity will have to be higher on that side to supply the needed blood. (Recall the equation of continuity: smaller area means larger velocity for the same flow rate, Eq. 13–7.) The increased blood velocity past the opening to the vertebral artery results in lower pressure (Bernoulli's principle). Thus, blood rising in the vertebral artery on the "good" side at normal pressure can be *diverted down* into the other vertebral artery because of the low pressure on that side (like the Venturi effect), instead of passing upward into the basilar artery and the brain. Hence the blood supply to the brain can be reduced due to "subclavian steal syndrome": the fast-moving blood in the subclavian artery "steals" the blood away from the brain. The resulting dizziness or weakness usually causes the person to stop the exertions, followed by a return to normal.

The sink where you brush your teeth contains a *trap*—a U-shaped pipe in the drain designed to retain water as a barrier against disagreeable odors (Fig. 13–29a). The air pressure on the two sides of the trap is the same, because of the *main vent*, which usually opens through the roof. However, the water in the trap could be pushed out of the trap, via the Bernoulli effect, if flush water from above passes at high speed—and low pressure—as shown in Fig. 13–29b. Building codes specify that drains of this type must have a separate vent (Fig. 13–29c) that maintains the atmospheric pressure on both sides of the trap, even when flush water passes through the system.

Bernoulli's equation ignores the effects of friction (viscosity) and the compressibility of the fluid. The energy that is transformed to internal (or potential) energy due to compression and to thermal energy by friction can be taken into account by adding terms to the right side of Eq. 13–8. These terms are difficult to calculate theoretically and are normally determined empirically. They do not significantly alter the explanations for the phenomena described above.

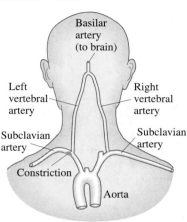

FIGURE 13–28 Rear of the head and shoulders showing arteries leading to the brain and to the arms. High blood velocity past the constriction in the left subclavian artery causes low pressure in the left vertebral artery, in which a reverse (downward) blood flow can then result: so-called "subclavian steal syndrome," resulting in a TIA (see text).

FIGURE 13–29
Making sure a trap works.

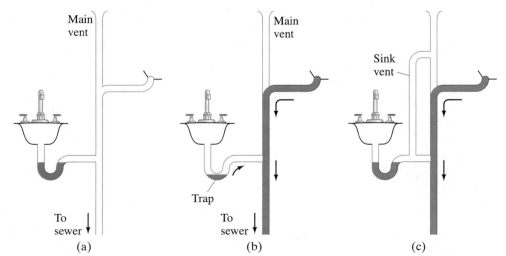

As already mentioned, real fluids have a certain amount of internal friction which is called **viscosity**. Viscosity exists in both liquids and gases, and is essentially a frictional force between adjacent layers of fluid as the layers move past one another. In liquids, viscosity is due to the cohesive forces between the molecules. In gases, it arises from collisions between the molecules.

Different fluids possess different amounts of viscosity: syrup is more viscous than water; grease is more viscous than engine oil; liquids in general are much more viscous than gases. The viscosity of different fluids can be expressed quantitatively by a *coefficient of viscosity*, η (the Greek lowercase letter eta), which is defined in the following way. A thin layer of fluid is placed between two flat plates. One plate is stationary and the other is made to move, Fig. 13–30. The fluid directly in contact with each plate is held to the surface by the adhesive force between the molecules of the liquid and those of the plate. Thus the upper surface of the fluid moves with the same speed v as the upper plate, whereas the fluid in contact with the stationary plate remains stationary. The stationary layer of fluid retards the flow of the layer just above it, which in turn retards the flow of the next layer, and so on. Thus the velocity varies continuously from 0 to v, as shown. The increase in velocity divided by the distance over which this change is made—equal to v/l—is called the *velocity gradient*. To move the upper plate requires a force, which you can verify by moving a flat plate across a puddle of syrup on a table. For a given fluid, it is found that the force required, F, is proportional to the area of fluid in contact with each plate, A, and to the speed, v, and is inversely proportional to the separation, l, of the plates: $F \propto vA/l$. For different fluids, the more viscous the fluid, the greater is the required force. Hence the proportionality constant for this equation is defined as the *coefficient of viscosity*, η:

$$F = \eta A \frac{v}{l}. \tag{13–11}$$

Solving for η, we find $\eta = Fl/vA$. The SI unit for η is $\text{N} \cdot \text{s/m}^2 = \text{Pa} \cdot \text{s}$ (pascal·second). In the cgs system, the unit is $\text{dyne} \cdot \text{s/cm}^2$ and this unit is called a *poise* (P). Viscosities are often given in centipoise $(1 \text{ cP} = 10^{-2} \text{ P})$. Table 13–3 lists the coefficient of viscosity for various fluids. The temperature is also specified, since it has a strong effect; the viscosity of liquids such as motor oil, for example, decreases rapidly as temperature increases.[†]

[†]The Society of Automotive Engineers assigns numbers to represent the viscosity of oils: 30-weight (SAE 30) is more viscous than 10-weight. Multigrade oils, such as 20–50, are designed to maintain viscosity as temperature increases; 20–50 means the oil is 20-wt when cool but is like a 50-wt pure oil when it is hot (engine running temperature).

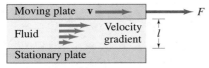

FIGURE 13–30
Determination of viscosity.

TABLE 13–3 Coefficient of Viscosity for Various Fluids

Fluid	Temperature (°C)	Coefficient of Viscosity, $\eta(\text{Pa} \cdot \text{s})$[†]
Water	0	1.8×10^{-3}
	20	1.0×10^{-3}
	100	0.3×10^{-3}
Whole blood	37	$\approx 4 \times 10^{-3}$
Blood plasma	37	$\approx 1.5 \times 10^{-3}$
Ethyl alcohol	20	1.2×10^{-3}
Engine oil (SAE 10)	30	200×10^{-3}
Glycerine	20	1500×10^{-3}
Air	20	0.018×10^{-3}
Hydrogen	0	0.009×10^{-3}
Water vapor	100	0.013×10^{-3}

[†]$1 \text{ Pa} \cdot \text{s} = 10 \text{ P} = 1000 \text{ cP}$

*13–11 Flow in Tubes: Poiseuille's Equation

If a fluid had no viscosity, it could flow through a level tube or pipe without a force being applied. Because of viscosity, a pressure difference between the ends of a tube is necessary for the steady flow of any real fluid, be it water or oil in a pipe, or blood in the circulatory system of a human, even when the tube is level.

The rate of flow of a fluid in a round tube depends on the viscosity of the fluid, the pressure difference, and the dimensions of the tube. The French scientist J. L. Poiseuille (1799–1869), who was interested in the physics of blood circulation (and after whom the "poise" is named), determined how the variables affect the flow rate of an incompressible fluid undergoing laminar flow in a cylindrical tube. His result, known as *Poiseuille's equation*, is:

$$Q = \frac{\pi R^4 (P_1 - P_2)}{8 \eta L},$$ (13–12)

Poiseuille's equation for flow rate in a tube

where R is the inside radius of the tube, L is its length, $P_1 - P_2$ is the pressure difference between the ends, η is the coefficient of viscosity, and Q is the volume rate of flow (volume of fluid flowing past a given point per unit time which in SI has units of m^3/s). Equation 13–12 applies only to laminar flow.

Poiseuille's equation tells us that the flow rate Q is directly proportional to the "pressure gradient," $(P_1 - P_2)/L$, and it is inversely proportional to the viscosity of the fluid. This is just what we might expect. It may be surprising, however, that Q also depends on the *fourth* power of the tube's radius. This means that for the same pressure gradient, if the tube radius is halved, the flow rate is decreased by a factor of 16! Thus the rate of flow, or alternately the pressure required to maintain a given flow rate, is greatly affected by only a small change in tube radius. Applying this to blood flow—although only approximately because of the presence of corpuscles and turbulence—we can see how reduction of artery radius by buildup of cholesterol and/or arteriosclerosis requires the heart to work much harder to maintain proper flow rates.

*13–12 Surface Tension and Capillarity

Up to now in this chapter, we have mainly been studying what happens to fluids as a whole. But the *surface* of a liquid at rest also behaves in an interesting way. A number of common observations suggest that the surface of a liquid acts like a stretched membrane under tension. For example, a drop of water on the end of a dripping faucet, or hanging from a thin branch in the early morning dew (Fig. 13–31), forms into a nearly spherical shape as if it were a tiny balloon filled with water. A steel needle can be made to float on the surface of water even though it is denser than the water. The surface of a liquid acts as if it is under tension, and this tension, acting parallel to the surface, arises from the attractive forces between the molecules. This effect is called *surface tension*. More specifically, a quantity called the *surface tension*, γ (the Greek letter gamma), is defined as the force F per unit length L that acts across any line in a surface, tending to pull the surface closed:

$$\gamma = \frac{F}{L}.$$ (13–13)

To understand this, consider the U-shaped apparatus shown in Fig. 13–32, which encloses a thin film of liquid. Because of surface tension, a force F is required to pull the movable wire and thus increase the surface area of the liquid. The liquid contained by the wire apparatus is a thin film having both a top and a bottom surface. Hence the length of the surface being increased is $2l$, and the surface tension is $\gamma = F/2l$. A delicate apparatus of this type can be used to measure the surface tension of various liquids. The surface tension of water is 0.072 N/m at 20°C. Table 13–4 gives the values for other liquids. Note that temperature has a considerable effect on the surface tension.

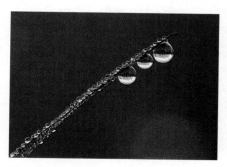

FIGURE 13–31 Spherical water droplets, dew on a grass blade.

FIGURE 13–32 U-shaped wire apparatus holding a film of liquid to measure surface tension ($\gamma = F/2l$).

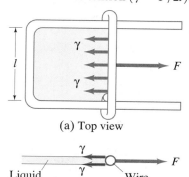

(a) Top view

(b) Edge view (magnified)

TABLE 13–4 Surface Tension of Some Substances

Substance	Surface Tension (N/m)
Mercury (20°C)	0.44
Blood, whole (37°C)	0.058
Blood, plasma (37°C)	0.073
Alcohol, ethyl (20°C)	0.023
Water (0°C)	0.076
(20°C)	0.072
(100°C)	0.059
Benzene (20°C)	0.029
Soap solution (20°C)	≈ 0.025
Oxygen (−193°C)	0.016

FIGURE 13–33 A water strider.

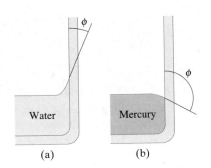

FIGURE 13–35 Water (a) "wets" the surface of glass, whereas (b) mercury does not "wet" the glass.

FIGURE 13–36 Capillarity.

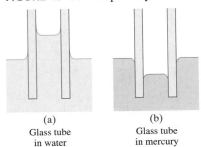

(a) Glass tube in water   (b) Glass tube in mercury

Because of surface tension, insects (Fig. 13–33) can walk on water; and objects more dense than water, such as a steel needle, can actually float on the surface. Figure 13–34a shows how the surface tension can support the weight w of an object. Actually, the object sinks slightly into the fluid, so w is the "effective weight" of that object—its true weight less the buoyant force. If the object is spherical in shape, the surface tension acts at all points around a horizontal circle of approximately radius r (Fig. 13–34a). Only the vertical component, $\gamma \cos \theta$, acts to balance w. We set the length L equal to the circumference of the circle, $L \approx 2\pi r$, so the net upward force due to surface tension is $F \approx (\gamma \cos \theta)L \approx 2\pi r \gamma \cos \theta$.

EXAMPLE 13–15 ESTIMATE Insect walks on water. The base of an insect's leg is approximately spherical in shape, with a radius of about 2.0×10^{-5} m. The 0.0030-g mass of the insect is supported equally by the six legs. Estimate the angle θ (see Fig. 13–34) for an insect on the surface of water. Assume the water temperature is 20°C.

SOLUTION Since the insect is in equilibrium, the upward surface tension force is equal to the effective pull of gravity downward on each leg:

$$2\pi r \gamma \cos \theta \approx w,$$

where w is one-sixth the weight of the insect (since it has six legs). Then

$$(6.28)(2.0 \times 10^{-5} \text{ m})(0.072 \text{ N/m}) \cos \theta \approx \tfrac{1}{6}(3.0 \times 10^{-6} \text{ kg})(9.8 \text{ m/s}^2)$$

$$\cos \theta \approx \frac{0.49}{0.90} = 0.54.$$

So $\theta \approx 57°$. Notice that if $\cos \theta$ were greater than 1, this would indicate that the surface tension would not be great enough to support the weight.

FIGURE 13–34 Surface tension acting on (a) a sphere, and (b) an insect leg. Example 13–15.

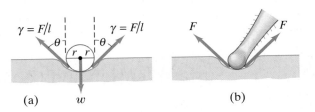

(a)   (b)

Soaps and detergents have the effect of lowering the surface tension of water. This is desirable for washing and cleaning since the high surface tension of pure water prevents it from penetrating easily between the fibers of material and into tiny crevices. Substances that reduce the surface tension of a liquid are called *surfactants*.

Surface tension plays a role in another interesting phenomenon, capillarity. It is a common observation that water in a glass container rises up slightly where it touches the glass, Fig. 13–35a. The water is said to "wet" the glass. Mercury, on the other hand, is depressed when it touches the glass, Fig. 13–35b; the mercury does not wet the glass. Whether or not a liquid wets a solid surface is determined by the relative strength of the cohesive forces between the molecules of the liquid compared to the adhesive forces between the molecules of the liquid and those of the container. (*Cohesion* refers to the force between molecules of the same type and *adhesion* to the force between molecules of different types.) Water wets glass because the water molecules are more strongly attracted to the glass molecules than they are to other water molecules. The opposite is true for mercury: the cohesive forces are stronger than the adhesive forces.

In tubes having very small diameters, liquids are observed to rise or fall relative to the level of the surrounding liquid. This phenomenon is called **capillarity**, and such thin tubes are called **capillaries**. Whether the liquid rises or falls (Fig. 13–36) depends on the relative strengths of the adhesive and cohesive forces. Thus water rises in a glass tube whereas mercury falls. The actual amount of rise (or fall) depends on the surface tension—which is what keeps the liquid surface from breaking apart.

*13–13 | Pumps, and the Heart

We conclude this chapter with a brief discussion of pumps of various types, including the heart. Pumps can be classified into categories according to their function. A *vacuum pump* is designed to reduce the pressure (usually of air) in a given vessel. A *force pump*, on the other hand, is a pump that is intended to increase the pressure—for example, to lift a liquid (such as water from a well) or to push a fluid through a pipe. Figure 13–37 illustrates the principle behind a simple reciprocating pump. It could be a vacuum pump, in which case the intake is connected to the vessel to be evacuated. A similar mechanism is used in some force pumps, and in this case the fluid is forced under increased pressure through the outlet. Other kinds of pumps are illustrated in Fig. 13–38. The centrifugal pump, or any force pump, can be used as a *circulating pump*—that is, to circulate a fluid around a closed path, such as the cooling water or lubricating oil in an automobile.

The heart of a human (and of other animals as well) is essentially a circulating pump. The action of a human heart is shown in Fig. 13–39. There are actually two separate paths for blood flow. The longer path takes blood to the parts of the body, via the arteries, bringing oxygen to body tissues and picking up carbon dioxide which it carries back to the heart via veins. This blood is then pumped into the lungs (the second path), where the carbon dioxide is released and oxygen is taken up. The oxygen-laden blood is returned to the heart, where it is again pumped to the tissues of the body.

➡ PHYSICS APPLIED

Pumps and the heart

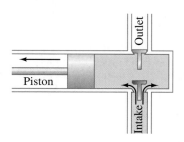

FIGURE 13–37 An example of one kind of pump. As the diagram indicates, the intake valve opens and air (or fluid that is being pumped) fills the empty space when the piston moves to the left. When the piston moves to the right (not shown), the outlet valve opens and fluid is forced out.

FIGURE 13–38 (a) Centrifugal pump: the rotating blades force fluid through the outlet pipe; this kind of pump is used in vacuum cleaners and as a water pump in automobiles. (b) Rotary oil-seal pump, used to obtain vacuums as low as 10^{-4} mm-Hg: gas (usually air) from the vessel to be evacuated diffuses into the space G via the intake pipe I; the rotating off-center cylinder C traps the gas in G and pushes it out the exhaust valve E, in the meantime allowing more gas to diffuse into G for the next cycle. The sliding valve V is kept in contact with C by a spring S, and this prevents the exhaust gas from returning to G. (c) Diffusion pump, used to obtain vacuums as low as 10^{-8} mm-Hg: air molecules from the vessel to be evacuated diffuse into the jet, where the rapidly moving jet of oil sweeps the molecules away. A "forepump" is needed, which is a mechanical pump, such as the rotary type (b), and acts as a first stage in reducing the pressure.

FIGURE 13–39 (a) In the diastole phase, the heart relaxes between beats. Blood moves into the heart; both atria are filled rapidly. (b) When the atria contract, the systole, or pumping, phase begins. The contraction pushes the blood through the mitral and tricuspid valves into the ventricles. (c) The contraction of the ventricles forces the blood through the semilunar valves into the pulmonary artery which leads to the lungs, and to the aorta (the body's largest artery) which leads to the arteries serving all the body. (d) When the heart relaxes, the semilunar valves close; blood fills the atria, beginning the cycle again.

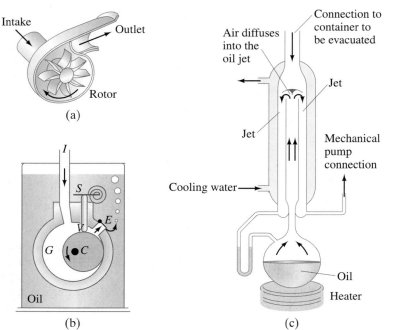

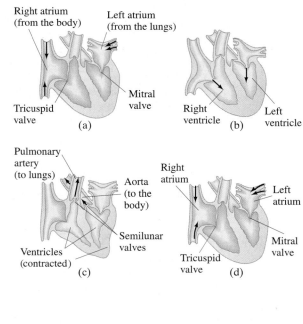

Summary

The three common phases of matter are **solid, liquid**, and **gas**. Liquids and gases are collectively called **fluids**, meaning they have the ability to flow. The **density** of a material is defined as its mass per unit volume. **Specific gravity** is the ratio of the density of the material to the density of water (at 4°C).

Pressure is defined as force per unit area. The pressure at a depth h in a liquid is given by

$$P = \rho g h,$$

where ρ is the density of the liquid and g is the acceleration due to gravity. If the density of a fluid is not uniform, the pressure P varies with height y as

$$\frac{dP}{dy} = -\rho g.$$

Pascal's principle says that an external pressure applied to a confined fluid is transmitted throughout the fluid.

Pressure is measured using a manometer or other type of gauge. A **barometer** is used to measure atmospheric pressure. Standard atmospheric pressure (average at sea level) is $1.013 \times 10^5 \, \text{N/m}^2$. **Gauge pressure** is the total pressure less atmospheric pressure.

Archimedes' principle states that an object submerged wholly or partially in a fluid is buoyed up by a force equal to the weight of fluid it displaces.

Fluid flow can be characterized either as **streamline** (sometimes called **laminar**), in which the layers of fluid move smoothly and regularly along paths called streamlines, or as **turbulent**, in which case the flow is not smooth and regular but is characterized by irregularly shaped whirlpools.

Fluid flow rate is the mass or volume of fluid that passes a given point per unit time. The **equation of continuity** states that for an incompressible fluid flowing in an enclosed tube, the product of the velocity of flow and the cross-sectional area of the tube remains constant:

$$Av = \text{constant}.$$

Bernoulli's principle tells us that where the velocity of a fluid is high, the pressure in it is low, and where the velocity is low, the pressure is high. **Bernoulli's equation** for steady laminar flow of an incompressible and nonviscous fluid is

$$P_1 + \tfrac{1}{2}\rho v_1^2 + \rho g y_1 = P_2 + \tfrac{1}{2}\rho v_2^2 + \rho g y_2.$$

for two points along the flow.

Viscosity refers to friction within a fluid that prevents the fluid from flowing freely and is essentially a frictional force between adjacent layers of fluid as they move past one another.

Questions

1. If one material has a higher density than another, does this mean the molecules of the first must be heavier than those of the second? Explain.

2. Airplane travelers often note that their cosmetics bottles and other containers have leaked after a trip. What might cause this?

3. The three containers in Fig. 13–40 are filled with water to the same height and have the same surface area at the base; hence the water pressure, and the total force on the base of each, is the same. Yet the total weight of water is different for each. Explain this "hydrostatic paradox."

FIGURE 13–40   Question 3.

4. Consider what happens when you push both a pin and the blunt end of a pen against your skin with the same force. Decide what determines whether your skin suffers a cut—the net force applied to it or the pressure.

5. It is often said that water seeks its own level. Explain.

6. A small amount of water is boiled in a one-gallon gasoline can. The can is removed from the heat and the lid put on. Shortly thereafter the can collapses. Explain.

7. Explain how the tube in Fig. 13–41, known as a **siphon**, can transfer liquid from one container to a lower one even though the liquid must flow uphill for part of its journey. (Note that the tube must be filled with liquid to start with.)

FIGURE 13–41
A siphon. Question 7.

8. An ice cube floats in a glass of water filled to the brim. What can you say about the density of ice? As the ice melts, will the glass overflow?

9. Will an ice cube float in a glass of alcohol? Why or why not?

10. A barge filled high with sand approaches a low bridge over the river and cannot quite pass under it. Should sand be added to, or removed from, the barge?

11. Will an empty balloon have precisely the same apparent weight on a scale as one that is filled with air? Explain.

12. Does the buoyant force on a diving bell deep beneath the ocean have precisely the same value as when the bell is just beneath the surface? Explain.

13. A small wooden boat floats in a swimming pool, and the level of the water at the edge of the pool is marked. Consider the following situations and determine whether the level of the water will rise, fall, or stay the same. (*a*) The boat is removed from the water. (*b*) The boat in the water holds an iron anchor which is removed from the boat and placed on the shore. (*c*) The iron anchor is removed from the boat and dropped in the pool.

14. Explain why helium weather balloons, which are used to measure atmospheric conditions at high altitude, are normally released while filled to only 10%–20% of their maximum volume.

15. Why do you float more easily in salt water than in fresh?

16. Roofs of houses are sometimes "blown" off (or are they pushed off?) during a tornado or hurricane. Explain, using Bernoulli's principle.

17. If you dangle two pieces of paper vertically, a few inches apart (Fig. 13–42), and blow between them, how do you think the papers will move? Try it and see. Explain.

FIGURE 13–42 Question 17.

18. Why does the canvas top of a convertible bulge out when the car is traveling at high speed?

19. Children are told to avoid standing too close to a rapidly moving train because they might get sucked under it. Is this possible? Explain.

20. Why does a sailboat need a keel? In a small sailboat, the keel (a vertical "board" that extends below the boat into the water) is removed when the boat is anchored. Why?

21. A tall Styrofoam cup is filled with water. Two holes are punched in the cup near the bottom, and water begins rushing out. If the cup is dropped so it falls freely, will the water continue to flow from the holes? Explain.

22. With a little effort, you can blow across a dime on a table and make it land in a cup as shown in Fig. 13–43 without touching either cup or dime. Explain (and try it).

FIGURE 13–43 Question 22.

23. Why do airplanes normally take off into the wind?

24. Why does the stream of water from a faucet become narrower as it falls (Fig. 13–44)?

FIGURE 13–44 Water coming from a faucet. Question 24 and Problem 93.

25. A baseball pitcher puts spin on the ball when throwing a curve. Use Bernoulli's principle to explain in detail why the ball curves. Explain why a spinning ball with a very smooth surface curves in the opposite direction to one with a rough surface (such as a baseball or tennis ball). [This is a challenging question. See "The Physics of Baseball," *Physics Today*, May 1995, p. 29.]

26. Two ships moving in parallel paths close to one another risk colliding. Why?

Problems

Section 13–1

1. (I) The approximate volume of the granite monolith known as El Capitan in Yosemite National Park (Fig. 13–45) is about $10^8 \, \text{m}^3$. What is its approximate mass?

FIGURE 13–45
Problem 1.

2. (I) What is the approximate mass of air in a living room $4.8 \, \text{m} \times 3.8 \, \text{m} \times 2.8 \, \text{m}$?

3. (I) If you tried to nonchalantly smuggle gold bricks by filling your backpack, whose dimensions are $60 \, \text{cm} \times 25 \, \text{cm} \times 15 \, \text{cm}$, what would its mass be?

4. (I) Estimate your volume. [*Hint*: Because you can swim on or just under the surface of the water in a swimming pool, you have a pretty good idea of your density.]

5. (II) A bottle has a mass of 35.00 g when empty and 98.44 g when filled with water. When filled with another fluid, the mass is 88.78 g. What is the specific gravity of this other fluid?

6. (II) If 5.0 L of antifreeze solution (specific gravity = 0.80) is added to 4.0 L of water to make a 9.0-L mixture what is the specific gravity of the mixture?

Sections 13–2 to 13–5

7. (I) Estimate the pressure exerted on a floor by (*a*) a pointed loudspeaker leg (60 kg on four legs) of area = $0.05 \, \text{cm}^2$, and compare it (*b*) to the pressure exerted by a 1500-kg elephant standing on one foot $(\text{area} = 800 \, \text{cm}^2)$.

8. (I) (*a*) Calculate the total force of the atmosphere acting on the top of a table that measures $1.6 \, \text{m} \times 2.9 \, \text{m}$. (*b*) What is the total force acting upward on the underside of the table?

9. (II) In a movie, Tarzan is shown evading his captors by hiding underwater for many minutes while breathing through a long thin reed. Assuming the maximum pressure difference his lungs can manage and still breathe is −80 mm-Hg, calculate the deepest he could have been.

10. (II) The gauge pressure in each of the four tires of an automobile is 240 kPa. If each tire has a "footprint" of $200 \, \text{cm}^2$, estimate the mass of the car.

11. (II) The maximum gauge pressure in a hydraulic lift is 17.0 atm. What is the largest size vehicle (kg) it can lift if the diameter of the output line is 24.5 cm?

12. (II) How high would the level be in an alcohol barometer at normal atmospheric pressure?

13. (II) What is the total force and the absolute pressure on the bottom of a swimming pool 22.0 m by 8.5 m whose uniform depth is 2.0 m? What will be the pressure against the *side* of the pool near the bottom?

14. (II) How high would the atmosphere extend if it were of uniform density, equal to that at sea level throughout?

15. (II) Water and then oil (which don't mix) are poured into a U-shaped tube, open at both ends. They come to equilibrium as shown in Fig. 13–46. What is the density of the oil? [*Hint*: Pressures at points a and b are equal. Why?]

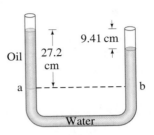

FIGURE 13–46 Problem 15.

16. (II) Determine the water gauge pressure at a house at the bottom of a hill fed by a full tank of water 5.0 m deep and connected to the house by a pipe that is 100 m long at an angle of 60° from the horizontal (Fig. 13–47). Neglect turbulence, and frictional and viscous effects. How high would the water shoot if it came vertically out of a broken pipe in front of the house?

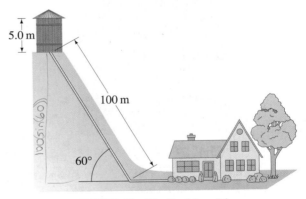

FIGURE 13–47 Problem 16.

17. (II) Estimate the air pressure on the summit of Mt. Everest (8850 m above sea level).

18. (II) Determine the minimum gauge pressure needed in the water pipe leading into a building if water is to come out of a faucet on the twelfth floor, 36.5 m above.

19. (II) An open-tube mercury manometer is used to measure the pressure in an oxygen tank. On a day when the atmospheric pressure is 1040 mbar, what is the absolute pressure (in Pa) in the tank if the height of the mercury in the open tube is (*a*) 28.0 cm higher, (*b*) 4.2 cm lower, than the mercury in the tube connected to the tank?

20. (II) A hydraulic press for compacting powdered samples has a large cylinder which is 10.0 cm in diameter, and a small cylinder with a diameter of 2.0 cm (Fig. 13–48). A lever is attached to the small cylinder as shown. The sample, which is placed on the large cylinder, has an area of 4.0 cm². What is the pressure on the sample if 300 N is applied to the lever?

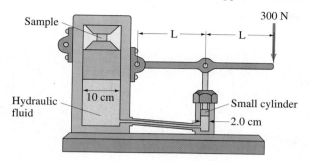

FIGURE 13–48 Problem 20.

21. (II) In working out his principle, Pascal showed dramatically how force can be multiplied with fluid pressure. He placed a long thin tube of 0.30-cm radius vertically into a 20-cm-radius wine barrel, Fig. 13–49. He found that when the barrel was filled with water and the tube filled to a height of 12 m, the barrel burst. Calculate (a) the mass of fluid in the tube, and (b) the net force on the lid of the barrel.

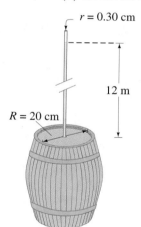

FIGURE 13–49
Problem 21 (Not to scale).

22. (III) A beaker of liquid accelerates from rest, on a horizontal surface, with acceleration a to the right. (a) Show that the surface of the liquid makes an angle $\theta = \tan^{-1} a/g$ with the horizontal. (b) Which edge of the water surface is higher? (c) How does the pressure vary with depth below the surface?

23. (III) Water stands at a height h behind a vertical dam of uniform width b. (a) Use integration to show that the total force of the water on the dam is $F = \frac{1}{2}\rho g h^2 b$. (b) Show that the torque about the base of the dam due to this force can be considered to act with a lever arm equal to h/3. (c) For a freestanding concrete dam of uniform thickness t and height h, what minimum thickness is needed to prevent overturning? Do you need to add in atmospheric pressure for this last part? Explain.

24. (III) Estimate the density of the water 6.0 km deep in the sea. (See Section 12–5 and Table 12–1.) By what fraction does it differ from the density at the surface?

25. (III) A cylindrical bucket of liquid (density ρ) is rotated about its symmetry axis, which is vertical. If the angular velocity is ω, show that the pressure at a distance r from the rotation axis is

$$P = P_0 + \frac{1}{2}\rho\omega^2 r^2,$$

where P_0 is the pressure at $r = 0$.

Section 13–6

26. (I) The hydrometer of Example 13–10 sinks to a depth of 22.9 cm when placed in a fermenting vat. What is the density of the brewing liquid?

27. (I) A geologist finds that a moon rock whose mass is 7.85 kg has an apparent mass of 6.18 kg when submerged in water. What is the density of the rock?

28. (I) What fraction of a piece of aluminum will be submerged when it floats in mercury?

29. (II) A spherically shaped balloon has a radius of 7.35 m, and is filled with helium. How large a cargo can it lift, assuming that the skin and structure of the balloon have a mass of 1000 kg? Neglect the buoyant force on the cargo volume itself.

30. (II) A 78-kg person has an apparent mass of 54 kg (because of buoyancy) when standing in water that comes up to the hips. Estimate the mass of each leg. Assume the body has SG = 1.00.

31. (II) What is the likely identity of a metal (see Table 13–1) if a sample has a mass of 63.5 g when measured in air and an apparent mass of 56.4 g when submerged in water?

32. (II) Calculate the true mass (in vacuum) of a piece of aluminum whose apparent mass is 2.0000 kg when weighed in air.

33. (II) An undersea research chamber for aquanauts is spherical with an external diameter of 6.0 m. The mass of the chamber, when occupied, is 75,000 kg. It is anchored to the sea bottom by a cable. What is (a) the buoyant force on the chamber, and (b) the tension in the cable?

34. (II) A scuba diver is diving off the shores of the Cayman Islands. The diver and her gear displace a volume of 65.0 L and have a total mass of 63.0 kg. (a) What is the buoyant force on the diver? (b) Will the diver sink or float?

35. (II) (a) Show that the buoyant force F_B on a partially submerged object such as a ship acts at the center of gravity of the fluid before it is displaced. This point is called the **center of buoyancy**. (b) For a ship to be stable, should its center of buoyancy be above, below, or at the same point as, its center of gravity? Explain (See Fig. 13–50.)

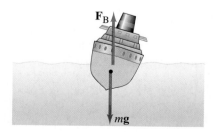

FIGURE 13–50 Problem 35.

36. (II) Archimedes' principle can be used not only to determine the specific gravity of a solid using a known liquid (Example 13–9). The reverse can be done as well. (a) As an example, a 3.40-kg aluminum ball has an apparent mass of 2.10 kg when submerged in a particular liquid: calculate the density of the liquid. (b) Derive a simple formula for determining the density of a liquid using this procedure.

37. (II) A 0.48-kg piece of wood floats in water but is found to sink in alcohol (specific gravity = 0.79) in which it has an apparent mass of 0.047 kg. What is the SG of the wood?

38. (II) The specific gravity of ice is 0.917, whereas that for seawater is 1.025. What fraction of an iceberg is above the surface of the water?

39. (III) A polar bear partially supports herself by pulling part of her body out of the water onto a rectangular slab of ice. The ice sinks down so that only half of what was once exposed now is exposed, and the bear has 70 percent of her volume (and weight) out of the water. Estimate the bear's mass, assuming that the total volume of the ice is 10 m^3, and the bear's specific gravity is 1.0.

40. (III) A 3.15-kg piece of wood (SG = 0.50) floats on water. What minimum mass of lead, hung from it by a string, will cause it to sink?

41. (III) If an object floats in water, its density can be determined by tying a sinker on it so that both the object and the sinker are submerged. Show that the specific gravity is given by $w/(w_1 - w_2)$, where w is the weight of the object alone in air, w_1 is the apparent weight when a sinker is tied to it and the sinker only is submerged, and w_2 is the apparent weight when both the object and the sinker are submerged.

Sections 13–7 to 13–9

42. (I) Using the data of Example 13–12, calculate the average speed of blood flow in the major arteries of the body which have a total cross-sectional area of about 2.0 cm^2.

43. (I) A 15-cm-radius air duct is used to replenish the air of a room 9.2 m × 5.0 m × 4.5 m every 12 min. How fast does the air flow in the duct?

44. (I) Show that Bernoulli's equation reduces to the hydrostatic variation of pressure with depth (Eq. 13–6) when there is no flow $(v_1 = v_2 = 0)$.

45. (I) How fast does water flow from a hole at the bottom of a very wide, 4.6-m-deep storage tank filled with water? Ignore viscosity.

46. (II) A $\frac{5}{8}$-inch (inside) diameter garden hose is used to fill a round swimming pool 6.1 m in diameter. How long will it take to fill the pool to a depth of 1.2 m if water issues from the hose at a speed of 0.33 m/s?

47. (II) What gauge pressure in the water mains is necessary if a firehose is to spray water to a height of 15 m?

48. (II) A 6.0-cm-diameter pipe gradually narrows to 4.0 cm. When water flows through this pipe at a certain rate, the gauge pressure in these two sections is 32 kPa and 24 kPa. What is the volume rate of flow?

49. (II) What is the volume rate of flow of water from a 1.85-cm-diameter faucet if the pressure head is 12.0 m?

50. (II) If wind blows at 25 m/s over your house, what is the net force on the roof if its area is 240 m^2?

51. (II) What is the lift (in newtons) due to Bernoulli's principle on a wing of area 86 m^2 if the air passes over the top and bottom surfaces at speeds of 340 m/s and 290 m/s, respectively?

52. (II) Estimate the air pressure inside a category 5 hurricane where the wind speed is 300 km/h. (See Fig. 13–51.)

FIGURE 13–51 Problem 52.

53. (II) Show that the power needed to drive a fluid through a pipe is equal to the volume rate of flow, Q, times the pressure difference, $P_1 - P_2$.

54. (II) Water at a gauge pressure of 3.8 atm at street level flows into an office building at a speed of 0.60 m/s through a pipe 5.0 cm in diameter. The pipes taper down to 2.6 cm in diameter by the top floor, 20 m above (Fig. 13–52). Calculate the flow velocity and the gauge pressure in such a pipe on the top floor. Assume no branch pipes and ignore viscosity.

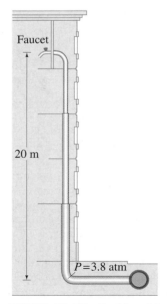

FIGURE 13–52 Problem 54.

55. (II) In Fig. 13–53, take into account the speed of the top surface of the tank and show that the speed of fluid leaving the opening at the bottom is

$$v_1 = \sqrt{2gh/(1 - A_1^2/A_2^2)},$$

where $h = y_2 - y_1$, and A_1 and A_2 are the areas of the opening and the top surface, respectively.

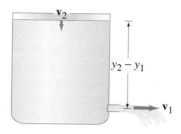

FIGURE 13–53 Problems 55, 56, 59, and 60.

56. (III) Suppose the top surface of the vessel in Fig. 13–53 is subjected to an external gauge pressure P_2. (a) Derive a formula for the speed, v_1, at which the liquid flows from the opening at the bottom into atmospheric pressure, P_A. Assume the velocity of the liquid surface, v_2, is approximately zero. (b) If $P_2 = 0.85$ atm and $y_2 - y_1 = 2.1$ m, determine v_1 for water.

57. (III) *Thrust of a rocket.* (a) Use Bernoulli's equation and the equation of continuity to show that the emission speed of the propelling gases of a rocket is

$$v = \sqrt{2(P - P_0)/\rho},$$

where ρ is the density of the gas, P is the pressure of the gas inside the rocket, and P_0 is atmospheric pressure just outside the exit orifice. Assume that the gas density stays approximately constant, and that the area of the exit orifice, A_0, is much smaller than the cross-sectional area, A, of the inside of the rocket (take it to be a large cylinder). Assume also that the gas speed is not so high that significant turbulence or non-steady flow sets in. (b) Show that the thrust force on the rocket due to the emitted gases is

$$F = 2A_0(P - P_0).$$

58. (III) (a) Show that the flow velocity measured by a venturi meter is given by the relation

$$v_1 = A_2 \sqrt{2(P_1 - P_2)/\rho(A_1^2 - A_2^2)}.$$

See Fig. 13–27. (b) A venturi tube is measuring the flow of water; it has a main diameter of 3.0 cm tapering down to a throat diameter of 1.0 cm; if the pressure difference is measured to be 18 mm-Hg, what is the velocity of the water?

59. (III) Suppose the opening in the tank of Fig. 13–53 is a height h_1 above the base and the liquid surface is a height h_2 above the base. The tank rests on level ground. (a) At what horizontal distance from the base of the tank will the fluid strike the ground? (b) At what other height, h_1', can a hole be placed so that the emerging liquid will have the same "range"?

60. (III) (a) In Fig. 13–53, show that the level of the liquid, $h = y_2 - y_1$, drops at a rate

$$\frac{dh}{dt} = -\sqrt{\frac{2ghA_1^2}{A_2^2 - A_1^2}},$$

where A_1 and A_2 are the areas of the opening and the top surface, and viscosity is ignored. (b) Determine h as a function of time by integrating. Let $h = h_0$ at $t = 0$. (c) How long will it take to empty a 9.4-cm-tall cylinder filled with 1.0 L of water if the opening is at the bottom and has a 0.50-cm diameter?

* Section 13–10

* 61. (II) A viscometer consists of two concentric cylinders, 10.20 cm and 10.60 cm in diameter. A particular liquid fills the space between them to a depth of 12.0 cm. The outer cylinder is fixed, and a torque of 0.024 m·N keeps the inner cylinder turning at a steady rotational speed of 62 rev/min. What is the viscosity of the liquid?

* Section 13–11

* 62. (I) A gardener feels it is taking him too long to water a garden with a $\frac{3}{8}$-in-diameter hose. By what factor will his time be cut if he uses a $\frac{5}{8}$-in-diameter hose? Assume nothing else is changed.

* 63. (I) Engine oil (assume SAE 10, Table 13–3) passes through a fine 1.80-mm-diameter tube in a prototype engine. The tube is 5.5 cm long. What pressure difference is needed to maintain a flow rate of 5.6 mL/min?

* 64. (II) What must be the pressure difference between the two ends of a 1.9-km section of pipe, 29 cm in diameter, if it is to transport oil $(\rho = 950 \text{ kg/m}^3, \eta = 0.20 \text{ Pa·s})$ at a rate of 450 cm³/s?

* 65. (II) What diameter must a 17.5-m-long air duct have if the ventilation and heating system is to replenish the air in a room 9.0 m × 12.0 × 4.0 m every 10 min? Assume the pump can exert a gauge pressure of 0.71×10^{-3} atm.

* 66. (II) Assuming a constant pressure gradient, by what factor does a blood vessel decrease in radius if the blood flow is reduced by 75 percent?

* 67. (II) Poiseuille's equation does not hold if the flow velocity is high enough that turbulence sets in. The onset of turbulence occurs when the so-called **Reynolds number**, Re, exceeds approximately 2000. Re is defined as

$$Re = \frac{2\bar{v}r\rho}{\eta},$$

where $\bar{v}$ is the average speed of the fluid, ρ is its density, η is its viscosity, and r is the radius of the tube in which the fluid is flowing. (a) Determine if blood flow through the aorta is laminar or turbulent, when the average speed of blood in the aorta $(r = 1.0 \text{ cm})$ during the resting part of the heart's cycle is about 30 cm/s. (b) During exercise, the blood-flow speed approximately doubles. Calculate the Reynolds number in this case and determine if the flow is laminar or turbulent.

* 68. (II) Water is to be pumped through a 10.0-cm-diameter pipe over a distance of 300 m. The far end of the pipe is 20 m above the pump and is at normal atmospheric pressure. What gauge pressure must the pump develop for there to be any flow at all?

* 69. (III) A patient is to be given a blood transfusion. The blood is to flow through a tube from a raised bottle to a needle inserted in the vein (Fig. 13–54). The inside diameter of the 4.0-cm-long needle is 0.40 mm and the required flow rate is 4.0 cm³ of blood per minute. How high should the bottle be placed above the needle? Obtain ρ and η from the Tables. Assume the blood pressure is 18 torr above atmospheric pressure.

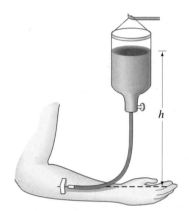

FIGURE 13–54 Problems 69 and 75.

* Section 13–12

* 70. (I) If the force F needed to move the wire in Fig. 13–32 is 5.1×10^{-3} N, calculate the surface tension γ of the enclosed fluid. Assume $l = 0.070$ m.

* 71. (I) Calculate the force needed to move the wire in Fig. 13–32 if it is immersed in a soapy solution and the wire is 18.2 cm long.

* 72. (II) If the base of an insect's leg has a radius of about 3.0×10^{-5} m and its mass is 0.016 g, would you expect the six-legged insect to remain on top of the water?

* 73. (II) The surface tension of a liquid can be determined by measuring the force F needed to just lift a circular platinum ring of radius r from the surface of the liquid. (a) Find a formula for γ in terms of F and r. (b) At 30°C, if $F = 8.40 \times 10^{-3}$ N and $r = 2.8$ cm, calculate γ for the tested liquid.

* 74. (III) Show that inside a soap bubble, there must be a pressure ΔP in excess of that outside equal to $\Delta P = 4\gamma/r$, where r is the radius of the bubble and γ is the surface tension. [*Hint*: Think of the bubble as two hemispheres in contact with each other; and remember that there are two surfaces to the bubble. Note that this result applies to any kind of membrane, where 2γ is the tension per unit length in that membrane.]

General Problems

75. Intravenous infusions are often made under gravity, as shown in Fig. 13–54. Assuming the fluid has a density of 1.00 g/cm³, at what height h should the bottle be placed so the liquid pressure is (a) 65 mm-Hg, (b) 550 mm-H₂O? (c) If the blood pressure is 18 mm-Hg above atmospheric pressure, how high should the bottle be placed so that the fluid just barely enters the vein?

76. A 2.4-N force is applied to the plunger of a hypodermic needle. If the diameter of the plunger is 1.3 cm and that of the needle 0.20 mm, (a) with what force does the fluid leave the needle? (b) What force on the plunger would be needed to push fluid into a vein where the gauge pressure is 18 mm-Hg? Answer for the instant just before the fluid starts too move.

77. A bicycle pump is used to inflate a tire. The initial tire pressure is 210 kPa (30 psi). At the end of the pumping process, the final pressure is 310 kPa (45 psi). If the diameter of the plunger in the cylinder of the pump is 3.0 cm, what is the range of the force that needs to be applied to the pump handle from beginning to end?

78. Estimate the pressure on the mountains underneath the Antarctic ice pack, which is typically 4 km thick.

79. What is the approximate difference in air pressure between the top and the bottom of the World Trade Center buildings in New York City? They are 410 m tall and are located at sea level. Express as a fraction of atmospheric pressure at sea level.

80. Giraffes are a wonder of cardiovascular engineering. Calculate the difference in pressure (in atmospheres) that the blood vessels in a giraffe's head have to accommodate as it lowers its head from a full upright position to ground level for a drink. The height of an average giraffe is about 6 meters.

81. When you drive up into the mountains, or descend rapidly from the mountains, your ears "pop," which means that the pressure behind the eardrum is being equalized to that outside. If this did not happen, what would be the approximate force on an eardrum of area 0.50 cm² if a change in altitude of 1000 m takes place?

82. One arm of a U-shaped tube (open at both ends) contains water and the other alcohol. If the two fluids meet exactly at the bottom of the U, and the alcohol is at a height of 18.0 cm, at what height will the water be?

83. A simple model (Fig. 13–55) considers a continent as a block (density = 2800 kg/m³) floating in the mantle rock around it (density = 3300 kg/m³). Assuming the continent is 35 km thick (the average thickness of the Earth's crust), estimate the height of the continent above the surrounding rock.

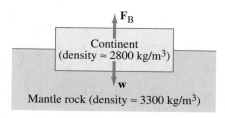

FIGURE 13–55
Problem 83.

84. The contraction of the left ventricle (chamber) of the heart pumps blood to the body. Assuming that the inner surface of the left ventricle has an area of 82 cm² and the maximum pressure in the blood is 120 mm-Hg, estimate the force exerted by the ventricle at maximum pressure.

85. Estimate the total mass of the Earth's atmosphere using the known value of atmospheric pressure at sea level.

86. Suppose a person can reduce the pressure in the lungs to −80 mm-Hg gauge pressure. How high can water then be sucked up a straw?

87. How high should the pressure head be if water is to come from a faucet at a speed of 7.2 m/s? Ignore viscosity.

88. A ship, carrying freshwater to a desert island in the Caribbean, has a horizontal cross-sectional area of 2650 m² at the waterline. When unloaded, the ship rises 8.50 m higher in the sea. How much water was delivered?

89. A raft is made of 10 logs lashed together. Each is 38 cm in diameter and has a length of 6.1 m. How many people can the raft hold before they start getting their feet wet, assuming the average person has a mass of 70 kg? Do *not* neglect the weight of the logs. Assume the specific gravity of wood is 0.60.

90. During each heartbeat, approximately 70 cm³ of blood is pushed from the heart at an average pressure of 105 mm-Hg. Calculate the power output of the heart, in watts, assuming 70 beats per minute.

91. A bucket of water is accelerated upward at 2.4 *g*. What is the buoyant force on a 3.0-kg granite rock (SG = 2.7) submerged in the water? Will the rock float? Why or why not?

92. The drinking fountain outside your classroom shoots water about 16 cm up in the air from a nozzle of diameter 0.60 cm. The pump at the base of the unit (1.1 m below the nozzle) pushes water into a 1.2-cm-diameter supply pipe that goes up to the nozzle. What gauge pressure does the pump have to provide? Ignore the viscosity; your answer will therefore be an underestimate.

93. The stream of water from a faucet decreases in diameter as it falls (Fig. 13–44). Derive an equation for the diameter of the stream as a function of the distance *y* below the faucet, given that the water has speed v_0 when it leaves the faucet, whose diameter is *D*.

94. Four lawn sprinkler heads are fed by a 1.9-cm-diameter pipe. The water comes out of the heads at an angle of 30° to the horizontal and covers a radius of 8.0 m. (*a*) What is the velocity of the water coming out of the sprinkler head? (Assume zero air resistance.) (*b*) If the output diameter of each head is 3.0 mm, how many liters of water do the four heads deliver per second? (*c*) How fast is the water flowing inside the 1.9-cm-diameter pipe?

95. You need to siphon water from a clogged sink. The sink has an area of 0.48 m² and is filled to a height of 4.0 cm. Your siphon tube rises 50 cm above the bottom of the sink and then descends 100 cm to a pail as shown in Fig. 13–56. The siphon tube has a diameter of 2.0 cm. (*a*) Assuming that the water enters the siphon tube with almost zero velocity, calculate its velocity when it enters the pail. (*b*) Estimate how long it will take to empty the sink.

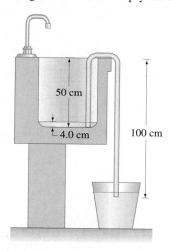

FIGURE 13–56
Problems 95 and 96.

96. Consider a siphon which transfers water (20°C) from one vessel to a second (lower) one, as in Fig. 13–56. Determine the rate of flow if the hose has a 1.2-cm-diameter and the difference in water levels of the two containers is 64 cm.

97. An airplane has a mass of 2.0×10^6 kg, and the air flows past the lower surface of the wings at 100 m/s. If the wings have a surface area of 1200 m², how fast must the air flow over the upper surface of the wing if the plane is to stay in the air? Consider only the Bernoulli effect.

98. A hydraulic lift is used to jack a 1000 kg car 10 cm off the floor. The diameter of the output piston is 15 cm and the input force is 250 N. (*a*) What is the area of the input piston? (*b*) What is the work done in lifting the car 10 cm? (*c*) If the travel for each stroke of the input piston is 12 cm, how high does the car move up for each stroke? (*d*) How many strokes are required to jack the car up 10 cm? (*e*) Show that energy is conserved.

The pendulum of a clock is an example of oscillatory motion. Many kinds of oscillatory motion are sinusoidal, or nearly so, and are referred to as being simple harmonic motion. Real systems generally have at least some friction, and the motion is damped. When an external sinusoidal force is exerted on a system able to oscillate, resonance occurs if the driving force is at or near the natural frequency of vibration.

CHAPTER 14

Oscillations

Many objects vibrate or oscillate—an object on the end of a spring, a tuning fork, the balance wheel of an old watch, a pendulum, a plastic ruler held firmly over the edge of a table and gently struck, the strings of a guitar or piano. Spiders detect prey by the vibrations of their webs, cars oscillate up and down when they hit a bump, buildings and bridges vibrate when heavy trucks pass or the wind is fierce. Indeed, because most solids are elastic (see Chapter 12), most material objects vibrate (at least briefly) when given an impulse. Electrical oscillations occur in radio and television sets. At the atomic level, atoms vibrate within a molecule, and the atoms of a solid vibrate about their relatively fixed positions. Because it is so common in everyday life and occurs in so many areas of physics, oscillatory (or vibrational) motion is of great importance. Vibrational motion is not really a "new" phenomenon because vibrations of mechanical systems are fully described on the basis of Newtonian mechanics.

14–1 Oscillations of a Spring

When a **vibration** or an **oscillation** repeats itself, back and forth, over the same path, the motion is **periodic**. The simplest form of periodic motion is represented by an object oscillating on the end of a coil spring. Because many other types of vibrational motion closely resemble this system, we will look at it in detail. We assume that the mass of the spring can be ignored, and that the spring is mounted horizontally (Fig. 14–1a), so that the object of mass m slides without friction on the horizontal surface. Any spring has a natural length at which the net force on the mass m is zero; the position of the mass at this point is called the **equilibrium position**. If the mass is moved either to the left, which compresses the spring, or to the right, which stretches it, the spring exerts a force on the mass that acts in the direction of returning the mass to the equilibrium position; hence it is called a "restoring force." The magnitude of the restoring force F is found to be directly proportional to the displacement x the spring has been stretched or compressed from the equilibrium position (Fig. 14–1b and c):

$$F = -kx. \qquad (14–1)$$

Note that the equilibrium position is at $x = 0$. Equation 14–1, which is often referred to as Hooke's law (see Sections 8–2 and 12–5), is accurate as long as the spring is not compressed to the point where the coils come close to touching, or stretched beyond the elastic region (see Fig. 12–19). The minus sign in Eq. 14–1 indicates that the restoring force is always in the direction opposite to the displacement x. For example, if we choose the positive direction to the right in Fig. 14–1, x is positive when the spring is stretched, but the direction of the restoring force is to the left (negative direction). If the spring is compressed, x is negative (to the left) but the force F acts toward the right (Fig. 14–1c).

The proportionality constant k in Eq. 14–1 is called the "spring constant." In order to stretch the spring a distance x, one has to exert an (external) force on the spring at least equal to $F = +kx$. The greater the value of k, the greater the force needed to stretch a spring a given distance.

Note that the force F in Eq. 14–1 is *not* a constant, but varies with position. Therefore the acceleration of the mass m is not constant, so we *cannot* use the equations for constant acceleration developed in Chapter 2.

Let us examine what happens when the spring is initially stretched a distance $x = A$, as shown in Fig. 14–2a, and then released. The spring exerts a force on the mass that pulls it toward the equilibrium position. But because the mass has been accelerated by the force, it passes the equilibrium position with considerable speed. Indeed, as the mass reaches the equilibrium position, the force on it decreases to zero, but its speed at this point is a maximum, Fig. 14–2b. As it moves farther to the left, the force on it acts to slow it down, and it stops momentarily at $x = -A$, Fig. 14–2c. It then begins moving back in the opposite direction, Fig. 14–2d, until it reaches the original starting point, $x = A$, Fig. 14–2e. It then repeats the motion, moving back and forth symmetrically between $x = A$ and $x = -A$.

To discuss vibrational motion, we need to define a few terms. The distance x of the mass from the equilibrium point at any moment is called the **displacement**. The maximum displacement—the greatest distance from the equilibrium point—is called the **amplitude**, A. One **cycle** refers to the complete to-and-fro motion from some initial point back to that same point, say from $x = A$ to $x = -A$ back to $x = A$. The **period**, T, is defined as the time required for one complete cycle. Finally, the **frequency**, f, is the number of complete cycles per second. Frequency is generally specified in hertz (Hz), where $1\,\text{Hz} = 1$ cycle per second (s^{-1}). It is easy to see, from their definitions, that frequency and period are inversely related:

$$f = \frac{1}{T} \quad \text{and} \quad T = \frac{1}{f}; \qquad (14–2)$$

for example, if the frequency is 5 cycles per second, then each cycle takes $\frac{1}{5}$ s.

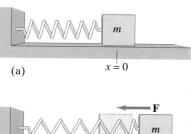

(a)

(b)

(c)

FIGURE 14–1 Mass vibrating at the end of a spring.

FIGURE 14–2 Force on, and velocity of, mass at different positions of its oscillation.

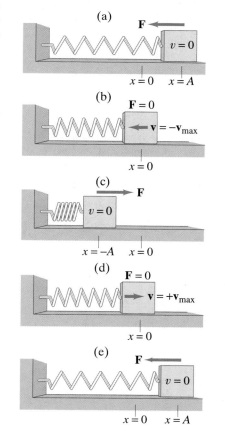

(a)

(b)

(c)

(d)

(e)

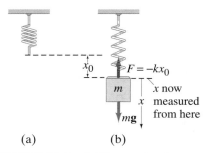

FIGURE 14–3 (a) Free spring, hung vertically. (b) Mass m attached to spring in new equilibrium position, which occurs when $\Sigma F = 0 = kx_0 - mg$.

FIGURE 14–4 Photo of a car's spring. (Also visible is the shock absorber—see Fig. 14–21.)

SHM

SHO

Equation of motion (SHM)

The oscillation of a spring hung vertically is essentially the same as that of a horizontal spring. Because of the force of gravity, the length of the vertical spring at equilibrium will be longer than when it is horizontal, as shown in Fig. 14–3. The spring is in equilibrium when $\Sigma F = 0 = kx_0 - mg$, so the spring stretches an extra amount $x_0 = mg/k$ to be in equilibrium. If x is measured from this new equilibrium position, Eq. 14–1 can be used directly with the same value of k.

EXAMPLE 14–1 **Car springs.** When a family of four people with a total mass of 200 kg step into their 1200-kg car, the car's springs compress 3.0 cm. (*a*) What is the spring constant of the car's springs (Fig. 14–4), assuming they act as a single spring? (*b*) How far will the car lower if loaded with 300 kg?

SOLUTION (*a*) The added force of $(200\ \text{kg})(9.8\ \text{m/s}^2) = 1960\ \text{N}$ causes the springs to compress $3.0 \times 10^{-2}\ \text{m}$. Therefore, by Eq. 14–1, the spring constant is

$$k = \frac{F}{x} = \frac{1960\ \text{N}}{3.0 \times 10^{-2}\ \text{m}} = 6.5 \times 10^4\ \text{N/m}.$$

(*b*) If the car is loaded with 300 kg,

$$x = \frac{F}{k} = \frac{(300\ \text{kg})(9.8\ \text{m/s}^2)}{(6.5 \times 10^4\ \text{N/m})} = 4.5 \times 10^{-2}\ \text{m},$$

or 4.5 cm. We could have obtained this answer without solving for k: since x is proportional to F, if 200 kg compresses the spring 3.0 cm, then 1.5 times the force will compress the spring 1.5 times as much, or 4.5 cm.

14–2 Simple Harmonic Motion

Any vibrating system for which the net restoring force is directly proportional to the negative of the displacement (as in Eq. 14–1, $F = -kx$) is said to exhibit **simple harmonic motion** (SHM). Such a system is often called a **simple harmonic oscillator** (SHO). We saw in Chapter 12 (Section 12–5) that most solid materials stretch or compress according to Eq. 14–1 as long as the displacement is not too great. Because of this, many natural vibrations are simple harmonic or close to it.

Let us now determine the position x as a function of time for a mass attached to the end of a simple spring with spring constant k. To do so, we make use of Newton's second law, $F = ma$. Since the acceleration $a = d^2x/dt^2$, we have

$$ma = \Sigma F$$
$$m\frac{d^2x}{dt^2} = -kx,$$

where m is the mass[†] which is oscillating. We rearrange this to obtain

$$\frac{d^2x}{dt^2} + \frac{k}{m}x = 0, \tag{14–3}$$

which is known as the **equation of motion** for the simple harmonic oscillator. Mathematically it is called a *differential equation*, since it involves derivatives. We want to determine what function of time, $x(t)$, satisfies this equation. We might guess the form of the solution by noting that if a pen were attached to a vibrating mass (Fig. 14–5) and a sheet of paper moved at a steady rate beneath it, the pen would trace the curve shown. The shape of this curve looks a lot like it might be

[†] In the case of a mass, m', on the end of a spring, the spring itself also oscillates and at least a part of its mass must be included. It can be shown—see the Problems—that approximately one-third the mass of the spring, m_s, must be included, so $m = m' + \frac{1}{3}m_s$ in our equation. Often m_s is small enough to be ignored.

sinusoidal (such as cosine or sine) as a function of time, and its height is the amplitude A. Let us then guess that the general solution to Eq. 14–3 can be written in a form such as

$$x = A \cos(\omega t + \phi),$$ **(14–4)**

General solution: position as functions of time

where we include the constant ϕ in the argument to be general.[†] Let us now put this (guessed at) solution into Eq. 14–3 and see if it really works. We need to differentiate the $x = x(t)$ twice:

$$\frac{dx}{dt} = \frac{d}{dt}[A \cos(\omega t + \phi)] = -\omega A \sin(\omega t + \phi)$$

$$\frac{d^2x}{dt^2} = -\omega^2 A \cos(\omega t + \phi).$$

We now put the latter into Eq. 14–3, along with Eq. 14–4 for x:

$$\frac{d^2x}{dt^2} + \frac{k}{m}x = 0$$

$$-\omega^2 A \cos(\omega t + \phi) + \frac{k}{m} A \cos(\omega t + \phi) = 0$$

or

$$\left(\frac{k}{m} - \omega^2\right) A \cos(\omega t + \phi) = 0.$$

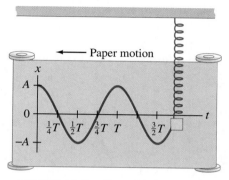

FIGURE 14–5 Sinusoidal nature of SHM as a function of time. In this case, $x = A \cos(2\pi t/T)$.

Our solution, Eq. 14–4, does indeed satisfy the equation of motion (Eq. 14–3) for any time t. But it does so only if $(k/m - \omega^2) = 0$; hence

$$\omega^2 = \frac{k}{m}.$$ **(14–5)**

Equation 14–4 is the general solution and it contains two arbitrary constants A and ϕ, which we should expect because the second derivative in Eq. 14–3 implies that two integrations are needed, each yielding a constant. They are "arbitrary" only in a calculus sense, in that they can be anything and still satisfy the differential equation, Eq. 14–3. In real physical situations, however, A and ϕ are determined by the **initial conditions**. Suppose, for example, the mass is started at its maximum displacement and is released from rest. This is, in fact, what is shown in Fig. 14–5, and for this case $x = A \cos \omega t$. Let us confirm it: we are given $v = 0$ at $t = 0$, where

Initial conditions

$$v = \frac{dx}{dt} = \frac{d}{dt}[A \cos(\omega t + \phi)] = -\omega A \sin(\omega t + \phi) = 0. \quad [\text{at } t = 0]$$

For v to be zero at $t = 0$, then $\sin(\omega t + \phi) = \sin(0 + \phi)$ is zero if $\phi = 0$ (ϕ could also be $\pi, 2\pi$, etc.), and when $\phi = 0$, then

$$x = A \cos \omega t,$$

as we expected. We see immediately that A is the amplitude of the motion, and it is determined initially by how far you pulled the mass m from equilibrium before releasing it.

[†] Another possible way to write the solution is the combination $x = a \cos \omega t + b \sin \omega t$, where a and b are constants. This is equivalent to Eq. 14–4 as can be seen using the trigonometric identity $\cos(A \pm B) = \cos A \cos B \mp \sin A \sin B$.

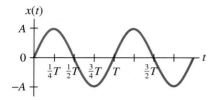

FIGURE 14–6 Special case of SHM where the mass m starts, at $t = 0$, at the equilibrium position $x = 0$ and has initial velocity toward positive values of x ($v > 0$ at $t = 0$).

FIGURE 14–7 A plot of $x = A\cos(\omega t + \phi)$ when $\phi < 0$.

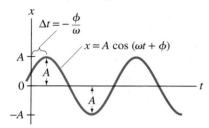

Position as function of time

Consider another interesting case: at $t = 0$, the mass m is at $x = 0$ and is struck, giving it an initial velocity toward increasing values of x. Then at $t = 0$, $x = 0$, so we can write $x = A\cos(\omega t + \phi) = A\cos\phi = 0$, which can happen only if $\phi = \pm\pi/2$ (or $\pm 90°$). Whether $\phi = +\pi/2$ or $-\pi/2$ depends on $v = dx/dt = -\omega A\sin(\omega t + \phi) = -\omega A\sin\phi$, which we are given as positive ($v > 0$ at $t = 0$); hence $\phi = -\pi/2[\sin(-90°) = -1]$. Thus our solution for this case is

$$x = A\cos\left(\omega t - \frac{\pi}{2}\right)$$
$$= A\sin\omega t,$$

where we used $\cos(\theta - \pi/2) = \sin\theta$. The solution in this case is a pure sine wave, Fig. 14–6, where A is still the amplitude.

Lots of other situations are possible, such as that shown in Fig. 14–7. The constant ϕ is called the **phase angle**, and it tells us how long after (or before) $t = 0$ the peak at $x = A$ is reached. Notice that the value of ϕ does not affect the shape of the $x(t)$ curve, but only affects the displacement at some arbitrary time, $t = 0$. Simple harmonic motion is thus always *sinusoidal*. Indeed, simple harmonic motion is *defined* as motion that is purely sinusoidal.

Since our oscillating mass repeats its motion after a time equal to its period T, it must be at the same position and moving in the same direction at $t = T$ as it was at $t = 0$. And since a sine or cosine function repeats itself after every 2π radians, then from Eq. 14–4, we must have

$$\omega T = 2\pi.$$

Hence

$$\omega = \frac{2\pi}{T} = 2\pi f,$$

where f is the frequency of the motion. We call ω the **angular frequency** (units are rad/s) to distinguish it from the frequency f (units are $s^{-1} = Hz$). Thus we can write Eq. 14–4 as

$$x = A\cos\left(\frac{2\pi t}{T} + \phi\right) \qquad \textbf{(14–6a)}$$

or

$$x = A\cos(2\pi f t + \phi), \qquad \textbf{(14–6b)}$$

where, because of Eq. 14–5

$$f = \frac{1}{2\pi}\sqrt{\frac{k}{m}}, \qquad \textbf{(14–7a)}$$

$$T = 2\pi\sqrt{\frac{m}{k}}. \qquad \textbf{(14–7b)}$$

f and T are independent of amplitude

Note that the *frequency and period do not depend on the amplitude*. Changing the amplitude of a simple harmonic oscillator does not affect its frequency. Equation 14–7a tells us that the greater the mass, the lower the frequency; and the stiffer the spring, the higher the frequency. This makes sense since a greater mass means more inertia and therefore slower response (or acceleration); and larger k means greater force and therefore quicker response. The frequency f (Eq. 14–7a) at which a SHO oscillates naturally is called its **natural frequency** (to distinguish it from a frequency at which it might be forced to oscillate by an outside force, as discussed in Section 14–8).

The simple harmonic oscillator is important in physics because whenever we have a net restoring force proportional to the displacement ($F = -kx$), which is at least a good approximation for a variety of systems, then the motion is simple harmonic—that is, sinusoidal.

EXAMPLE 14–2 Car springs again. What are the period and frequency of the car in Example 14–1 after hitting a bump? Assume the shock absorbers are poor, so the car really oscillates up and down.

➡ **PHYSICS APPLIED**

Car springs

SOLUTION From Eq. 14–7b,

$$T = 2\pi \sqrt{\frac{m}{k}} = 6.28 \sqrt{\frac{1400 \text{ kg}}{6.5 \times 10^4 \text{ N/m}}} = 0.92 \text{ s},$$

or slightly less than a second. The frequency $f = 1/T = 1.09$ Hz.

Let us continue our analysis of a simple harmonic oscillator. The velocity and acceleration of the oscillating mass can be obtained by differentiation of Eq. 14–4

$$v = \frac{dx}{dt} = -\omega A \sin(\omega t + \phi) \qquad (14\text{–}8)$$

$$a = \frac{d^2 x}{dt^2} = \frac{dv}{dt} = -\omega^2 A \cos(\omega t + \phi). \qquad (14\text{–}9)$$

The velocity and acceleration of a SHO also vary sinusoidally. In Fig. 14–8 we plot the displacement, velocity, and acceleration of a SHO as a function of time for the case when $\phi = 0$. As can be seen, the speed reaches its maximum

$$v_{\max} = \omega A = \sqrt{\frac{k}{m}} A$$

when the oscillating object is passing through its equilibrium point, $x = 0$. And the speed is zero at points of maximum displacement, $x = \pm A$. This is in accord with our discussion of Fig. 14–2. Similarly, the acceleration has its maximum value

$$a_{\max} = \omega^2 A = \frac{k}{m} A$$

which occurs where $x = \pm A$, and a is zero at $x = 0$, as we expect, since $ma = F = -kx$.

For the general case when $\phi \neq 0$, we can relate the constants A and ϕ to the initial values of x, v, and a by setting $t = 0$ in Eqs. 14–4, 14–8, and 14–9:

$$x_0 = x(0) = A \cos \phi$$

$$v_0 = v(0) = -\omega A \sin \phi = -v_{\max} \sin \phi$$

$$a_0 = a(0) = -\omega^2 A \cos \phi = -a_{\max} \cos \phi.$$

FIGURE 14–8 Displacement, x, velocity, dx/dt, and acceleration, d^2x/dt^2, of a simple harmonic oscillator when $\phi = 0$.

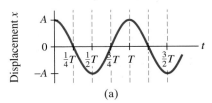

(a)

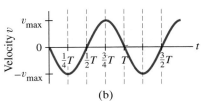

(b)

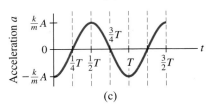

(c)

EXAMPLE 14–3 Loudspeaker cone. The cone of a loudspeaker vibrates in SHM at a frequency of 262 Hz ("middle C"). The amplitude at the center of the cone is $A = 1.5 \times 10^{-4}$ m, and at $t = 0$, $x = A$. (*a*) What is the equation describing the motion of the center of the cone? (*b*) What is its maximum velocity and maximum acceleration? (*c*) What is the position of the cone at $t = 1.00$ ms?

➡ **PHYSICS APPLIED**

Loudspeaker oscillations

SOLUTION (*a*) The amplitude $A = 1.5 \times 10^{-4}$ m and $\omega = 2\pi f = (6.28 \text{ rad})(262 \text{ s}^{-1}) = 1650$ rad/s. The motion begins ($t = 0$) with the cone at its maximum displacement ($x = A$ at $t = 0$), so we use the cosine function with $\phi = 0$:

$$x = A \cos \omega t = A \cos 2\pi f t = (1.5 \times 10^{-4} \text{ m}) \cos (1650 t).$$

(*b*) From Eq. 14–8,

$$v_{\max} = \omega A = 2\pi A f = 2\pi (1.5 \times 10^{-4} \text{ m})(262 \text{ s}^{-1}) = 0.25 \text{ m/s}.$$

From Eqs. 14–9 and 14–5:

$$a_{\max} = \omega^2 A = (2\pi f)^2 A$$

$$= 4\pi^2 (262 \text{ s}^{-1})^2 (1.5 \times 10^{-4} \text{ m}) = 410 \text{ m/s}^2,$$

which is more than 40 *g*'s.

(*c*) At $t = 1.00 \times 10^{-3}$ s,

$$x = (1.5 \times 10^{-4} \text{ m}) \cos[(1650 \text{ rad/s})(1.00 \times 10^{-3} \text{ s})]$$

$$= (1.5 \times 10^{-4} \text{ m}) \cos(1.65 \text{ rad}) = -1.2 \times 10^{-5} \text{ m}.$$

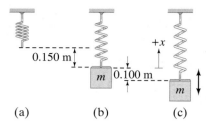

(a)   (b)   (c)

FIGURE 14–9 Example 14–4.

EXAMPLE 14–4 **Spring calculations.** In a testing device, a spring stretches 0.150 m when a 0.300-kg mass is hung from it (Fig. 14–9). The spring is then stretched an additional 0.100 m from this equilibrium point and released. Determine: (*a*) the values of the spring constant k and the angular frequency ω; (*b*) the amplitude of the oscillation A; (*c*) the maximum velocity, v_{max}; (*d*) the magnitude of the maximum acceleration of the mass; (*e*) the period T and frequency f; (*f*) the displacement x as a function of time; and (*g*) the velocity at $t = 0.150$ s.

SOLUTION (*a*) Since the spring stretches 0.150 m when 0.300 kg is hung from it, we find k from Eq. 14–1 to be

$$k = \frac{F}{x} = \frac{mg}{x} = \frac{(0.300 \text{ kg})(9.80 \text{ m/s}^2)}{0.150 \text{ m}} = 19.6 \text{ N/m}.$$

Also,

$$\omega = \sqrt{\frac{k}{m}} = \sqrt{\frac{19.6 \text{ N/m}}{0.300 \text{ kg}}} = 8.08 \text{ s}^{-1}.$$

(*b*) Since the spring is stretched 0.100 m from equilibrium (Fig. 14–9c) and is given no initial speed, $A = 0.100$ m.
(*c*) From Eq. 14–8, the maximum velocity is

$$v_{max} = \omega A = (8.08 \text{ s}^{-1})(0.100 \text{ m}) = 0.808 \text{ m/s}.$$

(*d*) Since $F = ma$, the maximum acceleration occurs where the force is greatest—that is, when $x = A = 0.100$ m. Thus

$$a_{max} = \frac{kA}{m} = \frac{(19.6 \text{ N/m})(0.100 \text{ m})}{0.300 \text{ kg}} = 6.53 \text{ m/s}^2.$$

(*e*) Equations 14–7a and b give

$$T = 2\pi \sqrt{\frac{m}{k}} = 6.28 \sqrt{\frac{0.300 \text{ kg}}{19.6 \text{ N/m}}} = 0.777 \text{ s}$$

$$f = \frac{1}{T} = 1.29 \text{ Hz}.$$

(*f*) The motion begins at a point of maximum displacement downward. If we take x positive upward, then at $t = 0$, $x = x_0 = -A = -0.100$ m. So we need a sinusoidal curve that has its maximum negative valve at $t = 0$; this is just a negative cosine:

$$x = -A \cos \omega t.$$

To write this in the form of Eq. 14–4 (no minus sign), recall that $\cos \theta = -\cos (\theta - \pi)$; then, putting in numbers,

$$x = -(0.100 \text{ m}) \cos 8.08t$$
$$= (0.100 \text{ m}) \cos (8.08t - \pi),$$

where t is in seconds and x is in meters. Note that the phase angle (Eq. 14–4) is $\phi = \pi$ or 180°.
(*g*) The velocity at any time t is (see part c)

$$v = \frac{dx}{dt} = A\omega \sin \omega t = (0.808 \text{ m/s}) \sin 8.08t.$$

At $t = 0.150$ s, $v = (0.808 \text{ m/s}) \sin(1.21 \text{ rad}) = 0.756 \text{ m/s}$, and is upward (+).

EXAMPLE 14–5 **Spring is started with a push.** Suppose the spring of Example 14–4 is stretched 0.100 m from equilibrium ($x_0 = -0.100$ m) but is given an upward shove of $v_0 = 0.400$ m/s. Determine (a) the phase angle ϕ, (b) the amplitude A, and (c) the displacement x as a function of time, $x(t)$.

SOLUTION (a) From Eq. 14–8, at $t = 0$, $v_0 = -\omega A \sin \phi$, and from Eq. 14–4, $x_0 = A \cos \phi$. Combining these, we get

$$\tan \phi = \frac{\sin \phi}{\cos \phi} = \frac{(v_0/-\omega A)}{(x_0/A)} = -\frac{v_0}{\omega x_0} = -\frac{0.400 \text{ m/s}}{(8.08 \text{ s}^{-1})(-0.100 \text{ m})} = 0.495.$$

A calculator gives the angle as 26.3°, but we note from this equation that both the sine and cosine are negative, so our angle is in the third quadrant. Hence

$$\phi = 26.3° + 180° = 206.3° = 3.60 \text{ rad}.$$

(b) Again using Eq. 14–4 at $t = 0$,

$$A = x_0/\cos \phi = (-0.100 \text{ m})/\cos (3.60 \text{ rad}) = 0.112 \text{ m}.$$

(c) $x = A \cos(\omega t + \phi) = 0.112 \cos(8.08t + 3.60)$.

14–3 | Energy in the Simple Harmonic Oscillator

When dealing with forces that are not constant, such as here with simple harmonic motion, it is often convenient and useful to use the energy approach.

For a simple harmonic oscillator, such as a mass m oscillating on the end of a massless spring, the restoring force is given by

$$F = -kx.$$

The potential energy function, as we have already seen in Chapter 8, is given by

$$U = -\int F \, dx = \tfrac{1}{2} kx^2,$$

where we set the constant of integration equal to zero so $U = 0$ at $x = 0$ (the equilibrium position).

The total mechanical energy is the sum of the kinetic and potential energies,

$$E = \tfrac{1}{2} mv^2 + \tfrac{1}{2} kx^2,$$

where v is the velocity of the mass m when it is a distance x from the equilibrium position. SHM can occur only if there is no friction, so the total mechanical energy E remains constant. As the mass oscillates back and forth, the energy continuously changes from potential energy to kinetic energy, and back again (Fig. 14–10). At the extreme points, $x = A$ and $x = -A$, all the energy is stored in the spring as potential energy (and is the same whether the spring is compressed or stretched to the full amplitude). At these extreme points, the mass stops momentarily as it changes direction, so $v = 0$ and:

$$E = \tfrac{1}{2} m(0)^2 + \tfrac{1}{2} kA^2 = \tfrac{1}{2} kA^2. \tag{14–10a}$$

Thus, the **total mechanical energy of a simple harmonic oscillator is proportional to the square of the amplitude.** At the equilibrium point, $x = 0$, all the energy is kinetic:

$$E = \tfrac{1}{2} mv^2 + \tfrac{1}{2} k(0)^2 = \tfrac{1}{2} mv_{\text{max}}^2, \tag{14–10b}$$

where v_{max} is the maximum velocity during the motion. At intermediate points the energy is part kinetic and part potential, and because energy is conserved

$$E = \tfrac{1}{2} mv^2 + \tfrac{1}{2} kx^2 = \tfrac{1}{2} kA^2 = \tfrac{1}{2} mv_{\text{max}}^2. \tag{14–10c}$$

We can confirm Eqs. 14–10a and b explicitly by inserting Eqs. 14–4 and 14–8 into this last relation:

$$E = \tfrac{1}{2} m\omega^2 A^2 \sin^2 (\omega t + \phi) + \tfrac{1}{2} kA^2 \cos^2 (\omega t + \phi).$$

Substituting with $\omega^2 = k/m$, or $kA^2 = m\omega^2 A^2 = mv_{\text{max}}^2$, and noting that $\sin^2 (\omega t + \phi) + \cos^2 (\omega t + \phi) = 1$, we obtain Eqs. 14–10a and b:

$$E = \tfrac{1}{2} kA^2 = \tfrac{1}{2} mv_{\text{max}}^2.$$

FIGURE 14–10 Energy changes from kinetic energy to potential energy and back again as the spring oscillates.

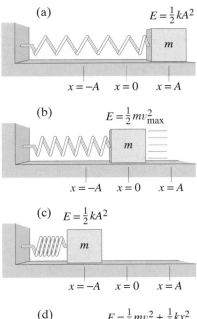

(a) $E = \tfrac{1}{2} kA^2$

$x = -A \quad x = 0 \quad x = A$

(b) $E = \tfrac{1}{2} mv_{\text{max}}^2$

$x = -A \quad x = 0 \quad x = A$

(c) $E = \tfrac{1}{2} kA^2$

$x = -A \quad x = 0 \quad x = A$

(d) $E = \tfrac{1}{2} mv^2 + \tfrac{1}{2} kx^2$

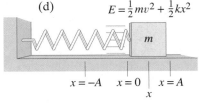

$x = -A \quad x = 0 \quad x = A$

x

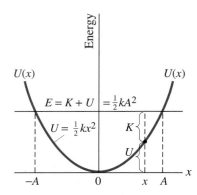

FIGURE 14–11 Graph of potential energy, $U = \frac{1}{2}kx^2$. $K + U = E = $ constant for any point x where $-A \leq x \leq A$. Values of K and U are indicated for an arbitrary point x.

We can now obtain an equation for the velocity v as a function of x by solving for v^2 in Eq. 14–10c:

$$v = \pm \sqrt{\frac{k}{m}(A^2 - x^2)} \qquad \textbf{(14–11a)}$$

or, since $v_{max} = A\sqrt{k/m}$,

$$v = \pm v_{max}\sqrt{1 - \frac{x^2}{A^2}}. \qquad \textbf{(14–11b)}$$

Again we see that v is a maximum at $x = 0$, and is zero at $x = \pm A$.

The potential energy, $U = \frac{1}{2}kx^2$, is plotted in Fig. 14–11. The upper horizontal line represents a particular value of the total energy $E = \frac{1}{2}kA^2$. The distance between the E line and the U curve represents the kinetic energy, K, and the motion[†] is restricted to x values between $-A$ and $+A$. These results are, of course, consistent with our full solution of the previous Section.

Energy conservation is a convenient way to obtain v, for example, if x is given (or vice versa), without having to deal with time t.

EXAMPLE 14–6 **Energy calculations.** For the simple harmonic oscillation of Example 14–4, determine (*a*) the total energy, (*b*) the kinetic and potential energies as a function of time, (*c*) the velocity when the mass is 0.050 m from equilibrium, (*d*) the kinetic and potential energies at half amplitude ($x = \pm A/2$).

SOLUTION (*a*) Since $k = 19.6\,\text{N/m}$ and $A = 0.100\,\text{m}$, the total energy E from Eq. 14–10a is

$$E = \tfrac{1}{2}kA^2 = \tfrac{1}{2}(19.6\,\text{N/m})(0.100\,\text{m})^2 = 9.80 \times 10^{-2}\,\text{J}.$$

(*b*) We have, from parts (*f*) and (*g*) of Example 14–4, $x = -(0.100\,\text{m})\cos 8.08t$ and $v = (0.808\,\text{m/s})\sin 8.08t$, so

$$U = \tfrac{1}{2}kx^2 = (9.80 \times 10^{-2}\,\text{J})\cos^2 8.08t$$
$$K = \tfrac{1}{2}mv^2 = (9.80 \times 10^{-2}\,\text{J})\sin^2 8.08t.$$

(*c*) We use Eq. 14–11b and find

$$v = v_{max}\sqrt{1 - x^2/A^2} = 0.70\,\text{m/s}.$$

(*d*) At $x = A/2 = 0.050\,\text{m}$, we have

$$U = \tfrac{1}{2}kx^2 = 2.5 \times 10^{-2}\,\text{J}$$
$$K = E - U = 7.3 \times 10^{-2}\,\text{J}.$$

CONCEPTUAL EXAMPLE 14–7 **Doubling the amplitude.** Suppose the spring in Fig. 14–10 is stretched twice as far (to $x = 2A$). What happens to (*a*) the energy of the system, (*b*) the maximum velocity, (*c*) the maximum acceleration?

RESPONSE (*a*) From Eq. 14–10a, the energy is related to the square of the amplitude, so stretching it twice as far quadruples the energy [You may protest: "I did work stretching the spring from $x = 0$ to $x = A$. Don't I do the same work stretching it from A to $2A$?" No. The force you have to exert for the second leg is more than for the first leg (because $F = -kx$), so the work done is more too.]
(*b*) From Eq. 14–10b, we can see that since the energy is quadrupled, the maximum velocity must be double what it was before.
(*c*) Since the force is twice as great when we stretch it twice as far, the acceleration here is also twice as great.

[†] See Section 8–9 for more discussion.

14–4 | Simple Harmonic Motion Related to Uniform Circular Motion

Simple harmonic motion has an interesting simple relationship to a particle rotating in a circle with uniform speed. Consider a mass m rotating in a circle of radius A with speed v_M on top of a table as shown in Fig. 14–12. As viewed from above, the motion is a circle. But a person who looks at the motion from the edge of the table, sees an oscillatory motion back and forth, and this corresponds precisely to SHM as we shall now see. What the person sees, and what we are interested in, is the projection of the circular motion onto the x axis, Fig. 14–12. To see that this motion is analogous to SHM, let us calculate the x component of the velocity v_M which is labeled v in Fig. 14–12. The two right triangles indicated in Fig. 14–12 are similar, so

$$\frac{v}{v_M} = \frac{\sqrt{A^2 - x^2}}{A}$$

or

$$v = v_M \sqrt{1 - \frac{x^2}{A^2}}.$$

This is exactly the equation for the speed of a mass oscillating with SHM, Eq. 14–11b, where $v_M = v_{max}$. Furthermore, we can see from Fig. 14–12 that if the angular displacement at $t = 0$ is ϕ, then after a time t the particle will have rotated through an angle $\theta = \omega t$, and so

$$x = A \cos(\theta + \phi) = A \cos(\omega t + \phi).$$

But what is ω here? The linear velocity v_M of our particle undergoing rotational motion is related to ω by $v_M = \omega A$ where A is the radius of the circle (see Eq. 10–4). To make one revolution requires a time T, so we also have $v_M = 2\pi A/T$ where $2\pi A$ is the circle's circumference. Hence

$$\omega = \frac{v_M}{A} = \frac{2\pi A/T}{A} = 2\pi/T = 2\pi f$$

where T is the time required for one rotation and f is the frequency. This corresponds precisely to the back-and-forth motion of a simple harmonic oscillator. Thus, the projection on the x axis of a particle rotating in a circle has the same motion as a mass undergoing SHM. Indeed, we can say that the projection of circular motion onto a straight line is SHM.

The projection of uniform circular motion onto the y axis is also simple harmonic. Thus uniform circular motion can be thought of as two simple harmonic motions operating at right angles.

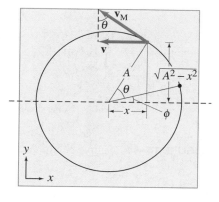

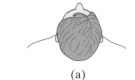

(a)

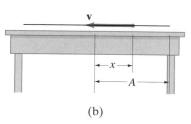

(b)

FIGURE 14–12 Analysis of simple harmonic motion as a side view (b) of circular motion (a).

14–5 | The Simple Pendulum

A **simple pendulum** consists of a small object (the pendulum bob) suspended from the end of a lightweight cord, Fig. 14–13. We assume that the cord doesn't stretch and that its mass can be ignored relative to that of the bob. The motion of a simple pendulum moving back and forth (Fig. 14–13) with negligible friction resembles simple harmonic motion: the pendulum oscillates along the arc of a circle with equal amplitude on either side of its equilibrium point (where it hangs vertically) and as it passes through the equilibrium point it has its maximum speed. But is it really undergoing SHM? That is, is the restoring force proportional to its displacement? Let us find out.

FIGURE 14–13 Strobe-light photo of an oscillating pendulum.

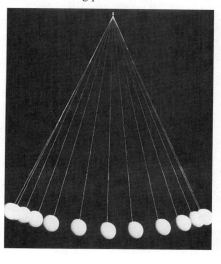

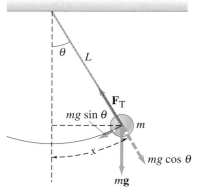

FIGURE 14–14 Simple pendulum.

The displacement of the pendulum along the arc is given by $x = L\theta$, where θ is the angle the cord makes with the vertical and L is the length of the cord, as shown in Fig. 14–14. If the restoring force is proportional to x or to θ, the motion will be simple harmonic. The restoring force is the component of the weight, mg, tangent to the arc:

$$F = -mg \sin\theta,$$

where the minus sign, as in Eq. 14–1, means the force is in the direction opposite to the angular displacement θ. Since F is proportional to the sine of θ and not to θ itself, the motion is *not* SHM. However, if θ is small, then $\sin\theta$ is very nearly equal to θ when the latter is specified in radians. This can be seen by looking at the series expansion† of $\sin\theta$ (or by looking at the trigonometry table inside the back cover), or by noting in Fig. 14–14 that the arc length x ($= L\theta$) is nearly the same length as the chord ($= L\sin\theta$) indicated by the straight dashed line, *if θ is small*. For angles less than 15°, the difference between θ (in radians) and $\sin\theta$ is less than 1 percent. Thus, to a very good approximation for small angles,

$$F = -mg \sin\theta \approx -mg\theta.$$

Using $x = L\theta$, we have

$$F \approx -\frac{mg}{L} x.$$

Thus, for small displacements, the motion is essentially simple harmonic, since this equation fits Hooke's law, $F = -kx$ where the effective force constant is $k = mg/L$. Thus we can write

$$\theta = \theta_{\max} \cos(\omega t + \phi)$$

where $\theta_{\max}$ is the maximum angular displacement and $\omega = 2\pi f = 2\pi/T$. To obtain ω we use Eq. 14–5, where for k we substitute mg/L: that is,‡ $\omega = \sqrt{(mg/L)/m}$, or

$$\omega = \sqrt{\frac{g}{L}}.$$ [θ small] **(14–12a)**

Then the frequency f is

$$f = \frac{\omega}{2\pi} = \frac{1}{2\pi}\sqrt{\frac{g}{L}},$$ [θ small] **(14–12b)**

and the period T is

$$T = \frac{1}{f} = 2\pi\sqrt{\frac{L}{g}}.$$ [θ small] **(14–12c)**

A surprising result is that the period does not depend on the mass of the pendulum bob! You may have noticed this if you pushed a small child and a large one on the same swing. The period *does* depend on the length L.

We saw in Section 14–2 that the period of an object undergoing SHM, including a simple pendulum, does not depend on the amplitude. Galileo is said to have first noted this fact while watching a swinging lamp in the cathedral at Pisa (Fig. 14–15). This discovery led to the pendulum clock, the first really precise time-piece, which became the standard for centuries.

Because a pendulum does not undergo *precisely* SHM, the period does depend slightly on the amplitude, the more so for large amplitudes. The accuracy

FIGURE 14–15 The swinging motion of this lamp, hanging by a very long cord from the ceiling of the cathedral at Pisa, is said to have been observed by Galileo and to have inspired him to the conclusion that the period of a pendulum does not depend on amplitude.

---

† $\sin\theta = \theta - \dfrac{\theta^3}{3!} + \dfrac{\theta^5}{5!} - \dfrac{\theta^7}{7!} + \cdots.$

‡ Be careful not to think that $\omega = d\theta/dt$ as in rotational motion. Here θ is the angle of the pendulum at any instant (Fig. 14–14), but we use ω now to represent *not* the rate this angle θ changes, but rather as a constant related to the period, $\omega = 2\pi f = \sqrt{g/L}$.

of a pendulum clock would be affected, after many swings, by the decrease in amplitude due to friction; but the mainspring in a pendulum clock (or the falling weight in a grandfather clock) supplies energy to compensate for the friction and to maintain the amplitude constant, so that the timing remains accurate.

EXAMPLE 14–8 **Measuring g.** For calibrating instruments that measure the acceleration of gravity, a geologist uses a simple pendulum whose length is 37.10 cm, and has a frequency of 0.8190 Hz at a particular location on the Earth. What is the acceleration of gravity at this location?

SOLUTION From Eq. 14–12b, we have

$$f = \frac{1}{2\pi}\sqrt{\frac{g}{L}}.$$

Solving for g, we obtain

$$g = (2\pi f)^2 L = (6.283 \times 0.8190\ \text{s}^{-1})^2(0.3710\ \text{m}) = 9.824\ \text{m/s}^2.$$

* 14–6 The Physical Pendulum and the Torsion Pendulum

Physical Pendulum

The term physical pendulum refers to any real extended body which oscillates back and forth, in contrast to the rather idealized simple pendulum where all the mass is assumed concentrated in the tiny pendulum bob. An example of a physical pendulum is a baseball bat suspended from the point O, as shown in Fig. 14–16. The force of gravity acts at the center of gravity (CG) of the body located a distance h from the pivot point O. The physical pendulum is best analyzed using the equations of rotational motion. The torque on a physical pendulum, calculated about point O, is

$$\tau = -mgh\sin\theta.$$

Newton's second law for rotational motion, Eq. 10–14, states that

$$\Sigma\tau = I\alpha = I\frac{d^2\theta}{dt^2},$$

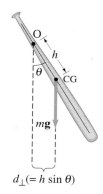

FIGURE 14–16 A physical pendulum suspended from point O.

where I is the moment of inertia of the body about the pivot point and $\alpha = d^2\theta/dt^2$ is the angular acceleration. Thus we have

$$I\frac{d^2\theta}{dt^2} = -mgh\sin\theta$$

or

$$\frac{d^2\theta}{dt^2} + \frac{mgh}{I}\sin\theta = 0,$$

where I is calculated about an axis through point O. For small angular amplitude, $\sin\theta \approx \theta$, so we have

$$\frac{d^2\theta}{dt^2} + \left(\frac{mgh}{I}\right)\theta = 0. \qquad \text{[small angular displacement]} \quad \textbf{(14–13)}$$

This is just the equation for SHM, Eq. 14–3, except that θ replaces x and mgh/I replaces k/m. Thus, for small angular displacements, a physical pendulum undergoes SHM, given by

$$\theta = \theta_{\max}\cos(\omega t + \phi),$$

where $\theta_{\max}$ is the maximum angular displacement and $\omega = 2\pi/T$. The period, T, is (see Eq. 14–7b, replacing m/k with I/mgh):

$$T = 2\pi\sqrt{\frac{I}{mgh}}. \qquad \text{[small angular displacement]} \quad \textbf{(14–14)}$$

Period of physical pendulum

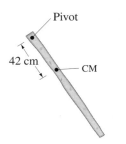

Pivot

42 cm — CM

FIGURE 14–17 Example 14–9.

FIGURE 14–18 A torsion pendulum. The disc oscillates in SHM between θ_{max} and $-\theta_{max}$.

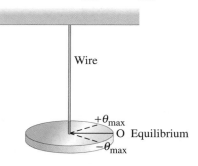

Wire

$+\theta_{max}$
O Equilibrium
$-\theta_{max}$

EXAMPLE 14–9 **Moment of inertia measurement.** An easy way to measure the moment of inertia of an object about any axis is to measure the period of oscillation about that axis. (*a*) Suppose a nonuniform 1.0-kg stick can be balanced at a point 42 cm from one end. If it is pivoted about that end (Fig. 14–17), it oscillates with a period of 3.0 s. What is its moment of inertia about this end? (*b*) What is its moment of inertia about an axis perpendicular to the stick through its center of mass?

SOLUTION (*a*) Given $T = 3.0$ s, and $h = 0.42$ m, we solve Eq. 14–14 for I:

$$I = mghT^2/4\pi^2 = 0.94 \text{ kg} \cdot \text{m}^2.$$

Since $I = \frac{1}{3}ML^2$ for a uniform stick of length L pivoted about one end (Fig. 10–21), do you think our stick is longer or shorter than 84 cm?
(*b*) We use the parallel-axis theorem (Section 10–8). The CM is where the stick balanced, 42 cm from the end, so from Eq. 10–17,

$$I_{CM} = I - Mh^2 = 0.94 \text{ kg} \cdot \text{m}^2 - (1.0 \text{ kg})(0.42 \text{ m})^2 = 0.76 \text{ kg} \cdot \text{m}^2.$$

Since an object does not oscillate about its CM, we can't measure I_{CM} directly, so the parallel-axis theorem provides a convenient method to determine I_{CM}.

Torsion Pendulum

Another type of oscillatory motion is a **torsion pendulum**, in which a disc (Fig. 14–18) or a bar (as in Cavendish's apparatus, Fig. 6–3) is suspended from a wire. The twisting (torsion) of the wire serves as the elastic force. The motion here will be SHM since the restoring torque is very closely proportional to the negative of the angular displacement,

$$\tau = -K\theta,$$

where K is a constant that depends of the properties of the system. Then

$$\omega = \sqrt{\frac{K}{I}}.$$

There is no small angle restriction here, as there is for the physical pendulum (where gravity acts), as long as the wire responds linearly in accordance with Hooke's law.

14–7 Damped Harmonic Motion

The amplitude of any real oscillating spring or swinging pendulum slowly decreases in time until the oscillations stop altogether. Figure 14–19 shows a typical graph of the displacement as a function of time. This is called **damped harmonic motion**. The damping[†] is generally due to the resistance of air and to internal friction within the oscillating system. The energy that is dissipated to thermal energy is reflected in a decreased amplitude of oscillation.

[†] To "damp" means to diminish, restrain, or extinguish, as to "dampen one's spirits."

FIGURE 14–19 Damped harmonic motion. The solid red curve represents a cosine times a decreasing exponential (the dashed curves).

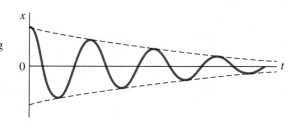

Since natural oscillating systems are damped in general, why do we even talk about (undamped) simple harmonic motion? The answer is that SHM is much easier to deal with mathematically. And if the damping is not large, the oscillations can be thought of as simple harmonic motion on which the damping is superposed, as represented by the dashed curves in Fig. 14–19. Although frictional damping does alter the frequency of vibration, the effect is usually small if the damping is small. Let us look at this in more detail.

The damping force depends on the speed of the oscillating object, and opposes the motion. In some simple cases the damping force can be approximated as being directly proportional to the speed:

$$F_{\text{damping}} = -bv,$$

where b is a constant.[†] For a mass oscillating on the end of a spring, the restoring force of the spring is $F = -kx$; so Newton's second law ($ma = \Sigma F$) becomes

$$ma = -kx - bv.$$

We bring all terms to the left side of the equation and substitute $v = dx/dt$ and $a = d^2x/dt^2$ and obtain

$$m\frac{d^2x}{dt^2} + b\frac{dx}{dt} + kx = 0, \qquad (14\text{–}15)$$

Equation of motion for damped harmonic motion

which is the equation of motion. To solve this equation, we guess at a solution and then check to see if it works. If the damping constant b is small, x as a function of t is as plotted in Fig. 14–19, which looks like a cosine function times a factor (represented by the dashed lines) that decreases in time. A simple function that does this is the exponential, $e^{-\alpha t}$, and the solution that satisfies Eq. 14–15 is

$$x = Ae^{-\alpha t}\cos\omega' t. \qquad (14\text{–}16)$$

General solution: x as function of t

where A, α, and ω' are assumed to be constants, and $x = A$ at $t = 0$. We have called the angular frequency ω' (and <u>not</u> ω) because it is not the same as the ω for SHM without damping ($\omega = \sqrt{k/m}$).

If we substitute Eq. 14–16 into Eq. 14–15 (we do this in the optional subsection below) we find that Eq. 14–15 is indeed a solution if α and ω' have the values

$$\alpha = \frac{b}{2m} \qquad (14\text{–}17)$$

$$\omega' = \sqrt{\frac{k}{m} - \frac{b^2}{4m^2}}. \qquad (14\text{–}18)$$

Thus x as a function of time t for a (lightly) damped harmonic oscillator is

$$x = Ae^{-(b/2m)t}\cos\omega' t. \qquad (14\text{–}19)$$

Of course a phase constant, ϕ, can be added to the argument of the cosine in Eq. 14–19. As it stands with $\phi = 0$, it is clear that the constant A in Eq. 14–19 is simply the initial displacement, $x = A$ at $t = 0$. The frequency f is

$$f = \frac{\omega'}{2\pi} = \frac{1}{2\pi}\sqrt{\frac{k}{m} - \frac{b^2}{4m^2}}. \qquad (14\text{–}20)$$

The frequency is lower, and the period longer, than for undamped SHM. (In many practical cases of light damping, however, ω' differs only slightly from $\omega = \sqrt{k/m}$.) This makes sense since we expect friction to slow down the motion. Equation 14–20 reduces to Eq. 14–7a, as it should, when there is no friction ($b = 0$). The constant $\alpha = b/2m$ is a measure of how quickly the oscillations decrease toward zero (Fig. 14–19). The time $t_L = 2m/b$ is the time taken for the oscillations to drop to $1/e$ of the original amplitude; t_L is called the "mean lifetime" of the oscillations. Note that the larger b is, the more quickly the oscillations die away.

[†] Such velocity-dependent forces were discussed in Section 5–5.

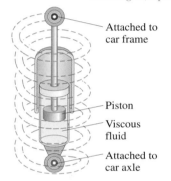

x

FIGURE 14–20 Underdamped (A), critically damped (B), and overdamped (C) motion.

Shock absorbers and building dampers

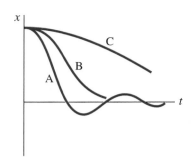

Attached to car frame

Piston

Viscous fluid

Attached to car axle

FIGURE 14–21 Automobile spring and shock absorber to provide damping so that car won't bounce up and down endlessly.

FIGURE 14–22 These huge dampers placed in a building structure look a lot like huge automobile shock absorbers, and they serve a similar purpose—to reduce the amplitude and the acceleration of movement when the shock of an earthquake hits.

The solution, Eq. 14–19, is not valid if b is so large that

$$b^2 > 4mk$$

since then ω' (Eq. 14–18) would become imaginary. In this case the system does not oscillate at all but returns directly to its equilibrium position, as we now discuss. Three common cases of *heavily damped* systems are shown in Fig. 14–20. Curve C represents the situation when the damping is so large $(b^2 \gg 4mk)$ that it takes a long time to reach equilibrium; the system is **overdamped**. Curve A represents an **underdamped** situation in which the system makes several swings before coming to rest $(b^2 < 4mk)$ and corresponds to a more heavily damped version of Eq. 14–19. Curve B represents **critical damping**: $b^2 = 4mk$; in this case equilibrium is reached in the shortest time. These terms all derive from the use of practical damped systems such as door closing mechanisms and shock absorbers in a car (Fig. 14–21) which are usually designed to give critical damping. But as they wear out, underdamping occurs: a door slams and a car bounces up and down several times whenever it hits a bump.

In many systems, the oscillatory motion is what counts, as in clocks and watches, and damping needs to be minimized. In other systems, oscillations are the problem, such as a car's springs, so a proper amount of damping (i.e., critical) is desired. Well-designed damping is needed for all kinds of applications. Large buildings, especially in California, are now built (or retrofitted) with huge dampers to reduce earthquake damage (Fig. 14–22).

EXAMPLE 14–10 **Simple pendulum with damping.** A simple pendulum has a length of 1.0 m (Fig. 14–23). It is set swinging with small-amplitude oscillations. After 5.0 minutes, the amplitude is only 50% of what it was initially. (*a*) What is the value of α for the motion? (*b*) By what factor does the frequency, ω', differ from ω, the undamped frequency?

SOLUTION (*a*) The equation of motion for damped harmonic motion is

$$x = Ae^{-\alpha t} \cos \omega' t, \quad \text{where } \alpha = \frac{b}{2m} \quad \text{and} \quad \omega' = \sqrt{\frac{k}{m} - \frac{b^2}{4m^2}},$$

for motion of a mass on the end of a spring. For the simple pendulum without damping, we saw in Section 14–5 that

$$F = -mg\theta.$$

Since $F = ma = mL\dfrac{d^2\theta}{dt^2}$, then

$$L\frac{d^2\theta}{dt^2} + g\theta = 0.$$

Introducing a damping term, $b(d\theta/dt)$, we have

$$L\frac{d^2\theta}{dt^2} + b\frac{d\theta}{dt} + g\theta = 0,$$

which is the same as Eq. 14–15 with θ replacing x, and L and g replacing m and k. Thus

$$\alpha = \frac{b}{2L} \quad \text{and} \quad \omega' = \sqrt{\frac{g}{L} - \frac{b^2}{4L^2}}.$$

At $t = 0$, Eq. 14–16 with θ replacing x is

$$\theta_0 = Ae^{-\alpha \cdot 0} \cos \omega' \cdot 0 = A.$$

Then at $t = 5.0\,\text{min} = 300\,\text{s}$, the amplitude has fallen to $0.50\,A$, so

$$Ae^{-\alpha(300\,\text{s})} = 0.50\,A.$$

We solve this for α and obtain $\alpha = \ln 2.0/(300\,\text{s}) = 2.3 \times 10^{-3}\,\text{s}^{-1}$.
(b) We have $L = 1.0\,\text{m}$, so $b = 2\alpha L = 4.6 \times 10^{-3}\,\text{m/s}$. Thus $(b^2/4L^2)$ is very much less than $g/L (= 9.8\,\text{s}^{-2})$, and the angular frequency of the motion remains almost the same as that of the undamped motion. Specifically,

$$\omega' = \sqrt{\frac{g}{L}}\left[1 - \frac{L}{g}\left(\frac{b^2}{4L^2}\right)\right]^{1/2} \approx \sqrt{\frac{g}{L}}\left[1 - \frac{1}{2}\frac{L}{g}\left(\frac{b^2}{4L^2}\right)\right]$$

where we used the binomial expansion. Then, with $\omega = \sqrt{g/L}$ (Eq. 14–12a),

$$\frac{\omega - \omega'}{\omega} \approx \frac{1}{2}\frac{L}{g}\left(\frac{b^2}{4L^2}\right) = 2.7 \times 10^{-7}.$$

So ω' differs from ω by less than one part in a million.

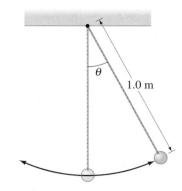

FIGURE 14–23 Example 14–10.

* Showing $x = Ae^{-\alpha t}\cos\omega' t$ is a Solution

We start with Eq. 14–16, to see if it is a solution to Eq. 14–15. First we take the first and second derivatives

$$\frac{dx}{dt} = -\alpha Ae^{-\alpha t}\cos\omega' t - \omega' Ae^{-\alpha t}\sin\omega' t$$

$$\frac{d^2x}{dt^2} = \alpha^2 Ae^{-\alpha t}\cos\omega' t + \alpha A\omega' e^{-\alpha t}\sin\omega' t + \omega'\alpha Ae^{-\alpha t}\sin\omega' t - \omega'^2 Ae^{-\alpha t}\cos\omega' t.$$

We next substitute these relations back into Eq. 14–15 and reorganize to obtain

$$Ae^{-\alpha t}\left[(m\alpha^2 - m\omega'^2 - b\alpha + k)\cos\omega' t + (2\omega'\alpha m - b\omega')\sin\omega' t\right] = 0. \qquad \textbf{(i)}$$

The left side of this equation must equal zero for all times t, but this can only be so for certain values of α and ω'. To determine α and ω', we choose two values of t that will make their evaluation easy. At $t = 0$, $\sin\omega' t = 0$, so the above relation reduces to $A(m\alpha^2 - m\omega'^2 - b\alpha + k) = 0$, which means[†] that

$$m\alpha^2 - m\omega'^2 - b\alpha + k = 0. \qquad \textbf{(ii)}$$

And at $t = \pi/2\omega'$, $\cos\omega' t = 0$ so Eq. (i) can be valid only if

$$2\alpha m - b = 0. \qquad \textbf{(iii)}$$

From Eq. (iii) we have

$$\alpha = \frac{b}{2m}$$

and from Eq. (ii)

$$\omega' = \sqrt{\alpha^2 - \frac{b\alpha}{m} + \frac{k}{m}} = \sqrt{\frac{k}{m} - \frac{b^2}{4m^2}}.$$

Thus we see that Eq. 14–16 is a solution to the equation of motion for the damped harmonic oscillator as long as α and ω' have these specific values (already given in Eqs. 14–17 and 14–18).

[†]It would also be satisfied by $A = 0$, but this gives the trivial and uninteresting solution $x = 0$ for all t—that is, no oscillation.

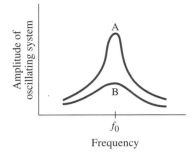

FIGURE 14–24 Resonance for lightly damped (A) and heavily damped (B) systems. (See Fig. 14–27 for a more detailed graph.)

FIGURE 14–25 This goblet breaks as it vibrates in resonance to a trumpet call.

FIGURE 14–26 (a) Large-amplitude oscillations of the Tacoma Narrows Bridge, due to gusty winds, led to its collapse (1940). (b) Collapse of freeway in California, due to the 1989 earthquake.

(a)

(b)

14–8 | Forced Vibrations; Resonance

When a vibrating system is set into motion, it vibrates at its natural frequency (Eqs. 14–7a and 14–12b). However, a system may have an external force applied to it that has its own particular frequency and then we have a **forced vibration**.

For example, we might pull the mass on the spring of Fig. 14–1 back and forth at a frequency f. The mass then vibrates at the frequency f of the external force, even if this frequency is different from the **natural frequency** of the spring, which we will now denote by f_0 where (see Eqs. 14–5 and 14–7a)

$$\omega_0 = 2\pi f_0 = \sqrt{\frac{k}{m}}.$$

In a forced vibration, the amplitude of vibration, and hence the energy transferred to the vibrating system, is found to depend on the difference between ω and ω_0 as well as on the amount of damping, reaching a maximum when the frequency of the external force equals the natural frequency of the system—that is, when $f = f_0$. The amplitude is plotted in Fig. 14–24 as a function of the external frequency f. Curve A represents light damping and curve B heavy damping. The amplitude can become large when the driving frequency f is near the natural frequency, $f \approx f_0$, as long as the damping is not too large. When the damping is small, the increase in amplitude near $f = f_0$ is very large (and often dramatic). This effect is known as **resonance**. The natural vibrating frequency f_0 of a system is called its **resonant frequency**.

A simple illustration of resonance is pushing a child on a swing. A swing, like any pendulum, has a natural frequency of oscillation that depends on its length L. If you push on the swing at a random frequency, the swing bounces around and reaches no great amplitude. But if you push with a frequency equal to the natural frequency of the swing, the amplitude increases greatly. The swing clearly illustrates that at resonance, relatively little effort is required to obtain a large amplitude.

The great tenor Enrico Caruso was said to be able to shatter a crystal goblet by singing a note of just the right frequency at full voice. This is an example of resonance, for the sound waves emitted by the voice act as a forced vibration on the glass. At resonance, the resulting vibration of the goblet may be large enough in amplitude that the glass exceeds its elastic limit and breaks (Fig. 14–25).

Since material objects are, in general, elastic, resonance is an important phenomenon in a variety of situations. It is particularly important in structural engineering, although the effects are not always foreseen. For example, it has been reported that a railway bridge collapsed because a nick in one of the wheels of a crossing train set up a resonant vibration in the bridge. Indeed, marching soldiers break step when crossing a bridge to avoid the possibility that their normal rhythmic march might match a resonant frequency of the bridge. The famous collapse of the Tacoma Narrows Bridge (Fig. 14–26a) in 1940 occurred as a result of strong gusting winds driving the span into large-amplitude oscillatory motion. The Oakland freeway collapse in the 1989 California earthquake (Fig. 14–26b) involved resonant oscillation which reached large amplitude on a section that was built on mudfill that readily transmitted that frequency.

We will meet important examples of resonance later. We will also see that vibrating objects often have not one, but many resonant frequencies.

*Equation of Motion and Its Solution

We now look at the equation of motion for a forced vibration, and its solution. Suppose the external force is sinusoidal and can be represented by

$$F_{ext} = F_0 \cos \omega t,$$

where $\omega = 2\pi f$ is the angular frequency applied externally to the oscillator. Then the equation of motion (with damping) is

$$ma = -kx - bv + F_0 \cos \omega t.$$

This can be written as

$$m \frac{d^2x}{dt^2} + b \frac{dx}{dt} + kx = F_0 \cos \omega t. \qquad \text{(14–21)}$$

Equation of motion for forced vibration

The external force, on the right of the equation, is the only term that does not involve x or one of its derivatives. It is left as an exercise (Problem 63) to show that

$$x = A_0 \sin(\omega t + \phi_0) \qquad \text{(14–22)}$$

General solution: x as function of t

is a solution to Eq. 14–21, by direct substitution, where

$$A_0 = \frac{F_0}{m \sqrt{\left(\omega^2 - \omega_0^2\right)^2 + b^2 \omega^2/m^2}} \qquad \text{(14–23)}$$

Amplitude

and

$$\phi_0 = \tan^{-1} \frac{\omega_0^2 - \omega^2}{\omega(b/m)}. \qquad \text{(14–24)}$$

Phase angle

Actually, the general solution to Eq. 14–21 is Eq. 14–22 plus another term of the form of Eq. 14–19 for the natural damped motion of the oscillator; this second term approaches zero in time, so in many cases we need to be concerned only with Eq. 14–22.

The amplitude of forced harmonic motion, A_0, depends strongly on the difference between the applied and the natural frequency. A plot of A_0 (Eq. 14–23) as a function of the applied frequency, ω, is shown in Fig. 14–27 (a more detailed version of Fig. 14–24) for three specific values of the damping constant b. Curve A $\left(b = \frac{1}{6} m\omega_0\right)$ represents light damping, curve B $\left(b = \frac{1}{2} m\omega_0\right)$ fairly heavy damping, and curve C $\left(b = \sqrt{2} m\omega_0\right)$ overdamped motion. The amplitude can become large when the driving frequency ω is near the natural frequency, $\omega \approx \omega_0$, as long as the damping is not too large. When the damping is small, the increase in amplitude near $\omega = \omega_0$ is very large and, as we saw, is known as *resonance*. The natural vibrating frequency ω_0 of a system is its *resonant frequency*.[†] If $b = 0$, resonance occurs at $\omega = \omega_0$ and the resonant peak $\left(\text{of } A_0\right)$ becomes infinite; in such a case, energy is being continuously transferred into the system and none is dissipated. For real systems, b is never precisely zero, and the resonant peak is finite. The peak does not occur precisely at $\omega = \omega_0$ (because of the term $b^2\omega^2/m^2$ in the denominator of Eq. 14–23), although it is quite close to ω_0 unless the damping is very large. If the damping is large, there is little or no peak (curve C in Fig. 14–27).

[†] Sometimes the resonant frequency is defined as the actual value of ω at which the amplitude has its maximum value, and this depends somewhat on the damping constant. Except for very heavy damping, this value is quite close to ω_0.

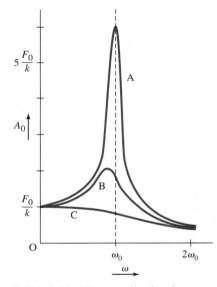

FIGURE 14–27 Amplitude of a forced harmonic oscillator as a function of ω. Curves A, B, and C correspond to light, heavy, and overdamped systems, respectively $\left(Q = m\omega_0/b = 6, 2, 0.71\right)$.

* Q value

The height and narrowness of a resonant peak is often specified by its **quality factor** or **Q value**, defined as

Quality factor or Q-value

$$Q = \frac{m\omega_0}{b}.$$

(14–25)

In Fig. 14–27, curve A has $Q = 6$, curve B has $Q = 2$, and curve C has $Q = 1/\sqrt{2}$. The smaller the damping constant b, the larger the Q value becomes, and the higher the resonance peak. The Q value is also a measure of the width of the peak. To see why, let ω_1 and ω_2 be the frequencies where the square of the amplitude A_0 has half its maximum value (we use the square because the power transferred to the system is proportional A_0^2—see the problems); then $\Delta\omega = \omega_1 - \omega_2$, which is called the *width* of the resonance peak, is related to Q by

$$\frac{\Delta\omega}{\omega_0} = \frac{1}{Q}.$$

(14–26)

(The proof of this relation, which is accurate only for weak damping, is the subject of Problem 67.) The larger the Q value, the narrower will be the resonance peak relative to its height. Thus a large Q value, representing a system of high quality, has a high, narrow resonance peak.

☐ Summary

A vibrating object undergoes **simple harmonic motion** (SHM) if the restoring force is proportional to the displacement,

$$F = -kx.$$

The maximum displacement is called the **amplitude**.

The **period**, T, is the time required for one complete cycle (back and forth), and the **frequency**, f, is the number of cycles per second; they are related by

$$f = \frac{1}{T}.$$

The period of vibration for a mass m on the end of an ideal massless spring is given by

$$T = 2\pi \sqrt{m/k}.$$

SHM is **sinusoidal**, which means that the displacement as a function of time follows a sine or cosine curve. The general solution can be written

$$x = A\cos(\omega t + \phi)$$

where A is the amplitude, ϕ is the **phase angle**, and

$$\omega = 2\pi f = \sqrt{\frac{k}{m}}.$$

The values of A and ϕ depend on the **initial** conditions (x and v at $t = 0$).

During SHM, the total energy $E = \frac{1}{2}mv^2 + \frac{1}{2}kx^2$ is continually changing from potential to kinetic and back again.

A swinging **simple pendulum** of length L approximates SHM if its amplitude is small and friction can be ignored. Its period is then given by (for small amplitudes)

$$T = 2\pi \sqrt{L/g},$$

where g is the acceleration of gravity.

When friction is present (for all real springs and pendulums), the motion is said to be **damped**. The maximum displacement decreases in time, and the mechanical energy is eventually all transformed to thermal energy. If the friction is very large, so no oscillations occur, the system is said to be **overdamped**. If the friction is small enough that oscillations occur, the system is **underdamped**, and the displacement is given by

$$x = Ae^{-\alpha t}\cos\omega't,$$

where α and ω' are constants. For a **critically damped** system, no oscillations occur and equilibrium is reached in the shortest time.

If an oscillating force is applied to a system capable of vibrating, the amplitude of vibration can be very large if the frequency of the applied force is near the **natural** (or **resonant**) **frequency** of the oscillator; this is called **resonance**.

Questions

1. Give some everyday examples of vibrating objects. Which follow SHM, at least approximately?
2. Contrast the equations for x, v, and a for uniformly accelerated linear motion (a = constant) with those for simple harmonic motion. Discuss their similarities and differences.
3. If a particle undergoes SHM with amplitude A, what is the total distance it travels in one period?
4. Real springs have mass. How will the true period and frequency differ from those given by the equations for a mass oscillating on the end of an idealized massless spring?
5. For a simple harmonic oscillator, when (if ever) are the displacement and velocity vectors in the same direction? When are the displacement and acceleration vectors in the same direction?
6. How could you double the maximum speed of a SHO?
7. A mass m hangs from a spring with stiffness constant k. The spring is cut in half and the same mass hung from it. Will the new arrangement have a higher or a lower stiffness constant than the original spring?
8. Two equal masses are attached to separate identical springs next to one another. One mass is pulled so its spring stretches 20 cm and the other pulled so its spring stretches only 10 cm. The masses are released simultaneously. Which mass reaches the equilibrium point first?
9. A 10-kg fish is attached to the hook of a vertical spring scale, and is then released. Describe the scale reading as a function of time.
10. Is the motion of a piston in an automobile engine simple harmonic? Explain.
11. If a pendulum clock is accurate at sea level, will it gain or lose time when taken to high altitude?
12. A tire swing hangs from a branch nearly to the ground. How could you estimate the height of the branch using only a stopwatch?
13. Does a car bounce on its springs faster when it is empty or when it is fully loaded?
14. What happens to the period of a playground swing if you rise up from sitting to a standing position?
15. Describe the possible motion of a solid object that is suspended so it is free to rotate about its center of gravity. Is it a physical pendulum?
16. A thin uniform rod of mass m is suspended from one end and oscillates with a frequency f. If a small sphere of mass $2m$ is attached to the other end, does the frequency increase or decrease? Explain.
17. Is the acceleration of a simple harmonic oscillator ever zero? If so, when? What about a damped harmonic oscillator?
18. A tuning fork of natural frequency 264 Hz sits on a table at the front of a room. At the back of the room, two tuning forks, one of natural frequency 260 Hz and one of 420 Hz are initially silent, but when the tuning fork at the front of the room is set into vibration, the 260-Hz fork spontaneously begins to vibrate but the 420-Hz fork does not. Explain.
19. Give several everyday examples of resonance.
20. Is a rattle in a car ever a resonance phenomenon? Explain.
21. Over the years, buildings have been able to be built out of lighter and lighter materials. How has this affected the natural vibration frequencies of buildings and the problems of resonance due to passing trucks, airplanes, or by wind and other natural sources of vibration?

Problems

Sections 14–1 and 14–2

1. (I) If a particle undergoes SHM with amplitude 0.15 m, what is the total distance it travels in one period?
2. (I) A fisherman's scale stretches 2.8 cm when a 3.7-kg fish hangs from it. (a) What is the spring constant? (b) What will be the amplitude and frequency of vibration if the fish is pulled down 2.5 cm more and released so that it vibrates up and down?
3. (I) When an 80-kg person climbs into a 1000-kg car, the car's springs compress vertically by 1.40 cm. What will be the frequency of vibration when the car hits a bump? (Ignore damping.)
4. (I) (a) What is the equation describing the motion of a spring that is stretched 8.8 cm from equilibrium and then released, and whose period is 0.75 s? (b) What will be its displacement after 1.8 s?
5. (II) A small fly of mass 0.60 g is caught in a spider's web. The web vibrates predominantly with a frequency of 10 Hz. (a) What is the value of the effective spring constant k for the web? (b) At what frequency would you expect the web to vibrate if an insect of mass 0.40 g were trapped?
6. (II) Determine the phase constant ϕ in Eq. 14–4 if, at $t = 0$, the oscillating mass is at (a) $x = -A$, (b) $x = 0$, (c) $x = A$, (d) $x = \frac{1}{2}A$, (e) $x = -\frac{1}{2}A$, (f) $x = A/\sqrt{2}$.
7. (II) A mass on the end of a spring is stretched a distance x_0 from equilibrium and released. At what distance from equilibrium will it have (a) velocity equal to half its maximum velocity and (b) acceleration equal to half its maximum acceleration?
8. (II) A balsa wood block of mass 50 g floats on a lake, bobbing up and down at a frequency of 2.5 Hz. (a) What is the value of the effective spring constant of the water? (b) A partially filled water bottle of mass 0.25 kg and almost the same size and shape of the balsa block is tossed into the water. At what frequency would you expect the bottle to bob up and down? Assume SHM.
9. (II) At what displacement from equilibrium is the speed of a SHO half the maximum value?

10. (II) A mass m at the end of a spring vibrates with a frequency of 0.88 Hz; when an additional 1.25 kg mass is added to m, the frequency is 0.48 Hz. What is the value of m?

11. (II) A block of mass m is supported by two parallel vertical springs, with spring constant k_1 and k_2 (Fig. 14–28). What will be the frequency of vibration?

FIGURE 14–28 Problem 11.

12. (II) The graph of displacement versus time for a small mass at the end of spring is shown in Fig. 14–29. At $t = 0$, $x = 0.43$ cm. (a) If $m = 14.3$ g, find the spring constant, k. (b) Write the equation for displacement x as a function of time.

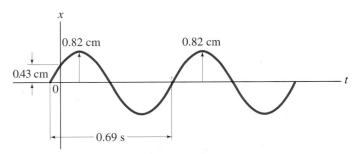

FIGURE 14–29 Problem 12.

13. (II) The position of a SHO as a function of time is given by $x = 3.8 \cos (7\pi t/4 + \pi/6)$ where t is in seconds and x in meters. Find (a) the period and frequency, (b) the position and velocity at $t = 0$, and (c) the velocity and acceleration at $t = 2.0$ s.

14. (II) A tuning fork vibrates at a frequency of 264 Hz and the tip of each prong moves 1.5 mm to either side of center. Calculate (a) the maximum speed and (b) the maximum acceleration of the tip of a prong.

15. (II) A spring vibrates with a frequency of 3.0 Hz when a weight of 0.50 kg is hung from it. What will its frequency be if only 0.35 kg hangs from it?

16. (II) (a) Show that

$$x = a \sin \omega t + b \cos \omega t$$

is a general solution of Eq. 14–3, and (b) determine the constants a and b in terms of the A and ϕ of Eq. 14–4.

17. (II) If a mass m hangs from a vertical spring, as shown in Fig. 14–3, show that $F = -kx$ holds for stretching or compression of the spring, where x is the displacement from the (vertical position) equilibrium point.

18. (II) A mass of 1.62 kg stretches a vertical spring 0.315 m. If the spring is stretched an additional 0.130 m and released, how long does it take to reach the (new) equilibrium position again?

19. (II) A spring of force constant 345 N/m vibrates with an amplitude of 22.0 cm when 0.250 kg hangs from it. (a) What is the equation describing this motion as a function of time? Assume that the mass passes downwards through the equilibrium point at $t = 0$. (b) At what times will the spring have its maximum and minimum extensions? Take y positive upwards.

20. (II) A 450-g object oscillates from a vertically hanging light spring once every 0.55 s. (a) Write down the equation giving its position y (+ upward) as a function of time t, assuming it started by being compressed 10 cm from the equilibrium position (where $y = 0$), and released. (b) How long will it take to get to the equilibrium position for the first time? (c) What will be its maximum speed? (d) What will be its maximum acceleration, and where will it first be attained?

21. (II) A uniform meter stick of mass M is pivoted on a hinge at one end and held horizontal by a spring with spring constant k attached at the other end (Fig. 14–30). If the stick oscillates up and down slightly, what is its frequency? [Hint: Write a torque equation about the hinge.]

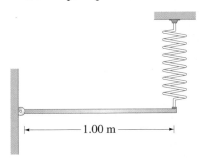

FIGURE 14–30
Problem 21.

22. (III) A mass m is at rest on the end of a spring of spring constant k. At $t = 0$ it is given an impulse J by a hammer. Write the formula for the subsequent motion in terms of $m, k, J,$ and t.

23. (III) A mass m is connected to two springs, with spring constants k_1 and k_2, in two different ways as shown in Fig. 14–31a and b. Show that the period for the configuration shown in part (a) is given by

$$T = 2\pi \sqrt{m \left(\frac{1}{k_1} + \frac{1}{k_2} \right)}$$

and for that in part (b) is given by

$$T = 2\pi \sqrt{\frac{m}{k_1 + k_2}}.$$

Ignore friction.

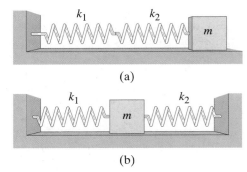

FIGURE 14–31
Problem 23.

24. (III) Two equal masses, m_1 and m_2, are connected by three identical springs of spring constant k as shown in Fig. 14–32. (a) Apply $\Sigma F = ma$ to each mass and obtain two differential equations for the displacements x_1 and x_2. (b) Determine the possible frequencies of vibration by assuming a solution of the form $x_1 = A_1 \cos \omega t$, $x_2 = A_2 \cos \omega t$.

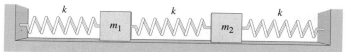

FIGURE 14–32 Problem 24.

25. (III) A spring with spring constant 250 N/m vibrates with an amplitude of 12.0 cm when 0.380 kg hangs from it. (a) What is the equation describing this motion as a function of time? Assume the mass passes through the equilibrium point, toward positive x (upward), at $t = 0.110$ s. (b) At what times will the spring have its maximum and minimum lengths? (c) What is the displacement at $t = 0$? (d) What is the force exerted by the spring at $t = 0$? (e) What is the maximum speed and when is it first reached after $t = 0$?

Section 14–3

26. (I) (a) At what displacement of a SHO is the energy half kinetic and half potential? (b) What fraction of the total energy of a SHO is kinetic and what fraction potential when the displacement is half the amplitude?

27. (I) A 2.00-kg mass vibrates according to the equation $x = 0.650 \cos 8.40t$ where x is in meters and t in seconds. Determine (a) the amplitude, (b) the frequency, (c) the total energy, and (d) the kinetic energy and potential energy when $x = 0.260$ m.

28. (II) A 0.35-kg mass at the end of a spring vibrates 3.0 times per second with an amplitude of 0.15 m. Determine (a) the velocity when it passes the equilibrium point, (b) the velocity when it is 0.10 m from equilibrium, (c) the total energy of the system, and (d) the equation describing the motion of the mass, assuming that at $t = 0$, x was a maximum.

29. (II) It takes a force of 95.0 N to compress the spring of a popgun 0.185 m to "load" a 0.200-kg ball. With what speed will the ball leave the gun?

30. (II) A 0.0125 kg bullet strikes a 0.300-kg block attached to a fixed horizontal spring whose spring constant is 2.25×10^3 N/m and sets it into vibration with an amplitude of 12.4 cm. What was the speed of the bullet if the two objects move together after impact?

31. (II) If one vibration has 10 times the energy of a second one of equal frequency, but the first's spring constant k is twice as large as the second's, how do their amplitudes compare?

32. (II) A mass of 240 g oscillates on a horizontal frictionless surface at a frequency of 3.5 Hz and with amplitude of 4.5 cm. (a) What is the effective spring constant for this motion? (b) How much energy is involved in this motion?

33. (II) A mass sitting on a horizontal, frictionless surface is attached to one end of a spring; the other end is fixed to a wall. 3.0 J of work is required to compress the spring by 0.12 m. If the mass is released from rest with the spring compressed, it experiences a maximum acceleration of 15 m/s². Find the value of (a) the spring constant and (b) the mass.

34. (II) An object with mass 2.1 kg is executing simple harmonic motion, attached to a spring with spring constant $k = 280$ N/m. When the object is 0.020 m from its equilibrium position, it is moving with a speed of 0.55 m/s. (a) Calculate the amplitude of the motion. (b) Calculate the maximum velocity attained by the object.

35. (II) Nikita devised the following method of measuring the muzzle velocity of a rifle (Fig. 14–33). She fires a bullet into a 6.023-kg wooden block resting on a smooth surface, and attached to a spring of spring constant $k = 142.7$ N/m. The bullet, whose mass is 7.870 g, remains embedded in the wooden block. She measures the distance that the block recoils and compresses the spring to be 9.460 cm. What is the speed v of the bullet?

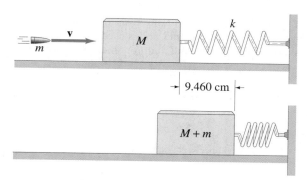

FIGURE 14–33 Problem 35.

36. (II) At $t = 0$, a 650-g mass at rest on the end of a horizontal spring ($k = 184$ N/m) is struck by a hammer which gives it an initial speed of 2.26 m/s. Determine (a) the period and frequency of the motion, (b) the amplitude, (c) the maximum acceleration, (d) the position as a function of time, (e) the total energy, and (f) the kinetic energy when $x = 0.40A$ where A is the amplitude.

37. (II) Obtain the displacement x as a function of time for the simple harmonic oscillator using the conservation of energy, Eq. 14–10. [Hint: Integrate Eq. 14–11a with $v = dx/dt$.]

Section 14–5

38. (I) A pendulum makes 42 vibrations in 50 s. What is its (a) period, and (b) frequency?

39. (I) How long must a simple pendulum be if it is to make exactly one swing per second? That is, one complete vibration takes exactly two seconds.

40. (I) What is the period of a simple pendulum on Mars, where the acceleration of gravity is about 0.37 that on Earth, if the pendulum has a period of 0.80 s on Earth?

41. (II) (a) Determine the length of a simple pendulum whose period is 1.00 s. (b) What would be the period of a 1.00-m-long simple pendulum?

42. (II) What is the period of a simple pendulum 73 cm long (a) on the Earth, and (b) when it is in a freely falling elevator?

43. (II) A simple pendulum is 0.30 m long. At $t = 0$ it is released starting at an angle of 14°. Ignoring friction, what will be the angular position of the pendulum at (a) $t = 0.65$ s, (b) $t = 1.95$ s, and (c) $t = 5.00$ s?

44. (II) Derive a formula for the maximum speed v_0 of a simple pendulum bob in terms of g, the length L, and the maximum angle of swing θ.

45. (II) A simple pendulum vibrates with an amplitude of $10.0°$. What fraction of the time does it spend between $+5.0$ and $-5.0°$? Assume SHM.

46. (II) The length of a simple pendulum is $0.68\,m$ and it is released at an angle of $12°$ to the vertical. (*a*) With what frequency does it vibrate? (*b*) What is the pendulum bob's speed when it passes through the lowest point of the swing?

* Section 14–6

* 47. (II) A plywood disk of radius $20.0\,cm$ and mass $3.00\,kg$ has a small hole drilled through it, $2.00\,cm$ from its edge (Fig. 14–34). The disk is hung from the wall by means of a metal pin through the hole, and is used as a pendulum. What is the period of this pendulum for small oscillations?

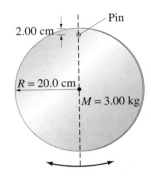

FIGURE 14–34 Problem 47.

* 48. (II) A pendulum consists of a tiny bob of mass M and a uniform cord of mass m and length L. (*a*) Determine a formula for the period. (*b*) What would be the fractional error if one used the formula for a simple pendulum, Eq. 14–12a?

* 49. (II) The balance wheel of a watch is a thin ring of radius $0.95\,cm$ and oscillates with a frequency of $3.10\,Hz$. If a torque of $1.1 \times 10^{-5}\,m \cdot N$ causes the wheel to rotate $60°$, calculate the mass of the balance wheel.

* 50. (II) (*a*) Determine the equation of motion (for θ as a function of time) for a torsion pendulum, Fig. 14–18, and show that the motion is simple harmonic. (*b*) Show that the period T is $T = 2\pi \sqrt{I/K}$. [The balance wheel of a mechanical watch is an example of a torsion pendulum in which the restoring torque is applied by a coil spring.]

* 51. (II) The human leg can be compared to a physical pendulum, with a "natural" swinging period at which walking is easiest. Consider the leg as two rods joined rigidly together at the knee; the axis for the leg is the hip joint. The length of each rod is about the same, $50\,cm$. The upper rod has a mass of $7.0\,kg$ and the lower rod has a mass of $4.0\,kg$. (*a*) Calculate the natural swinging period of the system. (*b*) Check your answer by standing on a chair and measuring the time for one or more complete back-and-forth swings. The effect of a shorter leg is, of course, a shorter swinging period, enabling a faster, although a shorter, "natural" stride.

* 52. (II) A student wants to use a meter stick as a pendulum. She plans to drill a small hole through the meter stick and suspend it from a smooth pin attached to the wall (Fig. 14–35). Where in the meter stick should she drill the hole to obtain the shortest possible period? How short an oscillation period can she obtain with a meter stick in this way?

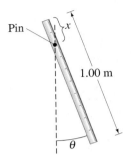

FIGURE 14–35 Problem 52.

* 53. (II) A meter stick is hung at its center from a thin wire (Fig. 14–36a). It is twisted and oscillates with a period of $6.0\,s$. The meter stick is sawed off to a length of $70.0\,cm$. This piece is again balanced at its center and set in oscillation (Fig. 14–36b). With what period does it oscillate?

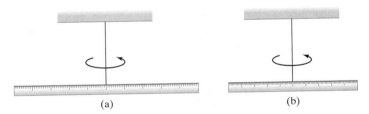

(a)   (b)

FIGURE 14–36 Problem 53.

* 54. (II) An aluminum disk, $12.5\,cm$ in diameter and $500\,g$ in mass, is mounted on a vertical shaft with an air bearing (Fig. 14–37). The disk also floats on an air bearing. One end of a flat coil spring is attached to the disk, the other end to the base of the apparatus. The disk is set into rotational oscillation and the frequency is $0.331\,Hz$. What is the torsional spring constant K ($\tau = -K\theta$)?

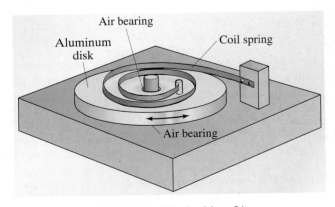

FIGURE 14–37 Problem 54.

Section 14–7

55. (I) A 750-g block oscillates on the end of a spring whose force constant is $k = 56.0 \text{ N/m}$. The mass moves in a fluid which offers a resistive force $F = -bv$, where $b = 0.162 \text{ N·s/m}$. (a) What is the period of the motion? (b) What is the fractional decrease in amplitude per cycle? (c) Write the displacement as a function of time if at $t = 0$, $x = 0$, and at $t = 1.00 \text{ s}$, $x = 0.120 \text{ m}$.

56. (II) A physical pendulum consists of an 80-cm-long, 300-g-mass, uniform wooden rod hung from a nail near one end (Fig. 14–38). The motion is damped because of friction in the pivot; the damping force is approximately proportional to $d\theta/dt$. The rod is set in oscillation by displacing it $15°$ from its equilibrium position and releasing it. After 8.0 seconds, the amplitude of the oscillation has been reduced to $5.5°$. If the angular displacement can be written as $\theta = Ae^{-\alpha t} \cos \omega' t$, find (a) α, (b) the approximate period of the motion, and (c) how long it takes for the amplitude to be reduced to $\frac{1}{2}$ of its original value.

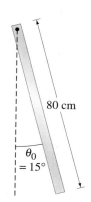

FIGURE 14–38 Problem 56.

57. (II) A damped harmonic oscillator loses 5.0 percent of its mechanical energy per cycle. (a) By what percentage does its frequency differ from the natural frequency $\omega_0 = \sqrt{k/m}$? (b) After how many periods will the amplitude have decreased to $1/e$ of its original value?

58. (III) (a) Show that the total mechanical energy, $E = \frac{1}{2}mv^2 + \frac{1}{2}kx^2$, as a function of time for a lightly damped harmonic oscillator is

$$E = \tfrac{1}{2}kA^2 e^{-(b/m)t} = E_0 e^{-(b/m)t},$$

where E_0 is the total mechanical energy at $t = 0$. (Assume $\omega' \gg b/2m$.) (b) Show that the fractional energy lost per period is

$$\frac{\Delta E}{E} = \frac{2\pi b}{m\omega_0} = \frac{2\pi}{Q},$$

where $\omega_0 = \sqrt{k/m}$ and $Q = m\omega_0/b$ is called the *quality factor* or *Q value* of the system. A larger Q value means the system can undergo oscillations for a longer time.

59. (III) A glider on an air track is connected by springs to either end of the track (Fig. 14–39). Both springs have the same spring constant, k, and the glider has mass M. (a) Determine the frequency of the oscillation, assuming no damping, if $k = 100 \text{ N/m}$ and $M = 200$ grams. (b) It is observed that after 55 oscillations, the amplitude of the oscillation has dropped to one-half of its initial value. Estimate the value of α, using Eq. 14–16. (c) How long does it take the amplitude to decrease to one-quarter of its initial value?

FIGURE 14–39 Problem 59.

Section 14–8

*60. (II) (a) At resonance $(\omega = \omega_0)$, what is the value of the phase angle ϕ_0? (b) What, then, is the displacement at a time when the driving force F_{ext} is a maximum, and at a time when $F_{\text{ext}} = 0$? (c) What is the phase difference (in degrees) between the driving force and the displacement in this case?

*61. (II) Construct an accurate resonance curve, from $\omega = 0$ to $\omega = 2\omega_0$, for $Q = 4.0$.

*62. (II) The amplitude of a driven harmonic oscillator reaches a value of $28.6 \, F_0/m$ at a resonant frequency of 382 Hz. What is the Q value of this system?

*63. (II) By direct substitution, show that Eq. 14–22, with Eqs. 14–23 and 14–24, is a solution of the equation of motion (Eq. 14–21) for the forced oscillator.

*64. (II) Differentiate Eq. 14–23 to show that the resonant amplitude peaks at

$$\omega = \sqrt{\omega_0^2 - b^2/2m^2}.$$

*65. (III) Consider a simple pendulum (point mass bob) 0.50 m long with a Q of 400. (a) How long does it take for the amplitude (assumed small) to decrease by two-thirds? (b) If the amplitude is 2.0 cm and the bob has mass 0.20 kg, what is the initial energy loss rate of the pendulum in watts? (c) If we are to stimulate resonance with a sinusoidal driving force, how close must the driving frequency be to the natural frequency of the pendulum?

*66. (III) *Power transferred to driven oscillator.* (a) Show that the power input to a forced oscillator due to the external force F_{ext} is

$$P = F_{\text{ext}} v = \frac{F_0^2 \omega \cos \phi_0 \cos^2 \omega t - \frac{1}{2}F_0^2 \omega \sin \phi_0 \sin 2\omega t}{m \sqrt{(\omega^2 - \omega_0^2)^2 + \omega^2 b^2/m^2}}.$$

(b) Show that the average power input, averaged over one (or many) full cycles, is

$$\bar{P} = \frac{\omega F_0^2 \cos \phi_0}{2m \sqrt{(\omega^2 - \omega_0^2)^2 + \omega^2 b^2/m^2}} = \tfrac{1}{2}\omega A_0 F_0 \cos \phi_0$$

or

$$\bar{P} = \tfrac{1}{2} F_0 v_{\max} \cos \phi_0,$$

where $v_{\max}$ is the maximum value of dx/dt. (c) Plot $\bar{P}$ versus ω from $\omega = 0$ to $\omega = 2\omega_0$ for $Q = 6.0$. Note that although the amplitude is not zero at $\omega = 0$, $\bar{P}$ is zero at $\omega = 0$.

*67. (III) Derive Eq. 14–26.

General Problems

68. A 52-kg person jumps from a window to a fire net 20.0 m below, which stretches the net 1.1 m. Assume that the net behaves like a simple spring, and calculate how much it would stretch if the same person were lying in it. How much would it stretch if the person jumped from 35 m?

69. The length of a simple pendulum is 0.63 m, the pendulum bob has a mass of 365 grams, and it is released at an angle of 15° to the vertical. (a) With what frequency does it vibrate? (b) What is the pendulum bob's speed when it passes through the lowest point of the swing? Assume SHM. (c) What is the total energy stored in this oscillation assuming no losses?

70. A 0.650-kg mass vibrates according to the equation $x = 0.25 \sin(5.50t)$ where x is in meters and t is in seconds. Determine (a) the amplitude, (b) the frequency, (c) the period, (d) the total energy, and (e) the kinetic energy and potential energy when x is 10 cm.

71. A bungee jumper with mass 72.0 kg jumps from a high bridge. After reaching his lowest point, he oscillates up and down, hitting a low point eight more times in 34.7 s. He finally comes to rest 25.0 m below the level of the bridge. Calculate the spring constant and the unstretched length of the bungee cord.

72. (a) A crane has hoisted a 1200-kg car at the junkyard. The steel crane cable is 20.0 m long and has a diameter of 6.4 mm. A breeze starts the car bouncing at the end of the cable. What is the period of the bouncing? [Hint: Refer to Table 12–1]. (b) What amplitude of bouncing will likely cause the cable to snap? (See Table 12–2, and assume Hooke's law holds all the way up to the breaking point.)

73. An energy-absorbing car bumper has a spring constant of 500 N/m. Find the maximum compression of the bumper if the car, with mass 1500 kg, collides with a wall at a speed of 2.0 m/s (approximately 5 mi/h).

74. A "seconds" pendulum has a period of exactly 2.0 seconds—each one-way swing takes 1.0 s. What is the length of a seconds pendulum in Austin, Texas, where $g = 9.793$ m/s²? If the pendulum is moved to Paris, where $g = 9.809$ m/s², by how many millimeters must we lengthen the pendulum? What is the length of a "seconds" pendulum on the Moon, where $g = 1.62$ m/s²?

75. A block of jello rests on a cafeteria plate as shown in Fig. 14–40 (which also gives the dimensions of the block). You push it sideways as shown, and then you let go. The jello springs back and begins to vibrate. In analogy to a mass vibrating on a spring, estimate the frequency of this vibration, given that the shear modulus of jello is 520 N/m² and its density is 1300 kg/m³.

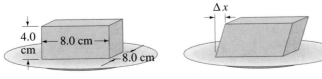

FIGURE 14–40 Problem 75.

76. A simple pendulum oscillates with frequency f. What is its frequency if it accelerates at $\frac{1}{2}g$ (a) upward, and (b) downward?

77. A 420-kg wooden raft floats on a lake. When a 75-kg man stands on the raft, it sinks 3.5 cm deeper into the water. When he steps off, the raft vibrates for a while. (a) What is the frequency of vibration? (b) What is the total energy of vibration (ignoring damping)?

78. A 5.0-kg box slides into a spring of spring constant 310 N/m (Fig. 14–41), compressing it 24 cm. (a) What is the incoming speed of the block? (b) How long is the box in contact with the spring before it bounces off in the opposite direction?

FIGURE 14–41 Problem 78.

79. A 1.60-kg table is supported on four springs. A 0.55-kg chunk of modeling clay is held above the table and dropped so that it hits the table with a speed of 1.65 m/s (Fig. 14–42). The clay makes an inelastic collision with the table and the table and clay oscillate up and down. After a long time the table comes to rest 6.0 cm below its original position. (a) What is the effective spring constant of all four springs taken together? (b) With what maximum amplitude does the platform oscillate?

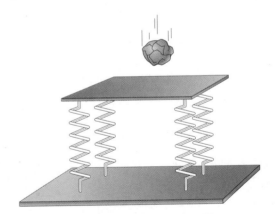

FIGURE 14–42 Problem 79.

80. A diving board oscillates with simple harmonic motion of frequency 5.0 cycles per second. What is the maximum amplitude with which the end of the board can vibrate in order that a pebble placed there (Fig. 14–43) does not lose contact with the board during the oscillation?

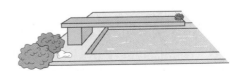

FIGURE 14–43 Problem 80.

81. A rectangular block of wood floats in a calm lake. Show that, if friction is ignored, when the block is pushed gently down into the water, it will then vibrate with SHM. Also, determine an equation for the force constant.

82. A mass m is gently placed on the end of a freely hanging spring. The mass then falls 22.0 cm before it stops and begins to rise. What is the frequency of the oscillation?

83. The water in a U-shaped tube is displaced an amount Δx from equilibrium. (The level in one side is $2\Delta x$ above the level in the other side.) If friction is neglected, will the water oscillate harmonically? Determine a formula for the equivalent spring constant k. Does k depend on the density of the liquid, the cross-section of the tube, or the length of the water column?

84. In some diatomic molecules, the force each atom exerts on the other can be approximated by $F = -C/r^2 + D/r^3$, where C and D are positive constants. (a) Graph F versus r from $r = 0$ to $r = 2D/C$. (b) Show that equilibrium occurs at $r = r_0 = D/C$. (c) Let $\Delta r = r - r_0$ be a small displacement from equilibrium, where $\Delta r \ll r_0$. Show that for such small displacements, the motion is approximately simple harmonic, and (d) determine the force constant. (e) What is the period of such motion? [Hint: Assume one atom is kept at rest.]

85. A mass m is connected to two springs with equal spring constants k (Fig. 14–44). In the horizontal position shown, each spring is stretched by an amount Δa. The mass is raised vertically and begins to oscillate up and down. Assuming that the displacement is small, and ignoring gravity, show that the motion is simple harmonic and find the period. [This kind of motion is called transverse oscillation. If we hooked many masses together like this, their transverse oscillations would be a model for a vibrating string in a musical instrument.]

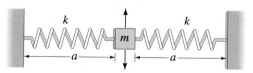

FIGURE 14–44 Problem 85.

86. Carbon dioxide is a linear molecule. The carbon–oxygen bonds in this molecule act very much like springs. Figure 14–45 shows one possible way the oxygen atoms in this molecule can vibrate: the oxygen atoms vibrate symmetrically in and out, while the central carbon atom remains at rest. Hence each oxygen atom acts like a simple harmonic oscillator with a mass equal to the mass of an oxygen atom. It is observed that this oscillation occurs with a frequency of $f = 2.83 \times 10^{13}$ Hz. What is the spring constant of the C–O bond?

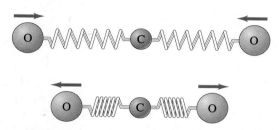

FIGURE 14–45 Problem 86, the CO_2 molecule.

87. A thin, straight, uniform rod of length $l = 1.00$ m and mass $m = 160$ g hangs from a pivot at one end. (a) What is its period for small-amplitude oscillations? (b) What is the length of a simple pendulum that will have the same period?

88. Imagine that a 10-cm-diameter circular hole were drilled all the way through the center of the Earth (Fig. 14–46). At one end of the hole, you drop an apple into the hole. Show that, if you assume that the Earth has a constant density, the subsequent motion of the apple is simple harmonic. How long will the apple take to return? Assume that we can ignore all frictional effects.

FIGURE 14–46 Problem 88.

Waves—such as water waves or waves traveling along a cord or slinky—travel away from their source. The waves on a stretched cord shown in these four photographs, however, are "standing waves." They seem not to be traveling; but each can be thought of as the sum of a wave traveling to the right interfering with its reflection traveling back toward the left. We can also think of these standing waves as oscillations of the stretched cord at resonance. Each of these standing waves occurs at a particular frequency. Can you guess how the frequency of each is related to the others?

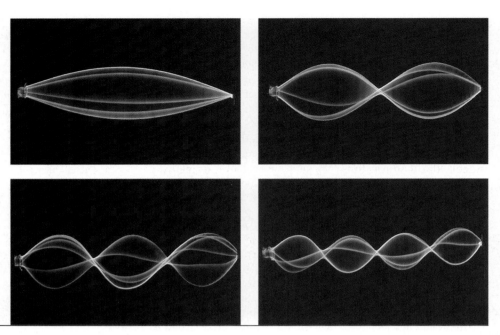

CHAPTER 15

Wave Motion

When you throw a stone into a lake or pool of water, circular waves form and move outward, Fig. 15–1. Waves will also travel along a cord (or a "slinky") that is stretched out straight on a table if you vibrate one end back and forth as shown in Fig. 15–2. Water waves and waves on a cord are two common examples of wave motion.

Vibrations and wave motion are intimately related subjects. Waves—whether ocean waves, waves on a string, earthquake waves, or sound waves in air—have as their source a vibration. In the case of sound, not only is the source a vibrating object, but so is the detector—the eardrum or the membrane of a microphone.

FIGURE 15–1 A stone thrown into a lake: water waves spread outward from the source.

15–3 Energy Transported by Waves

Waves transport energy from one place to another. As waves travel through a medium, the energy is transferred as vibrational energy from particle to particle of the medium. For a sinusoidal wave of frequency f, the particles move in SHM as a wave passes, and each particle has an energy $E = \frac{1}{2}kD_M^2$ where D_M is the maximum displacement (amplitude) of its motion, either transversely or longitudinally (see Eq. 14–10a, in which we have replaced A by D_M). Using Eq. 14–7a we can write $k = 4\pi^2 mf^2$, where m is the mass of a particle (or small volume) of the medium. Then in terms of the frequency

$$E = \tfrac{1}{2}kD_M^2 = 2\pi^2 mf^2 D_M^2.$$

For three-dimensional waves traveling in an elastic medium, the mass $m = \rho V$, where ρ is the density of the medium and V is the volume of a small slice of the medium. The volume $V = Al$ where A is the cross-sectional area through which the wave travels (Fig. 15–12), and we can write l as the distance the wave travels in a time t as $l = vt$, where v is the speed of the wave. Thus $m = \rho V = \rho Al = \rho Avt$ and

$$E = 2\pi^2 \rho Avtf^2 D_M^2. \qquad (15\text{–}5)$$

From this equation we have the important result that the **energy transported by a wave is proportional to the square of the amplitude, and to the square of the frequency**. The average *rate* of energy transferred is the average power P:

$$\overline{P} = \frac{E}{t} = 2\pi^2 \rho Avf^2 D_M^2. \qquad (15\text{–}6)$$

Finally, the **intensity**, I, of a wave is defined as the average power transferred across unit area perpendicular to the direction of energy flow:

$$I = \frac{\overline{P}}{A} = 2\pi^2 v\rho f^2 D_M^2. \qquad (15\text{–}7)$$

If a wave flows out from the source in all directions, it is a three-dimensional wave. Examples are sound traveling in the open air, earthquake waves, and light waves. If the medium is isotropic (same in all directions), the wave is said to be a *spherical wave* (Fig. 15–13). As the wave moves outward, the energy it carries is spread over a larger and larger area since the surface area of a sphere of radius r is $4\pi r^2$. Thus the intensity of a wave is

$$I = \frac{\overline{P}}{A} = \frac{\overline{P}}{4\pi r^2}.$$

If the power output $\overline{P}$ is constant, then the intensity decreases as the inverse square of the distance from the source:

$$I \propto \frac{1}{r^2}. \qquad \text{[spherical wave]} \quad (15\text{–}8a)$$

If we consider two points at distances r_1 and r_2 from the source, as in Fig. 15–13, then $I_1 = \overline{P}/4\pi r_1^2$ and $I_2 = \overline{P}/4\pi r_2^2$, so

$$\frac{I_2}{I_1} = \frac{r_1^2}{r_2^2}. \qquad (15\text{–}8b)$$

Thus, for example, when the distance doubles $(r_2/r_1 = 2)$, then the intensity is reduced to $\frac{1}{4}$ its earlier value: $I_2/I_1 = \left(\frac{1}{2}\right)^2 = \frac{1}{4}$.

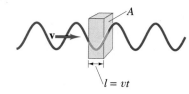

FIGURE 15–12 Calculating the energy carried by a wave moving with velocity v.

Wave energy $\propto$ (amplitude)^2
Wave energy $\propto (f)^2$

Intensity

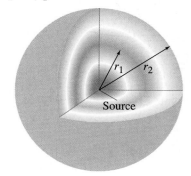

FIGURE 15–13 Wave traveling outward from source has spherical shape. Two different crests (or compressions) are shown, of radius r_1 and r_2.

Intensity $I \propto \dfrac{1}{r^2}$

The amplitude of a wave also decreases with distance. Since the intensity is proportional to the square of the amplitude $(I \propto D_M^2,$ Eq. 15–7), then the amplitude must decrease as $1/r$, so that $I \propto D_M^2$ will be proportional to $1/r^2$, Eq. 15–8a; that is,

Amplitude $\propto \dfrac{1}{r}$

$$D_M \propto \frac{1}{r}.$$

To see this directly from Eq. 15–6, we consider again two different distances from the source, r_1 and r_2. For constant power output, $A_1 D_{M1}^2 = A_2 D_{M2}^2$ where D_{M1} and D_{M2} are the amplitudes of the wave at r_1 and r_2, respectively. Since $A_1 = 4\pi r_1^2$ and $A_2 = 4\pi r_2^2$, we have $(D_{M1}^2 r_1^2) = (D_{M2}^2 r_2^2)$, or

$$\frac{D_{M2}}{D_{M1}} = \frac{r_1}{r_2}.$$

When the wave is twice as far from the source, the amplitude is half as large, as long as we can ignore damping due to friction.

EXAMPLE 15–3 **Earthquake intensity.** If the intensity of an earthquake P wave 100 km from the source is 1.0×10^6 W/m^2, what is the intensity 400 km from the source?

SOLUTION The intensity decreases as the square of the distance from the source. Therefore, at 400 km, the intensity will be $\left(\frac{1}{4}\right)^2 = \frac{1}{16}$ of its value at 100 km, or 6.2×10^4 W/m^2. Alternatively, Eq. 15–8b could be used: $I_2 = I_1 r_1^2/r_2^2 = (1.0 \times 10^6 \text{ W/m}^2)(100 \text{ km})^2/(400 \text{ km})^2 = 6.2 \times 10^4$ W/m^2.

The situation is different for a one-dimensional wave, such as a transverse wave on a string or a longitudinal wave pulse traveling down a thin uniform metal rod. The area remains constant, so the amplitude D_M also remains constant (ignoring friction). Thus the amplitude and the intensity do not decrease with distance.

In practice, frictional damping is generally present, and some of the energy is transformed into thermal energy. Thus the amplitude and intensity of a one-dimensional wave decrease with distance from the source, and for a three-dimensional wave the decrease will be greater than that discussed above, although the effect may often be small.

15–4 Mathematical Representation of a Traveling Wave

Let us now consider a one-dimensional wave traveling along the x axis. It could be, for example, a transverse wave on a cord or a longitudinal wave traveling in a rod or in a fluid-filled tube. Let us assume the wave shape is sinusoidal and has a particular wavelength λ and frequency f. At $t = 0$, suppose the wave shape is given by

$$D(x) = D_M \sin \frac{2\pi}{\lambda} x, \tag{15–9}$$

as shown by the solid curve in Fig. 15–14. $D(x)$ is the **displacement**[†] of the wave (be it a longitudinal or transverse wave) at position x, and D_M is the **amplitude**

[†] Some books use $y(x)$ in place of $D(x)$. To avoid confusion, we reserve y (and z) for the coordinate positions of waves in two or three dimensions. Our $D(x)$ can stand for pressure (in longitudinal waves), position displacement (transverse mechanical waves) or—as we will see later—electric or magnetic fields (for electromagnetic waves).

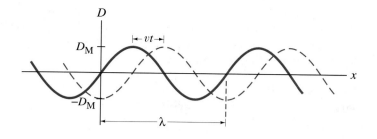

(maximum displacement) of the wave. This relation gives a shape that repeats itself every wavelength, which is what we want so that the displacement is the same at, for example, $x = 0$, $x = \lambda$, $x = 2\lambda$, and so on (since $\sin 4\pi = \sin 2\pi = \sin 0$).

Now suppose the wave is moving to the right with velocity v. Then, after a time t, each part of the wave (indeed, the whole wave "shape") has moved to the right a distance vt; see the dashed curve in Fig. 15–14. Consider any point on the wave at $t = 0$: say, a crest which is at some position x. After a time t, that crest will have traveled a distance vt so its new position is a distance vt greater than its old position. To describe this same point on the wave shape, the argument of the sine function must be the same, so we replace x in Eq. 15–9 by $(x - vt)$:

$$D(x, t) = D_M \sin\left[\frac{2\pi}{\lambda}(x - vt)\right].$$ **(15–10a)**

1-D wave, moving in positive x direction

Said another way, if you are riding on a crest, the argument of the sine function, $(2\pi/\lambda)(x - vt)$, remains the same $(= \pi/2, 5\pi/2$, and so on); as t increases, x must increase at the same rate so that $(x - vt)$ remains constant.

Equation 15–10a is the mathematical representation of a sinusoidal wave traveling along the x axis to the right (increasing x). It gives the displacement $D(x, t)$ of the wave at any chosen point x at any time t. The function $D(x, t)$ describes a curve that represents the actual shape of the wave in space at time t. Since $v = \lambda f$ (Eq. 15–1) we can write Eq. 15–10a in other ways that are often convenient:

$$D(x, t) = D_M \sin\left(\frac{2\pi x}{\lambda} - \frac{2\pi t}{T}\right),$$ **(15–10b)**

1-D wave, moving in positive x direction

where $T = 1/f = \lambda/v$ is the period; and

$$D(x, t) = D_M \sin(kx - \omega t),$$ **(15–10c)**

1-D wave, moving in positive x direction

where $\omega = 2\pi f = 2\pi/T$ is the angular frequency and

$$k = \frac{2\pi}{\lambda}$$ **(15–11)**

is called the **wave number**. (Do not confuse the wave number k with the spring constant k; they are very different quantities.) All three forms, Eq. 15–10a, b, and c, are equivalent; Eq. 15–10c is the simplest to write and is perhaps the most common. The quantity $(kx - \omega t)$, and its equivalent in the other two equations, is called the **phase** of the wave. The velocity v of the wave is often called the **phase velocity**, since it describes the velocity of the phase (or shape) of the wave and it can be written in terms of ω and k:

$$v = \lambda f = \left(\frac{2\pi}{k}\right)\left(\frac{\omega}{2\pi}\right) = \frac{\omega}{k}.$$ **(15–12)**

For a wave traveling along the x axis to the left (decreasing values of x), we start again with Eq. 15–9 and note that a particular point on the wave changes position by $-vt$ in a time t. So x in Eq. 15–9 must be replaced by $(x + vt)$. Thus, for a wave traveling to the left with velocity v,

1-D wave
moving in
negative x
direction

$$D(x, t) = D_M \sin\left[\frac{2\pi}{\lambda}(x + vt)\right] \tag{15–13a}$$

$$= D_M \sin\left(\frac{2\pi x}{\lambda} + \frac{2\pi t}{T}\right) \tag{15–13b}$$

$$= D_M \sin(kx + \omega t). \tag{15–13c}$$

In other words, we simply replace v in Eqs. 15–10 by $-v$.

Let us look for a moment at Eq. 15–13c (or, just as well, at Eq. 15–10c). At $t = 0$ we have

$$D(x, 0) = D_M \sin kx,$$

which is what we started with, a sinusoidal wave shape. If we look at the wave shape in space at a particular later time t_1, then we have

$$D(x, t_1) = D_M \sin(kx + \omega t_1).$$

That is, if we took a picture of the wave at $t = t_1$, we would see a sine wave with a phase constant ωt_1. Thus, for fixed $t = t_1$, the wave has a sinusoidal shape in space. On the other hand, if we consider a fixed point in space, say $x = 0$, we can see how the wave varies in time:

$$D(0, t) = D_M \sin \omega t$$

where we used Eq. 15–13c. This is just the equation for simple harmonic motion (see Section 14–2, Eq. 14–4). For any other fixed value of x, say $x = x_1$, $D = D_M \sin(\omega t + kx_1)$ which differs only by a phase constant kx_1. Thus, at any fixed point in space, the displacement undergoes the oscillations of simple harmonic motion in time. Equations 15–10 and 15–13 combine both these aspects to give us the representation for a **traveling sinusoidal wave** (also called a **harmonic wave**).

The argument of the sine in Eqs. 15–10 and 15–13 can in general contain a phase angle ϕ,

$$D(x, t) = D_M \sin(kx \pm \omega t + \phi),$$

to adjust for the position of the wave at $t = 0$, $x = 0$, just as in Section 14–2 (see Fig. 14–7). If the displacement is zero at $t = 0$, $x = 0$, as in Fig. 14–6, then $\phi = 0$.

Now let us consider a general wave (or wave pulse) of any shape. If frictional losses are small, experiment shows that the wave maintains its shape as it travels. Thus we can make the same arguments as we did right after Eq. 15–9. Suppose our wave has some shape at $t = 0$, given by

$$D(x, 0) = D(x)$$

where $D(x)$ is the displacement of the wave at x and is not necessarily sinusoidal. Then at some later time, if the wave is traveling to the right along the x axis, the wave will have the same shape but all parts will have moved a distance vt where v is the phase velocity of the wave. Hence we must replace x by $x - vt$ to obtain the amplitude at time t:

$$D(x, t) = D(x - vt). \tag{15–14}$$

Similarly, if the wave moves to the left, we must replace x by $x + vt$, so

$$D(x, t) = D(x + vt). \tag{15–15}$$

Thus, any wave traveling along the x axis must have the form of Eq. 15–14 or 15–15.

EXAMPLE 15-4 **A traveling wave.** The left-hand end of a long horizontal stretched cord oscillates transversely in SHM with frequency $f = 250\,\text{Hz}$ and amplitude 2.6 cm. The cord is under a tension of 140 N and has a linear density $\mu = 0.12\,\text{kg/m}$. At $t = 0$, the end of the cord has an upward displacement of 1.6 cm and is falling (Fig. 15–15). Determine (a) the wavelength of waves produced and (b) the equation for the traveling wave.

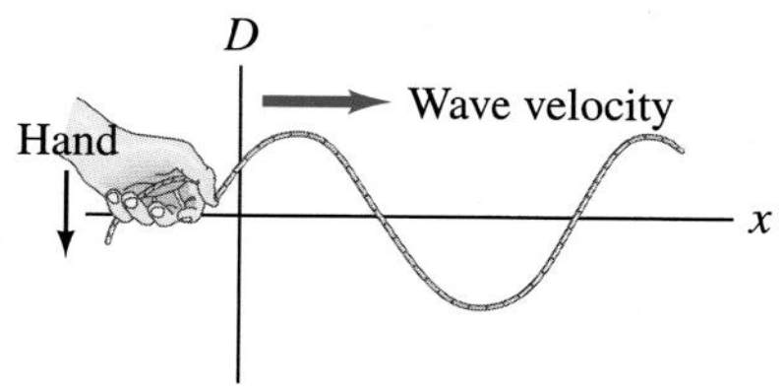

SOLUTION (a) The wave velocity is

$$v = \sqrt{\frac{F_T}{\mu}} = \sqrt{\frac{140\,\text{N}}{0.12\,\text{kg/m}}} = 34\,\text{m/s}.$$

Then

$$\lambda = \frac{v}{f} = \frac{34\,\text{m/s}}{250\,\text{Hz}} = 0.14\,\text{m} \quad \text{or} \quad 14\,\text{cm}.$$

(b) Let $x = 0$ at the left-hand end of the cord. The phase of the wave at $t = 0$ is not zero in general as was assumed in Eqs. 15–9, 10, and 13. The general form for a wave traveling to the right is then

$$D(x, t) = D_M \sin(kx - \omega t + \phi),$$

where ϕ is the phase angle. In our case, the amplitude $D_M = 2.6$ cm; and at $t = 0$, $x = 0$, we are given $D = 1.6$ cm. Thus

$$1.6 = 2.6 \sin \phi,$$

so $\phi = 38° = 0.66\,\text{rad}$. We also have $\omega = 2\pi f = 1570\,\text{s}^{-1}$ and $k = 2\pi/\lambda = 45\,\text{m}^{-1}$. Hence

$$D = 0.026 \sin(45x - 1570t + 0.66),$$

where D and x are in meters and t in seconds.

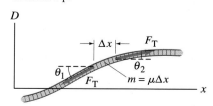

FIGURE 15–15 Example 15–4: The wave at $t = 0$ (the hand is falling). Not to scale.

The Wave Equation

Many types of waves satisfy an important general equation that is the equivalent of Newton's second law of motion for particles. This "equation of motion for a wave" is called the **wave equation**, and we derive it now for waves traveling on a stretched horizontal string.

We assume the amplitude of the wave is small compared to the wavelength so that each point on the string can be assumed to move only vertically and the tension in the string, F_T, does not vary during a vibration. We apply Newton's second law, $\Sigma F = ma$, to the vertical motion of a tiny section of the string as shown in Fig. 15–16. The amplitude of the wave is small, so the angles θ_1 and θ_2 that the string makes with the horizontal are small. The length of this section is then approximately Δx, and its mass is $\mu\,\Delta x$, where μ is the mass per unit length of the string. The net vertical force on this section of string is $F_T \sin\theta_2 - F_T \sin\theta_1$. So Newton's second law applied to the vertical (y) direction gives

$$\Sigma F_y = ma_y$$

$$F_T \sin\theta_2 - F_T \sin\theta_1 = (\mu\,\Delta x)\,\frac{\partial^2 D}{\partial t^2}.$$

FIGURE 15–16 Deriving the wave equation from Newton's second law: a segment of string under tension F_T.

We have written the acceleration as $a_y = \partial^2 D/\partial t^2$ since the motion is only vertical, and we use the partial derivative notation because the displacement D is a function of both x and t.

Because the angles θ_1 and θ_2 are assumed small, $\sin\theta \approx \tan\theta$ and $\tan\theta$ is equal to the slope s of the string at each point:

$$\sin\theta \approx \tan\theta = \frac{\partial D}{\partial x} = s.$$

Thus our equation at the bottom of the previous page becomes

$$F_T(s_2 - s_1) = \mu\,\Delta x\,\frac{\partial^2 D}{\partial t^2}$$

or

$$F_T\frac{\Delta s}{\Delta x} = \mu\,\frac{\partial^2 D}{\partial t^2}, \tag{a}$$

where $\Delta s = s_2 - s_1$ is the difference in the slope between the two ends of our tiny section. Now we take the limit of $\Delta x \to 0$, so that

$$F_T\lim_{\Delta x \to 0}\frac{\Delta s}{\Delta x} = F_T\frac{\partial s}{\partial x}$$

$$= F_T\frac{\partial}{\partial x}\left(\frac{\partial D}{\partial x}\right) = F_T\frac{\partial^2 D}{\partial x^2}$$

since the slope $s = \partial D/\partial x$, as we saw above. Substituting this into the equation labelled (a) above gives

$$F_T\frac{\partial^2 D}{\partial x^2} = \mu\,\frac{\partial^2 D}{\partial t^2}$$

or

$$\frac{\partial^2 D}{\partial t^2} = \frac{F_T}{\mu}\frac{\partial^2 D}{\partial t^2}.$$

We saw earlier in this chapter (Eq. 15–2) that the velocity of waves on a string is given by $v = \sqrt{F_T/\mu}$, so we can write this last equation as

Wave equation (1–D)
$$\frac{\partial^2 D}{\partial x^2} = \frac{1}{v^2}\frac{\partial^2 D}{\partial t^2}. \tag{15–16}$$

This is the **one-dimensional wave equation**, and it can describe not only small amplitude waves on a stretched string, but also small amplitude longitudinal waves (such as sound waves) in gases, liquids, and elastic solids, in which case D can refer to the pressure variations. In this case, the wave equation is a direct consequence of Newton's second law applied to a continuous elastic medium. The wave equation also describes electromagnetic waves for which D refers to the electric or magnetic field, as we shall see in Chapter 32. Equation 15–16 applies to waves traveling in one dimension only. For waves spreading out in three dimensions, the wave equation is the same, with the addition of $\partial^2 D/\partial y^2$ and $\partial^2 D/\partial z^2$ to the left side of Eq. 15–16.

The wave equation is a *linear* equation: the displacement D appears singly in each term. There are no terms that contain D^2, or $D(\partial D/\partial x)$, or the like in which D appears more than once. Thus, if $D_1(x, t)$ and $D_2(x, t)$ are two different solutions of a linear equation, such as the wave equation, then the linear combination

Superposition principle
$$D_3(x, t) = aD_1(x, t) + bD_2(x, t),$$

where a and b are constants, is also a solution. This is readily seen by direct substitution into the wave equation. This is the essence of the *superposition principle*, which we discuss in the next Section. Basically it says that if two waves pass through the same region of space at the same time, the actual displacement is the sum of the separate displacements. For waves on a string, or for sound waves, this is valid only for small-amplitude waves. If the amplitude is not small enough, the equations for wave propagation may become nonlinear and the principle of super-position would not hold and other new effects may occur.

EXAMPLE 15–5 **Wave equation solution.** Verify that the sinusoidal wave of Eq. 15–10c, $D(x, t) = D_M \sin(kx - \omega t)$, satisfies the wave equation.

SOLUTION We take the derivative of Eq. 15–10c twice with respect to t:

$$\frac{\partial D}{\partial t} = -\omega D_M \cos(kx - \omega t)$$

$$\frac{\partial^2 D}{\partial t^2} = -\omega^2 D_M \sin(kx - \omega t).$$

With respect to x, the derivatives are

$$\frac{\partial D}{\partial x} = k D_M \cos(kx - \omega t)$$

$$\frac{\partial^2 D}{\partial x^2} = -k^2 D_M \sin(kx - \omega t).$$

If we now divide the second derivatives we get

$$\frac{\partial^2 D / \partial t^2}{\partial^2 D / \partial x^2} = \frac{-\omega^2 D_M \sin(kx - \omega t)}{-k^2 D_M \sin(kx - \omega t)} = \frac{\omega^2}{k^2}.$$

From Eq. 15–12 we have $\omega^2/k^2 = v^2$, so we see that Eq. 15–10 does satisfy the wave equation (Eq. 15–16).

15–6 The Principle of Superposition

When two or more waves pass through the same region of space at the same time, it is found that for many waves *the actual displacement is the vector* (or *algebraic*) *sum of the separate displacements*. This is called the **principle of superposition**. It is valid for mechanical waves as long as the displacements are not too large and there is a linear relationship between the displacement and the restoring force of the oscillating medium.[†] If the amplitude of a mechanical wave, for example, is so large that it goes beyond the elastic region of the medium, and Hooke's law is no longer operative, the superposition principle is no longer accurate.[‡] For the most part, we will consider systems for which the superposition principle can be assumed to hold.

One result of the superposition principle is that if two waves pass through the same region of space, they continue to move independently of one another. You may have noticed, for example, that the ripples on the surface of water (two-dimensional waves) that form from two rocks striking the water at different places will pass through each other.

Figure 15–17, shows an example of the superposition principle. In this case there are three waves present, on a stretched string, each of different amplitude and frequency. At any time, such as at the instant shown, the actual amplitude at any position x is the algebraic sum of the amplitude of the three waves at that position. The actual wave is not a simple sinusoidal wave and is called a *composite* (or *complex*) *wave*. (Amplitudes are exaggerated in Fig. 15–17.)

It can be shown that any complex wave can be considered as being composed of many simple sinusoidal waves of different amplitudes, wavelengths, and frequencies. This is known as *Fourier's theorem*. A complex periodic wave of period T can be represented as a sum of pure sinusoidal terms whose frequencies are integral multiples of $f = 1/T$. If the wave is not periodic, the sum becomes an integral (called a *Fourier integral*). Although we will not go into the details here, we see the importance of considering sinusoidal waves (and simple harmonic motion): because any other wave shape can be considered a sum of such pure sinusoidal waves.

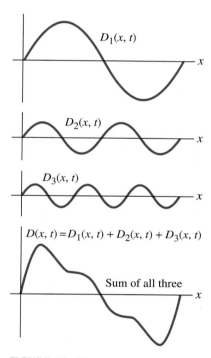

FIGURE 15–17 The superposition principle for one-dimensional waves. Composite wave formed from three sinusoidal waves of different amplitudes and frequencies $(f_0, 2f_0, 3f_0)$ at a certain instant in time. The amplitude of the composite wave at each point in space, at any time, is the algebraic sum of the amplitudes of the component waves. Amplitudes are shown exaggerated; for the superposition principle to hold, they must be small compared to the wavelengths.

[†] For electromagnetic waves in vacuum, Chapter 32, the superposition principle always holds.

[‡] Intermodulation distortion in high-fidelity equipment is an example of the superposition principle not holding when two frequencies do not combine linearly in the electronics.

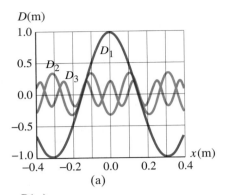

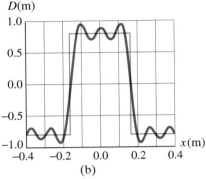

FIGURE 15–18 Example 15–6. Making a square wave.

FIGURE 15–19 Reflection of a wave pulse on a rope when the end of the rope is (a) fixed and (b) free.

EXAMPLE 15–6 **Making a square wave.** At $t = 0$, three waves are given by $D_1 = D_M \cos kx$, $D_2 = \frac{1}{3} D_M \cos 3kx$ and $D_3 = \frac{1}{5} D_M \cos 5kx$ where $D_M = 1.0$ m and $k = 10$ m^{-1}. Plot the sum of the three waves from $x = -0.4$ to $+0.4$ m. (These three waves are the first three Fourier components of a "square wave.")

SOLUTION The first wave, D_1 has amplitude of 1.0 m and wavelength $\lambda = 2\pi/k = 2\pi/10$ m $= 0.628$ m. The second wave, D_2 has amplitude of 0.33 m and wavelength $\lambda = 2\pi/3k = 2\pi/30$ m $= 0.209$ m. The third wave, D_3 has amplitude of 0.20 m and wavelength $\lambda = 2\pi/5k = 2\pi/50$ m $= 0.126$ m. Each wave is plotted in Fig. 15–18a. The sum of the three waves is shown in Fig. 15–18b. The sum begins to resemble a "square wave," shown in blue in Fig. 15–18b.

When the restoring force is not precisely proportional to the displacement for mechanical waves in some continuous medium, the speed of sinusoidal waves depends on the frequency. The variation of speed with frequency is called **dispersion**. The different sinusoidal waves that compose a complex wave will travel with slightly different speeds in such a case. Consequently, a complex wave will change shape as it travels if the medium is "dispersive." A pure sine wave will not change shape under these conditions, however, except by the influence of friction or dissipative forces. If there is no dispersion (or friction), even a complex linear wave doesn't change shape.

15–7 | Reflection and Transmission

When a wave strikes an obstacle, or comes to the end of the medium it is traveling in, at least a part of the wave is reflected. You have probably seen water waves reflect off of a rock or the side of a swimming pool. And you may have heard a shout reflected from a distant cliff—which we call an "echo."

A wave pulse traveling down a rope is reflected as shown in Fig. 15–19. You can observe this for yourself (try it with a rope lying on a table) and see that the reflected pulse is inverted as in Fig. 15–19a if the end of the rope is fixed; and returns right side up if the end is free as in Fig. 15–19b. When the end is fixed to a

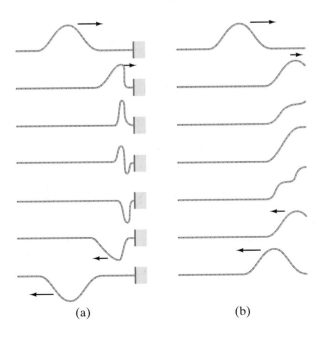

(a)   (b)

support, as in Fig. 15–19a, the pulse reaching that fixed end exerts a force (upward) on the support. The support exerts an equal but opposite force (Newton's third law) downward on the rope. This downward force on the rope is what "generates" the inverted reflected pulse. The inversion of the reflected pulse in part (a) is said to be a phase change of 180°. It is as if the phase shifted by $\frac{1}{2}\lambda$, or 180°, from a crest to a trough. In Fig. 15–19b, the free end is constrained by neither a support nor by additional rope. It therefore tends to overshoot—its displacement being momentarily greater than that of the traveling pulse. The overshooting end exerts an upward pull on the rope, and this is what generates the reflected pulse, which is not inverted (no phase change).

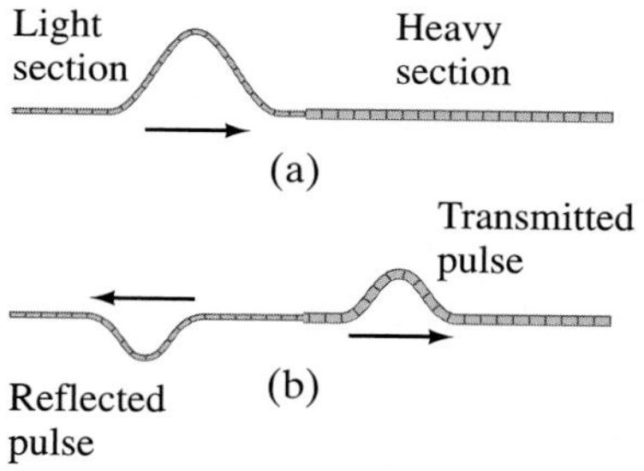

When the wave pulse on the rope in Fig. 15–19a reaches the wall, not all of the energy is reflected. Some of it is absorbed by the wall. Part of the absorbed energy is transformed into thermal energy, and part continues to propagate through the material of the wall. This is more clearly illustrated by considering a pulse that travels down a rope which consists of a light section and a heavy section, as shown in Fig. 15–20. When the wave reaches the boundary between the two sections, part of the pulse is reflected and part is transmitted, as shown. The heavier the second section, the less is transmitted; and when the second section is a wall or rigid support, very little is transmitted. For a periodic wave, the frequency does not change across the boundary since the boundary point oscillates at that frequency. Thus if the transmitted wave has a lower speed, its wavelength is also less ($\lambda = v/f$).

For a two- or three-dimensional wave, such as a water wave, we are concerned with **wave fronts**, by which we mean all the points along the wave forming the wave crest (what we usually refer to simply as a "wave" when we are at the seashore). A line drawn in the direction of motion, perpendicular to the wave front, is called a **ray**, as shown in Fig. 15–21. Note in Fig. 15–21b that wave fronts far from the source have lost almost all their curvature and are nearly straight, as ocean waves often are; they are then called **plane waves**.

For reflection of a two- or three-dimensional plane wave, as shown in Fig. 15–22, the angle that the incoming or *incident wave* makes with the reflecting surface is equal to the angle made by the reflected wave. This is the **law of reflection: the angle of reflection equals the angle of incidence**. The "angle of incidence" is defined as the angle the incident ray makes with the perpendicular to the reflecting surface (or the wave front makes with a tangent to the surface), and the "angle of reflection" is the corresponding angle for the reflected wave.

Wave fronts

Plane waves

Law of reflection

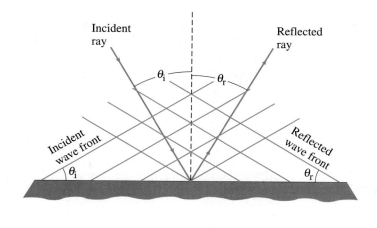

FIGURE 15–20 When a wave pulse traveling to the right (a) reaches a discontinuity, then (b) part is reflected and part is transmitted.

FIGURE 15–21 Rays, signifying the direction of motion, are always perpendicular to the wave fronts (wave crests). (a) Circular or spherical waves near the source. (b) Far from the source, the wave fronts are nearly straight or flat, and are called plane waves.

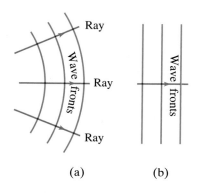

FIGURE 15–22 Law of reflection.

15–8 | Interference

Interference refers to what happens when two waves pass through the same region of space at the same time. It is an example of the superposition principle. Consider for example the two wave pulses on a string traveling toward each other as shown in Fig. 15–23. In part (a) the two pulses have the same amplitude but one is a crest and the other a trough; in part (b) they are both crests. In both cases, the waves meet and pass right by each other. However, when they overlap, the resultant displacement is the *algebraic sum of their separate displacements* (the principle of superposition). In Fig. 15–23a, the two wave amplitudes are opposite one another as they pass by and the result is called **destructive interference**. In Fig. 15–23b, the resultant displacement is greater than that of either pulse and the result is called **constructive interference**.

Destructive interference

Constructive interference

FIGURE 15–23 Two waves pulses pass each other. Where they overlap, interference occurs: (a) destructive; (b) constructive.

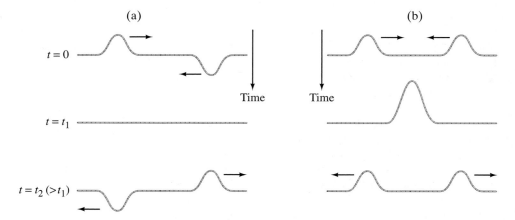

When two rocks are thrown into a pond simultaneously, the two sets of circular waves interfere with one another as shown in Fig. 15–24. In some areas of overlap, crests of one wave meet crests of the other (and troughs meet troughs); this is constructive interference and the water oscillates up and down with greater amplitude than either wave separately. In other areas, destructive interference occurs

FIGURE 15–24 Interference of water waves. Constructive interference occurs where one wave's maximum (a crest) meets the other's maximum. Destructive interference ("flat water") occurs where one wave's maximum (a crest) meets the other's miminum (a trough).

(a)

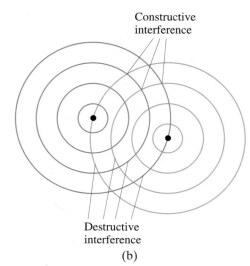

(b)

where the water actually does not move at all—this is where crests of one wave meet troughs of the other, and vice versa. In the first case, of constructive interference, the two waves are **in phase**, whereas in destructive interference the two waves are **out of phase** by one-half wavelength or 180°. Of course the relative phase of the two waves in most areas is intermediate between these two extremes, resulting in partially destructive interference. All three of these situations are shown in Fig. 15–25, where the amplitudes are plotted versus time at a given point in space. We will deal with interference in more detail when we discuss sound and light.

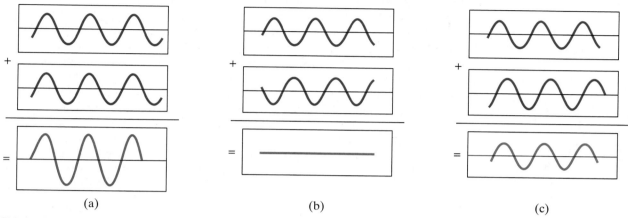

(a)   (b)   (c)

FIGURE 15–25 Two waves interfere: (a) constructively, (b) destructively, (c) partially destructively.

15–9 | Standing Waves; Resonance

If you shake one end of a cord (or slinky) and the other end is kept fixed, a continuous wave will travel down to the fixed end and be reflected back, inverted (Fig. 15–19a). As you continue to vibrate the cord, there will be waves traveling in both directions, and the wave traveling down the cord will interfere with the reflected wave coming back. Usually there will be quite a jumble. But if you vibrate the cord at just the right frequency, the two traveling waves will interfere in such a way that a large-amplitude **standing wave** will be produced, Fig. 15–26. It is called a "standing wave" because it doesn't appear to be traveling. The cord simply appears to have segments that oscillate up and down in a fixed pattern. The points of destructive interference, where the cord remains still at all times, are called **nodes**. Points of constructive interference, where the end oscillates with maximum amplitude, are called **antinodes**. The nodes and antinodes remain in fixed positions for a given frequency.

Standing waves can occur at more than one frequency. The lowest frequency of vibration that produces a standing wave gives rise to the pattern shown in Fig. 15–26a. The standing waves shown in parts (b) and (c) are produced at precisely twice and three times the lowest frequency, respectively, assuming the tension in the cord is the same. The cord can also vibrate with four loops at four times the lowest frequency, and so on.

The frequencies at which standing waves are produced are the **natural frequencies** or **resonant frequencies** of the cord, and the different standing wave patterns shown in Fig. 15–26 are different "resonant modes of vibration." Although a standing wave on a cord is the result of the interference of two waves traveling in opposite directions, it is also a vibrating object at resonance. Standing waves then represent the same phenomenon as the resonance of a vibrating spring or pendulum, which we discussed in Chapter 14. The only difference is that a spring or pendulum has only one resonant frequency, whereas the cord has an infinite number of resonant frequencies, each of which is a whole-number multiple of the lowest resonant frequency.

FIGURE 15–26 Standing waves corresponding to three resonant frequencies.

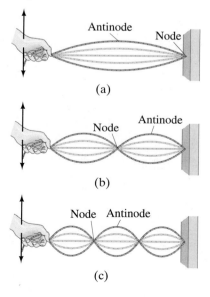

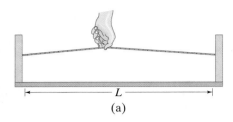

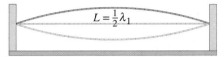

$L = \frac{1}{2}\lambda_1$

Fundamental or first harmonic, f_1

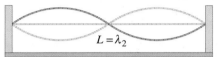

$L = \lambda_2$

First overtone or second harmonic, $f_2 = 2f_1$

$L = \frac{3}{2}\lambda_3$

Second overtone or third harmonic, $f_3 = 3f_1$

(b)

FIGURE 15–27 (a) A string is plucked. (b) Only standing waves corresponding to resonant frequencies persist for long.

Now let us consider a cord stretched between two supports that is plucked like a guitar or violin string, Fig. 15–27a. Waves of a great variety of frequencies will travel in both directions along the string, will be reflected at the ends, and will travel back in the opposite direction. Most of these waves interfere in a random way with each other and quickly die away. However, those waves that correspond to the resonant frequencies of the string will persist. The ends of the string, since they are fixed, will be nodes. There may be other nodes as well. Some of the possible resonant modes of vibration (standing waves) are shown in Fig. 15–27b. Generally, the motion will be a combination of these different resonant modes; but only those frequencies that correspond to a resonant frequency will be present.

To determine the resonant frequencies, we first note that the wavelengths of the standing waves bear a simple relationship to the length L of the string. The lowest frequency, called the **fundamental frequency**, corresponds to one antinode (or loop). And as can be seen in Fig. 15–27b, the whole length corresponds to one-half wavelength. Thus $L = \frac{1}{2}\lambda_1$, where λ_1 stands for the wavelength of the fundamental. The other natural frequencies are called **overtones**; when they are integral multiples of the fundamental (as they are for a simple string), they are also called **harmonics**, with the fundamental being referred to as the **first harmonic**.[†] The next mode after the fundamental has two loops and is called the **second harmonic** (or first overtone); the length of the string L at the second harmonic corresponds to one complete wavelength: $L = \lambda_2$. For the third and fourth harmonics, $L = \frac{3}{2}\lambda_3$, and $L = 2\lambda_4$, respectively, and so on. In general, we can write

$$L = \frac{n\lambda_n}{2}, \qquad \text{where } n = 1, 2, 3, \cdots.$$

The integer n labels the number of the harmonic: $n = 1$ for the fundamental, $n = 2$ for the second harmonic, and so on. We solve for λ_n and find

$$\lambda_n = \frac{2L}{n}, \qquad n = 1, 2, 3, \cdots. \tag{15–17}$$

In order to find the frequency f of each vibration we use Eq. 15–1, $f = v/\lambda$, and we see that

$$f_n = \frac{v}{\lambda_n} = \frac{nv}{2L} = nf_1,$$

where $f_1 = v/\lambda_1 = v/2L$ is the fundamental frequency. We see that each resonant frequency is an integer multiple of the fundamental frequency.

Because a standing wave is equivalent to two traveling waves moving in opposite directions, the concept of wave velocity still makes sense and is given by Eq. 15–2 in terms of the tension F_T in the string and its mass per unit length μ; that is, $v = \sqrt{F_T/\mu}$ for both traveling waves.

EXAMPLE 15–7 Piano string. A piano string is 1.10 m long and has a mass of 9.00 g. (a) How much tension must the string be under if it is to vibrate at a fundamental frequency of 131 Hz? (b) What are the frequencies of the first four harmonics?

SOLUTION (a) The wavelength of the fundamental is $\lambda = 2L = 2.20$ m (Eq. 15–17). The velocity is then $v = \lambda f = (2.20 \text{ m})(131 \text{ s}^{-1}) = 288$ m/s. Then, from Eq. 15–2, we have

$$F_T = \mu v^2$$
$$= \left(\frac{9.00 \times 10^{-3} \text{ kg}}{1.10 \text{ m}} \right)(288 \text{ m/s})^2 = 679 \text{ N}.$$

(b) The frequencies of the second, third, and fourth harmonics are two, three, and four times the fundamental frequency: 262, 393, and 524 Hz.

[†] The term "harmonic" comes from music, because such integral multiples of frequencies "harmonize."

A standing wave does appear to be standing in place (and a traveling wave appears to move). The term "standing" wave is also meaningful from the point of view of energy. Since the string is at rest at the nodes, no energy flows past these points. Hence the energy is not transmitted down the string but "stands" in place in the string.

Standing waves are produced not only on strings, but on any object that is set into vibration. Even when a rock or a piece of wood is struck with a hammer, standing waves are set up that correspond to the natural resonant frequencies of that object. In general, the resonant frequencies depend on the dimensions of the object, just as for a string they depend on its length. For example a small object does not have as low resonant frequencies as does a large object. All musical instruments depend on standing waves to produce their musical sounds, from stringed instruments to wind instruments (in which a column of air vibrates as a standing wave) to drums and other percussion instruments, as we will discuss in the next chapter.

Mathematical Representation of a Standing Wave

In Section 15–4 we saw how to write an equation for the displacement D of a linear traveling wave as a function of position x and time t. We can do the same for a standing wave on a string. We already discussed that a standing wave can be considered to consist of two traveling waves that move in opposite directions. These can be written (see Eqs. 15–10c and 15–13c)

$$D_1(x, t) = D_M \sin(kx - \omega t)$$
$$D_2(x, t) = D_M \sin(kx + \omega t)$$

since, assuming no damping, the amplitudes are equal as are the frequencies and wavelengths. The sum of these two traveling waves produces a standing wave which can be written mathematically as

$$D = D_1 + D_2 = D_M[\sin(kx - \omega t) + \sin(kx + \omega t)].$$

From the trigonometric identity $\sin \theta_1 + \sin \theta_2 = 2 \sin \frac{1}{2}(\theta_1 + \theta_2) \cos \frac{1}{2}(\theta_1 - \theta_2)$, we can rewrite this as

$$D = 2D_M \sin kx \cos \omega t. \qquad (15\text{--}18) \qquad \textit{Standing wave}$$

If we let $x = 0$ at the left-hand end of the string, then the right-hand end is at $x = L$ where L is the length of the string. Since the string is fixed at its two ends (Fig. 15–27), $D(x, t)$ must be zero at $x = 0$ and at $x = L$. Equation 15–18 already satisfies the first condition ($D = 0$ at $x = 0$) and satisfies the second condition if $\sin kL = 0$ which means

$$kL = \pi, 2\pi, 3\pi, \ldots, n\pi, \ldots$$

where n = integer, or, since $k = 2\pi/\lambda$,

$$\lambda = \frac{2L}{n}. \quad (n = \text{integer})$$

This is just Eq. 15–17.

Equation 15–18, with the condition $\lambda = 2L/n$, is the mathematical representation of a standing wave. We see that a particle at any position x vibrates in simple harmonic motion (because of the factor $\cos \omega t$). All particles of the string vibrate with the same frequency $f = \omega/2\pi$, but the amplitude depends on x and equals $2D_M \sin kx$. (Compare this to a traveling wave for which all particles vibrate with the same amplitude.) The amplitude has a maximum, equal to $2D_M$, when $kx = \pi/2, 3\pi/2, 5\pi/2$, and so on—that is, at

$$x = \frac{\lambda}{4}, \frac{3\lambda}{4}, \frac{5\lambda}{4}, \ldots.$$

These are, of course, the positions of the antinodes (see Fig. 15–27).

Wave forms. Two waves traveling in opposite directions on a string fixed at $x = 0$ are described by the functions

$$D_1 = (0.20 \text{ m}) \sin(2.0x - 4.0t)$$
$$D_2 = (0.20 \text{ m}) \sin(2.0x + 4.0t)$$

and they produce a standing wave pattern (x is in m, t is in s). (*a*) Determine the function for the standing wave. (*b*) What is the maximum amplitude at $x = 0.45$ m? (*c*) Where is the other end fixed ($x > 0$)? (*d*) What is the maximum amplitude, and where does it occur?

SOLUTION (*a*) The two waves are of the form $D = D_M \sin(kx \pm \omega t)$, so

$$k = 2.0 \text{ m}^{-1} \quad \text{and} \quad \omega = 4.0 \text{ s}^{-1}.$$

These combine to form a standing wave of the form of Eq. 15–18:

$$D = 2D_M \sin kx \cos \omega t = (0.40 \text{ m}) \sin(2.0x) \cos(4.0t).$$

(*b*) At $x = 0.45$ m,

$$D = (0.40 \text{ m}) \sin(0.90) \cos(4.0t) = (0.31 \text{ m}) \cos(4.0t).$$

The maximum amplitude at this point is $D = 0.31$ m and occurs when the cosine $= 1$.

(*c*) These waves make a standing wave pattern, so both ends of the string must be nodes. Nodes occur every half wavelength, which for our string is

$$\frac{\lambda}{2} = \frac{1}{2} \frac{2\pi}{k} = \frac{\pi}{2.0} \text{ m} = 1.57 \text{ m}.$$

If the string includes only one loop, its length is $L = 1.57$ m. But since we aren't given more information, it could be twice as long, including two loops, $L = 3.14$ m, or any integral number times 1.57 m, and still provide a standing wave pattern for these waves—see Fig. 15–28.

(*d*) The nodes occur at $x = 0$, $x = 1.57$ m, and, if the string is longer than $L = 1.57$ m, at $x = 3.14$ m, 4.71 m, and so on. The maximum amplitude (antinode) is 0.40 m [from part (*b*) above] and occurs midway between the nodes. For $L = 1.57$ m, there is only one antinode, at $x = 0.79$ m.

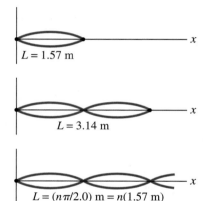

FIGURE 15–28 Example 15–8: possible lengths for the string.

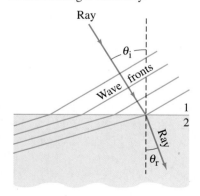

FIGURE 15–29 Refraction of waves crossing a boundary.

* 15–10 Refraction[†]

When any wave strikes a boundary, some of the energy is reflected and some is transmitted or absorbed. When a two- or three-dimensional wave traveling in one medium crosses a boundary into a medium where its velocity is different, the transmitted wave may move in a different direction than the incident wave, as shown in Fig. 15–29. This phenomenon is known as **refraction**. One example is a water wave; the velocity decreases in shallow water and the waves refract, Fig. 15–30. [When the wave velocity changes gradually, as in Fig. 15–30, without a sharp boundary, the waves change direction (refract) gradually.] In Fig. 15–29, the velocity of the wave in medium 2 is less than in medium 1. In this case, the direction of the wave bends so it travels more nearly perpendicular to the boundary. That is, the *angle of refraction*, θ_r, is less than the *angle of incidence*, θ_i. To see why this is so, and to help us get a quantitative relation between θ_r and θ_i, let us think of each wave front as a row of soldiers. The soldiers are marching from firm

[†]This Section and the next are covered in more detail in Chapters 33 to 36, on optics.

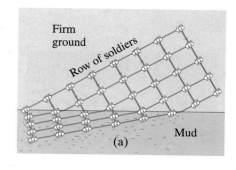

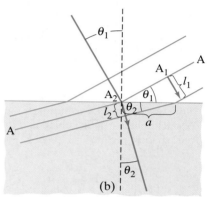

FIGURE 15–30 Water waves refracting as they approach the shore, where their velocity is less. There is no distinct boundary, as in Fig. 15–29, and the wave velocity changes gradually.

ground (medium 1) into mud (medium 2) and hence are slowed down. The soldiers that reach the mud first are slowed down first and the row bends as shown in Fig. 15–31a. Let us consider the wave front (or row of soldiers) labeled A in Fig. 15–31b. In the same time t that A_1 moves a distance $l_1 = v_1 t$, we see that A_2 moves a distance $l_2 = v_2 t$. The two triangles shown (one includes θ_1 and l_1, the other θ_2 and l_2) have the side labeled a in common. Thus

$$\sin\theta_1 = \frac{l_1}{a} = \frac{v_1 t}{a}$$

and

$$\sin\theta_2 = \frac{l_2}{a} = \frac{v_2 t}{a}.$$

Dividing these two equations, we find that

$$\frac{\sin\theta_2}{\sin\theta_1} = \frac{v_2}{v_1}. \tag{15–19}$$

Law of refraction

FIGURE 15–31 (a) Soldier analogy, to (b) derive law of refraction for waves.

Since θ_1 is the angle of incidence (θ_i), and θ_2 is the angle of refraction (θ_r), Eq. 15–19 gives the quantitative relation between the two. Of course, if the wave were going in the opposite direction, the argument would not be changed. Only θ_1 and θ_2 would change roles: θ_1 would be the angle of refraction and θ_2 the angle of incidence. Clearly then, if the wave travels into a medium where it can move faster, it will bend in the opposite way, $\theta_r > \theta_i$. We see from Eq. 15–19 that if the velocity increases, the angle increases, and vice versa.

Earthquake waves refract within the Earth as they travel through rock of different densities (and therefore the velocity is different) just as water waves do. Light waves refract as well, and when we discuss light we shall find Eq. 15–19 very useful.

EXAMPLE 15–9 **Refraction of earthquake wave.** An earthquake P wave passes across a boundary in rock where its velocity increases from 6.5 km/s to 8.0 km/s. If it strikes this boundary at 30°, what is the angle of refraction?

Earthquake wave refraction

SOLUTION Since $\sin 30° = 0.50$, Eq. 15–19 yields

$$\sin\theta_2 = \frac{(8.0\ \text{m/s})}{(6.5\ \text{m/s})}(0.50) = 0.62$$

so $\theta_2 = 38°$.

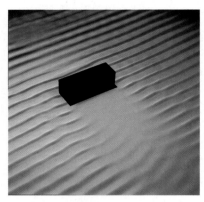

FIGURE 15–32 Wave diffraction. The waves come from the upper left. Note how the waves, as they pass the obstacle, bend around behind it, into the "shadow region."

Waves spread as they travel, and when they encounter an obstacle they bend around it somewhat and pass into the region behind as shown in Fig. 15–32 for water waves. This phenomenon is called **diffraction**.

The amount of diffraction depends on the wavelength of the wave and on the size of the obstacle, as shown in Fig. 15–33. If the wavelength is much larger than the object, as with the grass blades of Fig. 15–33a, the wave bends around them almost as if they were not there. For larger objects, parts (b) and (c), there is more of a "shadow" region behind the obstacle where we might not expect the waves to penetrate—but they do, at least a little. But notice in part (d), where the obstacle is the same as in part (c) but the wavelength is longer, that there is more diffraction into the shadow region. As a rule of thumb, *only if the wavelength is smaller than the size of the object will there be a significant shadow region.* It is worth noting that this rule applies to *reflection* from an obstacle as well. Very little of a wave is reflected unless the wavelength is smaller than the size of the obstacle.

A rough guide to the amount of diffraction is

$$\theta(\text{radians}) \approx \frac{\lambda}{L}$$

where θ is roughly the angular spread of waves after passing through an opening of width L or around an obstacle of width L.

That waves can bend around obstacles, and thus can carry energy to areas behind obstacles, is clearly different from energy carried by material particles. A clear example is the following: if you are standing around a corner on one side of a building, you can't be hit by a baseball thrown from the other side, but you can hear a shout or other sound because the sound waves diffract around the edges.

FIGURE 15–33 Water waves passing objects of various sizes. Note that the larger the wavelength compared to the size of the object, the more diffraction there is into the "shadow region."

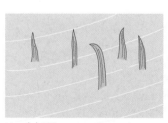 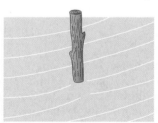  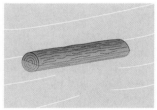

(a) Water waves passing blades of grass

(b) Stick in water

(c) Short-wavelength waves passing log

(d) Long-wavelength waves passing log

Summary

Vibrating objects act as sources of **waves** that travel outward from the source. Waves on water and on a string are examples. The wave may be a **pulse** (a single crest) or it may be continuous (many crests and troughs).

The **wavelength** of a continuous wave is the distance between two successive crests (or any two identical points on the wave shape).

The **frequency** is the number of full wavelengths (or crests) that pass a given point per unit time.

The **wave velocity** (how fast a crest moves) is equal to the product of wavelength and frequency,

$$v = \lambda f.$$

The **amplitude** of a wave is the maximum height of a crest, or depth of a trough, relative to the normal (or equilibrium) level.

In a **transverse wave**, the oscillations are perpendicular to the direction in which the wave travels. An example is a wave on a string.

In a **longitudinal wave**, the oscillations are along (parallel to) the line of travel; sound is an example.

The velocity of both longitudinal and transverse waves in matter is proportional to the square root of an elastic force factor divided by an inertia factor (or density).

Waves carry energy from place to place without matter being carried. The energy transported by a wave, the power (energy transported per unit time), and the **intensity** of a wave (energy transported across unit area per unit time) are all proportional to the square of the amplitude of the wave.

For a wave traveling outward in three dimensions from a point source, the intensity (ignoring damping) decreases with the square of the distance from the source,

$$I \propto \frac{1}{r^2}.$$

The amplitude decreases linearly with distance from the source.

A one-dimensional transverse wave traveling to the right along the x axis (x increasing) can be represented by a formula for its amplitude as a function of position and time as

$$D(x, t) = D_M \sin\left[\left(\frac{2\pi}{\lambda}\right)(x - vt)\right] = D_M \sin(kx - \omega t)$$

where

$$k = \frac{2\pi}{\lambda} \quad \text{and} \quad \omega = 2\pi f.$$

If a wave is traveling toward decreasing values of x,

$$D(x, t) = D_M \sin(kx + \omega t).$$

When two or more waves pass through the same region of space at the same time, the displacement at any given point will be the vector sum of the displacements of the separate waves. This is the **principle of superposition**. It is valid for mechanical waves if the amplitudes are small enough that the restoring force of the medium is proportional to displacement.

Waves reflect off objects in their path. When a wave strikes a boundary between two materials in which it can travel, part of the wave is reflected and part is transmitted.

When the **wave front** of a two or three dimensional wave strikes an object, the angle of reflection equals the angle of incidence.

When two waves pass through the same region of space at the same time, they **interfere**. From the superposition principle, the resultant displacement at any point and time is the sum of their separate displacements. This can result in **constructive interference, destructive interference,** or something in between depending on the amplitudes and relative phases of the waves.

Waves traveling on a cord (or other medium) of fixed length interfere with waves that have reflected off the end and are traveling back in the opposite direction. At certain frequencies, **standing waves** can be produced in which the waves seem to be standing still rather than traveling. The cord (or other medium) is vibrating as a whole. This is a resonance phenomenon and the frequencies at which standing waves occur are called **resonant frequencies**. The points of destructive interference (no vibration) are called **nodes**. Points of constructive interference (maximum amplitude of vibration) are called **antinodes**.

The wavelengths of standing waves are given by $\lambda_n = 2L/n$ where n is an integer.

Questions

1. Is the frequency of a simple periodic wave equal to the frequency of its source? Why or why not?

2. Explain the difference between the speed of a transverse wave traveling down a rope and the speed of a tiny piece of the rope.

3. Why do the strings used for the lowest-frequency notes on a piano normally have wire wrapped around them?

4. What kind of waves do you think will travel down a horizontal metal rod if you strike its end (a) vertically from above and (b) horizontally parallel to its length?

5. Since the density of air decreases with an increase in temperature, but the bulk modulus B is nearly independent of temperature, how would you expect the speed of sound waves in air to vary with temperature?

6. Give examples, other than those already mentioned, of one-, two-, and three-dimensional waves.

7. The speed of sound in most solids is somewhat greater than in air, yet the density of solids is much greater (10^3–10^4 times). Explain.

8. Give two reasons why circular water waves decrease in amplitude as they travel away from the source.

9. Two linear waves have the same amplitude and otherwise are identical, except one has half the wavelength of the other. Which transmits more energy? By what factor?

10. The intensity of a sound in real-life situations does not decrease precisely with the square of the distance from the source as we might expect from Eq. 15–8. Why not?

11. Will any function of ($x - vt$)—see Eq. 15–14—represent a wave motion? Why or why not? If not, give an example.

12. When a sinusoidal wave crosses the boundary between two sections of rope as in Fig. 15–20, the frequency does not change (although the wavelength and velocity do change). Explain why.

13. If a sinusoidal wave on a two-section string (Fig. 15–20) is inverted upon reflection, does the transmitted wave have a longer or shorter wavelength?

14. Is energy always conserved when two waves interfere? Explain.

15. If a string is vibrating in three segments, are there any places one can touch it with a knife blade without disturbing the motion?

16. When a standing wave exists on a string, the vibrations of incident and reflected waves cancel at the nodes. Does this mean that energy was destroyed? Explain.

17. Why can you make water slosh back and forth in a pan only if you shake the pan at a certain frequency?

18. Can the amplitude of the standing waves in Fig. 15–26 be greater than the amplitude of the vibrations that cause them (up and down motion of the hand)?

19. When a cord is vibrated as in Fig. 15–26 by hand or by a mechanical vibrator, the "nodes" are not quite true nodes (at rest). Explain. [*Hint*: Consider damping and energy flow from hand or vibrator.]

* 20. AM radio signals can usually be heard behind a hill, but FM often cannot. That is, AM signals bend more than FM. Explain. (Radio signals, as we shall see, are carried by electromagnetic waves whose wavelength for AM is typically 200 to 600 m and for FM about 3 m.)

Problems

Sections 15–1 and 15–2

1. (I) A fisherman notices that wave crests pass the bow of his anchored boat every 4.0 s. He measures the distance between two crests to be 9.0 m. How fast are the waves traveling?

2. (I) AM radio signals have frequencies between 550 kHz and 1600 kHz (kilohertz) and travel with a speed of 3.0×10^8 m/s. What are the wavelengths of these signals? On FM the frequencies range from 88 MHz to 108 MHz (megahertz) and travel at the same speed. What are their wavelengths?

3. (I) A sound wave in air has a frequency of 262 Hz and travels with a speed of 330 m/s. How far apart are the wave crests (compressions)?

4. (I) Calculate the speed of longitudinal waves in (a) water and (b) granite.

5. (I) Determine the wavelength of a 5000-Hz sound wave traveling along an iron rod.

6. (II) A rope of mass 0.65 kg is stretched between two supports 30 m apart. If the tension in the rope is 120 N, how long will it take a pulse to travel from one support to the other?

7. (II) A 0.40-kg rope is stretched between two supports, 4.8 m apart. When one support is struck by a hammer, a transverse wave travels down the rope and reaches the other support in 0.85 s. What is the tension in the rope?

8. (II) A sailor strikes the side of his ship just below the surface of the sea. He hears the echo of the wave reflected from the ocean floor directly 3.5 s later. How deep is the ocean at this point?

9. (II) A gondola is connected to the top of a hill by a steel cable of length 600 m and diameter 1.5 cm. As the gondola comes to the end of its run, it bumps into the side and sends a wave pulse along the cable. It is observed that it took 16 s for the pulse to return. (a) What is the speed of the pulse? (b) What is the tension in the cable?

10. (II) The wave on a string shown in Fig. 15–34 is moving to the right with a speed of 1.80 m/s. (a) Draw the shape of the string 1.00 s later and indicate which parts of the string are moving up and which down at that instant. (b) What is the vertical speed of point A on the string at the instant shown in the figure?

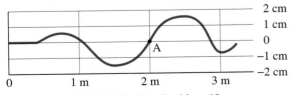

FIGURE 15–34 Problem 10.

11. (II) S and P waves from an earthquake travel at different speeds and this difference helps in the determination of the earthquake "epicenter" (where the disturbance took place). (a) Assuming typical speeds of 9.0 km/s and 5.5 km/s for P and S waves, respectively, how far away did the earthquake occur if a particular seismic station detects the arrival of these two types of waves exactly 94 s apart? (b) Is one seismic station sufficient to determine the position of the focus? Explain.

Section 15–3

12. (I) Two earthquake waves have the same frequency as they travel through the same portion of the Earth, but one is carrying twice the energy. What is the ratio of the amplitudes of the two waves?

13. (I) Compare (a) the intensities and (b) the amplitudes of an earthquake P wave as it passes two points 10 km and 20 km from the source.

14. (II) The intensity of a particular earthquake wave is measured to be 2.2×10^6 W/m² at a distance of 100 km from the source. (a) What was the intensity when it passed a point only 4.0 km from the source? (b) What was the total power passing through an area of 5.0 m² at a distance of 4.0 km?

15. (II) Show that if damping is ignored the amplitude D_M of circular water waves decreases as the square root of the distance r from the source: $D_M \propto 1/\sqrt{r}$.

16. (II) (a) Show that the intensity of a wave is equal to the energy density (energy per unit volume) in the wave times the wave speed. (b) What is the energy density 2.0 m from a 100-W light bulb? Light travels 3.0×10^8 m/s.

17. (II) A small steel wire of diameter 1.0 mm is connected to a vibrator and is under a tension of 4.5 N. The frequency of the vibrator is 60.0 Hz and it is observed that the amplitude of the wave on the steel wire is 0.50 cm. (a) What is the power output of the vibrator, assuming that the wave is not reflected back? (b) If the power output stays constant but the frequency is doubled, what is the amplitude of the wave?

18. (II) (a) Show that the average rate with which energy is transported along a cord by a mechanical wave of frequency f and amplitude D_M is

$$\overline{P} = 2\pi^2 \mu v f^2 D_M^2,$$

where v is the speed of the wave and μ is the mass per unit length of the cord. (b) If the cord is under a tension $F_T = 100\,N$ and has mass per unit length 0.10 kg/m, what power is required to transmit 120-Hz transverse waves of amplitude 2.0 cm?

Section 15–4

19. (I) Suppose at $t = 0$, a wave shape is represented by $D = D_M \sin(2\pi x/\lambda + \phi)$; that is, it differs from Eq. 15–9 by a constant phase factor ϕ. What then will be the equation for a wave traveling to the left along the x axis as a function of x and t?

20. (II) A transverse traveling wave on a cord is represented by $D = 0.48 \sin(5.6x + 84t)$ where D and x are in meters and t in seconds. For this wave determine (a) the wavelength, (b) frequency, (c) velocity (magnitude and direction), (d) amplitude, and (e) maximum and minimum speeds of particles of the cord.

21. (II) Consider a point on the string of Example 15–4 that is 1.00 m from the left-hand end. Determine (a) the maximum velocity of this point, and (b) its maximum acceleration. (c) What is its velocity and acceleration at $t = 2.0\,s$?

22. (II) Show, for a sinusoidal transverse wave traveling on a string, that the slope of the string at any point x is equal to the ratio of the transverse speed of the particle to the speed of the wave at that point.

23. (II) A transverse wave pulse travels to the right along a string with a speed $v = 2.0\,m/s$. At $t = 0$ the shape of the pulse is given by the function

$$D = 0.45 \cos(3.0x + 1.2),$$

where D and x are in meters. (a) Plot D versus x at $t = 0$. (b) Determine a formula for the wave pulse at any time t assuming there are no frictional losses. (c) Plot $D(x, t)$ versus x at $t = 1.0\,s$. (d) Repeat parts (b) and (c) assuming the pulse is traveling to the left.

24. (II) A 440-Hz longitudinal wave in air has a speed of 345 m/s. (a) What is the wavelength? (b) How much time is required for the phase to change by 90° at a given point in space? (c) At a particular instant, what is the phase difference (in degrees) between two points 4.4 cm apart?

25. (II) Write down the equation for the wave in Problem 24 if its amplitude is 0.020 cm and at $t = 0$, $x = 0$, and $D = -0.020$ cm.

26. (II) A sinusoidal wave traveling on a string in the negative x direction has amplitude 1.00 cm, wavelength 3.00 cm, and frequency 200 Hz. At $t = 0$, the particle of string at $x = 0$ is displaced a distance $D = 0.80$ cm above the origin and is moving upward. (a) Sketch the shape of the wave at $t = 0$ and (b) determine the function of x and t that describes the wave.

* Section 15–5

* 27. (II) Determine if the function $D = D_M \sin kx \cos \omega t$ is a solution of the wave equation.

* 28. (II) Show by direct substitution that the following functions satisfy the wave equation: (a) $D(x, t) = D_M \ln(x + vt)$; (b) $D(x, t) = (x - vt)^4$.

* 29. (II) Show that the wave forms of Eqs. 15–13 and 15–15 satisfy the wave equation, Eq. 15–16.

* 30. (II) Let two linear waves be represented by $D_1 = f_1(x, t)$ and $D_2 = f_2(x, t)$. If both these waves satisfy the wave equation (Eq. 15–16), show that any combination $D = C_1 D_1 + C_2 D_2$ does as well, where C_1 and C_2 are constants.

Section 15–7

31. (II) Consider a sine wave traveling down the stretched two-part cord of Fig. 15–20. Determine a formula (a) for the ratio of the speeds of the wave in the two sections, v_2/v_1, and (b) for the ratio of the wavelengths in the two sections. (The frequency is the same in both sections. Why?) (c) Is the wavelength larger in the heavier cord or the lighter?

32. (II) A cord has two sections with linear densities of 0.10 kg/m and 0.20 kg/m, Fig. 15–35. An incident wave, given by $D = (0.050\,m) \sin(6.0x - 12.0t)$, where x is in meters and t in seconds, travels along the lighter cord. (a) What is the wavelength on the lighter section of the cord? (b) What is the tension in the cord? (c) What is the wavelength when the wave travels on the heavier section?

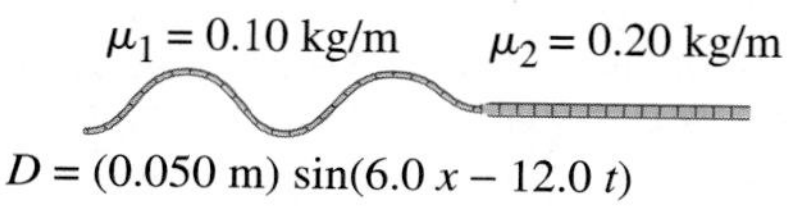

$$\mu_1 = 0.10 \text{ kg/m} \qquad \mu_2 = 0.20 \text{ kg/m}$$

$$D = (0.050 \text{ m}) \sin(6.0\,x - 12.0\,t)$$

FIGURE 15–35 Problem 32.

33. (III) A cord stretched to a tension F_T, consists of two sections (as in Fig. 15–21) whose linear densities are μ_1 and μ_2. Take $x = 0$ to be the point (a knot) where they are joined, with μ_1 referring to that section of cord to the left and μ_2 that to the right. A sinusoidal wave, $D = A \sin[k_1(x - v_1t)]$, starts at the left end of the cord. When it reaches the knot, part of it is reflected and part is transmitted. Let the equation of the reflected wave be $D = A_R \sin[k_1(x + v_1t)]$ and that for the transmitted wave be $D = A_T \sin[k_2(x - v_2t)]$. Since the frequency must be the same in both sections, we have $\omega_1 = \omega_2$ or $k_1 v_1 = k_2 v_2$. (a) Using the fact that the rope is continuous—so that a point an infinitesimal distance to the left of the knot has the same displacement at any moment (due to incident plus reflected waves) as a point just to the right of the knot (due to the transmitted wave)— show that $A = A_T + A_R$. (b) Assuming that the slope ($\partial D/\partial x$) of the string just to the left of the knot is the same as the slope just to the right of the knot, show that the amplitude of the reflected wave is given by

$$A_R = \left(\frac{v_1 - v_2}{v_1 + v_2}\right)A = \left(\frac{k_2 - k_1}{k_2 + k_1}\right)A.$$

(c) What is A_T in terms of A?

34. (I) The two pulses shown in Fig. 15–36 are moving toward each other. (a) Sketch the shape of the string at the moment they directly overlap. (b) Sketch the shape of the string a few moments later. (c) In Fig. 15–23a, at the moment the pulses pass each other, the string is straight. What has happened to the energy at this moment?

FIGURE 15–36 Problem 34.

35. (II) Suppose two linear waves of equal amplitude and frequency have a phase difference ϕ as they travel in the same medium. They can be represented by

$$D_1 = D_M \sin(kx - \omega t)$$
$$D_2 = D_M \sin(kx - \omega t + \phi).$$

(a) Use the trigonometric identity $\sin \theta_1 + \sin \theta_2 = 2 \sin \frac{1}{2}(\theta_1 + \theta_2) \cos \frac{1}{2}(\theta_1 - \theta_2)$ to show that the resultant wave is given by

$$D = \left(2D_M \cos \frac{\phi}{2}\right) \sin\left(kx - \omega t + \frac{\phi}{2}\right).$$

(b) What is the amplitude of this resultant wave? Is the wave purely sinusoidal, or not? (c) Show that constructive interference occurs if $\phi = 0, 2\pi, 4\pi$, and so on, and destructive interference occurs if $\phi = \pi, 3\pi, 5\pi$, etc. (d) Describe the resultant wave, by equation and in words, if $\phi = \pi/2$.

Section 15–9

36. (I) A violin string vibrates at 294 Hz when unfingered. At what frequency will it vibrate if it is fingered one-fourth of the way down from the end?

37. (I) If a violin string vibrates at 440 Hz as its fundamental frequency, what are the frequencies of the first four harmonics?

38. (I) A particular string resonates in four loops at a frequency of 264 Hz. Give at least three other frequencies at which it will resonate.

39. (I) In an earthquake, it is noted that a footbridge oscillated up and down in a one loop (fundamental standing wave) pattern once every 2.0 s. What other possible resonant periods of motion are there for this bridge? What frequencies do they correspond to?

40. (II) The velocity of waves on a string is 270 m/s. If the frequency of standing waves is 131 Hz, how far apart are the nodes?

41. (II) If two successive harmonics of a vibrating string are 280 Hz and 350 Hz, what is the frequency of the fundamental?

42. (II) A guitar string is 90.0 cm long and has a mass of 3.6 g. From the bridge to the support post ($= L$) is 60.0 cm and the string is under a tension of 520 N. What are the frequencies of the fundamental and first two overtones?

43. (II) Show that the frequency of standing waves on a string of length L and linear density μ, which is stretched to a tension F_T, is given by

$$f = \frac{n}{2L}\sqrt{\frac{F_T}{\mu}}$$

where n is an integer.

44. (II) One end of a horizontal string of linear density 4.8×10^{-4} kg/m is attached to a small amplitude mechanical 60-Hz vibrator. The string passes over a pulley, a distance $L = 1.40$ m away, and weights are hung from this end. What mass must be hung from this end of the string to produce (a) one loop, (b) two loops, and (c) five loops of a standing wave? Assume the string at the vibrator is a node, which is nearly true. Why can the amplitude of the standing wave be much greater than the vibrator amplitude?

45. (II) In Problem 44, the length of the string may be adjusted by moving the pulley. If the hanging mass is fixed at 0.080 kg, how many different standing wave patterns may be achieved by varying L between 10 cm and 1.5 m?

46. (II) The displacement of a standing wave is given by $D = 8.6 \sin(0.60x) \cos(58t)$, where x and D are in centimeters and t is in seconds. (a) What is the distance (cm) between nodes? (b) Give the amplitude, frequency, and speed of each of the component waves. (c) Find the speed of a particle of the string at $x = 3.20$ cm when $t = 2.5$ s.

47. (II) The displacement of a transverse wave traveling on a string is represented by $D = 4.2 \sin(0.71x - 47t + 2.1)$, where D and x are in cm and t in s. (a) Find an equation that represents a wave which, when traveling in the opposite direction, will produce a standing wave when added to this one. (b) What is the equation describing the standing wave?

48. (II) When you slosh the water back and forth in a tub at just the right frequency, the water alternately rises and falls at each end, remaining relatively calm at the center. Suppose the frequency to produce such a standing wave in a 60-cm-wide tub is 0.85 Hz. What is the speed of the water wave?

49. (II) A particular violin string plays at a frequency of 294 Hz. If the tension is increased 10 percent what will the new frequency be?

50. (II) Two traveling waves are described by the functions

$$D_1 = D_M \sin(kx - \omega t)$$
$$D_2 = D_M \sin(kx + \omega t),$$

where $D_M = 0.15$ m, $k = 3.5$ m^{-1}, and $\omega = 1.2$ s^{-1}. (a) Plot these two waves, from $x = 0$ to a point $x(>0)$ that includes one full wavelength. Choose $t = 1.0$ s. Plot the sum of the two waves and identify the nodes and antinodes in the plot, and compare to the analytic (mathematical) representation.

51. (II) Plot the two waves given in Problem 50, and their sum, as a function of time from $t = 0$ to $t = T$ (one period). Choose (a) $x = 0$ and (b) $x = \lambda/4$. Interpret your results.

52. (II) A standing wave on a 1.80-m-long horizontal string displays three loops when the string vibrates at 120 Hz. The maximum swing of the string (top to bottom) at the center of each loop is 12.0 cm. (a) What is the function describing the standing wave? (b) What are the functions describing the two equal-amplitude waves traveling in opposite directions that make up the standing wave?

* **53.** (I) An earthquake P wave traveling 8.0 km/s strikes a boundary within the Earth between two kinds of material. If it approaches the boundary at an incident angle of 50° and the angle of refraction is 31°, what is the speed in the second medium?

* **54.** (I) Water waves approach an underwater "shelf" where the velocity changes from 2.8 m/s to 2.5 m/s. If the incident wave crests make a 40° angle with the shelf, what will be the angle of refraction?

* **55.** (II) A longitudinal earthquake wave strikes a boundary between two types of rock at a 25° angle. As it crosses the boundary, the specific gravity of the rock changes from 3.7 to 2.8. Assuming that the elastic modulus is the same for both types of rock, determine the angle of refraction.

* **56.** (II) It is found for any type of wave, say an earthquake wave, that if it reaches a boundary beyond which its speed is increased, there is a maximum incident angle if there is to be a transmitted refracted wave. This maximum incident angle θ_{iM} corresponds to an angle of refraction equal to 90°. If $\theta_i > \theta_{iM}$, all the wave is reflected at the boundary and none is refracted (because this would correspond to $\sin\theta_r > 1$, where θ_r is the angle of refraction, which is impossible). This phenomenon is referred to as *total internal reflection*. (a) Find a formula for θ_{iM} using Eq. 15–19. (b) At what angles of incidence will there be only reflection and no transmission for an earthquake P wave traveling 7.5 km/s where it reaches a different kind of rock where its speed is 9.3 km/s?

* **57.** (II) A sound wave is traveling in warm air when it hits a layer of cold dense air. If the sound wave hits the cold air interface at an angle of 25°, what is the angle of refraction? Assume that the cold air temperature is −10°C and the warm air temperature is +10°C. The speed of sound as a function of temperature can be approximated by $v = (331 + 0.60\,T)$ m/s, where T is in °C.

General Problems

58. When you walk with a cup of coffee (diameter 8 cm) at just the right pace of about 1 step per second, the coffee builds up its "sloshing" until eventually, after a few steps, it starts to spill over the top. What is the speed of the waves in the coffee?

59. Two solid rods have the same bulk modulus but one is twice as dense as the other. In which rod will the speed of longitudinal waves be greater, and by what factor?

60. Two waves traveling along a stretched string have the same frequency, but one transports three times the power of the other. What is the ratio of the amplitudes of the two waves?

61. A bug on the surface of a pond is observed to move up and down a total vertical distance of 0.10 m, lowest to highest point, as a wave passes. (a) What is the amplitude of the wave? (b) If the ripples increase to 0.15 m, by what factor does the bug's maximum kinetic energy change?

62. A particular guitar string is supposed to vibrate at 200 Hz, but it is measured to actually vibrate at 205 Hz. By what percentage should the tension in the string be changed to get the frequency to the correct value?

63. An earthquake-produced surface wave can be approximated by a sinusoidal transverse wave. Assuming a frequency of 0.50 Hz (typical of earthquakes, which actually include a mixture of frequencies), what minimum amplitude will cause objects to leave contact with the ground?

64. A uniform cord of length L and mass m is hung vertically from a support. (a) Show that the speed of transverse waves in this cord is $\sqrt{gh}$, where h is the height above the lower end. (b) How long does it take for a pulse to travel upward from one end to the other?

65. A transverse wave pulse travels to the right along a string with a speed $v = 3.0$ m/s. At $t = 0$ the shape of the pulse is given by the function

$$D = \frac{4.0}{x^2 - 2.0},$$

where D and x are in meters. (a) Plot D versus x at $t = 0$. (b) Determine a formula for the wave pulse at any time t assuming there are no frictional losses. (c) Plot $D(x, t)$ versus x at $t = 0.50$ s. (d) Repeat parts (b) and (c) assuming the pulse is traveling to the left.

66. (a) Show that if the tension in a stretched string is changed by a small amount ΔF_T, the frequency of the fundamental is changed by an amount $\Delta f = \frac{1}{2}(\Delta F_T/F_T)f$. (b) By what percent must the tension in a piano string be increased or decreased to raise the frequency from 438 Hz to 442 Hz. (c) Does the formula in part (a) apply to the overtones as well?

67. Two strings on a musical instrument are tuned to play at 392 Hz (G) and 440 Hz (A). (a) What are the first two overtones for each string? (b) If the two strings have the same length and are under the same tension, what must be the ratio of their masses (M_G/M_A)? (c) If the strings, instead, have the same mass per unit length and are under the same tension, what is the ratio of their lengths (L_G/L_A)? (d) If their masses and lengths are the same, what must be the ratio of the tensions in the two strings?

68. A highway overpass was observed to resonate as one full loop $\left(\frac{1}{2}\lambda\right)$ when a small earthquake shook the ground vertically at 4.0 Hz. The highway department put a support at the center of the overpass, anchoring it to the ground as shown in Fig. 15–37. What resonant frequency would you now expect for the overpass? It is noted that earthquakes rarely do significant shaking above 5 or 6 Hz. Did the modifications do any good? Explain.

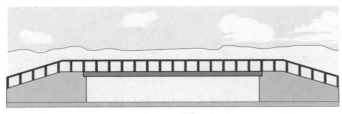

Before modification

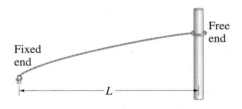

Added support

After modification

FIGURE 15–37   Problem 68.

69. A string can have a "free" end if that end is attached to a ring that can slide without friction on a vertical pole (Fig. 15–38). Determine the wavelengths of the resonant vibrations of such a string with one end fixed and the other free.

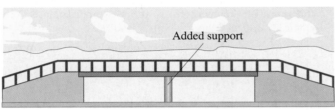

Fixed end

Free end

FIGURE 15–38
Problem 69.

70. A string fixed at two ends is pulled up at the center into the triangular shape shown in Fig. 15–39. Assuming that the tension F_T remains constant, calculate the energy of the vibrations of the string when it is released. [*Hint:* What work does it take to stretch the string up?]

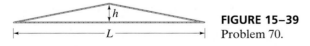

FIGURE 15–39
Problem 70.

71. Figure 15–40 shows the wave shape of a sinusoidal wave traveling to the right at two instants of time. What is the mathematical representation of this wave?

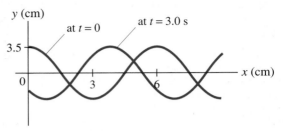

FIGURE 15–40   Problem 71.

72. Estimate the average power of a water wave when it hits the chest of an adult standing in the water at the seashore. Assume that the amplitude of the wave is 0.50 m, the wavelength is 2.5 m, and the period is 5.0 s.

73. Two wave pulses are traveling in opposite directions with the same speed of 5.0 cm/s as shown in Fig. 15–41. At $t = 0$, the leading edges of the two pulses are 15 cm apart. Sketch the wave pulses at $t = 1.0, 2.0$ and 3.0 s.

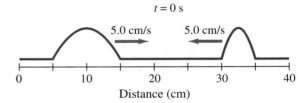

FIGURE 15–41   Problem 73.

74. For a spherical wave traveling uniformly away from a point source, show that the displacement can be represented by

$$D = \left(\frac{A}{r}\right)\sin(kr - \omega t),$$

where r is the radial distance from the source and A is a constant.

If music be the food of physics, play on.

This orchestra contains stringed instruments whose sound depends on transverse standing waves on strings, and wind instruments whose sound comes from longitudinal standing waves in an air column. Percussive instruments create more complicated standing waves. The human voice utilizes vocal chords (almost like strings) and the throat cavity (vibrating air) to produce its particular timbre.

We also study the decibel intensity scale, the ear's response, sound interference and the Doppler effect.

CHAPTER 16

Sound

Sound is associated with our sense of hearing and, therefore, with the physiology of our ears and the psychology of our brain which interprets the sensations that reach our ears. The term *sound* also refers to the physical sensation that stimulates our ears: namely, longitudinal waves.

We can distinguish three aspects of any sound. First, there must be a *source* for a sound; and as with any wave, the source of a sound wave is a vibrating object. Second, the energy is transferred from the source in the form of longitudinal sound *waves*. And third, the sound is *detected* by an ear or an instrument. Sound is an important factor in the design of buildings, especially theaters and auditoriums, but also in factories and other work places. We start this Chapter by looking at some aspects of sound waves themselves.

16–1 | Characteristics of Sound

We already saw in Chapter 15, Fig. 15–6, how a vibrating drumhead produces a sound wave in air. Indeed, we usually think of sound waves traveling in the air, for normally it is the vibrations of the air that force our eardrums to vibrate. But sound waves can also travel in other materials.

Two stones struck together under water can be heard by a swimmer beneath the surface, for the vibrations are carried to the ear by the water. When you put your ear flat against the ground, you can hear an approaching train or truck. In this case the ground does not actually touch your eardrum, but the longitudinal wave transmitted by the ground is called a sound wave just the same, for its vibrations cause the outer ear and the air within it to vibrate. Clearly, sound cannot travel in

TABLE 16–1   Speed of
Sound in Various Materials,
at 20°C and 1 atm

Material	Speed (m/s)
Air	343
Air (0°C)	331
Helium	1005
Hydrogen	1300
Water	1440
Seawater	1560
Iron and steel	≈ 5000
Glass	≈ 4500
Aluminum	≈ 5100
Hardwood	≈ 4000

➥ PHYSICS APPLIED

How far away is the lightning?

Loudness

Pitch

Audible frequency range

➥ PHYSICS APPLIED

Autofocusing camera

the absence of matter. For example, a bell ringing inside an evacuated jar cannot be heard, nor does sound travel through the empty reaches of outer space.

The **speed of sound** is different in different materials. In air at 0°C and 1 atm, sound travels at a speed of 331 m/s. We saw in Eq. 15–4 $\left(v = \sqrt{B/\rho}\right)$ that the speed depends on the elastic modulus, B, and the density, ρ, of the material. Thus for helium, whose density is much less than that of air but whose elastic modulus is not greatly different, the speed is about three times as great as in air. In liquids and solids, which are much less compressible and therefore have much greater elastic moduli, the speed is larger still. The speed of sound in various materials is given in Table 16–1. The values depend somewhat on temperature, but this is significant mainly for gases. For example, in air near room temperature, the speed increases approximately 0.60 m/s for each Celsius degree increase in temperature:

$$v \approx (331 + 0.60\,T)\,\text{m/s}, \qquad \text{[Speed of sound in air]}$$

where T is the temperature in °C. Unless stated otherwise, we will assume in this chapter that $T = 20°C$, so that $v = \left[331 + (0.60)(20)\right]\text{m/s} = 343\,\text{m/s}$.

CONCEPTUAL EXAMPLE 16–1 **Distance from a lightning strike.** A rule of thumb that tells how close lightning has hit is: "one mile for every five seconds before the thunder is heard." Justify, noting that the speed of light is so high $(3 \times 10^8\,\text{m/s})$ that the time for light to travel is negligible compared to the time for sound.

RESPONSE The speed of sound in air is about 340 m/s, so to travel 1 km = 1000 m takes about 3 seconds. One mile is about 1.6 kilometers, so the time for the thunder to travel a mile is about $(1.6)(3) \approx 5$ seconds.

Two aspects of any sound are immediately evident to a human listener. These are "loudness" and "pitch," and each refers to a sensation in the consciousness of the listener. But to each of these subjective sensations there corresponds a physically measurable quantity. **Loudness** is related to the energy in the sound wave, and we shall discuss it in Section 16–3.

The **pitch** of a sound refers to whether it is high, like the sound of a piccolo or violin, or low, like the sound of a bass drum or string bass. The physical quantity that determines pitch is the frequency, as was first noted by Galileo. The lower the frequency, the lower the pitch, and the higher the frequency, the higher the pitch.[†] The human ear responds to frequencies in the range from about 20 Hz to about 20,000 Hz. (Recall that 1 Hz is 1 cycle per second.) This is called the **audible range**. These limits vary somewhat from one individual to another. One general trend is that as people age, they are less able to hear the high frequencies, so that the high-frequency limit may be 10,000 Hz or less.

Sound waves whose frequencies are outside the audible range may reach the ear, but we are not generally aware of them. Frequencies above 20,000 Hz are called **ultrasonic** (do not confuse with *supersonic*, which is used for an object moving with a speed faster than the speed of sound). Many animals can hear ultrasonic frequencies; dogs, for example, can hear sounds as high as 50,000 Hz, and bats can detect frequencies as high as 100,000 Hz.

EXAMPLE 16–2 **Autofocusing with sound waves.** Autofocusing cameras emit a pulse of very high frequency (ultrasonic) sound that travels to the object being photographed, and include a sensor that detects the returning reflected sound, as shown in Fig. 16–1. To get an idea of the time sensitivity of the detector, calculate the travel time of the pulse for an object (*a*) 1.0 m away, (*b*) 20 m away.

[†] Although pitch is determined mainly by frequency, it also depends to a slight extent on loudness. For example, a very loud sound may seem slightly lower in pitch than a quiet sound of the same frequency.

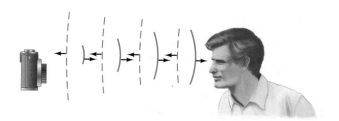

FIGURE 16–1 Example 16–2.
Autofocusing camera emits an
ultrasonic pulse. Solid blue lines
represent the moving wave front of
the outgoing wave pulse moving to
the right; dashed lines represent the
wave front of the pulse reflected off
the person's face, returning to the
camera. The time information allows
the camera mechanism to adjust the
lens to focus on the face.

SOLUTION We assume the temperature is about 20°C, so the speed of sound, as calculated above, is 343 m/s. (*a*) The pulse travels 1.0 m to the object and 1.0 m back, for a total of 2.0 m. Since speed = distance/time, we have

$$t = \frac{\text{distance}}{\text{speed}} = \frac{2.0 \text{ m}}{343 \text{ m/s}} = 0.0059 \text{ s} = 5.9 \text{ ms}.$$

(*b*) The total distance now is $2 \times 20 \text{ m} = 40 \text{ m}$, so

$$t = \frac{40 \text{ m}}{343 \text{ m/s}} = 0.12 \text{ s} = 120 \text{ ms}.$$

Sound waves whose frequencies are below the audible range (that is, less than 20 Hz) are called **infrasonic**. Sources of infrasonic waves include earthquakes, thunder, volcanoes, and waves produced by vibrating heavy machinery. This last source can be particularly troublesome to workers, for infrasonic waves—even though inaudible—can cause damage to the human body. These low-frequency waves act in a resonant fashion, causing considerable motion and irritation of internal organs of the body.

16–2 Mathematical Representation of Longitudinal Waves

In Section 15–4, we saw that a one-dimensional sinusoidal wave traveling along the *x* axis can be represented by the relation (Eq. 15–10c)

$$D = D_M \sin(kx - \omega t). \tag{16–1}$$

Here the wave number *k* is related to the wavelength λ by $k = 2\pi/\lambda$, and $\omega = 2\pi f$ where *f* is the frequency; *D* is the displacement at position *x* and time *t*, and D_M is its maximum value, the *amplitude*. For a transverse wave—such as a wave on a string—the displacement *D* is perpendicular to the direction of wave propagation along the *x* axis. But for a longitudinal wave the displacement *D* is *along the direction of wave propagation*. That is, *D* is parallel to *x* and represents the displacement of a tiny volume element of the medium from its equilibrium position.

Longitudinal (sound) waves can also be considered from the point of view of variations in pressure rather than displacement. Indeed, longitudinal waves are often called **pressure waves**. The pressure variation is usually easier to measure than the displacement (see Example 16–4). As can be seen in Fig. 16–2, in a wave "compression" (where molecules are closest together), the pressure is higher than normal, whereas in an expansion (or rarefaction) the pressure is less than normal. Figure 16–3 shows a graphical representation of a sound wave in air in terms of (a) displacement and (b) pressure. Note that the displacement wave is a quarter wavelength, or 90° ($\pi/2$ rad), out of phase with the pressure wave: where the pressure is a maximum or minimum, the molecules are momentarily at rest in equilibrium so the displacement from equilibrium is zero; and where the pressure variation is zero, the displacement is a maximum or minimum.

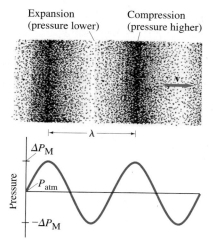

FIGURE 16–2 Longitudinal sound wave traveling to the right, and its graphical representation in terms of pressure.

FIGURE 16–3 Representation of a sound wave in terms of (a) displacement and (b) pressure.

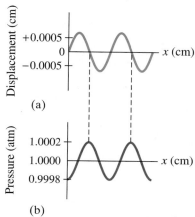

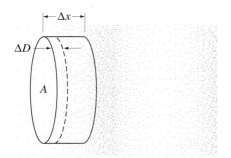

FIGURE 16–4 Longitudinal wave in a fluid moves to the right. A thin layer of fluid, in a thin cylinder of area A and thickness Δx, changes in volume as a result of pressure variation as the wave passes. At the moment shown, the pressure will increase as the wave moves to the right, so the thickness of our layer will decrease, by an amount ΔD.

Pressure Wave Derivation

Let us now derive the mathematical representation of the pressure variation in a traveling longitudinal wave. From the definition of the bulk modulus, B (Eq. 12–7a),

$$\Delta P = -B(\Delta V/V),$$

where $\Delta V/V$ is the fractional change in volume of the medium due to a pressure change ΔP, and ΔP represents the pressure difference from the normal pressure P_0 (no wave present). The negative sign reflects the fact that the volume decreases ($\Delta V < 0$) if the pressure is increased. Consider now a layer of fluid through which the longitudinal wave is passing (Fig. 16–4). If this layer has thickness Δx and area A, then its volume is $V = A\,\Delta x$. As a result of pressure variation in the wave, the volume will change by an amount $\Delta V = A\,\Delta D$ where ΔD is the change in thickness of this layer as it compresses or expands. (Remember that D represents the displacement of the medium.) Thus we have

$$\Delta P = -B\frac{A\,\Delta D}{A\,\Delta x}.$$

To be precise, we take the limit of $\Delta x \to 0$, so we obtain

$$\Delta P = -B\frac{\partial D}{\partial x}, \tag{16–2}$$

where we use the partial derivative notation since D is a function of both x and t. If the displacement D is sinusoidal as given by Eq. 16–1, then we have from Eq. 16–2 that

$$\Delta P = -(BD_M k)\cos(kx - \omega t). \tag{16–3}$$

Thus the pressure varies sinusoidally as well, but is out of phase from the displacement by 90° or a quarter wavelength; see Fig. 16–3. The quantity $BD_M k$ is called the **pressure amplitude**, ΔP_M. It represents the maximum and minimum amounts by which the pressure varies from the normal ambient pressure. We can thus write

$$\Delta P = -\Delta P_M \cos(kx - \omega t), \tag{16–4}$$

where, using $v = \sqrt{B/\rho}$ (Eq. 15–4), and $k = \omega/v = 2\pi f/v$ (Eq. 15–12), then

$$\begin{aligned}
\Delta P_M &= BD_M k \\
&= \rho v^2 D_M k \\
&= 2\pi\rho v D_M f. \tag{16–5}
\end{aligned}$$

16–3 | Intensity of Sound; Decibels

Intensity

Like pitch, **loudness** is a sensation in the consciousness of a human being. It too is related to a physically measurable quantity, the **intensity** of the wave. Intensity is defined as the energy transported by a wave per unit time across unit area perpendicular to the energy flow, and, as we saw in the previous chapter, is proportional to the square of the wave amplitude. Intensity has units of power per unit area, or watts/meter² (W/m^2).

The human ear can detect sounds with an intensity as low as $10^{-12}\,\text{W/m}^2$ and as high as $1\,\text{W/m}^2$ (and even higher, although above this it is painful). This is an incredibly wide range of intensity, spanning a factor of 10^{12} from lowest to highest. Presumably because of this wide range, what we perceive as loudness is not directly proportional to the intensity. To produce a sound that seems twice as loud requires a sound wave that has about ten times the intensity. This is roughly true at any sound level for frequencies near the middle of the audible range. For example, a sound wave of intensity $10^{-2}\,\text{W/m}^2$ sounds to an average human like it is about twice as loud as one whose intensity is $10^{-3}\,\text{W/m}^2$, and four times as loud as $10^{-4}\,\text{W/m}^2$.

Sound Level

Because of this relationship between the subjective sensation of loudness and the physically measurable quantity "intensity," it is usual to specify sound intensity levels using a logarithmic scale. The unit on this scale is a **bel**, after the inventor, Alexander Graham Bell, or much more commonly, the **decibel** (dB) which is $\frac{1}{10}$ bel ($10\,dB = 1$ bel). The **sound level**, β, of any sound is defined in terms of its intensity, I, as

$$\beta \text{ (in dB)} = 10 \log \frac{I}{I_0}, \qquad \text{(16–6)}$$

Sound level (decibels)

where I_0 is the intensity of some reference level and the logarithm is to the base 10. I_0 is usually taken as the minimum intensity audible to an average person, the "threshold of hearing," which is $I_0 = 1.0 \times 10^{-12}\,W/m^2$. Thus, for example, the sound level of a sound whose intensity $I = 1.0 \times 10^{-10}\,W/m^2$ will be

$$\beta = 10 \log \left(\frac{1.0 \times 10^{-10}\,W/m^2}{1.0 \times 10^{-12}\,W/m^2} \right) = 10 \log 100 = 20\,dB,$$

since log 100 is equal to 2.0. Notice that the sound level at the threshold of hearing is 0 dB. That is, $\beta = 10 \log 10^{-12}/10^{-12} = 10 \log 1 = 0$ since log 1 = 0. Notice too that an increase in intensity by a factor of 10 corresponds to a level increase of 10 dB. An increase in intensity by a factor of 100 corresponds to a level increase of 20 dB. Thus a 50-dB sound is 100 times more intense than a 30-dB sound, and so on.

The intensities and sound levels for a number of common sounds are listed in Table 16–2.

TABLE 16–2 Intensity of Various Sounds

Source of the Sound	Sound Level (dB)	Intensity (W/m^2)
Jet plane at 30 m	140	100
Threshold of pain	120	1
Loud rock concert	120	1
Siren at 30 m	100	1×10^{-2}
Auto interior, at 90 km/h	75	3×10^{-5}
Busy street traffic	70	1×10^{-5}
Conversation, at 50 cm	65	3×10^{-6}
Quiet radio	40	1×10^{-8}
Whisper	20	1×10^{-10}
Rustle of leaves	10	1×10^{-11}
Threshold of hearing	0	1×10^{-12}

EXAMPLE 16–3 **Loudspeaker response.** A high-quality loudspeaker is advertised to reproduce, at full volume, frequencies from 30 Hz to 18,000 Hz with uniform intensity ± 3 dB. That is, over this frequency range, the sound level does not vary by more than 3 dB from the average. By what factor does the intensity change for the maximum sound level change of 3 dB?

SOLUTION Let us call the average intensity I_1 and the average level β_1. Then the maximum intensity, I_2, corresponds to a level $\beta_2 = \beta_1 + 3$ dB. Thus

$$\beta_2 - \beta_1 = 10 \log \frac{I_2}{I_0} - 10 \log \frac{I_1}{I_0}$$

$$3\,dB = 10 \left(\log \frac{I_2}{I_0} - \log \frac{I_1}{I_0} \right)$$

$$= 10 \log \frac{I_2}{I_1}$$

because $(\log a - \log b) = \log a/b$. Then

$$\log \frac{I_2}{I_1} = 0.30, \qquad \text{or} \qquad \frac{I_2}{I_1} = 10^{0.30} = 2.0,$$

so ± 3 dB corresponds to a doubling (or halving) of the intensity.

➡ **PHYSICS APPLIED**

Loudspeaker response (± 3 dB)

It is worth noting that a sound level difference of 3 dB (which corresponds to a doubled intensity as we just saw) corresponds to only a very small change in the subjective sensation of apparent loudness. Indeed, the average human can distinguish a difference in level of only about 1 or 2 dB.

The Ear's Response

Sensitivity of the ear

The ear is not equally sensitive to all frequencies. To hear the same loudness for sounds of different frequencies requires different intensities. Studies averaged over large numbers of people have produced the curves shown in Fig. 16–5. On this graph, each curve represents sounds that seemed to be equally loud. The

Loudness (in "phons")

number labeling each curve represents the **loudness level** (the units are called *phons*), which is numerically equal to the sound level in dB at 1000 Hz. For example, the curve labeled 40 represents sounds that are heard by an average person to have the same loudness as a 1000-Hz sound with a sound level of 40 dB. From this 40-phon curve, we see that a 100-Hz tone must be at a level of about 62 dB to sound as loud as a 1000-Hz tone of only 40 dB. The lowest curve in Fig. 16–5 (labeled 0) represents the sound level, as a function of frequency, for the softest sound that is just audible by a very good ear (the "threshold of hearing"). Note that the ear is most sensitive to sounds of frequency between 2000 and 4000 Hz, which are common in speech and music. Note too that whereas a 1000-Hz sound is audible at a level of 0 dB, a 100-Hz sound must be at least 40 dB to be heard.

The top curve in Fig. 16–5, labeled 120, represents the "threshold of feeling or pain." Sounds above this level can actually be felt and cause pain.

Figure 16–5 shows that at lower intensity levels, our ears are less sensitive to the high and low frequencies relative to middle frequencies. The "loudness" control on stereo systems is intended to compensate for this. As the volume is turned down, the loudness control boosts the high and low frequencies relative to the middle frequencies so that the sound will have a more "normal-sounding" frequency balance. Many listeners, however, find the sound more pleasing or natural without the loudness control.

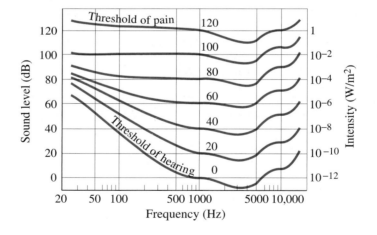

FIGURE 16–5 Sensitivity of the human ear as a function of frequency (see text). Note that the frequency scale is "logarithmic" in order to cover a wide range of frequencies.

Intensity Related to Amplitude

The intensity I is proportional to the square of the amplitude as we saw in Chapter 15. Indeed, using Eq. 15–7, $I = 2\pi^2 v\rho f^2 D_M^2$, we can relate the amplitude quantitatively to the intensity I or level β as the following Example shows.

EXAMPLE 16–4 **How tiny the displacement is.** (*a*) Calculate the maximum displacement of air molecules for a sound at the threshold of hearing, having a frequency of 1000 Hz. (*b*) Determine the maximum pressure variation in such a sound wave.

SOLUTION (*a*) The "threshold of hearing" at 1000 Hz is (Fig. 16–5) about 0 dB or 1.0×10^{-12} W/m². We use Eq. 15–7 of Chapter 15 and solve for D_M:

$$D_M = \frac{1}{\pi f} \sqrt{\frac{I}{2\rho v}}$$

$$= \frac{1}{(3.14)(1.0 \times 10^3 \text{ s}^{-1})} \sqrt{\frac{1.0 \times 10^{-12} \text{ W/m}^2}{(2)(1.29 \text{ kg/m}^3)(343 \text{ m/s})}} = 1.1 \times 10^{-11} \text{ m},$$

where we have taken the density of air to be 1.29 kg/m³ and the speed of sound in air (assumed 20°C) as 343 m/s. We see how incredibly sensitive the human ear is: It can detect displacements of air molecules which are actually less than the diameter of atoms (about 10^{-10} m).

Ear detects displacements smaller than size of atoms

(*b*) Now we are dealing with sound as a pressure wave (Section 16–2). From Eq. 16–5,

$$\Delta P_M = 2\pi \rho v D_M f = 3.1 \times 10^{-5} \text{ Pa}$$

or 3.1×10^{-10} atm. Again we see that the human ear is incredibly sensitive.

By combining Eqs. 15–7 and 16–5, we can write the intensity in terms of the pressure amplitude, ΔP_M:

$$I = 2\pi^2 v \rho f^2 D_M^2 = 2\pi^2 v \rho f^2 (\Delta P_M / 2\pi \rho v f)^2$$

$$I = \frac{(\Delta P_M)^2}{2v\rho}. \qquad \qquad \textbf{(16–7)}$$

Intensity related to pressure amplitude

The intensity, when given in terms of pressure amplitude, thus does not depend on frequency.

Normally, the loudness or intensity of a sound decreases as you get farther from the source of the sound. In enclosed rooms, this effect is altered because of absorption or reflection from the walls. However, if a source is in the open so that sound can radiate freely in all directions, the intensity decreases as the inverse square of the distance,

$$I \propto \frac{1}{r^2}$$

as we saw in Section 15–3, Eq. 15–8. Of course, if there is significant reflection from surrounding structures or the ground, the situation will be more complicated.

EXAMPLE 16–5 **Airplane roar.** The sound level of a jet plane at a distance of 30 m is 140 dB. What is the sound level 300 m away? (Ignore reflections from the ground.)

➡ **P H Y S I C S   A P P L I E D**

Jet plane noise

SOLUTION The intensity I at 30 m is found from Eq. 16–6:

$$140 \text{ dB} = 10 \log \left(\frac{I}{10^{-12} \text{ W/m}^2} \right).$$

FIGURE 16–6 Airport worker with sound-intensity reducing ear covers (headphones).

Reversing the log equation to solve for I we have:

$$10^{14} = \frac{I}{10^{-12} \text{ W/m}^2},$$

so $I = 10^2$ W/m². At 300 m, 10 times as far away, the intensity will be $\left(\frac{1}{10}\right)^2 = 1/100$ as much, or 1 W/m². Hence, the sound level is

$$\beta = 10 \log \left(\frac{1 \text{ W/m}^2}{10^{-12} \text{ W/m}^2} \right) = 120 \text{ dB}.$$

Even at 300 m, the sound is at the threshold of pain. This is why workers at airports wear ear covers to protect their ears from damage (Fig. 16–6).

TABLE 16–3 Equally Tempered Chromatic Scale[†]

Note	Frequency (Hz)
C	262
C♯ or D♭	277
D	294
D♯ or E♭	311
E	330
F	349
F♯ or G♭	370
G	392
G♯ or A♭	415
A	440
A♯ or B♭	466
B	494
C′	524

[†] Only one octave is included.

➡ **PHYSICS APPLIED**

Musical instrument

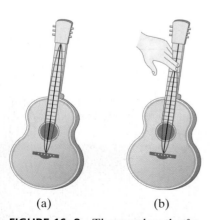

(a)   (b)

FIGURE 16–8 The wavelength of a fingered string (b) is shorter than that of an unfingered string (a). Hence, the frequency of the fingered string is higher. Only one string is shown on this guitar, and only the simplest standing wave, the fundamental, is shown.

16–4 | Sources of Sound: Vibrating Strings and Air Columns

The source of any sound is a vibrating object. Almost any object can vibrate and hence be a source of sound. We now discuss some simple sources of sound, particularly musical instruments. In musical instruments, the source is set into vibration by striking, plucking, bowing, or blowing. Standing waves are produced and the source vibrates at its natural resonant frequencies. The vibrating source is in contact with the air (or other medium) and pushes on it to produce sound waves that travel outward. The frequencies of the waves are the same as the source, but the speed and wavelengths can be different. A drum has a stretched membrane that vibrates. Xylophones and marimbas have metal or wood bars that can be set into vibration. Bells, cymbals, and gongs also make use of a vibrating metal. The most widely used instruments make use of vibrating strings, such as the violin, guitar, and piano, or make use of vibrating columns of air, such as the flute, trumpet, and pipe organ. We have already seen that the pitch of a pure sound is determined by the frequency. Typical frequencies for musical notes on the so-called "equally tempered chromatic scale" are given in Table 16–3 for the octave beginning with middle C. Note that one octave corresponds to a doubling of frequency. For example, middle C has frequency of 262 Hz whereas C′ (C above middle C) has twice the frequency, 524 Hz.

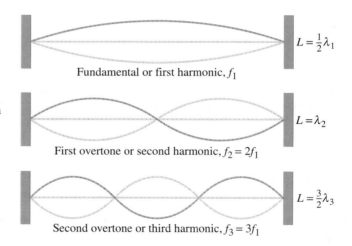

FIGURE 16–7 Standing waves on a string—only the lowest three frequencies are shown.

Fundamental or first harmonic, f_1 $L = \frac{1}{2}\lambda_1$

First overtone or second harmonic, $f_2 = 2f_1$ $L = \lambda_2$

Second overtone or third harmonic, $f_3 = 3f_1$ $L = \frac{3}{2}\lambda_3$

Stringed Instruments

We saw in Chapter 15, Fig. 15–27, how standing waves are established on a string, and we show this again here in Fig. 16–7. This is the basis for all stringed instruments. The pitch is normally determined by the lowest resonant frequency, the **fundamental**, which corresponds to nodes occurring only at the ends. The wavelength of the fundamental on the string is equal to twice the length of the string. Therefore, the fundamental frequency is $f = v/\lambda = v/2L$, where v is the velocity of the wave on the string. When a finger is placed on the string of, say, a guitar or violin, the effective length of the string is shortened. So its fundamental frequency, and pitch, is higher since the wavelength of the fundamental is shorter (Fig. 16–8). The strings on a guitar or violin are all the same length. They sound at a different pitch because the strings have different mass per unit length, μ, which affects the velocity as seen in Eq. 15–2, $v = \sqrt{F_T/\mu}$. Thus the velocity on a heavier string is less and the frequency will be less for the same wavelength. The tension may also be different; adjusting the tension is the means for tuning the instrument. In pianos and harps, the strings are each of different length. For the lower notes the strings are not only longer, but heavier as well, and the reason is illustrated in the following Example.

EXAMPLE 16–6 **Piano strings.** The highest key on a piano corresponds to a frequency about 150 times that of the lowest key. If the string for the highest note is 5.0 cm long, how long would the string for the lowest note have to be if it had the same mass per unit length and was under the same tension?

SOLUTION The velocity would be the same on each string, so the frequency is inversely proportional to the length L of the string ($f = v/\lambda = v/2L$). Thus

$$\frac{L_L}{L_H} = \frac{f_H}{f_L},$$

where the subscripts L and H refer to the lowest and highest notes, respectively. Thus $L_L = L_H(f_H/f_L) = (5.0 \text{ cm})(150) = 750 \text{ cm}$, or 7.5 m. This would be ridiculously long ($\approx 25 \text{ ft}$) for a piano. The longer lower strings are made heavier, so that even on grand pianos the strings are no longer than about 3 m.

EXAMPLE 16–7 **Frequencies and wavelengths in the violin.** A 0.32-m-long violin string is tuned to play A above middle C at 440 Hz. (*a*) What is the wavelength of the fundamental string vibration, and (*b*) what are the frequency and wavelength of the sound wave produced? (*c*) Why is there a difference?

SOLUTION (*a*) From Fig. 16–7 we see that the wavelength of the fundamental is

$$\lambda = 2L = 0.64 \text{ m} = 64 \text{ cm}.$$

This is the wavelength of the standing wave on the string.
(*b*) The sound wave that travels outward in the air (to reach our ears) has the same frequency, 440 Hz (why?). Its wavelength is

$$\lambda = \frac{v}{f} = \frac{343 \text{ m/s}}{440 \text{ Hz}} = 0.78 \text{ m} = 78 \text{ cm},$$

where v is the speed of sound in air (assumed at 20°C), Section 16–1.
(*c*) The wavelength of the sound wave is different from that of the standing wave on the string because the speed of sound in air (343 m/s at 20°C) is different from the speed of the wave on the string ($= f\lambda = 440 \text{ Hz} \times 0.64 \text{ m} = 280 \text{ m/s}$), which of course depends on the tension in the string and its mass per unit length.

(a)

(b)

FIGURE 16–9 (a) Sounding box (guitar); (b) sounding board (to which the strings are attached, inside of a piano).

Stringed instruments would not be very loud if they relied on their vibrating strings to produce the sound waves since the strings are simply too thin to compress and expand much air. Stringed instruments therefore make use of a kind of mechanical amplifier known as a *sounding board* (piano) or *sounding box* (guitar, violin), which acts to amplify the sound by putting a greater surface area in contact with the air (Fig. 16–9). When the strings are set into vibration, the sounding board or box is set into vibration as well. Since it has much greater area in contact with the air, it can produce a more intense sound wave. On an electric guitar, the sounding box is not so important since the vibrations of the strings are amplified electronically.

FIGURE 16–10 Wind instruments: clarinet (left) and flute.

Wind Instruments

Instruments such as woodwinds, the brasses, and the pipe organ produce sound from the vibrations of standing waves in a column of air within a tube or pipe (Fig. 16–10). Standing waves can occur in the air of any cavity, but the frequencies present are complicated for any but very simple shapes such as a long, narrow tube. In some instruments, a vibrating reed or the vibrating lip of the player helps to set up vibrations of the air column. In others, a stream of air is directed against one edge of the opening or mouthpiece, leading to turbulence which sets up the vibrations. Because of the disturbance, whatever its source, the air within the tube vibrates with a variety of frequencies; but only frequencies that correspond to standing waves will persist.

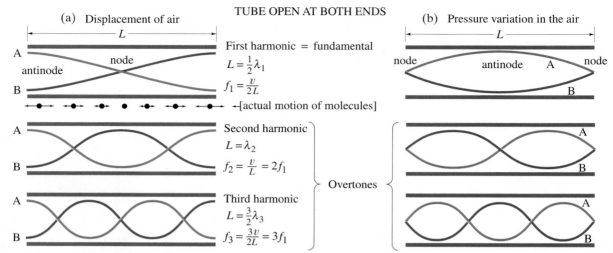

(a) Displacement of air

First harmonic = fundamental
$$L = \tfrac{1}{2}\lambda_1$$
$$f_1 = \frac{v}{2L}$$
←[actual motion of molecules]

Second harmonic
$$L = \lambda_2$$
$$f_2 = \frac{v}{L} = 2f_1$$

Third harmonic
$$L = \tfrac{3}{2}\lambda_3$$
$$f_3 = \frac{3v}{2L} = 3f_1$$

⎱ Overtones ⎰

(b) Pressure variation in the air

FIGURE 16–11 Modes of vibration (standing waves) for a tube open at both ends ("open tube"). The simplest modes of vibration are shown in (a), on the left, in terms of the motion of the air (displacement), and in (b), on the right, in terms of air pressure. These are graphs, shown placed within the tube, and are labeled A and B where B represents the wave form a half period after the moment when it has the shape labeled A. The actual motion of molecules for one case is shown just below the tube at upper left.

For a string fixed at both ends, Fig. 16–7, the standing waves have nodes (no movement) at the two ends, and one or more antinodes (large amplitude of vibration) in between; a node separates successive antinodes. The lowest-frequency standing wave, the *fundamental*, corresponds to a single antinode. The higher-frequency standing waves are called **overtones** or **harmonics**, as was discussed in Section 15–9. Specifically, the first harmonic is the fundamental, the second harmonic has twice the frequency of the fundamental,[†] and so on.

The situation is similar for a column of air, but we must remember that it is now air itself that is vibrating. We can describe the waves either in terms of the flow of the air—that is, in terms of the *displacement* of air—or in terms of the *pressure* in the air (see Figs. 16–2 and 16–3). In terms of displacement, the air at the closed end of a tube is a displacement node since the air is not free to move there, whereas near the open end of a tube there will be an antinode since the air can move freely. The air within the tube vibrates in the form of longitudinal standing waves. The possible modes of vibration for a tube open at both ends (called an **open tube**), are shown graphically in Fig. 16–11. They are shown for a tube that is open at one end but closed at the other (called a **closed tube**) in Fig. 16–12. [A tube closed at *both* ends, having no connection to the outside air, would be useless as an instrument.] The graphs in part (a) of each Figure (left sides) represent the displacement amplitude of the vibrating air in the tube. Note that these are graphs, and that the air molecules themselves oscillate *horizontally*, parallel to the tube length, as shown by the small arrows in the top diagram of Fig. 16–11a (on the left). The exact position of the antinode near the open end of a tube depends on the diameter of the tube, but if the diameter is small compared to the length, which is the usual case, the antinode occurs very close to the end as shown. We assume this is the case in what follows. (The position of the antinode may also depend slightly on the wavelength and other factors.)

Let us look in detail at the open tube, in Fig. 16–11a, which might be a flute. An open tube has displacement antinodes at both ends since the air is free to move at open ends. Notice that there must be at least one node within an open tube if there is to be a standing wave at all. A single node corresponds to the *fundamental frequency* of the tube. Since the distance between two successive nodes, or between two successive antinodes, is $\tfrac{1}{2}\lambda$, there is one-half a wavelength within the length of the tube for the simplest case of the fundamental (top diagram in Fig. 16–11a): $L = \tfrac{1}{2}\lambda$ or $\lambda = 2L$. So the fundamental frequency is $f_1 = v/\lambda = v/2L$, where v is the velocity of sound in air. The standing

---

[†] When the resonant frequencies above the fundamental (that is, the overtones) are integral multiples of the fundamental, they are called harmonics. But if the overtones are not integral multiples of the fundamental, as is the case for a vibrating drumhead, for example, they are not harmonics.

(a) Displacement of air

(b) Pressure variation in the air

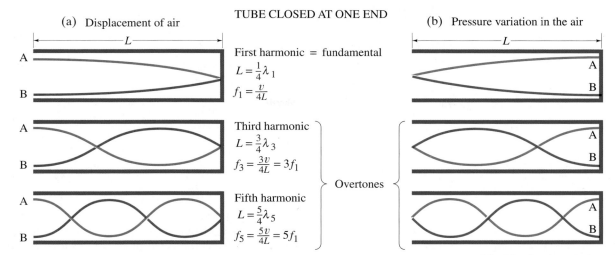

First harmonic = fundamental
$$L = \tfrac{1}{4}\lambda_1$$
$$f_1 = \frac{v}{4L}$$

Third harmonic
$$L = \tfrac{3}{4}\lambda_3$$
$$f_3 = \frac{3v}{4L} = 3f_1$$

Overtones

Fifth harmonic
$$L = \tfrac{5}{4}\lambda_5$$
$$f_5 = \frac{5v}{4L} = 5f_1$$

FIGURE 16–12 Modes of vibration (standing waves) for a tube closed at one end ("closed tube"). See caption for Fig. 16–11.

Open tubes produce all harmonics

wave with two nodes is the *first overtone* or *second harmonic* and has half the wavelength ($L = \lambda$) and twice the frequency. Indeed, the frequency of each overtone is an integral multiple of the fundamental frequency, as shown in Fig. 16–11a. This is just what is found for a string.

For a closed tube, shown in Fig. 16–12a, which could be a clarinet, there is always a displacement node at the closed end (because the air is not free to move) and an antinode at the open end (where the air can move freely). Since the distance between a node and the nearest antinode is $\tfrac{1}{4}\lambda$, we see that the fundamental in a closed tube corresponds to only one-fourth of a wavelength within the length of the tube: $L = \lambda/4$, and $\lambda = 4L$. The fundamental frequency is thus $f_1 = v/4L$, or half what it is for an open pipe of the same length. There is another difference, for as we can see from Fig. 16–12a, only the odd harmonics are present in a closed pipe: the overtones have frequencies equal to $3, 5, 7, \cdots$ times the fundamental frequency. There is no way for waves with $2, 4, 6, \cdots$ times the fundamental frequency to have a node at one end and an antinode at the other, and thus they cannot exist as standing waves in a closed tube.

If this description in terms of displacement seems hard to understand, or you want to understand it from another point of view, then consider a description in terms of the *pressure* in the air, shown in part (b) of Figs. 16–11 and 16–12 (right sides). Where the air in a wave is compressed, the pressure is higher, whereas in a wave expansion (or rarefaction), the pressure is less than normal. The open end of a tube is open to the atmosphere. Hence the pressure variation at an open end must be a *node*: the pressure doesn't alternate, but remains at the outside atmospheric pressure. If a tube has a closed end, the pressure at that closed end can readily alternate to be above or below atmospheric pressure. Hence there is a pressure *antinode* at a closed end of a tube. Of course there can be pressure nodes and antinodes within the tube, and some of the possible vibrational modes in terms of pressure for an open tube are shown in Fig. 16–11b, and for a closed tube are shown in Fig. 16–12b.

Pipe organs (Fig. 16–13) make use of both open and closed pipes. Notes of different pitch are sounded using different pipes with different lengths from a few centimeters to 5 m or more. Other musical instruments act either like a closed tube or an open tube. A flute, for example, is an open tube, for it is open not only where you blow into it, but also at the opposite end as well. The different notes on a flute and many other instruments are obtained by shortening the length of the tube—that is, by uncovering holes along its length. In a trumpet, on the other hand, pushing down on the valves opens additional lengths of tube. In all these instruments, the longer the length of the vibrating air column, the lower the frequency.

Closed tubes produce only odd harmonics

FIGURE 16–13 The pipe organ used by J. S. Bach in Leipzig, Germany.

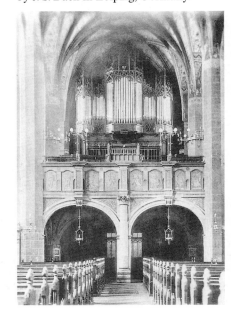

EXAMPLE 16–8 **Open and closed organ pipes.** What will be the fundamental frequency and first three overtones for a 26-cm-long organ pipe at 20°C if it is (a) open and (b) closed?

SOLUTION At 20°C, the speed of sound in air is 343 m/s (Section 16–1). (a) For the open pipe, Fig. 16–11, the fundamental frequency is

$$f_1 = \frac{v}{2L} = \frac{343 \text{ m/s}}{2(0.26 \text{ m})} = 660 \text{ Hz.}$$

The overtones, which include all harmonics, are 1320 Hz, 1980 Hz, 2640 Hz, and so on. (b) For a closed pipe, Fig. 16–12, we have

$$f_1 = \frac{v}{4L} = \frac{343 \text{ m/s}}{4(0.26 \text{ m})} = 330 \text{ Hz.}$$

But only the odd harmonics will be present, so the first three overtones will be 990 Hz, 1650 Hz, and 2310 Hz. (The closed pipe plays 330 Hz, which, from Table 16–3, is E above middle C, whereas the open pipe of the same length plays 660 Hz, an octave higher.)

EXAMPLE 16–9 **Flute.** A flute is designed to play middle C (262 Hz) as the fundamental frequency when all the holes are covered. Approximately how long should the distance be from the mouthpiece to the far end of the flute? (Note: This is only approximate since the antinode does not occur precisely at the mouthpiece.) Assume the temperature is 20°C.

SOLUTION The speed of sound in air at 20°C is 343 m/s. Because a flute is open at both ends, we use Fig. 16–11: the fundamental frequency f_1 is related to the length of the vibrating air column by $f = v/2L$. Solving for L, we find

$$L = \frac{v}{2f} = \frac{343 \text{ m/s}}{2(262 \text{ s}^{-1})} = 0.655 \text{ m.}$$

➡ P H Y S I C S   A P P L I E D

Temperature effect on staying in tune

EXAMPLE 16–10 **A cold flute.** If the temperature is only 10°C, what will be the frequency of the note played when all the openings are covered in the flute of Example 16–9?

SOLUTION The length L is still 65.5 cm. But now the velocity of sound is less since it changes by 0.60 m/s per each C°. For a drop of 10 C°, the velocity decreases by 6 m/s to 337 m/s. The frequency will be

$$f = \frac{v}{2L} = \frac{337 \text{ m/s}}{2(0.655 \text{ m})} = 257 \text{ Hz.}$$

We see why players of wind instruments take time to "warm up" their instruments so they will be in tune. The effect of temperature on stringed instruments is much smaller.

FIGURE 16–14 Example 16–11.

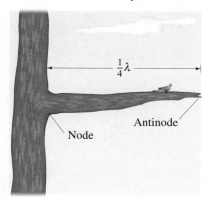

Node   Antinode $\frac{1}{4}\lambda$

CONCEPTUAL EXAMPLE 16–11 **Wind noise frequencies.** Wind can be noisy—it can "howl" in trees; it can "moan" in chimneys. Why is this so? What is really causing the noise, and about what range of frequencies would you expect to hear?

SOLUTION In each case, jets of air in the wind cause vibrations or oscillations, which produce the sound. The end of a tree limb fixed to the tree trunk is a node, whereas the other end is free to move and therefore is an antinode; the tree limb is thus about $\frac{1}{4}\lambda$ (Fig. 16–14). We estimate $v \approx 4000$ m/s for the speed of sound in wood (Table 16–1). Suppose that a tree limb has length $L \approx 2$ m; then $\lambda = 4L = 8$ m and $f = v/\lambda = (4000 \text{ m/s})/(8 \text{ m}) \approx 500$ Hz.

Wind can excite air oscillations in a chimney, much like in an organ pipe or flute. A chimney is a fairly long tube, perhaps 3 m in length, acting like a tube open at either one end or even both ends. If open at both ends ($\lambda = 2L$), with $v = 340$ m/s, we find $f_1 \approx v/2L \approx 56$ Hz, which is a fairly low note—no wonder chimneys "moan"!

* 16–5 Quality of Sound, and Noise

Whenever we hear a sound, particularly a musical sound, we are aware of its loudness, its pitch, and also of a third aspect called "quality." For example, when a piano and then a flute play a note of the same loudness and pitch (say middle C), there is a clear difference in the overall sound. We would never mistake a piano for a flute. This is what is meant by the **quality** of a sound. For musical instruments, the terms *timbre* or *tone color* are also used.

Just as loudness and pitch can be related to physically measurable quantities, so too can quality. The quality of a sound depends on the presence of overtones—their number and their relative amplitudes. Generally, when a note is played on a musical instrument, the fundamental as well as overtones are present simultaneously. We saw in Fig. 15–17 how the superposition of three wave forms, in that case the fundamental and first two overtones (with particular amplitudes), would combine to give a composite *waveform*. Of course, more than two overtones are usually present.

The relative amplitudes of the various overtones are different for different musical instruments, and this is what gives each instrument its characteristic quality or timbre. A graph showing the relative amplitudes of the harmonics produced by an instrument is called a "sound spectrum." Several typical examples for different instruments are shown in Fig. 16–15. Normally, the fundamental has the greatest amplitude and its frequency is what is heard as the pitch, although players can sometimes make the first overtone (for example) sound the loudest.

The manner in which an instrument is played strongly influences the sound quality. Plucking a violin string, for example, makes a very different sound than pulling a bow across it. The sound spectrum at the very start (or end) of a note, as when a hammer strikes a piano string, can be very different from the subsequent sustained tone. This too affects the subjective tone quality of an instrument.

An ordinary sound, like that made by striking two stones together, is a noise that has a certain quality, but a clear pitch is not discernible. A noise such as this is a mixture of many frequencies which bear little relation to one another. If a sound spectrum were made of this noise, it would not show discrete lines like those of Fig. 16–15. Instead it would show a continuous, or nearly continuous, spectrum of frequencies. Such a sound we call "noise" in comparison with the more harmonious sounds which contain frequencies that are simple multiples of the fundamental.

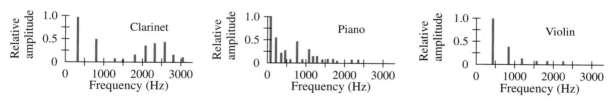

FIGURE 16–15 Sound spectra. The shapes of the spectra change as the instruments play different notes.

16–6 Interference of Sound Waves; Beats

Interference in Space

We saw in Section 15–8 that when two waves simultaneously pass through the same region of space, they interfere with one another. Since this can occur for any kind of wave, we should expect that interference will occur with sound waves, and indeed it does.

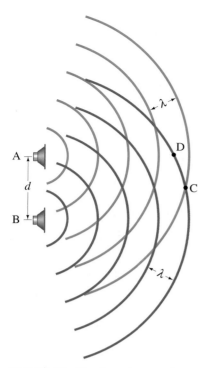

FIGURE 16–16 Sound waves from two loudspeakers interfere.

FIGURE 16–17 Sound waves of a single frequency from loudspeakers A and B (see Fig. 16–16) constructively interfere at C and destructively interfere at D.

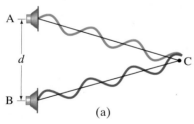

(a)

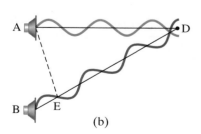

(b)

Beats

As a simple example, consider two large loudspeakers, A and B, a distance d apart on the stage of an auditorium as shown in Fig. 16–16. Let us assume the two speakers are emitting sound waves of the same single frequency and that they are in phase: that is, when one speaker is forming a compression, so is the other. (We ignore reflections from walls, floor, etc.) The curved lines in the diagram represent the crests of sound waves from each speaker. Of course, we must remember that for a sound wave, a crest is a compression in the air whereas a trough—which falls between two crests—is a rarefaction. A person or detector at a point such as C, which is the same distance from each speaker, will experience a loud sound because the interference will be constructive. On the other hand, at a point such as D in the diagram, little if any sound will be heard because destructive interference occurs—compressions of one wave meet rarefactions of the other and vice versa (see Fig. 15–24 and related discussion on water waves in Section 15–8).

An analysis of this situation is perhaps clearer if we graphically represent the wave forms as in Fig. 16–17. In Fig. 16–17a it can be seen that at point C, constructive interference occurs since both waves simultaneously have crests or simultaneously have troughs. In Fig. 16–17b we see the situation for point D. The wave from speaker B must travel a greater distance than the wave from A. Thus the wave from B lags behind that from A. In this diagram, point E is chosen so that the distance ED is equal to AD. Thus we see that if the distance BE is equal to precisely one-half the wavelength of the sound, the two waves will be exactly out of phase when they reach D, and destructive interference occurs. This then is the criterion for determining at what points destructive interference occurs: destructive interference occurs at any point whose distance from one speaker is greater than its distance from the other speaker by exactly one-half wavelength. Notice that if this extra distance (BE in Fig. 16–17b) is equal to a whole wavelength (or 2, 3, ··· wavelengths), then the two waves will be in phase and *constructive interference* occurs. If the distance BE equals $\frac{1}{2}, 1\frac{1}{2}, 2\frac{1}{2}, \cdots$ wavelengths, *destructive interference* occurs.

It is important to realize that a person sitting at point D hears nothing at all at this particular frequency, yet sound is coming from both speakers. Indeed, if one of the speakers is turned off, the sound from the other speaker would be clearly heard.

If a loudspeaker emits a whole range of frequencies, only specific ones will destructively interfere completely at a given point.

EXAMPLE 16–12 **Loudspeakers' interference.** Two loudspeakers are 1.00 m apart. A person stands 4.00 m from one speaker. How far must she be from the second speaker in order to detect destructive interference when the speakers emit 1150-Hz sound waves in phase with each other? Assume the temperature is 20°C.

SOLUTION The wavelength of this sound is

$$\lambda = \frac{v}{f} = \frac{343 \text{ m/s}}{1150 \text{ Hz}} = 0.30 \text{ m}.$$

For destructive interference to occur, the person must be one-half wavelength farther from one loudspeaker than from the other, or 0.15 m. Thus the person must be 4.15 m (or 3.85 m) from the second speaker. If the speakers are less than 0.15 m apart, there would be no point that was 0.15 m farther from one speaker than the other, and there would be no point where destructive interference would occur.

Beats—Interference in Time

We have been discussing interference of sound waves that takes place in space. An interesting and important example of interference that occurs in time is the phenomenon known as **beats**: two sources of sound—say, two tuning forks—are close in frequency but not exactly the same. Sound waves from the two sources interfere with each other and the sound level at a given position alternately rises and falls; the regularly spaced intensity changes are called beats.

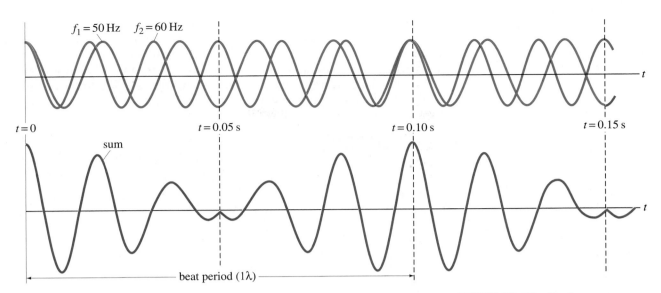

To see how beats arise, consider two equal-amplitude sound waves of frequency $f_1 = 50\,\text{Hz}$ and $f_2 = 60\,\text{Hz}$, respectively. In 1.00 s, the first source makes 50 vibrations whereas the second makes 60. We now examine the waves at one point in space equidistant from the two sources. The waveforms for each wave as a function of time are shown on the top graph of Fig. 16–18; the magenta line represents the 50 Hz wave, the blue line represents the 60 Hz wave. The lower graph in Fig. 16–18 shows the sum of the two waves. At time $t = 0$ the two waves are shown to be in phase and interfere constructively. Because the two waves vibrate at different rates, at time $t = 0.05\,\text{s}$ they are completely out of phase and destructive interference occurs as shown in the figure. At $t = 0.10\,\text{s}$, they are again in phase and the resultant amplitude again is large. Thus the resultant amplitude is large every 0.10 s and in between it drops drastically. This rising and falling of the intensity is what is heard as beats.[†] In this case the beats are 0.10 s apart. That is, the **beat frequency** is ten per second or 10 Hz. This result, that the beat frequency equals the difference in frequency of the two waves can be shown in general as follows.

Beat frequency = difference in the two wave frequencies

Let the two waves, of frequencies f_1 and f_2, be represented at a fixed point in space by

$$D_1 = D_M \sin 2\pi f_1 t$$

and

$$D_2 = D_M \sin 2\pi f_2 t.$$

The resultant displacement, by the principle of superposition, is

$$D = D_1 + D_2 = D_M(\sin 2\pi f_1 t + \sin 2\pi f_2 t).$$

Using the trigonometric identity $\sin A + \sin B = 2 \sin \frac{1}{2}(A + B) \cos \frac{1}{2}(A - B)$, we have

$$D = \left[2D_M \cos 2\pi \left(\frac{f_1 - f_2}{2} \right) t \right] \sin 2\pi \left(\frac{f_1 + f_2}{2} \right) t. \qquad \textbf{(16–8)}$$

We can interpret Eq. 16–8 as follows. The superposition of the two waves results in a wave that vibrates at the average frequency of the two components, $(f_1 + f_2)/2$. This vibration has an amplitude given by the expression in brackets, and this amplitude varies in time, from zero to a maximum of $2D_M$ (the sum of the separate amplitudes), with a frequency of $(f_1 - f_2)/2$. A beat occurs whenever $\cos 2\pi[(f_1 - f_2)/2]t$ equals $+1$ or -1 (see Fig. 16–18); that is, two beats occur per cycle, so the beat frequency is twice $(f_1 - f_2)/2$ which is just $f_1 - f_2$, the difference in frequency of the component waves.

[†] Beats will be heard even if the amplitudes are not equal, as long as the difference is not great.

The phenomenon of beats can occur with any kind of wave and is a very sensitive method for comparing frequencies. For example, to tune a piano, a piano tuner listens for beats produced between his standard tuning fork and that of a particular string on the piano, and knows it is in tune when the beats disappear. The members of an orchestra can tune up by listening for beats between their instruments and that of a standard tone (usually A above middle C at 440 Hz) produced by a piano or an oboe.

EXAMPLE 16–13 **Beats.** A tuning fork produces a steady 400-Hz tone. When this tuning fork is struck and held near a vibrating guitar string, twenty beats are counted in five seconds. What are the possible frequencies produced by the guitar string?

SOLUTION The beat frequency is

$$f_{\text{beat}} = 20 \text{ vibrations}/5 \text{ s} = 4 \text{ Hz}.$$

This is the difference of the frequencies of the two waves, and because one wave is known to be 400 Hz, the other must be either 404 Hz or 396 Hz.

16–7 | Doppler Effect

You may have noticed that the pitch of the siren on a speeding firetruck drops abruptly as it passes you. Or you may have noticed the change in pitch of a blaring horn on a fast-moving car as it passes by. The pitch of the sound from the engine of a race car changes as it passes an observer. When a source of sound is moving toward an observer, the pitch is higher than when the source is at rest; and when the source is traveling away from the observer, the pitch is lower. This phenomenon is known as the **Doppler effect**[†] and occurs for all types of waves. Let us now see why it occurs, and calculate the change in frequency for sound waves.

To be concrete, consider the siren of a firetruck at rest, which is emitting sound of a particular frequency in all directions as shown in Fig. 16–19a. The wave velocity depends only on the medium in which it is traveling, and is independent of the velocity of the source or observer. If our source, the firetruck, is moving, the siren emits sound at the same frequency as it does at rest. But the sound wavefronts it emits forward are closer together than when the firetruck is at rest, as shown in Fig. 16–19b. This is because the firetruck, as it moves, is "chasing after" the previously emitted wavefronts. Thus an observer on the sidewalk will detect more wave crests passing per second, so the frequency heard is higher. The wavefronts emitted behind the truck, on the other hand, are farther apart than when the truck is at rest because the truck is speeding away from them. Hence, fewer wave crests per second pass by an observer behind the truck and the pitch is lower.

[†] After J. C. Doppler (1803–1853).

FIGURE 16–19 (a) Both observers on the sidewalk hear the same frequency from the firetruck at rest. (b) Doppler effect: observer toward whom the firetruck moves hears a higher-frequency sound, and observer behind the firetruck hears a lower frequency.

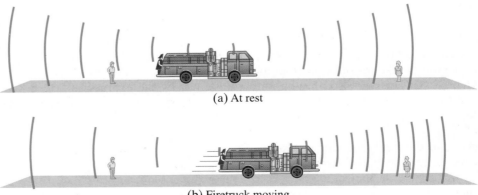

(a) At rest

(b) Firetruck moving

To calculate the change in frequency, we make use of Fig. 16–20, and we assume the air (or other medium) is at rest in our reference frame. In Fig. 16–20a, the source of the sound, shown as a dot, is at rest; two successive wave crests are shown, the second of which is just in the process of being emitted. The distance between these crests is λ, the wavelength. If the frequency of the source is f, then the time between emissions of wave crests is

Frequency change, moving source

$$T = \frac{1}{f}.$$

In Fig. 16–20b, the source is moving with a velocity v_S. In a time T (as just defined), the first wave crest has moved a distance $d = vT$, where v is the velocity of the sound wave in air (which is, of course, the same whether the source is moving or not). In this same time, the source has moved a distance $d_S = v_S T$. Then the distance between successive wave crests, which is the new wavelength λ', is (since $d = \lambda$)

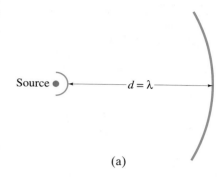

$$\lambda' = d - d_S$$
$$= \lambda - v_S T$$
$$= \lambda - v_S \frac{\lambda}{v} = \lambda\left(1 - \frac{v_S}{v}\right).$$

The change in wavelength, $\Delta\lambda$, is

$$\Delta\lambda = \lambda' - \lambda = -v_S \frac{\lambda}{v}.$$

So the shift in wavelength is directly proportional to the speed v_S of the source. The new frequency, on the other hand, is given by

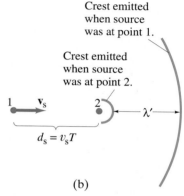

$$f' = \frac{v}{\lambda'} = \frac{v}{\lambda\left(1 - \dfrac{v_S}{v}\right)}$$

or, since $v/\lambda = f$,

$$f' = \frac{f}{\left(1 - \dfrac{v_S}{v}\right)}. \qquad \begin{bmatrix} \text{source moving toward} \\ \text{stationary observer} \end{bmatrix} \quad \textbf{(16–9a)}$$

FIGURE 16–20 Determination of frequency change in the Doppler effect (see text). Red dot is the source.

Because the denominator is less than 1, $f' > f$. For example, if a source emits a sound of frequency 400 Hz when at rest, then when the source moves toward a fixed observer with a speed of 30 m/s, the observer hears a frequency (at 20°C) of

$$f' = \frac{400\ \text{Hz}}{1 - \dfrac{30\ \text{m/s}}{343\ \text{m/s}}} = 438\ \text{Hz}.$$

For a source moving *away* from the observer at a speed v_S, the new wavelength will be

$$\lambda' = d + d_S,$$

and the change in wavelength will be

$$\Delta\lambda = \lambda' - \lambda = +v_S \frac{\lambda}{v}.$$

The frequency of the wave will be

$$f' = \frac{f}{\left(1 + \dfrac{v_S}{v}\right)}. \qquad \begin{bmatrix} \text{source moving away from} \\ \text{stationary observer} \end{bmatrix} \quad \textbf{(16–9b)}$$

In this case, if a source vibrating at 400 Hz is moving away from a fixed observer at 30 m/s, the observer hears a frequency of about 368 Hz.

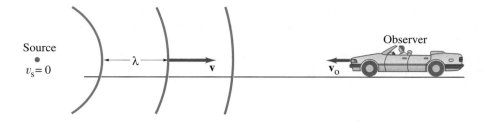

FIGURE 16–21 Observer moving with speed v_O toward a stationary source "sees" wave crests pass at speed $v' = v + v_O$ where v is the speed of the sound waves in air.

Frequency change, moving observer

The Doppler effect also occurs when the source is at rest and the observer is in motion. If the observer is traveling toward the source, the pitch is higher; and if the observer is traveling away from the source, the pitch is lower. Quantitatively the change in frequency is slightly different than for the case of a moving source. With a fixed source and a moving observer, the distance between wave crests, the wavelength λ, is not changed. But the velocity of the crests with respect to the observer is changed. If the observer is moving toward the source, Fig. 16–21, the speed of the waves relative to the observer is $v' = v + v_O$, where v is the velocity of sound in air (we assume the air is still) and v_O is the velocity of the observer. Hence, the new frequency is

$$f' = \frac{v'}{\lambda} = \frac{v + v_O}{\lambda}$$

or, since $\lambda = v/f$,

$$f' = \left(1 + \frac{v_O}{v}\right)f. \qquad \begin{bmatrix} \text{observer moving toward} \\ \text{stationary source} \end{bmatrix} \quad \textbf{(16–10a)}$$

If the observer is moving away from the source, the relative velocity is $v' = v - v_O$, so

$$f' = \left(1 - \frac{v_O}{v}\right)f. \qquad \begin{bmatrix} \text{observer moving away} \\ \text{from stationary source} \end{bmatrix} \quad \textbf{(16–10b)}$$

EXAMPLE 16–14 **A moving siren.** The siren of a police car at rest emits at a predominant frequency of 1600 Hz. What frequency will you hear if you are at rest and the police car moves at 25.0 m/s (*a*) toward you, (*b*) away from you?

SOLUTION (*a*) We use Eq. 16–9a:

$$f' = \frac{f}{\left(1 - \dfrac{v_S}{v}\right)} = \frac{1600 \text{ Hz}}{\left(1 - \dfrac{25.0 \text{ m/s}}{343 \text{ m/s}}\right)} = 1726 \text{ Hz}.$$

(*b*) We use Eq. 16–9b:

$$f' = \frac{f}{\left(1 + \dfrac{v_S}{v}\right)} = \frac{1600 \text{ Hz}}{\left(1 + \dfrac{25.0 \text{ m/s}}{343 \text{ m/s}}\right)} = 1491 \text{ Hz}.$$

When a sound wave is reflected from a moving obstacle, the frequency of the reflected wave will, because of the Doppler effect, be different from that of the incident wave. This is illustrated in the following Example.

EXAMPLE 16–15 **Two Doppler shifts.** A 5000-Hz sound wave is emitted by a stationary source toward an object moving 3.50 m/s toward the source (Fig. 16–22). What is the frequency of the wave reflected by the moving object as detected by a detector at rest near the source?

SOLUTION There are actually two Doppler shifts in this situation. First, the emitted wave strikes the moving object which is in effect a moving observer (Fig. 16–22a) that "detects" a sound wave of frequency (Eq. 16–10a):

$$f' = \left(1 + \frac{v_O}{v}\right)f = \left(1 + \frac{3.50 \text{ m/s}}{343 \text{ m/s}}\right)(5000 \text{ Hz}) = 5051 \text{ Hz}.$$

Second, the moving object takes this wave of frequency f' and reemits (or reflects) it, acting effectively as a moving source, so the frequency detected, f'', will be given by Eq. 16–9a:

$$f'' = \frac{f'}{\left(1 - \frac{v_S}{v}\right)} = \frac{5051 \text{ Hz}}{\left(1 - \frac{3.50 \text{ m/s}}{343 \text{ m/s}}\right)} = 5103 \text{ Hz}.$$

Thus the frequency shifts by 103 Hz.

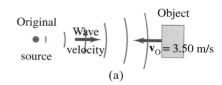

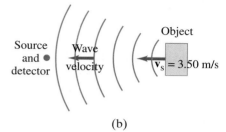

(b)

FIGURE 16–22 Example 16–15.

The incident wave and the reflected wave in Example 16–15, when mixed together (say, electronically), interfere with one another and beats are produced. The beat frequency is equal to the difference in the two frequencies, 103 Hz. This Doppler technique is used in a variety of medical applications, usually with ultrasonic waves in the megahertz frequency range. For example, ultrasonic waves reflected from red blood cells can be used to determine the velocity of blood flow. Similarly, the technique can be used to detect the movement of the chest of a young fetus and to monitor its heartbeat.

➡ **PHYSICS APPLIED**
Medical uses

For convenience, we can write Eqs. 16–9 and 16–10 as a single equation that covers all cases of both source and observer in motion:

$$f' = f\left(\frac{v \pm v_O}{v \mp v_S}\right). \qquad \textbf{(16–11)}$$

Source and observer moving

The upper signs apply if source and/or observer move toward each other; the lower signs apply if they are moving apart.

➡ **PROBLEM SOLVING**
Getting the signs right

Doppler Effect for Light

The Doppler effect occurs for other types of waves as well. Light and other types of electromagnetic waves (such as radar) exhibit the Doppler effect: although the formulas for the frequency shift are not identical to Eqs. 16–9 and 16–10, the effect is similar. One important application is for weather forecasting using radar. The time delay between the emission of radar pulses and their reception after being reflected off raindrops gives the position of precipitation. Measuring the Doppler shift in frequency (as in Example 16–15) tells how fast the storm is moving and in which direction.

➡ **PHYSICS APPLIED**
Doppler effect for EM waves and weather forecasting

Another important application is to astronomy where the velocities of distant galaxies can be determined from the Doppler shift. Light from such galaxies is shifted toward lower frequencies, indicating that the galaxies are moving away from us. This is called the **red shift** since red has the lowest frequency of visible light. The greater the frequency shift, the greater the velocity of recession. It is found that the farther the galaxies are from us, the faster they move away. This observation is the basis for the idea that the universe is expanding, and is one basis for the idea that the universe began with a great explosion, affectionately called the "Big Bang" (see Chapter 45).

➡ **PHYSICS APPLIED**
Red shift in cosmology

*16–8 Shock Waves and the Sonic Boom

An object such as an airplane traveling faster than the speed of sound is said to have a **supersonic speed**. Such a speed is often given as a **Mach**[†] **number**, which is defined as the ratio of the object's speed to that of sound in the medium at that location. For example, a plane traveling 600 m/s high in the atmosphere, where the speed of sound is only 300 m/s, has a speed of Mach 2.

[†]After the Austrian physicist Ernst Mach (1838–1916).

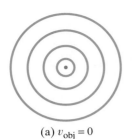
(a) $v_{obj} = 0$

(b) $v_{obj} < v_{snd}$

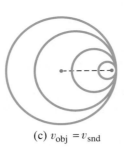
(c) $v_{obj} = v_{snd}$

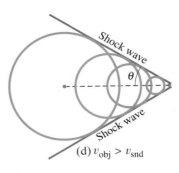
(d) $v_{obj} > v_{snd}$

FIGURE 16–23 Sound waves emitted by an object at rest (a) or moving (b, c, and d). If the object's velocity is less than the velocity of sound, the Doppler effect occurs (b); if its velocity is greater than the velocity of sound, a shock wave is produced (d).

Shock wave

Sonic boom

FIGURE 16–24 Bow waves produced by a boat.

FIGURE 16–25 (a) The (double) sonic boom has already been heard by person A on the left. It is just being heard by person B in the center. And it will shortly be heard by person C on the right. (b) Special photo of supersonic aircraft showing shock waves produced in the air. (Several closely spaced shock waves are produced by different parts of the aircraft.)

When a source of sound moves at subsonic speeds, the pitch of the sound is altered, as we have seen (the Doppler effect); see also Fig. 16–23a and b. But if a source of sound moves faster than the speed of sound, a more dramatic effect known as a **shock wave** occurs. In this case the source is actually "outrunning" the waves it produces. As shown in Fig. 16–23c, when the source is traveling at the speed of sound, the wave fronts it emits in the forward direction "pile up" directly in front of it. When the object moves at a supersonic speed, the wave fronts pile up on one another along the sides, as shown in Fig. 16–23d. The different wave crests overlap one another and form a single very large crest which is the shock wave. Behind this very large crest there is usually a very large trough. A shock wave is essentially the result of constructive interference of a large number of wave fronts. A shock wave in air is analogous to the bow wave of a boat traveling faster than the speed of the water waves it produces, Fig. 16–24.

When an airplane travels at supersonic speeds, the noise it makes and its disturbance of the air form into a shock wave containing a tremendous amount of sound energy. When the shock wave passes a listener, it is heard as a loud "sonic boom." A sonic boom lasts only a fraction of a second, but the energy it contains is often sufficient to break windows and cause other damage. It can be psychologically unnerving as well. Actually, a sonic boom is made up of two or more booms since major shock waves can form at the front and the rear of the aircraft, as well as at the wings, etc. (Fig. 16–25). Bow waves of a boat are also multiple, as can be seen in Fig. 16–24.

When an aircraft approaches the speed of sound, it encounters a barrier of sound waves in front of it (see Fig. 16–23c). In order to exceed the speed of sound, extra thrust is needed to pass through this "sound barrier." This is called "breaking the sound barrier." Once a supersonic speed is attained, this barrier no longer impedes the motion. It is sometimes erroneously thought that a sonic boom is produced only at the moment an aircraft is breaking through the sound barrier. Actually, a shock wave follows the aircraft at all times it is traveling at supersonic speeds. A series of observers on the ground will each hear a loud "boom" as the shock wave passes, Fig. 16–25. The shock wave consists of a cone whose apex is at the aircraft. The angle of this cone, θ, (see Fig. 16–23 d) is given by

$$\sin \theta = \frac{v_{snd}}{v_{obj}}, \qquad (16\text{–}12)$$

where v_{obj} is the velocity of the object (the aircraft) and v_{snd} is the velocity of sound in the medium (the proof is left as a Problem).

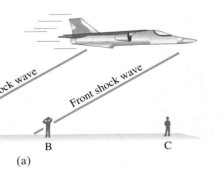

(a)

(b)

*16–9 Applications; Sonar, Ultrasound, and Ultrasound Imaging

The reflection of sound is used in many applications to determine distance. The **sonar**[†] or pulse-echo technique is used to locate underwater objects. A transmitter sends out a sound pulse through the water, and a detector receives its reflection, or echo, a short time later. This time interval is carefully measured, and from it the distance to the reflecting object can be determined since the speed of sound in water is known. The depth of the sea and the location of reefs, sunken ships, submarines, or schools of fish can be determined in this way. The interior structure of the Earth is studied in a similar way by detecting reflections of waves traveling through the Earth whose source was a deliberate explosion (called "soundings"). An analysis of waves reflected from various structures and boundaries within the Earth reveals characteristic patterns that are also useful in the exploration for oil and minerals.

Sonar generally makes use of **ultrasonic** frequencies: that is, waves whose frequencies are above 20 kHz, beyond the range of human detection. For sonar, the frequencies are typically in the range 20 kHz to 100 kHz. One reason for using ultrasound waves, other than the fact that they are inaudible, is that for shorter wavelengths there is less diffraction, so the beam spreads less and smaller objects can be detected.

The diagnostic use of ultrasound in medicine, in the form of images (sometimes called "sonograms") is an important and interesting application of physical principles. A **pulse-echo technique** is used, much like sonar. A high-frequency sound pulse is directed into the body, and its reflections from boundaries or interfaces between organs and other structures and lesions in the body are then detected. It is even possible to produce "real-time" ultrasound images, as if one were watching a movie of a section of the interior of the body.

The pulse-echo technique for medical imaging works as follows. A brief pulse of ultrasound is emitted by a transducer that transforms an electrical pulse into a sound-wave pulse. Part of the pulse is reflected at each interface surface in the body, and most (usually) continues on. The detection of the reflected pulses by the same transducer can then be displayed on the screen of a display terminal or monitor, as shown in Fig. 16–26a. The time elapsed from when the pulse is emitted to when each reflection (echo) is received is proportional to the distance to the reflecting surface. For example, if the distance from transducer to the vertebra is 25 cm, the pulse travels a round-trip distance of 2×25 cm $= 0.50$ m; the speed of sound in human tissue is about 1540 m/s (close to water), so the time taken is $t = d/v = (0.50\text{ m})/(1540\text{ m/s}) = 320\ \mu\text{s}$.

The *strength* of a reflected pulse depends mainly on the difference in density of the two materials on either side of the interface and can be displayed as a pulse or as a dot (Figs. 16–26b and c). Each echo dot (Fig. 16–26c) can be represented as a point, whose position is given by the time delay and whose brightness depends on the strength of the echo. A two-dimensional image can then be formed out of these dots from a series of scans. The transducer is moved, and at each position it sends out a pulse and receives echoes as shown in Fig. 16–27. Each trace can be plotted, spaced appropriately one below the other, to form an image on a display terminal as shown in Fig. 16–27b. Only 10 lines are shown in Fig. 16–27, so the image is crude. More lines give a more precise image. Photographs of ultrasound images are shown in Fig. 16–28.

[†] Sonar stands for "*so*und *na*vigation *r*anging."

→ PHYSICS APPLIED

Sonar and imaging

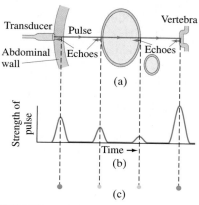

FIGURE 16–26 (a) Ultrasound pulse passes through abdomen, reflecting from surfaces in its path. (b) Reflected pulses plotted as a function of time when received by transducer. The vertical dashed lines point out which reflected pulse goes with which surface. (c) Dot display for the same echoes: brightness of each dot is related to signal strength.

FIGURE 16–27 (a) Ten traces are made across the abdomen by moving the transducer, or by using an array of transducers. (b) The echoes are plotted to produce the image. More closely spaced traces would give a more detailed image.

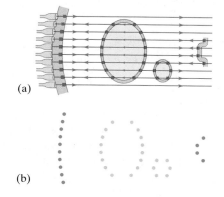

FIGURE 16–28 (a) Ultrasound image of a human fetus (with head at the left) within the uterus. (b) False-color high-resolution ultrasound image of fetus (different colors represent different intensities of reflected pulses).

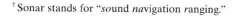
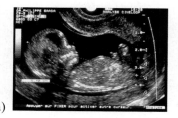

(a)   (b)

Summary

Sound travels as a longitudinal wave in air and other materials. In air, the speed of sound increases with temperature; at 20°C, it is about 343 m/s.

The **pitch** of a sound is determined by the frequency; the higher the frequency, the higher the pitch.

The **audible range** of frequencies for humans is roughly 20 Hz to 20,000 Hz (1 Hz = 1 cycle per second).

The **loudness** or **intensity** of a sound is related to the amplitude of the wave. Because the human ear can detect sound intensities from 10^{-12} W/m^2 to over 1 W/m^2, intensity levels are specified on a logarithmic scale. The **sound level** β, specified in decibels, is defined in terms of intensity I as

$$\beta = 10 \log (I/I_0),$$

where the reference intensity I_0 is usually taken to be 10^{-12} W/m^2.

Musical instruments are simple sources of sound in which *standing waves* are produced.

The strings of a stringed instrument may vibrate as a whole with nodes only at the ends; the frequency at which this occurs is called the **fundamental**. The string can also vibrate at higher frequencies, called **overtones** or **harmonics**, in which there are one or more additional nodes. The frequency of each harmonic is a whole-number multiple of the fundamental.

In wind instruments, standing waves are set up in the column of air within the tube.

The vibrating air in an **open tube** (open at both ends) has displacement antinodes at both ends. The fundamental frequency corresponds to a wavelength equal to twice the tube length: $\lambda_1 = 2L$. The harmonics have frequencies that are 2, 3, 4, ⋯ times the fundamental frequency, just as for strings.

For a **closed tube** (closed at one end), the fundamental corresponds to a wavelength four times the length of the tube: $\lambda_1 = 4L$. Only the odd harmonics are present, equal to 1, 3, 5, 7, ⋯ times the fundamental frequency.

Sound waves from different sources can interfere with each other. If two sounds are at slightly different frequencies, **beats** can be heard at a frequency equal to the difference in frequency of the two sources.

The **Doppler effect** refers to the change in pitch of a sound due to the motion either of the source or of the listener. If they are approaching each other, the pitch is higher; if they are moving apart, the pitch is lower.

Questions

1. What is the evidence that sound travels as a wave?

2. What is the evidence that sound is a form of energy?

3. Children sometimes play with a homemade "telephone" by attaching a string to the bottoms of two paper cups. When the string is stretched and a child speaks into one of the cups, the sound can be heard at the other cup. Explain clearly how the sound wave travels from one cup to the other. (See Fig. 16–29.)

FIGURE 16–29   Question 3.

4. When a sound wave passes from air into water, do you expect the frequency or wavelength to change?

5. What evidence can you give that the speed of sound in air does not depend significantly on frequency?

6. The voice of a person who has inhaled helium sounds very high-pitched. Why?

7. What is the main reason the speed of sound in hydrogen is greater than the speed of sound in air?

8. The molecules of a gas, such as air, move around randomly at fairly high speeds (Chapter 18). The distance between molecules, on the average, is many times their diameter. When a wave passes through a gas the impulse given to one molecule is given to another only when this distance is traveled and the two collide. Would you therefore expect the speed of sound in a gas to be limited by the average molecular speed?

9. Two tuning forks oscillate with the same amplitude, but one has twice the frequency. Which (if either) produces the more intense sound?

10. How does a rise in air temperature affect the loudness of sound coming from a source of fixed frequency and amplitude? (Assume that atmospheric pressure doesn't change.)

11. What is the reason that catgut strings on some musical instruments are wrapped with fine wire?

12. Explain how a tube might be used as a filter to reduce the amplitude of sounds in various frequency ranges. (An example is a car muffler.)

13. How will the air temperature in a room affect the pitch of organ pipes?

14. Why are the frets on a guitar spaced closer together as you move up the fingerboard toward the bridge?

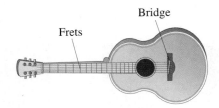

FIGURE 16–30   Question 14.

15. Standing waves can be said to be due to "interference in space," whereas beats can be said to be due to "interference in time." Explain.

16. In Fig. 16–16, if the frequency of the speakers were lowered, would the points D and C (where destructive and constructive interference occur) move farther apart or closer together?

17. Traditional methods of protecting the hearing of people who work in areas with very high noise levels have consisted mainly of efforts to block or reduce noise levels. With a relatively new technology, headphones are worn that do not block the ambient noise. Instead, a device is used which detects the noise, inverts it electronically, then feeds it to the headphones *in addition to* the ambient noise. How could adding *more* noise actually reduce the sound levels reaching the ears?

18. Suppose a source of sound moves at right angles to the line of sight of a listener at rest in still air. Will there be a Doppler effect? Explain.

19. If a wind is blowing, will this alter the frequency of the sound heard by a person at rest with respect to the source? Is the wavelength or velocity changed?

20. Figure 16–31 shows various positions of a child in motion on a swing. A monitor is blowing a whistle in front of the child on the ground. At which position will the child hear the highest frequency for the sound of the whistle? Explain your reasoning.

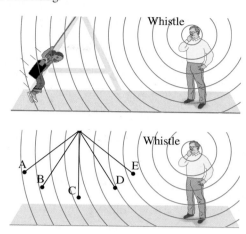

FIGURE 16–31 Question 20.

Problems

[Unless stated otherwise, assume $T = 20°C$ and $v_{sound} = 343$ m/s in air.]

Section 16–1

1. (I) A hiker determines the length of a lake by listening for the echo of her shout reflected by a cliff at the far end of the lake. She hears the echo 1.5 s after shouting. Estimate the length of the lake.

2. (I) (a) Calculate the wavelengths in air at 20°C for sounds in the maximum range of human hearing, 20 Hz to 20,000 Hz. (b) What is the wavelength of a 10-MHz ultrasonic wave?

3. (II) A person sees a heavy stone strike the concrete pavement. A moment later two sounds are heard from the impact: one travels in the air and the other in the concrete, and they are 1.4 s apart. How far away did the impact occur?

4. (II) A fishing boat is drifting just above a school of tuna on a foggy day. Without warning, an engine backfire occurs on another boat 1.0 km away (Fig. 16–32). How much time elapses before the backfire is heard (a) by the fish and (b) by the fishermen?

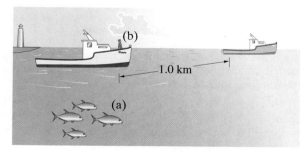

FIGURE 16–32 Problem 4.

5. (II) The sound from a very high burst of fireworks takes 4.5 s to arrive at your eardrums. The burst occurred 1500 m above you and traveled vertically through two stratified layers of air, the top one at 0°C and the bottom one at 20°C. How thick is each layer of air?

6. (II) If the camera of Example 16–2 focuses precisely at 20°C, what percent error will there be in distances when used at 0°C?

Section 16–2

7. (I) The pressure amplitude of a sound wave in air $(\rho = 1.29$ kg/m$^3)$ at 0°C is 3.0×10^{-3} Pa. What is the displacement amplitude if the frequency is (a) 100 Hz and (b) 10 kHz?

8. (I) What must be the pressure amplitude in a sound wave in air (0°C) if the air molecules undergo a maximum displacement equal to the diameter of an oxygen molecule, about 3×10^{-10} m? Assume a sound wave frequency of (a) 50 Hz and (b) 5.0 kHz.

9. (II) Write an expression that describes the pressure variation as a function of x and t for the waves described in Problem 8.

10. (II) The pressure variation in a sound wave is given by

$$\Delta P = 0.0025 \sin\left(\frac{\pi}{3} x - 1700\pi t\right),$$

where ΔP is in pascals, x in meters, and t in seconds. Determine (a) the wavelength, (b) the frequency, (c) the speed, and (d) the displacement amplitude of the wave. Assume the density of the medium to be $\rho = 2.7 \times 10^3$ kg/m^3.

11. (I) (a) What is the sound level of a sound whose intensity is 8.5×10^{-8} W/m²? (b) What is the intensity of a sound whose sound level is 25 dB?

12. (I) A 6000-Hz tone must have what sound level to seem as loud as a 100-Hz tone that has a 50-dB sound level? (See Fig. 16–5.)

13. (I) What are the lowest and highest frequencies that an average ear can detect when the sound level is 30 dB? (See Fig. 16–5)

14. (I) A stereo cassette player is said to have a signal-to-noise ratio of 63 dB. What is the ratio of intensities of the signal and the background noise?

15. (II) If the amplitude of a sound wave is tripled, (a) by what factor will the intensity increase? (b) By how many dB will the sound level increase?

16. (II) Human beings can typically detect a difference in sound intensity level of 2.0 dB. What is the ratio of the amplitudes of two sounds whose levels differ by this amount?

17. (II) Two sound waves have equal displacement amplitudes, but one has twice the frequency of the other. (a) Which has the greater pressure amplitude and by what factor is it greater? (b) What is the ratio of their intensities?

18. (II) A person standing a certain distance from an airplane with four equally noisy jet engines is experiencing a sound level bordering on pain, 120 dB. What sound level would this person experience if the captain shut down all but one engine?

19. (II) A 40-dB sound wave strikes an eardrum whose area is 5.0×10^{-5} m². (a) How much energy is absorbed by the eardrum per second? (b) At this rate, how long would it take your eardrum to receive a total energy of 1.0 J?

20. (II) (a) Estimate the power output of sound from a person speaking in normal conversation. Use Table 16–2. Assume the sound spreads roughly uniformly over a hemisphere in front of the mouth. (b) How many people would produce a total sound output of 100 W of ordinary conversation?

21. (II) If two firecrackers produce a sound level of 90 dB when fired simultaneously at a certain place, what will be the sound level if only one is exploded?

22. (II) What would be the sound level (in dB) of a sound wave in air that corresponds to a displacement amplitude of vibrating air molecules of 1.3 mm at 330 Hz?

23. (II) (a) Calculate the maximum displacement of air molecules when a 210-Hz sound wave passes whose intensity is at the threshold of pain (120 dB). (b) What is the pressure amplitude in this wave?

24. (II) Expensive stereo amplifier A is rated at 250 W per channel, while the more modest amplifier B is rated at 40 W per channel. (a) Estimate the sound level in decibels you would expect at a point 2.5 m from a loudspeaker connected in turn to each amp. (b) Will the expensive amp sound twice as loud as the cheaper one?

25. (II) At a rock concert, a dB meter registered 130 dB when placed 3.4 m in front of a loudspeaker on the stage. (a) What was the power output of the speaker, assuming uniform spherical spreading of the sound and neglecting absorption in the air? (b) How far away would the sound level be a somewhat reasonable 90 dB?

26. (II) A jet plane emits 5.0×10^5 J of sound energy per second. (a) What is the sound level 30 m away? Air absorbs sound at a rate of about 7.0 dB/km; calculate what the sound level will be (b) 1.0 km and (c) 5.0 km away from this jet plane, taking into account air absorption.

27. (II) (a) Show that the sound level, β, can be written in terms of the pressure amplitude, ΔP_M, as

$$\beta(\text{dB}) = 20 \log \frac{\Delta P_M}{\Delta P_{M0}},$$

where ΔP_{M0} is the pressure amplitude at some reference level. (b) The reference pressure amplitude ΔP_{M0} is often taken to be 3.0×10^{-5} N/m² corresponding to an intensity of 1.0×10^{-12} W/m². What would the sound level be if ΔP_M were 1 atm?

Section 16–4

28. (I) The G string on a violin has a fundamental frequency of 196 Hz. The length of the vibrating portion is 32 cm and has a mass of 0.68 g. Under what tension must the string be placed?

29. (I) (a) What resonant frequency would you expect from blowing across the top of an empty soda bottle that is 15 cm deep? (b) How would that change if it was one-third full of soda?

30. (I) How far from the end of the flute in Example 16–9 should the hole be that must be uncovered to play D above middle C at 294 Hz?

31. (I) If you were to build a pipe organ with open-tube pipes spanning the range of human hearing (20 Hz to 20 kHz), what would be the range of the lengths of pipes required?

32. (I) At a science museum there is a display called a sewer pipe symphony. It consists of many plastic pipes of various lengths, which are open on both ends. (a) If the pipes have lengths of 3.0 m, 2.5 m, 2.0 m, 1.5 m and 1.0 meters, what frequencies will be heard by a visitor's ear placed near the ends of the pipes? (b) Why does this display work better on a noisy day rather than a quiet day?

33. (I) An organ pipe is 78.0 cm long. What are the fundamental and first three audible overtones if the pipe is (a) closed at one end, and (b) open at both ends?

34. (II) An unfingered guitar string is 0.73 m long and is tuned to play E above middle C (330 Hz). (a) How far from the end of this string must the finger be placed to play A above middle C (440 Hz)? (b) What are the frequency and wavelength of the sound wave produced in air at 20°C?

35. (II) (a) Determine the length of an open organ pipe that emits middle C (262 Hz) when the temperature is 21°C. (b) What are the wavelength and frequency of the fundamental standing wave in the tube? (c) What are λ and f in the traveling sound wave produced in the outside air?

36. (II) The human ear canal is approximately 2.5 cm long. It is open to the outside and is closed at the other end by the tympanic membrane. Estimate the frequencies of the standing wave vibrations in the ear canal. What is the relationship of your answer to the information in the graph of Fig. 16–5?

37. (II) An organ is in tune at 20°C. By what percent will the frequency be off at 5.0°C?

38. (II) A particular organ pipe can resonate at 264 Hz, 440 Hz, and 616 Hz, but not at any other intermediate frequencies. (a) Is this an open or closed pipe? (b) What is the fundamental frequency of this pipe?

39. (II) (a) At $T = 15°C$, how long must an open organ pipe be if it is to have a fundamental frequency of 294 Hz? (b) If this pipe were filled with helium, what would its fundamental frequency be?

40. (II) A pipe in air at 20°C is to be designed to produce two successive harmonics at 240 Hz and 280 Hz. How long must the pipe be, and is it open or closed?

41. (II) A uniform narrow tube 1.95 m long is open at both ends. It resonates at two successive harmonics of frequency 275 Hz and 330 Hz. What is the speed of sound in the gas in the tube?

42. (II) How many overtones are present within the audible range for a 2.16 m-long organ pipe at 20°C (a) if it is open, and (b) if it is closed?

* Section 16–5

* 43. (II) Approximately what are the intensities of the first two overtones of a violin compared to the fundamental? How many decibels softer than the fundamental are the first and second overtones? (See Fig. 16–15.)

Section 16–6

44. (I) A piano tuner hears one beat every 2.0 s when trying to adjust two strings, one of which is sounding 440 Hz. How far off in frequency is the other string?

45. (I) A certain dog whistle operates at 23.5 kHz, while another (brand X) operates at an unknown frequency. If neither whistle can be heard by humans when played separately, but a shrill whine of frequency 5000 Hz occurs when they are played simultaneously, estimate the operating frequency of brand X.

46. (II) Two violin strings are tuned to the same frequency, 294 Hz. The tension in one string is then decreased by 2.0 percent. What will be the beat frequency heard when the two strings are played together?

47. (II) Two piano strings are supposed to be vibrating at 132 Hz, but a piano tuner hears three beats every 2.0 s when they are played together. (a) If one is vibrating at 132 Hz, what must be the frequency of the other (is there only one answer)? (b) By how much (in percent) must the tension be increased or decreased to bring them in tune?

48. (II) How many beats will be heard if two identical flutes each try to play middle C (262 Hz), but one is at 5.0°C and the other at 25.0°C?

49. (II) Two loudspeakers are 1.8 m apart. A person stands 3.0 m from one speaker and 3.5 m from the other. (a) What is the lowest frequency at which destructive interference will occur at this point? (b) Calculate two other frequencies that also result in destructive interference at this point (give the next two highest). Let $T = 20°C$.

50. (II) The two sources of sound in Fig. 16–16 face each other and emit sounds of equal amplitude and equal frequency (250 Hz) but 180° out of phase. For what minimum separation of the two speakers will there be some point at which (a) complete constructive interference occurs and (b) complete destructive interference occurs. (Assume $T = 20°C$)

51. (II) The two sources shown in Fig. 16–16 emit sound waves, in phase, each of wavelength λ and amplitude D_M. Consider a point such as C or D in the diagram, and let r_A and r_B be the distances of this point from source A and source B, respectively. Show that if r_A and r_B are nearly equal $(r_A - r_B \ll r_A)$ then the amplitude varies approximately with position as

$$\left(\frac{2D_M}{r_A}\right) \cos \frac{\pi}{\lambda}(r_A - r_B).$$

52. (II) Two loudspeakers are placed 3.00 m apart, as shown in Fig. 16–33. They emit 440-Hz sounds, in phase. A microphone is placed 3.20 m distant from a point midway between the two speakers, where an intensity maximum is recorded. (a) How far must the microphone be moved to the right to find the first intensity minimum? (b) Suppose the speakers are reconnected so that the 440-Hz sounds they emit are exactly out of phase. At what positions are the intensity maximum and minimum now?

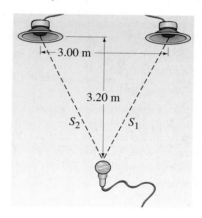

FIGURE 16–33
Problem 52.

53. (II) A guitar string produces 4 beats/s when sounded with a 350-Hz tuning fork and 9 beats/s when sounded with a 355-Hz tuning fork. What is the vibrational frequency of the string? Explain your reasoning.

54. (II) You have three tuning forks, A, B, and C. Fork B has a frequency of 440 Hz; when A and B are sounded together, a beat frequency of 3 Hz is heard. When B and C are sounded together, the beat frequency is 4 Hz. What are the possible frequencies of A and C? What beat frequencies are possible when A and C are sounded together?

55. (II) Show that the two speakers in Fig. 16–16 must be separated by at least a distance d equal to one-half the wavelength λ of sound if there is to be any place where complete destructive interference occurs. The speakers are in phase.

56. (II) A source emits sound of wavelengths 2.64 m and 2.76 m in air. (a) How many beats per second will be heard (assume $T = 20°C$)? (b) How far apart in space are the regions of maximum intensity?

57. (I) The predominant frequency of a certain police car's siren is 1550 Hz when at rest. What frequency do you detect if you move with a speed of 30.0 m/s (a) toward the car, and (b) away from the car?

58. (I) A bat at rest sends out ultrasonic sound waves at 50,000 Hz and receives them returned from an object moving radially away from it at 25.0 m/s. What is the received sound frequency?

59. (II) A bat flies toward a wall at a speed of 5.0 m/s. As it flies, the bat emits an ultrasonic sound wave with frequency 30,000 Hz. What frequency does the bat hear in the reflected wave?

60. (II) In one of the original Doppler experiments, one tuba was played on a moving train car at a frequency of 75 Hz, and a second identical tuba played the same tone while at rest in the railway station. What beat frequency was heard if the train car approached the station at a speed of 10.0 m/s?

61. (II) Two automobiles are equipped with the same single-frequency horn. When one is at rest and the other is moving toward an observer at 15 m/s, a beat frequency of 5.5 Hz is heard. What is the frequency the horns emit? Assume $T = 20°C$.

62. (II) Compare the shift in frequency if a 2000-Hz source is moving toward you at 15 m/s versus if you are moving toward it at 15 m/s. Are the two frequencies exactly the same? Are they close? Repeat the calculation for 150 m/s and then again for 300 m/s. What can you conclude about the asymmetry of the Doppler formulas? Show that at low speeds (relative to the speed of sound), the two formulas— source approaching and detector approaching—yield the same result.

63. (II) A Doppler flow meter uses ultrasound waves to measure blood-flow speeds. Suppose the device emits sound at 3.5 MHz, and the speed of sound in human tissue is taken to be 1540 m/s. What is the expected beat frequency if blood is flowing normally in large leg arteries at 2.0 cm/s directly away from the sound source?

64. (II) The Doppler effect using ultrasonic waves of frequency 2.25×10^6 Hz is used to monitor the heartbeat of a fetus. A (maximum) beat frequency of 500 Hz is observed. Assuming that the speed of sound in tissue is 1.54×10^3 m/s, calculate the maximum velocity of the surface of the beating heart.

65. (II) In Problem 64, the beat frequency is found to appear and then disappear 180 times per minute, which reflects the fact that the heart is beating and its surface changes speed. What is the heartbeat rate?

66. (II) (a) Use the binomial expansion to show that Eqs. 16–9a and 16–10a become essentially the same for small relative velocity between source and observer. (b) What percent error would result if Eq. 16–10a were used instead of Eq. 16–9a for a relative velocity of 22 m/s?

67. (III) A factory whistle emits sound of frequency 570 Hz. On a day when the wind velocity is 12.0 m/s from the north, what frequency will observers hear who are located, at rest, (a) due north, (b) due south, (c) due east, and (d) due west, of the whistle? What frequency is heard by a cyclist heading (e) north or (f) west, toward the whistle at 15.0 m/s? Assume $T = 20°C$.

* 68. (I) (a) How fast is an object moving on land if it is moving at Mach 0.33? (b) A high-flying Concorde passenger jet displays its Mach number on a screen while cruising at 3000 km/h to be 3.2. What is the speed of sound at that altitude?

* 69. (II) Show that the angle θ a sonic boom makes with the path of a supersonic object is given by Eq. 16–12.

* 70. (II) An airplane travels at Mach 2.3 where the speed of sound is 310 m/s. (a) What is the angle the shock wave makes with the direction of the airplane's motion? (b) If the plane is flying at a height of 7100 m, how long after it is directly overhead will a person on the ground hear the shock wave?

* 71. (II) A space probe enters the thin atmosphere of another planet where the speed of sound is only about 35 m/s. (a) What is the probe's Mach number if its initial speed is 15,000 km/h? (b) What is the apex angle of the shock wave it produces?

* 72. (II) A meteorite traveling 8000 m/s strikes the ocean. Determine the shock wave angle it produces (a) in the air just before entering the ocean, and (b) in the water just after entering. Assume $T = 20°C$.

* 73. (II) You look directly overhead and see a plane exactly 1.5 km above the ground flying faster than the speed of sound. By the time you hear the sonic boom, the plane has traveled a horizontal distance of 2.0 km. See Fig. 16–34. Determine (a) the angle of the shock cone, θ, and (b) the speed of the plane (the Mach number). Assume the speed of sound is 330 m/s.

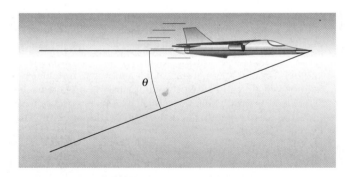

FIGURE 16–34 Problem 73.

* 74. (II) A supersonic jet traveling at Mach 1.8 at an altitude of 10,000 m passes directly over an observer on the ground. Where will the plane be relative to the observer when the latter hears the sonic boom? (See Fig. 16–35.)

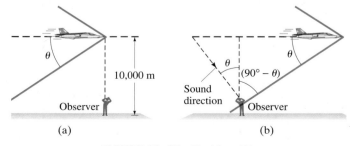

FIGURE 16–35 Problem 74.

General Problems

75. (II) A fish finder uses a sonar device that sends 20,000 Hz sound pulses downwards from the bottom of the boat, and then detects echoes. If the maximum depth for which it is designed to work is 200 meters, what is the minimum time between pulses?

76. Approximately how many octaves are there in the human audible range?

77. A stone is dropped from the top of a cliff. The splash it makes when striking the water below is heard 3.5 s later. How high is the cliff?

78. A single mosquito 5.0 m from a person makes a sound close to the threshold of human hearing (0 dB). What will be the sound level of 1000 such mosquitoes?

79. At the Indianapolis 500, you can estimate the speed of cars just by listening to the difference in pitch of the engine noise between approaching and receding cars. Suppose the sound of a certain car drops by a full octave as it goes by on the straightaway. How fast is it going?

80. A tight guitar string has a frequency of 540 Hz as its third harmonic. What will be its fundamental frequency if it is fingered at a length of only 60 percent of its original length?

81. Each string on a violin is tuned to a frequency $1\frac{1}{2}$ times that of its neighbor. If all the strings are to be placed under the same tension, what must be the mass per unit length of each string relative to that of the lowest string?

82. What is the resultant sound level when an 80-dB sound and an 85-dB sound are heard simultaneously?

83. The sound level 12.0 m from a loudspeaker, placed in the open, is 100 dB. What is the acoustic power output (W) of the speaker, assuming it radiates equally in all directions?

84. The A string of a violin is 32 cm long between fixed points with a fundamental frequency of 440 Hz and a linear density of 6.1×10^{-4} kg/m. (*a*) What are the wave speed and tension in the string? (*b*) What is the length of the tube of a simple wind instrument (say, an organ pipe) closed at one end whose fundamental is also 440 Hz if the speed of sound is 343 m/s in air? (*c*) What is the frequency of the first overtone of each instrument?

85. A stereo amplifier is rated at 150 W output at 1000 Hz. The power output drops by 10 dB at 15 kHz. What is the power output in watts at 15 kHz?

86. A tuning fork is set into vibration above a vertical open tube filled with water (Fig. 16–36). The water level is allowed to drop slowly. As it does so, the air in the tube above the water level is heard to resonate with the tuning fork when the distance from the tube opening to the water level is 0.125 m and again at 0.395 m. What is the frequency of the tuning fork?

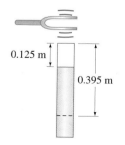

0.125 m

0.395 m

FIGURE 16–36
Problem 86.

87. Two loudspeakers face each other at opposite ends of a long corridor. They are connected to the same source which produces a pure tone of 280 Hz. A person walks from one speaker toward the other at a speed of 1.4 m/s. What "beat" frequency does the person hear?

88. Workers around jet aircraft typically wear protective devices over their ears. Assume that the intensity level of a jet airplane engine, at a distance of 30 m, is 140 dB, and that the average human ear has an effective radius of 2.0 cm. What would be the power intercepted by an unprotected ear at a distance of 30 m from a jet airplane engine?

89. The intensity at the "threshold of hearing" for the human ear at a frequency of about 1000 Hz is $I_0 = 1.0 \times 10^{-12}$ W/m², for which β, the sound level, is 0 dB. The "threshold of pain" at the same frequency is about 120 dB, or $I = 1.0$ W/m², corresponding to an increase of intensity by a factor of 10^{12}. By what factors do the displacement amplitude, D_M, and the pressure amplitude ΔP_M, vary?

90. As Fig. 16–5 shows, the human ear is not equally sensitive to all frequencies; the threshold of hearing is about 60 dB at 35 Hz, 0 dB at 1000 Hz and at 5000 Hz, and 20 dB at 15,000 Hz. Thus, near threshold, the ear is especially insensitive to low frequencies. Estimate the displacement amplitude for each of these four points. At which frequency is the ear most sensitive to displacement?

91. In audio and communications systems, the *gain*, β, in decibels is defined as $\beta = 10 \log(P_{\text{out}}/P_{\text{in}})$ where P_{in} is the power input to the system and P_{out} is the power output. A particular stereo amplifier puts out 100 W of power for an input of 1 mW. What is its gain in dB?

92. Two loudspeakers are at opposite ends of a railroad car as it moves past a stationary observer at 10.0 m/s, as shown in Fig. 16–37. If they have identical sound frequencies of 200 Hz, what is the beat frequency heard by the observer when (*a*) he listens from the position A, in front of the car, (*b*) he is between the speakers, at B, and (*c*) he hears the speakers after they have passed him, at C?

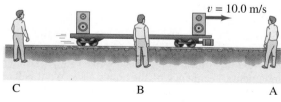

$v = 10.0$ m/s

C   B   A

FIGURE 16–37 Problem 92.

93. The frequency of a steam train whistle as it approaches you is 538 Hz. After it passes you, its frequency is measured as 486 Hz. How fast was the train moving (assume constant velocity)?

94. A 75-cm-long guitar string of mass 2.10 g is placed near a tube open at one end, and also 75 cm long. How much tension should be in the string if it is to produce resonance (in its fundamental mode) with the third harmonic in the tube?

95. If the velocity of blood flow in the aorta is normally about 0.32 m/s, what beat frequency would you expect if 5.50-MHz ultrasound waves were directed along the flow and reflected from the red blood cells? Assume that the waves travel with a speed of 1.54×10^3 m/s.

96. A source of sound waves (wavelength λ) is a distance l from a detector. Sound reaches the detector directly, and also by reflecting off an obstacle, as shown in Fig. 16–38. The obstacle is equidistant from source and detector. When the obstacle is a distance d to the right of the line of sight between source and detector, as shown, the two waves arrive in phase. How much farther to the right must the obstacle be moved if the two waves are to be out of phase by $\frac{1}{2}$ wavelength, so destructive interference occurs? (Assume $\lambda \ll l, d$.)

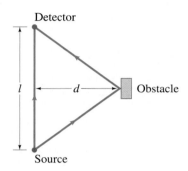

FIGURE 16–38 Problem 96.

97. A person hears a pure tone in the 500–1000 Hz range coming from two sources. The sound is loudest at points equidistant from the two sources. In order to determine exactly what the frequency is, the person moves about and finds that the sound level is minimal at a point 0.31 m farther from one source than the other. What is the frequency of the sound?

98. A bat flies toward a moth at speed 6.5 m/s while the moth is flying toward the bat at speed 5.0 m/s. The bat emits a sound wave of 51.35 kHz. What is the frequency of the wave detected by the bat after it reflects off the moth?

99. A dramatic demonstration, called "singing rods," involves a long, slender aluminum rod held in the hand near the rod's midpoint. The rod is stroked with the other hand. With a little practice, the rod can be made to "sing," or emit a clear, loud, ringing sound. For a 90-cm-long rod: (a) what is the fundamental frequency of the sound? (b) What is its wavelength in the rod, and (c) what is the traveling wavelength in air at 20°C?

100. Room acoustics for stereo listening can be compromised by the presence of standing waves, which can cause acoustic "dead spots" at the locations of the pressure nodes. Consider a living room with dimensions 5.0 m long, 4.0 m wide, and 2.8 m high. Calculate the fundamental frequencies for the standing waves in this room.

101. Assuming that the maximum displacement of the air molecules in a sound wave is about the same as that of the speaker cone that produces the sound (Fig. 16–39), estimate by how much a loudspeaker cone moves for a fairly loud (100 dB) sound of (a) 10 kHz, and (b) 40 Hz.

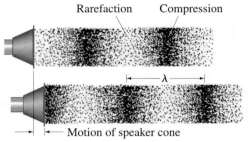

FIGURE 16–39 Problem 101.

102. A Doppler flow meter is used to measure the speed of blood flow. Transmitting and receiving elements are placed on the skin, as shown in Fig. 16–40. Typical sound-wave frequencies of about 5.0 MHz are used, which have a reasonable chance of being reflected from red blood cells. By measuring the frequency of the reflected waves, which are Doppler-shifted because the red blood cells are moving, the speed of the blood flow can be deduced. "Normal" blood flow speed is about 0.1 m/s. Suppose that an artery is partly constricted, so that the speed of the blood flow is increased, and the flow meter measures a Doppler shift of 900 Hz. What is the speed of blood flow in the constricted region? The effective angle between the sound waves (both transmitted and reflected) and the direction of blood flow is 45°. Assume the velocity of sound in tissue is 1540 m/s.

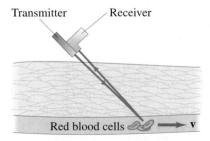

FIGURE 16–40 Problem 102.

The balloon in the foreground is about to leave on the first successful nonstop trip around the world in a balloon. To the rear, we can see the flame heating the air in another balloon. Heating the air inside a "hot-air" balloon raises the air's temperature and forces some of the air to escape from the opening at the bottom. This is an example of Charles's law: the volume of an enclosed gas at constant pressure increases directly with the Kelvin temperature. The balloon is open at the bottom, so the pressure is atmospheric. The volume increase caused by the heating forces some gas to escape through the opening. The reduced amount of gas inside means its density is lower, so there is a net buoyant force upward on the balloon. In this chapter we study temperature and its effects on matter: thermal expansion, thermal stresses, and the gas laws.

CHAPTER 17

Temperature, Thermal Expansion, and the Ideal Gas Law

The topics of temperature, heat and thermodynamics are the subject of the next four chapters, Chapters 17 through 20. Also included, and closely related, as a means of understanding, is the kinetic theory of gases.

We will often consider a particular **system**, by which we mean a particular object or set of objects; everything else in the universe is called the "environment." We can describe the **state** (or condition) of a particular system—such as a gas in a container—from either a microscopic or macroscopic point of view. A **microscopic** description would involve details of the motion of all the atoms or molecules making up the system, which could be very complicated. A **macroscopic** description is given in terms of quantities that are detectable directly by our senses, such as volume, mass, pressure, and temperature.

Microscopic vs. macroscopic properties

The description of processes in terms of macroscopic quantities is the field of **thermodynamics**. The number of macroscopic variables required to describe the state of a system at any time depends on the type of system. To describe the state of a pure gas in a container, for example, we need only three variables, which could be the volume, the pressure, and the temperature. Quantities such as these that can be used to describe the state of a system are called **state variables**.

The emphasis in this chapter is on the concept of temperature. We will begin, however, with a brief discussion of the theory that matter is made up of atoms and that these atoms are in continuous random motion. This theory is called the *kinetic theory* ("kinetic," you may remember, comes from the Greek word for "moving"), and we will discuss it in more detail in the next chapter.

445

17–1 | Atomic Theory of Matter

The idea that matter is made of atoms dates back to the ancient Greeks. According to the Greek philosopher Democritus, if a given substance—say, a piece of iron—were cut into smaller and smaller bits, eventually a smallest piece of that substance would be obtained which could not be divided further. This smallest piece was called an **atom**, which comes from the Greek *atomos*, "indivisible."[†] The only real alternative to the atomic theory of matter was the idea that matter is continuous and can be subdivided indefinitely.

Atomic and molecular masses

Today, we often speak of the relative masses of atoms and molecules—what we call the **atomic mass** or **molecular mass**, respectively.[‡] These are based on assigning the abundant carbon atom, ^{12}C, the value of exactly 12.0000 **unified atomic mass units** (u). In terms of kilograms,

$$1\,u = 1.6605 \times 10^{-27}\,kg.$$

The atomic mass of hydrogen is then 1.0078 u, and the values for other atoms are as listed in the periodic table inside the back cover of this book, and also in Appendix D. The molecular mass of a compound is the sum of the atomic masses of the atoms making up the molecule.

An important piece of evidence for the atomic theory is the so-called **Brownian movement**, named after the biologist Robert Brown, who is credited with its discovery in 1827. While he was observing tiny pollen grains suspended in water under his microscope, Brown noticed that the tiny grains moved about in tortuous paths (Fig. 17–1), even though the water appeared to be perfectly still. The atomic theory easily explains Brownian movement if the further reasonable assumption is made that the atoms of any substance are continually in motion. Then Brown's tiny pollen grains are jostled about by the vigorous barrage of rapidly moving molecules of water.

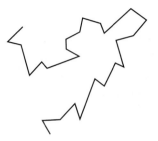

FIGURE 17–1 Path of a tiny particle (pollen grain, for example) suspended in water. The straight lines connect observed positions of the particle at equal time intervals. [The plot has the same general shape independent of the time interval—whether it be every 60 s or 0.1 s, a phenomenon referred to as *fractal* behavior.]

In 1905, Albert Einstein examined Brownian movement from a theoretical point of view and was able to calculate from the experimental data the approximate size and mass of atoms and molecules. His calculations showed that the diameter of a typical atom is about 10^{-10} m.

Phases of matter

At the start of Chapter 13, we distinguished the three common states of matter—solid, liquid, gas—based on macroscopic, or "large-scale," properties. Now let us see how these three phases of matter differ from the atomic, or microscopic, point of view. Clearly, atoms and molecules must exert attractive forces on each other. For how else could a brick or a piece of aluminum stay together in one piece? The attractive forces between molecules are of an electrical nature (more on this in later chapters). If the molecules come too close together, the force between them becomes repulsive (electric repulsion between their outer electrons). Thus molecules maintain a minimum distance from each other. In a solid material, the attractive forces are strong enough that the atoms or molecules are held in more or less fixed positions, often in an array known as a crystal lattice, as shown in Fig. 17–2a. The atoms or molecules in a solid are in motion—they vibrate about their nearly fixed positions. In a liquid, the atoms or molecules are moving more rapidly, or the forces between them are weaker, so that they are sufficiently free to pass over one another, as in Fig. 17–2b. In a gas, the forces are so weak, or the speeds so high, that the molecules do not even stay close together. They move rapidly every which way, Fig. 17–2c, filling any container and occasionally colliding with one another. On the average, the speeds are sufficiently high in a gas that

[†] Today, of course, we don't consider the atom as indivisible, but rather as consisting of a nucleus (containing protons and neutrons) and electrons.

[‡] The terms *atomic weight* and *molecular weight* are popularly used for these quantities, but properly speaking we are comparing masses.

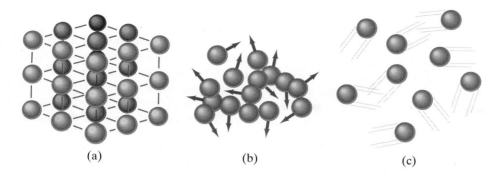

(a)　　　　　(b)　　　　　(c)

FIGURE 17–2 Atomic arrangements in (a) a crystalline solid, (b) a liquid, and (c) a gas.

when two molecules collide, the force of attraction is not strong enough to keep them close together and they fly off in new directions.

EXAMPLE 17–1 **ESTIMATE** **Distance between atoms.** The density of copper is $8.9 \times 10^3 \, \text{kg/m}^3$ and each copper atom has a mass of 63 u, where $1 \, \text{u} = 1.66 \times 10^{-27} \, \text{kg}$. Estimate the average distance between neighboring atoms.

SOLUTION The mass of 1 copper atom is $63 \times 1.66 \times 10^{-27} \, \text{kg} = 1.04 \times 10^{-25} \, \text{kg}$. This means that in a cube of copper 1 m on a side (volume $= 1 \, \text{m}^3$), there are

$$\frac{8.9 \times 10^3 \, \text{kg/m}^3}{1.04 \times 10^{-25} \, \text{kg/atom}} = 8.5 \times 10^{28} \, \text{atoms/m}^3.$$

The volume of a cube of side l is $V = l^3$, so on one edge of the 1-m-long cube there are $(8.5 \times 10^{28})^{\frac{1}{3}}$ atoms $= 4.4 \times 10^9$ atoms. Hence the distance between neighboring atoms is

$$\frac{1 \, \text{m}}{4.4 \times 10^9 \, \text{atoms}} = 2.3 \times 10^{-10} \, \text{m between atoms.}$$

17–2 Temperature and Thermometers

In everyday life, **temperature** is a measure of how hot or cold an object is. A hot oven is said to have a high temperature, whereas the ice of a frozen lake is said to have a low temperature.

Many properties of matter change with temperature. For example, most materials expand when heated.[†] An iron beam is longer when hot than when cold. Concrete roads and sidewalks expand and contract slightly according to temperature, which is why compressible spacers or expansion joints (Fig. 17–3) are placed at regular intervals. The electrical resistance of matter changes with temperature (see Chapter 25). So too does the color radiated by objects, at least at high temperatures: you may have noticed that the heating element of an electric stove glows with a red color when hot. At higher temperatures, solids such as iron glow orange or even white. The white light from an ordinary incandescent lightbulb comes from an extremely hot tungsten wire. The surface temperatures of the Sun and other stars can be measured by the predominant color (more precisely, wavelengths) of light they emit.

Instruments designed to measure temperature are called **thermometers**. There are many kinds of thermometers, but their operation always depends on some property of matter that changes with temperature. Most common thermometers

FIGURE 17–3 Expansion joint on a bridge.

Thermometers: measuring temperature

[†] Most materials expand when their temperature is raised, but not all. Water, for example, in the range 0°C to 4°C contracts with an increase in temperature (see Section 17–4).

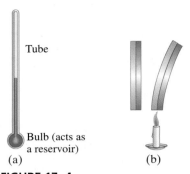

Tube

Bulb (acts as a reservoir)

(a)   (b)

FIGURE 17–4
(a) Mercury- or alcohol-in-glass thermometer; (b) bimetallic strip.

Temperature scales

FIGURE 17–5 Photograph of a thermometer using a coiled bimetallic strip.

rely on the expansion of a material with an increase in temperature. The first idea for a thermometer, by Galileo, made use of the expansion of a gas.

Common thermometers today consist of a hollow glass tube filled with mercury or with alcohol colored with a red dye. In such liquid-in-glass thermometers, the liquid expands more than the glass when the temperature is increased, so the liquid level rises in the tube (Fig. 17–4a). Although metals also expand with temperature, the change in length of a metal rod, say, is generally too small to measure accurately for ordinary changes in temperature. However, a useful thermometer can be made by bonding together two dissimilar metals whose rates of expansion are different (Fig. 17–4b). When the temperature is increased, the slightly different amounts of expansion cause the bimetallic strip to bend. Often the bimetallic strip is in the form of a coil, one end of which is fixed while the other is attached to a pointer, Fig. 17–5. This kind of thermometer is used as ordinary air thermometers, oven thermometers, automatic-off switch in electric coffeepots, and in room thermostats for determining when the heater or air conditioner should go on or off. Very precise thermometers make use of electrical properties (Chapter 25), such as resistance thermometers, thermocouples, and thermistors, which may have a digital readout.

In order to measure temperature quantitatively, some sort of numerical scale must be defined. The most common scale today is the **Celsius** scale, sometimes called the **centigrade** scale. In the United States, the **Fahrenheit** scale is also common. The most important scale in scientific work is the absolute, or Kelvin, scale, and it will be discussed later in this chapter.

One way to define a temperature scale is to assign arbitrary values to two readily reproducible temperatures. For both the Celsius and Fahrenheit scales these two fixed points are chosen to be the freezing point and the boiling point[†] of water, both taken at atmospheric pressure. On the Celsius scale, the freezing point of water is chosen to be 0°C ("zero degrees Celsius") and the boiling point 100°C. On the Fahrenheit scale, the freezing point is defined as 32°F and the boiling point 212°F. A practical thermometer is calibrated by placing it in carefully prepared environments at each of the two temperatures and marking the position of the mercury or pointer. For a Celsius scale, the distance between the two marks is then divided into one hundred equal intervals separated by small marks representing each degree between 0°C and 100°C (hence the name "centigrade scale" meaning "hundred steps"). For a Fahrenheit scale, the two points are labeled 32°F and 212°F and the distance between them is divided into 180 equal intervals. For temperatures below the freezing point of water and above the boiling point of water the scales can be extended using the same equally spaced intervals. However, ordinary thermometers can be used only over a limited temperature range because of their own limitations—for example, the mercury in a mercury-in-glass thermometer solidifies at some point, below which the thermometer will be useless. It is also rendered useless above temperatures where the fluid vaporizes. For very low or very high temperatures, specialized thermometers are required, some of which we will mention later.

Every temperature on the Celsius scale corresponds to a particular temperature on the Fahrenheit scale, Fig. 17–6. It is easy to convert from one to the other if you remember that 0°C corresponds to 32°F and that a range of 100° on the Celsius scale corresponds to a range of 180° on the Fahrenheit scale. Thus, one Fahrenheit degree (1 F°) corresponds to $100/180 = \frac{5}{9}$ of a Celsius degree (1 C°).

[†] The freezing point of a substance is defined as that temperature at which the solid and liquid phases coexist in equilibrium—that is, without any net liquid changing into the solid or vice versa. Experimentally, this is found to occur at only one definite temperature, for a given pressure. Similarly, the boiling point is defined as that temperature at which the liquid and gas coexist in equilibrium. Since these points vary with pressure, the pressure must be specified (usually it is 1 atm).

That is, $1 \, F° = \frac{5}{9} \, C°$. (Notice that when we refer to a specific temperature, we say "degrees Celsius," as in 20°C; but when we refer to a *change* in temperature or a temperature interval, to avoid misunderstanding we will say "Celsius degrees," as in "1 C°.") The conversion between the two temperature scales can be written

$$T(°C) = \tfrac{5}{9}[T(°F) - 32] \qquad \text{or} \qquad T(°F) = \tfrac{9}{5}T(°C) + 32.$$

Rather than memorizing these relations (it would be easy to confuse them), it is better to simply remember that $0°C = 32°F$ and $5 \, C° = 9 \, F°$ (or $100°C = 212°F$).

EXAMPLE 17–2 **Taking your temperature.** Normal body temperature is 98.6°F. What is this on the Celsius scale?

SOLUTION First we relate the given temperature to the freezing point of water (0°C). That is, 98.6°F is $98.6 - 32.0 = 66.6 \, F°$ above the freezing point of water. Since each F° is equal to $\frac{5}{9} \, C°$, this corresponds to $66.6 \times \frac{5}{9} = 37.0$ Celsius degrees above the freezing point. Since the freezing point is 0°C, the temperature is 37.0°C.

Different materials do not expand in quite the same way over a wide temperature range. Consequently, if we calibrate different kinds of thermometers exactly as described above, they will not usually agree precisely. Because of how we calibrated them, they will agree at 0°C and at 100°C. But because of different expansion properties, they may not agree precisely at intermediate temperatures (remember we arbitrarily divided the thermometer scale into 100 equal divisions between 0°C and 100°C). Thus a carefully calibrated mercury-in-glass thermometer might register 52.0°C, whereas a carefully calibrated thermometer of another type might read 52.6°C.

Because of this discrepancy, some standard kind of thermometer must be chosen so that these intermediate temperatures can be precisely defined. The chosen standard for this purpose is the so-called **constant-volume gas thermometer**. As shown in the simplified diagram of Fig. 17–7, this thermometer consists of a bulb filled with a dilute gas connected by a thin tube to a mercury manometer. The volume of the gas is kept constant by raising or lowering the right-hand tube of the manometer so that the mercury in the left tube coincides with the reference mark. An increase in temperature causes a proportional increase in pressure in the bulb or in volume of the gas. The tube must thus be lifted higher to keep the gas volume constant. The height of the mercury in the right-hand column is then a measure of the temperature. This thermometer can be calibrated and gives the same results for all gases in the limit of reducing the gas pressure in the bulb toward zero. The resulting scale is defined as the standard temperature scale (Section 17–10).

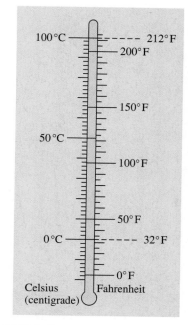

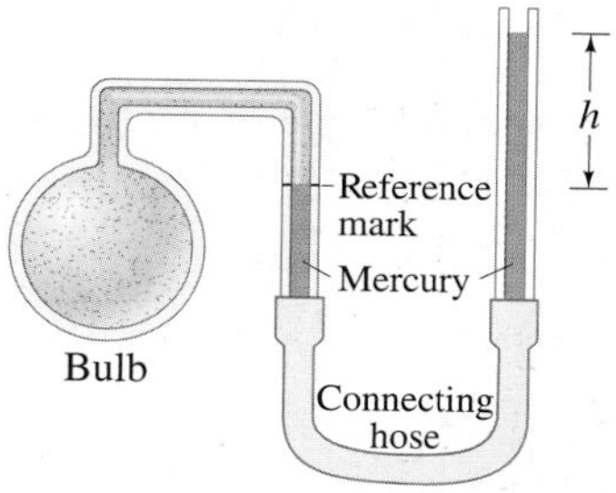

FIGURE 17–6 Celsius and Fahrenheit scales compared.

Constant-volume gas thermometer

FIGURE 17–7 Constant-volume gas thermometer.

17–3 Thermal Equilibrium and the Zeroth Law of Thermodynamics

We are all familiar with the fact that if two objects at different temperatures are placed in thermal contact (meaning thermal energy can pass from one to the other), the two objects will eventually reach the same temperature. They will then be said to be in **thermal equilibrium**. For example, an ice cube placed in a large glass of hot water melts to water, all of which eventually comes to the same temperature. If you put your hand in the water of an icy lake, you can *feel* the temperature of your hand drop. (It's best to pull your hand out before thermal equilibrium is reached!) Two objects are defined to be in *thermal equilibrium* if, when placed in thermal contact, their temperatures don't change.

Thermal equilibrium

Suppose you wanted to determine if two systems, A and B, are in thermal equilibrium, but without putting them in contact. You could do so by making use of a third system, C (which could be considered a thermometer). Suppose that C and A are in thermal equilibrium and that C and B are in thermal equilibrium. Does this imply that A and B are necessarily in thermal equilibrium with each other? Actually, it isn't completely obvious unless you do some experiments, and all experiments indicate that

Zeroth law of thermodynamics

if two systems are in thermal equilibrium with a third system, then they are in thermal equilibrum with each other.

This postulate is called the **zeroth law of thermodynamics**. It has this rather odd name because it was not until after the great first and second laws of thermodynamics (Chapters 19 and 20) were worked out that scientists realized that this apparently obvious postulate needed to be stated first.

Temperature is a property of a system that determines whether the system will be in thermal equilibrium with other systems. When two systems are in thermal equilibrium, their temperatures are, by definition, equal. This is consistent with our everyday notion of temperature, since when a hot body and a cold one are put into contact, they eventually come to the same temperature. Thus the importance of the zeroth law is that it allows a useful definition of temperature.

17–4 | Thermal Expansion

Most substances expand when heated and contract when cooled. However, the amount of expansion or contraction varies, depending on the material.

Linear Expansion

Experiments indicate that the change in length ΔL of almost all solids is, to a very good approximation, directly proportional to the change in temperature ΔT. As might be expected, the change in length is also proportional to the original length of the object, L_0, Fig. 17–8. That is, for the same temperature change, a 4-m-long iron rod will increase in length twice as much as a 2-m-long iron rod. We can write this proportionality as an equation:

$$\Delta L = \alpha L_0 \, \Delta T, \tag{17–1a}$$

where α, the proportionality constant, is called the *coefficient of linear expansion* for the particular material and has units of $(\text{C}°)^{-1}$. This equation can also be written as

$$L = L_0(1 + \alpha \Delta T), \tag{17–1b}$$

where L_0 is the length initially, at temperature T_0, and L is the length after heating or cooling to a temperature T. If the temperature change $\Delta T = T - T_0$ is negative, then $\Delta L = L - L_0$ is also negative, so the length decreases.

The values of α for various materials at 20°C are listed in Table 17–1. It should be noted that α does vary slightly with temperature (which is why thermometers made of different materials do not agree precisely). However, if the temperature range is not too great, the variation can usually be ignored.

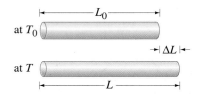

FIGURE 17–8 A thin rod of length L_0 at temperature T_0 is heated to a new uniform temperature T and acquires length L, where $L = L_0 + \Delta L$.

Linear expansion

➡ **PHYSICS APPLIED**

Expansion in structures

EXAMPLE 17–3 Bridge expansion. The steel bed of a suspension bridge is 200 m long at 20°C. If the extremes of temperature to which it might be exposed are −30°C to +40°C, how much will it contract and expand?

SOLUTION From Table 17–1, we find that $\alpha = 12 \times 10^{-6}(\text{C}°)^{-1}$. The increase in length when it is at 40°C will be

$$\Delta L = (12 \times 10^{-6}/\text{C}°)(200\,\text{m})(40°\text{C} - 20°\text{C}) = 4.8 \times 10^{-2}\,\text{m},$$

or 4.8 cm. When the temperature decreases to −30°C, $\Delta T = -50\,\text{C}°$. Then

$$\Delta L = (12 \times 10^{-6}/\text{C}°)(200\,\text{m})(-50\,\text{C}°) = -12.0 \times 10^{-2}\,\text{m},$$

or a decrease in length of 12 cm.

Experimentally, a solid rod gets longer when you raise its temperature, so we conclude that the average distance between atoms does increase. To understand this, let us look at a simplified potential-energy diagram as shown in Fig. 17–10, which represents the potential energy of two atoms versus their separation r. At large r, we assume the potential energy ≈ 0, and as r decreases, the potential energy decreases, indicating an attractive force as discussed in Section 8–9. For r less than r_0 (the equilibrium position) the potential-energy curve rises, indicating a repulsive force between atoms as they approach each other. The horizontal blue lines in Fig. 17–10 labeled E_2 and E_1 represent the total energy for two different temperatures, T_2 and T_1, where $T_2 > T_1$. The short vertical lines for E_1 and E_2 on the diagram represent the midpoints of the motion at these two temperatures. Because the potential-energy curve is not symmetrical, the average separation of atoms is greater for the higher temperature, as shown. Thus thermal expansion is due to the nonsymmetry of the potential-energy function. If the potential-energy curve were symmetrical, there would be no thermal expansion at all. Indeed, the fact that most substances expand when heated *implies* that the potential-energy curve must be asymmetrical, as in Fig. 17–10.

Anomalous Behavior of Water Below 4°C

Most substances expand more or less uniformly with an increase in temperature, as long as no phase change occurs. Water, however, does not follow the usual pattern. If water at 0°C is heated, it actually *decreases* in volume until it reaches 4°C. Above 4°C water behaves normally and expands in volume as the temperature is increased, Fig. 17–11. Water thus has its greatest density at 4°C. This anomalous behavior of water is of great importance for the survival of aquatic life during cold winters. When the water in a lake or river is above 4°C and begins to cool by contact with cold air, the water at the surface sinks because of its greater density and it is replaced by warmer water from below. This mixing continues until the temperature reaches 4°C. As the surface water cools further, it remains on the surface because it is less dense than the 4°C water below. Water then freezes first at the surface, and the ice remains on the surface since ice (specific gravity = 0.917) is less dense than water. The water at the bottom remains at 4°C until almost the whole body of water is frozen. If water were like most substances, becoming more dense as it cools, the water at the bottom of a lake would be frozen first. Lakes would freeze solid more easily since circulation would bring the warmer water to the surface to be efficiently cooled. The complete freezing of a lake would cause severe damage to its plant and animal life. Because of the unusual behavior of water below 4°C, it is rare for any large body of water to freeze completely, and this is helped by the layer of ice on the surface which acts as an insulator to reduce the flow of heat out of the water into the cold air above. Without this peculiar but wonderful property of water, life on this planet as we know it might not have been possible.

Not only does water expand as it cools from 4°C to 0°C, it expands even more as it freezes to ice. This is why ice cubes float in water and pipes break when water inside them freezes.

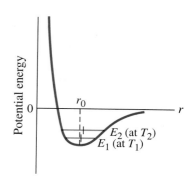

FIGURE 17–10 Typical curve of potential energy versus separation of atoms, r, for atoms in a crystal solid (simplified). Note that the midpoint (short vertical line) of the oscillatory motion of atoms is greater at the higher temperature T_2.

Water is unusual:
it expands when
cooled from 4°C to 0°C

➡ **PHYSICS APPLIED**

Life under ice

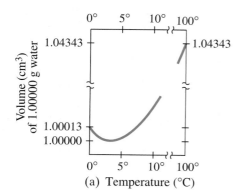

(a) Temperature (°C)

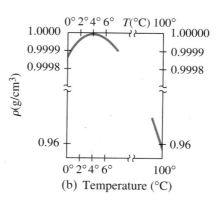

(b) Temperature (°C)

FIGURE 17–11 Behavior of water as a function of temperature near 4°C. (a) Volume, of 1.00000 gram of water, as a function of temperature. (b) Density vs. temperature. [Note the break in each axis.]

Thermal Stresses

In many situations, such as in buildings and roads, the ends of a beam or slab of material are rigidly fixed, which greatly limits expansion or contraction. If the temperature should change, large compressive or tensile stresses, called *thermal stresses*, will occur. The magnitude of such stresses can be calculated using the concept of elastic modulus developed in Chapter 12. To calculate the internal stress, we can think of this process as occurring in two steps. The rod expands (or contracts) by an amount ΔL given by Eq. 17–1; then a force is applied to compress (or expand) the material back to its original length. The force F required is given by Eq. 12–4:

$$\Delta L = \frac{1}{E} \frac{F}{A} L_0,$$

where E is Young's modulus for the material. To calculate the internal stress, F/A, we then set ΔL in Eq. 17–1a equal to ΔL in the equation above and find

$$\alpha L_0 \, \Delta T = \frac{1}{E} \frac{F}{A} L_0.$$

Hence, $F/A = \alpha E \, \Delta T$.

PHYSICS APPLIED

Highway buckling

EXAMPLE 17–7 **Stress in concrete on a hot day.** A highway is to be made of blocks of concrete 10 m long placed end to end with no space in between them to allow for expansion. If the blocks were placed at a temperature of 10°C, what force of compression would occur if the temperature reached 40°C? The contact area between each block is 0.20 m². Will fracture occur?

SOLUTION We solve for F in the equation above and use the value of E from Table 12–1:

$$F = \alpha \, \Delta T \, E \, A$$
$$= (12 \times 10^{-6}/\text{C}°)(30 \, \text{C}°)(20 \times 10^9 \, \text{N/m}^2)(0.20 \, \text{m}^2) = 1.4 \times 10^6 \, \text{N}.$$

The stress, F/A, is $(1.4 \times 10^6 \, \text{N})/(0.20 \, \text{m}^2) = 7.0 \times 10^6 \, \text{N/m}^2$. This is not far from the ultimate strength of concrete (Table 12–2) under compression and exceeds it for tension and shear. Hence, assuming the concrete is not perfectly aligned, part of the force will act in shear, and fracture is likely. How much space would you allow between blocks if you expected a temperature range of 0°F to 110°F?

17–6 The Gas Laws and Absolute Temperature

Equation 17–2 is not very useful for describing the expansion of a gas, partly because the expansion can be so great, and partly because gases generally expand to fill whatever container they are in. Indeed, Eq. 17–2 is meaningful only if the pressure is kept constant. The volume of a gas depends very much on the pressure as well as on the temperature. It is therefore valuable to determine a relation between the volume, the pressure, the temperature, and the mass of a gas. Such a relation is called an **equation of state**. (By the word *state*, we mean the physical condition of the system.)

If the state of a system is changed, we will always wait until the pressure and temperature have reached the same values throughout. We thus consider only **equilibrium states** of a system—when the variables that describe it (such as temperature and pressure) are the same throughout the system and are not changing in time. We also note that the results of this Section are accurate only for gases that are not too dense (the pressure is not too high, on the order of an atmosphere or so) and not close to the liquefaction (boiling) point.

For a given quantity of gas it is found experimentally that, to a good approximation, *the volume of a gas is inversely proportional to the pressure applied to it when the temperature is kept constant.* That is,

$$V \propto \frac{1}{P}.$$ [constant T]

where P is the absolute pressure (not "gauge pressure"—see Chapter 13). For example, if the pressure on a gas is doubled, the volume is reduced to half its original volume. This relation is known as **Boyle's law**, after Robert Boyle (1627–1691), who first stated it on the basis of his own experiments. A graph of P vs. V for a fixed temperature is shown in Fig. 17–12. Boyle's law can also be written

$$PV = \text{constant}.$$ [constant T]

That is, at constant temperature, if either the pressure or volume of the gas is allowed to vary, the other variable also changes so that the product PV remains constant.

Temperature also affects the volume of a gas, but a quantitative relationship between V and T was not found until more than a century after Boyle's work. The Frenchman Jacques Charles (1746–1823) found that when the pressure is not too high and is kept constant, the volume of a gas increases with temperature at a nearly constant rate, as in Fig. 17–13a. However, all gases liquefy at low temperatures (for example, oxygen liquefies at −183°C) and so the graph cannot be extended below the liquefaction point. Nonetheless, the graph is essentially a straight line and if projected to lower temperatures, as shown by the dashed line, it crosses the axis at about −273°C.

Such a graph can be drawn for any gas, and the straight line always projects back to −273°C at zero volume. This seems to imply that if a gas could be cooled to −273°C it would have zero volume, and at lower temperatures a negative volume, which makes no sense, of course. It could be argued that −273°C is the lowest temperature possible, and many other more recent experiments indicate that it is so. This temperature is called the **absolute zero** of temperature. Its value has been determined to be −273.15°C.

Absolute zero forms the basis of a temperature scale known as the **absolute** or **Kelvin scale**, and it is used extensively in scientific work. On this scale the temperature is specified as degrees Kelvin or, preferably, simply as kelvins (K) without the degree sign. The intervals are the same as for the Celsius scale, but the zero on this scale (0 K) is chosen as absolute zero itself. Thus the freezing point of water (0°C) is 273.15 K and the boiling point of water is 373.15 K. Indeed, any temperature on the Celsius scale can be changed to kelvins by adding 273.15 to it:

$$T(\text{K}) = T(\text{°C}) + 273.15.$$

Now let us look at Fig. 17–13b, where we see that the graph of the volume of a gas versus absolute temperature is a straight line that passes through the origin. Thus,

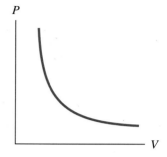

FIGURE 17–12 Pressure vs. volume of a gas at a constant temperature, showing the inverse relationship as given by Boyle's law: as the pressure increases, the volume decreases.

Boyle's law

Absolute zero

Kelvin scale

FIGURE 17–13 Volume of a gas as a function of (a) Celsius temperature, and (b) Kelvin temperature, when the pressure is kept constant.

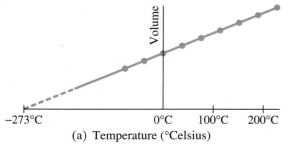

(a) Temperature (°Celsius)

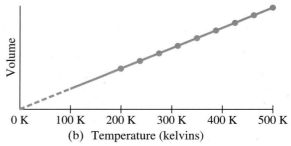

(b) Temperature (kelvins)

to a good approximation, *the volume of a given amount of gas is directly proportional to the absolute temperature when the pressure is kept constant.* This is known as **Charles's law**, and is written

Charles's law
$$V \propto T. \qquad \text{[constant } P\text{]}$$

A third gas law, known as **Gay-Lussac's law**, after Joseph Gay-Lussac (1778–1850), states that *at constant volume, the pressure of a gas is directly proportional to the absolute temperature*:

Gay-Lussac's law
$$P \propto T. \qquad \text{[constant } V\text{]}$$

A familiar example is that a closed jar, or aerosol can, thrown into a fire will explode due to the increase in gas pressure inside.

The laws of Boyle, Charles, and Gay-Lussac are not really laws in the sense that we use this term today (precise, deep, wide-ranging validity). They are really only approximations that are accurate for real gases only as long as the pressure and density of the gas are not too high, and the gas is not too close to condensation. The term *law* applied to these three relationships has become traditional, however, so we have stuck with that usage.

17–7 | The Ideal Gas Law

The gas laws of Boyle, Charles, and Gay-Lussac were obtained by means of a technique that is very useful in science: namely, to hold one or more variables constant in order to see clearly the effects of changing only one of the variables. These laws can now be combined into a single more general relation between the pressure, volume, and temperature of a fixed quantity of gas:

$$PV \propto T.$$

This relation indicates how any of the quantities P, V, or T will vary when the other two quantities change. This relation reduces to Boyle's, Charles's, or Gay-Lussac's law when either the temperature, the pressure, or the volume, respectively, is held constant.

Finally, we must incorporate the effect of the amount of gas present. Anyone who has blown up a balloon knows that the more air forced into the balloon, the bigger it gets (Figure 17–14). Indeed, careful experiments show that at constant temperature and pressure, the volume V of an enclosed gas increases in direct proportion to the mass m of gas present. Hence we write

$$PV \propto mT.$$

This proportion can be made into an equation by inserting a constant of proportionality. Experiment shows that this constant has a different value for different gases. However, the constant of proportionality turns out to be the same for all gases if, instead of the mass m, we use the number of moles.

Mole (unit)

One **mole** (abbreviated mol) is defined as the amount of substance that contains as many atoms or molecules as there are in 12.00 grams of carbon 12 (whose atomic mass is exactly 12 u). A simpler but equivalent definition is: 1 mol is that number of grams of a substance numerically equal to the molecular mass (Section 17–1) of the substance. For example, the molecular mass of hydrogen gas (H_2) is 2.0 u (since each molecule contains two atoms of hydrogen and each atom has an atomic mass of 1.0 u). Thus 1 mol of H_2 has a mass of 2.0 g. Similarly, 1 mol of neon gas has a mass of 20 g, and 1 mol of CO_2 has a mass of $[12 + (2 \times 16)] = 44$ g.

FIGURE 17–14 Blowing up a balloon means putting more air (more air molecules) into the balloon, which increases its volume. The pressure is nearly constant (atmospheric) except for the small effect of the balloon's elasticity.

The mole is the official unit in the SI system. In general, the number of moles, n, in a given sample of a pure substance is equal to its mass in grams divided by its molecular mass specified as grams per mole:

$$n \text{ (mol)} = \frac{\text{mass (grams)}}{\text{molecular mass (g/mol)}}.$$

For example, the number of moles in 132 g of CO_2 is

$$n = \frac{132 \text{ g}}{44 \text{ g/mol}} = 3.0 \text{ mol}.$$

We can now write the proportion discussed above as an equation:

$$PV = nRT, \qquad\qquad\qquad (17\text{–}3)$$

IDEAL GAS LAW

where n represents the number of moles and R is the constant of proportionality. R is called the **universal gas constant** because its value is found experimentally to be the same for all gases. The value of R, in several sets of units (only the first is the proper SI unit), is:

$$R = 8.315 \text{ J/(mol·K)} \qquad\qquad \text{[SI units]}$$
$$= 0.0821 \text{ (L·atm)/(mol·K)}$$
$$= 1.99 \text{ calories/(mol·K)}.^{\dagger}$$

Equation 17–3 is called the **ideal gas law**, or the **equation of state for an ideal gas**. We use the term "ideal" because real gases do not follow Eq. 17–3 precisely, particularly at high pressure (and density) or when the gas is near the liquefaction point (= boiling point). However, at pressures less than an atmosphere or so, and when T is not close to the liquefaction point of the gas, Eq. 17–3 is quite accurate and useful for real gases.

17–8 | Problem Solving with the Ideal Gas Law

The ideal gas law is an extremely useful tool, and we now consider some Examples. We will often refer to "standard conditions" or "standard temperature and pressure" (STP), which means $T = 273 \text{ K}$ (0°C) and $P = 1.00 \text{ atm} = 1.013 \times 10^5 \text{ N/m}^2 = 101.3 \text{ kPa}$.

STP = 273 K, 1 atm

EXAMPLE 17–8 **Volume of one mol at STP.** Determine the volume of 1.00 mol of any gas, assuming it behaves like an ideal gas, (a) at STP, (b) 20°C.

SOLUTION (a) We solve for V in Eq. 17–3:

$$V = \frac{nRT}{P} = \frac{(1.00 \text{ mol})(8.315 \text{ J/mol · K})(273 \text{ K})}{(1.013 \times 10^5 \text{ N/m}^2)}$$
$$= 22.4 \times 10^{-3} \text{ m}^3.$$

Since 1 liter is $1000 \text{ cm}^3 = 1 \times 10^{-3} \text{ m}^3$, 1 mol of any gas has a volume of 22.4 L at STP. [How big is 22.4 L? About the size of a cube one foot on a side; more precisely, $\sqrt[3]{22.4 \times 10^{-3} \text{ m}^3} = 0.28 \text{ m} = 28 \text{ cm}$ on a side.]
(b) At 20°C, $T = 293 \text{ K}$ and $V = 24.0 \text{ L}$.

1 mol of gas at STP has volume = 22.4 L

The value for the volume of 1 mol of an ideal gas at STP (22.4 L) is worth remembering, for it sometimes makes calculation simpler.

†Calories are defined in Chapter 19, and sometimes it is useful to use R as given in terms of calories.

Always remember, when using the ideal gas law, that temperatures must be given in kelvins (K) and that the pressure P must always be *absolute* pressure, not gauge pressure.

EXAMPLE 17–9 **ESTIMATE** **Mass of air in a room.** Estimate the mass of air in a room whose dimensions are 5 m × 3 m × 2.5 m high, at 20°C.

SOLUTION First we determine the number of moles n, and then we can multiply by the mass of one mole to get the total mass. Example 17–8 told us that 1 mol at 20°C has a volume of 24.0 L. The room's volume is 3 m × 5 m × 2.5 m, so

$$n = \frac{(3\,\text{m})(5\,\text{m})(2.5\,\text{m})}{24.0 \times 10^{-3}\,\text{m}^3} \approx 1600 \text{ mol}.$$

Air is a mixture of about 20% oxygen (O_2) and 80% nitrogen (N_2) whose atomic masses are 2 × 16 = 32 and 2 × 14 = 28 respectively, for an average of about 29. Thus, 1 mol of air has a mass of about 29 g = 0.029 kg, so our room has a mass of air

$$m \approx (1600 \text{ mol})(0.029 \text{ kg/mol}) \approx 50 \text{ kg}.$$

That is roughly 100 lbs of air!

Frequently, volume is specified in liters and pressure in atmospheres. Rather than convert these to SI units, we can instead use the value of R given above as 0.0821 L·atm/mol·K.

In many situations it is not necessary to use the value of R at all. For example, many problems involve a change in the pressure, temperature, and volume of a fixed amount of gas. In this case, $PV/T = nR$ = constant, since n and R remain constant. If we now let P_1, V_1, and T_1 represent the appropriate variables initially, and P_2, V_2, T_2 represent the variables after the change is made, then we can write

$$\frac{P_1 V_1}{T_1} = \frac{P_2 V_2}{T_2}.$$

If we know any five of the quantities in this equation, we can solve for the sixth. Or, if one of the three variables is constant $(V_1 = V_2$, or $P_1 = P_2$, or $T_1 = T_2)$ then we can use this equation to solve for one unknown when given the other three quantities.

EXAMPLE 17–10 **Check tires cold.** An automobile tire is filled to a gauge pressure of 200 kPa at 10°C. After driving 100 km, the temperature within the tire rises to 40°C. What is the pressure within the tire now?

SOLUTION Since the volume remains essentially constant, $V_1 = V_2$, and therefore

$$\frac{P_1}{T_1} = \frac{P_2}{T_2}.$$

This is, incidentally, a statement of Gay-Lussac's law. Since the pressure given is the gauge pressure (Section 13–3), we must add atmospheric pressure (= 101 kPa) to get the absolute pressure $P_1 = (200 \text{ kPa} + 101 \text{ kPa}) = 301$ kPa. And we convert temperatures to kelvin by adding 273:

$$P_2 = \frac{P_1}{T_1} T_2 = \frac{(3.01 \times 10^5 \text{ Pa})(313 \text{ K})}{(283 \text{ K})} = 333 \text{ kPa}.$$

Subtracting atmospheric pressure, we find the resulting gauge pressure to be 232 kPa, which is a 15 percent increase. This Example shows why car manuals suggest checking the pressure when the tires are cold.

17-9 Ideal Gas Law in Terms of Molecules: Avogadro's Number

The fact that the gas constant, R, has the same value for all gases is a remarkable reflection of simplicity in nature. It was first recognized, although in a slightly different form, by the Italian scientist Amedeo Avogadro (1776–1856). Avogadro stated that *equal volumes of gas at the same pressure and temperature contain equal numbers of molecules.* This is sometimes called **Avogadro's hypothesis**. That this is consistent with R being the same for all gases can be seen as follows. First of all, from Eq. 17–3 we see that for the same number of moles, n, and the same pressure and temperature, the volume will be the same for all gases as long as R is the same. Second, the number of molecules in 1 mole is the same for all gases.[†] Thus Avogadro's hypothesis is equivalent to R being the same for all gases.

Avogadro's hypothesis

The number of molecules in a mole is known as **Avogadro's number**, N_A. Although Avogadro conceived the notion, he was not able to actually determine the value of N_A. Indeed, precise measurements were not done until the twentieth century.

A number of methods have been devised to measure N_A and the accepted value today is

$$N_A = 6.02 \times 10^{23}. \qquad \text{[molecules/mole]}$$

Avogadro's number

Since the total number of molecules, N, in a gas is equal to the number per mole times the number of moles $(N = nN_A)$, the ideal gas law, Eq. 17–3, can be written in terms of the number of molecules present:

$$PV = nRT = \frac{N}{N_A} RT,$$

or

$$PV = NkT, \qquad\qquad (17\text{–}4)$$

IDEAL GAS LAW (IN TERMS OF MOLECULES)

where $k = R/N_A$ is called **Boltzmann's constant** and has the value

$$k = \frac{R}{N_A} = 1.38 \times 10^{-23}\,\text{J/K}.$$

EXAMPLE 17–11 **Hydrogen atom mass.** Use Avogadro's number to determine the mass of a hydrogen atom.

SOLUTION One mole of hydrogen (atomic mass = 1.008 u, Section 17–1) has a mass of 1.008×10^{-3} kg and contains 6.02×10^{23} atoms. Thus one atom has a mass

$$m = \frac{1.008 \times 10^{-3}\,\text{kg}}{6.02 \times 10^{23}} = 1.67 \times 10^{-27}\,\text{kg}.$$

Historically, the reverse process was one method used to obtain N_A: that is, from the measured mass of the hydrogen atom.

[†] For example, the molecular mass of H_2 gas is 2.0 atomic mass units (u), whereas that of O_2 gas is 32.0 u. Thus 1 mol of H_2 has a mass of 0.0020 kg and 1 mol of O_2 gas, 0.032 kg. The number of molecules in a mole is equal to the total mass M of a mole divided by the mass m of one molecule; since this ratio (M/m) is the same for all gases by definition of the mole, a mole of any gas must contain the same number of molecules.

It is important to have a very precisely defined temperature scale so that measurements of temperature made at different laboratories around the world can be accurately compared. We now discuss such a scale that has been accepted by the general scientific community.

The standard thermometer for this scale is the constant-volume gas thermometer already discussed in Section 17–2. The scale itself is called the **ideal gas temperature scale**, since it is based on the property of an ideal gas that the pressure is directly proportional to the absolute temperature (Gay-Lussac's law). A real gas, which would have to be used in any real constant-volume gas thermometer, approaches this ideal at low density. In other words, the temperature at any point in space is *defined* as being proportional to the pressure in the (nearly) ideal gas used in the thermometer. To set up a scale we need two fixed points. One fixed point will be $P = 0$ at $T = 0$ K. The second fixed point is chosen to be the **triple point** of water, which is that point where water in the solid, liquid, and gas states can coexist in equilibrium. This occurs only at a unique temperature and pressure,[†] and can be reproduced at different laboratories with great precision. The pressure at the triple point of water is 4.58 torr and the temperature is 0.01°C. This temperature corresponds to 273.16 K, since absolute zero is about −273.15°C. In fact, the triple point is now *defined* to be exactly 273.16 K.

The absolute or Kelvin temperature T at any point is then defined, using a constant-volume gas thermometer for an ideal gas, as

$$T = (273.16 \text{ K})\left(\frac{P}{P_{tp}}\right). \qquad \text{[ideal gas; constant volume]} \quad \textbf{(17–5a)}$$

In this relation, P_{tp} is the pressure in the thermometer at the triple point temperature of water, and P is the pressure in the thermometer when it is at the point where T is being determined. Note that if we let $P = P_{tp}$ in this relation, then $T = 273.16$ K, as it must.

The definition of temperature, Eq. 17–5a, with a constant-volume gas thermometer filled with a real gas is only approximate because we find that we get different results for the temperature depending on the type of gas that is used in the thermometer. Temperatures determined in this way also vary depending on the amount of gas in the bulb of the thermometer: for example, the boiling point of water at 1.00 atm is found from Eq. 17–5a to be 373.87 K when the gas is O_2 and $P_{tp} = 1000$ torr. If the amount of O_2 in the bulb is reduced so that at the triple point $P_{tp} = 500$ torr, the boiling point of water from Eq. 17–5a is then found to be 373.51 K. If H_2 gas is used instead, the corresponding values are 373.07 K and 373.11 K (see Fig. 17–15). But now suppose we use a particular real gas and make a series of measurements in which the amount of gas in the thermometer bulb is reduced to smaller and smaller amounts, so that P_{tp} becomes smaller and smaller. It is found experimentally that an extrapolation of such data to $P_{tp} = 0$ always gives the *same value* for the temperature of a given system (such as $T = 373.15$ K for the boiling point of water at 1.00 atm) as shown in Fig. 17–15. Thus the temperature T at any point in space, determined using a constant-volume gas thermometer containing a real gas, is defined using this limiting process:

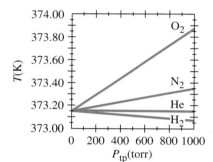

FIGURE 17–15 Temperature readings of a constant-volume gas thermometer for the boiling point of water at 1.00 atm are plotted, for different gases, as a function of the gas pressure in the thermometer at the triple point (P_{tp}). Note that as the amount of gas in the thermometer is reduced, so that $P_{tp} \to 0$, all gases give the same reading, 373.15 K. For pressure less than 0.10 atm (76 torr), the variation shown is less than 0.07 K.

Ideal gas temperature scale

$$T = (273.16 \text{ K})\lim_{P_{tp}\to 0}\left(\frac{P}{P_{tp}}\right). \qquad \text{[constant volume]} \quad \textbf{(17–5b)}$$

This defines the **ideal gas temperature scale**. One of the great advantages of this

---

[†] Liquid water and steam can coexist (the boiling point) at a range of temperatures depending on the pressure. Water boils at a lower temperature when the pressure is less, such as high in the mountains. The triple point represents a more precisely reproducible fixed point than does either the freezing point or boiling point of water at, say, 1 atm. See Section 18–3 for further discussion.

scale is that the value for T does not depend on the kind of gas used. But the scale does depend on the properties of gases in general. Helium has the lowest condensation point of all gases; at very low pressures it liquefies at about 1 K, so temperatures below this cannot be defined on this scale.

Summary

The atomic theory of matter postulates that all matter is made up of tiny entities called **atoms**, which are typically 10^{-10} m in diameter.

Atomic and **molecular masses** are specified on a scale where ordinary carbon (^{12}C) is arbitrarily given the value 12.0000 u (atomic mass units).

The distinction between solids, liquids, and gases can be attributed to the strength of the attractive forces between the atoms or molecules and to their average speed.

Temperature is a measure of how hot or cold a body is. **Thermometers** are used to measure temperature on the **Celsius** (°C), **Fahrenheit** (°F), and **Kelvin** (K) scales. Two standard points on each scale are the freezing point of water (0°C, 32°F, 273.15 K) and the boiling point of water (100°C, 212°F, 373.15 K). A one kelvin change in temperature equals a change of one Celsius degree or $\frac{9}{5}$ Fahrenheit degrees.

The change in length, ΔL, of a solid, when its temperature changes by an amount ΔT, is directly proportional to the temperature change and to its original length L_0. That is,

$$\Delta L = \alpha L_0 \, \Delta T,$$

where α is the *coefficient of linear expansion*.

The change in volume of most solids, liquids, and gases is proportional to the temperature change and to the original volume V_0: $\Delta V = \beta V_0 \, \Delta T$. The *coefficient of volume expansion*, β, is approximately equal to 3α for most solids.

Water is unusual because, unlike most materials whose volume increases with temperature, its volume actually decreases as the temperature increases in the range from 0°C to 4°C.

The **ideal gas law**, or **equation of state for an ideal gas**, relates the pressure P, volume V, and temperature T (in kelvins) of n moles of gas by the equation

$$PV = nRT,$$

where $R = 8.315$ J/mol·K for all gases. Real gases obey the ideal gas law quite accurately if they are not at too high a pressure or near their liquefaction point.

One **mole** of a substance is defined as the number of grams which is numerically equal to the atomic or molecular mass.

Avogadro's number, $N_A = 6.02 \times 10^{23}$, is the number of atoms or molecules in 1 mol of any pure substance.

The ideal gas law can be written in terms of the number of molecules N in the gas as

$$PV = NkT,$$

where $k = R/N_A = 1.38 \times 10^{-23}$ J/K is Boltzmann's constant.

Questions

1. Which has more atoms: 1 kg of iron or 1 kg of aluminum (see the Periodic Table or Appendix D)?
2. Name several properties of materials that could be exploited to make a thermometer.
3. Suppose system C is not in equilibrium with system A nor in equilibrium with system B. Does this imply that A and B are not in equilibrium? What can you infer regarding the temperatures of A, B, and C?
4. If system A is in equilibrium with system B, but B is not in equilibrium with system C, what can you say about the temperatures of A, B, and C?
5. A flat bimetallic strip consists of aluminum riveted to a strip of iron. When heated, which metal will be on the outside of the curve?
6. In the relation $\Delta L = \alpha L_0 \Delta T$, should L_0 be the initial length, the final length, or does it matter?
7. Why is it sometimes easier to remove the lid from a tightly closed jar after warming it under hot running water?
8. Long steam pipes often have a section in the shape of a U. Why?

9. Figure 17–16 shows a diagram of a simple *thermostat* used to control a furnace (or other heating or cooling system). The bimetallic strip consists of two strips of different metals bonded together. The electric switch is a glass vessel containing liquid mercury that conducts electricity when it can flow to touch both contact wires. Explain how this device controls the furnace and how it can be set at different temperatures.

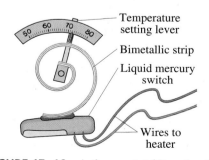

FIGURE 17–16 A thermostat (Question 9).

10. The units for the coefficients of expansion α are $(C°)^{-1}$, and there is no mention of a length unit such as meters. Would the expansion coefficient change if we used feet or millimeters instead of meters?

11. Explain why it is advisable to add water to an overheated automobile engine only slowly, and only with the engine running.

12. A glass container may break if one part of it is heated or cooled more rapidly than adjacent parts. Explain.

13. When a cold mercury-in-glass thermometer is first placed in a hot tub of water, the mercury initially descends a bit and then rises. Explain.

14. The principal virtue of Pyrex glass is that its coefficient of linear expansion is much smaller than that for ordinary glass (Table 17–1). Explain why this gives rise to the high "heat resistance" of Pyrex.

15. Will a grandfather clock, accurate at 20°C, run fast or slow on a hot day (30°C)? The clock uses a pendulum supported on a long thin brass rod.

16. Freezing a can of soda will cause its bottom and top to bulge so badly the can will not stand up. What has happened?

17. Why might you expect an alcohol-in-glass thermometer to be more precise than a mercury-in-glass thermometer?

18. Will the buoyant force on an aluminum sphere submerged in water increase or decrease if the temperature is increased from 20°C to 40°C?

19. A flat, uniform cylinder of lead floats in mercury at 0°C. Will the lead float higher or lower when the temperature is raised?

20. Which scale, Fahrenheit, Celsius, or Kelvin, might be considered most "natural" from a scientific point of view? Discuss.

21. If an atom is measured to have a mass of 6.7×10^{-27} kg, what atom do you think it is?

* 22. From a practical point of view, does it really matter what gas is used in a constant-volume gas thermometer? If so, explain. [*Hint*: See Fig. 17–15.]

Problems

Section 17–1

1. (I) How does the number of atoms in a 26.5-gram gold ring compare to the number in a silver ring of the same mass?

2. (I) How many atoms are there in a 3.4-gram copper penny?

Section 17–2

3. (I) (a) "Room temperature" is often taken to be 68°F; what is this on the Celsius scale? (b) The temperature of the filament in a lightbulb is about 1800°C; what is this on the Fahrenheit scale?

4. (I) (a) 15° below zero on the Celsius scale is what Fahrenheit temperature? (b) 15° below zero on the Fahrenheit scale is what Celsius temperature?

5. (I) In a foreign country, a thermometer tells you that you have a fever of 40.0°C. What is this in Fahrenheit?

6. (I) In an alcohol-in-glass thermometer, the alcohol column has length 11.82 cm at 0.0°C and length 22.85 cm at 100.0°C. What is the temperature if the column has length (a) 16.70 cm, and (b) 20.50 cm?

7. (II) At what temperature will the Fahrenheit and Centigrade scales yield the same numerical value?

Section 17–4

8. (I) A concrete highway is built of slabs 12 m long (20°C). How wide should the expansion cracks between the slabs be (at 20°C) to prevent buckling if the range of temperature is −30°C to +50°C?

9. (I) Super Invar, an alloy of iron and nickel, is a strong material with a very low coefficient of thermal expansion $[0.2 \times 10^{-6}(C°)^{-1}]$. A 2.0-m-long-tabletop of this alloy is used for sensitive laser measurements where extremely high tolerances are required. How much will this alloy table expand along its length if the temperature increases 5.0 C°? Compare to tabletops made of steel and marble.

10. (I) The Eiffel Tower (Fig. 17–17) is built of wrought iron approximately 300 m tall. Estimate how much its height changes between July (average temperature of 25°C) and January (average temperature of 2°C). Ignore the angles of the iron beams and treat the tower as a vertical beam.

FIGURE 17–17 Problem 10: The Eiffel Tower in Paris.

11. (II) To make a secure fit, rivets that are larger than the rivet hole are often used and the rivet is cooled (usually in dry ice) before it is placed in the hole. A steel rivet 1.871 cm in diameter is to be placed in a hole 1.869 cm in diameter. To what temperature must the rivet be cooled if it is to fit in the hole at 20°C?

12. (II) A uniform rectangular plate of length l and width w has coefficient of linear expansion α. Show that, if we neglect very small quantities, the change in area of the plate due to a temperature change ΔT is $\Delta A = 2\alpha l w\,\Delta T$. See Fig. 17–18.

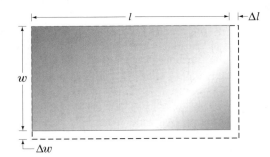

FIGURE 17–18 Rectangular plate is heated. Problem 12.

13. (II) An ordinary glass is filled to the brim with 350.0 mL of water at 100.0°C. If the temperature decreased to 20.0°C, how much water could be added to the glass?

14. (II) It is observed that 55.50 mL of water at 20°C completely fills a container to the brim. When the container and the water are heated to 60°C, 0.35 g of water is lost. (a) What is the coefficient of volume expansion of the container? (b) What is the most likely material of the container? Density of water at 60°C is 0.98324 g/mL.

15. (II) A quartz sphere is 8.75 cm in diameter. What will be its change in volume if it is heated from 30°C to 200°C?

16. (II) A brass plug is to be placed in a ring made of iron. At room temperature, the diameter of the plug is 8.753 cm and that of the inside of the ring is 8.743 cm. They must both be brought to what common temperature in order to fit?

17. (II) If a fluid is contained in a long narrow vessel so it can expand in essentially one direction only, show that the effective coefficient of linear expansion α is approximately equal to the coefficient of volume expansion β.

18. (II) (a) Show that the change in the density ρ of a substance, when the temperature changes by ΔT, is given by $\Delta\rho = -\beta\rho\,\Delta T$. (b) What is the fractional change in density of a lead sphere whose temperature decreases from 25°C to -40°C?

19. (III) The pendulum in a grandfather clock is made of brass and keeps perfect time at 17°C. How much time is gained or lost in a year if the clock is kept at 25°C? (Assume the frequency dependence on length for a simple pendulum applies.)

20. (III) (a) Determine a formula for the change in surface area of a uniform solid sphere of radius r if its coefficient of linear expansion is α (assumed constant) and its temperature is changed by ΔT. (b) What is the increase in area of a solid iron sphere of radius 60.0 cm if its temperature is raised from 20°C to 310°C?

21. (III) A 23.4-kg solid aluminum cylindrical wheel of radius 0.41 m is rotating about its axle in frictionless bearings with angular velocity $\omega = 32.8$ rad/s. If its temperature is now raised from 20.0°C to 75.0°C, what is the fractional change in ω?

* Section 17–5

*22. (I) At what temperature will the ultimate compressive strength of concrete be exceeded for the blocks discussed in Example 17–7?

*23. (I) An aluminum bar has the precisely desired length when at 15°C. How much stress is required to keep it at this length if the temperature increases to 35°C?

*24. (II) (a) A horizontal steel I-beam of cross-sectional area 0.041 m^2 is rigidly connected to two vertical steel girders. If the beam was installed when the temperature was 30°C, what stress is developed in the beam when the temperature drops to -30°C? (b) Is the ultimate strength of the steel exceeded? (c) What stress is developed if the beam is concrete and has a cross-sectional area of 0.13 m^2? Will it fracture?

*25. (III) A barrel of diameter 134.122 cm at 20°C is to be enclosed by an iron band. The circular band has an inside diameter of 134.110 cm at 20°C. It is 7.4 cm wide and 0.65 cm thick. (a) To what temperature must the band be heated so that it will fit over the barrel? (b) What will be the tension in the band when it cools to 20°C?

Section 17–6

26. (I) What are the following temperatures on the Kelvin scale: (a) 86°C, (b) 78°F, (c) -100°C, (d) 5500°C?

27. (I) Absolute zero is what temperature on the Fahrenheit scale?

28. (II) Typical temperatures in the interior of the Earth and Sun are about 4000°C and 15×10^6 °C, respectively. (a) What are these temperatures in kelvins? (b) What percent error is made in each case if a person forgets to change °C to K?

Sections 17–7 and 17–8

29. (I) If 3.00 m^3 of a gas initially at STP is placed under a pressure of 3.20 atm, the temperature of the gas rises to 38.0°C. What is the volume?

30. (I) In an internal combustion engine, air at atmospheric pressure and a temperature of about 20°C is compressed in the cylinder by a piston to $\frac{1}{9}$ of its original volume (compression ratio = 9.0). Estimate the temperature of the compressed air, assuming the pressure reaches 40 atm.

31. (II) Calculate the density of oxygen at STP using the ideal gas law.

32. (II) A storage tank contains 21.6 kg of nitrogen (N_2) at an absolute pressure of 3.65 atm. What will the pressure be if the nitrogen is replaced by an equal mass of CO_2?

33. (II) A storage tank at STP contains 18.5 kg of nitrogen (N_2). (a) What is the volume of the tank? (b) What is the pressure if an additional 15.0 kg of nitrogen is added without changing the temperature?

34. (II) If 18.75 mol of helium gas is at 10.0°C and a gauge pressure of 0.350 atm, calculate (a) the volume of the helium gas under these conditions, and (b) the temperature if the gas is compressed to precisely half the volume at a gauge pressure of 1.00 atm.

35. (II) What is the pressure inside a 35.0-L container holding 105.0 kg of argon gas at 20.0°C?

36. (II) A tank contains 30.0 kg of O_2 gas at a gauge pressure of 8.70 atm. If the oxygen is replaced by helium, how many kilograms of the latter will be needed to produce a gauge pressure of 7.00 atm?

37. (II) A hot-air balloon achieves its buoyant lift by heating the air inside the balloon which makes it less dense than the air outside. Suppose the volume of a balloon is 1800 m³ and the required lift is 2700 N (rough estimate of the weight of the equipment and passenger). Calculate the temperature of the air inside the balloon which will produce the required lift. Assume that the outside air temperature is 0° and that air is an ideal gas under these conditions. What factors limit the maximum altitude attainable by this method for a given load? (Neglect variables like wind.)

38. (II) A tire is filled with air at 15°C to a gauge pressure of 220 kPa. If the tire reaches a temperature of 38°C, what fraction of the original air must be removed if the original pressure of 220 kPa is to be maintained?

39. (II) If 61.5 L of oxygen at 18.0°C and an absolute pressure of 2.45 atm are compressed to 48.8 L and at the same time the temperature is raised to 50.0°C, what will the new pressure be?

40. (II) A child's helium-filled balloon escapes at sea level and 20.0°C. When it reaches an altitude of 3000 m, where the temperature is 5.0°C and the pressure only 0.70 atm, how will its volume compare to that at sea level?

41. (III) Compare the value for the density of water vapor at 100°C and 1 atm (Table 13–1) with the value predicted from the ideal gas law. Why would you expect a difference?

42. (III) An air bubble at the bottom of a lake 37.0 m deep has a volume of 1.00 cm³. If the temperature at the bottom is 5.5°C and at the top 21.0°C, what is the volume of the bubble just before it reaches the surface?

Section 17–9

43. (I) Calculate the number of molecules/m³ in an ideal gas at STP.

44. (I) How many moles of water are there in 1.000 L? How many molecules?

45. (I) Estimate the number of molecules you inhale in a 2.0-L breath.

46. (II) Estimate the number of (a) moles, and (b) molecules of water in all the Earth's oceans. Assume water covers 75 percent of the Earth to an average depth of 3 km.

47. (II) A cubic box of volume 5.1×10^{-2} m³ is filled with air at atmospheric pressure at 20°C. The box is closed and heated to 180°C. What is the net force on each side of the box?

48. (III) Estimate how many molecules of air are in each 2.0-L breath you inhale that were also in the last breath Galileo took. [*Hint:* Assume the atmosphere is about 10 km high and of constant density.]

* Section 17–10

* **49.** (I) At the boiling point of sulfur (444.6°C) the pressure in a constant-volume gas thermometer is 187 torr. Estimate (a) the pressure at the triple point of water, (b) the temperature when the pressure in the thermometer is 112 torr.

* **50.** (I) In a constant-volume gas thermometer, what is the limiting ratio of the pressure at the boiling point of water at 1 atm to that at the triple point? (Keep five significant figures.)

* **51.** (II) Use Fig. 17–15 to determine the inaccuracy of a constant-volume gas thermometer using oxygen if it reads a pressure $P = 268$ torr at the boiling point of water at 1 atm. Express answer (a) in kelvins and (b) as a percentage.

* **52.** (II) A constant-volume gas thermometer is being used to determine the temperature of the melting point of a substance. The pressure in the thermometer at this temperature is 218 torr; at the triple point of water, the pressure is 286 torr. Some gas is now released from the thermometer bulb so that the pressure at the triple point of water becomes 163 torr. At the temperature of the melting substance, the pressure is 128 torr. Estimate, as accurately as possible, the melting-point temperature of the substance.

General Problems

53. A precise steel tape measure has been calibrated at 20°C. At 34°C, (a) will it read high or low, and (b) what will be the percentage error?

54. A Pyrex measuring cup was calibrated at normal room temperature. How much error will be made in a recipe calling for 300 mL of cool water, if the water and the cup are hot, at 80°C, instead of at room temperature? Neglect the glass expansion.

55. The pressure in a helium gas cylinder is initially 35 atmospheres. After many balloons have been blown up, the pressure has decreased to 5 atm. What fraction of the original gas remains in the cylinder? Pressures are gauge pressures.

56. Write the ideal gas law in terms of the density of the gas.

57. Estimate the number of air molecules in a room of length 8.0 m, width 6.0 m, and height 4.2 m. Assume the temperature is 20°C. How many moles does that correspond to?

58. The lowest pressure attainable using the best available vacuum techniques is about 10^{-12} N/m². At such a pressure, how many molecules are there per cm³ at 0°C?

59. If a scuba diver fills his lungs to full capacity of 5.5 L when 10 m below the surface, to what volume would his lungs expand if he quickly rose to the surface? Is this advisable?

60. If a rod of original length L_1 has its temperature changed from T_1 to T_2, determine a formula for its new length L_2 in terms of T_1, T_2, and α. Assume (a) $\alpha = $ constant, (b) $\alpha = \alpha(T)$ is some function of temperature, and (c) $\alpha = \alpha_0 + bT$ where α_0 and b are constants.

61. A house has a volume of $770\ \text{m}^3$. (a) What is the total mass of air inside the house at $20°C$? (b) If the temperature drops to $-10°C$, what mass of air enters or leaves the house?

62. (a) Use the ideal gas law to show that, for an ideal gas at constant pressure, the coefficient of volume expansion is equal to $\beta = 1/T$ where T is the kelvin temperature. Compare to Table 17–1 for gases at $T = 293\ \text{K}$. (b) Show that the bulk modulus (Section 12–5) for an ideal gas held at constant temperature is $B = P$ where P is the pressure.

63. From the known value of atmospheric pressure at the surface of the Earth, estimate the total number of air molecules in the Earth's atmosphere.

64. (a) The tube of a mercury thermometer has an inside diameter of 0.140 mm. The bulb has a volume of $0.315\ \text{cm}^3$. How far will the thread of mercury move when the temperature changes from $11.5°C$ to $33.0°C$? Take into account expansion of the Pyrex glass. (b) Determine a formula for the length of the mercury column in terms of relevant variables.

65. What is the average distance between oxygen molecules at STP?

66. An iron cube floats in a bowl of liquid mercury at $0°C$. (a) If the temperature is raised to $25°C$, will the cube float higher or lower in the mercury? (b) By what percent will the fraction of volume submerged change?

67. If a steel band were to fit snugly around the Earth's equator at $20°C$, but then was heated to $35°C$, how high above the Earth would the band be (assume equal everywhere)?

68. Estimate the percent change in density of iron when it is still a solid, but deep in the Earth where the temperature is $2000°C$ and it is under 5000 atm of pressure. Take into account both thermal expansion and changes due to increased outside pressure. Assume both the bulk modulus and the volume coefficient of expansion do not vary with temperature and are the same as at normal room temperature. The bulk modulus for iron is about $90 \times 10^9\ \text{N/m}^2$.

69. A standard cylinder of oxygen used in a hospital has the following characteristics at room temperature (300 K): gauge pressure = 2000 psi (13,800 kPa), volume = 16 liters $(0.016\ \text{m}^3)$. How long will the cylinder last if the flow rate, measured at atmospheric pressure, is constant at 2.4 liters/min?

70. A helium party balloon, assumed to be a perfect sphere, has a radius of 18.0 cm. At room temperature ($20°C$), its internal pressure is 1.05 atm. Find the number of moles of helium in the balloon and the mass of helium needed to inflate the balloon to these values.

71. The density of gasoline at $0°C$ is $0.68 \times 10^3\ \text{kg/m}^3$. What is the density on a hot day, when the temperature is $32°C$?

72. A brass lid screws tightly onto a glass jar at $20°C$. To help open the jar, it can be placed into a bath of hot water. After this treatment, the temperatures of the lid and the jar are both $60°C$. The inside diameter of the lid is 8.0 cm. Find the size of the gap (difference in radius) that develops by this procedure.

73. The first length standard, adopted in the 18th century, was a platinum bar with two very fine marks separated by what was defined to be exactly one meter. If this standard bar was to be accurate to within $\pm 1.0\ \mu\text{m}$, how carefully would the trustees have needed to control the temperature? The coefficient of linear expansion is $9 \times 10^{-6}\ \text{C}°^{-1}$.

74. Reinforcing rods in concrete are made of steel, which has nearly the same coefficient of thermal expansion as concrete. What would happen if brass were used instead? Imagine a 2.5-cm diameter brass rod, embedded in a concrete matrix. If the temperature rises by 20 C°, the brass will be under compression and the concrete under tension. Will the concrete stay in one piece? (You will need data from Tables 12–1 and 12–2. Assume the same magnitude of stress in both materials.)

75. A helium balloon has volume V_0 and temperature T_0 at sea level where the pressure is P_0 and the air density is ρ_0. The balloon is allowed to float up in the air to altitude y where the temperature is T_1. (a) Show that the volume occupied by the balloon is then $V = V_0(T_1/T_0)e^{+cy}$ where $c = \rho_0 g/P_0 = 1.25 \times 10^{-4}\ \text{m}^{-1}$. (b) Show that the buoyant force does not depend on altitude y. Assume that the skin of the balloon maintains the helium pressure at a constant factor of 1.05 times greater than the outside pressure. [Hint: Assume that the pressure change with altitude is $P = P_0 e^{-cy}$, as in Example 13–4, Chapter 13.]

76. A scuba tank when fully charged has a pressure of 200 atmospheres at $20°C$. The volume of the tank is 11.3 liters. (a) What would the volume of the air be at 1.00 atmosphere and at the same temperature? (b) Before entering the water, a person consumes 2.0 liters of air in each breath, and breathes 12 times a minute. At this rate, how long would the tank last? (c) At a depth of 20.0 m of sea water and temperature of $10°C$, how long would the same tank last assuming the breathing rate does not change?

77. A temperature controller, designed to work in a steam environment, involves a bimetallic strip constructed of brass and steel, connected at their ends by rivets. Each of the metals is 2.0 mm thick. At $20°C$, the strip is 10.0 cm long and straight. Find the radius of curvature of the assembly at $100°C$. See Fig. 17–19.

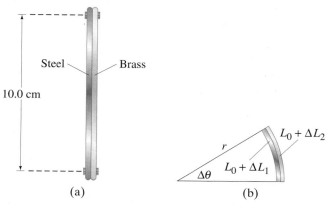

FIGURE 17–19 Problem 77.

Idaho's Salmon River, famous for rafting and fishing, in December. We see the three states of matter in the form of water as a liquid, as a solid (snow and ice), and as a gas (steam or fog). In this chapter we will examine the microscopic theory of matter as atoms or molecules that are always in motion, which we call kinetic theory. We will see that the temperature of a gas is directly related to the average kinetic energy of its molecules. We will consider ideal gases, but we'll also look at real gases and how they change phase including evaporation, vapor pressure, and humidity.

CHAPTER 18

Kinetic Theory of Gases

The analysis of matter in terms of atoms in continuous random motion is called the **kinetic theory**. We now investigate the properties of a gas from the point of view of kinetic theory, which is based on the laws of classical mechanics. But to apply Newton's laws to each one of the vast number of molecules in a gas $(>10^{25}/m^3$ at STP) is far beyond the capability of any present computer. Instead we take a statistical approach and determine averages of certain quantities, and these averages correspond to macroscopic variables. We will, of course, demand that our microscopic description correspond to the macroscopic properties of gases; otherwise our theory would be of little value. Most importantly, we will arrive at an important relation between the average kinetic energy of molecules in a gas and the absolute temperature.

18–1 | The Ideal Gas Law and the Molecular Interpretation of Temperature

We make the following assumptions about the molecules in a gas. These assumptions reflect a simple view of a gas, but nonetheless the results they predict correspond well to the essential features of real gases that are at low pressures and far from the liquefaction point. Under these conditions real gases follow the ideal gas law quite closely, and indeed the gas we now describe is referred to as an

ideal gas. The assumptions, which represent the basic postulates of the kinetic theory, are:

Postulates of kinetic theory

1. There are a large number of molecules, N, each of mass m, moving in random directions with a variety of speeds. This assumption is in accord with our observation that a gas fills its container and, in the case of air on the Earth, is kept from escaping only by the force of gravity.

2. The molecules are, on the average, far apart from one another. That is, their average separation is much greater than the diameter of each molecule.

3. The molecules are assumed to obey the laws of classical mechanics, and are assumed to interact with one another only when they collide. Although molecules exert weak attractive forces on each other between collisions, the potential energy associated with these forces is small compared to the kinetic energy, and we ignore it for now.

4. Collisions with another molecule or the wall of the vessel are assumed to be perfectly elastic, like the collisions of perfectly elastic billiard balls (Chapter 9). We assume the collisions are of very short duration compared to the time between collisions. Then we can ignore the potential energy associated with collisions in comparison to the kinetic energy between collisions.

We can see immediately how this kinetic view of a gas can explain Boyle's law (Section 17–6). The pressure exerted on a wall of a container of gas is due to the constant bombardment of molecules. If the volume is reduced by (say) half, the molecules are closer together and twice as many will be striking a given area of the wall per second. Hence we expect the pressure to be twice as great, in agreement with Boyle's law.

Boyle's law explained

Now let us calculate quantitatively the pressure in a gas based on kinetic theory. For purposes of argument, we imagine that the molecules are contained in a rectangular vessel whose ends have area A and whose length is l, as shown in Fig. 18–1a. The pressure exerted by the gas on the walls of its container is, according to our model, due to the collisions of the molecules with the walls. Let us focus our attention on the wall, of area A, at the left end of the container and examine what happens when one molecule strikes this wall, as shown in Fig. 18–1b. This molecule exerts a force on the wall, and the wall exerts an equal and opposite force back on the molecule. The magnitude of this force, according to Newton's second law, is equal to the molecule's rate of change of momentum, $F = dp/dt$. Assuming the collision is elastic, only the x component of the molecule's momentum changes, and it changes from $-mv_x$ (it is moving in the negative x direction) to $+mv_x$. Thus the change in momentum, $\Delta(mv)$, which is the final momentum minus the initial momentum, is

$$\Delta(mv) = mv_x - (-mv_x) = 2mv_x$$

for one collision. This molecule will make many collisions with the wall, each separated by a time Δt, which is the time it takes the molecule to travel across the box and back again, a distance (x component) equal to $2l$. Thus $2l = v_x \Delta t$ or

$$\Delta t = 2l/v_x.$$

The time Δt between collisions is very small, so the number of collisions per second is very large. Thus the average force—averaged over many collisions—will be equal to the force exerted during one collision divided by the time between collisions (Newton's second law):

$$F = \frac{\Delta(mv)}{\Delta t} = \frac{2mv_x}{2l/v_x} = \frac{mv_x^2}{l}. \qquad \text{[due to one molecule]}$$

During its passage back and forth across the container, the molecule may collide

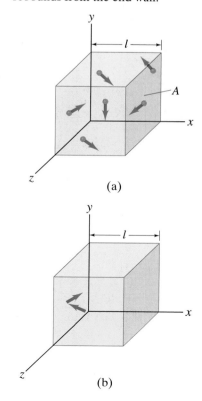

FIGURE 18–1 (a) Molecules of a gas moving about in a cubical container. (b) Arrows indicate the momentum of one molecule as it rebounds from the end wall.

(a)

(b)

with the tops and sides of the container, but this does not alter its x component of momentum and thus does not alter our result. It may also collide with other molecules, which may change its v_x. However, any loss (or gain) of momentum is acquired by the other molecule, and because we will eventually sum over all the molecules, this effect will be included. So our result above is not altered.

Of course, the actual force due to one molecule is intermittent, but because a huge number of molecules are striking the wall per second, the force is, on average, nearly constant. To calculate the force due to *all* the molecules in the box, we have to add the contributions of each. Thus the net force on the wall is

$$F = \frac{m}{l}\left(v_{x1}^2 + v_{x2}^2 + \cdots + v_{xN}^2\right),$$

where v_{x1} means v_x for molecule number 1 (we arbitrarily assign each molecule a number) and the sum extends over the total of N molecules. Now the average value of the square of the x component of velocity is

$$\overline{v_x^2} = \frac{v_{x1}^2 + v_{x2}^2 + \cdots + v_{xN}^2}{N}. \tag{18–1}$$

Thus we can write the force as

$$F = \frac{m}{l} N\overline{v_x^2}.$$

We know that the square of any vector is equal to the sum of the squares of its components (theorem of Pythagoras). Thus $v^2 = v_x^2 + v_y^2 + v_z^2$ for any velocity v. Taking averages, we obtain

$$\overline{v^2} = \overline{v_x^2} + \overline{v_y^2} + \overline{v_z^2}.$$

Since the velocities of the molecules in our gas are assumed to be random, there is no preference to one direction or another. Hence

$$\overline{v_x^2} = \overline{v_y^2} = \overline{v_z^2}.$$

Combining this relation with the one just above, we get

$$\overline{v^2} = 3\overline{v_x^2}.$$

We substitute this into the equation for net force F:

$$F = \frac{m}{l} N \frac{\overline{v^2}}{3}.$$

The pressure on the wall is then

$$P = \frac{F}{A} = \frac{1}{3}\frac{Nm\overline{v^2}}{Al}$$

or

Pressure in a gas

$$P = \frac{1}{3}\frac{Nm\overline{v^2}}{V}, \tag{18–2}$$

where $V = lA$ is the volume of the container. This is the result we were seeking, the pressure in a gas expressed in terms of molecular properties.

Equation 18–2 can be rewritten in a clearer form by multiplying both sides by V and slightly rearranging the right side:

$$PV = \tfrac{2}{3}N\left(\tfrac{1}{2}m\overline{v^2}\right). \tag{18–3}$$

The quantity $\tfrac{1}{2}m\overline{v^2}$ is the average kinetic energy $\left(\overline{K}\right)$ of the molecules in the gas. If we compare Eq. 18–3 with Eq. 17–4, the ideal gas law $PV = NkT$, we see that the two agree if

$$\tfrac{2}{3}\left(\tfrac{1}{2}m\overline{v^2}\right) = kT,$$

or

$$\overline{K} = \tfrac{1}{2}m\overline{v^2} = \tfrac{3}{2}kT. \qquad \text{[Ideal gas]} \quad \textbf{(18–5)}$$

TEMPERATURE RELATED TO AVERAGE KINETIC ENERGY OF MOLECULES

This equation tells us that

the average translational kinetic energy of molecules in an ideal gas is directly proportional to the absolute temperature.

The higher the temperature, according to kinetic theory, the faster the molecules are moving on the average. This relation is one of the triumphs of the kinetic theory.

EXAMPLE 18–1 **Molecular kinetic energy.** What is the average translational kinetic energy of molecules in an ideal gas at 37°C?

SOLUTION We use Eq. 18–4 and change 37°C to 310 K:

$$\overline{K} = \tfrac{3}{2}kT$$
$$= \tfrac{3}{2}(1.38 \times 10^{-23} \text{ J/K})(310 \text{ K}) = 6.42 \times 10^{-21} \text{ J}.$$

Note that a mole of molecules would have a total of translational kinetic energy equal to $(6.42 \times 10^{-21} \text{ J})(6.02 \times 10^{23}) = 3900 \text{ J}$, which equals the kinetic energy of a 1-kg stone traveling faster that 85 m/s.

Equation 18–4 holds not only for gases, but also applies reasonably accurately to liquids and solids. Thus the result of Example 18–1 would apply to molecules within living cells at body temperature (37°C).

We can use Eq. 18–4 to calculate how fast molecules are moving on the average. Notice that the average in Eqs. 18–1 through 18–4 is over the *square* of the velocity. The square root of $\overline{v^2}$ is called the **root-mean-square** velocity, v_{rms} (since we are taking the square *root* of the *mean* of the *square* of the velocity):

Root-mean-square (rms) velocity

$$v_{\text{rms}} = \sqrt{\overline{v^2}} = \sqrt{\frac{3kT}{m}}. \qquad \textbf{(18–5)}$$

rms speed of molecules

The **mean speed**, $\overline{v}$, is the average of the magnitudes of the speeds themselves; $\overline{v}$ is generally not equal to v_{rms}. To see the difference between the mean speed and the rms speed, consider the following Example.

Mean speed

EXAMPLE 18–2 **Mean speed and rms speed.** Eight particles have the following speeds, given in m/s: 1.0, 6.0, 4.0, 2.0, 6.0, 3.0, 2.0, 5.0. Calculate (*a*) the mean speed and (*b*) the rms speed.

SOLUTION (*a*) The mean speed is

$$\overline{v} = \frac{1.0 + 6.0 + 4.0 + 2.0 + 6.0 + 3.0 + 2.0 + 5.0}{8} = 3.6 \text{ m/s}.$$

(*b*) The rms speed is (Eq. 18–1):

$$v_{\text{rms}} = \sqrt{\frac{(1.0)^2 + (6.0)^2 + (4.0)^2 + (2.0)^2 + (6.0)^2 + (3.0)^2 + (2.0)^2 + (5.0)^2}{8}} \text{ m/s}$$
$$= 4.0 \text{ m/s}.$$

We see in this Example that $\overline{v}$ and v_{rms}, are not necessarily equal. In fact, for an ideal gas they differ by about 8 percent. We will see in the next Section how to calculate $\overline{v}$ for an ideal gas. We already have the tool to calculate v_{rms} (Eq. 18–5).

EXAMPLE 18–3 **Speeds of air molecules.** What is the rms speed of air molecules (O_2 and N_2) at room temperature (20°C)?

SOLUTION We must apply Eq. 18–5 to oxygen and nitrogen separately since they have different masses. The masses of one molecule of O_2 (molecular mass = 32 u) and N_2 (molecular mass = 28 u) are ($1\,u = 1.66 \times 10^{-27}$ kg)

$$m(O_2) = (32)(1.66 \times 10^{-27}\,\text{kg}) = 5.3 \times 10^{-26}\,\text{kg}$$
$$m(N_2) = (28)(1.66 \times 10^{-27}\,\text{kg}) = 4.7 \times 10^{-26}\,\text{kg}.$$

Thus, for oxygen

$$v_{\text{rms}} = \sqrt{\frac{3kT}{m}} = \sqrt{\frac{(3)(1.38 \times 10^{-23}\,\text{J/K})(293\,\text{K})}{(5.3 \times 10^{-26}\,\text{kg})}}$$
$$= 480\,\text{m/s},$$

and for nitrogen the result is $v_{\text{rms}} = 510$ m/s. These speeds are more than 1500 km/h or 1000 mi/h.

Equation 18–4, $\overline{K} = \frac{3}{2}kT$, implies that as the temperature approaches absolute zero the kinetic energy of molecules approaches zero. Modern quantum theory, however, tells us this is not quite so. Instead, as absolute zero is approached, the kinetic energy approaches a very small nonzero minimum value. Even though all real gases become liquid or solid near 0 K, molecular motion does not cease, even at absolute zero.

18–2 | Distribution of Molecular Speeds

The Maxwell Distribution

The molecules in a gas are assumed to be in random motion, which means that many molecules have speeds less than the average speed and others have speeds greater than the average. In 1859, James Clerk Maxwell (1831–1879) worked out a formula for the most probable distribution of speeds in a gas containing N molecules. We will not give a derivation here but merely quote his result:

$$f(v) = 4\pi N\left(\frac{m}{2\pi kT}\right)^{\frac{3}{2}} v^2 e^{-\frac{1}{2}\frac{mv^2}{kT}}. \tag{18–6}$$

Maxwell distribution

FIGURE 18–2 Distribution of speeds of molecules in an ideal gas. Note that $\overline{v}$ and v_{rms} are not at the peak of the curve (that speed is called the "most probable speed," v_{p}). This is because the curve is skewed to the right: it is not symmetrical.

$f(v)$ is called the **Maxwell distribution of speeds**, and is plotted in Fig. 18–2. The quantity $f(v)\,dv$ represents the number of molecules that have speed between v and $v + dv$. Notice that $f(v)$ does not give the number of molecules with speed v; $f(v)$ must be multiplied by dv to give the number of molecules (clearly the number of molecules must depend on the "width" or "range" of velocities, dv). In the formula for $f(v)$, m is the mass of a single molecule, T is the absolute temperature, and k is Boltzmann's constant. Since N is the total number of molecules in the gas, when we sum over all the molecules in the gas we must get N; thus we must have

$$\int_0^\infty f(v)\,dv = N.$$

(Problem 15 is an exercise to show that this is true.)

Experiments to determine the distribution of speeds in real gases, starting in the 1920s, confirmed with considerable accuracy the Maxwell distribution (for gases at not too high a pressure) and the direct proportion between average kinetic energy and absolute temperature, Eq. 18–4.

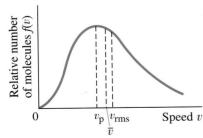

The Maxwell distribution for a given gas depends only on the absolute temperature. Figure 18–3 shows the distributions for two different temperatures. Just as v_{rms} increases with temperature, so the whole distribution curve shifts to the right at higher temperatures.

Figure 18–3 illustrates how kinetic theory can be used to explain why many chemical reactions, including those in biological cells, take place more rapidly as the temperature increases. Most chemical reactions take place in a liquid solution, and the molecules in a liquid have a distribution of speeds close to the Maxwell distribution. Two molecules may chemically react only if their kinetic energy is great enough so that when they collide, they penetrate into each other somewhat. The minimum energy required is called the *activation energy*, E_A, and it has a specific value for each chemical reaction. The molecular speed corresponding to a kinetic energy of E_A for a particular reaction is indicated in Fig. 18–3. The relative number of molecules with energy greater than this value is given by the area under the curve to the right of $v(E_A)$. In Fig. 18–3, the respective areas for two different temperatures are indicated by the two different shadings in the figure. It is clear that the number of molecules that have kinetic energies in excess of E_A increases greatly for only a small increase in temperature. The rate at which a chemical reaction occurs is proportional to the number of molecules with energy greater than E_A, and thus we see why reaction rates increase rapidly with increased temperature.

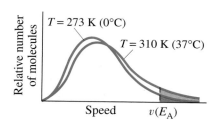

FIGURE 18–3 Distribution of molecular speeds for two different temperatures.

➡ **PHYSICS APPLIED**

How chemical reactions depend on temperature

*Calculations Using the Maxwell Distribution

Let us see how the Maxwell distribution can be used to obtain some interesting results.

EXAMPLE 18–4 **Determining $\bar{v}$ and v_p.** Determine formulas for (*a*) the average speed, $\bar{v}$, and (*b*) the most probable speed, v_p, of molecules in an ideal gas at temperature T.

SOLUTION (*a*) The average value of any quantity is found by multiplying each possible value of the quantity (say, speed) by the number of molecules that have that value, and then summing all these numbers and dividing by N (the total number). We are given a continuous distribution of speeds (Eq. 18–6), so the sum becomes an integral over the product of v and the number $f(v)\,dv$ that have speed v:

$$\bar{v} = \frac{\int_0^\infty v f(v)\,dv}{N} = 4\pi \left(\frac{m}{2\pi kT}\right)^{\frac{3}{2}} \int_0^\infty v^3 e^{-\frac{1}{2}\frac{mv^2}{kT}}\,dv.$$

We can look up the definite integral in the tables, or integrate by parts, and obtain

$$\bar{v} = 4\pi \left(\frac{m}{2\pi kT}\right)^{\frac{3}{2}} \left(\frac{2k^2 T^2}{m^2}\right) = \sqrt{\frac{8}{\pi}\frac{kT}{m}} \approx 1.60 \sqrt{\frac{kT}{m}}.$$

Average speed, $\bar{v}$

(*b*) The *most probable speed* is that speed which occurs more than any others, and thus is that speed where $f(v)$ has its maximum value. Since $df(v)/dv = 0$ at the maximum of the curve, we have

$$\frac{df(v)}{dv} = 4\pi \left(\frac{m}{2\pi kT}\right)^{\frac{3}{2}} \left(2ve^{-\frac{mv^2}{2kT}} - \frac{2mv^3}{2kT} e^{-\frac{mv^2}{2kT}}\right) = 0.$$

Solving for v, we get

$$v_p = \sqrt{\frac{2kT}{m}} \approx 1.41 \sqrt{\frac{kT}{m}}.$$

Most probable speed, v_p

(Another solution is $v = 0$, but this corresponds to a minimum, not a maximum.)

In summary,

$$v_p = \sqrt{2\frac{kT}{m}} \approx 1.41\sqrt{\frac{kT}{m}} \qquad \textbf{(18–7a)}$$

$$\bar{v} = \sqrt{\frac{8}{\pi}\frac{kT}{m}} \approx 1.60\sqrt{\frac{kT}{m}} \qquad \textbf{(18–7b)}$$

and from Eq. 18–5

$$v_{rms} = \sqrt{3\frac{kT}{m}} \approx 1.73\sqrt{\frac{kT}{m}}.$$

These are all indicated in Fig. 18–2. From Eq. 18–6 and Fig. 18–2, it is clear that the speeds of molecules in a gas vary from zero up to many times the average speed, but as can be seen from the graph, most molecules have speeds that are not far from the average. Less than 1 percent of the molecules exceed four times v_{rms}.

EXAMPLE 18–5 **ESTIMATE** **Using $f(v)$.** Suppose a sample of helium gas at temperature $T = 300\,\text{K}$ contains $N = 10^6$ atoms, each of mass $m_{He} = 6.65 \times 10^{-27}\,\text{kg}$. (*a*) What is the most probable speed, v_p, of a helium atom in the sample? (*b*) Estimate how many atoms in the sample have speeds in the range between v_p and $v_p + 40\,\text{m/s}$. (*c*) Now let $v = 10\,v_p$, and calculate the number of molecules with speeds between v and $v + 40\,\text{m/s}$.

SOLUTION (*a*) The most probable speed v_p is (Eq. 18–7a):

$$v_p = \sqrt{2\frac{kT}{m}} = \sqrt{\frac{(2)(1.38 \times 10^{-23}\,\text{J/K})(300\,\text{K})}{(6.65 \times 10^{-27}\,\text{kg})}} = 1.12 \times 10^3\,\text{m/s}.$$

(*b*) The number of molecules having a speed between v and $v + dv$ is equal to $f(v)\,dv$, as we discussed just after Eq. 18–6. For our estimate over a small finite range, $\Delta v = 40\,\text{m/s}$, we replace the differential dv by Δv and write

$$\Delta N = f(v)\,\Delta v.$$

As shown in Fig. 18–4, ΔN is equal to the area under the $f(v)$ curve between v and $v + \Delta v$. We approximate this area by a rectangle $f(v)$ high by Δv wide. A better estimate is obtained using the midpoint of the range between $v = 1120\,\text{m/s}$ and $v + \Delta v = 1160\,\text{m/s}$, namely $1140\,\text{m/s}$, although it doesn't make a great deal of difference here. Thus

$$\Delta N = f(v)\Delta v = 4\pi N\left(\frac{m}{2\pi kT}\right)^{\frac{3}{2}}v^2 e^{-\frac{mv^2}{2kT}}\,\Delta v$$

$$= (4\pi)(10^6)\left(\frac{6.65 \times 10^{-27}\,\text{kg}}{2\pi(1.38 \times 10^{-23}\,\text{J/K})(300\,\text{K})}\right)^{\frac{3}{2}}$$

$$\times (1.14 \times 10^3\,\text{m/s})^2 e^{-\frac{(6.65\times10^{-27}\,\text{kg})(1.14\times10^3\,\text{m/s})^2}{2(1.38\times10^{-23}\,\text{J/K})(300\,\text{K})}}\,(40\,\text{m/s})$$

$$\approx 30{,}000$$

or about 3% of all molecules.

(*c*) If we now replace $v = v_p$ by $v = 10\,v_p = 1.12 \times 10^4\,\text{m/s}$, and repeat the calculation with all else the same, we obtain $\Delta N = f(v)\,\Delta v = 4 \times 10^{-39}$, which means essentially that *no* molecules have speeds this high.

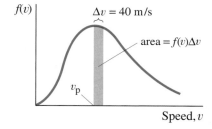

FIGURE 18–4 Calculating the number of molecules ΔN having speeds between v_p and $v_p + 40\,\text{m/s}$. $\Delta N = f(v)\,\Delta v$ is equal to the area shown shaded.

18–3 | Real Gases and Changes of Phase

The ideal gas law

$$PV = NkT$$

is an accurate description of the behavior of a real gas as long as the pressure is not too high and as long as the temperature is far from the liquefaction point. But what happens when these two criteria are not satisfied? First we discuss real gas behavior, and then we will look at how kinetic theory can help us understand this behavior.

Let us look at a graph of pressure plotted against volume for a given amount of gas. On such a "PV diagram," Fig. 18–5, each point represents an equilibrium state of the given substance. The various curves (labeled A, B, C, and D) show how the pressure varies as the volume is changed at constant temperature for several different values of the temperature. The dashed curve A′ represents the behavior of a gas as predicted by the ideal gas law; that is, PV = constant. The solid curve A represents the behavior of a real gas at the same temperature. Notice that at high pressure, the volume of a real gas is less than that predicted by the ideal gas law. The curves B and C in Fig. 18–5 represent the gas at successively lower temperatures, and we see that the behavior deviates even more from the curves predicted by the ideal gas law (for example, B′), and the deviation is greater the closer the gas is to liquefying.

To explain this, we note that at higher pressure we expect the molecules to be closer together. And, particularly at lower temperatures, the potential energy associated with the attractive forces between the molecules (which we ignored before) is no longer negligible compared to the now reduced kinetic energy of the molecules. These attractive forces tend to pull the molecules closer together so that at a given pressure, the volume is less than expected from the ideal gas law. At still lower temperatures, these forces cause liquefaction, and the molecules become very close together.

Curve D represents the situation when liquefaction occurs. At low pressure on curve D (on the right in Fig. 18–5), the substance is a gas and occupies a large volume. As the pressure is increased, the volume decreases until point b is reached. Beyond b, the volume decreases with no change in pressure; the substance is gradually changing from the gas to the liquid phase. At point a, all of the substance has changed to liquid. Further increase in pressure reduces the volume only slightly— liquids are nearly incompressible—so the curve is very steep as shown. The area which is colored yellow in Fig. 18–5, within the orange dashed line, represents the region where the gas and liquid phases exist together in equilibrium.

Curve C in Fig. 18–5 represents the behavior of the substance at its **critical temperature**; and the point c (the one point where this curve is horizontal) is called the **critical point**. At temperatures less than the critical temperature (and this is the definition of the term), a gas will change to the liquid phase if sufficient pressure is applied. Above the critical temperature, no amount of pressure can cause a gas to change phase and become a liquid. What happens instead is that the gas becomes denser and denser as the pressure is increased and gradually it acquires properties resembling a liquid, but no liquid surface forms. The critical temperatures for various gases are given in Table 18–1. Scientists tried for many years to liquefy oxygen without success. Only after the discovery of the behavior of substances associated with the critical point was it realized that oxygen can be liquefied only if first cooled below its critical temperature of −118°C.

Often a distinction is made between the terms "gas" and "vapor": a substance below its critical temperature in the gaseous state is called a **vapor**; when above the critical temperature, it is called a **gas**. This is indicated in Fig. 18–5.

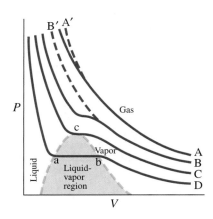

FIGURE 18–5 *PV* diagram for a real substance. Curves A, B, C, and D represent the same gas at different temperatures.

PV diagram

**TABLE 18–1
Critical Temperatures and Pressures**

Substance	Critical Temperature °C	K	Critical Pressure (atm)
Water	374	647	218
CO_2	31	304	72.8
Oxygen	−118	155	50
Nitrogen	−147	126	33.5
Hydrogen	−239.9	33.3	12.8
Helium	−267.9	5.3	2.3

Vapor vs. gas

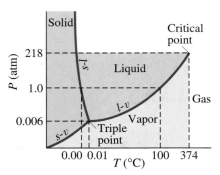

FIGURE 18–6 Phase diagram for water (note that the scales are not linear).

FIGURE 18–7 Phase diagram for carbon dioxide (CO_2).

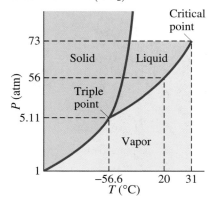

The behavior of a substance can be diagrammed not only on a *PV* diagram but also on a *PT* diagram. A *PT* diagram, often called a **phase diagram**, is particularly convenient for comparing the different phases of a substance. Figure 18–6 is the phase diagram for water. The curve labeled *l-v* represents those points where the liquid and vapor phases are in equilibrium—it is thus a graph of the boiling point versus pressure. Note that the curve correctly shows that at a pressure of 1 atm the boiling point of water is 100°C and that the boiling point is lowered for a decreased pressure. The curve *s-l* represents points where solid and liquid exist in equilibrium, and thus is a graph of the freezing point versus pressure. At 1 atm, the freezing point of water is, of course, 0°C, as shown. Notice also in Fig. 18–6 that at a pressure of 1 atm, the substance is in the liquid phase if the temperature is between 0°C and 100°C, but is in the solid or vapor phase if the temperature is below 0°C or above 100°C. The curve labeled *s-v* is the *sublimation point* versus pressure curve. **Sublimation** refers to the process whereby at low pressures a solid changes directly into the vapor phase without passing through the liquid phase. For water, this occurs at pressures less than 0.0060 atm. Carbon dioxide (CO_2), which in the solid phase is called dry ice, sublimates even at atmospheric pressure.

The intersection of the three curves (in Fig. 18–6) is the **triple point**, and it is only at this point that the three phases can exist together in equilibrium. Because the triple point corresponds to a unique value of temperature and pressure $(273.16 \text{ K}, 6.03 \times 10^{-3} \text{ atm, for water})$, it is precisely reproducible and is often used as a point of reference (Section 17–10).

Notice that the *s-l* curve for water slopes upward to the left. This is true only of substances that *expand* upon freezing; for at a higher pressure, a lower temperature is needed to cause the liquid to freeze. More commonly, substances contract upon freezing and the *s-l* curve slopes upward to the right, as shown for CO_2 in Fig. 18–7.

The phase transitions we have been discussing are the common ones. Some substances, however, can exist in several forms in the solid phase. A transition from one of these phases to another occurs at a particular temperature and pressure, just like ordinary phase changes. For example, ice has been observed in at least eight different modifications at very high pressure. Ordinary helium has two distinct liquid phases, called helium I and II. They exist only at temperatures within a few degrees of absolute zero. Helium II exhibits very unusual properties referred to as **superfluidity**. It has essentially zero viscosity and exhibits strange properties such as actually climbing up the sides of an open container.

18–4 Vapor Pressure and Humidity

Evaporation

If a glass of water is left out overnight, the water level will have dropped by morning. We say the water has evaporated, meaning that some of the water has changed to the vapor or gas phase.

This process of **evaporation** can be explained on the basis of kinetic theory. The molecules in a liquid move past one another with a variety of speeds that follow, approximately, the Maxwell distribution. There are strong attractive forces between these molecules, which is what keeps them close together in the liquid phase. A molecule in the upper regions of the liquid may, because of its speed, leave the liquid momentarily. But just as a rock thrown into the air returns to the Earth, so the attractive forces of the other molecules can pull the vagabond molecule back to the liquid surface—that is, if its velocity is not too large. A molecule with a high enough velocity, however, will escape from the liquid entirely, like a rocket escaping the Earth, and become part of the gas phase. Only those molecules that have kinetic energy above a particular value can escape to the gas phase. We have already seen that kinetic theory predicts that the relative number of molecules with kinetic energy above a particular value (such as E_A in Fig. 18–3) increases with temperature. This is in accord with the well-known observation that the evaporation rate is greater at higher temperatures.

Because it is the fastest molecules that escape from the surface, the average speed of those remaining is less. When the average speed is less, the absolute temperature is less. Thus kinetic theory predicts that *evaporation is a cooling process*. You have no doubt noticed this effect when you stepped out of a warm shower and felt cold as the water on your body began to evaporate; and after working up a sweat on a hot day, even a slight breeze makes you feel cool through evaporation.

Vapor Pressure

Air normally contains water vapor (water in the gas phase) and it comes mainly from evaporation. To look at this process in a little more detail, consider a closed container that is partially filled with water (it could just as well be any other liquid) and from which the air has been removed (Fig. 18–8). The fastest moving molecules quickly evaporate into the space above. As they move about, some of these molecules strike the liquid surface and again become part of the liquid phase: this is called **condensation**. The number of molecules in the vapor increases for a time, until a point is reached where the number returning to the liquid equals the number leaving in the same time interval. Equilibrium then exists, and the space is said to be *saturated*. The pressure of the vapor when it is saturated is called the **saturated vapor pressure** (or sometimes simply the vapor pressure).

The saturated vapor pressure does not depend on the volume of the container. If the volume above the liquid were reduced suddenly, the density of molecules in the vapor phase would be increased temporarily. More molecules would then be striking the liquid surface per second. There would be a net flow of molecules back to the liquid phase until equilibrium was again reached, and this would occur at the same value of the saturated vapor pressure, as long as the temperature had not changed.

The saturated vapor pressure of any substance depends on the temperature. At higher temperatures, more molecules have sufficient kinetic energy to break from the liquid surface into the vapor phase. Hence equilibrium will be reached at a higher pressure. The saturated vapor pressure of water at various temperatures is given in Table 18–2. Notice that even solids—for example, ice—have a measurable saturated vapor pressure.

In everyday situations, evaporation from a liquid takes place into the air above it rather than into a vacuum. This does not materially alter the discussion above relating to Fig. 18–8. Equilibrium will still be reached when there are sufficient molecules in the gas phase that the number reentering the liquid equals the number leaving. The concentration of particular molecules (such as water) in the gas phase is not affected by the presence of air, although collisions with air molecules may lengthen the time needed to reach equilibrium. Thus equilibrium occurs at the same value of the saturated vapor pressure as if air weren't there.

Of course, if the container is large or is not closed, all the liquid may evaporate before saturation is reached. And if the container is not sealed—as, for example, a room in your house—it is not likely that the air will become saturated with water vapor (unless it is raining outside).

Boiling

The saturated vapor pressure of a liquid increases with temperature. When the temperature is raised to the point where the saturated vapor pressure at that temperature equals the external pressure, **boiling** occurs (Fig. 18–9). As the boiling point is approached, tiny bubbles tend to form in the liquid, which indicate a change from the liquid to the gas phase. However, if the vapor pressure inside the bubbles is less than the external pressure, the bubbles immediately are crushed. As the temperature is increased, the saturated vapor pressure inside a bubble eventually becomes equal to or exceeds the external air pressure. The bubble will then not collapse but will increase in size and rise to the surface. Boiling has then begun. *A liquid boils when its saturated vapor pressure equals the external pressure.* This occurs for water under 1 atm (760 torr) of pressure at 100°C, as can be seen from Table 18–2.

➡ PHYSICS APPLIED

Evaporation cools

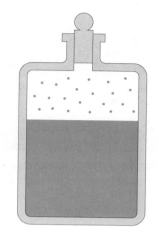

FIGURE 18–8 Vapor appears above a liquid in a closed container.

TABLE 18–2 Saturated Vapor Pressure of Water

Temperature (°C)	Saturated Vapor Pressure	
	torr (= mm-Hg)	Pa (= N/m²)
−50	0.030	4.0
−10	1.95	2.60×10^2
0	4.58	6.11×10^2
5	6.54	8.72×10^2
10	9.21	1.23×10^3
15	12.8	1.71×10^3
20	17.5	2.33×10^3
25	23.8	3.17×10^3
30	31.8	4.24×10^3
40	55.3	7.37×10^3
50	92.5	1.23×10^4
60	149	1.99×10^4
70	234	3.12×10^4
80	355	4.73×10^4
90	526	7.01×10^4
100	760	1.01×10^5
120	1489	1.99×10^5
150	3570	4.76×10^5

At boiling, saturated vapor pressure equals external pressure

FIGURE 18–9 Boiling: bubbles of water vapor float upward from the bottom of the pot (where the temperature is highest).

Relative humidity

The boiling point of a liquid clearly depends on the external pressure. At high elevations, the boiling point of water is somewhat less than at sea level since the air pressure is less. For example, on the summit of Mt. Everest (8850 m) the air pressure is about one-third of what it is at sea level, and from Table 18–2 we can see that water will boil at about 70°C. Cooking food by boiling takes longer at high elevations, since the temperature is less. Pressure cookers, however, reduce cooking time, because they build up a pressure as high as 2 atm, allowing higher boiling temperatures to be attained.

Partial Pressure and Humidity

When we refer to the weather as being dry or humid, we are referring to the water vapor content of the air. In a gas such as air, which is a mixture of several types of gases, the total pressure is the sum of the *partial pressures* of each gas present.[†] By **partial pressure**, we mean the pressure each gas would exert if it alone were present. The partial pressure of water in the air can be as low as zero and can vary up to a maximum equal to the saturated vapor pressure of water at the given temperature. Thus, at 20°C, the partial pressure of water cannot exceed 17.5 torr (see Table 18–2). The **relative humidity** is defined as the ratio of the partial pressure to the saturated vapor pressure at a given temperature. It is usually expressed as a percentage:

$$\text{Relative humidity} = \frac{\text{partial pressure of } H_2O}{\text{saturated vapor pressure of } H_2O} \times 100\%.$$

Thus, when the humidity is close to 100 percent, the air holds nearly all the water vapor it can.

EXAMPLE 18–6 Relative humidity. On a particular hot day, the temperature is 30°C and the partial pressure of water vapor in the air is 21.0 torr. What is the relative humidity?

SOLUTION From Table 18–2, the saturated vapor pressure of water at 30°C is 31.8 torr. Hence the relative humidity is

$$\frac{21.0 \text{ torr}}{31.8 \text{ torr}} \times 100\% = 66\%.$$

➡ **PHYSICS APPLIED**

Weather

Air is saturated with water vapor when the partial pressure of water in the air is equal to the saturated vapor pressure at that temperature. If the partial pressure of water exceeds the saturated vapor pressure, the air is said to be **supersaturated**. This situation can occur when a temperature decrease occurs. For example, suppose the temperature is 30°C and the partial pressure of water is 21 torr, which represents a humidity of 66 percent as we saw above. Suppose now that the temperature falls to, say, 20°C, as might happen at nightfall. From Table 18–2 we see that the saturated vapor pressure of water at 20°C is 17.5 torr. Hence the relative humidity would be greater than 100 percent, and the supersaturated air cannot hold this much water. The excess water condenses and appears as dew—or perhaps instead as fog or rain.

When air containing a given amount of water is cooled, a temperature is reached where the partial pressure of water equals the saturated vapor pressure. This is called the **dew point**. Measurement of the dew point is the most accurate means of determining the relative humidity. One method uses a polished metal surface in contact with air, which is gradually cooled down. The temperature at which moisture begins to appear on the surface is the dew point, and the partial pressure of water can then be obtained from saturated vapor pressure tables. If, for example, on a given day the temperature is 20°C, and the dew point is 5°C, then the partial pressure of water (Table 18–2) in the original air was 6.54 torr, whereas its saturated vapor pressure was 17.5 torr; hence the relative humidity was 6.54/17.5 = 37 percent.

[†] For example, 78 percent (by volume) of air molecules are nitrogen and 21 percent oxygen, with much smaller amounts of water vapor, argon, and other gases. At an air pressure of 1 atm, oxygen exerts a partial pressure of 0.21 atm and nitrogen 0.78 atm.

Van der Waals Equation of State

In Section 18–3, we discussed how real gases deviate from ideal gas behavior, particularly at high densities or when near condensing to a liquid. We would like to understand these deviations using a microscopic (molecular) point of view. J. D. van der Waals (1837–1923) analyzed this problem and in 1873 arrived at an equation of state which fits real gases more accurately than the ideal gas law. His analysis is based on kinetic theory but takes into account: (1) the finite size of molecules (we previously neglected the actual volume of the molecules themselves, compared to the total volume of the container, and this assumption becomes poorer as the density increases and molecules become closer together); (2) the range of the forces between molecules may be greater than the size of the molecules (we previously assumed that intermolecular forces act only during collision, when the molecules are "in contact"). Let us now look at this analysis and derive the van der Waals equation of state.

Assume the molecules in a gas are spherical with radius r. If we assume these molecules behave like hard spheres, then two molecules collide and bounce off one another if the distance between their centers (Fig. 18–10) gets as small as $2r$. Thus the actual volume in which the molecules can move about is somewhat less than the volume V of the container holding the gas. The amount of "unavailable volume" depends on the number of molecules and on their size. Let b represent the "unavailable volume per mole" of gas. Then in the ideal gas law we replace V by $(V - nb)$, where n is the number of moles, and we obtain

$$P(V - nb) = nRT.$$

If we divide through by n

$$P\left(\frac{V}{n} - b\right) = RT. \qquad \textbf{(18–8)}$$

This relation (sometimes called the **Clausius equation of state**) predicts that for a given temperature T and volume V, the pressure P will be greater than for an ideal gas. This makes sense since the reduced "available" volume means the number of collisions with the walls is increased.

Next we consider the effects of attractive forces between molecules, which are responsible for holding molecules in the liquid and solid states at lower temperatures. These forces are electrical in nature and although they act even when molecules are not touching, we assume their range is small—that is, they act mainly between nearest neighbors. Molecules at the edge of the gas, headed toward a wall of the container, are slowed down by a net force pulling them back into the gas. Thus these molecules will exert less force and less pressure on the wall than if there were no attractive forces. The reduced pressure will be proportional to the density of molecules in the layer of gas at the surface, and also to the density in the next layer, which exerts the inward force.[†] Therefore we expect the pressure to be reduced by a factor proportional to the density squared $(n/V)^2$, here written as moles per volume. If the pressure P is given by Eq. 18–8, then we should reduce this by an amount $a(n/V)^2$ where a is a proportionality constant. Thus we have

$$P = \frac{RT}{(V/n) - b} - \frac{a}{(V/n)^2}$$

or

$$\left(P + \frac{a}{(V/n)^2}\right)\left(\frac{V}{n} - b\right) = RT, \qquad \textbf{(18–9)}$$

which is the **van der Waals equation of state**.

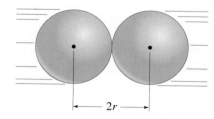

FIGURE 18–10 Molecules, of radius r, colliding.

Van der Waals equation of state

[†]This is similar to the gravitational force in which the force on mass m_1 due to mass m_2 is proportional to the product of their masses (Newton's law of universal gravitation, Chapter 6).

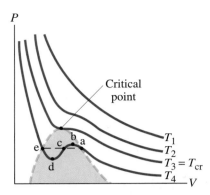

FIGURE 18–11 *PV* diagram for a van der Waals gas, shown for four different temperatures. For T_1, T_2, and T_3 (T_3 is chosen equal to the critical temperature), the curves fit experimental data very well for most gases. The curve labeled T_4, a temperature below the critical point, passes through the liquid–vapor region. The maximum (point b) and minimum (point d) would seem to be artifacts, since we usually see constant pressure, as indicated by the horizontal dashed line. However, for very pure supersaturated vapors or supercooled liquids, the sections ab and ed, respectively, have been observed. (The section bd would be unstable and has not been observed.)

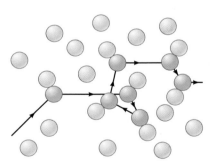

FIGURE 18–12 Zigzag path of a molecule colliding with other molecules.

FIGURE 18–13 Molecule at left moves to the right with speed $\bar{v}$. It collides with any molecule whose center is within the cylinder of radius $2r$.

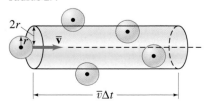

The constants a and b in the van der Waals equation are different for different gases and are determined by fitting to experimental data for each gas. For CO_2 gas, the best fit is obtained for $a = 3.6 \times 10^{-3} \, \text{N} \cdot \text{m}^4/\text{mol}^2$ and $b = 4.2 \times 10^{-5} \, \text{m}^3/\text{mol}$. Figure 18–11 shows a typical *PV* diagram for Eq. 18–9 (a "van der Waals gas") for four different temperatures, with detailed caption, and it should be compared to Fig. 18–5 for real gases.

Neither the van der Waals equation of state nor the many other equations of state that have been proposed are accurate for all gases under all conditions. Yet Eq. 18–9 is a very useful relation. And because it is quite accurate for many situations, its derivation gives us further insight into the nature of gases at the microscopic level. Note that at low densities $a/(V/n)^2 \ll P$ and $b \ll V/n$, so that the van der Waals equation reduces to the equation of state for an ideal gas, $PV = nRT$.

* 18–6 Mean Free Path

If gas molecules were truly point particles, they would never collide with one another. Thus, when a person opened a perfume bottle, you would be able to smell it almost instantaneously across the room, since molecules travel hundreds of meters per second. In fact it takes some time before you detect an odor, and according to kinetic theory, this must be due to collisions between molecules of nonzero size.

If we were to follow the path of a particular molecule, we would expect to see it follow a zigzag path as shown in Fig. 18–12. Between each collision the molecule would move in a straight-line path. (Not quite true if we take account of the small intermolecular forces that act between collisions.) An important parameter for a given situation is the **mean free path**, which is defined as the average distance a molecule travels between collisions. We would expect that the greater the gas density, and the larger the molecules, the shorter the mean free path would be. We now determine the nature of this relationship for an ideal gas.

Suppose our gas is made up of molecules which are hard spheres of radius r. A collision will occur whenever the centers of two molecules come within a distance $2r$ of one another. Let us follow a molecule as it traces a straight-line path. In Fig. 18–13, the dashed line represents the path of our particle if it made no collisions. Also shown is a cylinder of radius $2r$. If the center of another molecule lies within this cylinder, a collision will occur. (Of course, when a collision occurs the particle's path would change direction, as would our imagined cylinder, but our result won't be altered by unbending a zigzag cylinder into a straight one for purposes of calculation.) Assume our molecule is an average one, moving at the mean speed $\bar{v}$ in the gas. For the moment, let us assume that the other molecules are not moving, and that the concentration of molecules (number per unit volume) is N/V. Then the number of molecules whose center lies within the cylinder of Fig. 18–13 is N/V times the volume of this cylinder, and this also represents the number of collisions that will occur. In a time Δt, our molecule travels a distance $\bar{v}\Delta t$, so the length of the cylinder is $\bar{v}\Delta t$ and its volume is $\pi(2r)^2\bar{v}\Delta t$. Hence the number of collisions that occur in a time Δt is $(N/V)\pi(2r)^2\bar{v}\Delta t$. We define the **mean free path**, l_M, as the average distance between collisions. This distance is equal to the distance traveled ($\bar{v}\Delta t$) in a time Δt divided by the number of collisions made in time Δt:

$$l_M = \frac{\bar{v}\Delta t}{(N/V)\pi(2r)^2\bar{v}\Delta t} = \frac{1}{4\pi r^2(N/V)}. \qquad \textbf{(18–10a)}$$

Thus we see that l_M is inversely proportional to the cross-sectional area ($=\pi r^2$) of the molecules and to their concentration (number/volume), N/V. However, Eq. 18–10a is not fully correct since we assumed the other molecules are all at rest. In fact, they are moving, and the number of collisions in a time Δt must depend on the *relative* speed of the colliding molecules, rather than on $\bar{v}$. Hence the number of collisions per second is $(N/V)\pi(2r)^2 v_{rel}\,\Delta t$ (rather than $(N/V)\pi(2r)^2\bar{v}\,\Delta t$), where v_{rel} is the average relative speed of colliding molecules. A careful calculation shows that for a Maxwellian distribution of speeds $v_{rel} = \sqrt{2}\bar{v}$. Hence the mean free path is

$$l_M = \frac{1}{4\pi\sqrt{2}r^2(N/V)}.$$
(18–10b)   *Mean free path*

EXAMPLE 18–7 **ESTIMATE** **Mean free path of air molecules at STP.** Estimate the mean free path of air molecules at STP, standard temperature and pressure (0°C, 1 atm). The diameter of O_2 and N_2 molecules is about 3×10^{-10} m.

SOLUTION We saw in Example 17–8 that 1 mol of an ideal gas occupies a volume of 22.4×10^{-3} m^3 at STP. Hence

$$\frac{N}{V} = \frac{6.02 \times 10^{23} \text{ molecules}}{22.4 \times 10^{-3} \text{ m}^3} = 2.69 \times 10^{25} \text{ molecules/m}^3.$$

Then

$$l_M = \frac{1}{4\pi\sqrt{2}(1.5 \times 10^{-10} \text{ m})^2(2.7 \times 10^{25} \text{ m}^{-3})} \approx 9 \times 10^{-8} \text{ m}$$

which is about 300 times the diameter of a molecule.

At very low densities, such as in an evacuated vessel, the concept of mean free path loses meaning since collisions with walls of the container may occur more frequently than collisions with other molecules. For example, in a cubical box 20 cm on a side containing air at 10^{-7} torr, the mean free path is about 700 m, which means many more collisions are made with the walls than with other molecules. (Note, nonetheless, that the box contains over 10^{12} molecules.) If the concept of mean free path were to include any type of collision, it would be closer to 0.2 m than to the 700 m calculated from Eq. 18–10.

*18–7 | Diffusion

If you carefully place a few drops of food coloring in a container of water as in Fig. 18–14, you will find that the color spreads throughout the water. The process may take some time (assuming you don't shake the glass), but eventually the color will become uniform. This mixing, known as **diffusion**, takes place because of the random movement of the molecules. Diffusion occurs in gases too. Common examples include perfume or smoke (or the odor of something cooking on the stove) diffusing in air, although convection (moving air currents) often plays a greater role in spreading odors than does diffusion. Diffusion depends on concentration, by which we mean the number of molecules or moles per unit volume. In general, *the diffusing substance moves from a region where its concentration is high to one where its concentration is low.*

Diffusion occurs from high toward low concentration

FIGURE 18–14 A few drops of food coloring spreads slowly throughout the water, eventually becoming uniform.

(a)   (b)   (c)

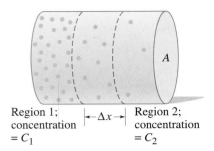

Region 1;
concentration
$= C_1$

$\leftarrow \Delta x \rightarrow$

Region 2;
concentration
$= C_2$

A

FIGURE 18–15 Diffusion occurs
from a region of high concentration
to one of lower concentration (only
one type of molecule is shown).

**TABLE 18–3   Diffusion
Constants, D (20°C, 1 atm)**

Diffusing Molecules	Medium	$D(m^2/s)$
H_2	Air	6.3×10^{-5}
O_2	Air	1.8×10^{-5}
O_2	Water	100×10^{-11}
Blood hemoglobin	Water	6.9×10^{-11}
Glycine (an amino acid)	Water	95×10^{-11}
DNA (mass 6×10^6 u)	Water	0.13×10^{-11}

Diffusion can be readily understood on the basis of kinetic theory and the
random motion of molecules. Consider a tube of cross-sectional area A containing
molecules in a higher concentration on the left than on the right, Fig. 18–15. We
assume the molecules are in random motion. Yet there will be a net flow of molecules to the right. To see why this is true, let us consider the small section of tube of
length Δx as shown. Molecules from both regions 1 and 2 cross into this central section as a result of their random motion. The more molecules there are in a region,
the more will strike a given area or cross a boundary. Since there is a greater concentration of molecules in region 1 than in region 2, more molecules cross into the
central section from region 1 than from region 2. There is, then, a net flow of molecules from left to right, from high concentration toward low concentration. The
flow stops only when the concentrations become equal.

You might expect that the greater the difference in concentration, the greater
the flow rate. This is indeed the case. In 1855, the physiologist Adolf Fick
(1829–1901) determined experimentally that the rate of diffusion (J) is directly proportional to the change in concentration per unit distance $(C_1 - C_2)/\Delta x$ (which is
called the **concentration gradient**), and to the cross-sectional area A (see Fig. 18–15):

$$J = DA \frac{C_1 - C_2}{\Delta x},$$

or, in terms of derivatives

$$J = DA \frac{dC}{dx}. \tag{18–11}$$

*Diffusion equation
(Fick's law)*

D is a constant of proportionality called the **diffusion constant**. Equation 18–11 is
known as the **diffusion equation**, or **Fick's law**. If the concentrations are given in
mol/m^3, then J is the number of moles passing a given point per second; if the concentrations are given in kg/m^3, then J is the mass movement per second (kg/s).

The diffusion constant D will depend on the properties of the diffusing molecules and on those of the substance in which it is immersed (often water or air),
and also on the temperature and the external pressure. The values of D for a variety of substances are given in Table 18–3.

EXAMPLE 18–8   ESTIMATE   Diffusion of ammonia in air. To get an idea of
the time required for diffusion, estimate how long it might take for ammonia (NH_3)
to be detected 10 cm from a bottle after it is opened, assuming only diffusion.

SOLUTION This will be an order-of-magnitude calculation. The rate of diffusion J can be set equal to the number of molecules N diffusing across area A in a
time t: $J = N/t$. Then $t = N/J$ and using Eq. 18–11,

$$t = \frac{N}{J} = \frac{N}{AD} \frac{\Delta x}{\Delta C}.$$

The average concentration (midway between bottle and nose) can be approximated
by $\overline{C} \approx N/V$, where V is the volume over which the molecules move and is roughly
of the order of $V \approx A \Delta x$ where Δx is 10 cm. We substitute $N = \overline{C}A \Delta x$ into the
above equation:

$$t \approx \frac{(\overline{C} A \Delta x)\Delta x}{AD \Delta C} = \frac{\overline{C}}{\Delta C} \frac{(\Delta x)^2}{D}.$$

The concentration of ammonia is high near the bottle and low near the detecting
nose, so $\overline{C} \approx \Delta C/2$ or $(\overline{C}/\Delta C) \approx \frac{1}{2}$. Since NH_3 molecules have a size somewhere
between H_2 and O_2, from Table 18–3, we can estimate $D \approx 4 \times 10^{-5}\ m^2/s$. Then

$$t \approx \frac{1}{2} \frac{(0.10\ m)^2}{(4 \times 10^{-5}\ m^2/s)} \approx 100\ s.$$

or about a minute or two. This seems rather long from experience, suggesting that
air currents (convection) are more important than diffusion for transmitting odors.

Summary

According to the **kinetic theory** of gases, which is based on the idea that a gas is made up of molecules that are moving rapidly and at random, the average kinetic energy of the molecules is proportional to the Kelvin temperature T:

$$\overline{K} = \tfrac{1}{2}m\overline{v^2} = \tfrac{3}{2}kT$$

where k is Boltzmann's constant.

At any moment, there exists a wide distribution of molecular speeds within a gas. The **Maxwell distribution of speeds** is derived from simple kinetic theory assumptions, and is in good accord with experiment for gases at not too high a pressure.

The behavior of real gases at high pressure, and/or when near their liquefaction point, deviates from the ideal gas law. These deviations are due to the finite size of molecules and to the attractive forces between molecules.

Below the **critical temperature**, a gas can change to a liquid if sufficient pressure is applied; but if the temperature is higher than the critical temperature, no amount of pressure will cause a liquid surface to form.

The **triple point** of a substance is that unique temperature and pressure at which all three phases—solid, liquid, and gas—can coexist in equilibrium. Because of its precise reproducibility, the triple point of water is often taken as a standard reference point.

Evaporation of a liquid is the result of the fastest moving molecules escaping from the surface. Because the average molecular velocity is less after the fastest molecules escape, the temperature decreases when evaporation takes place.

Saturated vapor pressure refers to the pressure of the vapor above a liquid when the two phases are in equilibrium. The vapor pressure of a substance (such as water) depends strongly on temperature and is equal to atmospheric pressure at the boiling point.

Relative humidity of air at a given place is the ratio of the partial pressure of water vapor in the air to the saturated vapor pressure at that temperature; it is usually expressed as a percentage.

Questions

1. Why doesn't the size of different molecules enter into the gas laws?

2. When a gas is rapidly compressed (say, by pushing down a piston) its temperature increases. When a gas expands against a piston, it cools. Explain these changes in temperature using the kinetic theory, in particular noting what happens to the momentum of molecules when they strike the moving piston.

3. In Section 18–1 we assumed the gas molecules made perfectly elastic collisions with the walls of the container. This assumption is not necessary as long as the walls are at the same temperature as the gas. Why?

4. Explain in words how Charles's law follows from kinetic theory and the relation between average kinetic energy and the absolute temperature.

5. Explain in words how Gay-Lussac's law follows from kinetic theory.

6. As you go higher in the Earth's atmosphere, the ratio of N_2 molecules to O_2 molecules increases. Why?

7. Can you determine the temperature of a vacuum?

8. Is temperature a macroscopic or microscopic variable?

9. Discuss why the Maxwell distribution of speeds (Fig. 18–2) is not a symmetric curve.

10. Explain why the peak of the curve for 310 K in Fig. 18–3 is not as high as for 273 K. (Assume the total number of molecules is the same for both.)

11. Explain, using the Maxwell distribution of speeds, (a) why the Moon has very little atmosphere, (b) why hydrogen, if at one time in the Earth's atmosphere, would probably have escaped.

12. Explain why putting food in the freezer retards spoilage.

13. Would the average kinetic energy of molecules in an ideal gas correspond to $\overline{v}$, v_{rms}, v_p, or to some other value?

14. If the pressure in a gas is doubled while its volume is held constant, by what factor do (a) v_{rms} and (b) $\overline{v}$ change?

15. If a container of gas is at rest, the average velocity of molecules must be zero. Yet the average speed is not zero. Explain.

16. Draw, roughly, the Maxwellian distribution of *velocities*. Discuss what the values would be for mean, rms, and most probable velocities. Would the curve be symmetric?

17. What everyday observation would tell you that not all molecules in a material have the same speed?

18. We saw that the saturated vapor pressure of a liquid (say, water) does not depend on the external pressure. Yet the temperature of boiling does depend on the external pressure. Is there a contradiction? Explain.

19. Alcohol evaporates more quickly than water at room temperature. What can you infer about the molecular properties of one relative to the other?

20. Explain why a hot humid day is far more uncomfortable than a hot dry day at the same temperature.

21. Is it possible to boil water at room temperature (20°C) without heating it? Explain.

22. What exactly does it mean when we say that oxygen boils at −183°C?

23. A length of thin wire is placed over a block of ice (or an ice cube) at 0°C and weights are hung from the ends of the wire. It is found that the wire cuts its way through the ice cube, but leaves a solid block of ice behind it. This process is called *regelation*. Explain how this happens by inferring how the freezing point of water depends on pressure.

24. How do a gas and a vapor differ?

25. (*a*) At suitable temperatures and pressures, can ice be melted by applying pressure? (*b*) At suitable temperatures and pressures, can carbon dioxide be melted by applying pressure?

26. Why does dry ice not last long at room temperature?

27. Under what conditions can liquid CO_2 exist? Be specific. Can it exist as a liquid at normal room temperature?

* **28.** Name several ways to reduce the mean free path in a gas.

* **29.** Discuss why sound waves can travel in a gas only if their wavelength is somewhat larger than the mean free path.

Problems

Section 18–1

1. (I) (*a*) What is the average kinetic energy of an oxygen molecule at STP? (*b*) What is the total translational kinetic energy of 2.0 mol of O_2 molecules at 20°C?

2. (I) Calculate the rms speed of helium atoms near the surface of the sun at a temperature of about 6000 K.

3. (I) By what factor will the rms speed of gas molecules increase if the temperature is increased from 0°C to 100°C?

4. (I) A gas is at 20°C. To what temperature must it be raised to double the rms speed of its molecules?

5. (I) Twelve molecules have the following speeds, given in arbitrary units: 6, 2, 4, 6, 0, 4, 1, 8, 5, 3, 7, and 8. Calculate (*a*) the mean speed, and (*b*) the rms speed.

6. (II) The rms speed of molecules in a gas at 20.0°C is to be increased by 1.0%. To what temperature must it be raised?

7. (II) If the pressure in a gas is doubled while its volume is held constant, by what factor does v_{rms} change?

8. (II) Show that the rms speed of molecules in a gas is given by $v_{rms} = \sqrt{3P/\rho}$, where P is the pressure in the gas, and ρ is the gas density.

9. (II) Show that for a mixture of two gases at the same temperature, the ratio of their rms speeds is equal to the inverse ratio of the square roots of their molecular masses.

10. (II) What is the rms speed of nitrogen molecules contained in an 8.5-m³ volume at 2.1 atm if the total amount of nitrogen is 1300 mol?

11. (II) Calculate (*a*) the rms speed of an oxygen molecule at 0°C and (*b*) determine how many times per second it would move back and forth across a 7.0-m-long room on the average, assuming it made very few collisions with other molecules.

12. (II) What is the average distance between nitrogen molecules at STP?

13. (II) The two isotopes of uranium, ^{235}U and ^{238}U (the superscripts refer to their atomic masses), can be separated by a gas diffusion process by combining them with fluorine to make the gaseous compound UF_6. Calculate the ratio of the rms speeds of these molecules for the two isotopes.

Section 18–2

14. (I) A group of 25 particles have the following speeds: two have speed 10 m/s, seven have 15 m/s, four have 20 m/s, three have 25 m/s, six have 30 m/s, one has 35 m/s, and two have 40 m/s. Determine (*a*) the average speed, (*b*) the rms speed, and (*c*) the most probable speed.

* **15.** (II) Starting from the Maxwell distribution of speeds, Eq. 18–6, show (*a*)$\int_0^\infty f(v)\, dv = N$, and (*b*)

$$\int_0^\infty v^2 f(v)\, dv/N = 3kT/m.$$

Section 18–3

16. (I) (*a*) At atmospheric pressure, in what phases can CO_2 exist? (*b*) For what range of pressures and temperatures can CO_2 be a liquid? Refer to Fig. 18–7.

17. (I) CO_2 exists in what phase when the pressure is 30 atm and the temperature is 30°C (Fig. 18–7)?

18. (I) Water is in which phase when the pressure is 0.01 atm and the temperature is (*a*) 90°C, (*b*) −20°C?

19. (II) You have a sample of water and are able to control temperature and pressure arbitrarily. (*a*) Using Fig. 18–6, describe the phase changes you would see if you started at a temperature of 100°C, a pressure of 220 atm, and decreased the pressure down to 0.004 atm while keeping the temperature fixed. (*b*) Repeat part (*a*) with the temperature at 0.0°C. Assume that you held the system at the starting conditions long enough for the system to stabilize before making further changes.

Section 18–4

20. (I) What is the dew point (approximately) if the humidity is 50 percent on a day when the temperature is 25°C?

21. (I) What is the air pressure at a place where water boils at 90°C?

22. (I) If the air pressure at a particular place in the mountains is 0.85 atm, estimate the temperature at which water boils.

23. (I) What is the temperature on a day when the partial pressure of water is 530 Pa and the relative humidity is 40 percent?

24. (I) What is the partial pressure of water on a day when the temperature is 25°C and the relative humidity is 35%?

25. (II) What is the approximate pressure inside a pressure cooker if the water is boiling at a temperature of 120°C? Assume no air escaped during the heating process, which started at 20°C.

26. (II) If the humidity in a room of volume 240 m³ at 25°C is 80 percent, what mass of water can still evaporate from an open pan?

27. (II) An *autoclave* is a device used to sterilize laboratory instruments. It is essentially a high-pressure steam boiler and operates on the same principle as a pressure cooker. However, because hot steam under pressure is more effective in killing microorganisms than moist air at the same temperature and pressure, the air is removed and replaced by steam. Typically, the gauge pressure inside the autoclave is 1.0 atm. What is the temperature of the steam? Assume the steam is in equilibrium with boiling water.

28. (III) Air that is at its dew point of 5°C is drawn into a building where it is heated to 25°C. What will be the relative humidity at this temperature? Assume constant pressure of 1.0 atm. Take into account the expansion of the air.

* Section 18–5

*29. (II) For oxygen gas, the van der Waals equation of state achieves its best fit for $a = 0.14 \, N \cdot m^4/mol^2$ and $b = 3.2 \times 10^{-5} \, m^3/mol$. Determine the pressure in 1.0 mole of the gas at 0°C if its volume is 0.40 L, calculated using (a) the van der Waals equation, (b) the ideal gas law.

*30. (II) In the van der Waals equation of state, the constant b represents the amount of "unavailable volume" occupied by the molecules themselves. Thus V is replaced by $(V - nb)$, where n is the number of moles. For oxygen, b is about $3.2 \times 10^{-5} \, m^3/mol$. Estimate the diameter of an oxygen molecule.

*31. (III) (a) From the van der Waals equation of state, show that the critical temperature and pressure are given by

$$T_{cr} = \frac{8a}{27bR}, \qquad P_{cr} = \frac{a}{27b^2}.$$

[*Hint*: Use the fact that the P versus V curve has an inflection point at the critical point so that the first and second derivatives are zero.] (b) Determine a and b for CO_2 from the measured values of $T_{cr} = 304$ K and $P_{cr} = 72.8$ atm.

*32. (III) (a) Show that a collision between two spherical molecules each of radius r, is equivalent to a collision between a point particle and a sphere of radius $2r$, and thus the center of one molecule cannot penetrate a volume equal to that of a sphere of radius $2r$ when two molecules collide. (b) Show that the total unavailable volume per mole is $b = 16\pi r^3 N_A/3$, where N_A is Avogadro's number. [*Hint*: In summing the total excluded volume, multiply by $\frac{1}{2}$ to avoid counting each pair of molecules twice.] Note that the unavailable volume is four times the actual molecular volume. (c) Estimate the diameter of a CO_2 molecule, for which $b = 4.2 \times 10^{-5} \, m^3/mol$.

* Section 18–6

*33. (II) At about what pressure would the mean free path of air molecules be (a) 1.0 m and (b) equal to the diameter of air molecules, $\approx 3 \times 10^{-10}$ m?

*34. (II) (a) The mean free path of CO_2 molecules at STP is measured to be about 5.6×10^{-8} m. Estimate the diameter of a CO_2 molecule. (b) Do the same for He gas for which $l_M \approx 25 \times 10^{-8}$ at STP.

*35. (II) A cubic box of volume $4.4 \times 10^{-3} \, m^3$ contains 70 marbles, each 1.5 cm in diameter. What is the mean free path for a marble (a) when the box is shaken vigorously and (b) when it is slightly shaken?

*36. (II) A cubic box 1.20 m on a side is evacuated so the pressure of air inside is 10^{-6} torr. Estimate how many molecular collisions there are per each collision with a wall (0°C).

*37. (II) A very small amount of hydrogen gas is released into the air. If the air is at 1.0 atm and 25°C, estimate the mean free path for a H_2 molecule. What assumptions did you make?

*38. (II) Estimate the maximum allowable pressure in a 38-cm-long cathode ray tube if 98 percent of all electrons must hit the screen without first striking an air molecule.

*39. (II) (a) Show that the number of collisions a molecule makes per second, called the *collision frequency*, f, is given by $f = \bar{v}/l_M$, and thus $f = 4\sqrt{2}\,\pi r^2 \bar{v} N/V$. (b) What is the collision frequency for N_2 molecules in air at $T = 20°C$ and $P = 1.0 \times 10^{-2}$ atm?

*40. (II) We saw in Example 18–7 that the mean free path of air molecules at STP, l_M, is about 9×10^{-8} m. Estimate the collision frequency f, the number of collisions per unit time.

*41. (II) Suppose that a gas contains two types of molecules in concentrations n_1 and n_2. Their radii are r_1 and r_2. Use arguments similar to those leading to Eq. 18–10a to derive the following relation for the mean free path for type 1 molecules

$$l_{M1} = \frac{1}{4\pi r_1^2 n_1 + \pi(r_1 + r_2)^2 n_2}.$$

*42. (III) At some instant, suppose we have N_0 identical molecules. Show that the number N of molecules that travel a distance x or more before the next collision is given by $N = N_0 e^{-x/l_M}$, where l_M is the mean free path. This is called the *survival equation*.

* Section 18–7

*43. (I) Approximately how long would it take for the ammonia of Example 18–8 to be detected 1.5 m from the bottle after it is opened? What does this suggest about the relative importance of diffusion and convection for carrying odors?

*44. (II) What is the time needed for a glycine molecule (see Table 18–3) to diffuse a distance of 15 μm in water at 20°C if its concentration varies over that distance from 1.00 mol/m³ to 0.40 mol/m³? Compare this "speed" to its rms (thermal) speed. The molecular mass of glycine is about 75 u.

*45. (II) Oxygen diffuses from the surface of insects to the interior through tiny tubes called tracheae. An average trachea is about 2 mm long and has cross-sectional area of $2 \times 10^{-9} \, m^2$. Assuming the concentration of oxygen inside is half what it is outside in the atmosphere, (a) show that the concentration of oxygen in the air (assume 21 percent is oxygen) at 20°C is about 8.7 mol/m³, then (b) calculate the diffusion rate J, and (c) estimate the average time for a molecule to diffuse in. Assume the diffusion constant is $1 \times 10^{-5} \, m^2/s$.

General Problems

46. What is the rms speed of nitrogen molecules contained in a 12.8 m³ volume at 3.42 atm if the total amount of nitrogen is 1800 mol.

47. In outer space the density of matter is about one atom per cm³, mainly hydrogen atoms, and the temperature is about 2.7 K. Calculate the rms speed of these hydrogen atoms, and the pressure (in atmospheres).

48. Calculate approximately the total translational kinetic energy of all the molecules in an *E. coli* bacterium of mass 2.0×10^{-15} kg at 37°C. Assume 70 percent of the cell, by weight, is water, and the other molecules have an average molecular mass on the order of 10^5 u.

49. (*a*) Calculate the approximate rms speed of an amino acid, whose molecular mass is 89 u, in a living cell at 37°C. (*b*) What would be the rms speed of a protein of molecular mass 50,000 u at 37°C?

50. The escape speed from the Earth is 1.12×10^4 m/s, so that a gas molecule travelling away from Earth near the outer boundary of the Earth's atmosphere would, at this speed, be able to escape from the Earth's gravitational field and be lost to the atmosphere. At what temperature is the average speed of (*a*) oxygen molecules, and (*b*) helium atoms equal to 1.12×10^4 m/s?

51. The second postulate of kinetic theory is that the molecules are, on the average, far apart from one another. That is, their average separation is much greater than the diameter of each molecule. Is this assumption reasonable? To check, calculate the average volume occupied by one molecule of a gas at STP, and compare it to the size of the molecule (diameter of a typical gas molecule is about 0.2 nm). If the molecules were the diameter of ping-pong balls, say 4 cm, how far away, on average, would the next ping-pong ball be?

52. Estimate how many times per second a collision of a molecule of oxygen gas occurs with one of the walls of its container, which is a cube with sides 1.0 m. There are 2.0 mol of oxygen inside at 20°C. [*Hint:* Let $\bar{v}_x \approx \sqrt{\overline{v_x^2}}$.]

53. Consider a container of oxygen gas at a temperature of 20°C that is 0.50 m tall. Compare the gravitational potential energy of a molecule at the top of the container (assuming the potential energy is zero at the bottom) with the average kinetic energy of the molecules. Is it reasonable to neglect the potential energy?

54. In humid climates, people constantly *dehumidify* their cellars in order to prevent rot and mildew. If the cellar in a house (kept at 20°C) has 85 m² of floor space and a ceiling height of 2.8 m, what is the mass of water that must be removed from it in order to drop the humidity from 95 percent to a more reasonable 30 percent?

55. The temperature of an ideal gas is increased from 120°C to 290°C while the volume and the number of moles stay constant. By what factor does the pressure change? By what factor does v_{rms} change?

56. A scuba tank has a volume of 2800 cm³. For very deep dives, the tank is filled with 50 percent (by volume) pure oxygen and 50 percent pure helium. (*a*) How many molecules are there of each type in the tank if it is filled at 20°C to a gauge pressure of 10 atm? (*b*) What is the ratio of the average kinetic energies of the two types of molecule? (*c*) What is the ratio of the rms speeds of the two types of molecule?

57. A space vehicle returning from the Moon enters the atmosphere at a speed of about 40,000 km/h. Molecules (assume nitrogen) striking the nose of the vehicle with this speed correspond to what temperature? (Because of this high temperature, the nose of a space vehicle must be made of special materials; indeed, part of it does vaporize, and this is seen as a bright blaze upon reentry.)

58. At room temperature, it takes approximately 2.45×10^3 J to evaporate 1.00 g of water. Estimate the average speed of evaporating molecules. What multiple of v_{rms} (at 20°C) for water molecules is this? (Assume Eq. 18–4 holds.)

59. Calculate the total water vapor pressure in the air on the following two days: (*a*) a hot summer day, with the temperature 30°C and the relative humidity at 40%; (*b*) a cold winter day, with the temperature 5°C and the relative humidity at 80%.

* 60. At 300 K, an 8.50 mol sample of carbon dioxide occupies a volume of 0.200 m³. Calculate the gas pressure, first by assuming the ideal gas law, and then by using the van der Waals equation of state. (The values for *a* and *b* are given in the text following Eq. 18–9.) In this range of pressure and volume, the van der Waals equation is very accurate. What percentage error did you make in assuming ideal-gas-law behavior?

* 61. The density of atoms, mostly hydrogen, in interstellar space is about one per cubic centimeter. Estimate the mean free path of the hydrogen atoms, assuming an atomic diameter of 10^{-10} m.

* 62. Using the ideal gas law, find an expression for the mean free path l_M that involves pressure and temperature instead of (N/V). Use this expression to find the mean free path for nitrogen molecules at a pressure of 10 atm and 300 K.

When it is cold, warm clothes act as insulators to reduce heat loss from the body to the outside by conduction and convection.

Heat radiation from a campfire can warm you, and the fire can also transfer energy directly by heat conduction to a cooking pot.

Heat, like work, represents a transfer of energy. Heat is defined as a transfer of energy due to a difference of temperature. Work is a transfer of energy by mechanical means, not due to a temperature difference. The first law of thermodynamics links the two in a general statement of energy conservation: the heat Q added to a system minus the net work W done by the system equals the change in internal energy ΔU of the system: $\Delta U = Q - W$. Internal energy U is the sum total of all the energy of the molecules of the system.

Heat and the First Law of Thermodynamics

When a pot of cold water is placed on a hot burner of a stove, the temperature of the water increases. We say that heat flows from the hot burner to the cold water. When two objects at different temperatures are put in contact, heat spontaneously flows from the hotter one to the colder one. The spontaneous flow of heat is in the direction tending to equalize the temperature. If the two objects are kept in contact long enough for their temperatures to become equal, the two bodies are said to be in thermal equilibrium, and there is no further heat flow between them. For example, when the mercury in a fever thermometer is still rising, heat is flowing from the patient's mouth to the thermometer; when the mercury stops, the thermometer is then in equilibrium with the person's mouth, and they are at the same temperature.

Heat and temperature are often confused. They are very different concepts and we will make the clear distinction between them. We begin this Chapter by defining and using the concept of heat. We will also begin our discussion of thermodynamics, which is the name we give to the study of processes in which energy is transferred as heat and as work.

19–1 | Heat as Energy Transfer

We use the term "heat" in everyday life as if we knew what we meant. But the term is often used inconsistently, so it is important for us to define heat clearly, and to clarify the phenomena and concepts related to heat.

We commonly speak of the "flow" of heat—heat flows from a stove burner to a pot of coffee, from the Sun to the Earth, from a person's mouth into a fever thermometer. Heat flows spontaneously from an object at higher temperature to one at lower temperature. Indeed, an eighteenth-century model of heat pictured heat flow as movement of a fluid substance called *caloric*. However, the caloric fluid was never able to be detected. In the nineteenth century, it was found that the various phenomena associated with heat could be described consistently without the need to use the fluid model. The new model viewed heat as being akin to work and energy, as we will discuss in a moment. First we note that a common unit for heat, still in use today, is named after caloric. It is called the **calorie** (cal) and is defined as *the amount of heat necessary to raise the temperature of 1 gram of water by 1 Celsius degree.* [To be precise, the particular temperature range from 14.5°C to 15.5°C is specified because the heat required is very slightly different at different temperatures. The difference is less than 1 percent over the range 0°C to 100°C and we will ignore it for most purposes.] More often used than the calorie is the **kilocalorie** (kcal), which is 1000 calories. Thus *1 kcal is the heat needed to raise 1 kg of water by 1 C°.* Sometimes a kilocalorie is called a **Calorie** (with a capital C), and it is by this unit that the energy value of food is specified (we might also call it the "dietary calorie"). In the British system of units, heat is measured in British thermal units (Btu). One Btu is defined as the heat needed to raise the temperature of 1 lb of water by 1 F°. It can be shown (Problem 4) that 1 Btu = 0.252 kcal = 1055 J.

calorie (unit)

Kilocalorie (= dietary Calorie)

Btu

The idea that heat is related to energy was pursued by a number of scientists in the 1800s, particularly by an English brewer, James Prescott Joule (1818–1889). Joule performed a number of experiments that were crucial for establishing our present-day view that heat, like work, represents a transfer of energy. One of Joule's experiments is shown (simplified) in Fig. 19–1. The falling weight causes the paddle wheel to turn and do work on the water: the friction between the water and the paddle wheel causes the temperature of the water to rise slightly (barely measurable, in fact, by Joule). Of course, the same temperature rise could also be obtained by heating the water on a hot stove. In this and a great many other experiments (some involving electrical energy), Joule determined that a given amount of work done was always equivalent to a particular amount of heat input. Quantitatively, 4.186 joules (J) of work was found to be equivalent to 1 calorie (cal) of heat. This is known as the **mechanical equivalent of heat**:

$$4.186 \text{ J} = 1 \text{ cal}$$
$$4.186 \times 10^3 \text{ J} = 1 \text{ kcal}.$$

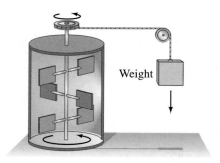

FIGURE 19–1 Joule's experiment on the mechanical equivalent of heat.

Weight

Mechanical equivalent of heat

As a result of these and other experiments, scientists came to interpret heat not as a substance, and not even as a form of energy. Rather, heat refers to a *transfer of energy*: when heat flows from a hot object to a cooler one, it is energy that is being transferred from the hot to the cold object. Thus, **heat** is *energy that is transferred from one body to another because of a difference in temperature*. In SI units, the unit for heat, as for any form of energy, is the joule. Nonetheless, calories and kcal are still sometimes used. Today the calorie is *defined* in terms of the joule (via the mechanical equivalent of heat, above), rather than in terms of the properties of water, as given previously. The latter is still handy to remember: 1 cal raises 1 g of water by 1 C°, or 1 kcal raises 1 kg of water by 1 C°.

Heat is energy transferred because of a ΔT

➡ PHYSICS APPLIED

Working off Calories

EXAMPLE 19–1 **ESTIMATE** **Working off the extra calories.** A young couple throw caution to the wind one afternoon, eating too much ice cream and cake. They realize that they each overate by about 500 Calories, and to compensate they want to do an equivalent amount of work climbing stairs or a mountain. How much vertical height must each person walk up? Each has a mass of 60 kg.

SOLUTION 500 Calories is 500 kcal, which in joules is

$$(500 \, \text{kcal})(4.186 \times 10^3 \, \text{J/kcal}) = 2.1 \times 10^6 \, \text{J}.$$

The work done to climb a vertical height h is $W = mgh$. We want to solve for h given that $W = 2.1 \times 10^6 \, \text{J}$:

$$h = \frac{W}{mg} = \frac{2.1 \times 10^6 \, \text{J}}{(60 \, \text{kg})(9.80 \, \text{m/s}^2)} = 3600 \, \text{m}.$$

They need to climb a very high mountain (over 11,000 ft) or many flights of stairs. [The human body does not transform energy with 100 percent efficiency—more like 20 percent—just as no engine does. As we'll see in the next chapter, some energy is always "wasted," so the couple would actually have to climb only about $(0.2)(3600 \, \text{m}) \approx 700 \, \text{m}$.]

19-2 | Internal Energy

We introduce the concept of internal energy now since it will help clarify ideas about heat. The sum total of all the energy of all the molecules in an object is called its **thermal energy** or **internal energy**. (We will use the two terms interchangeably.) Occasionally, the term "heat content" of a body is used for this purpose, but it is not a good term because it can be confused with heat itself. Heat, as we have seen, is not the energy a body contains, but rather refers to the amount of energy transferred from one body to another at a different temperature.

Internal energy

Distinguishing Temperature, Heat, and Internal Energy

Using the kinetic theory, we can make a clear distinction between temperature, heat, and internal energy. Temperature (in kelvins) is a measure of the *average* kinetic energy of individual molecules. Thermal energy and internal energy refer to the *total* energy of all the molecules in the object. (Thus two equal-mass hot ingots of iron may have the same temperature, but two of them have twice as much thermal energy as one does.) Heat, finally, refers to a *transfer* of energy (such as thermal energy) from one object to another because of a difference in temperature.

Heat vs. internal energy vs. temperature

Notice that the direction of heat flow between two objects depends on their temperatures, not on how much internal energy each has. Thus, if 50 g of water at 30°C is placed in contact (or mixed) with 200 g of water at 25°C, heat flows *from* the water at 30°C *to* the water at 25°C even though the internal energy of the 25°C water is much greater because there is so much more of it.

Direction of heat flow depends on temperature

Internal Energy of an Ideal Gas

Let us calculate the internal energy of n moles of an ideal monatomic (one atom per molecule) gas. The internal energy, U, is the sum of the translational kinetic energies of all the atoms. This sum is just equal to the average kinetic energy per molecule times the total number of molecules, N:

$$U = N(\tfrac{1}{2} m \overline{v^2}).$$

Using Eq. 18–4, $\overline{K} = \tfrac{1}{2} m \overline{v^2} = \tfrac{3}{2} kT$, we can write this as

$$U = \tfrac{3}{2} NkT$$

or

$$U = \tfrac{3}{2} nRT, \qquad \text{[monatomic ideal gas]} \quad \textbf{(19–1)}$$

Internal energy of an ideal monatomic gas

where n is the number of moles. Thus, the internal energy of an ideal gas depends only on temperature and the number of moles of gas.

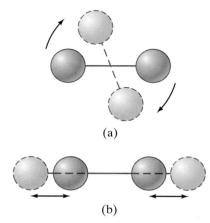

(a)

(b)

FIGURE 19–2 Molecules can have (a) rotational and (b) vibrational energy, as well as translational energy.

Relation between heat flow and temperature change

Specific heat

If the gas molecules contain more than one atom, then the rotational and vibrational energy of the molecules (Fig. 19–2) must also be taken into account. The internal energy will be greater at a given temperature than for a monatomic gas, but it will still be a function only of temperature for an ideal gas.

The internal energy of real gases also depends mainly on temperature, but where they deviate from ideal gas behavior, the internal energy also depends somewhat on pressure and volume.

The internal energy of liquids and solids is quite complicated, for it includes electrical potential energy associated with the forces (or "chemical" bonds) between atoms and molecules.

19–3 Specific Heat

If heat is put into an object, its temperature rises. But how much does the temperature rise? That depends. As early as the eighteenth century, experimenters had recognized that the amount of heat Q required to change the temperature of a given material is proportional to the mass m of the material and to the temperature change ΔT. This remarkable simplicity in nature can be expressed in the equation

$$Q = mc\,\Delta T, \tag{19–2}$$

where c is a quantity characteristic of the material called its **specific heat**. Because $c = Q/m\Delta T$, specific heat is specified in units of $J/kg\cdot C°$ (the proper SI unit) or $kcal/kg\cdot C°$. (The value of c in $cal/gm\cdot C°$ is the same as in $kcal/kg\cdot C°$.) For water at 15°C and a constant pressure of 1 atm, $c = 1.00\,kcal/kg\cdot C°$ or $4.19 \times 10^3\,J/kg\cdot C°$, since, by definition of the cal and the joule, it takes 1 kcal of heat to raise the temperature of 1 kg of water by 1 C°. Table 19–1 gives the values of specific heat for many substances at 20°C. The values of c depend to some extent on temperature (as well as slightly on pressure), but for temperature changes that are not too great, c can often be considered constant.[†]

TABLE 19–1 Specific Heats (at 1 atm constant pressure and 20°C unless otherwise stated)

| Substance | Specific Heat, c | |
	kcal/kg·C° (= cal/gm·C°)	J/kg·C°
Aluminum	0.22	900
Alcohol (ethyl)	0.58	2400
Copper	0.093	390
Glass	0.20	840
Iron or steel	0.11	450
Lead	0.031	130
Marble	0.21	860
Mercury	0.033	140
Silver	0.056	230
Wood	0.4	1700
Water		
Ice (−5°C)	0.50	2100
Liquid (15°C)	1.00	4186
Steam (110°C)	0.48	2010
Human body (average)	0.83	3470
Protein	0.4	1700

EXAMPLE 19–2 How heat depends on specific heat. (a) How much heat is required to raise the temperature of an empty 20-kg vat made of iron from 10°C to 90°C? (b) What if the vat is filled with 20 kg of water?

SOLUTION (a) From Table 19–1, the specific heat of iron is 450 J/kg·C°. The change in temperature is $(90°C − 10°C) = 80\,C°$. Thus,

$$Q = mc\,\Delta T = (20\,kg)(450\,J/kg\cdot C°)(80\,C°) = 7.2 \times 10^5\,J = 720\,kJ.$$

(b) The water alone would require

$$Q = mc\,\Delta T = (20\,kg)(4186\,J/kg\cdot C°)(80\,C°) = 6.7 \times 10^6\,J = 6700\,kJ$$

or almost 10 times what an equal mass of iron requires. The total, for the vat plus the water, is $720\,kJ + 6700\,kJ = 7400\,kJ$.

If the iron vat in part (a) had been *cooled* from 90°C to 10°C, 720 kJ of heat would have flowed *out* of the iron. In other words, Eq. 19–2 is valid for heat flow either in or out, with a corresponding increase or decrease in temperature. We saw in part (b) of Example 19–2 that water requires almost 10 times as much heat as an equal mass of iron to make the same temperature change. Water has one of the highest specific heats of all substances, which makes it an ideal substance for hot-water space-heating systems and other uses that require a minimal drop in temperature for a given amount of heat transfer. It is the water content, too, that causes the apples rather than the crust in hot apple pie to burn our tongues, through heat transfer.

[†] To take into account the dependence of c on T, we can write Eq. 19–2 in differential form: $dQ = mc(T)\,dT$. Then the heat Q required to change the temperature from T_1 to T_2 is

$$Q = \int_{T_1}^{T_2} mc(T)\,dT,$$

where $c(T)$ is a function of temperature.

19–4 Calorimetry—Solving Problems

When different parts of an isolated system are at different temperatures, but are placed in thermal contact, heat will flow from the part at higher temperature to the part at lower temperature. If the system is completely isolated, no energy can flow into or out of it. Again, the *conservation of energy* plays an important role for us: the heat lost by one part of the system is equal to the heat gained by the other part:

heat lost = heat gained.

Energy conservation

Let us take an Example.

EXAMPLE 19–3 **The cup cools the tea.** If 200 cm^3 of tea at 95°C is poured into a 150-g glass cup initially at 25°C (Fig. 19–3), what will be the final temperature T of the mixture when equilibrium is reached. assuming no heat flows to the surroundings?

SOLUTION Since tea is mainly water, its specific heat is 4186 J/kg·C° (Table 19–1) and its mass m is its density times its volume ($V = 200$ cm^3 = 200×10^{-6} m^3): $m = \rho V = (1.0 \times 10^3$ kg/m$^3)(200 \times 10^{-6}$ m$^3) = 0.20$ kg. Applying conservation of energy, we can set

heat lost by tea = heat gained by cup

$$m_{tea} c_{tea}(95°C - T) = m_{cup} c_{cup}(T - 25°C)$$

where T is the as yet unknown final temperature. Putting in numbers and using Table 19–1, we solve for T, and find

$$(0.20 \text{ kg})(4186 \text{ J/kg·C°})(95°C - T) = (0.15 \text{ kg})(840 \text{ J/kg·C°})(T - 25°C)$$
$$79{,}400 \text{ J} - (836 \text{ J/C°})T = (126 \text{ J/C°})T - 6300 \text{ J}$$
$$T = 89°C.$$

The tea drops in temperature by 6 C° by coming into equilibrium with the cup.

Alternate Solution: We can set up this Example (and others) by a different approach. We can write that the total heat transferred into or out of the isolated system is zero:

$$\Sigma Q_i = 0.$$

Then each term is written as $Q = mc(T_i - T_f)$ where T_i and T_f are the initial and final temperatures. In the present Example:

$$\Sigma Q = m_{tea} c_{tea}(95°C - T) + m_{cup} c_{cup}(25°C - T) = 0.$$

Note that the second term is negative because T will be greater than 25°C. Solving the algebra gives the same result.

The exchange of energy, as exemplified in Example 19–3, is the basis for a technique known as **calorimetry**, which is the quantitative measurement of heat exchange. To make such measurements, a **calorimeter** is used; a simple water calorimeter is shown in Fig. 19–4. It is very important that the calorimeter be well insulated so that only a minimal amount of heat is exchanged with the outside. One important use of the calorimeter is in the determination of specific heats of substances. In the technique known as the "method of mixtures," a sample of the substance is heated to a high temperature, which is accurately measured, and then quickly placed in the cool water of the calorimeter. The heat lost by the sample will be gained by the water and the calorimeter. By measuring the final temperature of the mixture, the specific heat can be calculated, as illustrated in the following Example.

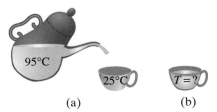

FIGURE 19–3 Example 19–3.

➡ **PROBLEM SOLVING**

Alternate approach

FIGURE 19–4
Simple water calorimeter.

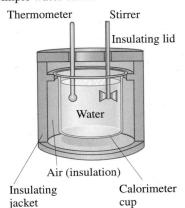

EXAMPLE 19–4 **Unknown specific heat determined by calorimetry.** An engineer wishes to determine the specific heat of a new alloy. A 0.150-kg sample of the alloy is heated to 540°C. It is then quickly placed in 400 g of water at 10.0°C, which is contained in a 200-g aluminum calorimeter cup. (We do not need to know the mass of the insulating jacket since we assume the air space between it and the cup insulates it well, so that its temperature does not change significantly.) The final temperature of the mixture is 30.5°C. Calculate the specific heat of the alloy.

SOLUTION Again we apply the conservation of energy and write that the heat lost equals the heat gained:

$$\begin{pmatrix} \text{heat lost} \\ \text{by sample} \end{pmatrix} = \begin{pmatrix} \text{heat gained} \\ \text{by water} \end{pmatrix} + \begin{pmatrix} \text{heat gained by} \\ \text{calorimeter cup} \end{pmatrix}$$

$$m_s c_s \, \Delta T_s = m_w c_w \, \Delta T_w + m_{cal} c_{cal} \, \Delta T_{cal}$$

where the subscripts s, w, and cal refer to the sample, water, and calorimeter, respectively. When we put in values and use Table 19–1, this equation becomes

$$(0.150 \, \text{kg})(c_s)(540°C - 30.5°C) = (0.40 \, \text{kg})(4186 \, \text{J/kg·C°})(30.5°C - 10.0°C)$$
$$+ (0.20 \, \text{kg})(900 \, \text{J/kg·C°})(30.5°C - 10.0°C)$$
$$76.4 \, c_s = (34,300 + 3700) \, \text{J/kg·C°}$$
$$c_s = 500 \, \text{J/kg·C°}.$$

In making this calculation, we have ignored any heat transferred to the thermometer and the stirrer (which is needed to quicken the heat transfer process and thus reduce heat loss to the outside). It can be taken into account by adding additional terms to the right side of the above equation and will result in a slight correction to the value of c_s (see Problem 14). It should be noted that the quantity $m_{cal} c_{cal}$ is often called the **water equivalent** of the calorimeter—that is, $m_{cal} c_{cal}$ is numerically equal to the mass of water (in kilograms) that would absorb the same amount of heat.

➡ **PHYSICS APPLIED**

Measuring Calorie content

A **bomb calorimeter** is used to measure the heat released when a substance burns. Important applications are the burning of foods to determine their Calorie content, and burning of seeds and other substances to determine their "energy content," or heat of combustion. A carefully weighed sample of the substance, together with an excess amount of oxygen at high pressure, is placed in a sealed container (the "bomb"). The bomb is placed in the water of the calorimeter and a fine wire passing into the bomb is then heated briefly, which causes the mixture to ignite.

19–5 Latent Heat

When a material changes phase from solid to liquid, or from liquid to gas (see also Section 18–3), a certain amount of energy is involved in this **change of phase**. For example, let us trace what happens when a 1.0-kg block of ice at −40°C has heat added to it at a steady rate until all the ice has changed to water, then the (liquid) water is heated to 100°C and changed to steam above 100°C, all at 1 atm pressure. As shown in the graph of Fig. 19–5, as heat is added to the ice, its temperature rises at a rate of about 2 C°/kcal of heat added (since for ice, $c \approx 0.50 \, \text{kcal/kg·C°}$). However, when 0°C is reached, the temperature stops increasing even though heat is still being added. Now as heat is added, the ice gradually changes to water in the liquid state without any change in temperature. After about 40 kcal have been added at 0°C, half the ice remains and half has changed to water. After about 80 kcal, or 330 kJ, has been added, all the ice has changed to water, still at 0°C. Further addition of heat causes the water's temperature to again increase, now at a rate of 1 C°/kcal. When 100°C is reached, the temperature again remains constant as the heat added changes the liquid water to vapor (steam). About 540 kcal (2260 kJ) is

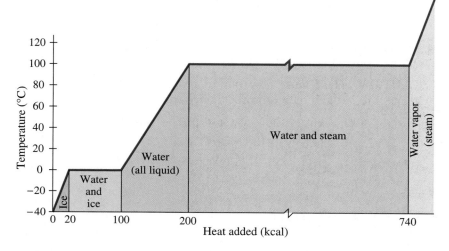

FIGURE 19–5 Temperature as a function of the heat added to bring 1.0 kg of ice at −40°C to steam above 100°C. Note the scale break between 200 and 740 kcal (the page isn't wide enough to fit it).

required to change the 1.0 kg of water completely to steam, after which the curve rises again, indicating that the temperature of the steam now rises as heat is added.

The heat required to change 1.0 kg of a substance from the solid to the liquid state is called the **heat of fusion**; it is denoted by L_F. The heat of fusion of water at 0°C is 79.7 kcal/kg or, in proper SI units, 333 kJ/kg $(= 3.33 \times 10^5 \, \text{J/kg})$. The heat required to change a substance from the liquid to the vapor phase is called the **heat of vaporization**, L_V, and for water at 100°C it is 539 kcal/kg or 2260 kJ/kg. Other substances follow graphs similar to Fig. 19–5, although the melting-point and boiling-point temperatures are different, as are the specific heats and heats of fusion and vaporization. Values for the heats of fusion and vaporization, which are also called the **latent heats**, are given in Table 19–2 for a number of substances.

Heat of fusion

Heat of vaporization

The heats of vaporization and fusion also refer to the amount of heat *released* by a substance when it changes from a gas to a liquid, or from a liquid to a solid. Thus, steam releases 2260 kJ/kg when it changes to water, and water releases 333 kJ/kg when it becomes ice.

Of course, the heat involved in a change of phase depends not only on the latent heat, but also on the total mass of the substance. That is,

$$Q = mL,$$

where L is the latent heat of the particular process and substance, m is the mass of the substance, and Q is the heat required or given off during the phase change. For example, when 5.00 kg of water freezes at 0°C, $(5.00 \, \text{kg})(3.33 \times 10^5 \, \text{J/kg}) = 1.67 \times 10^6 \, \text{J}$ of energy is released.

TABLE 19–2 Latent Heats (at 1 atm)

Substance	Melting Point (°C)	Heat of Fusion kcal/kg[†]	J/kg	Boiling Point (°C)	Heat of Vaporization kcal/kg[†]	J/kg
Oxygen	−218.8	3.3	0.14×10^5	−183	51	2.1×10^5
Nitrogen	−210.0	6.1	0.26×10^5	−195.8	48	2.00×10^5
Ethyl alcohol	−114	25	1.04×10^5	78	204	8.5×10^5
Ammonia	−77.8	8.0	0.33×10^5	−33.4	33	1.37×10^5
Water	0	79.7	3.33×10^5	100	539	22.6×10^5
Lead	327	5.9	0.25×10^5	1750	208	8.7×10^5
Silver	961	21	0.88×10^5	2193	558	23×10^5
Iron	1808	69.1	2.89×10^5	3023	1520	63.4×10^5
Tungsten	3410	44	1.84×10^5	5900	1150	48×10^5

[†] Numerical values in kcal/kg are the same in cal/g.

Calorimetry sometimes involves a change of state as the following Examples show. Indeed, latent heats are often measured using calorimetry.

EXAMPLE 19–5 **Making ice.** How much energy does a refrigerator have to remove from 1.5 kg of water at 20°C to make ice at −12°C?

SOLUTION Heat must flow out to reduce the water from 20°C to 0°C, to change it to ice, and then to lower the ice from 0°C to −12°C,

$$Q = mc_w(20°C - 0°C) + mL_F + mc_{ice}[0° - (-12°C)]$$
$$= (1.5\,kg)(4186\,J/kg \cdot C°)(20\,C°) + (1.5\,kg)(3.33 \times 10^5\,J/kg)$$
$$+ (1.5\,kg)(2100\,J/kg \cdot C°)(12\,C°)$$
$$= 6.6 \times 10^5\,J = 660\,kJ.$$

EXAMPLE 19–6 **Will all the ice melt?** At a party, a 0.50-kg chunk of ice at −10°C is placed in 3.0 kg of "iced" tea at 20°C. At what temperature and in what phase will the final mixture be? The tea can be considered as water.

➡ **PROBLEM SOLVING**

First determine (or estimate) the final state

SOLUTION In this situation, before we can write down an equation, we must first check to see if the final state will be all ice, a mixture of ice and water at 0°C, or all water. To bring the 3.0 kg of water at 20°C down to 0°C would require an energy release of

$$m_w c_w(20°C - 0°C) = (3.0\,kg)(4186\,J/kg \cdot C°)(20\,C°) = 250\,kJ.$$

On the other hand, to raise the ice from −10°C to 0°C would require

$$m_{ice} c_{ice}[0°C - (-10°C)] = (0.50\,kg)(2100\,J/kg \cdot C°)(10\,C°) = 10.5\,kJ,$$

and to change the ice to water at 0°C would require

$$m_{ice} L_F = (0.50\,kg)(333\,kJ/kg) = 167\,kJ,$$

for a total of 10.5 kJ + 167 kJ = 177 kJ. This is not enough energy to bring the 3.0 kg of water at 20°C down to 0°C, so we know that the mixture must end up all water, somewhere between 0°C and 20°C. Now we can determine the final temperature T by applying the conservation of energy and writing

Then determine the final temperature

$$\begin{pmatrix} \text{heat to raise} \\ \text{0.50 kg of ice} \\ \text{from } -10°C \\ \text{to } 0°C \end{pmatrix} + \begin{pmatrix} \text{heat to change} \\ \text{0.50 kg} \\ \text{of ice} \\ \text{to water} \end{pmatrix} + \begin{pmatrix} \text{heat to raise} \\ \text{0.50 kg of water} \\ \text{from } 0°C \\ \text{to } T \end{pmatrix} = \begin{pmatrix} \text{heat lost by} \\ \text{3.0 kg of} \\ \text{water cooling} \\ \text{from } 20°C \text{ to } T \end{pmatrix}$$

so

$$10.5\,kJ + 167\,kJ + (0.50\,kg)(4186\,J/kg \cdot C°)(T)$$
$$= (3.0\,kg)(4186\,J/kg \cdot C°)(20°C - T).$$

Solving for T gives $T = 5.1°C$.

We can make use of kinetic theory to see why energy is needed to melt or vaporize a substance. At the melting point, the latent heat of fusion does not increase the kinetic energy (and the temperature) of the molecules in the solid, but instead is used to overcome the potential energy associated with the forces between the molecules. That is, work must be done against these attractive forces to break the molecules loose from their relatively fixed positions in the solid so they can freely roll over one another in the liquid phase. Similarly, energy is required for molecules held close together in the liquid phase to escape into the gaseous phase. This process is a more violent reorganization of the molecules than is melting (the average distance between the molecules is greatly increased) and hence the heat of vaporization is generally much greater than the heat of fusion for a given substance.

Calorimetry

1. Be sure you have enough information to apply energy conservation. Ask yourself: is the system isolated (or very nearly so, enough to get a good estimate)? Do we know or can we calculate all significant sources of heat energy flow?

2. Apply conservation of energy. One approach is to write

heat gained = heat lost.

For each substance in the system, a heat (energy) term will appear on either the left or right of this equation.

3. If no phase changes occur, each term in the energy conservation equation (above) will have the form

$$Q(\text{gain}) = mc(T_f - T_i) \quad \text{or} \quad Q(\text{lost}) = mc(T_i - T_f)$$

where T_i and T_f are the initial and final temperatures

of the substance, and m and c are its mass and specific heat.

4. If phase changes do or might occur, there may be terms in the energy conservation equation of the form $Q = mL$, where L is the latent heat. But *before* applying energy conservation, determine (or estimate) in which phase the final state will be, as we did in Example 19–6 by calculating the different contributing values for heat Q.

5. Be sure each term appears on the correct side of the energy equation (heat gained or heat lost) and that each ΔT is positive.

6. Note that when the system reaches thermal equilibrium, the final temperature of each substance will have the *same* value. There is only one T_f.

7. Solve your energy equation for the unknown.

The latent heat to change a liquid to a gas is needed not only at the boiling point. Water can change from the liquid to the gas phase even at room temperature. This process is called **evaporation** (see also Section 18–4). The value of the heat of vaporization increases slightly with a decrease in temperature: at 20°C, for example, it is 2450 kJ/kg (585 kcal/kg) compared to 2260 kJ/kg (539 kcal/kg) at 100°C. When water evaporates, it cools, since the energy required (the latent heat of vaporization) comes from the water itself. So its internal energy, and therefore its temperature, must drop.[†]

19–6 | The First Law of Thermodynamics

Up to now in this chapter we have discussed internal energy and heat. But work too is often involved in thermodynamic processes.

In Chapter 8 we saw that work is done when energy is transferred from one body to another by mechanical means. In Section 19–1 we saw that heat is a transfer of energy from one body to a second body at a lower temperature. Thus, heat is much like work. To distinguish them, *heat* is defined as a *transfer of energy due to a difference in temperature*, whereas work is a transfer of energy that is not due to a temperature difference.

Heat distinguished from work

In discussing thermodynamics, we shall often refer to particular systems. A **system** is any object or set of objects that we wish to consider. Everything else in the universe we will refer to as its "environment." There are several categories of systems. A **closed system** is one for which no mass enters or leaves (but energy may be exchanged with the environment). In an **open system**, mass may enter or leave (as well as energy). Many (idealized) systems we study in physics are closed systems. But many systems, including plants and animals, are open systems since they exchange materials (food, oxygen, waste products) with the environment. A closed system is said to be **isolated** if no energy in any form passes across its boundaries; otherwise it is not isolated.

Open and closed systems

[†] According to kinetic theory, evaporation is a cooling process because it is the fastest-moving molecules that escape from the surface. Hence the average speed of the remaining molecules is less, so by Eq. 18–4 the temperature is less.

In Section 19–2, we defined the internal energy of a system as the sum total of all the energy of the molecules of the system. We would expect that the internal energy of a system would be increased if work were done on the system, or if heat were added to it. Similarly the internal energy would be decreased if heat flowed out of the system or if work were done by the system on something else.

Thus, from conservation of energy, it is reasonable to propose an important law: the change in internal energy of a closed system, ΔU, will be equal to the heat added to the system minus the work done by the system; in equation form:

FIRST LAW OF THERMODYNAMICS

$$\Delta U = Q - W, \tag{19–3}$$

where Q is the net heat *added* to the system and W is the net work done *by* the system. We must be careful and consistent in following the sign conventions for Q and W. Because W in Eq. 19–3 is the work done *by* the system, then if work is done *on* the system, W will be negative and U will increase. [Of course, we could have defined W as the work done *on* the system, in which case there would be a plus sign in Eq. 19–3; but it is conventional to define W and Q as we have done.] Similarly, Q is positive for heat added to the system, so if heat leaves the system, Q is negative. Equation 19–3 is known as the **first law of thermodynamics**. It is one of the great laws of physics, and its validity rests on experiments (such as Joule's) in which no exceptions have been seen. Since Q and W represent energy transferred into or out of the system, the internal energy changes accordingly. Thus, the first law of thermodynamics is a great and broad statement of the *law of conservation of energy*. It is worth noting that the conservation of energy law was not formulated until the nineteenth century, for it depended on the interpretation of heat as a transfer of energy.

Heat added is +
Heat lost is −
Work on system is −
Work by system is +

First law of thermodynamics is conservation of energy

Equation 19–3 applies to a closed system. It also applies to an open system if we take into account the change in internal energy due to the increase or decrease in the amount of matter. For an isolated system, no work is done and no heat enters or leaves the system, so $W = Q = 0$, and hence $\Delta U = 0$.

Internal energy is a property of a system; work and heat are not

A given system, in a particular state, can be said to have a certain amount of internal energy, U. This cannot be said of heat or work. A system in a given state does not "have" a certain amount of heat or work. Rather, when work is done on a system (such as compressing a gas), or when heat is added or removed from a system, the state of the system *changes*. Thus, work and heat are involved in *thermodynamic processes* that can change the system from one state to another; they are not characteristic of the state itself, as are pressure P, volume V, temperature T, internal energy U, and the mass m or number of moles n. Because U is a *state variable*, which we saw in Chapter 17 is a variable that depends only on the state of the system and not on how the system arrived in that state, we can write

$$\Delta U = U_2 - U_1 = Q - W,$$

where U_1 and U_2 represent the internal energy of the system in states 1 and 2, and Q and W are the heat added to the system and work done by the system in going from state 1 to state 2.

It is sometimes useful to write the first law of thermodynamics in differential form:

$$dU = dQ - dW.$$

Here, dU represents an infinitesimal change in internal energy when an infinitesimal amount[†] of heat dQ is added to the system, and the system does an infinitesimal amount of work dW.

[†] The differential form of the first law is often written

$$dU = đQ - đW,$$

where the bars on the differential sign ($đ$) are used to remind us that W and Q are not functions of the state variables (such as P, V, T, n). Internal energy, U, is a function of the state variables, and so dU represents the differential (called an *exact differential*) of some function U. The differentials $đW$ and $đQ$ are not exact differentials (they are not the differential of some mathematical function); they thus only represent infinitesimal amounts. This issue won't really be of concern in this book.

Using the first law. (*a*) An amount of heat equal to 2500 J is added to a system, and 1800 J of work is done on the system. What is the change in internal energy of the system?

SOLUTION We use the first law of thermodynamics, Eq. 19–3. The heat added to the system is $Q = 2500$ J. The work done by the system W is -1800 J, with a minus sign because 1800 J done *on* the system equals -1800 J done *by* the system. Hence

$$\Delta U = 2500 \, J - (-1800 \, J) = 2500 \, J + 1800 \, J = 4300 \, J.$$

You may have intuitively thought that the 2500 J and the 1800 J would need to be added together, since both refer to energy transferred into the system. You would have been right. We did this exercise in detail to emphasize the importance of keeping careful track of signs.

19–7 Applying the First Law of Thermodynamics; Calculating the Work

Let us analyze some simple processes in the light of the first law of thermodynamics.

Isothermal Processes ($\Delta T = 0$)

First we consider an idealized process that is carried out at constant temperature. Such a process is called an **isothermal** process (from the Greek meaning "same temperature"). If the system is an ideal gas, then $PV = nRT$ (Eq. 17–3), so for a fixed amount of gas kept at constant temperature, $PV = $ constant. Thus the process follows a curve like AB on the PV diagram shown in Fig. 19–6, which is a curve for $PV = $ constant. Each point on the curve, such as point A, represents the state of the system at a given moment—that is, its pressure P and volume V. At a lower temperature, another isothermal process would be represented by a curve like A'B' in Fig. 19–6 (the product $PV = nRT = $ constant is less when T is less). The curves shown in Fig. 19–6 are referred to as *isotherms*.

Let us assume that the gas is enclosed in a container fitted with a movable piston, Fig. 19–7, and that the gas is in contact with a **heat reservoir** (a body whose mass is so large that, ideally, its temperature does not change significantly when heat is exchanged with our system). We also assume that the process of compression (volume decreases) or expansion (volume increases) is done **quasistatically** ("almost statically"), by which we mean extremely slowly, so that all of the gas stays in equilibrium at the same constant temperature. If the gas is initially in a state represented by point A in Fig. 19–6 and an amount of heat Q is added to the system, the system will move to another point, B, on the diagram. If the temperature is to remain constant, the gas must expand and do an amount of work W on the environment (it exerts a force on the piston and moves it through a distance). The temperature and mass are kept constant so, from Eq. 19–1, the internal energy does not change: $\Delta U = \frac{3}{2} nR\Delta T = 0$. Hence, by the first law of thermodynamics, Eq. 19–3, $\Delta U = Q - W = 0$, so $W = Q$: the work done by the gas in an isothermal process equals the heat added to the gas.

Adiabatic Processes ($Q = 0$)

An **adiabatic** process is one in which no heat is allowed to flow into or out of the system: $Q = 0$. This situation can occur if the system is extremely well insulated, or the process happens so quickly that heat—which flows slowly—has no time to flow in or out. The very rapid expansion of gases in an internal combustion engine

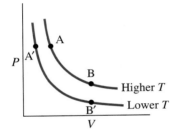

FIGURE 19–6 *PV* diagram for an ideal gas undergoing isothermal processes at two different temperatures.

Heat reservoir

FIGURE 19–7 An ideal gas in a cylinder fitted with a movable piston.

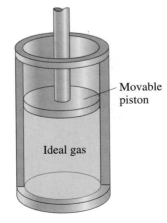

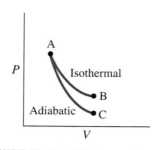

FIGURE 19–8 *PV* diagram for adiabatic (AC) and isothermal (AB) processes on an ideal gas.

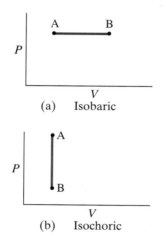

(a) Isobaric

(b) Isochoric

FIGURE 19–9 (a) Isobaric ("same pressure") process. (b) Isochoric ("same volume") process.

FIGURE 19–10 The work done by a gas when its volume increases by $dV = A\,dl$ is $dW = P\,dV$.

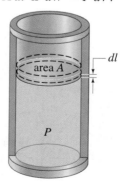

is one example of a process that is very nearly adiabatic. A slow adiabatic expansion of an ideal gas follows a curve like that labeled AC in Fig. 19–8. Since $Q = 0$, we have from Eq. 19–3 that $\Delta U = -W$. That is, the internal energy decreases if the gas expands; hence the temperature decreases as well (because $\Delta U = \frac{3}{2}nR\Delta T$). Thus an adiabatic *PV* curve is steeper than an isotherm: in Fig. 19–8 the product $PV(= nRT)$ is less at point C than at point B (curve AB is for an isothermal process, for which $\Delta U = 0$ and $\Delta T = 0$). In an adiabatic compression (going from C to A, for example), work is done *on* the gas, and hence the internal energy increases and the temperature rises. In a diesel engine, the rapid adiabatic compression reduces the volume by a factor of 15 or more; the temperature rise is so great that the air-fuel mixture ignites spontaneously.

| **CONCEPTUAL EXAMPLE 19–8** | **Simple adiabatic process.** Here is an adiabatic process that you can do with just a rubber band. Hold the rubber band loosely with two hands and gauge the temperature with your lips. Stretch the rubber band suddenly and again touch it lightly to your lips. You should notice an increase in temperature. Explain clearly why the temperature increases.

RESPONSE Stretching the rubber band *suddenly* makes the process adiabatic because there is no time for heat to enter or leave the system, so $Q = 0$. You do work on the system, representing an energy input, so by the first law of thermodynamics the internal energy increases, corresponding to an increase in temperature (Eq. 19–1); so the rubber band heats up.

Isobaric and Isochoric Processes

Isothermal and adiabatic processes are just two possible processes that can occur. Two other simple thermodynamic processes are illustrated on the *PV* diagram of Fig. 19–9: (a) an **isobaric** process is one in which the pressure is kept constant, so the process is represented by a horizontal straight line on the *PV* diagram, Fig. 19–9a; (b) an **isochoric** or isovolumetric process is one in which the volume does not change (Fig. 19–9b). In these, and in all other processes, the first law of thermodynamics holds.

Work Done in Volume Changes

We often want to calculate the work done in a process. Suppose we have a gas confined to a cylindrical container fitted with a movable piston (Fig. 19–10). We must always be careful to define exactly what our system is. In this case we choose our system to be the gas; so the container's walls and the piston are parts of the environment. Now let us calculate the work done by the gas when it expands quasistatically, so that P and T are defined for the system at all instants.[†] The gas expands against the piston, whose area is A. The gas exerts a force $F = PA$ on the piston, where P is the pressure in the gas. The work done by the gas to move the piston an infinitesimal displacement dl is

$$dW = \mathbf{F} \cdot d\mathbf{l} = PA\,dl = P\,dV \tag{19-4}$$

since the infinitesimal increase in volume is $dV = A\,dl$. If the gas were *compressed* so that dl pointed into the gas, the volume would decrease and $dV < 0$. The work done by the gas in this case would then be negative, which is equivalent to saying that positive work was done *on* the gas, not by it. For a finite change in volume from V_A to V_B, the work W done by the gas will be

$$W = \int dW = \int_{V_A}^{V_B} P\,dV. \tag{19-5}$$

Equations 19–4 and 19–5 are valid for the work done in any volume change—by a gas, a liquid, or a solid—as long as it is done quasistatically.

[†] If the gas expanded or were compressed quickly, there would be turbulence, and different parts would be at different pressure (and temperature).

In order to integrate Eq. 19–5, we need to know how the pressure varies during the process, and this depends on the type of process. Let us first consider a quasistatic isothermal expansion of an ideal gas. This process is represented by the curve between points A and B on the PV diagram of Fig. 19–11. The work done by the gas in this process, according to Eq. 19–5, is just the area between the PV curve and the V axis, and is shown shaded in Fig. 19–11. We can do the integral in Eq. 19–5 for an ideal gas by using the ideal gas law, $P = nRT/V$. The work done is

$$W = \int_{V_A}^{V_B} P\, dV = nRT \int_{V_A}^{V_B} \frac{dV}{V} = nRT \ln \frac{V_B}{V_A}. \qquad \left[\begin{array}{l}\text{isothermal process;} \\ \text{ideal gas}\end{array}\right] \quad (19\text{–}6)$$

Let us next consider a different way of taking an ideal gas between the same states A and B. This time, let us lower the pressure in the gas from P_A to P_B, as indicated by the line AD in Fig. 19–12. (In this *isochoric* process, heat must be allowed to flow out of the gas so its temperature drops.) Then let the gas expand from V_A to V_B at constant pressure $(= P_B)$, which is indicated by the line DB in Fig. 19–12. (In this *isobaric* process, heat must be added to the gas to raise its temperature.) No work is done in the isochoric process AD, since $dV = 0$:

$$W = 0. \qquad \text{[isochoric process; ideal gas]}$$

In the isobaric process DB the pressure remains constant, so

$$W = \int_{V_A}^{V_B} P\, dV = P_B(V_B - V_A) = P\,\Delta V. \qquad \left[\begin{array}{l}\text{isobaric process;} \\ \text{ideal gas}\end{array}\right] \quad (19\text{–}7a)$$

The work done is again represented by the area between the curve (ADB) on the PV diagram, and the V axis as indicated by the shading in Fig. 19–12. Using the ideal gas law, we can also write

$$W = P_B(V_B - V_A) = nRT_B\left(1 - \frac{V_A}{V_B}\right). \qquad \left[\begin{array}{l}\text{isobaric process;} \\ \text{ideal gas}\end{array}\right] \quad (19\text{–}7b)$$

As can be seen from the shaded areas in Figs. 19–11 and 19–12, or by putting in numbers in Eqs. 19–6 and 19–7 (try it for $V_B = 2V_A$), the work done in these two processes is different. This is a general result. *The work done in taking a system from one state to another depends not only on the initial and final states but also on the type of process (or "path").*

This result reemphasizes the fact that work cannot be considered a property of a system. The same is true of heat. The heat input required to change the gas from state A to state B depends on the process; for the isothermal process of Fig. 19–11, the heat input turns out to be greater than for the process ADB of Fig. 19–12. In general, *the amount of heat added or removed in taking a system from one state to another depends not only on the initial and final states but also on the path or process.*

EXAMPLE 19–9 **First law in isobaric and isochoric processes.** An ideal gas is slowly compressed at a constant pressure of 2.0 atm from 10.0 L to 2.0 L, path B to D in Fig. 19–12. (In this process, some heat flows out and the temperature drops.) Heat is then added to the gas, holding the volume constant, and the pressure and temperature are allowed to rise until the temperature reaches its original value, process D to A in Fig. 19–12. Calculate (*a*) the total work done by the gas in the process BDA, and (*b*) the total heat flow into the gas.

SOLUTION (*a*) Work is done only in the first part, the compression (BD)

$$W = P\,\Delta V = (2.0 \times 10^5\,\text{N/m}^2)(2.0 \times 10^{-3}\,\text{m}^3 - 10.0 \times 10^{-3}\,\text{m}^3) = -1.6 \times 10^3\,\text{J}.$$

From D to A no work is done ($\Delta V = 0$); so the total work done by the gas is -1.6×10^3 J, where the minus means that $+1.6 \times 10^3$ J of work is done *on* the gas. (*b*) Since the temperature at the beginning and at the end of the process is the same, there is no change in internal energy of our ideal gas: $\Delta U = 0$. From the first law of thermodynamics we have

$$0 = \Delta U = Q - W,$$

so $Q = W = -1.60 \times 10^3$ J. Since Q is negative, 1600 J of heat flows out of the gas for the whole process, BDA.

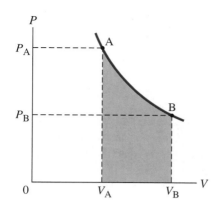

FIGURE 19–11 Work done by an ideal gas in an isothermal process equals the area under the PV curve. Shaded area equals the work done by the gas when it expands from V_A to V_B.

FIGURE 19–12 Process ADB consists of an isochoric (AD) and an isobaric (DB) process.

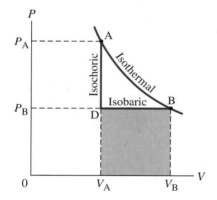

EXAMPLE 19–10 **Work done in an engine.** In an engine, 0.25 moles of gas in the cylinder expands rapidly and adiabatically against the piston. In the process, the temperature drops from 1150 K to 400 K. How much work does the gas do? Assume the gas is ideal.

SOLUTION Instead of calculating the work directly, we can more simply use the first law of thermodynamics, if we can determine ΔU since we know $Q = 0$ because the process is adiabatic. We determine ΔU from Eq. 19–1 for the internal energy of an ideal monatomic gas:

$$\Delta U = U_f - U_i = \tfrac{3}{2}nR(T_f - T_i)$$
$$= \tfrac{3}{2}(0.25 \text{ mol})(8.315 \text{ J/mol·K})(400 \text{ K} - 1150 \text{ K}) = -2300 \text{ J}.$$

Then, from Eq. 19–3, $W = Q - \Delta U = 0 - (-2300 \text{ J}) = 2300 \text{ J}$.

Free Expansion

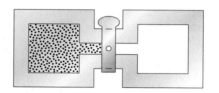

FIGURE 19–13 Free expansion.

For a free expansion
$Q = 0, W = 0, \Delta U = 0$

One type of adiabatic process is a so-called **free expansion** in which a gas is allowed to expand in volume adiabatically without doing any work. The apparatus to accomplish a free expansion is shown in Fig. 19–13. It consists of two well-insulated compartments (to ensure no heat flow in or out) connected by a valve or stopcock. One compartment is filled with gas, the other is empty. When the valve is opened, the gas expands to fill both containers. No heat flows in or out ($Q = 0$), and no work is done because the gas does not move any other object. Thus $Q = W = 0$ and by the first law of thermodynamics, $\Delta U = 0$. *The internal energy of a gas does not change in a free expansion.* For an ideal gas, $\Delta T = 0$ also, since U depends only on T (Section 19–2). Experimentally, the free expansion has been used to determine if the internal energy of *real gases* depends only on T. The experiments are very difficult to do accurately, but it has been found that the temperature of a real gas drops very slightly in a free expansion. Thus the internal energy of real gases does depend, a little, on pressure or volume as well as on temperature.

Note, incidentally, that a free expansion could not be plotted on a *PV* diagram, because the process is rapid, not quasistatic. The intermediate states are not equilibrium states, and hence the pressure (and even the volume at some instants) is not clearly defined.

19–8 Molar Specific Heats for Gases, and the Equipartition of Energy

In Section 19–3 we discussed the concept of specific heat and applied it to solids and liquids. Much more than for solids and liquids, the values of the specific heat for gases depends on how the process is carried out. Two important processes are those in which either the volume or the pressure is kept constant. Although for solids and liquids it matters little, Table 19–3 shows that the specific heats of gases at constant volume (c_V) and at constant pressure (c_P) are quite different.

Molar Specific Heats for Gases

The difference in specific heats for gases is nicely explained in terms of the first law of thermodynamics and kinetic theory. Indeed, the values of the specific heats can be calculated using the kinetic theory, and the results are in close agreement with experiment. Our discussion is simplified if we use **molar specific heats**, C_V and C_P, which are defined as the heat required to raise 1 mol of the gas by 1 C° at constant volume and at constant pressure, respectively. That is, in analogy to Eq. 19–2, the heat Q needed to raise the temperature of n moles of gas by ΔT is

Molar specific heats

$$Q = nC_V \Delta T \qquad \text{[volume constant]} \quad \textbf{(19–8a)}$$
$$Q = nC_P \Delta T. \qquad \text{[pressure constant]} \quad \textbf{(19–8b)}$$

TABLE 19–3 Specific heats of gases at 15°C

Gas	Specific heats (kcal/kg·K)		Molar specific heats (cal/mol·K)		$C_P - C_V$ (cal/mol·K)	$\gamma = \dfrac{C_P}{C_V}$
	c_V	c_P	C_V	C_P		
Monatomic						
He	0.75	1.15	2.98	4.97	1.99	1.67
Ne	0.148	0.246	2.98	4.97	1.99	1.67
Diatomic						
N_2	0.177	0.248	4.96	6.95	1.99	1.40
O_2	0.155	0.218	5.03	7.03	2.00	1.40
Triatomic						
CO_2	0.153	0.199	6.80	8.83	2.03	1.30
H_2O (100°C)	0.350	0.482	6.20	8.20	2.00	1.32
Polyatomic						
C_2H_6	0.343	0.412	10.30	12.35	2.05	1.20

It is clear from the definition of molar specific heat (or by comparing Eqs. 19–2 and 19–8) that

$$C_V = Mc_V$$
$$C_P = Mc_P,$$

where M is the molecular mass of the gas ($M = m/n$ in grams/mol). The values for molar specific heats are included in Table 19–3, and we see that the values are nearly the same for different gases that have the same number of atoms per molecule.

Let us now make use of the kinetic theory and see first why the specific heats of gases are higher for constant-pressure processes than for constant-volume processes. Let us imagine that an ideal gas is slowly heated via these two different processes—first at constant volume, and then at constant pressure. In both of these processes, we let the temperature increase by the same amount, ΔT. In the process done at constant volume, no work is done since $\Delta V = 0$. Thus, according to the first law of thermodynamics, the heat added (which we now denote by Q_V) all goes into increasing the internal energy of the gas:

$$Q_V = \Delta U.$$

In the process carried out at constant pressure, work is done, and hence the heat added, Q_P, must not only increase the internal energy but also is used to do the work $W = P\Delta V$. Thus, more heat must be added in this process than in the first process at constant volume. For the process at constant pressure, we have from the first law of thermodynamics

$$Q_P = \Delta U + P\Delta V.$$

Since ΔU is the same in the two processes (ΔT was chosen to be the same), we can combine the two above equations:

$$Q_P - Q_V = P\Delta V.$$

From the ideal gas law, $V = nRT/P$, so for a process at constant pressure we have $\Delta V = nR\Delta T/P$. Putting this into the above equation and using Eqs. 19–8, we find

$$nC_P \, \Delta T - nC_V \, \Delta T = P\left(\frac{nR\Delta T}{P}\right)$$

or, after cancellations,

$$C_P - C_V = R. \tag{19–9}$$

Since the gas constant $R = 8.315 \, \text{J/mol·K} = 1.99 \, \text{cal/mol·K}$, our prediction is that C_P will be larger than C_V by about 1.99 cal/mol·K. Indeed, this is very close to what is obtained experimentally, as can be seen in the next to last column in Table 19–3.

Let us now calculate the molar specific heat of a monatomic gas using the kinetic theory model of gases. First we consider a process carried out at constant volume. Since no work is done in this process the first law of thermodynamics tells us that if heat Q is added to the gas, the internal energy of the gas changes by

$$\Delta U = Q.$$

For an ideal monatomic gas, the internal energy, U, is the total kinetic energy of all the molecules,

$$U = N(\tfrac{1}{2}m\overline{v^2}) = \tfrac{3}{2}nRT$$

as we saw in Section 19–2. Then, using Eq. 19–8a, we can write $\Delta U = Q$ in the form

$$\Delta U = \tfrac{3}{2}nR\,\Delta T = nC_V\,\Delta T \tag{19–10}$$

or

$$C_V = \tfrac{3}{2}R. \tag{19–11}$$

Since $R = 8.315\,\text{J/mol·K} = 1.99\,\text{cal/mol·K}$, kinetic theory predicts that $C_V = 2.98\,\text{cal/mol·K}$ for an ideal monatomic gas. This is very close to the experimental values for monatomic gases such as helium and neon (Table 19–3). From Eq. 19–9, C_P is predicted to be about $4.97\,\text{cal/mol·K}$, also in agreement with experiment.

Equipartition of Energy

The measured molar specific heats for more complex gases (Table 19–3), such as diatomic (two atoms) and triatomic (three atoms) gases, increase with the increased number of atoms per molecule. We can explain this by assuming that the internal energy includes not only translational kinetic energy but other forms of energy as well. Take, for example, a diatomic gas. As shown in Fig. 19–14 the two atoms can rotate about two different axes (but rotation about a third axis passing through the two atoms would give rise to very little energy since the moment of inertia is so small). The molecules can have rotational as well as translational kinetic energy. It is useful to introduce the idea of **degrees of freedom**, by which we mean the number of independent ways molecules can possess energy. For example, a monatomic gas is said to have three degrees of freedom, since an atom can have velocity along the x axis, the y axis, and the z axis. These are considered to be three independent motions because a change in any one of the components would not affect any of the others. A diatomic molecule has the same three degrees of freedom associated with translational kinetic energy plus two more degrees of freedom associated with rotational kinetic energy, for a total of five degrees of freedom. A quick look at Table 19–3 indicates that the C_V for diatomic gases is about $\tfrac{5}{3}$ times as great as for a monatomic gas—that is, in the same ratio as their degrees of freedom. This led nineteenth-century physicists to an important idea, the **principle of equipartition of energy**. This principle states that energy is shared equally among the active degrees of freedom, and in particular each active degree of freedom of a molecule has on the average an energy equal to $\tfrac{1}{2}kT$. Thus, the average energy for a molecule of a monatomic gas would be $\tfrac{3}{2}kT$ (which we already knew) and of a diatomic gas $\tfrac{5}{2}kT$. Hence the internal energy of a diatomic gas would be $U = N(\tfrac{5}{2}kT) = \tfrac{5}{2}nRT$, where n is the number of moles. Using the same argument we did for monatomic gases, we see that for diatomic gases the molar specific heat at constant volume would be $\tfrac{5}{2}R = 4.97\,\text{cal/mol·K}$, in accordance with measured values. More complex molecules have even more degrees of freedom and thus greater molar specific heats.

The situation was complicated, however, by measurements that showed that for diatomic gases at very low temperatures, C_V has a value of only $\tfrac{3}{2}R$, as if it had only three degrees of freedom. And at very high temperatures, C_V was about $\tfrac{7}{2}R$, as if there were seven degrees of freedom. The explanation is that at low temperatures, nearly all molecules have only translational kinetic energy. That is, no energy goes into rotational energy, so only three degrees of freedom are "active." At very high temperatures, on the other hand, all five degrees of freedom are active plus

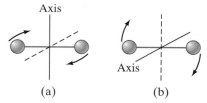

FIGURE 19–14 A diatomic molecule can rotate about two different axes.

Degrees of freedom

Equipartition of energy

two additional ones. We can interpret the two new degrees of freedom as being associated with the two atoms vibrating as if they were connected by a spring, as shown in Fig. 19–15. One degree of freedom comes from the kinetic energy of the vibrational motion, and the second comes from the potential energy of vibrational motion $(\frac{1}{2}kx^2)$. At room temperature, these two degrees of freedom are apparently not active. See Fig. 19–16. Just why fewer degrees of freedom are "active" at lower temperatures was eventually explained by Einstein using the quantum theory. [According to quantum theory, energy does not take on continuous values but is quantized—it can have only certain values, and there is a certain minimum energy. The minimum rotational and vibrational energies are higher than for simple translational kinetic energy, so at lower temperatures and lower translational kinetic energy, there is not enough energy to excite the rotational or vibrational kinetic energy.] Calculations based on kinetic theory and the principle of equipartition of energy (as modified by the quantum theory) give numerical results in accord with experiment.

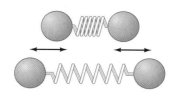

FIGURE 19–15 A diatomic molecule can vibrate, as if the two atoms were connected by a spring. Of course they are not connected by a spring, but rather exert forces on each other that are electrical in nature, but of a form that resembles a spring force.

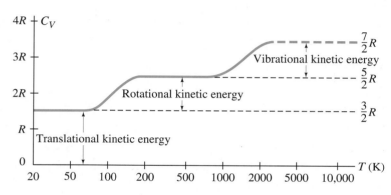

FIGURE 19–16 Molar specific heat C_V as a function of temperature for hydrogen molecules (H_2). As the temperature is increased, some of the translational kinetic energy can be transferred in collisions into rotational kinetic energy and, at still higher temperature, into vibrational kinetic energy. [Note: H_2 dissociates into two atoms at about 3200 K, so the last part of the curve is shown dashed.]

Solids

The principle of equipartition of energy can be applied to solids as well. The molar specific heat of any solid, at high temperature, is close to $3R$ (6.0 cal/mol·K), Fig. 19–17. This is called the Dulong and Petit value after the scientists who first measured it in 1819. (Note that Table 19–1 gave the specific heats per kilogram, not per mole.) At high temperatures, each atom apparently has six degrees of freedom, although some are not active at low temperatures. Each atom in a crystalline solid can vibrate about its equilibrium position as if it were connected by springs to each of its neighbors (Fig. 19–18). Thus it can have three degrees of freedom for kinetic energy and three more associated with potential energy of vibration in each of the x, y, and z directions, which is in accord with the measured values.

FIGURE 19–17 Molar heat capacities of solids as a function of temperature.

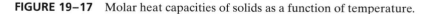

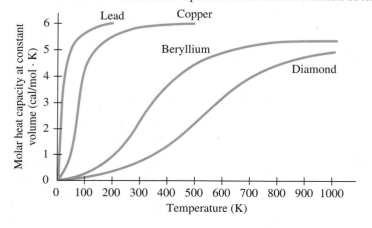

FIGURE 19–18 The atoms in a crystalline solid can vibrate about their equilibrium positions as if they were connected to their neighbors by springs. (The forces between atoms are actually electrical in nature.)

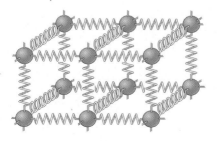

19–9 | Adiabatic Expansion of a Gas

The PV curve for the quasistatic (slow) adiabatic expansion ($Q = 0$) of an ideal gas was shown in Fig. 19–8 (curve AC). It is somewhat steeper than for an isothermal process ($\Delta T = 0$), which indicates that for the same change in volume the change in pressure will be greater. Hence the temperature of the gas must drop during an adiabatic expansion. Conversely, the temperature rises during an adiabatic compression.

We can derive the relation between the pressure P and the volume V of an ideal gas that is allowed to slowly expand adiabatically. We begin with the first law of thermodynamics, written in differential form:

$$dU = dQ - dW = -dW = -P\, dV,$$

since $dQ = 0$ for an adiabatic process. Equation 19–10 gives us a relation between ΔU and C_V, which is valid for any ideal gas process since U is a function only of T for an ideal gas. We write this in differential form:

$$dU = nC_V\, dT.$$

When we combine these last two equations, we obtain

$$nC_V\, dT + P\, dV = 0.$$

We next take the differential of the ideal gas law, $PV = nRT$, allowing P, V, and T to vary:

$$P\, dV + V\, dP = nR\, dT.$$

We solve for dT in this relation and substitute it into the previous relation and get

$$nC_V\left(\frac{P\, dV + V\, dP}{nR}\right) + P\, dV = 0$$

or

$$(C_V + R)P\, dV + C_V V\, dP = 0.$$

We note from Eq. 19–9 that $C_V + R = C_P$, so we have

$$C_P P\, dV + C_V V\, dP = 0,$$

or

$$\frac{C_P}{C_V} P\, dV + V\, dP = 0.$$

We define

$$\gamma = \frac{C_P}{C_V} \tag{19–12}$$

so that our last equation becomes

$$\frac{dP}{P} + \gamma \frac{dV}{V} = 0.$$

This is integrated to become

$$\ln P + \gamma \ln V = \text{constant}.$$

This simplifies (using the rules for addition and multiplication of logarithms) to

$$PV^\gamma = \text{constant}. \qquad \begin{bmatrix}\text{quasistatic adiabatic} \\ \text{process; ideal gas}\end{bmatrix} \tag{19–13}$$

This is the relation between P and V for a quasistatic adiabatic expansion or contraction. We will find it very useful when we discuss heat engines in the next chapter. Table 19–3 gives values of γ for some real gases. Figure 19–8 compares an adiabatic expansion (Eq. 19–13) in curve AC to an isothermal expansion

(PV = constant) in curve AB. It is important to remember that the ideal gas law, $PV = nRT$, continues to hold even for an adiabatic expansion (PV^γ = constant); clearly PV is not constant, meaning T is not constant.

EXAMPLE 19–11 **Adiabatic vs. isothermal expansion.** An ideal monatomic gas is allowed to expand slowly until its pressure is reduced to exactly half its original value. By what factor does the volume change if the process is (a) adiabatic; (b) isothermal?

SOLUTION (a) From Eq. 19–13, $P_1 V_1^\gamma = P_2 V_2^\gamma$; hence

$$\frac{V_2}{V_1} = \left(\frac{P_1}{P_2}\right)^{1/\gamma} = (2)^{3/5} = 1.52$$

since $\gamma = C_P/C_V = (5/2)/(3/2) = 5/3$.

(b) Equation 19–13 does not apply for an isothermal process. But the ideal gas law, $PV = nRT$, always applies for an ideal gas. Hence, since $T_1 = T_2$, then $P_1 V_1 = P_2 V_2$ and

$$\frac{V_2}{V_1} = \frac{P_1}{P_2} = 2.0.$$

19–10 Heat Transfer: Conduction, Convection, Radiation

Heat is transferred from one place or body to another in three different ways: by *conduction*, *convection*, and *radiation*. We now discuss each of these in turn; but in practical situations, any two or all three may be operating at the same time. We start with conduction.

Three methods of heat transfer

Conduction

When a metal poker is put in a hot fire, or a silver spoon is placed in a hot bowl of soup, the exposed end of the poker or spoon soon becomes hot as well, even though it is not directly in contact with the source of heat. We say that heat has been conducted from the hot end to the cold end.

Heat **conduction** in many materials can be visualized as the result of molecular collisions. As one end of the object is heated, the molecules there move faster and faster. As they collide with their slower-moving neighbors, they transfer some of their energy to these molecules whose speeds thus increase. These in turn transfer some of their energy by collision with molecules still farther along the object. Thus the energy of thermal motion is transferred by molecular collision along the object. In metals, according to modern theory, it is collisions of free electrons within the metal with each other and with the metal lattice atoms that are visualized as being mainly responsible for conduction.

Heat conduction takes place only if there is a difference in temperature. Indeed, it is found experimentally that the rate of heat flow through a substance is proportional to the difference in temperature between its ends. The rate of heat flow also depends on the size and shape of the object. To investigate this quantitatively, let us consider the heat flow through a uniform object, as illustrated in Fig. 19–19. It is found experimentally that the heat flow ΔQ per time interval Δt is given by

$$\frac{\Delta Q}{\Delta t} = kA \frac{T_1 - T_2}{l}, \tag{19–14a}$$

where A is the cross-sectional area of the object, l is the distance between the two ends, which are at temperatures T_1 and T_2, and k is a proportionality constant called the **thermal conductivity**, which is characteristic of the material.

FIGURE 19–19 Heat conduction between areas at temperatures T_1 and T_2. If T_1 is greater than T_2, the heat flows to the right; the rate is given by Eq. 19–14a.

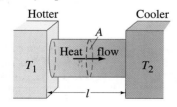

Rate of heat flow by conduction

TABLE 19–4
Thermal Conductivities

| Substance | Thermal Conductivity, k | |
	kcal/(s·m·C°)	J/(s·m·C°)
Silver	10×10^{-2}	420
Copper	9.2×10^{-2}	380
Aluminum	5.0×10^{-2}	200
Steel	1.1×10^{-2}	40
Ice	5×10^{-4}	2
Glass	2.0×10^{-4}	0.84
Brick	2.0×10^{-4}	0.84
Concrete	2.0×10^{-4}	0.84
Water	1.4×10^{-4}	0.56
Human tissue	0.5×10^{-4}	0.2
Wood	0.3×10^{-4}	0.1
Fiberglass	0.12×10^{-4}	0.048
Cork	0.1×10^{-4}	0.042
Wool	0.1×10^{-4}	0.040
Goose down	0.06×10^{-4}	0.025
Polyurethane	0.06×10^{-4}	0.024
Air	0.055×10^{-4}	0.023

FIGURE 19–20 Example 19–12.

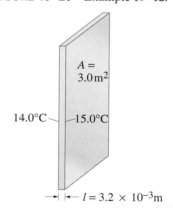

$A = 3.0\,m^2$

14.0°C — 15.0°C

$l = 3.2 \times 10^{-3}\,m$

Wind can cause much greater heat loss

➡ **PHYSICS APPLIED**

Thermal windows

In some cases (such as when k or A cannot be considered constant) we need to consider the limit of an infinitesimally thin slab of thickness dx. Then Eq. 19–14a becomes

$$\frac{dQ}{dt} = -kA\frac{dT}{dx}, \tag{19–14b}$$

where dT/dx is the temperature gradient and the negative sign is included since the heat flow is in the direction opposite to the temperature gradient.[†]

The thermal conductivities, k, for a variety of substances are given in Table 19–4. Substances for which k is large conduct heat rapidly and are said to be good **conductors**. Most metals fall in this category, although there is a wide range even among them as you may observe by holding the ends of a silver spoon and a stainless-steel spoon immersed in the same hot cup of soup. Substances for which k is small, such as fiberglass and down, are poor conductors of heat and are therefore good **insulators**. The relative magnitudes of k can explain simple phenomena such as why a tile floor feels much colder on the feet than a rug-covered floor at the same temperature. Tile is a better conductor of heat than the rug. Heat transferred from your foot to the rug is not conducted away rapidly, so the rug's surface quickly heats up to the temperature of your foot. But the tile conducts the heat away rapidly and thus can take more heat from your foot quickly, so your foot's surface temperature drops.

EXAMPLE 19–12 **Heat loss through windows.** A major source of heat loss from a house is through the windows. Calculate the rate of heat flow through a glass window 2.0 m × 1.5 m in area and 3.2 mm thick, if the temperatures at the inner and outer surfaces are 15.0°C and 14.0°C, respectively (Fig. 19–20).

SOLUTION Since $A = (2.0\,\text{m})(1.5\,\text{m}) = 3.0\,\text{m}^2$, $l = 3.2 \times 10^{-3}\,\text{m}$, and using Table 19–4 to get k, we have from Eq. 19–14a

$$\frac{\Delta Q}{\Delta t} = kA\frac{T_1 - T_2}{l} = \frac{(0.84\,\text{J/s}\cdot\text{m}\cdot\text{C°})(3.0\,\text{m}^2)(15.0°\text{C} - 14.0°\text{C})}{(3.2 \times 10^{-3}\,\text{m})}$$

$$= 790\,\text{J/s}.$$

This is equivalent to $(790\,\text{J/s})/(4.19 \times 10^3\,\text{J/kcal}) = 0.19\,\text{kcal/s}$, or $(0.19\,\text{kcal/s})\cdot(3600\,\text{s/h}) = 680\,\text{kcal/h}$.

You might notice in this Example that 15°C is not very warm for the living room of a house. The room itself may indeed be much warmer, and the outside might be colder than 14°C. But the temperatures of 15°C and 14°C were specified as those at the window surfaces, and there is usually a considerable drop in temperature of the air in the vicinity of the window both on the inside and the outside. That is, the layer of air on either side of the window acts as an insulator, and normally the major part of the temperature drop between the inside and outside of the house takes place across the air layer. If there is a heavy wind, the air outside a window will constantly be replaced with cold air; the temperature gradient across the glass will be greater and there will be a much greater rate of heat loss. Increasing the width of the air layer, such as using two panes of glass separated by an air gap, will reduce the heat loss more than simply increasing the glass thickness, since the thermal conductivity of air is much less than for glass.

[†]This is quite similar to the relations describing diffusion (Chapter 18) and the flow of fluids through a tube (Chapter 13). In those cases, the flow of matter was found to be proportional to the concentration gradient or to the pressure gradient. This close similarity is one reason we speak of the "flow" of heat. Yet we must keep in mind that no substance is flowing in this case—it is energy that is being transferred.

The insulating properties of clothing come from the insulating properties of air. Without clothes, our bodies would heat the air in contact with the skin and would soon become reasonably comfortable because air is a very good insulator. But since air moves—there are breezes and drafts, and people themselves move about—the warm air would be replaced by cold air, thus increasing the temperature difference and the heat loss from the body. Clothes keep us warm by holding air so it cannot move readily. It is not the cloth that insulates us, but the air that the cloth traps. Down is a very good insulator because even a small amount of it fluffs up and traps a great amount of air. On this basis, can you see one reason why drapes in front of a window reduce heat loss from a house?

➡ PHYSICS APPLIED

How clothes keep us warm

Clothing traps air

For practical purposes, the thermal properties of building materials, particularly when considered as insulation, are commonly specified by R-values (or "thermal resistance"), defined for a given thickness l of material as:

$$R = \frac{l}{k}.$$

The R-value of a given piece of material thus combines the thickness l and the thermal conductivity k in one number. In the United States, R-values are given in British units (although often not stated at all!), as $\text{ft}^2 \cdot \text{h} \cdot \text{F}°/\text{Btu}$. Table 19–5 gives R-values for some common building materials: note that R-values increase directly with material thickness. For example, 2 inches of fiberglass has $R = 6 \, \text{ft}^2 \cdot \text{h} \cdot \text{F}°/\text{Btu}$, half that for 4 inches.

TABLE 19–5 R-values

Material	Thickness	R-value ($\text{ft}^2 \cdot \text{h} \cdot \text{F}°/\text{Btu}$)
Glass	$\frac{1}{8}$ inch	1
Brick	$3\frac{1}{2}$ inch	0.6–1
Plywood	$\frac{1}{2}$ inch	0.6
Fiberglass insulation	4 inches	12

Convection

Although liquids and gases are generally not very good conductors of heat, they can transfer heat quite rapidly by convection. **Convection** is the process whereby heat is transferred by the mass movement of molecules from one place to another. Whereas conduction involves molecules (and/or electrons) moving only over small distances and colliding, convection involves the movement of molecules over large distances.

A forced-air furnace, in which air is heated and then blown by a fan into a room, is an example of *forced convection*. *Natural convection* occurs as well, and one familiar example is that hot air rises. For instance, the air above a radiator (or other type of heater) expands as it is heated, and hence its density decreases; because its density is less, it rises, just as a log submerged in water floats upward because its density is less than that of water. Warm or cold ocean currents, such as the balmy Gulf Stream, represent natural convection on a large scale. Wind is another example of convection, and weather in general is a result of convective air currents.

When a pot of water is heated (Fig. 19–21), convection currents are set up as the heated water at the bottom of the pot rises because of its reduced density and is replaced by cooler water from above. This principle is used in many heating systems, such as the hot-water radiator system shown in Fig. 19–22. Water is heated in the furnace and as its temperature increases, it expands and rises, as shown. This causes the water to circulate in the system. Hot water then enters the radiators, heat is transferred by conduction to the air, and the cooled water returns to the furnace. Thus, the water circulates because of convection; pumps are sometimes used to improve circulation. The air throughout the room also becomes heated as a result of convection. The air heated by the radiators rises and is replaced by cooler air, resulting in convective air currents, as shown.

Other types of furnaces also depend on convection. Hot-air furnaces with registers (openings) near the floor often do not have fans but depend on natural convection, which can be appreciable. In other systems, a fan is used. In either case, it is important that cold air can return to the furnace so that convective currents circulate throughout the room if the room is to be uniformly heated.

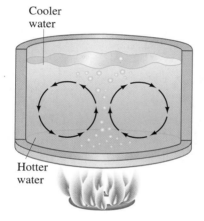

Cooler water

Hotter water

FIGURE 19–21 Convection currents in a pot of water being heated on a stove.

FIGURE 19–22 Convection plays a role in heating a house. The circular green arrows show the convective air currents in the rooms.

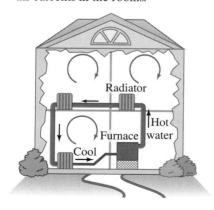

Radiator

Hot water

Furnace

Cool

Radiation

Convection and conduction require the presence of matter as a medium to carry the heat from the hotter to the colder region. But a third type of heat transfer occurs without any medium at all. All life on Earth depends on the transfer of energy from the Sun, and this energy is transferred to the Earth over empty (or nearly empty) space. This form of energy transfer is heat—since the Sun's temperature is much higher (6000 K) than Earth's—and is referred to as **radiation**. The warmth we receive from a fire is mainly radiant energy (most of the air heated by a fire rises by convection up the chimney and does not reach us).

As we shall see in later chapters, radiation involves electromagnetic waves. Suffice it to say for now that radiation from the Sun consists of visible light plus many other wavelengths that the eye is not sensitive to, including infrared (IR) radiation which is mainly responsible for heating the Earth.

The rate at which an object radiates energy has been found to be proportional to the fourth power of the Kelvin temperature, T. That is, a body at 2000 K as compared to one at 1000 K radiates energy at a rate $2^4 = 16$ times greater. The rate of radiation is also proportional to the area A of the emitting object, so the rate at which energy leaves the object, $\Delta Q / \Delta t$, is

Radiation $\propto T^4$

$$\frac{\Delta Q}{\Delta t} = e\sigma A T^4. \tag{19-15}$$

This is called the **Stefan-Boltzmann equation**, and σ is a universal constant called the **Stefan-Boltzmann constant** which has the value

Stefan-Boltzmann constant

$$\sigma = 5.67 \times 10^{-8} \, \text{W/m}^2 \cdot \text{K}^4.$$

The factor e, called the **emissivity**, is a number between 0 and 1 that is characteristic of the material. Very black surfaces, such as charcoal, have emissivity close to 1, whereas bright shiny surfaces have e close to zero and thus emit correspondingly less radiation. The value of e depends somewhat on the temperature of the body.

Not only do light shiny surfaces emit less radiation, but they absorb little of the radiation that falls upon them (most is reflected). Black and very dark objects, on the other hand, absorb nearly all the radiation that falls on them—which is why light-colored clothing is usually preferable to dark clothing on a hot day. Thus, **a good absorber is also a good emitter**.

Good absorber is good emitter

Any object not only emits energy by radiation, but also absorbs energy radiated by other bodies. If an object of emissivity e and area A is at a temperature T_1, it radiates energy at a rate $e\sigma A T_1^4$. If the object is surrounded by an environment at temperature T_2 and high emissivity (≈ 1), the rate the surroundings radiate energy is proportional to T_2^4, and the rate that energy is absorbed by the object is proportional to T_2^4. The *net* rate of radiant heat flow from the object is given by the equation

Net flow rate of heat radiation

$$\frac{\Delta Q}{\Delta t} = e\sigma A \big(T_1^4 - T_2^4\big), \tag{19-16}$$

where A is the surface area of the object, T_1 its temperature and e its emissivity (at temperature T_1), and T_2 is the temperature of the surroundings. Notice in this equation that the rate of heat absorption by an object was taken to be $e\sigma A T_2^4$. That is, the proportionality constant is the same for both emission and absorption. This must be true to correspond with the experimental fact that equilibrium between object and surroundings is reached when they come to the same temperature. That is, $\Delta Q / \Delta t$ must equal zero when $T_1 = T_2$, so the coefficients of emission and absorption terms must be the same. This confirms the idea that a good emitter is a good absorber.

Because both the object and its surroundings radiate energy, there is a net transfer of energy from one to the other unless everything is at the same temperature. From Eq. 19–16 it is clear that if $T_1 > T_2$, the net flow of heat is from the body to the surroundings, so the body cools. But if $T_1 < T_2$, the net heat flow is from the surroundings into the body, and its temperature rises. If different parts of the surroundings are at different temperatures, Eq. 19–16 becomes more complicated.

EXAMPLE 19–13 **ESTIMATE** **Two teapots.** A ceramic teapot ($e = 0.70$) and a shiny one ($e = 0.10$) each hold 0.75 L of tea at 95°C. (*a*) Estimate the rate of heat loss from each, and (*b*) estimate the temperature drop after 30 min for each. Consider only radiation and assume the surroundings are at 20°C.

SOLUTION (*a*) A teapot that holds 0.75 L can be approximated by a cube 10 cm on a side, with five sides exposed, so its surface area would be about $5 \times 10^{-2} \text{ m}^2$. The rate of heat loss would be about

$$\frac{\Delta Q}{\Delta t} = e\sigma A(T_1^4 - T_2^4)$$
$$= e(5.67 \times 10^{-8} \text{ W/m}^2 \cdot \text{K}^4)(5 \times 10^{-2} \text{ m}^2)[(368 \text{ K})^4 - (293 \text{ K})^4]$$
$$= e(30) \text{ W},$$

or about 20 W for the ceramic pot ($e = 0.70$) and 3 W for the shiny one ($e = 0.10$). (*b*) To estimate the temperature drop, we use the concept of specific heat and ignore the contribution of the pots compared to that of the 0.75 L of water. Then, using Eq. 19–2,

$$\frac{\Delta T}{\Delta t} = \frac{\Delta Q/\Delta t}{mc} = \frac{e(30) \text{ J/s}}{(0.75 \text{ kg})(4.19 \times 10^3 \text{ J/kg} \cdot \text{C}°)} = e(0.010) \text{ C}°\text{/s},$$

which for 30 min (1800 s) represents about 12 C° for the ceramic pot and about 2 C° for the shiny one. The shiny one clearly has an edge, at least as far as radiation is concerned. However, convection and conduction could play a greater role than radiation.

Heating of an object by radiation from the Sun cannot be calculated using Eq. 19–16 since this equation assumes a uniform temperature, T_2, of the environment surrounding the object, whereas the Sun is essentially a point source. Hence the Sun must be treated as a separate source of energy. Heating by the Sun is calculated using the fact that about 1350 J of energy strikes the atmosphere of the Earth from the Sun per second per square meter of area at right angles to the Sun's rays. This number, 1350 W/m², is called the **solar constant**. The atmosphere may absorb as much as 70 percent of this energy before it reaches the ground, depending on the cloud cover. On a clear day, about 1000 W/m² reaches the Earth's surface. An object of emissivity e with area A facing the Sun absorbs heat at a rate, in watts, of about

$$\frac{\Delta Q}{\Delta t} = (1000 \text{ W/m}^2)eA \cos\theta,$$

where θ is the angle between the Sun's rays and a line perpendicular to the area A (Fig. 19–23). That is, $A \cos\theta$ is the "effective" area, at right angles to the Sun's rays. The explanations for the seasons and the polar ice caps (Fig. 19–24), and why the Sun heats the Earth more at midday than at sunrise or sunset, are also related to this $\cos\theta$ factor.

Notice that if a person wears light-colored clothing, e is much smaller, so the energy absorbed is less.

An interesting application of thermal radiation to diagnostic medicine is **thermography**. A special instrument, the thermograph, scans the body, measuring the intensity of radiation from many points and forming a picture that resembles an X-ray (Fig. 19–25). Areas where metabolic activity is high, such as in tumors, can often be detected on a thermogram as a result of their higher temperature and consequent increased radiation.

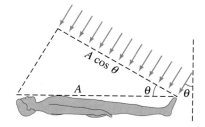

FIGURE 19–23 Radiant energy striking a body at an angle θ.

Radiation from the Sun

FIGURE 19–24 June sun makes an angle of about 23° with the equator. Thus (a) θ in the southern United States is near 0° (direct summer sun), whereas (b) in the southern hemisphere, θ is 50° or 60°, and less heat can be absorbed—hence it is winter. (c) At the poles, there is never strong direct sun; $\cos\theta$ varies from about $\frac{1}{2}$ in summer to 0 in winter; with so little heating, ice can form.

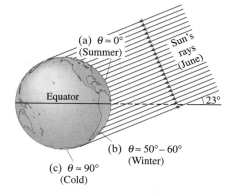

FIGURE 19–25 Thermograms of a healthy person's arms and hands (a) before smoking, (b) after smoking a cigarette, showing temperature decrease due to impaired blood circulation associated with smoking. The thermograms have been color-coded according to temperature; the scale on the right goes from blue (cold) to white (hot).

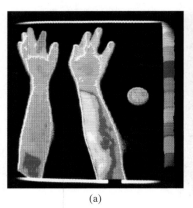
(a)

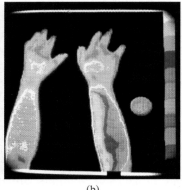
(b)

➡ PHYSICS APPLIED

Astronomy—size of a star

EXAMPLE 19–14 ESTIMATE Star radius. The giant star Betelgeuse emits radiant energy at a rate 10^4 times greater than our Sun, whereas its surface temperature is only half (2900 K) that of our Sun. Estimate the radius of Betelgeuse, assuming $e = 1$. The Sun's radius is $r_S = 7 \times 10^8$ m.

SOLUTION We assume both Betelgeuse and the Sun are spherical, with surface area $4\pi r^2$. We solve Eq. 19–15 for A:

$$4\pi r^2 = A = \frac{(\Delta Q / \Delta t)}{e\sigma T^4}.$$

Then

$$\frac{r_B^2}{r_S^2} = \frac{(\Delta Q / \Delta t)_B}{(\Delta Q / \Delta t)_S} \cdot \frac{T_S^4}{T_B^4} = (10^4)(2^4) = 16 \times 10^4.$$

Hence $r_B = \sqrt{16 \times 10^4}\, r_S = (400)(7 \times 10^8\,\text{m}) \approx 3 \times 10^{11}$ m. If Betelgeuse were our Sun, it would envelop us $\left(\text{Earth is } 1.5 \times 10^{11}\,\text{m from the Sun}\right)$.

Summary

Thermal energy, or **internal energy**, U, refers to the total energy of all the molecules in a body. For an ideal monatomic gas

$$U = \tfrac{3}{2} NkT = \tfrac{3}{2} nRT.$$

Heat refers to the transfer of energy from one body to another because of a difference of temperature. Heat is thus measured in energy units, such as joules.

Heat and thermal energy are also sometimes specified in calories or kilocalories, where

$$1\,\text{cal} = 4.186\,\text{J}$$

is the amount of heat needed to raise the temperature of 1 g of water by 1 C°.

The **specific heat**, c, of a substance is defined as the energy (or heat) required to change the temperature of unit mass of substance by 1 degree; as an equation,

$$Q = mc\,\Delta T,$$

where Q is the heat absorbed or given off, ΔT is the temperature increase or decrease, and m is the mass of the substance.

When heat flows within an isolated system, conservation of energy tells us that the heat gained by one part of the system is equal to the heat lost by the other part of the system; this is the basis of **calorimetry**, which is the quantitative measurement of heat exchange.

Exchange of energy occurs, without a change in temperature, whenever a substance changes phase. The **heat of fusion** is the heat required to melt 1 kg of a solid into the liquid phase; it is also equal to the heat given off when the substance changes from liquid to solid. The **heat of vaporization** is the energy required to change 1 kg of a substance from the liquid to the vapor phase; it is also the energy given off when the substance changes from vapor to liquid.

The **first law of thermodynamics** states that the change in internal energy, ΔU, of a system is equal to the heat added to the system, Q, minus the work, W, done by the system:

$$\Delta U = Q - W.$$

This important law is a broad restatement of the conservation of energy and is found to hold for all processes.

Two simple thermodynamic processes are **isothermal**, which is a process carried out at constant temperature, and **adiabatic**, a process in which no heat is exchanged.

The work done by (or on) a gas to change its volume by dV is $dW = P\,dV$, where P is the pressure.

Work and heat are not functions of the state of a system (as are P, V, T, n, and U) but depend on the type of process that takes a system from one state to another.

The **molar specific heat** of an ideal gas at constant volume, C_V, and at constant pressure, C_P, are related by $C_P - C_V = R$, where R is the gas constant. For a monatomic ideal gas, $C_V = \frac{3}{2}R$.

For ideal gases made up of diatomic or more complex molecules, C_V is equal to $\frac{1}{2}R$ times the number of **degrees of freedom** of the molecule. Unless the temperature is very high, some of the degrees of freedom may not be active and so do not contribute. According to the **principle of equipartition of energy**, energy is shared equally among the active degrees of freedom in an amount $\frac{1}{2}kT$ per molecule on average.

When an ideal gas expands (or contracts) adiabatically ($Q = 0$), the relation $PV^\gamma = $ constant holds, where $\gamma = C_P/C_V$.

Heat is transferred from one place (or body) to another in three different ways. In **conduction**, energy is transferred from higher-kinetic energy molecules or electrons to lower-kinetic energy neighbors when they collide.

Convection is the transfer of energy by the mass movement of molecules over considerable distances.

Radiation, which does not require the presence of matter, is energy transfer by electromagnetic waves, such as from the Sun. All bodies radiate energy in an amount that is proportional to the fourth power of their Kelvin temperature (T^4) and to their surface area. The energy radiated (or absorbed) also depends on the nature of the surface (dark surfaces absorb and radiate more than bright shiny ones), which is characterized by the emissivity, e.

☐ Questions

1. Is heat really involved in the Joule experiment, Fig. 19–1?
2. What happens to the work done when a jar of orange juice is vigorously shaken?
3. When a hot object warms a cooler object, does temperature flow between them? Are the temperature changes of the two objects equal?
4. (a) If two objects of different temperature are placed in contact, will heat naturally flow from the object with higher internal energy to the object with lower internal energy? (b) Is it possible for heat to flow even if the internal energies of the two objects are the same? Explain.
5. In warm regions where tropical plants grow but the temperature may drop below freezing a few times in the winter, the destruction of sensitive plants due to freezing can be reduced by watering them in the evening. Explain.
6. The specific heat of water is quite large. Explain why this fact makes water particularly good for heating systems (that is, hot-water radiators).
7. Why does water in a canteen stay cooler if the cloth jacket surrounding the canteen is kept moist?
8. Explain why burns caused by steam on the skin are often more severe than burns caused by water at 100°C.
9. Explain, using the concepts of latent heat and internal energy, why water cools (its temperature drops) when it evaporates.
10. Will potatoes cook faster if the water is boiling faster?
11. The temperature very high in the Earth's atmosphere can be 700°C. Yet an animal there would freeze to death rather than roast. Explain.
12. What happens to the internal energy of water vapor in the air that condenses on the outside of a cold glass of water? Is work done or heat exchanged? Explain.
13. Use the conservation of energy to explain why the temperature of a gas increases when it is compressed—say, by pushing down on a cylinder—whereas the temperature decreases when the gas expands.
14. In an isothermal process, 3700 J of work is done by an ideal gas. Is this enough information to tell how much heat has been added to the system? If so, how much?
15. One liter of air is cooled at constant pressure until its volume is halved. Then it is allowed to expand isothermally back to its original volume. Draw the process on a PV diagram.
16. Is it possible for the temperature of a system to remain constant even though heat flows into or out of it? If so, give examples.
17. Discuss how the first law of thermodynamics can apply to metabolism in humans. In particular, note that a person does work W, but very little heat Q is added to the body (rather, it tends to flow out). Why then doesn't the internal energy drop drastically in time?
18. Explain in words why C_P is greater than C_V.
19. Explain why the temperature of a gas increases when it is adiabatically compressed.
20. An ideal monatomic gas is allowed to expand slowly to twice its volume (1) isothermally; (2) adiabatically; (3) isobarically. Plot each on a PV diagram. In which process is ΔU the greatest, and in which is ΔU the least? In which is W the greatest and the least? In which is Q the greatest and the least?
21. Ceiling fans are sometimes reversible, so that they drive the air down in one season and pull it up in another season. Which way should you set the fan for summer? For winter?
22. Down sleeping bags and parkas are often specified as so many inches or centimeters of *loft*, the actual thickness of the garment when it is fluffed up. Explain.
23. Microprocessor chips nowadays have a "heat sink" glued on top that looks like a series of fins. Why is it shaped like that?
24. Sea breezes are often encountered on sunny days at the shore of a large body of water. Explain in light of the fact that the temperature of the land rises more rapidly than that of the nearby water.
25. The Earth cools off at night much more quickly when the weather is clear than when cloudy. Why?
26. Explain why air-temperature readings are always taken with the thermometer in the shade.
27. A premature baby in an incubator can be dangerously cooled even when the air temperature in the incubator is warm. Explain.

28. Heat loss occurs through windows by the following processes: (1) ventilation around edges; (2) through the frame, particularly if it is metal; (3) through the glass panes and (4) radiation. (a) For the first three, what is (are) the mechanism(s): conduction, convection, or radiation? (b) Heavy curtains reduce which of these heat losses? Explain in detail.

29. A piece of wood lying in the Sun absorbs more heat than a piece of shiny metal. Yet the wood feels less hot than the metal when you pick it up. Explain.

30. Explain why cities situated on the ocean tend to have less extreme temperatures than inland cities at the same latitude.

31. Early in the day, after the sun has reached the slope of a mountain, there tends to be a gentle upward movement of air. Later, after a slope goes into shadow, there is a gentle downdraft. Explain.

32. In the Northern Hemisphere the amount of heat required to heat a room where the windows face north is much higher than that required where the windows face south. Explain.

Problems

Section 19–1

1. (I) How much heat (joules) is required to raise the temperature of 30.0 kg of water from 15°C to 95°C?

2. (I) To what temperature will 7700 J of heat raise 3.0 kg of water initially at 10.0°C?

3. (II) When a 3.0-g bullet, traveling with a speed of 400 m/s, passes through a tree, its speed is reduced to 200 m/s. How much heat Q is produced and shared by the bullet and the tree?

4. (II) A British thermal unit (Btu) is a unit of heat in the British system of units. One Btu is defined as the heat needed to raise 1 pound of water by 1 F°. Show that

$$1 \text{ Btu} = 0.252 \text{ kcal} = 1055 \text{ J}.$$

5. (II) A water heater can generate 7200 kcal/h. How much water can it heat from 15°C to 50°C per hour?

6. (II) A small immersion heater is rated at 350 W. Estimate how long it will take to heat a cup of soup (assume this is 250 mL of water) from 20°C to 60°C.

7. (II) How many kilocalories of heat are generated when the brakes are used to bring a 1000-kg car to rest from a speed of 95 km/h?

Sections 19–3 and 19–4

8. (I) What is the specific heat of a metal substance if 135 kJ of heat is needed to raise 5.1 kg of the metal from 20°C to 30°C?

9. (I) An automobile cooling system holds 16 L of water. How much heat does it absorb if its temperature rises from 20°C to 90°C?

10. (II) A 35-g glass thermometer reads 21.6°C before it is placed in 135 mL of water. When the water and thermometer come to equilibrium, the thermometer reads 39.2°C. What was the original temperature of the water?

11. (II) The 1.20-kg head of a hammer has a speed of 6.5 m/s just before it strikes a nail (Fig. 19–26) and is brought to rest. Estimate the temperature rise of a 14-g iron nail generated by ten such hammer blows done in quick succession. Assume the nail absorbs all the energy.

12. (II) What will be the equilibrium temperature when a 245-g block of copper at 300°C is placed in a 150-g aluminum calorimeter cup containing 820 g of water at 12.0°C?

13. (II) A hot iron horseshoe (mass = 0.40 kg) which has just been forged, is dropped into 1.35 L of water in a 0.30-kg iron pot initially at 20°C. If the final equilibrium temperature is 25°C, estimate the initial temperature of the hot horseshoe.

14. (II) When 215 g of a substance is heated to 330°C and then plunged into a 100-g aluminum calorimeter cup containing 150 g of water at 12.5°C, the final temperature, as registered by a 17-g glass thermometer, is 35.0°C. What is the specific heat of the substance?

15. (II) How long does it take a 750-W coffeepot to bring to a boil 0.75 L of water initially at 8.0°C? Assume that the part of the pot which is heated with the water is made of 360 g of aluminum, and that no water boils away.

16. (II) Estimate the Calorie content of 100 g of Mullaney's fudge brownie from the following measurements. A 10-g sample of the brownie is allowed to dry before putting it in a bomb calorimeter. The aluminum bomb has a mass of 0.615 kg and is placed in 2.00 kg of water contained in an aluminum calorimeter cup of mass 0.524 kg. The initial temperature of the mixture is 15.0°C and its temperature after ignition is 36.0°C.

17. (II) (a) Show that if the specific heat varies as a function of temperature, $c(T)$, the heat needed to raise the temperature of a substance from T_1 to T_2 is given by

$$Q = \int_{T_1}^{T_2} mc(T) \, dT.$$

(b) Suppose $c(T) = c_0(1 + aT)$ for some substance, where $a = 2.0 \times 10^{-3} \text{ C°}^{-1}$ and T is the Celsius temperature. Determine the heat required to raise the temperature from T_1 to T_2. (c) What is the mean value of c over the range T_1 to T_2 for part (b), expressed in terms of c_0, the heat capacity at 0°C?

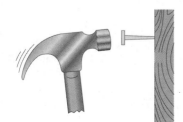

FIGURE 19–26
Problem 11.

18. (I) How much heat is needed to melt 15.50 kg of silver that is initially at 20°C?

19. (I) During exercise, a person may give off 180 kcal of heat in 30 min by evaporation of water from the skin. How much water has been lost?

20. (I) If 2.80×10^5 J of energy is supplied to a flask of liquid oxygen at −183°C, how much oxygen will evaporate?

21. (II) What will be the final result when equal amounts of ice at 0°C and steam at 100°C are mixed together?

22. (II) A 40-g ice cube at its melting point is dropped into an insulated container of liquid nitrogen. How much nitrogen evaporates if it is at its boiling point of 77 K and has a latent heat of vaporization of 200 kJ/kg? Assume for simplicity that the specific heat of ice is a constant and is equal to its value near its melting point.

23. (II) A cube of ice is taken from the freezer at −8.5°C and placed in a 75-g aluminum calorimeter filled with 300 g of water at room temperature of 20°C. The final situation is observed to be all water at 17°C. What was the mass of the ice cube?

24. (II) An iron boiler of mass 230 kg contains 760 kg of water at 20°C. A heater supplies energy at the rate of 52,000 kJ/h. How long does it take for the water (a) to reach the boiling point, and (b) to all have changed to steam?

25. (II) On a hot day's race, a bicyclist consumes 8.0 L of water over the span of four hours. Making the approximation that all of the cyclist's energy goes into evaporating this water as sweat, how much energy in kcal did the rider use during the ride? (Since the efficiency of the rider is only about 20 percent, most of the energy consumed does go to heat, so our approximation is not far off.)

26. (II) What mass of steam at 100°C must be added to 1.00 kg of ice at 0°C to yield liquid water at 20°C?

27. (II) The specific heat of mercury is 138 J/kg·C°. Determine the latent heat of fusion of mercury using the following calorimeter data: 1.00 kg of solid Hg at its melting point of −39.0°C is placed in a 0.620-kg aluminum calorimeter with 0.430 kg of water at 12.80°C; the resulting equilibrium temperature is 5.06°C.

28. (II) A 54.0-kg ice-skater moving at 4.8 m/s glides to a stop. Assuming the ice is at 0°C and that 50 percent of the heat generated by friction is absorbed by the ice, how much ice melts?

29. (II) At a crime scene, the forensic investigator notes that the 8.2-g lead bullet that was stopped in a door-frame apparently melted completely on impact. Assuming the bullet was fired at room temperature (20°C), what does the investigator calculate the *minimum* muzzle velocity of the gun was?

Sections 19–6 and 19–7

30. (I) In Example 19–9, if the heat lost from the gas in the process BD is 2.78×10^3 J, what is the change in internal energy of the gas?

31. (I) An ideal gas expands isothermally, performing 5.00×10^3 J of work in the process. Calculate (a) the change in internal energy of the gas, and (b) the heat absorbed during this expansion.

32. (I) 1.0 L of air initially at 6.0 atm of (absolute) pressure is allowed to expand isothermally until the pressure is 1.0 atm. It is then compressed at constant pressure to its initial volume and lastly is brought back to its original pressure by heating at constant volume. Draw the process on a PV diagram, including numbers and labels for the axes.

33. (I) Sketch a PV diagram of the following process: 2.0 L of ideal gas at atmospheric pressure are cooled at constant pressure to a volume of 1.0 L, and then expanded isothermally back to 2.0 L, whereupon the pressure is increased at constant volume until the original pressure is reached.

34. (I) A gas is enclosed in a cylinder fitted with a light frictionless piston and maintained at atmospheric pressure. When 1400 kcal of heat is added to the gas, the volume is observed to increase slowly from 12.0 m³ to 18.2 m³. Calculate (a) the work done by the gas and (b) the change in internal energy of the gas.

35. (II) An ideal gas has its pressure cut in half slowly, while being kept in a rigid wall container. In the process, 1300 kJ of heat left the gas. (a) How much work was done during this process? (b) What was the change in internal energy of the gas during this process?

36. (II) In an engine, an ideal gas is compressed adiabatically to half its volume. In doing so, 2350 J of work is done on the gas. (a) How much heat flows into or out of the gas? (b) What is the change in internal energy of the gas? (c) Does its temperature rise or fall?

37. (II) An ideal gas expands at a constant pressure of 5.0 atm from 400 mL to 710 mL. Heat then flows out of the gas, at constant volume, and the pressure and temperature are allowed to drop until the temperature reaches its original value. Calculate (a) the total work done by the gas in the process, and (b) the total heat flow into the gas.

38. (II) Consider the following two-step process. Heat is allowed to flow out of an ideal gas at constant volume so that its pressure drops from 2.2 atm to 1.5 atm. Then the gas expands at constant pressure, from a volume of 6.8 L to 10.0 L, where the temperature reaches its original value. See Fig. 19–27. Calculate (a) the total work done by the gas in the process, (b) the change in internal energy of the gas in the process, and (c) the total heat flow into or out of the gas.

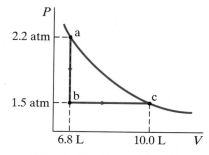

FIGURE 19–27 Problem 38.

39. (II) Suppose 2.00 mol of an ideal gas of volume $V_1 = 3.50$ m³ at $T_1 = 300$ K is allowed to expand isothermally to $V_2 = 7.00$ m³ at $T_2 = 300$ K. Determine (a) the work done by the gas, (b) the heat added to the gas, and (c) the change in internal energy of the gas.

40. (II) Determine (a) the work done and (b) the change in internal energy of 1.00 kg of water when it is all boiled to steam at 100°C. Assume a constant pressure of 1.00 atm.

41. (II) How much work is done to slowly compress, isothermally, 2.50 L of nitrogen at 0°C and 1 atm to 1.50 L at 0°C?

42. (II) The PV diagram in Fig. 19–28 shows two possible states of a system containing 1.45 moles of a monatomic ideal gas. ($P_1 = P_2 = 450 \text{ N/m}^2$, $V_1 = 2.00 \text{ m}^3$, $V_2 = 8.00 \text{ m}^3$). (a) Draw the process which depicts an isobaric expansion from state 1 to state 2 and label this process (A). (b) Find the work done by the gas and the change in internal energy of the gas in process (A). (c) Draw the process which depicts an isothermal expansion from state 1 to the volume V_2 followed by an isochoric increase in temperature to state 2 and label this two-step process (B). (d) Find the change in internal energy of the gas for the two-step process (B).

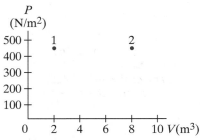

FIGURE 19–28 Problem 42.

43. (II) Show that Eqs. 19–4 and 19–5 are valid for any shape volume that changes. To do so, draw an arbitrary closed curve to represent the boundary of the volume; then draw a slightly larger curve to represent an increase in volume; choose a small section of the original boundary, of area ΔA, and show that $dW = P\,\Delta A\,dl = P\,dV$ for this section; then integrate over the whole boundary, and finally integrate over a finite volume.

44. (II) When a gas is taken from a to c along the curved path in Fig. 19–29, the work done by the gas is $W = -35 \text{ J}$ and the heat added to the gas is $Q = -63 \text{ J}$. Along path abc, the work done is $W = -48 \text{ J}$. (a) What is Q for path abc? (b) If $P_c = \frac{1}{2}P_b$, what is W for path cda? (c) What is Q for path cda? (d) What is $U_a - U_c$? (e) If $U_d - U_c = 5 \text{ J}$, what is Q for path da?

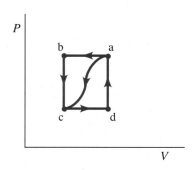

FIGURE 19–29 Problems 44, 45, and 46.

45. (III) In the process of taking a gas from state a to state c along the curved path shown in Fig. 19–29, 80 J of heat leave the system and 55 J of work are done *on* the system. (a) Determine the change in internal energy, $U_a - U_c$. (b) When the gas is taken along the path cda, the work done by the gas is $W = 38 \text{ J}$. How much heat Q is added to the gas in the process cda? (c) If $P_a = 2.5P_d$, how much work is done by the gas in the process abc? (d) What is Q for path abc? (e) If $U_a - U_b = 10 \text{ J}$, what is Q for the process bc? Here is a summary of what is given:

$$Q_{a\to c} = -80 \text{ J} \qquad U_a - U_b = 10 \text{ J}$$
$$W_{a\to c} = -55 \text{ J} \qquad P_a = 2.5P_d$$
$$W_{cda} = 38 \text{ J}$$

46. (III) Suppose a gas is taken clockwise around the rectangular cycle shown in Fig. 19–29, starting at b, then to a, to d, to c, and returning to b. Using the values given in Problem 45. (a) describe each leg of the process, and then calculate (b) the net work done during the cycle, (c) the net heat flow during the cycle, and (d) the total internal energy change during the cycle. (e) What percentage of the *intake* heat was turned into usable work: i.e., how efficient is this "rectangular" cycle (give as a percentage)?

* 47. (III) Determine the work done by 1.00 mol of a van der Waals gas (Section 18–5) when it expands from volume V_1 to V_2 isothermally.

Section 19–8

48. (I) What is the internal energy of 4.50 mol of an ideal diatomic gas at 600 K, assuming all degrees of freedom are active?

49. (I) If a heater supplies $1.8 \times 10^6 \text{ J/h}$ to a room $6.5 \text{ m} \times 4.6 \text{ m} \times 3.0 \text{ m}$ containing air at 20°C and 1.0 atm, by how much will the temperature rise in one hour, assuming no heat losses to the outside?

50. (I) Show that if the molecules of a gas have n degrees of freedom, then theory predicts $C_V = (n/2)R$ and $C_P = [(n + 2)/2]R$.

51. (I) Estimate the molar specific heat and specific heat at both constant pressure and constant volume for hydrogen gas (H_2) at room temperature.

52. (II) Show that the work done by n moles of an ideal gas when it expands adiabatically is $W = nC_V(T_1 - T_2)$, where T_1 and T_2 are the initial and final temperatures, and C_V is the molar specific heat at constant volume.

53. (II) A certain gas has specific heat $c_V = 0.0356 \text{ kcal/kg} \cdot \text{C}°$, which changes little over a wide temperature range. What is the atomic mass of this gas? What gas is it?

54. (II) The specific heat at constant volume of a particular gas is $0.182 \text{ kcal/kg} \cdot \text{K}$ at room temperature, and its molecular mass is 34. (a) What is its specific heat at constant pressure? (b) What do you think is the molecular structure of this gas?

55. (II) An audience of 2500 fills a concert hall of volume 30,000 m³. If there were no ventilation, by how much would the temperature rise over a period of 2.0 h due to the metabolism of the people (70 W/person)?

56. (II) A sample of 770 mol of nitrogen gas is maintained at a constant pressure of 1.00 atm in a flexible container. The gas is heated from 40°C to 180°C. Calculate (a) the heat added to the gas, (b) the work done by the gas, and (c) the change in internal energy.

57. (II) One mole of N_2 gas at 0°C is heated to 100°C at constant pressure (1.00 atm). Determine (a) the change in internal energy, (b) the work the gas does, and (c) the heat added to it.

58. (III) A 1.00-mol sample of an ideal diatomic gas at a pressure of 1.00 atm and temperature of 490 K undergoes a process in which its pressure increases linearly with temperature. The final temperature and pressure are 720 K and 1.60 atm. Determine (a) the change in internal energy, (b) the work done by the gas, and (c) the heat added to the gas. (Assume 5 active degrees of freedom.)

Section 19–9

59. (I) A 1.00-mol sample of an ideal diatomic gas, originally at 1.00 atm and 20°C, expands adiabatically to twice its volume. What are the final pressure and temperature for the gas? (Assume no molecular vibration.)

60. (II) Show, using Eqs. 19–4 and 19–13, that the work done by a gas that slowly expands adiabatically from pressure P_1 and volume V_1, to P_2 and V_2, is given by $W = (P_1V_1 - P_2V_2)/(\gamma - 1)$.

61. (II) 1.0 mol of an ideal monatomic gas at 300 K and 3.0 atm expands adiabatically to a final pressure of 1.0 atm. How much work does the gas do in the expansion?

62. (II) An ideal gas at 400 K is expanded adiabatically to 4.2 times its original volume. Determine its resulting temperature if the gas is (a) monatomic; (b) diatomic (no vibrations); (c) diatomic (molecules do vibrate).

63. (II) A 4.65-mol sample of an ideal diatomic gas expands adiabatically from a volume of 0.1210 m³ to 0.750 m³. Initially the pressure was 1.00 atm. Determine: (a) the initial and final temperatures; (b) the change in internal energy; (c) the heat lost by the gas; (d) the work done on the gas. (Assume no molecular vibration.)

64. (II) An ideal monatomic gas, consisting of 2.6 mol of volume 0.084 m³, expands adiabatically. The initial and final temperatures are 25°C and −68°C. What is the final volume of the gas?

65. (III) A 1.00-mol sample of an ideal monatomic gas, originally at a pressure of 1.00 atm, undergoes a three-step process: (1) it is expanded adiabatically from $T_1 = 550$ K to $T_2 = 389$ K; (2) it is compressed at constant pressure until its temperature reaches T_3; (3) it then returns to its original pressure and temperature by a constant-volume process. (a) Plot these processes on a PV diagram. (b) Determine T_3. (c) Calculate the change in internal energy, the work done by the gas, and the heat added to the gas for each process, and (d) for the complete cycle.

Section 19–10

66. (I) Calculate the rate of heat flow by conduction in Example 19–12 assuming there are strong gusty winds and that the external temperature is −5°C.

67. (I) (a) How much power is radiated by a tungsten sphere (emissivity $e = 0.35$) of radius 18.0 cm at a temperature of 25°C? (b) If the sphere is enclosed in a room whose walls are kept at −5°C, what is the net flow rate of energy out of the sphere?

68. (I) Over what distance must there be heat flow by conduction from the blood capillaries beneath the skin to the surface if the temperature difference is 0.50 C°? Assume 200 W must be tranferred through the whole body's surface area of 1.5 m².

69. (I) What is the rate of energy absorption from the Sun by a person lying flat on the beach on a clear day if the Sun makes a 30° angle with the vertical? Assume that $e = 0.70$, the area of the body exposed to the Sun is 0.80 m², and that 1000 W/m² reaches the Earth's surface.

70. (II) Two rooms, each a cube 4.0 m on a side, share a 15-cm thick brick wall. Because of a number of 100-W light-bulbs in one room, the air is at 30°C, while in the other room it is at 10°C. How many of the 100-W light-bulbs are needed to maintain the temperature difference across the wall?

71. (II) How long does it take the Sun to melt a block of ice at 0°C with an area of 1.0 m² and thickness 1.6 cm? Assume that the Sun's rays make an angle of 30° with the normal to the area, and that the emissivity of ice is 0.050.

72. (II) A copper rod and an aluminum rod of the same length and cross-sectional area are attached end to end (Fig. 19–30). The copper end is placed in a furnace which is maintained at a constant temperature of 250°C. The aluminum end is placed in an ice bath held at constant temperature of 0.0°C. Calculate the temperature at the point where the two rods are joined.

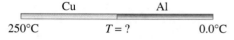

FIGURE 19–30 Problem 72.

73. (II) (a) Estimate, using the solar constant, the rate at which the whole Earth receives energy from the Sun. (b) Assume the Earth radiates an equal amount back into space (that is, the Earth is in equilibrium). Then, assuming the Earth is a perfect emitter ($e = 1.0$), estimate its average surface temperature.

74. (II) Suppose the insulating qualities of the wall of a house come mainly from a 4.0-inch layer of brick and an R-19 layer of insulation, as shown in Fig. 19–31. What is the total rate of heat loss through such a wall, if its total area is 240 ft^2 and the temperature difference across it is 10 F°?

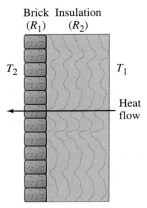
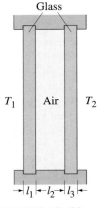

FIGURE 19–31 Two layers of insulation of a wall of a building. Problem 74.

FIGURE 19–32 Problem 75.

75. (II) A double-glazed window is one with two panes of glass separated by an air space, Fig. 19–32. (*a*) Show that the rate of heat flow by conduction is given by

$$\frac{\Delta Q}{\Delta t} = \frac{A(T_2 - T_1)}{l_1/k_1 + l_2/k_2 + l_3/k_3},$$

where $k_1, k_2,$ and k_3 are the thermal conductivities for glass, air, and glass, respectively. (*b*) Generalize this expression for any number of materials placed next to one another.

76. (III) Approximately how long should it take 8.50 kg of ice at 0°C to melt when it is placed in a carefully sealed Styrofoam icebox of dimensions 25 cm × 35 cm × 50 cm whose walls are 1.5 cm thick? Assume that the conductivity of Styrofoam is double that of air and that the outside temperature is 30°C.

77. (III) A house thermostat is normally set to 22°C, but at night it is turned down to 12°C for 7.0 h. Estimate how much more heat would be needed (state as a percentage of daily usage) if the thermostat were not turned down at night. Assume that the outside temperature averages 0°C for the 7.0 h at night and 8°C for the remainder of the day, and that the heat loss from the house is proportional to the difference in temperature inside and out. To obtain an estimate from the data, you will have to make other simplifying assumptions; state what these are.

78. (III) A cylindrical pipe has inner radius R_1 and outer radius R_2. The interior of the pipe carries hot water at temperature T_1. The temperature outside is T_2 $(< T_1)$. (*a*) Show that the rate of heat loss for a length L of pipe is

$$\frac{dQ}{dt} = \frac{2\pi k(T_1 - T_2)L}{\ln(R_2/R_1)},$$

where k is the thermal conductivity of the pipe. (*b*) Suppose the pipe is steel with $R_1 = 3.0$ cm, $R_2 = 4.0$ cm, and $T_2 = 30$°C. If the pipe holds still water at $T_1 = 71$°C, what will be the initial rate of change of its temperature? (*c*) Suppose water at 71°C enters the pipe and moves at a speed of 8.0 cm/s. What will be its temperature drop per centimeter of travel?

General Problems

79. To get an idea of how much thermal energy is contained in the world's oceans, estimate the heat liberated when a cube, 1 km on a side, of ocean water is cooled by 1 K. (Approximate the ocean water by pure water for this estimate.)

80. A 15-g lead bullet is tested by firing it into a fixed block of wood with a mass of 0.92 kg. If the block and imbedded bullet together absorb all the heat energy generated and, after thermal equilibrium has been reached, the system has a temperature rise measured as 0.020 C°, estimate the entering speed of the bullet.

81. The specific heat of mercury is 0.033 kcal/kg·C°. When 1.00 kg of solid mercury at its melting point of −39°C is placed in a 0.50-kg aluminum calorimeter filled with 1.20 kg of water at 20.0°C, the final temperature of the mixture is found to be 16.5°C. What is the heat of fusion of mercury in kcal/kg?

82. A scuba diver releases a 3.00-cm-diameter (spherical) bubble of air from a depth of 14.0 m. Assume the temperature is constant at 298 K, and that the air behaves as a perfect gas. (*a*) How large is the bubble when it reaches the surface? (*b*) Sketch a PV diagram for the process. (*c*) Apply the first law of thermodynamics to the bubble, and find the work done by the air in rising to the surface, the change in its internal energy, and the heat added or removed from the air in the bubble as it rises. Take the density of water to be 1000 kg/m^3.

83. A reciprocating compressor is a device that compresses air by a back-and-forth straight-line motion, like a piston in a cylinder. Consider a reciprocating compressor running at 400 rpm. During a compression stroke, 1.00 mol of air is compressed. The initial temperature of the air is 390 K, the engine of the compressor is supplying 10 kW of power to compress the air, and heat is being removed at the rate of 1.5 kW. Calculate the temperature change per compression stroke.

84. Suppose 2.0 mol of neon (an ideal monatomic gas) at STP are compressed slowly and isothermally to $\frac{1}{3}$ the original volume. The gas is then allowed to expand quickly and adiabatically back to its original volume. Find the highest and lowest temperatures and pressures attained by the gas, and show on a PV diagram where these values occur.

85. At very low temperatures, the molar specific heat of many substances varies as the cube of the absolute temperature:

$$C = k\frac{T^3}{T_0^3},$$

which is sometimes called Debye's law. For rock salt, $T_0 = 281$ K and $k = 1940$ J/mol·K. Determine the heat needed to raise 3.5 mol of salt from 22.0 K to 55.0 K.

86. (*a*) Find the total power radiated into space by the Sun, assuming it to be a perfect emitter at $T = 5500$ K. The Sun's radius is 7.0×10^8 m. (*b*) From this, determine the power per unit area arriving at the Earth, 1.5×10^{11} m away (Fig. 19–33).

$$r = 1.5 \times 10^{11} \text{ m}$$

Sun ———————————— Earth   **FIGURE 19–33**
Problem 86.

87. During light activity, a 70-kg person may generate 200 kcal/h. Assuming that 20 percent of this goes into useful work and the other 80 percent is converted to heat, calculate the temperature rise of the body after 1.00 h if none of this heat were transferred to the environment.

88. The *heat capacity*, C, of an object is defined as the amount of heat needed to raise its temperature by 1 C°. Thus, to raise the temperature by ΔT requires heat Q given by

$$Q = C\,\Delta T.$$

(*a*) Write the heat capacity C in terms of the specific heat, c, of the material. (*b*) What is the heat capacity of 1.0 kg of water? (*c*) Of 25 kg of water?

89. A very long 2.00-cm-diameter lead rod absorbs 410 kJ of heat. By how much does its length change? What would happen if the rod were only 2.0 cm long?

90. A mountain climber wears down clothing 3.5 cm thick with total surface area 1.9 m². The temperature at the surface of the clothing is -20°C and at the skin is 34°C. Determine the rate of heat flow by conduction through the clothing (*a*) assuming it is dry and that the thermal conductivity, k, is that of down, and (*b*) assuming the clothing is wet, so that k is that of water and the jacket has matted down to 0.50 cm thickness.

91. A marathon runner has an average metabolism rate for the race of about 1000 kcal/h. If the runner has a mass of 65.0 kg, how much water would the runner lose to evaporation from the skin for a race that lasts 2.5 h?

92. Estimate the rate at which heat can be conducted from the interior of the body to the surface. Assume that the thickness of tissue is 4.0 cm, that the skin is at 34°C and the interior at 37°C, and that the surface area is 15 m². Compare this to the measured value of about 230 W that must be dissipated by a person working lightly. This clearly shows the necessity of convective cooling by the blood.

93. *Newton's law of cooling* states that for small temperature differences, if a body at a temperature T_1 is in surroundings at a temperature T_2, the body cools at a rate given by

$$\frac{\Delta Q}{\Delta t} = K(T_1 - T_2)$$

where K is a constant. It includes the effects of conduction, convection, and radiation. That this linear relationship should hold is obvious if only conduction is considered. Show that it is also approximately true for radiation by showing that Eq. 19–16 reduces to

$$\Delta Q/\Delta t = 4\sigma e A T_2^3 (T_1 - T_2)$$
$$= \text{constant} \times (T_1 - T_2)$$

if $(T_1 - T_2)$ is small.

94. A house has well-insulated walls 15.5 cm thick (assume conductivity of air) and area 410 m², a roof of wood 6.5 cm thick and area 280 m², and plain glass windows 0.65 cm thick and total area 33 m². (*a*) Assuming that the heat loss is only by conduction, calculate the rate at which heat must be supplied to this house to maintain its temperature at 23°C if the outside temperature is -10°C. (*b*) If the house is initially at 10°C, estimate how much heat must be supplied to raise the temperature to 23°C within 30 min. Assume that only the air needs to be heated and that its volume is 750 m³. Take the specific heat of air to be 0.24 kcal/kg·C°.

95. A leaf of area 40 cm² and mass 4.5×10^{-4} kg directly faces the Sun on a clear day. The leaf has an emissivity of 0.85 and a specific heat of 0.80 kcal/kg·K. (*a*) Estimate the rate of rise of the leaf's temperature. (*b*) Calculate the temperature the leaf would reach if it lost all its heat by radiation (the surroundings are at 20°C). (*c*) In what other ways can the heat be dissipated by the leaf?

96. An iron meteorite melts when it enters the Earth's atmosphere. If its initial temperature, outside of the atmosphere, was -125°C, calculate the minimum velocity the meteorite must have had before it entered Earth's atmosphere.

97. Write a formula for the density of a gas when it is allowed to expand (*a*) as a function of temperature when the pressure is kept constant and (*b*) as a function of pressure when temperature is kept constant.

98. If 2.0 moles of an ideal monatomic gas expand adiabatically, performing 7500 J of work in the process, what is the change in temperature of the gas?

99. When 5.30×10^4 J of heat are added to a gas enclosed in a cylinder fitted with a light frictionless piston maintained at atmospheric pressure, the volume is observed to increase from 2.2 m³ to 4.1 m³. Calculate (*a*) the work done by the gas, and (*b*) the change in internal energy of the gas. (*c*) Graph this process on a *PV* diagram.

100. A diesel engine accomplishes ignition without a spark plug by an adiabatic compression of air to a temperature above the ignition temperature of the diesel fuel, which is injected into the cylinder at the peak of the compression. Suppose air is taken into the cylinder at 300 K and volume V_1 and is compressed adiabatically to 560°C (≈ 1000°F) and volume V_2. Assuming that the air behaves as an ideal gas whose ratio of C_P to C_V is 1.4, calculate the compression ratio V_1/V_2 of the engine.

101. An ice sheet forms on a lake. The air above the sheet is at -15°C, whereas the water is at 0°C. Assume that the heat of fusion of the water freezing on the lower surface is conducted through the sheet to the air above. How much time will it take to form a sheet of ice 25 cm thick?

102. The temperature within the Earth's crust increases about 1.0 C° for each 30 m of depth. The thermal conductivity of the crust is 0.80 W/C°·m. (*a*) Determine the heat transferred from the interior to the surface for the entire Earth in one day. (*b*) Compare this heat to the amount of energy incident on the Earth in one day due to radiation from the Sun.

103. A 100-W lightbulb generates 95 W of heat, which is dissipated through a glass bulb that has a radius of 3.0 cm and is 1.0 mm thick. What is the difference in temperature between the inner and outer surfaces of the glass?

Two uses for a heat engine: a modern coal-burning power plant, and an old steam locomotive. Both produce steam which does work—on turbines to generate electricity, and on a piston that moves linkage to turn locomotive wheels. The efficiency of any engine—no matter how carefully engineered—is limited by nature as described in the second law of thermodynamics. This great law is best stated in terms of a quantity called entropy, which is unlike any other. Entropy is *not* conserved, but instead is constrained always to increase in any real process. Entropy is a measure of disorder. The second law of thermodynamics tells us that as time moves forward, the disorder in the universe increases.

We also discuss many practical matters such as heat engines, heat pumps, and refrigeration.

CHAPTER 20

Second Law of Thermodynamics

In this final chapter on heat and thermodynamics, we discuss the famous second law of thermodynamics, and the quantity "entropy" that arose from this fundamental law and is its quintessential expression. We also discuss heat engines—the engines that transform heat into work in power plants, trains, and motor vehicles—because they first showed us that a new law was needed. Finally, we briefly discuss the third law of thermodynamics.

20–1 The Second Law of Thermodynamics— Introduction

The first law of thermodynamics states that energy is conserved. There are, however, many processes we can imagine that conserve energy but are not observed to occur in nature. For example, when a hot object is placed in contact with a cold object, heat flows from the hotter one to the colder one, never spontaneously the reverse. If heat were to leave the colder object and pass to the hotter one, energy could still be conserved. Yet it doesn't happen spontaneously.[†] As a second example, consider what happens when you drop a rock and it hits the ground. The initial

---

[†] By spontaneously, we mean by itself without input of work of some sort. (A refrigerator does move heat from a cold environment to a warmer one, but only by doing work.)

potential energy of the rock changes to kinetic energy as the rock falls, and when the rock hits the ground this energy in turn is transformed into internal energy (thermal energy) of the rock and the ground in the vicinity of the impact; the molecules move faster and the temperature rises slightly. But have you seen the reverse happen—a rock at rest on the ground suddenly rise up in the air because the thermal energy of molecules is transformed into kinetic energy of the rock as a whole? Energy could be conserved in this process, yet we never see it happen.

The first law of thermodynamics, conservation of energy, would not be violated if either of these processes occurred in reverse. To explain this lack of reversibility, scientists in the latter half of the nineteenth century came to formulate a new principle known as the **second law of thermodynamics**. This law is a statement about which processes occur in nature and which do not. It can be stated in a variety of ways, all of which are equivalent. One statement, due to R. J. E. Clausius (1822–1888), is that

heat flows naturally from a hot object to a cold object; heat will not flow spontaneously from a cold object to a hot object.

SECOND LAW OF THERMODYNAMICS (Clausius statement)

Since this statement applies to one particular process, it is not obvious how it applies to other processes. A more general statement is needed that will include other possible processes in a more obvious way.

The development of a general statement of the second law was based partly on the study of heat engines. A **heat engine** is any device that changes thermal energy into mechanical work, such as steam engines and automobile engines. We now examine heat engines, both from a practical point of view and to show their importance in developing the second law of thermodynamics.

20–2 | Heat Engines

It is easy to produce thermal energy by doing work—for example, by simply rubbing your hands together briskly, or indeed by any frictional process. But to get work from thermal energy is more difficult, and the invention of a practical device to do this came only about 1700 with the development of the steam engine.

The basic idea behind any heat engine is that mechanical energy can be obtained from thermal energy only when heat is allowed to flow from a high temperature to a low temperature. In the process, some of the heat can then be transformed to mechanical work, as diagrammed schematically in Fig. 20–1. That is, a heat input $|Q_H|$ at a high temperature T_H is partly transformed into work $|W|$ and partly exhausted as heat $|Q_L|$ at a lower temperature T_L. By conservation of energy, $|Q_H| = |W| + |Q_L|$. The high and low temperatures, T_H and T_L, are called the **operating temperatures** of the engine. We will be interested only in engines that run in a repeating *cycle* (that is, the system returns repeatedly to its starting point) and thus can run continuously.

We use absolute value signs around Q_L, Q_H and W because we are interested only in the magnitudes, and thus avoid worrying about the sign conventions established in Section 19–6. These quantities are thus positive, and arrows on diagrams (as in Fig. 20–1) tell us the direction of energy transfers.

The temperatures T_H and T_L may or may not be precisely constant, depending on the cycle.

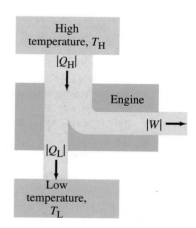

FIGURE 20–1 Schematic diagram of energy transfers for a heat engine.

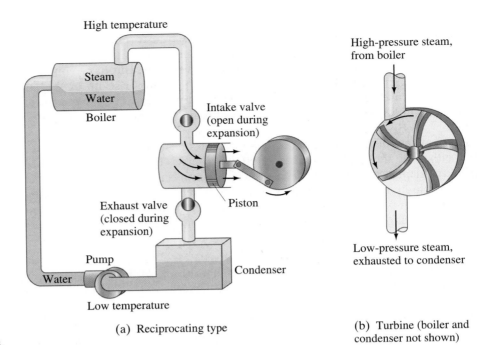

(a) Reciprocating type

(b) Turbine (boiler and condenser not shown)

FIGURE 20–2 Steam engines.

Steam Engine and Internal Combustion Engine

The operation of two practical engines, the steam engine and the internal combustion engine (used in most automobiles), is illustrated in Figs. 20–2 and 20–3. Steam engines are of two main types, each making use of steam heated by combustion of coal, oil, or gas (or nuclear energy). In the so-called reciprocating type, Fig. 20–2a, the heated steam passes through the open intake valve and expands against a piston, forcing it to move. When the piston reaches the end of its stroke and begins to return to its original position, the exhaust valve opens and the piston forces the gases out. In a steam turbine, Fig. 20–2b, everything is essentially the same, except that the reciprocating piston is replaced by a rotating turbine that resembles a paddlewheel with many sets of blades. Most of our electricity today is generated using

FIGURE 20–3 Four-cycle internal combustion engine: (a) the gasoline–air mixture flows into the cylinder as the piston moves down; (b) the piston moves upward and compresses the gas; (c) firing of the spark plug ignites the gasoline–air mixture, raising it to a high temperature; (d) the gases, now at high temperature and pressure, expand against the piston in this, the power stroke; (e) the burned gases are pushed out to the exhaust pipe; the intake valve then opens, and the whole cycle repeats.

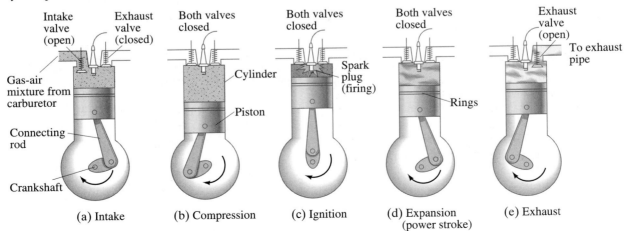

(a) Intake

(b) Compression

(c) Ignition

(d) Expansion (power stroke)

(e) Exhaust

steam turbines.[†] The material that is heated and cooled, steam in this case, is called the **working substance**. In a steam engine, the high temperature is obtained by burning coal, oil, or other fuel to heat the steam. In an internal combustion engine, the high temperature is achieved by burning the gasoline–air mixture in the cylinder itself (ignited by the spark plug), Fig. 20–3.

Why a ΔT is needed to drive a heat engine

To see why a *temperature difference* is required to run an engine, let us examine the steam engine. In the reciprocating engine, for example (Fig. 20–2a), suppose there were no condenser or pump, and that the steam was at the same temperature throughout the system. This would mean that the pressure of the gas being exhausted would be the same as that on intake. Thus, although work would be done by the gas *on* the piston when it expanded, an equal amount of work would have to be done *by* the piston to force the steam out the exhaust; hence, no net work would be done. In a real engine, the exhausted gas is cooled to a lower temperature and condensed so that the exhaust pressure is less than the intake pressure. Thus, although the piston must do work on the gas to expel it on the exhaust stroke, it is less than the work done by the gas on the piston during the intake. So a net amount of work can be obtained—but only if there is a difference of temperature. Similarly, in the gas turbine if the gas were not cooled, the pressure on each side of the blades would be the same; by cooling the gas on the exhaust side, the pressure on the front side of the blade is greater and hence the turbine turns.

The **efficiency**, e, of any heat engine can be defined as the ratio of the work it does, $|W|$, to the heat input at the high temperature, $|Q_H|$ (Fig. 20–1):

$$e = \frac{|W|}{|Q_H|}.$$

This is a sensible definition since $|W|$ is the output (what you get from the engine), whereas $|Q_H|$ is what you put in and pay for in fuel that burns. Because energy is conserved, the heat input $|Q_H|$ must equal the work done plus the heat that flows out at the low temperature ($|Q_L|$):

$$|Q_H| = |W| + |Q_L|.$$

Thus $|W| = |Q_H| - |Q_L|$, and the efficiency of an engine is

$$e = \frac{|W|}{|Q_H|} = \frac{|Q_H| - |Q_L|}{|Q_H|} = 1 - \frac{|Q_L|}{|Q_H|}. \qquad \textbf{(20–1)}$$

Efficiency of any heat engine

EXAMPLE 20–1 **Car efficiency.** An automobile engine has an efficiency of 20 percent and produces an average of 23,000 J of mechanical work per second during operation. How much heat is discharged from this engine per second?

SOLUTION The output heat is $|Q_L|$. We are given $e = 0.20$, so from Eq. 20–1 we find that

$$\frac{|Q_L|}{|Q_H|} = 1 - e = 0.80.$$

We also know that $e = |W|/|Q_H|$, by definition, so in one second

$$|Q_H| = \frac{|W|}{e} = \frac{23{,}000 \text{ J}}{0.20} = 1.15 \times 10^5 \text{ J}.$$

Thus

$$|Q_L| = 0.80 \, |Q_H| = (0.80)(1.15 \times 10^5 \text{ J}) = 9.2 \times 10^4 \text{ J}.$$

The engine discharges 9.2×10^4 J/s = 92,000 watts.

[†] Even nuclear power plants utilize steam turbines; the nuclear fuel—uranium—merely serves as fuel to heat the steam.

It is clear from Eq. 20–1 that the efficiency of an engine will be greater if $|Q_L|$ can be made small. However, from experience with a wide variety of systems, it has not been found possible to reduce $|Q_L|$ to zero. If $|Q_L|$ could be reduced to zero we would have a 100 percent efficient engine, as diagrammed in Fig. 20–4. That such a perfect engine (running continuously in a cycle) is not possible is another way of expressing the second law of thermodynamics. This can be stated formally as follows:

No device is possible whose sole effect is to transform a given amount of heat completely into work.

This is known as the **Kelvin-Planck statement of the second law of thermodynamics**. Said another way, *there can be no perfect (100 percent efficient) heat engine* such as that diagrammed in Fig. 20–4.

If the second law were not true, so that a perfect engine could be built, some rather remarkable things could happen. For example, if the engine of a ship did not need a low-temperature reservoir to exhaust heat into, the ship could sail across the ocean using the vast resources of the internal energy of the ocean water. Indeed, we would have no fuel problems at all!

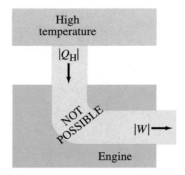

FIGURE 20–4 Schematic diagram of a hypothetical perfect heat engine in which all the heat input would be used to do work.

20–3 Reversible and Irreversible Processes; Carnot Engine

In the early nineteenth century, the French scientist N. L. Sadi Carnot (1796–1832) studied in detail the process of transforming heat into mechanical energy. His aim had been to determine how to increase the efficiency of heat engines, but his studies soon led him to investigate the foundations of thermodynamics itself.

As an aid in this pursuit, Carnot, in 1824, invented (on paper) an idealized type of engine which we now call the "Carnot engine." The importance of the Carnot engine is not really as a practical engine, but rather as an aid to the understanding of heat engines in general, and also because Carnot and his engine contributed to the establishment and understanding of the second law of thermodynamics.

Reversible and Irreversible Processes

Reversible process

The Carnot engine involves *reversible processes*, so before we discuss it we must discuss what is meant by reversible and irreversible processes. A **reversible process** is one that is carried out infinitely slowly, so that the process can be considered as a series of equilibrium states, and the whole process could be done in reverse with no change in magnitude of the work done or heat exchanged. For example, a gas contained in a cylinder fitted with a tight, movable, but frictionless piston could be compressed isothermally in a reversible way if done infinitely slowly. Not all very slow (quasistatic) processes are reversible, however. If there is friction present, for example (as between the movable piston and cylinder just mentioned), the work done in one direction (going from some state A to state B) will not be the negative of the work done in the reverse direction (state B to state A). Such a process would not be considered reversible. Of course a perfectly reversible process is not possible in reality since it would require an infinite time; reversible processes can be approached arbitrarily closely, however, and they are very important theoretically.

Irreversible process

All real processes are **irreversible**: they are not done infinitely slowly. There could be turbulence in the gas, friction would be present, and so on; any process could not be done precisely in reverse since the heat lost to friction would not reverse itself, the turbulence would be different, and so on. For any given volume there would not be a well-defined pressure P and temperature T since the system would not always be in an equilibrium state. Thus a real, irreversible, process cannot be plotted on a PV diagram (except insofar as it may approach an ideal reversible process). But a reversible process (since it is a quasistatic series of equilibrium states) always can be plotted on a *PV* diagram; and a reversible process, when done in reverse, retraces the same path on a *PV* diagram. Although all real processes are irreversible, reversible processes are conceptually important, just as the concept of an ideal gas is.

Carnot's Engine

Now let us look at Carnot's idealized engine. The Carnot engine makes use of a **reversible cycle**, by which we mean a series of reversible processes that take a given substance (the *working substance*) from an initial equilibrium state through many other equilibrium states and returns it again to the same initial state. In particular, the Carnot engine utilizes the **Carnot cycle**, which is illustrated in Fig. 20–5, with the working substance assumed to be an ideal gas. Let us take point a as the initial state. The gas is first expanded isothermally, and reversibly, path ab, at temperature T_H. To do so, we can imagine the gas to be in contact with a heat reservoir at a constant temperature T_H which delivers heat $|Q_H|$ to our working substance. Next the gas is expanded adiabatically and reversibly, path bc; no heat is exchanged and the temperature of the gas is reduced to T_L. The third step is a reversible isothermal compression, path cd, in contact with a heat reservoir at a constant low temperature, T_L, during which heat $|Q_L|$ flows out of the working substance. Finally, the gas is compressed adiabatically, path da, back to its original state. Thus a Carnot cycle consists of two isothermal and two adiabatic processes.

It is easy to see that the net work done in one cycle by a Carnot engine (or any other type of engine using a reversible cycle) is equal to the area enclosed by the curve representing the cycle on the *PV* diagram, the curve abcd in Fig. 20–5. (See Section 19–7.)

Carnot cycle

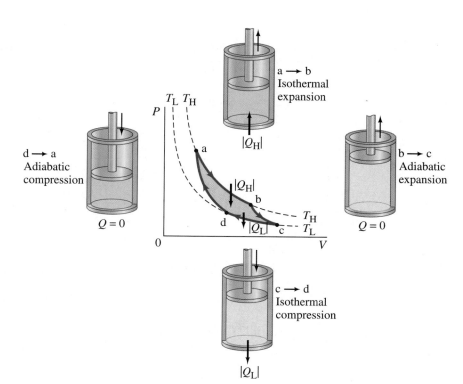

FIGURE 20–5 The Carnot cycle. Heat engines work in a cycle, and the cycle for the Carnot engine begins at point a on this *PV* diagram. (1) The gas is first expanded isothermally, with the addition of heat $|Q_H|$, along the path ab at temperature T_H. (2) Next the gas expands adiabatically from b to c—no heat is exchanged, but the temperature drops to T_L. (3) The gas is then compressed at constant temperature T_L, path c to d, and heat $|Q_L|$ flows out. (4) Finally, the gas is compressed adiabatically, path da, back to its original state. No Carnot engine actually exists, but as a theoretical engine it played an important role in the development of thermodynamics.

Carnot Efficiency and the Second Law of Thermodynamics

The efficiency of a Carnot engine, like any heat engine, is given by Eq. 20–1:

$$e = 1 - \frac{|Q_L|}{|Q_H|}.$$

For a Carnot engine, however, we can show that the efficiency depends only on the temperatures of the heat reservoirs, T_H and T_L. In the first isothermal process, ab in Fig. 20–5, the work done by the gas is (see Eq. 19–6)

$$W_{ab} = nRT_H \ln \frac{V_b}{V_a},$$

where n is the number of moles of the ideal gas used as working substance. Since the internal energy of an ideal gas does not change when the temperature remains constant, the first law of thermodynamics tells us that the heat added to the gas equals the work done by the gas:

$$|Q_H| = nRT_H \ln \frac{V_b}{V_a}.$$

Similarly, the heat lost by the gas in the isothermal process cd is

$$|Q_L| = nRT_L \ln \frac{V_c}{V_d}.$$

Since paths bc and da are adiabatic, we have (see Eq. 19–13)

$$P_b V_b^{\gamma} = P_c V_c^{\gamma} \quad \text{and} \quad P_d V_d^{\gamma} = P_a V_a^{\gamma}.$$

Also, from the ideal gas law,

$$\frac{P_b V_b}{T_H} = \frac{P_c V_c}{T_L} \quad \text{and} \quad \frac{P_d V_d}{T_L} = \frac{P_a V_a}{T_H}.$$

When we divide these last equations, term by term, into the corresponding set of equations on the line above, we obtain

$$T_H V_b^{\gamma-1} = T_L V_c^{\gamma-1} \quad \text{and} \quad T_L V_d^{\gamma-1} = T_H V_a^{\gamma-1}.$$

Next we divide the equation on the left by the one on the right and get

$$\left(\frac{V_b}{V_a}\right)^{\gamma-1} = \left(\frac{V_c}{V_d}\right)^{\gamma-1}.$$

Hence

$$\frac{V_b}{V_a} = \frac{V_c}{V_d}$$

or

$$\ln \frac{V_b}{V_a} = \ln \frac{V_c}{V_d}.$$

Using this result in our equations for $|Q_H|$ and $|Q_L|$ above we have

$$\frac{|Q_L|}{|Q_H|} = \frac{T_L}{T_H}. \qquad \text{[Carnot cycle]} \quad \textbf{(20–2)}$$

Hence the efficiency of a reversible Carnot engine is

Carnot (ideal) efficiency

$$e_{\text{ideal}} = 1 - \frac{|Q_L|}{|Q_H|} = 1 - \frac{T_L}{T_H}. \qquad \left[\begin{array}{l}\text{Carnot efficiency;}\\\text{Kelvin temperatures}\end{array}\right] \quad \textbf{(20–3)}$$

The temperatures T_L and T_H are the absolute or Kelvin temperatures as measured on the ideal gas temperature scale. Thus the efficiency of a Carnot engine depends only on the temperatures T_L and T_H.

We could imagine other possible reversible cycles that could be used for an ideal reversible engine. According to a theorem stated by Carnot:

All reversible engines operating between the same two constant temperatures T_H and T_L have the same efficiency. Any irreversible engine operating between the same two fixed temperatures will have an efficiency less than this.

Carnot's theorem

This is known as **Carnot's theorem**.[†] It tells us that Eq. 20–3, $e = 1 - (T_L/T_H)$, applies to any ideal reversible engine with fixed input and exhaust temperatures, T_H and T_L, and that this equation represents a maximum possible efficiency for a real (i.e., irreversible) engine.

In practice, the efficiency of real engines is always less than the Carnot efficiency. Well-designed engines reach perhaps 60 to 80 percent of Carnot efficiency.

EXAMPLE 20–2 **Steam engine efficiency.** A steam engine operates between 500°C and 270°C. What is the maximum possible efficiency of this engine?

SOLUTION We must first change the temperature to kelvins. Thus, $T_H = 773$ K and $T_L = 543$ K. Then from Eq. 20–3,

$$e_{ideal} = 1 - \frac{543}{773} = 0.30.$$

Thus, the maximum (or Carnot) efficiency is 30 percent. Realistically, an engine might attain 0.70 of this value or 21 percent. Note in this Example that the exhaust temperature is still rather high, 270°C. Steam engines are often arranged in series so that the exhaust of one engine is used as intake by a second or third engine.

EXAMPLE 20–3 **A phony claim?** An engine manufacturer makes the following claims: The heat input per second of the engine is 9.0 kJ at 475 K. The heat output per second is 4.0 kJ at 325 K. Do you believe these claims?

SOLUTION The efficiency of the engine is (Eq. 20–1)

$$e = \frac{|Q_H| - |Q_L|}{|Q_H|} = \frac{9.0 \text{ kJ} - 4.0 \text{ kJ}}{9.0 \text{ kJ}} = 0.56.$$

However, the maximum possible efficiency is given by the Carnot efficiency, Eq. 20–3:

$$e_{ideal} = \frac{T_H - T_L}{T_H} = \frac{475 \text{ K} - 325 \text{ K}}{475 \text{ K}} = 0.32.$$

So the manufacturer's claims violate the second law of thermodynamics and cannot be believed.

It is clear from Eq. 20–3 that a 100 percent efficient engine is not possible. Only if the exhaust temperature, T_L, were at absolute zero would 100 percent efficiency be obtainable. But reaching absolute zero is a practical (as well as theoretical) impossibility.[‡] Thus we can state, as we already did in Section 20–2, that **no device is possible whose sole effect is to transform a given amount of heat completely into work**. As we saw in Section 20–2, this is known as the *Kelvin-Planck statement of the second law of thermodynamics*. It tells us that there can be no perfect (100 percent efficient) heat engine such as the one diagrammed in Fig. 20–4.

SECOND LAW OF THERMODYNAMICS (Kelvin-Planck statement)

[†] Carnot's theorem can be shown to follow directly from either the Clausius or Kelvin-Planck statements of the second law of thermodynamics.

[‡] This result is known as the *third law of thermodynamics*, as discussed in Section 20–10.

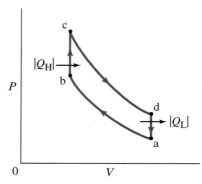

FIGURE 20–6 The Otto cycle.

* The Otto Cycle

The operation of an automobile internal combustion engine (see Fig. 20–3) can be approximated by a reversible cycle known as the *Otto cycle*, whose *PV* diagram is shown in Fig. 20–6. Unlike the Carnot cycle, the input and exhaust temperatures of the Otto cycle are *not* constant. Paths ab and cd are adiabatic, and paths bc and da are at constant volume. The gas (gasoline-air mixture) enters the cylinder at point a and is compressed adiabatically (compression stroke) to point b. At b ignition occurs (spark plug) and the burning of the gas adds heat $|Q_H|$ to the system at constant volume (approximately in a real engine). The temperature and pressure rise, and then in the power stroke, ca, the gas expands adiabatically. In the exhaust stroke, da, heat $|Q_L|$ is ejected to the environment (in a real engine, the gas leaves the engine and is replaced by a new mixture of air and fuel).

EXAMPLE 20–4 The Otto cycle. (*a*) For an ideal gas as working substance, show that the efficiency of an Otto cycle engine is

$$e = 1 - \left(\frac{V_a}{V_b}\right)^{1-\gamma}$$

where γ is the ratio of specific heats (Sections 19–8 and 19–9) and V_a/V_b is the *compression ratio*. (*b*) Calculate the efficiency for a compression ratio $V_a/V_b = 8.0$ assuming a diatomic gas like O_2 and N_2.

SOLUTION The heat exchanges take place at constant volume in the ideal Otto cycle, so from Eq. 19–8a:

$$|Q_H| = nC_V(T_c - T_b) \quad \text{and} \quad |Q_L| = nC_V(T_d - T_a).$$

Then from Eq. 20–1,

$$e = 1 - \frac{|Q_L|}{|Q_H|} = 1 - \left[\frac{T_d - T_a}{T_c - T_b}\right].$$

To get this in terms of the compression ratio, V_a/V_b, we use the result from Section 19–9, Eq. 19–13, $PV^\gamma = $ constant during the adiabatic processes ab and cd. Thus

$$P_a V_a^\gamma = P_b V_b^\gamma \quad \text{and} \quad P_c V_c^\gamma = P_d V_d^\gamma.$$

We use the ideal gas law, $P = nRT/V$, and substitute P into these two equations

$$T_a V_a^{\gamma-1} = T_b V_b^{\gamma-1} \quad \text{and} \quad T_c V_c^{\gamma-1} = T_d V_d^{\gamma-1}.$$

Then the efficiency (see above) is

$$e = 1 - \left[\frac{T_d - T_a}{T_c - T_b}\right] = 1 - \left[\frac{T_c(V_c/V_d)^{\gamma-1} - T_b(V_b/V_a)^{\gamma-1}}{T_c - T_b}\right].$$

But processes bc and da are at constant volume, so $V_c = V_b$ and $V_d = V_a$. Hence $V_c/V_d = V_b/V_a$ and

$$e = 1 - \left[\frac{(V_b/V_a)^{\gamma-1}(T_c - T_b)}{T_c - T_b}\right] = 1 - \left(\frac{V_b}{V_a}\right)^{\gamma-1} = 1 - \left(\frac{V_a}{V_b}\right)^{1-\gamma}.$$

(*b*) For diatomic molecules (Section 19–8), $\gamma = C_P/C_V = 1.4$ so

$$e = 1 - (8.0)^{1-\gamma} = 1 - (8.0)^{-0.4} = 0.56.$$

Real engines do not reach this high efficiency because they do not follow perfectly the Otto cycle, plus there is friction, turbulence, heat loss and incomplete combustion of the gases.

(a)

(b)

(c)

FIGURE 20–7 (a) An array of mirrors focuses sunlight on a boiler to produce steam at a solar energy installation. (b) A fossil-fuel steam plant. (c) Large cooling towers at an electric generating plant.

* Thermal Pollution

Much of the energy we utilize in everyday life—from motor vehicles to most of the electricity produced by power plants—make use of a heat engine. Electricity produced by falling water at dams, by windmills, or by solar cells (Fig. 20–7a) do not involve a heat engine. But over 90 percent of the electric energy produced in the U.S. is done at fossil-fuel steam plants (Fig. 20–7b), and they make use of a heat engine (essentially steam engines). In electric power plants, the steam drives the turbines and generators whose output is electric energy. Even nuclear power plants use nuclear fuel to run a steam engine. The heat output $|Q_L|$ from every heat engine, from power plants to cars, is referred to as **thermal pollution** because this heat $|Q_L|$ must be absorbed by the environment—such as by water from rivers or lakes or via large cooling towers (Fig. 20–7c). This heat raises the temperature of the cooling water, altering the natural ecology of aquatic life (largely because warmer water holds less oxygen); or, in the case of air cooling towers, the output heat raises the temperature of the atmosphere which affects the weather. Air pollution—by which we mean the chemicals released in the burning of fossil fuels in cars, power plants, and industrial furnaces—gives rise to smog and other problems $\left(CO_2 \text{ buildup in the atmosphere, causing the greenhouse effect and global warming}\right)$ that can be controlled to some extent and hopefully more so in the coming years. But thermal pollution is unavoidable. Engineers can try to design and build engines that are more efficient, but they cannot surpass the Carnot efficiency and must live with T_L being at best the ambient temperature of water or air. The second law of thermodynamics tells us the limit imposed by nature. What we can do, in the light of the second law of thermodynamics, is use less energy and conserve our fuel resources.

➡ **PHYSICS APPLIED**

*Heat engines
and thermal pollution*

FIGURE 20–8 Schematic diagram of energy transfers for a refrigerator or air conditioner.

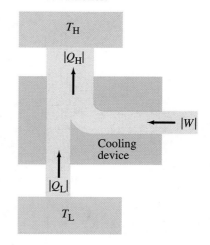

20–4 Refrigerators, Air Conditioners, and Heat Pumps

The operating principle of refrigerators, air conditioners, and heat pumps is just the reverse of a heat engine. Each operates to transfer heat out of a cool environment into a warm environment. As diagrammed in Fig. 20–8, by doing work $|W|$, heat is taken from a low-temperature region, T_L (inside a refrigerator, say), and a greater amount of heat is exhausted at a high temperature, T_H (the room). You can often feel this heated air blowing out beneath a refrigerator. The

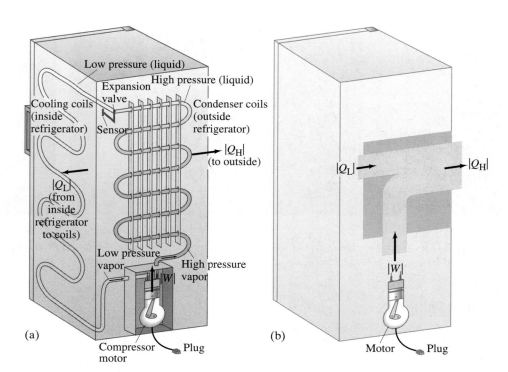

FIGURE 20–9 (a) Typical refrigerator system. The compressor motor forces a gas at high pressure through a heat exchanger (condenser) on the outside walls of the refrigerator where $|Q_H|$ is given off and the gas cools to become liquid. The liquid passes from a high-pressure region, via a valve, to low-pressure tubes on the inside walls of the refrigerator; the liquid evaporates at this lower pressure and thus absorbs heat ($|Q_L|$) from the inside of the refrigerator. The fluid returns to the compressor where the cycle begins again. (b) Schematic diagram, like Fig. 20–8.

work $|W|$ is usually done by a compressor motor which compresses a fluid, as illustrated in Fig. 20–9.

A perfect refrigerator—one in which no work is required to take heat from the low-temperature region to the high-temperature region—is not possible. This is **the Clausius statement of the second law of thermodynamics**, already mentioned in Section 20–1, which can be stated formally as

> **No device is possible whose sole effect is to transfer heat from one system at one temperature into a second system at a higher temperature.**

To make heat flow from a low-temperature body (or system) to one at a higher temperature, work must be done. Thus, *there can be no perfect refrigerator.*

Coefficient of performance (CP)

The **coefficient of performance (CP)** of a refrigerator is defined as the heat $|Q_L|$ removed from the low-temperature area (inside a refrigerator) divided by the work $|W|$ done to remove the heat (Fig. 20–8 or 20–9b):

$$CP = \frac{|Q_L|}{|W|}. \qquad \left[\begin{array}{c} \text{Refrigerator and} \\ \text{air conditioner} \end{array} \right] \quad \textbf{(20–4a)}$$

This makes sense since the more heat, $|Q_L|$, that can be removed from the inside of the refrigerator for a given amount of work, the better (more efficient) the refrigerator. Energy is conserved, so from the first law of thermodynamics we can write (see Fig. 20–8 or 20–9b) $|Q_L| + |W| = |Q_H|$, or $|W| = |Q_H| - |Q_L|$. Then Eq. 20–4a becomes

CP for

$$CP = \frac{|Q_L|}{|W|} = \frac{|Q_L|}{|Q_H| - |Q_L|}. \qquad \textbf{(20–4b)}$$

refrigerator

For an ideal refrigerator (not a perfect one, which is impossible), the best one could do would be

and

air conditioner

$$CP_{ideal} = \frac{T_L}{T_H - T_L}, \qquad \textbf{(20–4c)}$$

as for an ideal (Carnot) engine (Eq. 20–3).

➡ **PHYSICS APPLIED**

Air conditioner

An air conditioner works very much like a refrigerator, although the actual construction details are different because an air conditioner takes heat $|Q_L|$ from

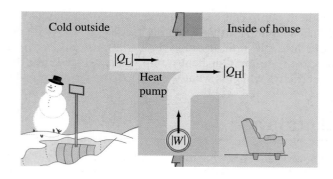

FIGURE 20–10 A heat pump "pumps" heat from the cold outside to the warm inside of a house.

inside a room or building at a low temperature, and deposits heat $|Q_H|$ outside to the environment at a higher temperature. Equations 20–4 also describe the coefficient of performance for an air conditioner.

Heat naturally flows from high temperature to low temperature. Refrigerators and air conditioners do work to accomplish the opposite: to make heat flow from cold to hot. We might say they "pump" heat from cold areas to hotter areas, against the natural tendency of heat to flow from hot to cold, just as water can be pumped uphill, against the natural tendency to flow downhill. The term **heat pump** is usually reserved for a device that can heat a house in winter by taking heat $|Q_L|$ from the outside at low temperature and delivering heat $|Q_H|$ to the warmer inside of the house, by doing work $|W|$; see Fig. 20–10. As in a refrigerator, there is an indoor and an outdoor heat exchanger (coils of the refrigerator) and a compressor. The operating principle is like that for a refrigerator or air conditioner; but the objective of a heat pump is to heat (deliver $|Q_H|$), rather than to cool (remove $|Q_L|$). Thus, the coefficient of performance of a heat pump is defined differently than for an air conditioner because it is the heat $|Q_H|$ delivered to the inside of the house that is important now:

$$ CP = \frac{|Q_H|}{|W|}. \qquad \text{[heat pump]} \quad \textbf{(20–5)} $$

➥ **PHYSICS APPLIED**

Heat pump

CP for heat pump

Most heat pumps can, however, be "turned around" and used as air conditioners in the summer.

EXAMPLE 20–5 **Heat pump.** A heat pump has a coefficient of performance of 3.0 and is rated to do work at 1500 W. (*a*) How much heat can it add to a room per second? (*b*) If the heat pump were turned around to act as an air conditioner in the summer, what would you expect its coefficient of performance to be, assuming all else stays the same?

SOLUTION (*a*) We use Eq. 20–5 for the heat pump and, since our device does 1500 J of work per second, it can pour heat into the room at a rate of

$$ |Q_H| = CP \times |W| = 3.0 \times 1500\,\text{J} = 4500\,\text{J} $$

per second, or at a rate of 4500 W.
(*b*) If our device is turned around in summer, it can take heat $|Q_L|$ from inside the house and do 1500 J of work per second, and then dumps $|Q_H|$ = 4500 J per second to the hot outside. Energy is conserved, so $|Q_L| + |W| = |Q_H|$ (see Fig. 20–10, but reverse inside and outside of house), so

$$ |Q_L| = |Q_H| - |W| = 4500\,\text{J} - 1500\,\text{J} = 3000\,\text{J}. $$

The coefficient of performance as an air conditioner would thus be (Eq. 20–4a)

$$ CP = \frac{|Q_L|}{|W|} = \frac{3000\,\text{J}}{1500\,\text{J}} = 2.0. $$

A good heat pump can be a money saver and an energy saver. Compare, for example, our heat pump in this Example to, say, a 1500-watt electric heater. We plug the latter into the wall, it draws 1500 watts of electricity and delivers 1500 watts of heat to the room. Our heat pump when plugged into the wall also draws 1500 W of electricity (which is what we pay for), but it delivers 4500 W of heat!

20–5 | Entropy

We have seen several aspects of the second law of thermodynamics; but we have not yet arrived at a general statement of it. Both the Clausius and Kelvin-Planck statements deal with rather specific situations. Yet, as mentioned at the beginning of this chapter, there are a great many processes that simply are not observed in nature, even though they would not violate the first law of thermodynamics if they did occur. To cover all these processes, a more general statement of the second law of thermodynamics is needed. This general statement can be made in terms of a quantity, introduced by Clausius in the 1860s, called *entropy*, which we now discuss.

In our study of the Carnot cycle we found (Eq. 20–2) that $|Q_L|/|Q_H| = T_L/T_H$. We rewrite this as

$$\frac{|Q_H|}{T_H} = \frac{|Q_L|}{T_L}.$$

In this relation, both $|Q_H|$ and $|Q_L|$ are positive since they are absolute values. Let us now remove the absolute value signs and recall our original convention as used in the first law (see Section 19–6), that Q is positive when it represents a heat flow into the system (as Q_H) and negative for a heat flow out of the system (as Q_L). Then this relation becomes

$$\frac{Q_H}{T_H} + \frac{Q_L}{T_L} = 0. \qquad \text{[Carnot cycle]} \quad \textbf{(20–6)}$$

Now consider *any* reversible cycle, as represented by the smooth (oval-shaped) curve in Fig. 20–11. Any reversible cycle can be approximated as a series of Carnot cycles. Figure 20–11 shows only six—the isotherms (dashed lines) are connected by adiabatic paths for each—and the approximation becomes better and better if we increase the number of Carnot cycles. Equation 20–6 is valid for each of these cycles, so we can write

$$\Sigma \frac{Q}{T} = 0 \qquad \text{[Carnot cycles]} \quad \textbf{(20–7)}$$

for the sum of all these cycles. But note that the heat output Q_L of one cycle is approximately equal to the negative of the heat input, Q_H, of the cycle below it (actual equality in the limit of an infinite number of infinitely thin Carnot cycles). Hence the heat flows on the inner paths of all these Carnot cycles cancel out, so the net heat transferred, and the work done, is the same for the series of Carnot cycles as for the original cycle. Hence, in the limit of infinitely many Carnot cycles, Eq. 20–7 applies to any reversible cycle. In this case Eq. 20–7 becomes

$$\oint \frac{dQ}{T} = 0, \qquad \text{[reversible cycle]} \quad \textbf{(20–8)}$$

where dQ represents an infinitesimal heat flow.[†] The symbol $\oint$ means take the integral around a closed path; the integral can be started at any point on the path such

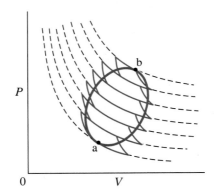

FIGURE 20–11 Any reversible cycle can be approximated as a series of Carnot cycles. (The dashed lines represent isotherms.)

[†] dQ is often written $đQ$: see footnote at end of Section 19–6.

as at a or b in Fig. 20–11, and proceed in either direction. Let us divide the cycle of Fig. 20–11 into two parts as indicated in Fig. 20–12. Then we rewrite Eq. 20–8 as

$$\int_{a}^{b}\frac{dQ}{T} + \int_{b}^{a}\frac{dQ}{T} = 0.$$
$$\quad\;_{I}\qquad\quad\;_{II}$$

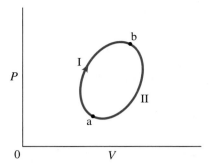

P

The first term is the integral from point a to point b along path I in Fig. 20–12, and the second term is the integral from b back to a along path II. (Path I plus path II is the whole cycle.) If one path is taken in reverse, say path II, dQ at each point becomes $-dQ$, since the path is reversible. We can therefore write

$$\int_{a}^{b}\frac{dQ}{T} = \int_{a}^{b}\frac{dQ}{T}.$$
$$\quad\;_{I}\qquad\quad\;_{II}$$
[reversible paths]   **(20–9)**

0 $\qquad\qquad V$

FIGURE 20–12 The integral, $\oint dS$, of the entropy for a reversible cycle is zero. Hence the difference in entropy between states a and b, $S_b - S_a = \int_a^b dS$, is the same for path I as for path II.

Since our cycle is arbitrary, Eq. 20–9 tells us that the integral of dQ/T between any two equilibrium states, a and b, does not depend on the path of the process. We therefore define a new quantity, called the **entropy**, S, by the relation

$$dS = \frac{dQ}{T}.$$
(20–10)

Entropy defined

From Eq. 20–8, we have that

$$\oint dS = 0.$$
[reversible cycle]   **(20–11)**

The quantity ΔS, where

$$\Delta S = S_b - S_a = \int_a^b dS = \int_a^b \frac{dQ}{T},$$
[reversible processes]   **(20–12)**

is seen from Eq. 20–9 to be *independent of the path between the two points a and b*. This is an important result. It tells us that the difference in entropy, $S_b - S_a$, between two equilibrium states of a system does not depend on how you get from one state to the other. Thus *entropy is a state variable*—its value depends only on the state of the system, and not on the process or the past history of how it got there.[†] This is in clear distinction to Q and W which are *not* state variables; their values do depend on the processes undertaken.

Entropy is a state variable

Entropy and the Second Law of Thermodynamics

We have defined a new quantity, S, the entropy, which can be used to describe the state of the system, along with P, T, V, U, and n. But what does this rather abstract quantity have to do with the second law of thermodynamics? To answer this, let us take some examples in which we calculate the entropy changes during particular processes. But note first that Eq. 20–12 can be applied only to reversible processes. How then do we calculate $\Delta S = S_b - S_a$ for a real process that is irreversible? What we can do is this: we figure out some other *reversible* process that takes the system between the same two states, and calculate ΔS for this reversible process. This will equal ΔS for the irreversible process since ΔS depends only on the initial and final states of the system.

How to determine ΔS (using a reversible process)

[†] Equation 20–12 says nothing about the absolute value of S; it only gives the change in S. This is much like potential energy (Chapter 8). However, one form of the so-called *third law of thermodynamics* (see also Section 20–10) states that as $T \to 0, S \to 0$.

EXAMPLE 20–6 **Entropy change in melting.** A 1.00-kg piece of ice at 0°C melts very slowly to water at 0°C. Assume the ice is in contact with a heat reservoir whose temperature is only infinitesimally greater than 0°C. Determine the entropy change of (a) the ice cube and (b) the heat reservoir.

SOLUTION (a) The process is carried out at a constant temperature $T = 273$ K and is done reversibly, so we can use Eq. 20–12:

$$\Delta S_{ice} = \int \frac{dQ}{T} = \frac{1}{T} \int dQ = \frac{Q}{T}.$$

Since the heat needed to melt the ice is $Q = mL$, where the heat of fusion $L = 79.7$ kcal/kg $= 3.33 \times 10^5$ J/kg, we have

$$\Delta S_{ice} = \frac{mL}{T} = \frac{(1.00 \text{ kg})(79.7 \text{ kcal/kg})}{273 \text{ K}} = 0.292 \text{ kcal/K},$$

or 1220 J/K.

(b) Heat $Q = mL$ is *removed* from the heat reservoir, so (since $T = 273$ K and is constant)

$$\Delta S_{res} = -\frac{Q}{T} = -0.292 \text{ kcal/K}.$$

Note that the *total* entropy change, $\Delta S_{ice} + \Delta S_{res}$, is zero.

EXAMPLE 20–7 **ESTIMATE** **Entropy change when mixing water.** A sample of 50.0 kg of water at 20.0°C is mixed with 50.0 kg of water at 24.0°C. Estimate the change in entropy without using calculus.

SOLUTION The final temperature of the mixture will be 22.0°C, since we started with equal amounts of water. A quantity of heat,

$$Q = mc \, \Delta T = (50.0 \text{ kg})(1.00 \text{ kcal/kg} \cdot \text{K})(2.0 \text{ C°}) = 100 \text{ kcal},$$

flows out of the hot water as it cools down from 24°C to 22°C, and this heat flows into the cold water as it warms from 20°C to 22°C. The total change in entropy, ΔS, will be the sum of the changes in entropy of the hot water, ΔS_H, and that of the cold water, ΔS_C:

$$\Delta S = \Delta S_H + \Delta S_C.$$

We estimate entropy changes by writing $\Delta S = Q/T_{av}$, where T_{av} is an "average" temperature for each process, which ought to give a reasonable estimate since the temperature change is small. For the hot water we use an average temperature of 23°C (296 K), and for the cold water an average temperature of 21°C (294 K). Thus

$$\Delta S_H \approx -\frac{100 \text{ kcal}}{296 \text{ K}} = -0.338 \text{ kcal/K}$$

and

$$\Delta S_C \approx \frac{100 \text{ kcal}}{294 \text{ K}} = 0.340 \text{ kcal/K}.$$

Note that the entropy of the hot water (S_H) decreases since heat flows out of the hot water. But the entropy of the cold water (S_C) increases by a greater amount. The total change in entropy is

$$\Delta S = \Delta S_H + \Delta S_C \approx -0.338 \text{ kcal/K} + 0.340 \text{ kcal/K}$$
$$\approx +0.002 \text{ kcal/K}.$$

We see that although the entropy of one part of the system decreased, the entropy of the other part increased by a greater amount so that the net change in entropy of the whole system is positive.

We can now easily show in general that for an isolated system of two bodies, the flow of heat from the higher-temperature (T_H) body to the lower-temperature (T_L) body always results in an increase in the total entropy. The two bodies eventually come to some intermediate temperature, T_M. The heat lost by the hotter body $(Q_H = -Q$ where Q is positive$)$ is equal to the heat gained by the colder one $(Q_L = Q)$, so the total change in entropy is

$$\Delta S = \Delta S_H + \Delta S_L = -\frac{Q}{T_{HM}} + \frac{Q}{T_{LM}},$$

where T_{HM} is some intermediate temperature between T_H and T_M for the hot body as it cools from T_H to T_M, and T_{LM} is the counterpart for the cold body; since the temperature of the hot body is, at all times during the process, greater than that of the cold body, then $T_{HM} > T_{LM}$. Hence

$$\Delta S = Q\left(\frac{1}{T_{LM}} - \frac{1}{T_{HM}}\right) > 0.$$

One body decreases in entropy, while the other gains in entropy, but the *total* change is positive.

EXAMPLE 20–8 **Entropy changes in a free expansion.** Consider the *adiabatic free expansion* of n moles of an ideal gas from volume V_1 to volume V_2, where $V_2 > V_1$, as was discussed in Section 19–7, Fig. 19–13. Calculate the change in entropy (a) of the gas and (b) of the surrounding environment. (c) Evaluate ΔS for 1.00 mole, with $V_2 = 2.00\ V_1$.

SOLUTION As we saw in Section 19–7, the gas is initially in a closed container of volume V_1, and, with the opening of a valve, adiabatically it expands into a previously empty container; the total volume of the two containers is V_2. The whole apparatus is thermally insulated from the surroundings, so no heat flows into the gas, $Q = 0$. The gas does no work, $W = 0$, so there is no change in internal energy, $\Delta U = 0$, and the temperature of the initial and final states is the same, $T_2 = T_1 = T$.
(a) The process takes place very quickly, and so is irreversible. So we cannot apply Eq. 20–12 to this process. Instead we must think of a reversible process that will take the gas from volume V_1 to V_2 at the same temperature, and use Eq. 20–12 on this reversible process to get ΔS. A reversible isothermal process will do the trick; in such a process, the internal energy does not change, so (from the first law)

$$dQ = dW = P\,dV.$$

Then

$$\Delta S_{gas} = \int \frac{dQ}{T} = \frac{1}{T}\int_{V_1}^{V_2} P\,dV;$$

from the ideal gas law $P = nRT/V$, so

$$\Delta S_{gas} = \frac{nRT}{T}\int_{V_1}^{V_2} \frac{dV}{V} = nR\ln\frac{V_2}{V_1}.$$

Since $V_2 > V_1$, $\Delta S_{gas} > 0$.
(b) Since no heat is transferred to the surrounding environment, there is no change of the state of the environment due to this process. Hence $\Delta S_{env} = 0$. Note that the total change in entropy, $\Delta S_{gas} + \Delta S_{env}$ is greater than zero.
(c) Since $n = 1.00$ and $V_2 = 2.00\ V_1$, then $\Delta S_{gas} = R\ln 2.00 = 5.76\ \text{J/K}$.

EXAMPLE 20–9 **Heat conduction.** A red-hot 2.0-kg piece of iron at temperature $T_1 = 880$ K is thrown into a huge lake whose temperature is $T_2 = 280$ K. Assume the lake is so large that its temperature rise is insignificant. Determine the change in entropy (a) of the iron and (b) of the surrounding environment (the lake).

SOLUTION (a) The process is irreversible, but the same entropy change will occur for a reversible process, and we use the concept of specific heat, Eq. 19–2. We assume the specific heat of the iron is constant at $c = 0.11$ kcal/kg·K. Then $dQ = mc\ dT$ and in a quasistatic reversible process

$$\Delta S_{iron} = \int \frac{dQ}{T} = mc \int_{T_1}^{T_2} \frac{dT}{T} = mc \ln \frac{T_2}{T_1} = -mc \ln \frac{T_1}{T_2}.$$

Putting in numbers, we find

$$\Delta S_{iron} = -(2.0 \text{ kg})(0.11 \text{ kcal/kg·K}) \ln \frac{880 \text{ K}}{280 \text{ K}} = -0.25 \text{ kcal/K}.$$

or $\Delta S = -1100$ J/K.
(b) The initial and final temperatures of the lake are the same, $T = 280$ K. The lake receives from the iron an amount of heat

$$Q = mc(T_2 - T_1) = (2.0 \text{ kg})(0.11 \text{ kcal/kg·K})(880 \text{ K} - 280 \text{ K}) = 130 \text{ kcal}.$$

Strictly speaking, this is an irreversible process (the lake heats up locally before equilibrium is reached), but is equivalent to a reversible isothermal transfer of heat $Q = 130$ kcal at $T = 280$ K. Hence

$$\Delta S_{env} = \frac{130 \text{ kcal}}{280 \text{ K}} = 0.46 \text{ kcal/K}$$

or 1900 J/K. Thus, although the entropy of the iron actually decreases, the *total* change in entropy of iron plus environment is positive: 0.46 kcal/K − 0.25 kcal/K = +0.21 kcal/K, or 800 J/K.

In each of these examples, the entropy of our system plus that of the environment (or surroundings) either stayed constant or increased. For any *reversible* process, such as that in Example 20–6, the total entropy change is zero. This can be seen in general as follows: any reversible process can be considered as a series of quasistatic isothermal transfers of heat ΔQ between a system and the environment, which differ in temperature only by an infinitesimal amount. Hence the change in entropy of either the system or environment is $\Delta Q/T$ and that of the other is $-\Delta Q/T$, so the total is

$$\Delta S = \Delta S_{syst} + \Delta S_{env} = 0. \qquad \text{[any reversible process]}$$

In Examples 20–7, 20–8, and 20–9, we found that the total entropy of system plus environment increases. Indeed, it has been found that for all real (irreversible) processes, the total entropy increases. No exceptions have been found. We can thus make the *general statement of the* **second law of thermodynamics** as follows:

The entropy of an isolated system never decreases. It either stays constant (reversible processes) or increases (irreversible processes).

Since all real processes are irreversible, we can equally well state the second law as:

The total entropy of any system plus that of its environment increases as a result of any natural process:

$$\Delta S = \Delta S_{syst} + \Delta S_{env} > 0. \qquad (20\text{–}13)$$

Although the entropy of one part of the universe may decrease in any process (see the Examples above), the entropy of some other part of the universe always increases by a greater amount, so the total entropy always increases.

Now that we finally have a quantitative general statement of the second law of thermodynamics, we can see that it is an unusual law. It differs considerably from other laws of physics, which are typically equalities (such as $F = ma$) or conservation laws (such as for energy and momentum). The second law introduces a new quantity, the entropy, S, but does not tell us it is conserved. Quite the opposite. Entropy always increases in time.

Entropy increases in all real processes

The second law of thermodynamics summarizes, very succinctly, which processes are observed in nature, and which are not. Or, said another way, it tells us about the *direction* processes go. For the reverse of any of the processes in the last few Examples, the entropy would decrease; and we never observe them. For example, we never observe heat flowing spontaneously from a cold body to a hot body, the reverse of Example 20–9. Nor do we ever observe a gas spontaneously compressing itself into a smaller volume, the reverse of Example 20–8 (gases always expand to fill their containers). Nor do we see thermal energy transform into kinetic energy of a rock so the rock rises spontaneously from the ground. Any of these processes would be consistent with the first law of thermodynamics (conservation of energy). It is the second law of thermodynamics they are not consistent with, and this is why we need the second law. If you were to see a movie run backward, you would probably realize it immediately because you would see odd occurrences—such as rocks rising spontaneously from the ground, or air rushing in from the atmosphere to fill an empty balloon (the reverse of free expansion). When watching a movie or video, we are tipped off to a reversal of time by observing whether entropy is increasing or decreasing. Hence entropy has been called **time's arrow**, for it tells us in which direction time is going.

Time's arrow

Is the general statement of the second law of thermodynamics—"the principle of entropy increase"—consistent with the Clausius and Kelvin-Planck statements? Yes, it is. This is easy to see, for if a process occurred in which heat flowed spontaneously out of a low-temperature (T_L) reservoir (so its entropy decreased), and all of it flowed into a high-temperature (T_H) reservoir (which increased in entropy), in violation of the Clausius statement, then the total change in entropy $\Delta S = Q/T_H - Q/T_L$ would be less than zero, since $T_L < T_H$. Thus the principle of entropy increase implies the Clausius statement. Can you show the equivalence of the entropy principle to the Kelvin-Planck statement?

Equivalence of second law statements

20–7 | Order to Disorder

The concept of entropy, as we have discussed it so far, may seem rather abstract. But we can relate it to the more ordinary concepts of *order* and *disorder*. In fact, the entropy of a system can be considered a *measure of the disorder of the system*. Then the second law of thermodynamics can be stated simply as:

Natural processes tend to move toward a state of greater disorder.

SECOND LAW OF THERMODYNAMICS (general statement)

Exactly what we mean by disorder may not always be clear, so we now consider a few examples. Some of these will show us how this very general statement of the second law actually applies beyond what we usually consider as thermodynamics.

Let us look at some very simple processes. First, a jar containing separate layers of salt and pepper is more orderly than when the salt and pepper are all mixed up. Shaking a jar containing separate layers results in a mixture, and no amount of shaking brings the layers back again. The natural process is from a state of relative order (layers) to one of relative disorder (a mixture), not the reverse. That is, disorder increases. Second, a solid coffee cup is a more "orderly"

(a)

(b)

(c)

FIGURE 20–13 Have you ever observed this process, a broken cup spontaneously reassembling and rising up to a table?

object than the pieces of a broken cup. Cups break when they fall, but they do not spontaneously mend themselves (Fig. 20–13). Again, the normal course of events is an increase of disorder.

Now let us consider some processes for which we have actually calculated the entropy change, and see that an increase in entropy results in an increase in disorder (or vice versa). When ice melts to water at 0°C, the entropy of the water increases, as we saw in Example 20–6. Intuitively, we can think of solid water, ice, as being more ordered than the less orderly fluid state which can flow all over the place. This change from order to disorder can be seen more clearly from the molecular point of view: the orderly arrangement of water molecules in an ice crystal has changed to the disorderly and somewhat random motion of the molecules in the fluid state.

When a hot object is put in contact with a cold object, heat flows from the high temperature to the low until the two objects reach the same intermediate temperature. At the beginning of the process we can distinguish two classes of molecules—those with a high average kinetic energy and those with a low average kinetic energy. After the process, all the molecules are in one class with the same average kinetic energy, and we no longer have the more orderly arrangement of molecules in two classes. Order has gone to disorder. Furthermore, note that the separate hot and cold objects could serve as the hot- and cold-temperature regions of a heat engine, and thus could be used to obtain useful work. But once the two objects are put in contact and reach the same temperature, no work can be obtained. Disorder has increased, since a system that has the ability to perform work must surely be considered to have a higher order than a system no longer able to do work.

An interesting example of the increase in entropy relates to the theory of biological evolution and to the growth of organisms. Clearly, a human being is a highly ordered organism. The process of evolution from the early macromolecules and simple forms of life to *Homo sapiens* represents increasing order. So, too, the development of an individual from a single cell to a grown person is a process of increasing order. Do these processes violate the second law of thermodynamics? No, they do not. In the processes of evolution and growth, and even during the life of an individual, waste products are eliminated. These small molecules that remain as a result of metabolism are simple molecules without much order in comparison to the macromolecules of life such as DNA and proteins. Thus they represent relatively higher disorder or entropy. Indeed, the total entropy of the molecules cast aside by organisms during the processes of evolution and growth is greater than the decrease in entropy associated with the order of the growing individual or evolving species.

20–8 | Energy Availability; Heat Death

In the process of heat conduction from a hot body to a cold one, we have seen that entropy increases and that order goes to disorder. The separate hot and cold objects could serve as the high- and low-temperature regions for a heat engine and thus could be used to obtain useful work. But after the two objects are put in contact with each other and reach the same uniform temperature, no work can be obtained from them. With regard to being able to do useful work, order has gone to disorder in this process.

The same can be said about a falling rock that comes to rest upon striking the ground. Just before hitting the ground, all the kinetic energy of the rock could have been used to do useful work. But once the rock's mechanical kinetic energy becomes thermal energy, this is no longer possible.

Both these examples illustrate another important aspect of the second law of thermodynamics—*in any natural process*, *some energy becomes unavailable to do useful work*. In any process, no energy is ever lost (it is always conserved). Rather, it becomes less useful—it can do less useful work. As time goes on, **energy is degraded**, in a sense; it goes from more orderly forms (such as mechanical) eventually to the least orderly form, internal or thermal energy. Entropy is a factor here because the amount of energy that becomes unavailable to do work is proportional to the change in entropy during any process.[†]

Degradation of energy

A natural outcome of this is the prediction that as time goes on, the universe will approach a state of maximum disorder. Matter will become a uniform mixture, heat will have flowed from high-temperature regions to low-temperature regions until the whole universe is at one temperature. No work can then be done. All the energy of the universe will have become degraded to thermal energy. All change will cease. This, the so-called **heat death** of the universe, has been much discussed by philosophers. This final state seems an inevitable consequence of the second law of thermodynamics, although it lies very far in the future. Yet it is based on the assumption that the universe is finite, which cosmologists are not fully sure of. The answers are not all in yet; and that makes science interesting.

"Heat death"

*20-9 Statistical Interpretation of Entropy and the Second Law

The ideas of entropy and disorder are made clearer with the use of a statistical or probabilistic analysis of the molecular state of a system. This statistical approach, which was first applied toward the end of the nineteenth century by Ludwig Boltzmann (1844–1906), makes a distinction between the "macrostate" and the "microstate" of a system. The **microstate** of a system would be specified when the position and velocity of every particle (or molecule) is given. The **macrostate** of a system is specified by giving the macroscopic properties of the system—the temperature, pressure, number of moles, and so on. In reality, we can know only the macrostate of a system. There are generally far too many molecules in a system to be able to know the velocity and position of every one at a given moment. Nonetheless, it is important to recognize that a great many different microstates can correspond to the *same* macrostate.

Microstates and macrostates

Let us take a simple example. Suppose you repeatedly shake four coins in your hand and drop them on the table. Specifying the number of heads and the number of tails that appear on a given throw is the macrostate of this system. Specifying each coin as being a head or a tail is the microstate of the system. In the following table we see the number of microstates that correspond to each macrostate:

Macrostate	Possible microstates (H = heads, T = tails)	Number of microstates
4 heads	H H H H	1
3 heads, 1 tail	H H H T, H H T H, H T H H, T H H H	4
2 heads, 2 tails	H H T T, H T H T, T H H T, H T T H, T H T H, T T H H	6
1 head, 3 tails	T T T H, T T H T, T H T T, H T T T	4
4 tails	T T T T	1

[†] It can be shown that the amount of energy that becomes unavailable to do useful work is equal to $T_L \Delta S$, where T_L is the lowest available temperature and ΔS is the total increase in entropy during the process.

A basic principle behind the statistical approach is that *each microstate is equally probable.* Thus the number of microstates that give the same macrostate corresponds to the relative probability of that macrostate occurring. The macrostate of two heads and two tails is the most probable one in our case of tossing four coins. Out of the total of 16 possible microstates, six correspond to two heads and two tails, so the probability of throwing two heads and two tails is 6 out of 16, or 38 percent. The probability of throwing one head and three tails is 4 out of 16, or 25 percent. The probability of four heads is only 1 in 16, or 6 percent. Of course if you threw the coins 16 times, you might not find that two heads and two tails appears exactly 6 times, or four tails exactly once. These are only probabilities or averages. But if you made 1600 throws, very nearly 38 percent of them would be two heads and two tails. The greater the number of tries, the closer are the percentages to the calculated probabilities.

If we consider tossing more coins, say 100, the relative probability of throwing all heads (or all tails) is greatly reduced. There is only one microstate corresponding to all heads. For 99 heads and 1 tail, there are 100 microstates since each of the coins could be the one tail. The relative probabilities for other macrostates are given in Table 20–1. There are a total of about 10^{30} microstates possible.[†] Thus the relative probability of finding all heads is 1 in 10^{30}, an incredibly unlikely event! The probability of obtaining 50 heads and 50 tails (see Table 20–1) is $1.0 \times 10^{29}/10^{30} = 0.10$ or 10 percent. The probability of obtaining between 45 and 55 heads is 90 percent.

Thus we see that as the number of coins increases, the probability of obtaining an orderly arrangement (all heads or all tails) becomes extremely unlikely. The least orderly arrangement (half heads, half tails) is the most probable; and the probability of being within a certain percentage (say 5 percent) of the most probable arrangement greatly increases as the number of coins increases. These same ideas can be applied to the molecules of a system. For example the most probable state of a gas (say the air in a room) is one in which the molecules take up the whole space and move about randomly; this corresponds to the Maxwellian distribution, Fig. 20–14a (see Section 18–2). On the other hand, the very orderly arrangement of all the molecules located in one corner of the room and all moving with the same velocity (Fig. 20–14b) is extremely unlikely.

From these examples, it is clear that probability is directly related to disorder and hence to entropy. That is, the most probable state is the one with greatest entropy or greatest disorder and randomness. Boltzmann showed that, consistent with Clausius's definition ($dS = dQ/T$) that the entropy of a system in a given (macro) state can be written

$$S = k \ln \mathcal{W}, \tag{20–14}$$

where k is Boltzmann's constant $(k = R/N_A = 1.38 \times 10^{-23}\,\text{J/K})$ and $\mathcal{W}$ is the number of microstates corresponding to the given macrostate. That is, $\mathcal{W}$ is

TABLE 20–1 Probabilities of various macrostates for 100 coin tosses

Macrostate		Number of microstates
Heads	Tails	$\mathcal{W}$
100	0	1
99	1	1.0×10^2
90	10	1.7×10^{13}
80	20	5.4×10^{20}
60	40	1.4×10^{28}
55	45	6.1×10^{28}
50	50	1.0×10^{29}
45	55	6.1×10^{28}
40	60	1.4×10^{28}
20	80	5.4×10^{20}
10	90	1.7×10^{13}
1	99	1.0×10^2
0	100	1

[†]Each coin has two possibilities, heads or tails. Then the possible number of microstates is $2 \times 2 \times 2 \times \ldots = 2^{100} = 1.27 \times 10^{30}$ (using a calculator or logarithms).

FIGURE 20–14 (a) Most probable distribution of molecular speeds in a gas (Maxwellian, or random); (b) orderly, but highly unlikely, distribution of speeds in which all molecules have nearly the same speed.

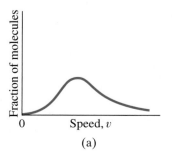

(a)

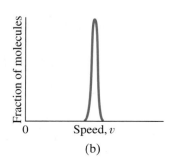

(b)

proportional to the probability of occurrence of that state. $\mathcal{W}$ is called the **thermodynamic probability**, or, sometimes, the disorder parameter.

EXAMPLE 20–10 **Free expansion—statistical determination of entropy.**
Use Eq. 20–14 to determine the change in entropy for the adiabatic *free expansion* of a gas, a calculation we did macroscopically in Example 20–8. Assume $\mathcal{W}$, the number of microstates for each macrostate, is the number of possible positions.

SOLUTION Assume the number of moles is $n = 1$ and then the number of molecules is $N = nN_A = N_A$, and assume the volume doubles, just as in Example 20–8. Because the volume doubles, the number of possible positions for each molecule doubles. Since there are N_A molecules, the total number of microstates increases by a factor of 2^{N_A} when the volume doubles. That is

$$\frac{\mathcal{W}_2}{\mathcal{W}_1} = 2^{N_A}.$$

The change in entropy is, from Eq. 20–14,

$$\Delta S = S_2 - S_1 = k(\ln \mathcal{W}_2 - \ln \mathcal{W}_1) = k \ln \frac{\mathcal{W}_2}{\mathcal{W}_1} = k \ln 2^{N_A} = kN_A \ln 2 = R \ln 2$$

which is the same result we got in Example 20–8.

In terms of probability, the second law of thermodynamics—which tells us that entropy increases in any process—reduces to the statement that those processes occur which are most probable. The second law thus becomes a trivial statement. However there is an additional element now. The second law in terms of probability does not *forbid* a decrease in entropy. Rather, it says the probability is extremely low. It is not impossible that salt and pepper should separate spontaneously into layers, or that a broken tea cup should mend itself. It is even possible that a lake should freeze over on a hot summer day (that is, heat flow out of the cold lake into the warmer surroundings). But the probability for such events occurring is extremely small. In our coin examples above, we saw that increasing the number of coins from 4 to 100 reduced drastically the probability of large deviations from the average, or most probable, arrangement. In ordinary systems we are dealing with incredibly large numbers of molecules: in 1 mol alone there are 6×10^{23} molecules. Hence the probability of deviation far from the average is incredibly tiny. For example, it has been calculated that the probability that a stone resting on the ground should transform 1 cal of thermal energy into mechanical energy and rise up into the air is much less likely than the probability that a group of monkeys typing randomly would by chance produce the complete works of Shakespeare.

*20–10 Thermodynamic Temperature Scale; Absolute Zero, and the Third Law of Thermodynamics

In Section 20–3 we discussed the Carnot engine and other (ideal) reversible engines. We saw that the efficiency of any reversible engine operating between two heat reservoirs depends only on the temperatures of these two reservoirs; and the efficiency does not depend on the working substance—it could be helium, water, or something else, and the efficiency would be the same. The efficiency is given by

$$e = 1 - \frac{|Q_L|}{|Q_H|},$$

where $|Q_H|$ is the heat absorbed from the high-temperature reservoir and $|Q_L|$ is the heat exhausted to the low-temperature reservoir. Since $|Q_L/Q_H|$ is the same for any reversible engine operating between the same two temperatures, Kelvin suggested using this fact to define an absolute temperature scale. That is, the ratio of the temperatures of the two reservoirs, T_H and T_L, is defined as the ratio of the heats, $|Q_H|$ and $|Q_L|$, exchanged with them by a Carnot or other reversible engine:

$$\frac{T_H}{T_L} = \frac{|Q_H|}{|Q_L|}. \tag{20–15}$$

This is the basis for the **Kelvin** or **thermodynamic temperature scale**.

In Section 20–3, we saw that this same relation, $T_H/T_L = |Q_H/Q_L|$, holds for a Carnot engine (Eq. 20–2) when the temperatures are based on the ideal gas temperature scale (Section 17–10) which we have been using up to now. Indeed, to complete the definition of the thermodynamic scale, we assign the value $T_{tp} = 273.16\,\text{K}$ to the triple point of water so that

$$T = (273.16\,\text{K})\left(\frac{|Q|}{|Q_{tp}|}\right),$$

where $|Q|$ and $|Q_{tp}|$ are the heats exchanged by a Carnot engine with reservoirs at temperatures T and T_{tp}. This corresponds precisely to the definition of the ideal gas scale. Thus, the thermodynamic scale is identical to the ideal gas scale over the range of validity of the latter (below about 1 K, no substance is a gas, so the ideal gas scale cannot be used). The thermodynamic scale is now considered the standard scale since it can be used over the entire range of possible temperatures, and it is also independent of the substance used. From a practical point of view, it is especially useful at very low temperatures.

Very low temperatures are difficult to obtain experimentally. In fact, it is found experimentally that the closer the temperature is to absolute zero, the more difficult it is to reduce the temperature further. And it is generally accepted that it is not possible to actually reach absolute zero in any finite number of processes. *Third law of thermodynamics* This last statement is one way to state[†] the **third law of thermodynamics**. Since the maximum efficiency that any heat engine can have is the Carnot efficiency

$$e = 1 - \frac{T_L}{T_H},$$

and since T_L can never be zero, we see that a 100-percent efficient heat engine is not possible.

[†] See also the statement in the footnote on page 529.

PROBLEM SOLVING Thermodynamics

1. Define the system you are dealing with; be careful to distinguish the system under study from its surroundings.

2. Be careful of signs associated with work and heat. In the first law work done *by* the system is positive; work done *on* the system is negative. Heat added to the system is positive, but heat removed from it is negative. With heat engines, we usually consider heat and work as positive and write energy conservation equations with + and − signs taking into account directions.

3. Watch the units used for work and heat; work is most often expressed in joules, and heat in calories or kilocalories. Be consistent: choose only one unit for use throughout a given problem.

4. Temperatures must generally be expressed in kelvins; temperature **differences** may be expressed in C° or K.

5. Efficiency (or coefficient of performance) is a ratio of two energy transfers: useful output divided by required input. Efficiency (but *not* coefficient of performance) is always less than 1 in value, and hence is often stated as a percentage.

6. The entropy of a system increases when heat is added to the system, and decreases when heat is removed. Because entropy is a state variable, the change in entropy ΔS for an irreversible process can be determined by calculating ΔS for a reversible process between the same two states.

Summary

A **heat engine** is a device for changing thermal energy, by means of heat flow, into useful work.

The **efficiency** of a heat engine is defined as the ratio of the work W done by the engine to the heat input $|Q_H|$. Because of conservation of energy, the work output equals $|Q_H| - |Q_L|$, where $|Q_L|$ is the heat exhausted to the environment; hence the efficiency

$$e = \frac{|W|}{|Q_H|} = 1 - \frac{|Q_L|}{|Q_H|}.$$

Carnot's (idealized) engine consists of two isothermal and two adiabatic processes in a reversible cycle. For a **Carnot engine**, or any reversible engine operating between two temperatures, T_H and T_L, the efficiency is

$$e_{\text{ideal}} = 1 - \frac{T_L}{T_H}.$$

Irreversible (real) engines always have an efficiency less than this.

All heat engines give rise to thermal pollution because they exhaust heat to the environment.

The operation of **refrigerators** and **air conditioners** is the reverse of that of a heat engine: work is done to extract heat from a cool region and exhaust it to a region at a higher temperature. A **heat pump** does work to bring heat from a cold exterior into a warmer interior.

The **second law of thermodynamics** can be stated in several equivalent ways:

(a) heat flows spontaneously from a hot object to a cold one, but not the reverse;

(b) there can be no 100 percent efficient heat engine—that is, one that can change a given amount of heat completely into work;

(c) natural processes tend to move toward a state of greater disorder or greater **entropy**.

Statement (c) is the most general statement of the second law of thermodynamics, and can be restated as: the total entropy, S, of any system plus that of its environment increases as a result of any natural process:

$$\Delta S > 0.$$

Entropy, which is a state variable, is a quantitative measure of the disorder of a system. The change in entropy of a system during a reversible process is given by $\Delta S = \int dQ/T$.

The second law of thermodynamics tells us in which direction processes tend to proceed; hence entropy is called "time's arrow."

As time goes on, energy is degraded to less useful forms—that is, it is less available to do useful work.

Questions

1. Is it possible to cool down a room on a hot summer day by leaving the refrigerator door open? Explain.

2. Can mechanical energy ever be transformed completely into heat or internal energy? Can the reverse happen? In each case, if your answer is no, explain why not; if yes, give examples.

3. Would a definition of heat engine efficiency as $e = |W|/|Q_L|$ be a useful one? Explain.

4. What plays the role of high-temperature and low-temperature reservoirs in (a) an internal combustion engine, (b) a steam engine? Are they, strictly speaking, heat reservoirs?

5. Which will give the greater improvement in the efficiency of a Carnot engine, a 10°C increase in the high-temperature reservoir, or a 10°C decrease in the low-temperature reservoir?

6. The oceans contain a tremendous amount of thermal energy. Why, in general, is it not possible to put this energy to useful work?

7. Discuss the factors that keep real engines from reaching Carnot efficiency.

8. The expansion valve in a refrigeration system, Fig. 20–9, is crucial for cooling the fluid. Explain how the cooling occurs.

9. Describe a process in nature that is nearly reversible.

10. (a) Describe how heat could be added to a system reversibly. (b) Could you use a stove burner to add heat to a system reversibly? Explain.

11. Powdered milk is very slowly (quasistatically) added to water while being stirred. Is this a reversible process?

12. Two identical systems are taken from state a to state b by two different *irreversible* processes. Will the change in entropy for the system be the same for each process? For the environment? Answer carefully and completely.

13. It can be said that the *total change in entropy during a process is a measure of the irreversibility of the process.* Discuss why this is valid, starting with the fact that $\Delta S = 0$ for a reversible process.

14. Use arguments, other than the principle of entropy increase, to show that for an adiabatic process, $\Delta S = 0$ if it is done reversibly and $\Delta S > 0$ if done irreversibly.

15. A gas is allowed to expand (a) adiabatically and (b) isothermally. In each process, does the entropy increase, decrease, or stay the same?

16. Entropy is often called "time's arrow" because it tells us in which direction natural processes occur. If a movie film were run backward, name some processes that you might see that would tell you that time was "running backward."

17. Give three examples, other than those mentioned in this chapter, of naturally occurring processes in which order goes to disorder. Discuss the observability of the reverse process.

18. Which do you think has the greater entropy, 1 kg of solid iron or 1 kg of liquid iron? Why?

19. What happens if you remove the lid of a bottle containing chlorine gas? Does the reverse process ever happen? Why or why not?

20. Think up several processes (other than those already mentioned in the text) that would obey the first law of thermodynamics, but, if they actually occurred, would violate the second law.

21. Describe how a free expansion, for which the entropy increases, can be considered as a process in which order goes to disorder. [*Hint:* consider a stack of papers dropped into a large box versus dropped all over the floor.]

22. Suppose you collect a lot of papers strewn all over the floor and put them in a neat stack; does this violate the second law of thermodynamics? Explain.

23. The first law of thermodynamics is sometimes whimsically stated as, "You can't get something for nothing," and the second law as, "You can't even break even." Explain how these statements could be equivalent to the formal statements.

24. Give three examples of naturally occurring processes that illustrate the degradation of usable energy into internal energy.

25. Living organisms convert relatively simple food molecules into a complex structure as they grow. Is this a violation of the second law of thermodynamics? Explain.

Problems

Section 20–2

1. (I) A heat engine exhausts 8500 J of heat while performing 2700 J of useful work. What is the efficiency of this engine?

2. (I) A heat engine does 8200 J of work in each cycle while absorbing 18.0 kcal of heat from a high-temperature reservoir. What is the efficiency of this engine?

3. (I) A typical power plant puts out 500 MW of electric power. Estimate the heat discharged per second, assuming that the plant has an efficiency of 38 percent.

4. (II) A four-cylinder gasoline engine has an efficiency of 0.25 and delivers 180 J of work per cycle per cylinder. The engine fires at 25 cycles per second. (*a*) Determine the work done per second. (*b*) What is the total heat input per second from the gasoline? (*c*) If the energy content of gasoline is 130 MJ per gallon, how long does one gallon last?

5. (II) The burning of gasoline in a car releases about 3.0×10^4 kcal/gal. If a car averages 38 km/gal when driving 90 km/h, which requires 20 hp, what is the efficiency of the engine under those conditions?

6. (II) Figure 20–15 is a *PV* diagram for a reversible heat engine in which 1.0 mol of argon, a nearly ideal monatomic gas, is initially at STP (point a). Points b and c are on an isotherm at $T = 423$ K. Process ab is at constant volume, process ac at constant pressure. (*a*) Is the path of the cycle carried out clockwise or counterclockwise? (*b*) What is the efficiency of this engine?

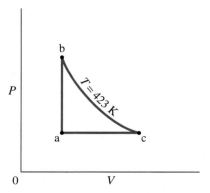

FIGURE 20–15 Problem 6.

7. (II) A 38-percent-efficient power plant puts out 810 MW (megawatts) of power (electrical). Cooling towers are used to take away the exhaust heat. If the air temperature is allowed to rise 7.5 C°, what volume of air (km^3) is heated per day? Will the local climate be heated significantly? If the heated air were to form a layer 200 m thick, how large an area would it cover for 24 h of operation? (The heat capacity of air is about 7.0 cal/mol·C° at constant pressure.)

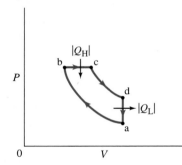

FIGURE 20–16
Problem 8.

8. (III) The operation of a *Diesel engine* can be idealized by the cycle shown in Fig. 20–16. Air is drawn into the cylinder during the intake stroke (not part of the idealized cycle). The air is compressed adiabatically, path ab. At point b diesel oil is injected into the cylinder which immediately burns since the temperature is very high. Combustion is slow, and during the first part of the power stroke, the gas expands at (nearly) constant pressure, path bc. After burning, the rest of the power stroke is adiabatic, path cd. Path da corresponds to the exhaust stroke. (*a*) Show that, for a quasistatic reversible engine undergoing this cycle using an ideal gas, the ideal efficiency is

$$e = 1 - \frac{(V_a/V_c)^{-\gamma} - (V_a/V_b)^{-\gamma}}{\gamma[(V_a/V_c)^{-1} - (V_a/V_b)^{-1}]},$$

where V_a/V_b is the "compression ratio", V_a/V_c is the "expansion ratio", and γ is defined by Eq. 19–12. (*b*) If $V_a/V_b = 15$ and $V_a/V_c = 5.0$, calculate the efficiency assuming the gas is diatomic (like N_2 and O_2) and ideal.

Section 20–3

9. (I) What is the maximum efficiency of a heat engine whose operating temperatures are 530°C and 305°C?

10. (I) The exhaust temperature of a heat engine is 220°C. What must be the high temperature if the Carnot efficiency is to be 36 percent?

11. (II) (a) Show that the work done by a Carnot engine is equal to the area enclosed by the Carnot cycle on a PV diagram, Fig. 20–5. (See Section 19–7.) (b) Generalize this to any reversible cycle.

12. (II) A heat engine exhausts its heat at 360°C and has a Carnot efficiency of 35 percent. What exhaust temperature would enable it to achieve a Carnot efficiency of 50 percent?

13. (II) A nuclear power plant operates at 75 percent of its maximum theoretical (Carnot) efficiency between temperatures of 660°C and 360°C. If the plant produces electric energy at the rate of 1.1 GW, how much exhaust heat is discharged per hour?

14. (II) An engine that operates at half its theoretical (Carnot) efficiency operates between 525°C and 280°C while producing work at the rate of 850 kW. How much heat is wasted per hour?

15. (II) Assume that a hiker needs 4000 kcal of energy to supply a day's worth of metabolism. Estimate the maximum height the person can climb in one day, using only this amount of energy. As a rough prediction, treat the person as an isolated heat engine, operating between the internal temperature of 37°C (98.6°F) and the ambient air temperature of 20°C.

16. (II) A Carnot engine performs work at the rate of 570 kW while using 1350 kcal of heat per second. If the temperature of the heat source is 580°C, at what temperature is the waste heat exhausted?

17. (II) A heat engine utilizes a heat source at 580°C and has a Carnot efficiency of 29 percent. To increase the efficiency to 35 percent, what must be the temperature of the heat source?

18. (II) At a steam power plant, steam engines work in pairs, the heat output of the first one being the approximate heat input of the second. The operating temperatures of the first are 680°C and 430°C, and of the second 415°C and 280°C. If the heat of combustion of coal is 2.8×10^7 J/kg, at what rate must coal be burned if the plant is put out 900 MW of power? Assume the efficiency of the engines is 65 percent of the ideal (Carnot) efficiency.

19. (II) Water is used to cool the power plant in Problem 18. If the water temperature is allowed to increase by no more than 5.5 C°, estimate how much water must pass through the plant per hour.

20. (II) Show that if two different adiabatic paths intersected at a single point on a PV diagram, they could be connected by an isotherm to form a cycle, and that an engine run on this cycle would violate the second law of thermodynamics. What do you conclude about the crossing of adiabatic lines?

21. (III) A Carnot cycle, shown in Fig. 20–5, has the following conditions: $V_a = 6.0$ L, $V_b = 15.0$ L, $T_H = 470°C$, and $T_L = 290°C$. The gas used in the cycle is 0.50 mole of a diatomic gas, $\gamma = 1.4$. Calculate (a) the pressures at a and b; (b) the volumes at c and d. (c) What is the work done along

process ab? (d) What is the heat lost along process cd? (e) Calculate the net work done for the whole cycle. (f) What is the efficiency of the cycle, using the definition $e = |W|/|Q_H|$? Show that this is the same as given by Eq. 20–3.

22. (III) One mole of monatomic gas undergoes a Carnot cycle with $T_H = 350°C$ and $T_L = 210°C$. The initial pressure is 10 atm. During the isothermal expansion, the volume doubles. (a) Find the values of the pressure and volume at the points a, b, c, and d (see Fig. 20–5). (b) Determine Q, W, and ΔU for each segment of the cycle. (c) Calculate the efficiency of the cycle.

Section 20–4

23. (I) The low temperature of a freezer cooling coil is −15°C and the discharge temperature is 30°C. What is the maximum theoretical coefficient of performance?

24. (I) If an ideal refrigerator keeps its contents at −15°C when the house temperature is 22°C, what is its coefficient of performance?

25. (I) A restaurant refrigerator has a coefficient of performance of 5.0. If the temperature in the kitchen outside the refrigerator is 29°C, what is the lowest temperature that could be obtained inside the refrigerator if it were ideal?

26. (II) A heat pump is used to keep a house warm at 22°C. How much work is required of the pump to deliver 2800 J of heat into the house if the outdoor temperature is (a) 0°C, (b) −15°C? Assume ideal (Carnot) behavior.

27. (II) An ideal engine has an efficiency of 35 percent. If it were run backward as a heat pump, what would be its coefficient of performance?

28. (II) The engine of Example 20–2 is run in reverse. How long would it take to freeze a tray of a dozen 40-g compartments of liquid water at room temperature (20°C) into a dozen ice cubes at the freezing point, assuming that it takes 450 W of input electric power to run it? Assume ideal (Carnot) behavior.

29. (II) A "Carnot" refrigerator (reverse of a Carnot engine) absorbs heat from the freezer compartment at a temperature of −17°C and exhausts it into the room at 25°C. (a) How much work must be done by the refrigerator to change 0.50 kg of water at 25°C into ice at −17°C? (b) If the compressor output is 200 W, what minimum time is needed to take 0.50 kg of 25°C water and freeze it at 0°C?

30. (II) (a) Given that the coefficient of performance of a refrigerator is defined (Eq. 20–4a) as

$$CP = \frac{|Q_L|}{|W|},$$

show that for an ideal (Carnot) refrigerator,

$$CP_{ideal} = \frac{T_L}{T_H - T_L}.$$

(b) Write the CP in terms of the efficiency e of the reversible heat engine obtained by running the refrigerator backward. (c) What is the coefficient of performance for an ideal refrigerator that maintains a freezer compartment at −16°C when the condenser's temperature is 22°C?

31. (II) What volume of water at 0°C can a freezer make into ice cubes in one hour, if the coefficient of performance of the cooling unit is 7.0 and the power input is 1.0 kilowatt?

32. (II) A central heat pump operating as an air conditioner draws 36,000 Btu per hour from a building and operates between the temperatures of 24°C and 38°C. (*a*) If its coefficient of performance is 24% of that of a Carnot air conditioner, what is the effective coefficient of performance? (*b*) What is the power required of the compressor motor? (*c*) What is the power in terms of hp?

Sections 20–5 and 20–6

33. (I) A 10.0-kg box having an initial speed of 3.0 m/s slides along a rough table and comes to rest. Estimate the total change in entropy of the universe. Assume all objects are at room temperature (293 K).

34. (I) What is the change in entropy of 1.00 m^3 of water at 0°C when it is frozen to ice at 0°C?

35. (II) If the water in Problem 34 were frozen by being in contact with a great deal of ice at −10°C, what would be the total change in entropy of the process?

36. (II) If 1.00 kg of water at 100°C is changed by a reversible process to steam at 100°C, determine the change in entropy of (*a*) the water, (*b*) the surroundings, and (*c*) the universe as a whole. (*d*) How would your answers differ if the process were irreversible?

37. (II) An aluminum rod conducts 7.50 cal/s from a heat source maintained at 240°C to a large body of water at 27°C. Calculate the rate at which entropy increases in this process.

38. (II) A 3.8-kg piece of aluminum at 30°C is placed in 1.0 kg of water in a Styrofoam container at room temperature (20°C). Estimate the net change in entropy of the system.

39. (II) What is the total change in entropy when 2.5 kg of water at 0°C is frozen to ice at 0°C by being in contact with 450 kg of ice at −15°C?

40. (II) When 2.0 kg of water at 20°C is mixed with 3.0 kg of water at 80°C in a well-insulated container, what is the change in entropy of the system?

41. (II) Show that the principle of entropy increase is equivalent to the Kelvin-Planck statement of the second law of thermodynamics.

42. (II) The temperature of 2.0 mol of an ideal diatomic gas goes from 25°C to 45°C at a constant volume. What is the change in entropy?

43. (II) (*a*) Calculate the change in entropy of 1.00 kg of water when it is heated from 0°C to 100°C. (*b*) Does the entropy of the surroundings change? If so, by how much?

44. (II) A 150-g insulated aluminum cup at 20°C is filled with 240 g of water at 100°C. (*a*) What is the final temperature of the mixture? (*b*) What is the total change in entropy as a result of the mixing process?

45. (II) Two samples of an ideal gas are initially at the same temperature and pressure; they are each compressed reversibly from a volume V to volume $V/2$, one isothermally, the other adiabatically. (*a*) In which sample is the final pressure greater? (*b*) Determine the change in entropy of the gas for each process. (*c*) What is the entropy change of the environment for each process?

46. (II) One mole of nitrogen (N_2) gas and one mole of argon (Ar) gas are in separate, equal-sized, insulated containers at the same temperature. The containers are then connected and the gases (assumed ideal) allowed to mix. (*a*) What is the change in entropy of the system and (*b*) of the environment? (*c*) Repeat part (*a*) but assume one container is twice as large as the other.

47. (II) (*a*) Why would you expect the total entropy change in a Carnot cycle to be zero? (*b*) Do a calculation to show that it is zero.

48. (III) The specific heat per mole of potassium at low temperatures is given by $C_V = aT + bT^3$, where $a = 2.08 \text{ mJ/mol} \cdot \text{K}^2$ and $b = 2.57 \text{ mJ/mol} \cdot \text{K}^4$. Calculate the entropy change of 0.25 mol of potassium when its temperature is lowered from 3.0 K to 1.0 K.

Section 20–8

49. (III) A general theorem states that the amount of energy that becomes unavailable to do useful work in any process is equal to $T_L \Delta S$, where T_L is the lowest temperature available and ΔS is the total change in entropy during the process. Show that this is valid in the specific cases of (*a*) a falling rock that comes to rest when it hits the ground; (*b*) the free adiabatic expansion of an ideal gas; and (*c*) the conduction of heat, Q, from a high-temperature (T_H) reservoir to a low-temperature (T_L) reservoir. [*Hint*: In part (*c*) compare to a Carnot engine.]

50. (III) Determine the work available in a 5.0 kg block of copper at 420 K if the surroundings are at 290 K. Use results of Problem 49.

* Section 20–9

* 51. (II) Suppose you repeatedly shake six coins in your hand and drop them on the table. Construct a table showing the number of microstates that correspond to each macrostate. What is the probability of obtaining (*a*) three heads and three tails and (*b*) six heads?

* 52. (II) Calculate the relative probabilities, when you throw two dice, of obtaining (*a*) a 7, (*b*) an 11, (*c*) a 5.

* 53. (II) Rank the following five-card hands in order of increasing probability: (*a*) four aces and a king; (*b*) six of hearts, eight of diamonds, queen of clubs, three of hearts, jack of spades; (*c*) two jacks, two queens, and an ace; and (*d*) any hand having no two equal-value cards. Explain your ranking using microstates and macrostates.

* 54. (II) (*a*) Suppose you have four coins, all with tails up. You now rearrange them so two heads and two tails are up. What was the change in entropy of the coins? (*b*) Suppose your system is the 100 coins of Table 20–1; what is the change in entropy of the coins if they are mixed randomly initially, 50 heads and 50 tails, and you arrange them so all 100 are heads? (*c*) Compare these entropy changes to ordinary thermodynamic entropy changes, such as Examples 20–6, 7, 8 and 9.

General Problems

55. It is not necessary that a heat engine's hot environment be hotter than ambient temperature. Liquid nitrogen (90 K) is about as cheap as bottled water. What would be the efficiency of an engine that made use of heat transferred from air at room temperature (293 K) to the liquid nitrogen "fuel"?

56. It has been suggested that a heat engine could be developed that made use of the temperature difference between water at the surface of the ocean and that several hundred meters deep. In the tropics, the temperatures may be 27°C and 4°C, respectively. What is the maximum efficiency such an engine could have? Why might such an engine be feasible in spite of the low efficiency? Can you imagine any adverse environmental effects that might occur?

57. Two 1100-kg cars are traveling 95 km/h in opposite directions when they collide and are brought to rest. Estimate the change in entropy of the universe as a result of this collision. Assume $T = 20°C$.

58. A 120-g insulated aluminum cup at 15°C is filled with 210 g of water at 50°C. After a few minutes, equilibrium is reached. Determine (a) the final temperature, and (b) the total change in entropy.

59. (a) What is the coefficient of performance of an ideal heat pump that extracts heat from 6°C air outside and deposits heat inside your house at 24°C? (b) If this heat pump operates on 1000 W of electrical energy, what is the maximum heat it can deliver into your house each hour?

60. An inventor claims to have designed and built an engine that produces 1.50 MW (megawatts) of usable work while taking in 3.00 MW of thermal energy at 425 K, and rejecting 1.50 MW of thermal energy at 215 K. Is there anything fishy about his claim? Explain.

61. Suppose a power plant delivers energy at 900 MW using steam turbines. The steam goes into the turbines superheated at 600 K and deposits its unused heat in river water at 285 K. Assume that the turbine operates as an ideal Carnot engine. (a) If the river flow rate is 37 m³/s, calculate the average temperature increase of the river water downstream from the power plant. (b) What is the entropy increase per kilogram of the downstream river water in J/kg·K?

62. A 100-hp car engine operates at about 15 percent efficiency. Assume the engine's water temperature of 85°C is its cold-temperature (exhaust) reservoir and 500°C is its thermal "intake" temperature (the temperature of the exploding gas/air mixture). (a) Calculate its efficiency relative to its maximum possible (Carnot) efficiency. (b) Estimate how much power (in watts) goes into moving the car, and how much heat, in joules and in kcal, is exhausted to the air in 1.0 h.

63. A falling rock has kinetic energy K just before striking the ground and coming to rest. What is the total change in entropy of the rock plus environment as a result of this collision?

64. An aluminum can, with negligible heat capacity, is filled with 500 g of water at 0°C and then is brought into thermal contact with a similar can filled with 500 g of water at 50°C. Find the change in entropy of the system if no heat is allowed to exchange with the surroundings.

65. (II) Thermodynamic processes can be represented not only on PV and PT diagrams; another useful one is a TS (temperature-entropy) diagram. (a) Draw a TS diagram for a Carnot cycle. (b) What does the area within the curve represent?

66. (II) A real heat engine working between heat reservoirs at 400 K and 850 K produces 600 J of work per cycle for a heat input of 1600 J. (a) Compare the efficiency of this real engine to that of a Carnot engine. (b) Calculate the total entropy change of the universe for each cycle of the real engine. (c) Calculate the total entropy change of the universe for a Carnot engine operating between the same two temperatures. (d) Show that the difference in work done by these two engines per cycle is $T_L \Delta S$, where T_L is the temperature of the low-temperature reservoir (400 K) and ΔS is the entropy increase per cycle of the real engine. (See also Problem 49 and Section 20–8.)

67. The *Stirling cycle*, shown in Fig. 20–17, is useful to describe external combustion engines as well as solar-power systems. Find the efficiency of the cycle in terms of the parameters shown, assuming a monatomic gas as the working substance. The processes ab and cd are isothermal whereas bc and da are at constant volume. How does it compare to the Carnot efficiency?

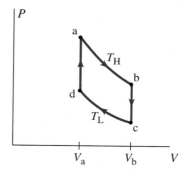

FIGURE 20–17 Problem 67.

(Problems continue on the next page.)

68. One mole of an ideal monatomic gas at STP first undergoes an isothermal expansion so that the volume at b is 2.5 times the volume at a (Fig. 20–18). Next, heat is extracted at a constant volume so that the pressure drops. The gas is then compressed adiabatically back to the original state. (a) Calculate the pressures at b and c. (b) Determine the temperature at c. (c) Determine the work done, heat input or extracted, and the change in entropy for each process. (d) What is the efficiency of this cycle?

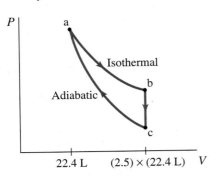

FIGURE 20–18 Problem 68.

69. As a mechanical engineer, you must design a cooling unit for a new freezer. The enclosure has an inner surface area of 6.0 m² and is bounded by walls that are 10 cm thick, with a thermal conductivity of 0.050 W/m·K. The inside must be kept at −10 C° in a room that is at 20°C. The motor for the cooling unit must run no more than 15% of the time. What is the minimum power requirement of the cooling motor?

70. A gas turbine operates under the *Brayton cycle*, which is depicted in the *PV* diagram of Fig. 20–19. In process ab the air-fuel mixture undergoes an adiabatic compression. This is followed, in process bc, with an isobaric (constant pressure) heating, by combustion. Process cd is an adiabatic expansion with expulsion of the products to the atmosphere. The return step, da, takes place at constant pressure. If the working gas behaves like an ideal gas, show that the efficiency of the Brayton Cycle is

$$e = 1 - \left(\frac{P_b}{P_a}\right)^{\frac{1-\gamma}{\gamma}} .$$

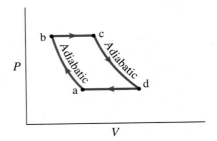

FIGURE 20–19 Problem 70.

* **71.** Estimate the probability that a bridge player will be dealt (a) all four aces (among 13 cards), and (b) all 13 cards of one suit.

APPENDIX A

Mathematical Formulas

A–1 Quadratic Formula

If

$$ax^2 + bx + c = 0$$

then

$$x = \frac{-b \pm \sqrt{b^2 - 4ac}}{2a}$$

A–2 Binomial Expansion

$$(1 \pm x)^n = 1 \pm nx + \frac{n(n-1)}{2!}x^2 \pm \frac{n(n-1)(n-2)}{3!}x^3 + \cdots$$

$$(x + y)^n = x^n\left(1 + \frac{y}{x}\right)^n = x^n\left(1 + n\frac{y}{x} + \frac{n(n-1)}{2!}\frac{y^2}{x^2} + \cdots\right)$$

A–3 Other Expansions

$$e^x = 1 + x + \frac{x^2}{2!} + \frac{x^3}{3!} + \cdots$$

$$\ln(1 + x) = x - \frac{x^2}{2} + \frac{x^3}{3} - \frac{x^4}{4} + \cdots$$

$$\sin\theta = \theta - \frac{\theta^3}{3!} + \frac{\theta^5}{5!} - \cdots$$

$$\cos\theta = 1 - \frac{\theta^2}{2!} + \frac{\theta^4}{4!} - \cdots$$

$$\tan\theta = \theta + \frac{\theta^3}{3} + \frac{2}{15}\theta^5 + \cdots \qquad |\theta| < \frac{\pi}{2}$$

In general:

$$f(x) = f(0) + \left(\frac{df}{dx}\right)_0 x + \left(\frac{d^2f}{dx^2}\right)_0 \frac{x^2}{2!} + \cdots$$

A–4 Areas and Volumes

Object	Surface area	Volume
Circle, radius r	πr^2	—
Sphere, radius r	$4\pi r^2$	$\frac{4}{3}\pi r^3$
Right circular cylinder, radius r, height h	$2\pi r^2 + 2\pi rh$	$\pi r^2 h$
Right circular cone, radius r, height h	$\pi r^2 + \pi r\sqrt{r^2 + h^2}$	$\frac{1}{3}\pi r^2 h$

A–5 Plane Geometry

1.

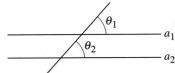

FIGURE A–1

If line a_1 is parallel to line a_2, then $\theta_1 = \theta_2$.

2.

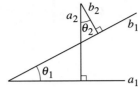

FIGURE A–2

If $a_1 \perp a_2$ and $b_1 \perp b_2$, then $\theta_1 = \theta_2$.

3. The sum of the angles in any plane triangle is 180°.

4. *Pythagorean theorem:*

FIGURE A–3

In any right triangle (one angle = 90°) of sides a, b, and c:

$$a^2 + b^2 = c^2$$

where c is the length of the hypotenuse (opposite the 90° angle).

A–6 Trigonometric Functions and Identities

(See Fig. A–4.)

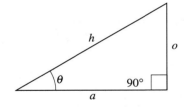

FIGURE A–4

$$\sin\theta = \frac{o}{h} \qquad\qquad \csc\theta = \frac{1}{\sin\theta} = \frac{h}{o}$$

$$\cos\theta = \frac{a}{h} \qquad\qquad \sec\theta = \frac{1}{\cos\theta} = \frac{h}{a}$$

$$\tan\theta = \frac{o}{a} = \frac{\sin\theta}{\cos\theta} \qquad\qquad \cot\theta = \frac{1}{\tan\theta} = \frac{a}{o}$$

$$a^2 + o^2 = h^2 \qquad\qquad \text{[Pythagorean theorem]}.$$

Figure A–5 shows the signs (+ or −) that cosine, sine, and tangent take on for angles θ in the four quadrants (0° to 360°). Note that angles are measured counterclockwise from the x axis as shown; negative angles are measured from *below* the x axis, clockwise: for example, −30° = +330°, and so on.

FIGURE A–5

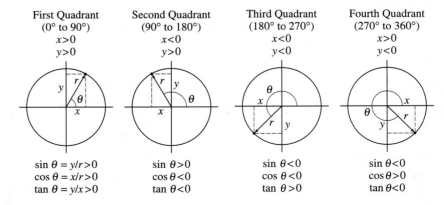

First Quadrant
(0° to 90°)
$x > 0$
$y > 0$

Second Quadrant
(90° to 180°)
$x < 0$
$y > 0$

Third Quadrant
(180° to 270°)
$x < 0$
$y < 0$

Fourth Quadrant
(270° to 360°)
$x > 0$
$y < 0$

$\sin\theta = y/r > 0$
$\cos\theta = x/r > 0$
$\tan\theta = y/x > 0$

$\sin\theta > 0$
$\cos\theta < 0$
$\tan\theta < 0$

$\sin\theta < 0$
$\cos\theta < 0$
$\tan\theta > 0$

$\sin\theta < 0$
$\cos\theta > 0$
$\tan\theta < 0$

The following are some useful identities among the trigonometric functions:

$$\sin^2\theta + \cos^2\theta = 1, \quad \sec^2\theta - \tan^2\theta = 1, \quad \csc^2\theta - \cot^2\theta = 1$$

$$\sin 2\theta = 2\sin\theta\cos\theta$$

$$\cos 2\theta = \cos^2\theta - \sin^2\theta = 2\cos^2\theta - 1 = 1 - 2\sin^2\theta$$

$$\tan 2\theta = \frac{2\tan\theta}{1 - \tan^2\theta}$$

$$\sin(A \pm B) = \sin A\cos B \pm \cos A\sin B$$

$$\cos(A \pm B) = \cos A\cos B \mp \sin A\sin B$$

$$\tan(A \pm B) = \frac{\tan A \pm \tan B}{1 \mp \tan A\tan B}$$

$$\sin(180° - \theta) = \sin\theta$$

$$\cos(180° - \theta) = -\cos\theta$$

$$\sin(90° - \theta) = \cos\theta$$

$$\cos(90° - \theta) = \sin\theta$$

$$\cos(-\theta) = \cos\theta$$

$$\sin(-\theta) = -\sin\theta$$

$$\tan(-\theta) = -\tan\theta$$

$$\sin\tfrac{1}{2}\theta = \sqrt{\frac{1 - \cos\theta}{2}}, \quad \cos\tfrac{1}{2}\theta = \sqrt{\frac{1 + \cos\theta}{2}}, \quad \tan\tfrac{1}{2}\theta = \sqrt{\frac{1 - \cos\theta}{1 + \cos\theta}}$$

$$\sin A \pm \sin B = 2\sin\left(\frac{A \pm B}{2}\right)\cos\left(\frac{A \mp B}{2}\right).$$

For any triangle (see Fig. A–6):

$$\frac{\sin\alpha}{a} = \frac{\sin\beta}{b} = \frac{\sin\gamma}{c} \qquad \text{[Law of sines]}$$

$$c^2 = a^2 + b^2 - 2ab\cos\gamma. \qquad \text{[Law of cosines]}$$

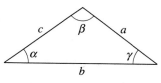

FIGURE A–6

A–7 Logarithms

The following identities apply to common logs (base 10), natural logs (base e) which are often abbreviated ln, or logs to any other base.

$$\log(ab) = \log a + \log b$$

$$\log\left(\frac{a}{b}\right) = \log a - \log b$$

$$\log a^n = n\log a.$$

A–8 Vectors

Vector addition is covered in Sections 3–2 to 3–5.
Vector multiplication is covered in Sections 3–3, 7–2 and 11–1.

Derivatives and Integrals

B–1 Derivatives: General Rules

(See also Section 2–3.)

$$\frac{dx}{dx} = 1$$

$$\frac{d}{dx}[af(x)] = a\frac{df}{dx} \qquad (a = \text{constant})$$

$$\frac{d}{dx}[f(x) + g(x)] = \frac{df}{dx} + \frac{dg}{dx}$$

$$\frac{d}{dx}[f(x)g(x)] = \frac{df}{dx}g + f\frac{dg}{dx}$$

$$\frac{d}{dx}[f(y)] = \frac{df}{dy}\frac{dy}{dx} \qquad \text{[chain rule]}$$

$$\frac{dx}{dy} = \frac{1}{\left(\dfrac{dy}{dx}\right)} \qquad \text{if } \frac{dy}{dx} \neq 0.$$

B–2 Derivatives: Particular Functions

$$\frac{da}{dx} = 0 \qquad (a = \text{constant})$$

$$\frac{d}{dx}x^n = nx^{n-1}$$

$$\frac{d}{dx}\sin ax = a\cos ax$$

$$\frac{d}{dx}\cos ax = -a\sin ax$$

$$\frac{d}{dx}\tan ax = a\sec^2 ax$$

$$\frac{d}{dx}\ln ax = \frac{1}{x}$$

$$\frac{d}{dx}e^{ax} = ae^{ax}$$

B–3 Indefinite Integrals: General Rules

(See also Section 7–3.)

$$\int dx = x$$

$$\int af(x)\,dx = a\int f(x)\,dx \qquad (a = \text{constant})$$

$$\int [f(x) + g(x)]\,dx = \int f(x)\,dx + \int g(x)\,dx$$

$$\int u\,dv = uv - \int v\,du \qquad \text{[integration by parts]}$$

B–4 Indefinite Integrals: Particular Functions

(An arbitrary constant can be added to the right side of each equation.)

$$\int a\,dx = ax \qquad (a = \text{constant})$$

$$\int x^m\,dx = \frac{1}{m+1}x^{m+1} \qquad (m \neq -1)$$

$$\int \sin ax\,dx = -\frac{1}{a}\cos ax$$

$$\int \cos ax\,dx = \frac{1}{a}\sin ax$$

$$\int \tan ax\,dx = \frac{1}{a}\ln|\sec ax|$$

$$\int \frac{1}{x}\,dx = \ln x$$

$$\int e^{ax}\,dx = \frac{1}{a}e^{ax}$$

$$\int \frac{dx}{x^2 + a^2} = \frac{1}{a}\tan^{-1}\frac{x}{a}$$

$$\int \frac{dx}{x^2 - a^2} = \frac{1}{2a}\ln\left(\frac{x-a}{x+a}\right) \qquad (x^2 > a^2)$$

$$= -\frac{1}{2a}\ln\left(\frac{a+x}{a-x}\right) \qquad (x^2 < a^2)$$

$$\int \frac{dx}{\sqrt{x^2 \pm a^2}} = \ln(x + \sqrt{x^2 \pm a^2})$$

$$\int \frac{dx}{(x^2 \pm a^2)^{\frac{3}{2}}} = \frac{\pm x}{a^2\sqrt{x^2 \pm a^2}}$$

$$\int \frac{x\,dx}{(x^2 \pm a^2)^{\frac{3}{2}}} = \frac{-1}{\sqrt{x^2 \pm a^2}}$$

Gravitational Force due to a Spherical Mass Distribution

In Chapter 6 (Section 6–1), we stated that the gravitational force exerted by or on a uniform sphere acts as if all the mass of the sphere were concentrated at its center. In other words, the gravitational force that a uniform sphere exerts on a particle outside it is

$$F = G\frac{mM}{r^2}, \qquad \text{[}m \text{ outside sphere of mass } M\text{]}$$

where m is the mass of the particle, M the mass of the sphere, and r the distance of m from the center of the sphere. Now we will derive this result. We will use the concepts of infinitesimally small quantities and integration.

First we consider a very thin, uniform spherical shell (like a thin-walled basketball) of mass M whose thickness t is small compared to its radius R (Fig. C–1). The force on a particle of mass m at a distance r from the center of the shell can be calculated as the vector sum of the forces due to all the particles of the shell. We imagine the shell divided up into thin (infinitesimal) circular strips so that all points on a strip are equidistant from our particle m. One of these circular strips, labeled AB, is shown in Fig. C–1. It is $Rd\theta$ wide, t thick, and has a radius $R\sin\theta$. The force on our particle m due to a tiny piece of the strip at point A is represented by the vector $\mathbf{F}_A$ shown. The force due to a tiny piece of the strip at point B, which is diametrically opposite A, is the force $\mathbf{F}_B$. We take the two pieces at A and B to be of equal mass, so $F_A = F_B$. The horizontal components of $\mathbf{F}_A$ and $\mathbf{F}_B$ are each equal to

$$F_A\cos\phi$$

and point toward the center of the shell. The vertical components of $\mathbf{F}_A$ and $\mathbf{F}_B$ are of equal magnitude and point in opposite directions, and so cancel. Since for every point on the strip there is a corresponding point diametrically opposite (as with A and B), we see that the net force due to the entire strip points toward the center of the shell. Its magnitude will be

$$dF = G\frac{m\,dM}{l^2}\cos\phi,$$

where dM is the mass of the entire circular strip and l is the distance from all

FIGURE C–1 Calculating the gravitational force on a particle of mass m due to a uniform spherical shell of radius R and mass M.

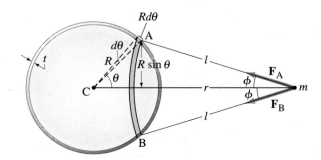

points on the strip to m, as shown. We write dM in terms of the density ρ; by density we mean the mass per unit volume (Section 13–1). Hence, $dM = \rho\, dV$, where dV is the volume of the strip and equals $(2\pi R \sin\theta)(t)(R\, d\theta)$. Then the force dF due to the circular strip shown is

$$dF = G\frac{m\rho 2\pi R^2 t \sin\theta\, d\theta}{l^2}\cos\phi. \tag{C–1}$$

To get the total force F that the entire shell exerts on the particle m, we must integrate over all the circular strips: that is, from $\theta = 0°$ to $\theta = 180°$. But our expression for dF contains l and ϕ, which are functions of θ. From Fig. C–1 we can see that

$$l\cos\phi = r - R\cos\theta.$$

Furthermore, we can write the law of cosines for triangle CmA:

$$\cos\theta = \frac{r^2 + R^2 - l^2}{2rR}. \tag{C–2}$$

With these two expressions we can reduce our three variables (l, θ, ϕ) to only one, which we take to be l. We do two things with Eq. C–2: (1) We put it into the equation for $l\cos\phi$ above:

$$\cos\phi = \frac{1}{l}(r - R\cos\theta) = \frac{r^2 + l^2 - R^2}{2rl};$$

and (2) we take the differential of both sides of Eq. C–2 (because $\sin\theta\, d\theta$ appears in the expression for dF, Eq. C–1):

$$-\sin\theta\, d\theta = -\frac{2l\, dl}{2rR} \quad\text{or}\quad \sin\theta\, d\theta = \frac{l\, dl}{rR},$$

since r and R are considered constants when summing over the strips. Now we insert these into Eq. C–1 for dF and find

$$dF = Gm\rho\pi t\,\frac{R}{r^2}\left(1 + \frac{r^2 - R^2}{l^2}\right)dl.$$

Now we integrate to get the net force on our thin shell of radius R. To integrate over all the strips ($\theta = 0°$ to $180°$), we must go from $l = r - R$ to $l = r + R$ (see Fig. C–1). Thus,

$$F = Gm\rho\pi t\,\frac{R}{r^2}\left[l - \frac{r^2 - R^2}{l}\right]_{l=r-R}^{l=r+R}$$

$$= Gm\rho\pi t\,\frac{R}{r^2}(4R).$$

The volume V of the spherical shell is its area $(4\pi R^2)$ times the thickness t. Hence the mass $M = \rho V = \rho 4\pi R^2 t$, and finally

$$F = G\frac{mM}{r^2}. \qquad \left[\begin{array}{c}\text{particle of mass } m \text{ outside a}\\ \text{thin uniform spherical shell of mass } M\end{array}\right]$$

This result gives us the force a thin shell exerts on a particle of mass m a distance r from the center of the shell, and *outside* the shell. We see that the force is the same as that between m and a particle of mass M at the center of the shell. In other words, for purposes of calculating the gravitational force exerted on or by a uniform spherical shell, we can consider all its mass concentrated at its center.

What we have derived for a shell holds also for a solid sphere, since a solid sphere can be considered as made up of many concentric shells, from $R = 0$ to $R = R_0$, where R_0 is the radius of the solid sphere. Why? Because if each shell has mass dM, we write for each shell, $dF = Gm\,dM/r^2$, where r is the distance from the center C to mass m and is the same for all shells. Then the total force equals the sum or integral over dM, which gives the total mass M. Thus the result

$$F = G\frac{mM}{r^2} \qquad \begin{bmatrix} \text{particle of mass } m \text{ outside} \\ \text{solid sphere of mass } M \end{bmatrix} \quad \text{(C–3)}$$

is valid for a solid sphere of mass M even if the density varies with distance from the center. (It is not valid if the density varies within each shell—that is, depends not only on R.) Thus the gravitational force exerted on or by spherical objects, including nearly spherical objects like the Earth, Sun, and Moon, can be considered to act as if the objects were point particles.

This result, Eq. C–3, is true only if the mass m is outside the sphere. Let us next consider a point mass m that is located inside the spherical shell of Fig. C–1. Here, r would be less than R, and the integration over l would be from $l = R - r$ to $l = R + r$, so

$$\left[l - \frac{r^2 - R^2}{l} \right]_{R-r}^{R+r} = 0.$$

Thus the force on any mass inside the shell would be zero. This result has particular importance for the electrostatic force, which is also an inverse square law. For the gravitational situation, we see that at points within a solid sphere, say 1000 km below the earth's surface, only the mass up to that radius contributes to the net force. The outer shells beyond the point in question contribute no net gravitational effect.

The results we have obtained here can also be reached using the gravitational analog of Gauss's law for electrostatics (Chapter 22).

Selected Isotopes

(1) Atomic Number Z	(2) Element	(3) Symbol	(4) Mass Number A	(5) Atomic Mass†	(6) % Abundance (or Radioactive Decay Mode)	(7) Half-life (if radioactive)
0	(Neutron)	n	1	1.008665	β^-	10.4 min
1	Hydrogen	H	1	1.007825	99.985%	
	Deuterium	D	2	2.014102	0.015%	
	Tritium	T	3	3.016049	β^-	12.33 yr
2	Helium	He	3	3.016029	0.000137%	
			4	4.002603	99.999863%	
3	Lithium	Li	6	6.015122	7.5%	
			7	7.016004	92.5%	
4	Beryllium	Be	7	7.016929	EC, γ	53.12 days
			9	9.012182	100%	
5	Boron	B	10	10.012937	19.9%	
			11	11.009305	80.1%	
6	Carbon	C	11	11.011434	β^+, EC	20.39 min
			12	12.000000	98.90%	
			13	13.003355	1.10%	
			14	14.003242	β^-	5730 yr
7	Nitrogen	N	13	13.005739	β^+	9.965 min
			14	14.003074	99.63%	
			15	15.000108	0.37%	
8	Oxygen	O	15	15.003065	β^+, EC	122.24 s
			16	15.994915	99.76%	
			18	17.999160	0.20%	
9	Fluorine	F	19	18.998403	100%	
10	Neon	Ne	20	19.992440	90.48%	
			22	21.991386	9.25%	
11	Sodium	Na	22	21.994437	β^+, EC, γ	2.6019 yr
			23	22.989770	100%	
			24	23.990963	β^-, γ	14.9590 h
12	Magnesium	Mg	24	23.985042	78.99%	
13	Aluminum	Al	27	26.981538	100%	

† The masses given in column (5) are those for the neutral atom, including the Z electrons.

(1) Atomic Number Z	(2) Element	(3) Symbol	(4) Mass Number A	(5) Atomic Mass†	(6) % Abundance (or Radioactive Decay Mode)	(7) Half-life (if radioactive)
14	Silicon	Si	28	27.976927	92.23%	
			31	30.975363	β^-, γ	157.3 min
15	Phosphorus	P	31	30.973762	100%	
			32	31.973907	β^-	14.262 days
16	Sulfur	S	32	31.972071	95.02%	
			35	34.969032	β^-	87.32 days
17	Chlorine	Cl	35	34.968853	75.77%	
			37	36.965903	24.23%	
18	Argon	Ar	40	39.962383	99.600%	
19	Potassium	K	39	38.963707	93.2581%	
			40	39.963999	0.0117%	
					β^-, EC, γ, β^+	1.28×10^9 yr
20	Calcium	Ca	40	39.962591	96.941%	
21	Scandium	Sc	45	44.955910	100%	
22	Titanium	Ti	48	47.947947	73.8%	
23	Vanadium	V	51	50.943964	99.750%	
24	Chromium	Cr	52	51.940512	83.79%	
25	Manganese	Mn	55	54.938049	100%	
26	Iron	Fe	56	55.934942	91.72%	
27	Cobalt	Co	59	58.933200	100%	
			60	59.933822	β^-, γ	5.2714 yr
28	Nickel	Ni	58	57.935348	68.077%	
			60	59.930791	26.233%	
29	Copper	Cu	63	62.929601	69.17%	
			65	64.927794	30.83%	
30	Zinc	Zn	64	63.929147	48.6%	
			66	65.926037	27.9%	
31	Gallium	Ga	69	68.925581	60.108%	
32	Germanium	Ge	72	71.922076	27.66%	
			74	73.921178	35.94%	
33	Arsenic	As	75	74.921596	100%	
34	Selenium	Se	80	79.916522	49.61%	
35	Bromine	Br	79	78.918338	50.69%	
36	Krypton	Kr	84	83.911507	57.0%	
37	Rubidium	Rb	85	84.911789	72.17%	
38	Strontium	Sr	86	85.909262	9.86%	
			88	87.905614	82.58%	
			90	89.907737	β^-	28.79 yr
39	Yttrium	Y	89	88.905848	100%	
40	Zirconium	Zr	90	89.904704	51.45%	
41	Niobium	Nb	93	92.906377	100%	
42	Molybdenum	Mo	98	97.905408	24.13%	

†The masses given in column (5) are those for the neutral atom, including the Z electrons.

(1) Atomic Number Z	(2) Element	(3) Symbol	(4) Mass Number A	(5) Atomic Mass†	(6) % Abundance (or Radioactive Decay Mode)	(7) Half-life (if radioactive)
43	Technetium	Tc	98	97.907216	β^-, γ	4.2×10^6 yr
44	Ruthenium	Ru	102	101.904349	31.6%	
45	Rhodium	Rh	103	102.905504	100%	
46	Palladium	Pd	106	105.903483	27.33%	
47	Silver	Ag	107	106.905093	51.839%	
			109	108.904756	48.161%	
48	Cadmium	Cd	114	113.903358	28.73%	
49	Indium	In	115	114.903878	95.7%; β^-, γ	4.41×10^{14} yr
50	Tin	Sn	120	119.902197	32.59%	
51	Antimony	Sb	121	120.903818	57.36%	
52	Tellurium	Te	130	129.906223	33.80%	7.9×10^{20} yr
53	Iodine	I	127	126.904468	100%	
			131	130.906124	β^-, γ	8.0207 days
54	Xenon	Xe	132	131.904154	26.9%	
			136	135.907220	8.9%	
55	Cesium	Cs	133	132.905446	100%	
56	Barium	Ba	137	136.905821	11.23%	
			138	137.905241	71.70%	
57	Lanthanum	La	139	138.906348	99.9098%	
58	Cerium	Ce	140	139.905434	88.48%	
59	Praseodymium	Pr	141	140.907647	100%	
60	Neodymium	Nd	142	141.907718	27.13%	
61	Promethium	Pm	145	144.912744	EC, γ, α	17.7 yr
62	Samarium	Sm	152	151.919728	26.7%	
63	Europium	Eu	153	152.921226	52.2%	
64	Gadolinium	Gd	158	157.924101	24.84%	
65	Terbium	Tb	159	158.925343	100%	
66	Dysprosium	Dy	164	163.929171	28.2%	
67	Holmium	Ho	165	164.930319	100%	
68	Erbium	Er	166	165.930290	33.6%	
69	Thulium	Tm	169	168.934211	100%	
70	Ytterbium	Yb	174	173.938858	31.8%	
71	Lutecium	Lu	175	174.940767	97.4%	
72	Hafnium	Hf	180	179.946549	35.100%	
73	Tantalum	Ta	181	180.947996	99.988%	
74	Tungsten (wolfram)	W	184	183.950933	30.67%	
75	Rhenium	Re	187	186.955751	62.60%; β^-	4.35×10^{10} yr
76	Osmium	Os	191	190.960927	β^-, γ	15.4 days
			192	191.961479	41.0%	
77	Iridium	Ir	191	190.960591	37.3%	
			193	192.962923	62.7%	
78	Platinum	Pt	195	194.964774	33.8%	

† The masses given in column (5) are those for the neutral atom, including the Z electrons.

(1) Atomic Number Z	(2) Element	(3) Symbol	(4) Mass Number A	(5) Atomic Mass†	(6) % Abundance (or Radioactive Decay Mode)	(7) Half-life (if radioactive)
79	Gold	Au	197	196.966551	100%	
80	Mercury	Hg	199	198.968262	16.87%	
			202	201.970625	29.86%	
81	Thallium	Tl	205	204.974412	70.476%	
82	Lead	Pb	206	205.974449	24.1%	
			207	206.975880	22.1%	
			208	207.976635	52.4%	
			210	209.984173	β^-, γ, α	22.3 yr
			211	210.988731	β^-, γ	36.1 min
			212	211.991887	β^-, γ	10.64 h
			214	213.999798	β^-, γ	26.8 min
83	Bismuth	Bi	209	208.980383	100%	
			211	210.987258	α, γ, β^-	2.14 min
84	Polonium	Po	210	209.982857	α, γ	138.376 days
			214	213.995185	α, γ	164.3 μs
85	Astatine	At	218	218.008681	α, β^-	1.5 s
86	Radon	Rn	222	222.017570	α, γ	3.8235 days
87	Francium	Fr	223	223.019731	β^-, γ, α	21.8 min
88	Radium	Ra	226	226.025402	α, γ	1600 yr
89	Actinium	Ac	227	227.027746	β^-, γ, α	21.773 yr
90	Thorium	Th	228	228.028731	α, γ	1.9116 yr
			232	232.038050	100%; α, γ	1.405×10^{10} yr
91	Protactinium	Pa	231	231.035878	α, γ	3.276×10^4 yr
92	Uranium	U	232	232.037146	α, γ	68.9 yr
			233	233.039628	α, γ	1.592×10^5 yr
			235	235.043923	0.720%, α, γ	7.038×10^8 yr
			236	236.045561	α, γ	2.342×10^7 yr
			238	238.050782	99.2745%; α, γ	4.468×10^9 yr
			239	239.054287	β^-, γ	23.45 min
93	Neptunium	Np	237	237.048166	α, γ	2.144×10^6 yr
			239	239.052931	β^-, γ	2.3565 d
94	Plutonium	Pu	239	239.052157	α, γ	24,110 yr
			244	244.064197	α	8.08×10^7 yr
95	Americium	Am	243	243.061373	α, γ	7370 yr
96	Curium	Cm	247	247.070346	α, γ	1.56×10^7 yr
97	Berkelium	Bk	247	247.070298	α, γ	1380 yr
98	Californium	Cf	251	251.079580	α, γ	898 yr
99	Einsteinium	Es	252	252.082972	α, EC, γ	472 d
100	Fermium	Fm	257	257.095099	α, γ	101 d
101	Mendelevium	Md	258	258.098425	α, γ	51.5 d
102	Nobelium	No	259	259.10102	α, EC	58 min
103	Lawrencium	Lr	262	262.10969	α, EC, fission	216 min

† The masses given in column (5) are those for the neutral atom, including the Z electrons.

(1) Atomic Number Z	(2) Element	(3) Symbol	(4) Mass Number A	(5) Atomic Mass†	(6) % Abundance (or Radioactive Decay Mode)	(7) Half-life (if radioactive)
104	Rutherfordium	Rf	261	261.10875	α	65 s
105	Dubnium	Db	262	262.11415	α, fission, EC	34 s
106	Seaborgium	Sg	266	266.12193	α, fission	21 s
107	Bohrium	Bh	264	264.12473	α	0.44 s
108	Hassium	Hs	269	269.13411	α	9 s
109	Meitnerium	Mt	268	268.13882	α	0.07 s
110			271	271.14608	α	0.06 s
111			272	272.15348	α	1.5 ms
112			277	277	α	0.24 ms
114			289	289	α	20 s
116			289	289	α	0.6 ms
118			293	293	α	0.1 ms

†The masses given in column (5) are those for the neutral atom, including the Z electrons.

Answers to Odd-Numbered Problems

1. (a) 1×10^{10} yr; (b) 3×10^{17} s.

3. (a) 1.156×10^3; (b) 2.18×10^1; (c) 6.8×10^{-3}; (d) 2.7635×10^1; (e) 2.19×10^{-1}; (f) 2.2×10^1.

5. 7.7%.

7. (a) 4%; (b) 0.4%; (c) 0.07%.

9. 1.0×10^5 s.

11. 9%.

13. (a) 0.286 6 m; (b) 0.000 085 V; (c) 0.000 760 kg; (d) 0.000 000 000 060 0 s; (e) 0.000 000 000 000 022 5 m; (f) 2,500,000,000 volts.

15. 1.8 m.

17. (a) 0.111 yd^2; (b) 10.76 ft^2.

19. (a) 3.9×10^{-9} in; (b) 1.0×10^8 atoms.

21. (a) 0.621 mi/h; (b) 1 m/s = 3.28 ft/s; (c) 0.278 m/s.

23. (a) 9.46×10^{15} m; (b) 6.31×10^4 AU; (c) 7.20 AU/h.

25. (a) 10^3; (b) 10^4; (c) 10^{-2}; (d) 10^9.

27. $\approx 20\%$.

29. 1×10^5 cm^3.

31. (a) ≈ 600 dentists.

33. $\approx 3 \times 10^8$ kg/yr.

35. 51 km.

37. $A = \left[L/T^4\right] = $ m/s^4, $B = \left[L/T^2\right] = $ m/s^2.

39. (a) 0.10 nm; (b) 1.0×10^5 fm; (c) 1.0×10^{10} Å; (d) 9.5×10^{25} Å.

41. (a) 3.16×10^7 s; (b) 3.16×10^{16} ns; (c) 3.17×10^{-8} yr.

43. (a) 1,000 drivers.

45. 1×10^{11} gal/yr.

47. 9 cm.

49. 4×10^5 t.

51. ≈ 4 yr.

53. 1.9×10^2 m.

55. (a) 3%, 3%; (b) 0.7%, 0.2%.

1. 5.0 h.

3. 61 m.

5. 0.78 cm/s (toward $+x$).

7. ≈ 300 m/s.

9. (a) 10.1 m/s; (b) +3.4 m/s, away from trainer.

11. (a) 0.28 m/s; (b) 1.2 m/s; (c) 0.28 m/s; (d) 1.6 m/s; (e) -1.0 m/s.

13. (a) 13.4 m/s; (b) +4.5 m/s, away from master.

15. 24 s.

17. 55 km/h, 0.

19. 6.73 m/s.

21. 5.2 s

23. -7.0 m/s^2, 0.72.

25. (a) 4.7 m/s^2; (b) 2.2 m/s^2; (c) 0.3 m/s^2; (d) 1.6 m/s^2.

27. $v = (6.0 \text{ m/s}) + (17 \text{ m/s}^2)t$, $a = 17$ m/s^2.

29. 1.5 m/s^2, 99 m.

31. 1.7×10^2 m.

33. 4.41 m/s^2, $t = 2.61$ s.

35. 55.0 m.

37. (a) 2.3×10^2 m; (b) 31 s; (c) 15 m, 13 m.

39. (a) 103 m; (b) 64 m.

41. 31 m/s.

43. (b) 3.45 s.

45. 32 m/s (110 km/h).

47. 2.83 s.

49. (a) 8.81 s; (b) 86.3 m/s.

51. 1.44 s.

53. 15 m/s.

55. 5.44 s.

59. 0.035 s.

61. 1.8 m above the top of the window.

63. 52 m.

65. 19.8 m/s, 20.0 m.

67. (a) $v = (g/k)\left(1 - e^{-kt}\right)$; (b) $v_{\text{term}} = g/k$.

69. $6h_{\text{Earth}}$.

71. 1.3 m.

73. (b) $H_{50} = 9.8$ m; (c) $H_{100} = 39$ m.

75. (a) 1.3 m; (b) 6.1 m/s; (c) 1.1 s.

77. (a) 3.88 s; (b) 73.9 m; (c) 38.0 m/s, 48.4 m/s.

79. (a) 52 min; (b) 31 min.

81. (a) $v_0 = 26$ m/s; (b) 35 m; (c) 1.2 s; (d) 4.1 s.

83. (a) 4.80 s; (b) 37.0 m/s; (c) 75.2 m.

85. She should decide to stop!

87. $\Delta v_{0\text{down}} = 0.8$ m/s, $\Delta v_{0\text{up}} = 0.9$ m/s.

89. 29.0 m.

CHAPTER 3

1. 263 km, 13° S of W.

3. $\mathbf{V}_{\text{wrong}} = \mathbf{V}_2 - \mathbf{V}_1$.

5. 13.6 m, 18° N of E,

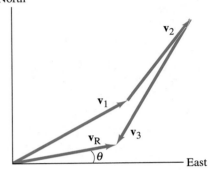

7. (a)

(b) $V_x = -11.7$, $V_y = 8.16$; (c) 14.3, 34.9° above $- x$-axis.

9. (a) $V_N = 476$ km/h, $V_W = 421$ km/h; (b) $d_N = 1.43 \times 10^3$ km, $d_W = 1.26 \times 10^3$ km.

11. (a) 4.2, 45° below $+x$-axis; (b) 5.1, 79° below $+x$-axis.

13. (a) 53.7, 1.40° above $-x$-axis; (b) 53.7, 1.40° below $+x$-axis.

15. (a) 94.5, 11.8° below $-x$-axis; (b) 226, 35.3° below $+x$-axis.

17. (a) $A_x = \pm 82.9$; (b) 166.6, 12.1° above $-x$-axis.

19. $(7.60 \text{ m/s})\mathbf{i} - (4.00 \text{ m/s})\mathbf{k}$; 8.59 m/s.

21. (a) Unknown; (b) 7.74 m/s^2, 33.2° north of east; (c) unknown.

23. (a) $\mathbf{v} = (4.0\,\text{m/s}^2)t\mathbf{i} + (3.0\,\text{m/s}^2)t\mathbf{j}$;
(b) $(5.0\,\text{m/s}^2)t$;
(c) $\mathbf{r} = (2.0\,\text{m/s}^2)t^2\mathbf{i} + (1.5\,\text{m/s}^2)t^2\mathbf{j}$;
(d) $\mathbf{v} = (8.0\,\text{m/s})\mathbf{i} + (6.0\,\text{m/s})\mathbf{j}$,
$\quad |\mathbf{v}| = 10.0\,\text{m/s}$,
$\quad \mathbf{r} = (8.0\,\text{m})\mathbf{i} + (6.0\,\text{m})\mathbf{j}$.

25. (a) $-(18.0\,\text{m/s})\sin(3.0\,\text{s}^{-1})t\mathbf{i}$
$\quad + (18.0\,\text{m/s})\cos(3.0\,\text{s}^{-1})t\mathbf{j}$;
(b) $-(54.0\,\text{m/s}^2)\cos(3.0\,\text{s}^{-1})t\mathbf{i}$
$\quad - (54.0\,\text{m/s}^2)\sin(3.0\,\text{s}^{-1})t\mathbf{j}$;
(c) circle; (d) $a = (9.0\,\text{s}^{-2})r, 180°$.

27. 44 m, 6.3 m.

29. 38° and 52°.

31. 1.40 s.

33. 22 m.

35.
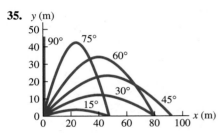

37. 5.71 s.

39. (a) 65.7 m; (b) 7.32 s; (c) 267 m;
(d) 42.2 m/s, 30.1° above the horizontal.

43. Unsuccessful, 34.7 m.

45. (a) $v_0 = 3.42\,\text{m/s}, 47.5°$ above the horizontal; (b) 5.32 m above the water; (c) $v_f = 10.5\,\text{m/s}, 77°$ below the horizontal.

47. $\theta = \tan^{-1}(gt/v_0)$ below the horizontal.

49. $\theta = \frac{1}{2}\tan^{-1}(-\cot\phi)$.

51. $7.29g$ up.

53. $5.9 \times 10^{-3}\,\text{m/s}^2$ toward the Sun.

55. $0.94g$.

59. 2.7 m/s, 22° from the river bank.

61. 23.1 s.

63. 1.41 m/s.

65. (a) 1.82 m/s; (b) 3.22 m/s.

67. (a) 60 m; (b) 75 s.

69. 58 km/h, 31°, 58 km/h opposite to $\mathbf{v}_{12}$.

71. $0.0889\,\text{m/s}^2$.

73. $D_x = 60\,\text{m}, D_y = -35\,\text{m}$,
$D_z = -12\,\text{m}; 70\,\text{m}; \theta_h = 30°$ from the x-axis toward the $-y$-axis,
$\theta_v = 9.8°$ below the horizontal.

75. 7.0 m/s.

77. $\pm28.5°, \pm25.2$.

79. 170 km/h, 41.5° N of E.

81. $1.6\,\text{m/s}^2$.

83. 2.7 s, 1.9 m/s.

85. (a) $Dv/(v^2 - u^2)$;
(b) $D/(v^2 - u^2)^{1/2}$.

87. 54.6° below the horizontal.

89. (a) 464 m/s; (b) 355 m/s.

91. Row at an angle of 23° upstream and run 243 m in a total time of 20.7 min.

93. 1.8×10^3 rev/day.

CHAPTER 4

3. 6.9×10^2 N.

5. (a) 5.7×10^2 N; (b) 99 N;
(c) 2.1×10^2 N; (d) 0.

7. 107 N.

9. -9.3×10^5 N, 25% of the weight of the train.

11. $m > 1.9\,\text{kg}$.

13. 2.1×10^2 N.

15. $-1.40\,\text{m/s}^2$ (down).

17. a (downward) $\geq 1.2\,\text{m/s}^2$.

19. $a_{\max} = 0.557\,\text{m/s}^2$.

21. (a) $2.2\,\text{m/s}^2$; (b) 18 m/s; (c) 93 s.

23. 3.0×10^3 N downward.

25. (a) 1.4×10^2 N; (b) 14.5 m/s.

27. Southwesterly direction.

29.
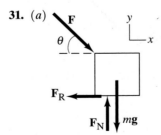
(a)　　(b)

31. (a)
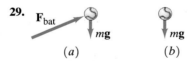
(b) 55.2 N; (c) 183 N; (d) 89 N.

33. (a) 8.98×10^4 N; (b) 1.24×10^4 N.

35. (a) 2.8×10^2 N; (b) $0.98\,\text{m/s}^2$.

37. $F_{T1} = 114\,\text{N}, F_{T2} = 132\,\text{N}$,
$F_{T3} = 229\,\text{N}$.

39. (a) $3.67\,\text{m/s}^2$; (b) 9.39 m/s.

41. (a) 2.2 m up the plane; (b) 2.2 s.

43. $\frac{5}{2}(F_0/m)t_0^2$.

47. (a)

(b) $a = m_2g/(m_1 + m_2)$,
$\quad F_T = m_1 m_2 g/(m_1 + m_2)$.

49. $a = [m_2 + m_C(\ell_2/\ell)]g/$
$\quad (m_1 + m_2 + m_C)$.

51. $1.74\,\text{m/s}^2, F_{T1} = 22.6\,\text{N}$,
$\quad F_{T2} = 20.9\,\text{N}$.

53. $(m + M)g\tan\theta$.

55. $F_{T1} = [4m_1 m_2 m_3/$
$\quad (m_1 m_3 + m_2 m_3 + 4m_1 m_2)]g$,
$F_{T3} = [8m_1 m_2 m_3/$
$\quad (m_1 m_3 + m_2 m_3 + 4m_1 m_2)]g$,
$a_1 = [(m_1 m_3 - 3m_2 m_3 + 4m_1 m_2)/$
$\quad (m_1 m_3 + m_2 m_3 + 4m_1 m_2)]g$,
$a_2 = [(-3m_1 m_3 + m_2 m_3 + 4m_1 m_2)/$
$\quad (m_1 m_3 + m_2 m_3 + 4m_1 m_2)]g$,
$a_3 = [(m_1 m_3 + m_2 m_3 - 4m_1 m_2)/$
$\quad (m_1 m_3 + m_2 m_3 + 4m_1 m_2)]g$.

57. $v = \{[2m_2\ell_2 + m_C(\ell_2^2/\ell)]g/$
$\quad (m_1 + m_2 + m_C)\}^{1/2}$.

59. 2.0×10^{-2} N.

61. 4.3 N.

63. 1.5×10^4 N.

65. 1.9 s.

67. (a) $2.45\,\text{m/s}^2$ (up the incline);
(b) 0.50 kg; (c) 7.35 N, 4.9 N.

69. 1.3×10^2 N.

71. 8.8°.

73. 82 m/s (300 km/h).

75. (a) $F = \frac{1}{2}Mg$;
(b) $F_{T1} = F_{T2} = \frac{1}{2}Mg, F_{T3} = \frac{3}{2}Mg$,
$\quad F_{T4} = Mg$.

77. -8.3×10^2 N.

79. (a) $0.606\,\text{m/s}^2$; (b) 150 kN.

CHAPTER 5

1. 35 N, no force.

3. (*a*)

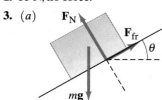

(*b*)

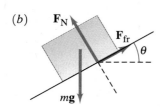

(*c*)

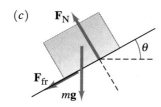

5. 0.20.

7. 69 N, $\mu_k = 0.54$.

9. 8.0 kg.

11. 1.3 m.

13. 1.3×10^3 N.

15. (*a*) 0.58; (*b*) 5.7 m/s; (*c*) 15 m/s.

17. (*a*) 1.4 m/s^2; (*b*) 5.4×10^2 N; (*c*) 1.41 m/s^2, 2.1×10^2 N.

19. (*a*) 86 cm up the plane; (*b*) 1.5 s.

21. (*a*) 2.8 m/s^2; (*b*) 2.1 N.

23. $a = \{\sin\theta - [(\mu_1 m_1 + \mu_2 m_2)/ (m_1 + m_2)]\cos\theta\}g,$
$F_T = [m_1 m_2(\mu_2 - \mu_1)/ (m_1 + m_2)]g\cos\theta.$

25. (*a*) $\mu_k = (v_0^2/2gd\cos\theta) - \tan\theta$; (*b*) $\mu_s \geq \tan\theta$.

27. (*a*) 2.0 m/s^2 up the plane; (*b*) 5.4 m/s^2 up the plane.

29. $\mu_k = 0.40$.

31. (*a*) $c = 14$ kg/m; (*b*) 5.7×10^2 N.

33. $F_{\min} = (m + M)g(\sin\theta + \mu\cos\theta)/ (\cos\theta - \mu_s\sin\theta).$

35. $v_{\max} = 21$ m/s, independent of the mass.

37. (*a*) 0.25 m/s^2 toward the center; (*b*) 6.3 N toward the center.

39. Yes, $v_{\text{top, min}} = (gR)^{1/2}$.

41. 0.34.

43. 2.1×10^2 N.

45. 5.91°, 14.3 N.

47. (*a*) 5.8×10^3 N; (*b*) 4.1×10^2 N; (*c*) 31 m/s.

49. $F_T = 2\pi m R f^2$.

51. 66 km/h $< v <$ 123 km/h.

53. (*a*) $(1.6 \text{ m/s}^2)\mathbf{i}$; (*b*) $(0.98 \text{ m/s}^2)\mathbf{i} - (1.7 \text{ m/s}^2)\mathbf{j}$; (*c*) $-(4.9 \text{ m/s}^2)\mathbf{i} - (1.6 \text{ m/s}^2)\mathbf{j}$.

55. (*a*) 9.0 m/s^2; (*b*) 15 m/s^2.

57. $\tau = m/b$.

59. (*a*) $v = (mg/b) + [v_0 - (mg/b)]e^{-bt/m}$; (*b*) $v = -(mg/b) + [v_0 + (mg/b)]e^{-bt/m}$, $v \geq 0$.

61. (*b*) 1.8°.

63. 10 m.

65. $\mu_s = 0.41$.

67. 2.3.

69. 101 N, $\mu_k = 0.719$.

71. (*b*) Will slide.

73. Emerges with a speed of 13 m/s.

75. 27.6 m/s, 0.439 rev/s.

77. $\Sigma F_{\text{tan}} = 3.3 \times 10^3$ N, $\Sigma F_R = 2.0 \times 10^3$ N.

79. (*a*) $F_{NC} > F_{NB} > F_{NA}$; (*b*) heaviest at C, lightest at A; (*c*) $v_{A\max} = (gR)^{1/2}$.

81. (*a*) 1.23 m/s; (*b*) 3.01 m/s.

83. $\phi = 31°$.

85. (*a*) $r = v^2/g\cos\theta$; (*b*) 92 m.

87. (*a*) 59 s; (*b*) greater normal force.

89. 29.2 m/s.

91. 302 m, 735 m.

93. $g(1 - \mu_s\tan\phi)/4\pi^2 f^2(\tan\phi + \mu_s)$ $< r < g(1 + \mu_s\tan\phi)/ 4\pi^2 f^2(\tan\phi - \mu_s)$.

CHAPTER 6

1. 1.52×10^3 N.

3. 1.6 m/s^2.

5. $g_h = 0.91 g_{\text{surface}}$.

7. 1.9×10^{-8} N toward center of square.

9. $Gm^2\{(2/x_0^2) + [3x_0/(x_0^2 + y_0^2)^{3/2}]\}\mathbf{i} + Gm^2\{[3y_0/(x_0^2 + y_0^2)^{3/2}] + (4/y_0^2)\}\mathbf{j}.$

11. 1.26.

13. 3.46×10^8 m from Earth's center.

15. (*b*) g decreases with an increase in height; (*c*) 9.493 m/s^2.

19. 7.56×10^3 m/s.

21. 2.0 h.

23. (*a*) 56 kg; (*b*) 56 kg; (*c*) 75 kg; (*d*) 38 kg; (*e*) 0.

25. (*a*) 22 N (toward the Moon); (*b*) -1.7×10^2 N (away from the Moon).

27. (*a*) Gravitational force provides required centripetal acceleration; (*b*) 9.6×10^{26} kg.

29. 7.9×10^3 m/s.

31. $v = (Gm/L)^{1/2}$.

33. 0.0587 days (1.41 h).

35. 1.6×10^2 yr.

37. 2×10^8 yr.

39. $r_{\text{Europa}} = 6.71 \times 10^5$ km, $r_{\text{Ganymede}} = 1.07 \times 10^6$ km, $r_{\text{Callisto}} = 1.88 \times 10^6$ km.

41. 9.0 Earth-days.

43. (*a*) 2.1×10^2 A.U. $(3.1 \times 10^{13}$ m); (*b*) 4.2×10^2 A.U.; (*c*) 4.2×10^2.

45. (*a*) 5.9×10^{-3} N/kg; (*b*) not significant.

47. 2.7×10^3 km.

49. 6.7×10^{12} m/s^2.

51. 4.4×10^7 m/s^2.

53. $G' = 1 \times 10^{-4}$ N·m^2/kg$^2 \approx 10^6 G$.

55. 5 h 35 min, 19 h 50 min.

57. (*a*) 10 h; (*b*) 6.5 km; (*c*) 4.2×10^{-3} m/s^2.

59. 5.4×10^{12} m, in the Solar System, Pluto.

61. $2.3 g_{\text{Earth}}$.

63. $m_P = g_P r^2/G$.

67. 7.9×10^3 m/s.

CHAPTER 7

1. 6.86×10^3 J.

3. 1.27×10^4 J.

5. 8.1×10^3 J.

7. $1 \text{ J} = 1 \times 10^7$ erg = 0.738 ft·lb.

9. 1.0×10^4 J.

13. (*a*) 3.6×10^2 N; (*b*) -1.3×10^3 J; (*c*) -4.6×10^3 J; (*d*) 5.9×10^3 J; (*e*) 0.

15. $W_{FN} = W_{mg} = 0$, $W_{FP} = -W_{\text{fr}} = 2.0 \times 10^2$ J.

21. (*a*) -16.1; (*b*) -238; (*c*) -3.9.

23. $\mathbf{C} = -1.3\mathbf{i} + 1.8\mathbf{j}$.

25. $\theta_x = 42.7°, \theta_y = 63.8°, \theta_z = 121°$.

27. 95°, $-35°$ from x-axis.

31. 0.089 J.

33. 2.3×10^3 J.

35. 2.7×10^3 J.

37. $(kX^2/2) + (kX^4/4) + (kX^5/5)$.

39. (a) 5.0×10^{10} J.

41. (a) $\sqrt{3}$; (b) $\frac{1}{4}$.

43. -5.02×10^5 J.

45. 3.0×10^2 N in the direction of the motion of the ball.

47. 24 m/s (87 km/h or 54 mi/h), the mass cancels.

49. (a) 72 J; (b) -35 J; (c) 37 J.

51. 10.2 m/s.

53. $\mu_k = F/2mg$.

55. (a) 6.5×10^2 J; (b) -4.9×10^2 J; (c) 0; (d) 4.0 m/s.

57. (a) 1.66×10^5 J; (b) 21.0 m/s; (c) 2.13 m.

59. $v_p = 2.0 \times 10^7$ m/s, $v_{pc} = 2.0 \times 10^7$ m/s; $v_e = 2.9 \times 10^8$ m/s, $v_{ec} = 8.4 \times 10^8$ m/s.

61. 1.74×10^3 J.

63. (a) 15 J; (b) 4.2×10^2 J; (c) -1.8×10^2 J; (d) -2.5×10^2 J; (e) 0; (f) 12 J.

65. (a) 12 J; (b) 10 J; (c) -2.1 J.

67. 86 kJ, $\theta = 42°$.

69. $(A/k)e^{-(0.10 \text{ m})k}$.

71. 1.5 N.

73. 5.0×10^3 N/m.

75. (a) 6.6°; (b) 10.3°.

CHAPTER 8

1. 0.924 m.

3. 2.2×10^3 J.

5. (a) 51.7 J; (b) 15.1 J; (c) 51.7 J.

7. (a) Conservative; (b) $\frac{1}{2}kx^2 - \frac{1}{4}ax^4 - \frac{1}{5}bx^5 + $ constant.

9. (a) $\frac{1}{2}k(x^2 - x_0^2)$; (b) same.

11. 45.4 m/s.

13. 6.5 m/s.

15. (a) 1.0×10^2 N/m; (b) 22 m/s².

17. (a) 8.03 m/s; (b) 3.44 m.

19. (a) $v_{max} = [v_0^2 + (kx_0^2/m)]^{1/2}$; (b) $x_{max} = [x_0^2 + (mv_0^2/k)]^{1/2}$.

21. (a) 2.29 m/s; (b) 1.98 m/s; (c) 1.98 m/s; (d) $F_{Ta} = 0.87$ N, $F_{Tb} = 0.80$ N, $F_{Tc} = 0.80$ N; (e) $v_a = 2.59$ m/s, $v_b = 2.31$ m/s, $v_c = 2.31$ m/s.

23. $k = 12Mg/h$.

25. 4.5×10^6 J.

27. (a) 22 m/s; (b) 2.9×10^2 m.

29. 13 m/s.

31. 0.23.

33. 0.40.

35. (a) 0.13 m; (b) 0.77; (c) 0.46 m/s.

37. (a) $K = GM_E m_S/2r_S$; (b) $U = -GM_E m_S/r_S$; (c) $-\frac{1}{2}$.

39. (a) 6.2×10^5 m/s; (b) 4.2×10^4 m/s, $v_{esc}/v_{orbit} = \sqrt{2}$.

45. (a) 1.07×10^4 m/s; (b) 1.17×10^4 m/s; (c) 1.12×10^4 m/s.

47. (a) $dv_{esc}/dr = -\frac{1}{2}(2GM_E/r^3)^{1/2}$ $= -v_{esc}/2r$; (b) 1.09×10^4 m/s.

49. $GmM_E/12r_E$.

51. 1.1×10^4 m/s.

55. 5.4×10^2 N.

57. (a) 1.0×10^3 J; (b) 1.0×10^3 W.

59. 2.1×10^4 W, 28 hp.

61. 4.8×10^2 W.

63. 1.2×10^3 W.

65. 1.8×10^6 W.

67. (a) -25 W; (b) $+4.3 \times 10^3$ W; (c) $+1.5 \times 10^3$ W.

69. (a) 80 J; (b) 60 J; (c) 80 J; (d) 5.7 ms at $x = 0$; (e) 32 m/s² at $x = \pm x_0$.

71. $a^2/4b$.

73. 8.0 m/s.

75. 32.5 hp.

77. (a) 28 m/s; (b) 1.2×10^2 m.

79. (a) $(2gL)^{1/2}$; (b) $(1.2gL)^{1/2}$.

81. (a) 1.1×10^6 J; (b) 60 W (0.081 hp); (c) 4.0×10^2 W (0.54 hp).

83. (a) 40 m/s; (b) 2.6×10^5 W.

87. (a) 29°; (b) 6.4×10^2 N; (c) 9.2×10^2 N.

89. (a) $-\dfrac{U_0}{r}\left(\dfrac{r_0}{r} + 1\right)e^{-r/r_0}$; (b) 0.030; (c) $F(r) = -C/r^2$, 0.11.

91. 6.7 hp.

93. (a) 2.8 m; (b) 1.5 m; (c) 1.5 m.

95. 61 hp.

97. (a) 5.00×10^3 m/s; (b) 2.89×10^3 m/s.

CHAPTER 9

1. 6.0×10^7 N, up.

3. (a) 0.36 kg·m/s; (b) 0.12 kg·m/s.

5. $(26 \text{ N·s})\mathbf{i} - (28 \text{ N·s})\mathbf{j}$.

7. (a) $(8h/g)^{1/2}$; (b) $(2gh)^{1/2}$; (c) $-(8m^2 gh)^{1/2}$ (up); (d) mg (down), a surprising result.

9. 3.4×10^4 kg.

11. 4.4×10^3 m/s.

13. -0.667 m/s (opposite to the direction of the package).

15. 2, lesser kinetic energy has greater mass.

17. $\frac{3}{2}v_0\mathbf{i} - v_0\mathbf{j}$.

19. 1.1×10^{-22} kg·m/s, 36° from the direction opposite to the electron's.

21. (a) $(100 \text{ m/s})\mathbf{i} + (50 \text{ m/s})\mathbf{j}$; (b) 3.3×10^5 J.

23. 130 N, not large enough.

25. 1.1×10^3 N.

27. (a) $2mv/\Delta t$; (b) $2mv/t$.

29. (a)

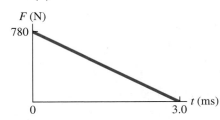

F (N)

780

t (ms)

0   3.0

(b) 1.2 N·s; (c) 1.2 N·s; (d) 3.9 g.

31. (a) $(0.84 \text{ N}) + (1.2 \text{ N/s})t$; (b) 18.5 N; (c) $(0.12 \text{ kg/s})\{[(49 \text{ m}^2/\text{s}^2) - (1.12 \text{ m}^2/\text{s}^3)t]^{1/2} + (9.80 \text{ m/s}^2)t\}$, 18.3 N.

33. $v_1' = -1.40$ m/s (rebound), $v_2' = 2.80$ m/s.

35. (a) 2.7 m/s; (b) 0.84 kg.

37. 3.2×10^3 m/s.

39. (a) 1.00; (b) 0.89; (c) 0.29; (d) 0.019.

41. (a) 0.32 m; (b) -3.1 m/s (rebound), 4.9 m/s; (c) Yes.

43. (a) $+M/(m + M)$; (b) 0.964.

45. 141°.

47. (b) $e = (h'/h)^{1/2}$.

49. (a) $v_1' = v_2' = 1.9$ m/s; (b) $v_1' = -1.6$ m/s, $v_2' = 7.9$ m/s; (c) $v_1' = 0$, $v_2' = 5.2$ m/s; (d) $v_1' = 3.1$ m/s, $v_2' = 0$; (e) $v_1' = -4.0$ m/s, $v_2' = 12$ m/s; result for (c) is reasonable, result for (d) is not reasonable, result for (e) is not reasonable.

51. 61° from first eagles's direction, 6.8 m/s.

53. (a) 30°; (b) $v_2' = v/\sqrt{3}$, $v_1' = v/\sqrt{3}$; (c) $\frac{2}{3}$.

55. $\theta_1' = 76°$, $v_n' = 5.1 \times 10^5$ m/s,
$v_{He}' = 1.8 \times 10^5$ m/s.

59. 6.5×10^{-11} m from the carbon atom.

61. 0.030 nm above center of H triangle.

63. $x_{CM} = 1.10$ m (East),
$y_{CM} = -1.10$ m (South).

65. $x_{CM} = 0$, $y_{CM} = 2r/\pi$.

67. $x_{CM} = 0$, $y_{CM} = 0$,
$z_{CM} = 3h/4$ above the point.

69. (a) 4.66×10^6 m.

71. (a) $x_{CM} = 4.6$ m; (b) 4.3 m;
(c) 4.6 m.

73. $mv/(M + m)$ up, balloon will also stop.

75. 55 m.

77. 0.899 hp.

79. (a) 2.3×10^3 N; (b) 2.8×10^4 N;
(c) 1.1×10^4 hp.

81. A "scratch shot".

83. 1.4×10^4 N, 43.3°.

85. 5.1×10^2 m/s.

87. $m_2 = 4.00m$.

89. 50%.

91. (a) No; (b) $v_1/v_2 = -m_2/m_1$;
(c) m_2/m_1; (d) does not move;
(e) center of mass will move.

93. 8.29 m/s.

95. (a) 2.5×10^{-13} m/s; (b) 1.7×10^{-17};
(c) 0.19 J.

97. $m \leq M/3$.

99. 29.6 km/s.

101. (a) 2.3 N·s; (b) 4.5×10^2 N.

103. (a) Inelastic collision; (b) 0.10 s;
(c) -1.4×10^5 N.

105. 0.28 m, 1.1 m.

CHAPTER 10

1. (a) $\pi/6$ rad = 0.524 rad;
(b) $19\pi/60 = 0.995$ rad;
(c) $\pi/2 = 1.571$ rad;
(d) $2\pi = 6.283$ rad;
(e) $7\pi/3 = 7.330$ rad.

3. 2.3×10^3 m.

5. (a) 0.105 rad/s;
(b) 1.75×10^{-3} rad/s;
(c) 1.45×10^{-4} rad/s; (d) zero.

7. (a) 464 m/s; (b) 185 m/s;
(c) 355 m/s.

9. (a) 262 rad/s;
(b) 46 m/s, 1.2×10^4 m/s² radial.

11. 7.4 cm.

13. (a) 1.75×10^{-4} rad/s²;
(b) $a_R = 1.17 \times 10^{-2}$ m/s²,
$a_{tan} = 7.44 \times 10^{-4}$ m/s².

15. (a) 0.58 rad/s2; (b) 12 s.

17. (a) $(1.67 \text{ rad/s}^4)t^3 - (1.75 \text{ rad/s}^3)t^2$;
(b) $(0.418 \text{ rad/s}^4)t^4 - (0.583 \text{ rad/s}^3)t^3$;
(c) 6.4 rad/s, 2.0 rad.

19. (a) ω_1 is in the $-x$-direction,
ω_2 is in the $+z$-direction;
(b) $\omega = 61.0$ rad/s, 35.0° above
$-x$-axis;
(c) $-(1.75 \times 10^3 \text{ rad/s}^2)\mathbf{j}$.

21. (a) 35 m·N; (b) 30 m·N.

23. 1.2 m·N (clockwise).

25. 3.5×10^2 N, 2.0×10^3 N.

27. 53 m·N.

29. (a) 3.5 kg·m²; (b) 0.024 m·N.

31. 2.25×10^3 kg·m², 8.8×10^3 m·N.

33. 9.5×10^4 m·N.

35. 10 m/s.

37. (a)

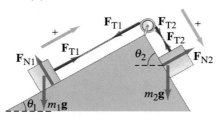

(b) $F_{T1} = 47$ N, $F_{T2} = 75$ N;
(c) 7.0 m·N, 1.7 kg·m².

39. Thin hoop (through center):
$k = R_0$;
Thin hoop (through diameter):
$k = \left[(MR_0^2/2) + (MW^2)/12\right]^{1/2}$;
Solid cylinder (through center):
$k = R/\sqrt{2}$;
Hollow cylinder (through center):
$k = \left[(R_1^2 + R_2^2)/2\right]^{1/2}$;
Uniform sphere (through center):
$k = (2r_0^2/5)^{1/2}$;
Rod (through center): $k = \ell/\sqrt{12}$;
Rod (through end): $k = \ell/\sqrt{3}$;
Plate (through center):
$k = \left[(\ell^2 + W^2)/12\right]^{1/2}$.

41. (a) 4.18 rad/s²; (b) 8.37 m/s²;
(c) 421 m/s²; (d) 3.07×10^3 N;
(e) 1.14°.

43. (a) $I_a = Ms^2/12$; (b) $I_b = Ms^2/12$.

45. (a) $5.30MR_0^2$; (b) -15%.

47. (a) $9MR_0^2/16$; (b) $MR_0^2/4$;
(c) $5MR_0^2/4$.

51. (b) $M\ell^2/12$, $Mw^2/12$.

53. 0.38 rev/s.

55. (a) As moment of inertia increases,
angular velocity must decrease;
(b) 1.6.

57. (a) 7.1×10^{33} kg·m²/s;
(b) 2.7×10^{40} kg·m²/s.

59. 0.45 rad/s, 0.80 rad/s.

61. 2.33×10^4 J.

63. 5×10^9, loss of gravitational
potential energy.

65. 1.4 m/s.

67. (a) 2.5 kg·m²; (b) 0.58 kg·m²;
(c) 0.35 s; (d) -72 J; (e) rotating.

69. 12.4 m/s.

71. 1.4×10^2 J.

73. (a) 4.48 m/s; (b) 1.21 J;
(c) $\mu_s \geq 0.197$.

75. $v = \left[10g(R_0 - r_0)/7\right]^{1/2}$.

77. (a) 4.5×10^5 J; (b) 0.18 (18%);
(c) 1.71 m/s²; (d) 6.4%.

79. (a) $12v_0^2/49 \mu_k g$;
(b) $v = 5v_0/7$, $\omega = 5v_0/7R$.

81. (a) 4.5 m/s², 19 rad/s²; (b) 5.8 m/s;
(c) 15.3 J; (d) 1.4 J;
(e) $K = 16.7$ J, $\Delta E = 0$;
(f) $a = 4.5$ m/s², $v = 5.8$ m/s, 14.1 J.

83. $\theta_{Sun} = 9.30 \times 10^{-3}$ rad (0.53°),
$\theta_{Moon} = 9.06 \times 10^{-3}$ rad (0.52°).

85. $\omega_1/\omega_2 = R_2/R_1$.

87. $\ell/2$, $\ell/2$.

89. (a) $-(I_W/I_P)\omega_W$ (down);
(b) $-(I_W/2I_P)\omega_W$ (down);
(c) $(I_W/I_P)\omega_W$ (up); (d) 0.

91. (a) $\omega_R/\omega_F = N_F/N_R$; (b) 4.0;
(c) 1.5.

93. (a) 1.5×10^2 rad/s²;
(b) 1.2×10^3 N.

95. (a) 0.070 rad/s²; (b) 40 rpm.

97. 7.9 N.

99. (b) 2.2×10^3 rad/s; (c) 24 min.

101. (a) 2.9 m; (b) 3.6 m.

103. (a) 1.2 rad/s; (b) 2.0×10^3 J,
1.2×10^3 J, loss of 8.0×10^2 J,
decrease of 40%.

105. (a) 1.7 m/s; (b) 0.84 m/s.

107. (a) $h_{min} = 2.7R_0$;
(b) $h_{min} = 2.7R_0 - 1.7r_0$.

109. (a) 0.84 m/s; (b) 0.96.

CHAPTER 11

7. (a) $-7.0\mathbf{i} - 14.0\mathbf{j} + 19.3\mathbf{k}$; (b) 164°.

11. $-(30.0$ m·N$)\mathbf{k}$ (in $-z$-direction).

13. $(18$ m·N$)\mathbf{i} \pm (14$ m·N$)\mathbf{j}$
$\mp (19$ m·N$)\mathbf{k}$.

19. $(55\mathbf{i} - 90\mathbf{j} + 42\mathbf{k})\,\text{kg}\cdot\text{m}^2/\text{s}.$

21. (a) $[(7m/9) + (M/6)]\ell^2\omega^2;$
(b) $[(14m/9) + (M/3)]\ell^2\omega.$

23. $2.30\,\text{m/s}^2.$

25. (a) $L = [R_0M_1 + R_0M_2 + (I/R_0)]v;$
(b) $a = M_2g/[M_1 + M_2 + (I/R_0^2)].$

27. Rod rotates at 7.8 rad/s about the center of mass, which moves with constant velocity of 0.21 m/s.

31. $F_1 = [(d + r\cos\phi)/2d]m_1r\omega^2\sin\phi,$
$F_2 = [(d - r\cos\phi)/2d]m_1r\omega^2\sin\phi.$

33. $16\,\text{N}, -7.5\,\text{N}.$

35. $3m^2v^2/g(3m + 4M)(m + M).$

37. $(1 - 4.7 \times 10^{-13})\omega_E.$

39. (a) 14 m/s; (b) 6.8 rad/s.

41. $1.02 \times 10^{-3}\,\text{kg}\cdot\text{m}^2.$

43. 2.2 rad/s (0.35 rev/s).

45. $\tan^{-1}(r\omega^2/g).$

47. (a) g, along a radial line; (b) $0.998g$, $0.0988°$ south from a radial line; (c) $0.997g$, along a radial line.

49. North or south direction.

51. (a) South; (b) $\omega D^2\sin\lambda/v_0;$ (c) 0.46 m.

53. (a) $(-9.0\mathbf{i} + 12\mathbf{j} - 8.0\mathbf{k})\,\text{kg}\cdot\text{m}^2/\text{s};$ (b) $(9.0\mathbf{j} - 6.0\mathbf{k})\,\text{m}\cdot\text{N}.$

55. (a) Turn in the direction of the lean; (b) $\Delta L = 0.98\,\text{kg}\cdot\text{m}^2/\text{s}$, $\Delta L = 0.18L_0.$

57. (a) $1.8 \times 10^3\,\text{kg}\cdot\text{m}^2/\text{s}^2;$ (b) $1.8 \times 10^3\,\text{m}\cdot\text{N};$ (c) $2.1 \times 10^3\,\text{W}.$

59. $v_{CM} = (3g\ell/4)^{1/2}.$

61. $(19\,\text{m/s})(1 - \cos\theta)^{1/2}.$

63. (a) $2.3 \times 10^{12}\,\text{rev/s};$ (b) $5.7 \times 10^{11}\,\text{rev/s}.$

CHAPTER 12

1. 379 N, 141°.

3. $1.6 \times 10^3\,\text{m}\cdot\text{N}.$

5. 6.52 kg.

7. 2.84 m from the adult.

9. 0.32 m.

11. $F_{T1} = 3.4 \times 10^3\,\text{N},$
$F_{T2} = 3.9 \times 10^3\,\text{N}.$

13. $F_1 = -2.94 \times 10^3\,\text{N}$ (down),
$F_2 = 1.47 \times 10^4\,\text{N}.$

15. Top hinge: $F_{Ax} = 55.2\,\text{N},$
$F_{Ay} = 63.7\,\text{N}$; bottom hinge:
$F_{Bx} = -55.2\,\text{N}, F_{By} = 63.7\,\text{N}.$

17. (a)

(b) $1.5 \times 10^4\,\text{N};$ (c) $6.7 \times 10^3\,\text{N}.$

19. $F_T = 1.4 \times 10^3\,\text{N}$ (up),
$F_{bone} = 2.1 \times 10^3\,\text{N}$ (down).

21. $2.7 \times 10^3\,\text{N}.$

23. 89.5 cm from the feet.

25. $F_1 = 5.8 \times 10^3\,\text{N}, F_2 = 5.6 \times 10^3\,\text{N}.$

27. (a) $2.1 \times 10^2\,\text{N};$ (b) $2.0 \times 10^3\,\text{N}.$

29. $7.1 \times 10^2\,\text{N}.$

31. $F_T = 2.5 \times 10^2\,\text{N},$
$F_{AH} = 2.5 \times 10^2\,\text{N},$
$F_{AV} = 2.0 \times 10^2\,\text{N}.$

33. (a) 1.00 N; (b) 1.25 N.

35. $\theta_{max} = 40°$, same.

37. (a) $F_T = 182\,\text{N};$ (b) $F_{N1} = 352\,\text{N},$
$F_{N2} = 236\,\text{N};$ (c) $F_B = 298\,\text{N}, 52.4°.$

39. $1.0 \times 10^2\,\text{N}.$

41. (a) $1.2 \times 10^5\,\text{N/m}^2;$ (b) $2.4 \times 10^{-6}.$

43. (a) $1.3 \times 10^5\,\text{N/m}^2;$ (b) $6.5 \times 10^{-7};$ (c) 0.0062 mm.

45. $9.6 \times 10^6\,\text{N/m}^2.$

47. $9.0 \times 10^7\,\text{N/m}^2, 9.0 \times 10^2\,\text{atm}.$

49. $1.0 \times 10^9\,\text{N},$

51. (a) $1.1 \times 10^2\,\text{m}\cdot\text{N};$ (b) wall; (c) all three.

53. $3.9 \times 10^2\,\text{N}$, thicker strings, maximum strength is exceeded.

55. (a) $4.4 \times 10^{-5}\,\text{m}^2;$ (b) 2.7 mm.

57. 1.2 cm.

61. (a) $F_T = 129\,\text{kN};$
$F_A = 141\,\text{kN}, 23.5°;$
(b) $F_{DE} = 64.7\,\text{kN}$ (tension),
$F_{CE} = 32.3\,\text{kN}$ (compression),
$F_{CD} = 64.7\,\text{kN}$ (compression),
$F_{BD} = 64.7\,\text{kN}$ (tension),
$F_{BC} = 64.7\,\text{kN}$ (tension),
$F_{AC} = 97.0\,\text{kN}$ (compression),
$F_{AB} = 64.7\,\text{kN}$ (compression).

63. (a) $4.8 \times 10^{-2}\,\text{m}^2;$ (b) $6.8 \times 10^{-2}\,\text{m}^2.$

65. $F_{AB} = 5.44 \times 10^4\,\text{N}$ (compression),
$F_{ACx} = 2.72 \times 10^4\,\text{N}$ (tension),
$F_{BC} = 5.44 \times 10^4\,\text{N}$ (tension),
$F_{BD} = 5.44 \times 10^4\,\text{N}$ (compression),
$F_{CD} = 5.44 \times 10^4\,\text{N}$ (tension),
$F_{CE} = 2.72 \times 10^4\,\text{N}$ (tension),
$F_{DE} = 5.44 \times 10^4\,\text{N}$ (compression).

67. 12 m.

69. $M_C = 0.191\,\text{kg}, M_D = 0.0544\,\text{kg},$
$M_A = 0.245\,\text{kg}.$

71. (a) $Mg[h/(2R - h)]^{1/2};$
(b) $Mg[h(2R - h)]^{1/2}/(R - h).$

73. $\theta_{max} = 29°.$

75. 6, 2.0 m apart.

77. 3.8.

79. $5.0 \times 10^5\,\text{N}, 3.2\,\text{m}.$

81. (a) 600 N; (b) $F_A = 0, F_B = 1200\,\text{N};$
(c) $F_A = 150\,\text{N}, F_B = 1050\,\text{N};$
(d) $F_A = 750\,\text{N}, F_B = 450\,\text{N}.$

83. $6.5 \times 10^2\,\text{N}.$

85. 0.67 m.

87. Right end is safe, left end is not safe, 0.10 m.

89. (a) $F_L = 3.3 \times 10^2\,\text{N}$ up,
$F_R = 2.3 \times 10^2\,\text{N}$ down;
(b) 65 cm from right hand;
(c) 123 cm from right hand.

91. $\theta \geq 40°.$

93. (b) beyond the table;
(c) $D = L\sum_{i=1}^{n}\frac{1}{2i};$ (d) 32 bricks.

95. $F_{TB} = 134\,\text{N}, F_{TA} = 300\,\text{N}.$

97. $2.6w, 31°$ above horizontal.

CHAPTER 13

1. $3 \times 10^{11}\,\text{kg}.$

3. $4.3 \times 10^2\,\text{kg}.$

5. 0.8477.

7. (a) $2.9 \times 10^7\,\text{N/m}^2;$
(b) $1.8 \times 10^5\,\text{N/m}^2.$

9. 1.1 m.

11. $8.28 \times 10^3\,\text{kg}.$

13. $1.2 \times 10^5\,\text{N/m}^2,$
$2.3 \times 10^7\,\text{N}$ (down),
$1.2 \times 10^5\,\text{N/m}^2.$

15. $6.54 \times 10^2\,\text{kg/m}^3.$

17. $3.36 \times 10^4\,\text{N/m}^2$ (0.331 atm).

19. (a) $1.41 \times 10^5\,\text{Pa};$ (b) $9.8 \times 10^4\,\text{Pa}.$

21. (a) 0.34 kg; (b) $1.5 \times 10^4\,\text{N}$ (up).

23. (c) $\geq 0.38h$, no.

27. $4.70 \times 10^3\,\text{kg/m}^3.$

29. $8.5 \times 10^2\,\text{kg}.$

31. Copper.

33. (a) $1.14 \times 10^6\,\text{N};$ (b) $4.0 \times 10^5\,\text{N}.$

35. (b) Above the center of gravity.

37. 0.88.

39. $7.9 \times 10^2\,\text{kg}.$

43. 4.1 m/s.

45. 9.5 m/s.

47. $1.5 \times 10^5\,\text{N/m}^2 = 1.5\,\text{atm}.$

49. $4.11 \times 10^{-3}\,\text{m}^3/\text{s}.$

51. $1.7 \times 10^6\,\text{N}.$

59. (a) $2[h_1(h_2 - h_1)]^{1/2}$;
(b) $h_1' = h_2 - h_1$.

61. $0.072\,\text{Pa}\cdot\text{s}$.

63. $4.0 \times 10^3\,\text{Pa}$.

65. 11 cm.

67. (a) Laminar; (b) 3200, turbulent.

69. 1.9 m.

71. $9.1 \times 10^{-3}\,\text{N}$.

73. (a) $\gamma = F/4\pi r$; (b) 0.024 N/m.

75. (a) 0.88 m; (b) 0.55 m; (c) 0.24 m.

77. $1.5 \times 10^2\,\text{N} \le F \le 2.2 \times 10^2\,\text{N}$.

79. 0.051 atm.

81. 0.63 N.

83. 5 km.

85. $5.3 \times 10^{18}\,\text{kg}$.

87. 2.6 m.

89. 39 people.

91. 37 N, not float.

93. $d = D[v_0^2/(v_0^2 + 2gy)]^{1/4}$.

95. (a) 3.2 m/s; (b) 20 s.

97. $1.9 \times 10^2\,\text{m/s}$.

CHAPTER 14

1. 0.60 m.

3. 1.19 Hz.

5. (a) 2.4 N/m; (b) 12 Hz.

7. (a) $0.866\,x_{max}$; (b) $0.500\,x_{max}$.

9. $0.866\,A$.

11. $[(k_1 + k_2)/m]^{1/2}/2\pi$.

13. (a) 8/7 s, 0.875 Hz; (b) 3.3 m, -10.4 m/s; (c) $+18$ m/s, -57 m/s^2.

15. 3.6 Hz.

19. (a) $y = -(0.220\,\text{m})\sin[(37.1\,\text{s}^{-1})t]$;
(b) maximum extensions at 0.0423 s, 0.211 s, 0.381 s,...; minimum extensions at 0.127 s, 0.296 s, 0.465 s,....

21. $f = (3k/M)^{1/2}/2\pi$.

25. (a) $x = (12.0\,\text{cm})\cos[(25.6\,\text{s}^{-1})t + 1.89\,\text{rad}]$;
(b) $t_{max} = 0.294$ s, 0.539 s, 0.784 s,...;
$t_{min} = 0.171$ s, 0.416 s, 0.661 s,...;
(c) -3.77 cm; (d) $+13.1$ N (up);
(e) 3.07 cm/s, 0.110 s.

27. (a) 0.650 m; (b) 1.34 Hz; (c) 29.8 J; (d) $K = 25.0\,\text{J}, U = 4.8\,\text{J}$.

29. 9.37 m/s.

31. $A_1 = 2.24A_2$.

33. (a) $4.2 \times 10^2\,\text{N/m}$; (b) 3.3 kg.

35. 352.6 m/s.

39. 0.9929 m.

41. (a) 0.248 m; (b) 2.01 s.

43. (a) $-12°$; (b) $+2.1°$; (c) $-13°$.

45. $\frac{1}{3}$.

47. 1.08 s.

49. 0.31 g.

51. (a) 1.6 s.

53. 3.5 s.

55. (a) 0.727 s; (b) 0.0755;
(c) $x = (0.189\,\text{m})e^{-(0.108/s)t}$
$\sin[(8.64\,\text{s}^{-1})t]$.

57. (a) $8.3 \times 10^{-4}\%$; (b) 39 periods.

59. (a) 5.03 Hz; (b) $0.0634\,\text{s}^{-1}$;
(c) 110 oscillations.

61.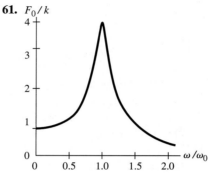

65. (a) 198 s; (b) $8.7 \times 10^{-6}\,\text{W}$;
(c) $8.8 \times 10^{-4}\,\text{Hz}$ on either side of f_0.

69. (a) 0.63 Hz; (b) 0.65 m/s; (c) 0.077 J.

71. 151 N/m, 20.3 m.

73. 0.11 m.

75. 3.6 Hz.

77. (a) 1.1 Hz; (b) 13 J.

79. (a) 90 N/m; (b) 8.9 cm.

81. $k = \rho_{\text{water}}\,gA$.

83. Block will oscillate with SHM, $k = 2\rho gA$, the density and the cross section.

85. $T = 2\pi(ma/2k\,\Delta a)^{1/2}$.

87. (a) 1.64 s; (b) 0.67 m.

CHAPTER 15

1. 2.3 m/s.

3. 1.26 m.

5. 0.72 m.

7. 2.7 N.

9. (a) 75 m/s; (b) $7.8 \times 10^3\,\text{N}$.

11. (a) $1.3 \times 10^3\,\text{km}$;
(b) cannot be determined.

13. (a) 0.25; (b) 0.50.

17. (a) 0.295 W; (b) 0.25 cm.

19. $D = D_M \sin[2\pi(x/\lambda + t/T) + \phi]$.

21. (a) 41 m/s; (b) $6.4 \times 10^4\,\text{m/s}^2$;
(c) 41 m/s, $8.2 \times 10^3\,\text{m/s}^2$.

23. (a, c)

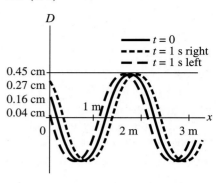

(b) $D = (0.45\,\text{m})\sin[(3.0\,\text{m}^{-1})x - (6.0\,\text{s}^{-1})t + 1.2]$.
(d) $D = (0.45\,\text{m})\sin[(3.0\,\text{m}^{-1})x + (6.0\,\text{s}^{-1})t + 1.2]$.

25. $D = -(0.020\,\text{cm})\cos[(8.01\,\text{m}^{-1})x - (2.76 \times 10^3\,\text{s}^{-1})t]$.

27. The function is a solution.

31. (a) $v_2/v_1 = (\mu_1/\mu_2)^{1/2}$;
(b) $\lambda_2/\lambda_1 = v_2/v_1 = (\mu_1/\mu_2)^{1/2}$;
(c) lighter cord.

33. (c) $A_T = [2k_1/(k_2 + k_1)]A = [2v_2/(v_1 + v_2)]A$.

35. (b) $2D_M \cos(\frac{1}{2}\phi)$, purely sinusoidal;
(d) $D = \sqrt{2}\,D_M \sin(kx - \omega t + \pi/4)$.

37. 440 Hz, 880 Hz, 1320 Hz, 1760 Hz.

39. $f_n = n(0.50\,\text{Hz}), n = 1, 2, 3,...$;
$T_n = (2.0\,\text{s})/n, n = 1, 2, 3,...$.

41. 70 Hz.

45. 4.

47. (a) $D_2 = (4.2\,\text{cm})\sin[(0.71\,\text{cm}^{-1})x + (47\,\text{s}^{-1})t + 2.1]$;
(b) $D_{\text{resultant}} = (8.4\,\text{cm})$
$\sin[(0.71\,\text{cm}^{-1})x + 2.1]$
$\cos[(47\,\text{s}^{-1})t]$.

49. 308 Hz.

51. (a)

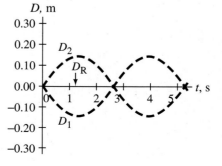

(b)

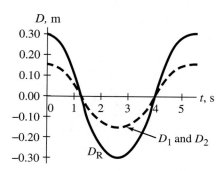

53. 5.4 km/s.

55. 29°.

57. 24°.

59. Speed will be greater in the less dense rod by a factor of $\sqrt{2}$.

61. (a) 0.050 m; (b) 2.3.

63. 0.99 m.

65. (a)

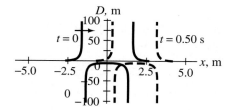

(b) $D = (4.0\,\text{m})/\{[x - (3.0\,\text{m/s})t]^2 - 2.0\,\text{m}^2\}$;

(c)

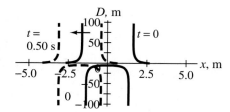

(d) $D = (4.0\,\text{m})/\{[x + (3.0\,\text{m/s})t]^2 - 2.0\,\text{m}^2\}$.

67. (a) 784 Hz, 1176 Hz, 880 Hz, 1320 Hz; (b) 1.26; (c) 1.12; (d) 0.794.

69. $\lambda_n = 4L/(2n - 1)$, $n = 1, 2, 3, \ldots$.

71. $y = (3.5\,\text{cm})\cos[(1.05\,\text{cm}^{-1})x - (1.39\,\text{s}^{-1})t]$.

73.

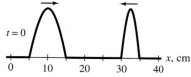

$t = 0$

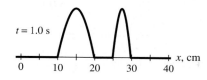

$t = 1.0\,\text{s}$

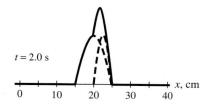

$t = 2.0\,\text{s}$

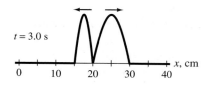

$t = 3.0\,\text{s}$

CHAPTER 16

1. 2.6×10^2 m.

3. 5.4×10^2 m.

5. 1200 m, 300 m.

7. (a) 1.1×10^{-8} m; (b) 1.1×10^{-10} m.

9. (a) $\Delta P = (4 \times 10^{-5}\,\text{Pa})$
$\sin[(0.949\,\text{m}^{-1})x - (315\,\text{s}^{-1})t]$;
(b) $\Delta P = (4 \times 10^{-3}\,\text{Pa})$
$\sin[(94.9\,\text{m}^{-1})x - (3.15 \times 10^4\,\text{s}^{-1})t]$.

11. (a) 49 dB; (b) 3.2×10^{-10} W/m².

13. 150 Hz to 20,000 Hz.

15. (a) 9; (b) 9.5 dB.

17. (a) Higher frequency is greater by a factor of 2; (b) 4.

19. (a) 5.0×10^{-13} W; (b) 6.3×10^4 yr.

21. 87 dB.

23. (a) 5.10×10^{-5} m; (b) 29.8 Pa.

25. (a) 1.5×10^3 W; (b) 3.4×10^2 m.

27. (b) 190 dB.

29. (a) 570 Hz; (b) 860 Hz.

31. 8.6 mm $< L <$ 8.6 m.

33. (a) 110 Hz, 330 Hz, 550 Hz, 770 Hz; (b) 220 Hz, 440 Hz, 660 Hz, 880 Hz.

35. (a) 0.66 m; (b) 262 Hz, 1.32 m; (c) 1.32 m, 262 Hz.

37. −2.6%.

39. (a) 0.578 m; (b) 869 Hz.

41. 215 m/s.

43. 0.64, 0.20, −2 dB, −7 dB.

45. 28.5 kHz.

47. (a) 130.5 Hz, or 133.5 Hz; (b) ±2.3%.

49. (a) 343 Hz; (b) 1030 Hz, 1715 Hz.

53. 346 Hz.

57. (a) 1690 Hz; (b) 1410 Hz.

59. 30,890 Hz.

61. 120 Hz.

63. 91 Hz.

65. 90 beats/min.

67. (a) 570 Hz; (b) 570 Hz; (c) 570 Hz; (d) 570 Hz; (e) 594 Hz; (f) 595 Hz.

71. (a) 120; (b) 0.96°.

73. (a) 37°; (b) 1.7.

75. 0.278 s.

77. 55 m.

79. 410 km/h (257 mi/h).

81. 1, 0.444, 0.198, 0.0878, 0.0389.

83. 18.1 W.

85. 15 W.

87. 2.3 Hz.

89. $\Delta P_M/\Delta P_{M0} = D_M/D_{M0} = 10^6$.

91. 50 dB.

93. 17.5 m/s.

95. 2.29 kHz.

97. 550 Hz.

99. (a) 2.8×10^3 Hz; (b) 1.80 m; (c) 0.12 m.

101. (a) 2.2×10^{-7} m; (b) 5.4×10^{-8} m.

CHAPTER 17

1. 0.548.

3. (a) 20°C; (b) ≈3300°F.

5. 104.0°F.

7. −40°F = −40°C.

9. $\Delta L_{\text{Invar}} = 2.0 \times 10^{-6}$ m, $\Delta L_{\text{steel}} = 1.2 \times 10^{-4}$ m, $\Delta L_{\text{marble}} = 2.5 \times 10^{-5}$ m.

11. −69°C.

13. 5.1 mL.

15. 0.060 cm³.

19. −40 min.

21. -2.8×10^{-3} (0.28%).

23. 3.5×10^7 N/m².

25. (a) 27°C; (b) 4.3×10^3 N.

27. −459.7°F.

29. 1.07 m³.

31. 1.43 kg/m³.

33. (a) 14.8 m³; (b) 1.81 atm.
35. 1.80 × 10³ atm.
37. 37°C.
39. 3.43 atm.
41. 0.588 kg/m³, water vapor is not an ideal gas.
43. 2.69 × 10²⁵ molecules/m³.
45. 4.8 × 10²² molecules.
47. 7.7 × 10³ N.
49. (a) 71.2 torr; (b) 157°C.
51. (a) 0.19 K; (b) 0.051%.
53. (a) Low; (b) 0.017%.
55. 1/6.
57. 5.1 × 10²⁷ molecules, 8.4 × 10³ mol.
59. 11 L, not advisable.
61. (a) 9.3 × 10² kg; (b) 1.0 × 10² kg.
63. 1.1 × 10⁴⁴ molecules.
65. 3.3 × 10⁻⁷ cm.
67. 1.1 × 10³ m.
69. 15 h.
71. 0.66 × 10³ kg/m³.
73. ±0.11 C°.
77. 3.6 m.

CHAPTER 18

1. (a) 5.65 × 10⁻²¹ J; (b) 7.3 × 10³ J.
3. 1.17.
5. (a) 4.5; (b) 5.2.
7. $\sqrt{2}$.
11. (a) 461 m/s; (b) 19 s⁻¹.
13. 1.00429.
17. Vapor.
19. (a) Gas, liquid, vapor;
   (b) gas, liquid, solid, vapor.
21. 0.69 atm.
23. 11°C.
25. 1.96 atm.
27. 120°C.
29. (a) 5.3 × 10⁶ Pa; (b) 5.7 × 10⁶ Pa.
31. (b) b = 4.28 × 10⁻⁵ m³/mol,
   a = 0.365 N·m⁴/mol².
33. (a) 10⁻⁹ atm; (b) 300 atm.
35. (a) 6.3 cm; (b) 0.58 cm.
37. 2 × 10⁻⁷ m.
39. (b) 4.7 × 10⁷ s⁻¹.
43. 7.8 h.
45. (b) 4 × 10⁻¹¹ mol/s; (c) 0.6 s.
47. 2.6 × 10² m/s,
   4 × 10⁻¹⁷ N/m² ≈ 4 × 10⁻²² atm.
49. (a) 2.9 × 10² m/s; (b) 12 m/s.
51. Reasonable, 70 cm.

53. $mgh = 4.3 \times 10^{-5}(\frac{1}{2}mv_{rms}^2)$, reasonable.
55. $P_2/P_1 = 1.43$, $T_2/T_1 = 1.20$.
57. 1.4 × 10⁵ K.
59. (a) 1.7 × 10³ Pa; (b) 7.0 × 10² Pa.
61. 2 × 10¹³ m.

CHAPTER 19

1. 1.0 × 10⁷ J.
3. 1.8 × 10² J.
5. 2.1 × 10² kg/h.
7. 83 kcal.
9. 4.7 × 10⁶ J.
11. 40 C°.
13. 186°C.
15. 7.1 min.
17. (b) $mc_0\left[(T_2 - T_1) + a(T_2^2 - T_1^2)/2\right]$;
   (c) $c_{mean} = c_0\left[1 + \frac{1}{2}a(T_2 + T_1)\right]$.
19. 0.334 kg (0.334 L).
21. $\frac{2}{3}m$ steam and $\frac{4}{3}m$ water at 100°C.
23. 9.4 g.
25. 4.7 × 10³ kcal.
27. 1.22 × 10⁴ J/kg.
29. 360 m/s.
31. (a) 0; (b) 5.00 × 10³ J.
33.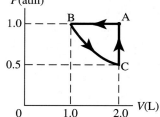
35. (a) 0; (b) −1300 kJ.
37. (a) 1.6 × 10² J; (b) +1.6 × 10² J.
39. W = 3.46 × 10³ J, ΔU = 0,
   Q = +3.46 × 10³ J (into the gas).
41. +129 J.
45. (a) +25 J; (b) +63 J; (c) −95 J;
   (d) −120 J; (e) −15 J.
47. $W = RT \ln\left(\frac{V_2 - b}{V_1 - b}\right) + a\left(\frac{1}{V_2} - \frac{1}{V_1}\right)$.
49. 22°C/h.
51. 4.98 cal/mol·K, 2.49 kcal/kg·K;
   6.97 cal/mol·K, 3.48 kcal/kg·K.
53. 83.7 g/mol, krypton.
55. 46 C°.
57. (a) 2.08 × 10³ J; (b) 8.32 × 10² J;
   (c) 2.91 × 10³ J.
59. 0.379 atm, −51°C.
61. 1.33 × 10³ J.

63. (a) T₁ = 317 K, T₂ = 153 K;
   (b) −1.59 × 10⁴ J; (c) −1.59 × 10⁴ J;
   (d) Q = 0.
65. (a)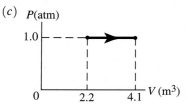
   (b) 231 K;
   (c) $Q_{1\to2} = 0$,
   $\Delta U_{1\to2} = -2.01 \times 10^3$ J,
   $W_{1\to2} = +2.01 \times 10^3$ J;
   $W_{2\to3} = -1.34 \times 10^3$ J,
   $\Delta U_{2\to3} = -1.97 \times 10^3$ J,
   $Q_{2\to3} = -3.31 \times 10^3$ J;
   $W_{3\to1} = 0$,
   $\Delta U_{3\to1} = +3.98 \times 10^3$ J,
   $Q_{3\to1} = +3.98 \times 10^3$ J;
   (d) $W_{cycle} = +0.67 \times 10^3$ J,
   $Q_{cycle} = +0.67 \times 10^3$ J,
   $\Delta U_{cycle} = 0$.
67. (a) 64 W; (b) 22 W.
69. 4.8 × 10² W.
71. 31 h.
73. (a) 1.7 × 10¹⁷ W; (b) 280K (7°C).
75. (b) $\Delta Q/\Delta t = A(T_2 - T_1)/\Sigma(\ell_i/k_i)$.
77. 22%.
79. 4 × 10¹⁵ J.
81. 2.8 kcal/kg.
83. 31 C°.
85. 682 J.
87. 2.8 C°.
89. 2.58 cm, rod vaporizes.
91. 4.3 kg.
95. (a) 2.3 C°/s; (b) 84°C;
   (c) convection, conduction, evaporation.
97. (a) $\rho = m/V = (mP/nR)/T$;
   (b) $\rho = (m/nRT)P$.
99. (a) 1.9 × 10⁵ J; (b) −1.4 × 10⁵ J;
   (c)
101. 4.98 × 10⁵ s = 5.8 d.
103. 10 C°.

CHAPTER 20

1. 24%.

3. 816 MW.

5. 18%.

7. 13 km^3/day, 63 km^2.

9. 28.0%.

13. 1.2×10^{13} J/h.

15. 1.4×10^3 m/day.

17. 659°C.

19. 3.7×10^8 kg/h.

21. (*a*) $P_a = 5.15 \times 10^5$ Pa,
$P_b = 2.06 \times 10^5$ Pa;
(*b*) $V_c = 30.0$ L, $V_d = 12.0$ L;
(*c*) 2.83×10^3 J; (*d*) -2.14×10^3 J;
(*e*) 0.69×10^3 J; (*f*) 24%.

23. 5.7.

25. −21°C.

27. 2.9.

29. (*a*) 3.9×10^4 J; (*b*) 3.0 min.

31. 76 L.

33. 0.15 J/K.

35. +11 kcal/K.

37. +0.0104 cal/K · s.

39. +172 J/K.

43. (*a*) 0.312 kcal/K;
(*b*) > -0.312 kcal/K.

45. (*a*) Adiabatic process;
(*b*) $\Delta S_i = -nR \ln 2, \Delta S_a = 0$;
(*c*) $\Delta S_{\text{surr,i}} = nR \ln 2, \Delta S_{\text{surr,a}} = 0$.

47. (*a*) Entropy is a state function.

51. (*a*) 5/16; (*b*) 1/64.

53. (*b*), (*a*), (*c*), (*d*).

55. 69%.

57. 2.6×10^3 J/K.

59. (*a*) 17; (*b*) 6.1×10^7 J/h.

61. (*a*) 5.3 C°; (*b*) +77 J/kg · K.

63. $\Delta S = K/T$.

65. (*a*)

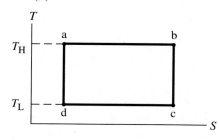

(*b*) area $= Q_{\text{net}} = W_{\text{net}}$.

67. $e_{\text{Stirling}} = (T_{\text{H}} - T_{\text{L}}) \ln(V_b/V_a)/$
$[T_{\text{H}} \ln(V_b/V_a)$
$+ \frac{3}{2}(T_{\text{H}} - T_{\text{L}})]$,
$\quad e_{\text{Stirling}} < e_{\text{Carnot}}$.

69. 0.091 hp.

71. (*a*) 1/379; (*b*) $1/1.59 \times 10^{11}$.

Index

Photo Credits

Periodic Table of the Elements §

Transition Elements

Key:
Symbol — Cl 17 — Atomic Number
Atomic Mass§ — 35.4527
$3p^5$ — Electron Configuration (outer shells only)

Group I	Group II				Transition Elements							Group III	Group IV	Group V	Group VI	Group VII	Group VIII
H 1 1.00794 $1s^1$																	**He** 2 4.002602 $1s^2$
Li 3 6.941 $2s^1$	**Be** 4 9.012182 $2s^2$											**B** 5 10.811 $2p^1$	**C** 6 12.0107 $2p^2$	**N** 7 14.00674 $2p^3$	**O** 8 15.9994 $2p^4$	**F** 9 18.9984032 $2p^5$	**Ne** 10 20.1797 $2p^6$
Na 11 22.989770 $3s^1$	**Mg** 12 24.3050 $3s^2$											**Al** 13 26.981538 $3p^1$	**Si** 14 28.0855 $3p^2$	**P** 15 30.973761 $3p^3$	**S** 16 32.066 $3p^4$	**Cl** 17 35.4527 $3p^5$	**Ar** 18 39.948 $3p^6$

Transition row labels: Sc Ti V Cr Mn Fe Co Ni Cu Zn

Sc 21 44.955910 $3d^14s^2$	**Ti** 22 47.867 $3d^24s^2$	**V** 23 50.9415 $3d^34s^2$	**Cr** 24 51.9961 $3d^54s^1$	**Mn** 25 54.938049 $3d^54s^2$	**Fe** 26 55.845 $3d^64s^2$	**Co** 27 58.933200 $3d^74s^2$	**Ni** 28 58.6934 $3d^84s^2$	**Cu** 29 63.546 $3d^104s^1$	**Zn** 30 65.39 $3d^104s^2$

K 19 — 39.0983 — $4s^1$; **Ca** 20 — 40.078 — $4s^2$; **Ga** 31 — 69.723 — $4p^1$; **Ge** 32 — 72.61 — $4p^2$; **As** 33 — 74.92160 — $4p^3$; **Se** 34 — 78.96 — $4p^4$; **Br** 35 — 79.904 — $4p^5$; **Kr** 36 — 83.80 — $4p^6$

Y 39 88.90585 $4d^15s^2$	**Zr** 40 91.224 $4d^25s^2$	**Nb** 41 92.90638 $4d^45s^1$	**Mo** 42 95.94 $4d^55s^1$	**Tc** 43 (98) $4d^55s^2$	**Ru** 44 101.07 $4d^75s^1$	**Rh** 45 102.90550 $4d^85s^1$	**Pd** 46 106.42 $4d^105s^0$	**Ag** 47 107.8682 $4d^105s^1$	**Cd** 48 112.411 $4d^105s^2$

Rb 37 — 85.4678 — $5s^1$; **Sr** 38 — 87.62 — $5s^2$; **In** 49 — 114.818 — $5p^1$; **Sn** 50 — 118.710 — $5p^2$; **Sb** 51 — 121.760 — $5p^3$; **Te** 52 — 127.60 — $5p^4$; **I** 53 — 126.90447 — $5p^5$; **Xe** 54 — 131.29 — $5p^6$

Hf 72 178.49 $5d^26s^2$	**Ta** 73 180.9479 $5d^36s^2$	**W** 74 183.84 $5d^46s^2$	**Re** 75 186.207 $5d^56s^2$	**Os** 76 190.23 $5d^66s^2$	**Ir** 77 192.217 $5d^76s^2$	**Pt** 78 195.078 $5d^96s^1$	**Au** 79 196.96655 $5d^106s^1$	**Hg** 80 200.59 $5d^106s^2$

Cs 55 — 132.90545 — $6s^1$; **Ba** 56 — 137.327 — $6s^2$; 57–71 † ; **Tl** 81 — 204.3833 — $6p^1$; **Pb** 82 — 207.2 — $6p^2$; **Bi** 83 — 208.98038 — $6p^3$; **Po** 84 — (209) — $6p^4$; **At** 85 — (210) — $6p^5$; **Rn** 86 — (222) — $6p^6$

Rf 104 (261) $6d^27s^2$	**Db** 105 (262) $6d^37s^2$	**Sg** 106 (266) $6d^47s^2$	**Bh** 107 (264)	**Hs** 108 (269)	**Mt** 109 (268)	110 (271)	111 (272)	112 (277)

Fr 87 — (223) — $7s^1$; **Ra** 88 — (226) — $7s^2$; 89–103 ‡ ; 114 (289) ; 116 (289) ; 118 (293)

---

†Lanthanide Series

La 57 138.9055 $5d^16s^2$	**Ce** 58 140.115 $4f^15d^16s^2$	**Pr** 59 140.90765 $4f^35d^06s^2$	**Nd** 60 144.24 $4f^45d^06s^2$	**Pm** 61 (145) $4f^55d^06s^2$	**Sm** 62 150.36 $4f^65d^06s^2$	**Eu** 63 151.964 $4f^75d^06s^2$	**Gd** 64 157.25 $4f^75d^16s^2$	**Tb** 65 158.92534 $4f^95d^06s^2$	**Dy** 66 162.50 $4f^105d^06s^2$	**Ho** 67 164.93032 $4f^115d^06s^2$	**Er** 68 167.26 $4f^125d^06s^2$	**Tm** 69 168.93421 $4f^135d^06s^2$	**Yb** 70 173.04 $4f^145d^06s^2$	**Lu** 71 174.967 $4f^145d^16s^2$

‡Actinide Series

Ac 89 (227.02775) $6d^17s^2$	**Th** 90 232.0381 $6d^27s^2$	**Pa** 91 231.03588 $5f^26d^17s^2$	**U** 92 238.0289 $5f^36d^17s^2$	**Np** 93 (237) $5f^46d^17s^2$	**Pu** 94 (244) $5f^66d^07s^2$	**Am** 95 (243) $5f^76d^07s^2$	**Cm** 96 (247) $5f^76d^17s^2$	**Bk** 97 (247) $5f^96d^07s^2$	**Cf** 98 (251) $5f^106d^07s^2$	**Es** 99 (252) $5f^116d^07s^2$	**Fm** 100 (257) $5f^126d^07s^2$	**Md** 101 (258) $5f^136d^07s^2$	**No** 102 (259) $5f^146d^07s^2$	**Lr** 103 (262) $5f^146d^17s^2$

§ Atomic mass values averaged over isotopes in percentages they occur on Earth's surface. For many unstable elements, mass of the longest-lived known isotope is given in parentheses. 1999 revisions. (See also Appendix D.)